The Microbiology Place

Log on.

Tune in.

Succeed.

To help you succeed in microbiology, your professor has arranged for you to enjoy access to great media resources, on an interactive CD-ROM and on The Microbiology Place website. You'll find that these resources that accompany your textbook will enhance your course materials.

Home   Search   Site Requirements   FAQs   Help   AW   Credits   Profile

the Microbiology place

1: The Microbial World and You

Microbiology: An Introduction, Eighth Edition

Introduction
Objectives
Activities
Practice Tests
Chapter Quizzes
Web Links
Flashcards
Lab Supplement
Content In the News
Research Navigator
Online Tutor
Careers
Instructor Resources

Welcome to The Microbiology Place, accompanying *Microbiology: An Introduction,* 8th edition by Tortora/Funke/Case. To begin, choose a chapter from the pull-down menu above and then select a feature on the left.

You will find an abundance of resources on this site:

- **Objectives** help you focus on the key concepts to master in each chapter.
- **Activities** include interactive tutorials, simulations, and case studies designed to review and test your understanding of concepts in microbiology. Play the role of an AIDS researcher, try your hand at sewage treatment, grow microbes in a bioreactor, and more!
- **Practice Tests** provide a convenient way to test yourself on the material in each chapter. Instant feedback is provided for each answer. There are 20 multiple-choice questions per chapter.
- **Chapter Quizzes** are made up of 40 quiz questions for each chapter of your textbook. The quizzes are in true/false, multiple-choice, and fill-in-the-blank format. Chapter Quizzes may be used as practice tests and/

Looking for micrographs? The Center for Disease Control's **Public Health Image Library** is a searchable database of images that may be used for educational purposes. From the image library home page, type in "micrograph" as your key word and then click "go."

What your system needs to use these media resources:

WINDOWS™
266 MHz Intel Pentium processor or greater
Windows 98 or higher
64 MB Minimum RAM
800 X 600 screen resolution
Browser:Internet Explorer 5.0 or higher; Netscape 4.7, 6.2, or higher
Plug-Ins: Shockwave Player 8, Flash Player 5, QuickTime 4
Internet Connection: 56k modem minimum for website
NOTE: Use of Netscape 6.0 is not recommended due to a known compatibility issue between Netscape 6.0 and the Flash and Shockwave plug-ins.

MACINTOSH™
266 MHz Minimum CPU
OS 8.6 or higher
64 MB Minimum RAM
800 x 600 screen resolution,thousands of colors
Browser: OS 8-9: Netscape 4.7, 6.2, or higher; (Web only: Internet Explorer 5 or higher); OS X: Internet Explorer 5.2 or higher
Plug-Ins: Shockwave Player 8, Flash Player 5, QuickTime 4
Internet Connection: 56k modem minimum for website
NOTE: Use of Netscape 6.0 is not recommended due to a known compatibility issue between Netscape 6.0 and the Flash and Shockwave plug-ins. Also, Mac OS X users should run this CD-ROM on Internet Explorer only due to known compatibility issues between OS X and Netscape.

Got technical questions?

For technical support, please visit www.aw.com/techsupport and complete the appropriate online form. You can also call our tech support hotline at 800-6-Pro-Desk (800-677-6337) Monday-Friday, 8 a.m. to 5 p.m. CST.

Here's your personal ticket to success:

How to log on to www.microbiologyplace.com

1. Go to www.microbiologyplace.com.
2. Click on *Microbiology: An Introduction 8th ed. by Tortora, Funke & Case.*
3. Click "Register"
4. Scratch off the silver foil coating below to reveal your pre-assigned access code.
5. Enter your pre-assigned access code exactly as it appears below.
6. Complete the online registration form to create your own personal user Login Name and Password.
7. Once your personal Login Name and Password are confirmed by email, go back to www.microbiologyplace.com, type in your new Login Name and Password, and click "Log In"

Your Access Code is:

Record your new login name and password on the back of this card.

Cut out this card and keep it handy. It's your ticket to valuable information. Students who purchase a used copy of this book may not have a valid access code. Students may purchase a subscription to this website at www.microbiologyplace.com.

Important: Please read the License Agreement located on the log in screen before using **The Microbiology Place™**. By using the website, you indicate that you have read, understood, and accepted the terms of this agreement.

Microbiology
AN INTRODUCTION
BRIEF EDITION

GERARD J. TORTORA

BERGEN COMMUNITY COLLEGE

BERDELL R. FUNKE

NORTH DAKOTA STATE UNIVERSITY

CHRISTINE L. CASE

SKYLINE COLLEGE

PEARSON

Benjamin Cummings

San Francisco Boston New York
Cape Town Hong Kong London Madrid Mexico City
Montreal Munich Paris Singapore Sydney Tokyo Toronto

Publisher: Daryl Fox
Executive Editor: Leslie Berriman
Senior Project Editor: Barbara Yien
Editorial Assistant: Blythe Robbins
Managing Editor: Wendy Earl
Production Editor: Sharon Montooth
Art and Photo Supervisor: Bradley Burch
Photo Editors: Kathleen Olson and Maureen Spuhler
Text and Cover Designer: Kathleen Cunningham
Copy Editor: Sally Peyrefitte
Proofreader: Martha Ghent
Illustrations: Precision Graphics and The GTS Companies, Inc.
Compositor: The GTS Companies, Inc.
Manufacturing Supervisor: Stacey Weinberger
Marketing Manager: Lauren Harp

Cover Photograph: Electron micrograph of conidiospores of *Aspergillus,* a common mold.
Magnification: 1600×. © Dennis Kunkel/Phototake, Inc.

Text, art, and photo credits appear following the Glossary.

Library of Congress Cataloging-in-Publication Data

Tortora, Gerard J.
  Microbiology: an introduction / Gerard J. Tortora, Berdell R. Funke,
Christine L. Case.–Brief ed.
    p. cm.
  Includes index.
  ISBN 0-8053-7753-0
  1. Microbiology. I. Funke, Berdell R. II. Case, Christine L., 1948–
III.
Title.
  QR41.2.T672 2004
  579–dc22

                   2004052729

ISBN 0-8053-7753-0

2 3 4 5 6 7 8 9 10–VHC–07 06 05
www.aw.com/bc

About the Authors

Gerard J. Tortora Jerry Tortora is a professor of biology and teaches microbiology, human anatomy, and physiology at Bergen Community College in Paramus, New Jersey. He received his M.A. in Biology from Montclair State College in 1965. He belongs to numerous biology/microbiology organizations, such as the American Society of Microbiology (ASM), Human Anatomy and Physiology Society (HAPS), American Association for the Advancement of Science (AAAS), National Education Association (NEA), New Jersey Educational Association (NJEA), and the Metropolitan Association of College and University Biologists (MACUB). Jerry is the author of a number of biological science textbooks. In 1995, he was selected as one of the finest faculty scholars at Bergen Community College and was named Distinguished Faculty Scholar. In 1996, Jerry received a National Institute for Staff and Organizational Development (NISOD) excellence award from the University of Texas and was selected to represent Bergen Community College in a campaign to increase awareness of the contributions of community colleges to higher education.

Berdell R. Funke Bert Funke received his Ph.D., M.S., and B.S. in microbiology from Kansas State University. He has spent his professional years as a professor of microbiology at North Dakota State University. He taught introductory microbiology, including laboratory sections, general microbiology, food microbiology, soil microbiology, clinical parasitology, and pathogenic microbiology. As a research scientist in the Experiment Station at North Dakota State, he has published numerous papers in soil microbiology and food microbiology.

Christine L. Case Chris Case is a registered microbiologist and a professor of microbiology at Skyline College in San Bruno, California, where she has taught for the past 32 years. She received her Ed.D. in curriculum and instruction from Nova Southeastern University and her M.A. in microbiology from San Francisco State University. She was Director for the Society for Industrial Microbiology (SIM) and is an active member of the ASM and Northern California SIM. She received the ASM and California Hayward outstanding educator awards. In addition to teaching, Chris contributes regularly to the professional literature, develops innovative educational methodologies, and maintains a personal and professional commitment to conservation and the importance of science in society. Chris is also an avid photographer, and many of her photographs appear in this book.

Preface

In recent years, more and more instructors have requested a shorter, paperback version of our hardcover textbook, *Microbiology: An Introduction* (now in its eighth edition). In response to these requests, we have developed this Brief Edition of *Microbiology: An Introduction*. The Brief Edition is essentially an alternative version of the hardcover eighth edition. The twenty chapters in this book are identical to the first twenty chapters of the hardcover edition. The key distinction is that the hardcover edition includes eight chapters focusing on microbial disease and environmental microbiology, whereas the Brief Edition omits these chapters. We made this decision after learning that some instructors do not have sufficient time to assign these chapters, or prefer covering these topics with handouts and student projects instead. The Brief Edition has been produced with the needs of these instructors and students in mind.

Like the hardcover edition, the Brief Edition provides the fundamentals of microbiology for students in allied health sciences, biotechnology technician training, and liberal arts. It is a beginning text, assuming no previous study of biology or chemistry.

During the 21 years since the publication of the first hardcover edition, hundreds of thousands of students have used *Microbiology: An Introduction* at more than 1000 colleges and universities, making it the best-selling introductory text in the world. We have been gratified to hear from instructors and students alike that the book has become a favorite among their textbooks—a learning tool that is both effective and enjoyable.

Features of the Brief Edition

The hallmark features of this text include:

■ **An appropriate balance between microbiological fundamentals and applications, and between medical applications and other applied areas of microbiology.** We have provided the solid grounding in fundamental facts and principles necessary to understand and adapt to the rapid developments in microbiology. At the same time, we have integrated applications throughout the text because we know that beginning students benefit from seeing the relevance of microbiology to their respective programs. We have made the text especially comprehensive in medically important areas of microbiology. For example, the principles and applications of immunity have been particulary emphasized. We believe this emphasis is deserved in the light of the tremendous importance of immunology, both as a basic science and as the source of many new valuable tools and techniques for microbiology and the health sciences.

■ **Straightforward presentation of complex topics.** Each chapter of the text was written with the student in mind, and maintains the clarity of explanation for which *Microbiology: An Introduction* has become known.

■ **Integrated learning objectives and end-of-chapter questions** help students check their understanding of key chapter concepts and learn critical problem-solving skills, needed in clinical and industrial situations.

■ **Applications and discovery-oriented boxes** focus on modern, practical uses of microbiology and biotechnology and emphasize the process of scientific discovery. The application boxes show real people doing science to provide students with examples of career opportunities in microbiology.

■ **An exceptional art program.** Well-developed, full-color illustrations throughout the book support and enhance the text. Key concepts and questions appear in figure legends to encourage students to think critically about illustrations. Key features of the illustration program include:

Consistent use of symbols and colors. Symbols for molecules such as phosphate groups (Ⓟ) and ATP ( ) are the same color and shape throughout the book, enabling students to progress from familiar parts of illustrated processes to unfamiliar ones with confidence.

Orientation Diagrams, miniature versions of overview illustrations with appropriate parts highlighted, help students keep the "big picture" in mind.

Step-by-step descriptions, in both the narration and illustrations, walk students through important microbiological processes and help them visualize the order of events. Color-coded numbers link text and figure legends to corresponding art.

Micrograph icons appear throughout the book to identify the types of microscopes used in the micrographs. **SEM** (scanning electron micrographs) icons and **TEM** (transmission electron micrographs) icons indicate the use of electron micrographs. The **LM** icon indicates light micrographs. A red icon **SEM** indicates a colorized micrograph.

Scope and Sequence

The Brief Edition is divided into three parts. Part One, Fundamentals of Microbiology, consists of chapters 1–9 and covers chemistry, microscopy, cell structure and function, metabolism, microbial growth, genetics, and biotechnology. Part Two, A Survey of the Microbial World, consists of chapters 10–13 and covers classification of microorganisms as well as detailed introductions to prokaryotes, eukaryotes, and viruses. Part Three, Interaction Between Microbe and Host, consists of chapters 14–20 and covers epidemiology, pathogenicity, immunology, and antimicrobial drugs. We have organized the book in what we think is a useful fashion, while recognizing that the material might be effectively presented in a number of other sequences. For those who wish to use a different order, we have made each chapter as independent as possible and have included numerous cross-references. The Instructor's Guide provides detailed guidelines for organizing the material in several other ways.

A Comprehensive Teaching and Learning Package

The Brief Edition is fully supported by a comprehensive suite of print and media supplements for both instructors and students.

Supplementary Materials for the Student

- **The Microbiology Place Website and CD-ROM** includes tutorials, interactive activities, flashcards of key terms, and simulations covering microbiology's toughest topics. Interactive activities cover the sulfur and phosphorous cycles, photosynthesis, serological tests, the use of PCR to identify microbes in environmental samples, the nitrogen cycle, gene cloning practice, host and pathogen responses during infection, and virus identification. Three unique simulations enhance understanding of the effects of sewage treatment on waterways, industrial fermentation in a bioreactor, and enzyme inhibitor drugs.

- **Student Study Guide,** by Berdell Funke (0-8053-7620-8). The study guide includes concise explanations of key concepts, definitions of important terms, art labeling exercises, critical thinking problems, and a variety of self-test questions with answers. Students can master key concepts and earn better grades with the help of clear, concise writing and creative and thought-provoking exercises.

- *Microbiology: A Photographic Atlas for the Laboratory* by Steve K. Alexander and Dennis Strete (0-8053-2732-0). Tailored for the introductory microbiology laboratory, this atlas contains approximately 400 high-quality, color photographs that demonstrate the results of laboratory procedures and the morphology of important microorganisms.

- *Laboratory Experiments in Microbiology,* Seventh Edition by Ted Johnson and Christine Case (0-8053-7673-9). This text promotes good laboratory practice and thinking skills with 57 classroom-tested experiments that meet ASM recommendations for the core curriculum considered essential to teach in every introductory microbiology laboratory. These experiments reinforce lecture concepts and promote critical thinking about metabolism, growth, genetics, and immunology. A *Preparation Guide* (0-8053-7674-7) is also available for instructors to aid in the preparation of lab sessions.

- *Microbiology Coloring Book* by I. Edward Alcamo and Lawrence Elson (0-06-041925-3). This unique study tool contains 105 plates, including text and illustrations to color, which enhance understanding and comprehension of important microbiological concepts.

Supplementary Materials for the Instructor

All supplemental teaching materials are available to qualified adopting instructors. Please contact your local Benjamin Cummings sales consultant or call our Customer Service at 1-800-922-0579. Some package items may not be available to adopters outside of the United States.

- **Instructor's Art and Photo Presentation CD-ROM** Included on the CD-ROM are electronic files of art, photos, line art and tables (approximately 700 in all, including all images found in the Brief Edition text, as well as many additional images of microbial diseases).

Images are identified by figure number and are included in both jpeg and PowerPoint® format. All the images in a chapter appear as small thumbnail slides in web-browser format—a user-friendly interface that allows the instructor to quickly locate a figure or table by sight and open the file with a click.

- **Prepared PowerPoint® Lecture Slides** These prepared lecture slides, written by Christine L. Case and included with the Instructor's Art and Photo Presentation CD-ROM, outline the contents of each chapter alongside images from the text. Instructors can adapt the PowerPoint® slides to their specific course without the need to assemble presentations from scratch.

- **Instructor's Guide and Test Bank** (0-8053-7619-4). Written and revised by Christine L. Case, the Instructor's Guide contains teaching tips, alternative course outlines, ideas for using special features, and answers to test questions. The Test Bank features over 1200 multiple-choice questions with answers as well as three to five essay questions per chapter.

- **Computerized Test Bank CD** (TestGen 3.0 0-8053-7617-8). This easy-to-use testing program allows you to view and edit electronic questions from the Test Bank, create multiple tests, and print them in a variety of formats.

- **Transparency Acetates** (0-8053-7618-6). Includes approximately 350 illustrations, including all illustrations from the text, as well as many additional illustrations of microbial disease processes.

- **Course Management Systems, including WebCT, Blackboard, and eCollege** Useful for on-line course management and distance learning courses, content selected from the Microbiology Place and the entire computerized Test Bank is available in the three leading course management systems. Please contact your local Benjamin Cummings representative for more details.

Acknowledgments

In preparation for this textbook, we have benefited from the guidance and advice of a large number of microbiology instructors across the country. The reviewers and focus group participants listed on the next page provided constructive criticism and valuable suggestions at various stages of the revision. We gratefully acknowledge our debt to these individuals.

We also thank the staff at Benjamin Cummings for their dedication to excellence. Publisher Daryl Fox and executive editor Leslie Berriman supported the idea of publishing this Brief Edition, demonstrating their commitment to responding to instructors' needs. Sally Peyrefitte's careful attention to continuity and detail in her copyedit of both text and art served to keep concepts and information clear throughout. Barbara Yien, Sharon Montooth, and Wendy Earl expertly guided the text through the production process. Bradley Burch effectively managed the large art program. The photo researchers, Kathleen Olson and Maureen Spuhler, made sure we had clear and striking images throughout the book. Kathleen Cunningham created the interior design and did a wonderful job with the cover. The GTS Companies did their usual outstanding job moving this book quickly and beautifully through composition; the skilled team was led by Ruth Sakata Corley. Stacey Weinberger guided the book through the manufacturing process. Leslie Austin expertly guided the production of the supplements, with support from Ziki Dekel and Ryan Shaw.

We would all like to acknowledge our spouses and families, who have provided invaluable support throughout the writing process.

Finally, we have an enduring appreciation for our students, whose comments and suggestions provide insight and remind us of their needs. This text is for them.

Gerard J. Tortora
Berdell R. Funke
Christine L. Case

Brief Contents

Contents

Chapter 5 continues

Part Two
A Survey of the Microbial World

Chapter 10 continues

Chapter 11

The Prokaryotes: Domains Bacteria and Archaea 304

Chapter 12

The Eukaryotes: Fungi, Algae, Protozoa, and Helminths 334

Chapter 13

Viruses, Viroids, and Prions 376

Part Three

Interaction Between Microbe and Host

Chapter 14
Principles of Disease and Epidemiology 408

Chapter 15
Microbial Mechanisms of Pathogenicity 437

Chapter 15 continues

Chapter 19
Disorders Associated with the Immune System 529

Chapter 20
Antimicrobial Drugs 559

Chapter 1

The Microbial World and You

NORMAL MICROBIOTA OF THE HUMAN MOUTH. *Normal microbiota in and on the human body can provide protection against harmful microorganisms.*

The overall theme of this textbook is the relationship between microbes and our lives. This relationship involves not only the familiar harmful effects of certain microorganisms, such as disease and food spoilage, but also their many beneficial effects. In this chapter we will introduce you to some of the many ways microbes affect our lives. They have been fruitful subjects of study for a number of years, as you will see in the short history of microbiology that opens the chapter. We then discuss the incredible diversity of microorganisms and their ecological importance in maintaining balance in the environment by recycling chemical elements such as carbon and nitrogen between the soil and the atmosphere. We will also examine how microbes are used in commercial and industrial applications to produce foods, chemicals, and drugs (such as penicillin) and to treat sewage, control pests, and clean up pollutants. Finally, we will discuss microbes as the cause of such diseases as West Nile encephalitis, AIDS, mad cow disease, diarrhea, and hemorrhagic fever.

Microbes in Our Lives

Learning Objective

• List several ways in which microbes affect our lives.

For many people, the words *germ* and *microbe* bring to mind a group of tiny creatures that do not quite fit into any of the categories in that old question, "Is it animal, vegetable, or mineral?" Microbes, also called microorganisms, are minute living things that individually are usually too small to be seen with the unaided eye. The group includes bacteria (Chapter 11), fungi (yeasts and molds), protozoa, and microscopic algae (Chapter 12). It also includes viruses, those noncellular entities sometimes regarded as being at the border between life and nonlife (Chapter 13). You will be introduced to each of these groups of microbes shortly.

We tend to associate these small organisms only with major diseases such as AIDS, uncomfortable infections, or such common inconveniences as spoiled food. However, the majority of microorganisms make crucial contributions to the welfare of the world's inhabitants by helping to maintain the balance of living organisms and chemicals in our environment. Marine and freshwater microorganisms form the basis of the food chain in oceans, lakes, and rivers. Soil microbes help break down wastes and incorporate nitrogen gas from the air into organic compounds, thereby recycling chemical elements in the soil, water, and air. Certain microbes play important roles in photosynthesis, a food- and oxygen-generating process that is critical to life on Earth. Humans and many other animals depend on the microbes in their intestines for digestion and the synthesis of some vitamins that their bodies require, including some B vitamins for metabolism and vitamin K for blood clotting.

Microorganisms also have many commercial applications. They are used in the synthesis of such chemical products as acetone, organic acids, enzymes, alcohols, and many drugs. The process by which microbes produce acetone and butanol was discovered in 1914 by Chaim Weizmann, a Russian-born chemist working in England. When World War I broke out in August of that year, the production of acetone was very important for making cordite (gunpowder) for use in the manufacture of munitions. Weizmann's discovery played a significant role in determining the outcome of the war.

The food industry also uses microbes in producing vinegar, sauerkraut, pickles, alcoholic beverages, green olives, soy sauce, buttermilk, cheese, yogurt, and bread (see the box on page 3). In addition, enzymes from microbes can now be manipulated such that the microbes produce substances they normally do not synthesize. These substances include cellulose, digestive aids, and drain cleaner, plus important therapeutic substances such as insulin.

Though only a minority of microorganisms are **pathogenic** (disease-producing), practical knowledge of microbes is necessary for medicine and the related health sciences. For example, hospital workers must be able to protect patients from common microbes that are normally harmless but pose a threat to the sick and injured.

Today we understand that microorganisms are found almost everywhere. Yet not long ago, before the invention of the microscope, microbes were unknown to scientists. Thousands of people died in devastating epidemics, the causes of which were not understood. Food spoilage often could not be controlled, and entire families died because vaccinations and antibiotics were not available to fight infections.

We can get an idea of how our current concepts of microbiology developed by looking at a few of the historic milestones in microbiology that have changed our lives. Before doing this, however, we will first take a look at the major groups of microbes and how they are named and classified.

Naming and Classifying Microorganisms

Nomenclature

Learning Objective

• Recognize the system of scientific nomenclature that uses two names: a genus and a specific epithet.

The system of nomenclature (naming) for organisms in use today was established in 1735 by Carolus Linnaeus. Scientific names are latinized because Latin was the language traditionally used by scholars. Scientific nomenclature assigns each organism two names—the **genus** (plural: **genera**) is the first name and is always capitalized; the **specific epithet** (**species** name) follows and is not capitalized. The organism is referred to by both the genus and the specific epithet, and both names are underlined or italicized. By custom, after a scientific name has been mentioned once, it can be abbreviated with the initial of the genus followed by the specific epithet.

Scientific names can, among other things, describe an organism, honor a researcher, or identify the habitat of a species. For example, consider *Staphylococcus aureus* (staf-i-lō-kok′kus ô′rē-us), a bacterium commonly found on human skin. *Staphylo-* describes the clustered arrangement of the cells; *coccus* indicates that they are shaped like spheres. The specific epithet, *aureus*, is Latin for golden, the color of many colonies of this bacterium. The genus of the bacterium *Escherichia coli* (esh-ėr-i′kē-ä kō′lī or kō′lē) is named for a scientist, Theodor Escherich, whereas its specific epithet, *coli*, reminds us that *E. coli* live in the colon, or large intestine.

Types of Microorganisms

Learning Objective

• Differentiate among the major characteristics of each group of microorganisms.

Bacteria

Bacteria (singular: *bacterium*) are relatively simple, single-celled (unicellular) organisms. Because their genetic material is not enclosed in a special nuclear membrane, bacterial cells are called **prokaryotes** (prō-kar′e-ōts), from Greek words meaning prenucleus. Prokaryotes include both the bacteria and the archaea.

Bacterial cells generally appear in one of several shapes. *Bacillus* (bä-sil′lus) (rodlike), illustrated in Figure 1.1a, *coccus* (spherical or ovoid), and *spiral* (corkscrew or curved) are among the most common shapes, but some bacteria are star-shaped or square (see Figures 4.1–4.5 on pages 76–78). Individual bacteria may form pairs, chains, clusters, or other groupings; such formations are usually characteristic of a particular genus or species of bacteria.

Bacteria are enclosed in cell walls that are largely composed of a carbohydrate and protein complex called *peptidoglycan*. (By contrast, cellulose is the main substance of plant and algal cell walls.) Bacteria generally reproduce by dividing into two equal cells; this process is called *binary fission*. For nutrition, most bacteria use organic chemicals, which in nature can be derived from either dead or living organisms. Some bacteria can manufacture their own food

APPLICATIONS OF MICROBIOLOGY

What Makes Sourdough Bread Different?

Imagine being a miner during the California gold rush. You've just made bread dough from your last supplies of flour and salt when someone yells, "Gold!" Temporarily forgetting your hunger, you run off to the gold fields. Many hours later you return. The dough has been rising longer than usual, but you are too cold, tired, and hungry to start a new batch. Later you find that your bread tastes different from previous batches; it is slightly sour. During the gold rush, miners baked so many sour loaves that they were nicknamed "sourdoughs."

Conventional bread is made from flour, water, sugar, salt, shortening, and a living microbe, yeast. The yeast belongs to the Kingdom Fungi and is named *Saccharomyces cerevisiae*. When flour is mixed with water, an enzyme in the flour breaks its starch into two sugars, maltose and glucose. After the ingredients for the bread are mixed, the yeast metabolizes the sugars and produces alcohol (ethanol) and carbon dioxide as waste products. This metabolic process is called *fermentation*. The dough rises as carbon dioxide bubbles get trapped in the sticky matrix. The alcohol, which evaporates during baking, and the carbon dioxide gas form spaces that remain in the bread.

Originally, breads were leavened by wild yeast from the air, which had been trapped in the dough. Later, bakers kept a starter culture of yeast—dough from the last batch of bread—to leaven each new batch of dough. Sourdough bread is made with a special sourdough starter culture that is added to flour, water, and salt. Perhaps the most famous sourdough bread made today comes from San Francisco, where a handful of bakeries have continuously cultivated their

starters for more than 150 years and have meticulously maintained the starters to keep out unwanted microbes that can produce different and undesirable flavors. After bakeries in other areas made several unsuccessful attempts to match the unique flavor of San Francisco sourdough, rumors attributed the taste to a unique local climate or contamination from bakery walls. Ted F. Sugihara and Leo Kline from the United States Department of Agriculture (USDA) set out to debunk these rumors and to determine the microbiological basis for the bread's different taste so that it could be made in other areas.

The USDA workers found that sourdough is eight to ten times more acidic than conventional bread because of the presence of lactic and acetic acids. These acids account for the sour flavor of the bread. The workers isolated and identi-

fied the yeast in the starter as *Saccharomyces exiguus*, a unique yeast that does not ferment maltose and thrives in the acidic environment of this dough. The sourdough question was not answered, however, because the yeast did not produce the acids and did not use maltose. Sugihara and Kline searched the starter for a second agent capable of fermenting maltose and producing the acids (see the figure). The bacterium they isolated, so carefully guarded all those years, was classified into the genus *Lactobacillus*. Many members of this genus are used in dairy fermentations and are found naturally on humans and other mammals. Analyses of cell structure and genetic composition showed that the sourdough bacterium is genetically different from other previously characterized lactobacilli. It has been given the name *Lactobacillus sanfrancisco*.

FIGURE 1.1 **Types of microorganisms. (a)** The rod-shaped bacterium *Haemophilus influenzae,* one of the bacterial causes of pneumonia. **(b)** *Mucor,* a common bread mold, is a type of fungus. When released from sporangia (round structures), spores land on a favorable surface and germinate into a network of hyphae (threads) that absorb nutrients. **(c)** An amoeba, a protozoan, approaching food. **(d)** The pond alga, *Volvox.* **(e)** Several human immuno-deficiency viruses (HIVs), the causative agents of AIDS (red spheres), budding from a CD4 lymphocyte.

- The classification of organisms on the basis of cellular structure includes Bacteria, Archaea, and Eukarya (Protists, Fungi, Plants, and Animals).

by photosynthesis, and some can derive nutrition from inorganic substances. Many bacteria can "swim" by using moving appendages called *flagella*. (For a complete discussion of bacteria, see Chapter 11.)

Archaea

Like bacteria, **archaea** (är′kē-ä) consist of prokaryotic cells, but if they have cell walls, the walls lack peptidoglycan. Archaea, often found in extreme environments, are divided into three main groups. The *methanogens* produce methane as a waste product from respiration. The *extreme halophiles* live in extremely salty environments such as the Great Salt Lake and the Dead Sea. The *extreme thermophiles* live in hot sulfurous water such as hot springs at Yellowstone National Park. Archaea are not known to cause disease in humans.

Fungi

Fungi (singular: **fungus**) are **eukaryotes** (yū-kar′ē-ōts), organisms whose cells have a distinct nucleus containing the cell's genetic material (DNA), surrounded by a special envelope called the nuclear membrane. Organisms in the Kingdom Fungi may be unicellular or multicellular (see Chapter 12, page 334). Large multicellular fungi, such as mushrooms, may look somewhat like plants, but they cannot carry out photosynthesis, as most plants can. True fungi have cell walls composed primarily of a substance called *chitin*. The unicellular forms of fungi, *yeasts*, are oval microorganisms that are larger than bacteria. The most typical fungi are *molds* (Figure 1.1b). Molds form visible masses called *mycelia*, which are composed of long filaments (*hyphae*) that branch and intertwine. The cottony growths sometimes found on bread and fruit are mold mycelia. Fungi can reproduce sexually or asexually. They obtain nourishment by absorbing solutions of organic material from their environment—whether soil, seawater, fresh water, or an animal or plant host. Organisms called *slime molds* have characteristics of both fungi and amoeba. They are discussed in detail in Chapter 12.

Protozoa

Protozoa (singular: **protozoan**) are unicellular, eukaryotic microbes (see Chapter 12, page 352). Protozoa move by pseudopods, flagella, or cilia. Amoebas (Figure 1.1c) move by using extensions of their cytoplasm called *pseudopods* (false feet). Other protozoa have long *flagella* or numerous shorter appendages for locomotion called *cilia*. Protozoa have a variety of shapes and live either as free entities or as parasites (organisms that derive nutrients from living hosts) that absorb or ingest organic compounds from their environment. Protozoa can reproduce sexually or asexually.

Algae

Algae (singular: **alga**) are photosynthetic eukaryotes with a wide variety of shapes and both sexual and asexual reproductive forms (Figure 1.1d). The algae of interest to microbiologists are usually unicellular (see Chapter 12, page 347). The cell walls of many algae, like those of plants, are composed of *cellulose*. Algae are abundant in fresh and salt water, in soil, and in association with plants. As photosynthesizers, algae need light and air for food production and growth, but they do not generally require organic compounds from the environment. As a result of photosynthesis, algae produce oxygen and carbohydrates, which are utilized by other organisms, including animals. Thus, they play an important role in the balance of nature.

Viruses

Viruses (Figure 1.1e) are very different from the other microbial groups mentioned here. They are so small that most can be seen only with an electron microscope, and they are acellular (not cellular). Structurally very simple, a virus particle contains a core made of only one type of nucleic acid, either DNA or RNA. This core is surrounded by a protein coat. Sometimes the coat is encased by an additional layer, a lipid membrane called an envelope. All living cells have RNA *and* DNA, can carry out chemical reactions, and can reproduce as self-sufficient units. Viruses can reproduce only by using the cellular machinery of other organisms. Thus, on the one hand viruses are considered to be living when they multiply within host cells they infect. In this sense, viruses are parasites of other forms of life. On the other hand, viruses are not considered to be living because outside of living hosts they are inert. (Viruses will be discussed in detail in Chapter 13.)

Multicellular Animal Parasites

Although multicellular animal parasites are not strictly microorganisms, they are of medical importance and therefore will be discussed in this text. The two major groups of parasitic worms are the flatworms and the roundworms, collectively called **helminths** (see Chapter 12, page 361). During some stages of their life cycle, helminths are microscopic in size. Laboratory identification of these organisms includes many of the same techniques used for the identification of microbes.

Classification of Microorganisms

Before the existence of microbes was known, all organisms were grouped into either the animal kingdom or the plant kingdom. When microscopic organisms with characteristics of animals or plants were discovered late in the seventeenth century, a new system of classification was needed. Still, biologists could not agree on the criteria for classifying the new organisms they were seeing until the late 1960s.

In 1978, Carl Woese devised a system of classification based on the cellular organization of organisms. It groups all organisms in three domains as follows:

1. Bacteria (cell walls contain peptidoglycan)
2. Archaea (cell walls, if present, lack peptidoglycan)
3. Eukarya, which includes the following:

   - Protists (slime molds, protozoa, and algae)
   - Fungi (unicellular yeasts, multicellular molds, and mushrooms)
   - Plants (includes mosses, ferns, conifers, and flowering plants)
   - Animals (includes sponges, worms, insects, and vertebrates)

Classification will be discussed in more detail in Chapters 10–13.

A Brief History of Microbiology

The science of microbiology dates back only a few hundred years, yet the recent discovery of *Mycobacterium tuberculosis* (mī-kō-bak-ti′rē-um tü-bėr-kū-lō′sis) DNA in 3000-year-old Egyptian mummies reminds us that microorganisms have been around for much longer. In fact, bacterial ancestors were the first living cells to appear on Earth. While we know relatively little about what earlier people thought about the causes, transmission, and treatment of disease, the history of the past few hundred years is better known. Let's look now at some key developments in microbiology that have helped the field progress to its current high-technology state.

The First Observations

Learning Objective

- *Explain the importance of observations made by Hooke and van Leeuwenhoek.*

One of the most important discoveries in the history of biology occurred in 1665 with the help of a relatively crude microscope. An Englishman, Robert Hooke, after observing a thin slice of cork reported to the world that life's smallest structural units were "little boxes," or "cells," as he called them. Using his improved version of a compound microscope (one that uses two sets of lenses), Hooke was able to see individual cells. Hooke's discovery marked the beginning of the **cell theory**—the theory that *all living things are composed of cells*. Subsequent investigations into the structure and functions of cells were based on this theory.

Though Hooke's microscope was capable of showing cells, he lacked the staining techniques that would have allowed him to see microbes clearly. The Dutch merchant and amateur scientist Antoni van Leeuwenhoek was probably the first to actually observe live microorganisms through magnifying lenses. Between 1673 and 1723, he wrote a series of letters to the Royal Society of London describing the "animalcules" he saw through his simple, single-lens microscope. Van Leeuwenhoek made detailed drawings of "animalcules" in rainwater, in liquid in which peppercorns had soaked, and in material scraped from his teeth. These drawings have since been identified as representations of bacteria and protozoa (Figure 1.2).

The Debate Over Spontaneous Generation

Learning Objectives

- *Compare spontaneous generation and biogenesis.*
- *Identify the contributions to microbiology made by Needham, Spallanzani, Virchow, and Pasteur.*

After van Leeuwenhoek discovered the previously "invisible" world of microorganisms, the scientific community of the time became interested in the origins of these tiny living things. Until the second half of the nineteenth century, many scientists and philosophers believed that some forms of life could arise spontaneously from nonliving matter; they called this hypothetical process **spontaneous generation.** Not much more than 100 years ago, people commonly believed that toads, snakes, and mice could be born of moist soil; that flies could emerge from manure; and that maggots, the larvae of flies, could arise from decaying corpses.

Evidence Pro and Con

A strong opponent of spontaneous generation, the Italian physician Francesco Redi, set out in 1668 (even before van Leeuwenhoek's discovery of microscopic life) to demonstrate that maggots did not arise spontaneously from decaying meat. Redi filled three jars with decaying meat and sealed them tightly. Then he arranged three other jars similarly but left them open. Maggots appeared in the open vessels after flies entered the jars and laid their eggs, but the sealed containers showed no signs of maggots. Still, Redi's antagonists were not convinced; they claimed that fresh air was needed for spontaneous generation. So Redi set up a second experiment, in which three jars were covered with a fine net instead of being sealed. No larvae appeared in the gauze-covered jars, even though air was present. Maggots appeared only when flies were allowed to leave their eggs on the meat.

(a) Van Leeuwenhoek using his microscope

Location of specimen

Lens

(b) Microscope replica

(c) Drawings of bacteria

FIGURE 1.2 Antoni van Leeuwenhoek's microscopic observations. (a) By holding his microscope toward a source of light, van Leeuwenhoek was able to observe living organisms too small to be seen with the unaided eye. **(b)** The specimen was placed on the tip of the adjustable point and viewed from the other side through the tiny, nearly spherical lens. The highest magnification possible with his microscopes was about 300× (times). **(c)** Some of van Leeuwenhoek's drawings of bacteria, made in 1683. The letters represent various shapes of bacteria. C–D represents a path of motion he observed.

■ Van Leeuwenhoek was the first person to view microorganisms that we now call bacteria and protozoa.

Redi's results were a serious blow to the long-held belief that large forms of life could arise from nonlife. However, many scientists still believed that small organisms, such as van Leeuwenhoek's "animalcules," were simple enough to be generated from nonliving materials.

The case for spontaneous generation of microorganisms seemed to be strengthened in 1745 when John Needham, an Englishman, found that even after he heated nutrient fluids (chicken broth and corn broth) before pouring them into covered flasks, the cooled solutions were soon teeming with microorganisms. Needham claimed that microbes developed spontaneously from the fluids. Twenty years later, Lazzaro Spallanzani, an Italian scientist, suggested that microorganisms from the air probably had entered Needham's solutions after they were boiled. Spallanzani showed that nutrient fluids heated *after* being sealed in a flask did not develop microbial growth. Needham responded by claiming the "vital force" necessary for spontaneous generation had been destroyed by the heat and was kept out of the flasks by the seals.

This intangible "vital force" was given all the more credence shortly after Spallanzani's experiment, when Laurent Lavoisier showed the importance of oxygen to life. Spallanzani's observations were criticized on the grounds that there was not enough oxygen in the sealed flasks to support microbial life.

The Theory of Biogenesis

The issue was still unresolved in 1858, when the German scientist Rudolf Virchow challenged spontaneous generation with the concept of **biogenesis,** the claim that living cells can arise only from preexisting living cells. Arguments about spontaneous generation continued until 1861, when the issue was resolved by the French scientist Louis Pasteur.

With a series of ingenious and persuasive experiments, Pasteur demonstrated that microorganisms are present in the air and can contaminate sterile solutions, but air itself does not create microbes. He filled several short-necked flasks with beef broth and then boiled their contents. Some were then left open and allowed to cool. In a few days, these

FIGURE 1.3 **Pasteur's experiment disproving the theory of sponta-neous generation.** ❶ Pasteur first poured beef broth into a long-necked flask. ❷ Next he heated the neck of the flask and bent it into an S-shaped curve; then he boiled the broth for several minutes. ❸ Microorganisms did not appear in the cooled solution, even after long periods, as you can see in this recent photograph of an actual flask used by Pasteur in a similar experiment.

■ Pasteur's discoveries form the basis for aseptic techniques, procedures that prevent contamination by unwanted organisms.

flasks were found to be contaminated with microbes. The other flasks, sealed after boiling, were free of microorganisms. From these results, Pasteur reasoned that microbes in the air were the agents responsible for contaminating non-living matter such as the broths in Needham's flasks.

Pasteur next placed broth in open-ended long-necked flasks and bent the necks into S-shaped curves (Figure 1.3). The contents of these flasks were then boiled and cooled. The broth in the flasks did not decay and showed no signs of life, even after months. Pasteur's unique design allowed air to pass into the flask, but the curved neck trapped any airborne microorganisms that might contaminate the broth. (Some of these original vessels, which were later sealed, are on display at the Pasteur Institute in Paris. Like the flask shown in Figure 1.3, they still show no sign of contamination more than 100 years later.)

Pasteur showed that microorganisms can be present in nonliving matter—on solids, in liquids, and in the air. Furthermore, he demonstrated conclusively that microbial life can be destroyed by heat and that methods can be devised to block the access of airborne microorganisms to nutrient environments. These discoveries form the basis of **aseptic techniques,** techniques that prevent contamination by

unwanted microorganisms, which are now the standard practice in laboratory and many medical procedures. Modern aseptic techniques are among the first and most important things that a beginning microbiologist learns.

Pasteur's work provided evidence that microorganisms cannot originate from mystical forces present in nonliving materials. Rather, any appearance of "spontaneous" life in nonliving solutions can be attributed to microorganisms that were already present in the air or in the fluids themselves. Scientists now believe that a form of spontaneous generation probably did occur on the primitive Earth when life first began, but they agree that this does not happen under today's environmental conditions.

The Golden Age of Microbiology

Learning Objectives

- Identify the importance of Koch's postulates.
- Explain how Pasteur's work influenced Lister and Koch.
- Identify the importance of Jenner's work.

For about 60 years, beginning with the work of Pasteur, there was an explosion of discoveries in microbiology. The period from 1857 to 1914 has been appropriately named the Golden Age of Microbiology. During this period, rapid advances, spearheaded mainly by Pasteur and Robert Koch, led to the establishment of microbiology as a science. Discoveries during these years included both the agents of many diseases and the role of immunity in the prevention and cure of disease. During this productive period, microbiologists studied the chemical activities of microorganisms, improved the techniques for performing microscopy and culturing microorganisms, and developed vaccines and surgical techniques. Some of the major events that occurred during the Golden Age of Microbiology are listed in Figure 1.4.

Fermentation and Pasteurization

One of the key steps that established the relationship between microorganisms and disease occurred when a group of French merchants asked Pasteur to find out why wine and beer soured. They hoped to develop a method that would prevent spoilage when those beverages were shipped long distances. At the time, many scientists believed that air converted the sugars in these fluids into alcohol. Pasteur found instead that microorganisms called yeasts convert the sugars to alcohol in the absence of air. This process, called **fermentation** (see Chapter 5, page 132), is used to make wine and beer. Souring and spoilage are caused by different microorganisms called bacteria. In the presence of air, bacteria change the alcohol in the beverage into vinegar (acetic acid).

Pasteur's solution to the spoilage problem was to heat the beer and wine just enough to kill most of the bacteria that caused the spoilage; the process, called **pasteurization,** is now commonly used to kill spoilage and potentially harmful bacteria in milk as well as in some alcoholic drinks. Showing the connection between spoilage of food and microorganisms was a major step toward establishing the relationship between disease and microbes.

The Germ Theory of Disease

As we have seen, the fact that many kinds of diseases are related to microorganisms was unknown until relatively recently. Before the time of Pasteur, effective treatments for many diseases were discovered by trial and error, but the causes of the diseases were unknown.

The realization that yeasts play a crucial role in fermentation was the first link between the activity of a microorganism and physical and chemical changes in organic materials. This discovery alerted scientists to the possibility that microorganisms might have similar relationships with plants and animals—specifically, that microorganisms might cause disease. This idea was known as the **germ theory of disease.**

The germ theory was a difficult concept for many people to accept at that time because for centuries disease was believed to be punishment for an individual's crimes or misdeeds. When the inhabitants of an entire village became ill, people often blamed the disease on demons appearing as foul odors from sewage or on poisonous vapors from swamps. Most people born in Pasteur's time found it inconceivable that "invisible" microbes could travel through the air to infect plants and animals, or remain on clothing and bedding to be transmitted from one person to another. But gradually scientists accumulated the information needed to support the new germ theory.

In 1865, Pasteur was called upon to help fight silkworm disease, which was ruining the silk industry throughout Europe. Years earlier, in 1835, Agostino Bassi, an amateur microscopist, had proved that another silkworm disease was caused by a fungus. Using data provided by Bassi, Pasteur found that the more recent infection was caused by a protozoan, and he developed a method for recognizing afflicted silkworm moths.

In the 1860s, Joseph Lister, an English surgeon, applied the germ theory to medical procedures. Lister was aware that in the 1840s, the Hungarian physician Ignaz Semmelweis had demonstrated that physicians, who at the time did not disinfect their hands, routinely transmitted infections (puerperal, or childbirth, fever) from one obstetrical patient to another. Lister had also heard of Pasteur's work connecting microbes to animal diseases. Disinfectants were not used at the time, but Lister knew that phenol (carbolic acid) kills bacteria, so he began treating surgical wounds with a phenol solution. The practice so reduced the incidence of infections and deaths that other surgeons quickly adopted it. Lister's technique was one of the earliest medical attempts to control infections caused by microorganisms. In fact, his findings proved that microorganisms cause surgical wound infections.

The first proof that bacteria actually cause disease came from Robert Koch in 1876. Koch, a German physician, was Pasteur's young rival in the race to discover the cause of anthrax, a disease that was destroying cattle and sheep in Europe. Koch discovered rod-shaped bacteria now known as *Bacillus anthracis* (bä-sil'lus an-thrā'sis) in the blood of cattle that had died of anthrax. He cultured the bacteria on nutrients and then injected samples of the culture into healthy animals. When these animals became sick and died, Koch isolated the bacteria in their blood and compared them with the bacteria originally isolated. He found that the two sets of blood cultures contained the same bacteria.

Koch thus established a sequence of experimental steps for directly relating a specific microbe to a specific disease. These steps are known today as **Koch's postulates** (see Figure 14.3 on page 414). During the past 100 years, these

1665	Hooke—First observation of cells
1673	van Leeuwenhoek—First observation of live microorganisms
1735	Linnaeus—Nomenclature for organisms
1798	Jenner—First vaccine
1835	Bassi—Silkworm fungus
1840	Semmelweis—Childbirth fever
1853	DeBary—Fungal plant disease

Louis Pasteur (1822–1895)

1857	Pasteur—Fermentation
1861	Pasteur—Disproved spontaneous generation
1864	Pasteur—Pasteurization
1867	Lister—Aseptic surgery
1876	*Koch—Germ theory of disease
1879	Neisser—*Neisseria gonorrhoeae*
1881	*Koch—Pure cultures
	Finley—Yellow fever
1882	*Koch—*Mycobacterium tuberculosis*
	Hess—Agar (solid) media

GOLDEN AGE OF MICROBIOLOGY

1883	*Koch—*Vibrio cholerae*
1884	*Metchnikoff—Phagocytosis
	Gram—Gram-staining procedure
	Escherich—*Escherichia coli*
1887	Petri—Petri dish
1889	Kitasato—*Clostridium tetani*
1890	*von Bering—Diphtheria antitoxin
	*Ehrlich—Theory of immunity
1892	Winogradsky—Sulfur cycle
1898	Shiga—*Shigella dysenteriae*
1910	Chagas—*Trypanosoma cruzi*; *Ehrlich—Syphilis

Robert Koch (1843–1910)

1928	*Fleming, Chain, Florey—Penicillin
	Griffith—Transformation in bacteria
1934	Lancefield—Streptococcal antigens
1935	*Stanley, Northrup, Sumner—Crystallized virus
1941	Beadle and Tatum—Relationship between genes and enzymes
1943	*Delbrück and Luria—Viral infection of bacteria
1944	Avery, MacLeod, McMarty—Genetic material is DNA
1946	Lederberg and Tatum—Bacterial conjugation
1953	*Watson and Crick—DNA structure
1957	*Jacob and Monod—Protein synthesis regulation
1959	Stewart—Viral cause of cancer
1962	*Edelman and Porter—Antibodies
1964	Epstein, Achong, Barr—Epstein-Barr virus as cause of human cancer
1971	*Nathans, Smith, Arber—Restriction enzymes (used for genetic engineering)
1973	Berg, Boyer, Cohen—Genetic engineering
1975	Dulbecco, Temin, Baltimore—Reverse transcriptase
1978	Woese—Archaea; *Arber, Smith, Nathans—Restriction endonucleases
	*Mitchell—Chemiosmotic mechanism
1981	Margulis—Origin of eukaryotic cells
1982	*Klug—Structure of tobacco mosaic virus
1983	*McClintock—Transposons

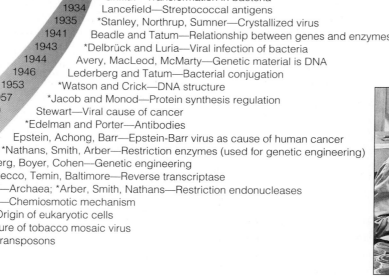

Rebecca C. Lancefield (1895–1981)

1988	*Deisenhofer, Huber, Michel—Bacterial photosynthesis pigments
1994	Cano—Reported to have cultured 40-million-year-old bacteria
1997	*Prusiner—Prions

FIGURE 1.4 Milestones in microbiology, highlighting those that occurred during the Golden Age of Microbiology. An asterisk indicates a Nobel laureate.

■ The Golden Age of Microbiology is so named because during this period, numerous discoveries led to the establishment of microbiology as a science.

same criteria have been invaluable in investigations proving that specific microorganisms cause many diseases. Koch's postulates, their limitations, and their application to disease will be discussed in greater detail in Chapter 14.

Vaccination

Often a treatment or preventive procedure is developed before scientists know why it works. The smallpox vaccine is an example of this. On May 4, 1796, almost 70 years before Koch established that a specific microorganism causes anthrax, Edward Jenner, a young British physician, embarked on an experiment to find a way to protect people from smallpox.

Smallpox epidemics were greatly feared. The disease periodically swept through Europe, killing thousands, and it wiped out 90% of the Native Americans on the East Coast when European settlers first brought the infection to the New World.

When a young milkmaid informed Jenner that she couldn't get smallpox because she already had been sick from cowpox—a much milder disease—he decided to put the girl's story to the test. First Jenner collected scrapings from cowpox blisters. Then he inoculated a healthy 8-year-old volunteer with the cowpox material by scratching the person's arm with a pox-contaminated needle. The scratch turned into a raised bump. In a few days, the volunteer became mildly sick but recovered and never again contracted either cowpox or smallpox. The process was called *vaccination*, from the Latin word *vacca*, meaning cow. Pasteur gave it this name in honor of Jenner's work. The protection from disease provided by vaccination (or by recovery from the disease itself) is called **immunity.** We will discuss the mechanisms of immunity in Chapter 17.

Years after Jenner's experiment, in about 1880, Pasteur discovered why vaccinations work. He found that the bacterium that causes fowl cholera lost its ability to cause disease (lost its *virulence*, or became *avirulent*) after it was grown in the laboratory for long periods. However, it—and other microorganisms with decreased virulence—was able to induce immunity against subsequent infections by its virulent counterparts. The discovery of this phenomenon provided a clue to Jenner's successful experiment with cowpox. Both cowpox and smallpox are caused by viruses. Even though cowpox virus is not a laboratory-produced derivative of smallpox virus, it is so closely related to the smallpox virus that it can induce immunity to both viruses. Pasteur used the term *vaccine* for cultures of avirulent microorganisms used for preventive inoculation.

Jenner's experiment marked the first time in a Western culture that a living viral agent—the cowpox virus—was used to produce immunity. Physicians in China had immunized patients by removing scales from drying pustules of a person suffering from a mild case of smallpox, grinding the scales to a fine powder, and inserting the powder into the nose of the person to be protected.

Some vaccines are still produced from avirulent microbial strains that stimulate immunity to the related virulent strain. Other vaccines are made from killed virulent microbes, from isolated components of virulent microorganisms, or by genetic engineering techniques.

The Birth of Modern Chemotherapy: Dreams of a "Magic Bullet"

Learning Objective

- *Identify the contributions to microbiology made by Ehrlich and Fleming.*

After the relationship between microorganisms and disease was established, medical microbiologists next focused on the search for substances that could destroy pathogenic microorganisms without damaging the infected animal or human. Treatment of disease by using chemical substances is called **chemotherapy.** (The term also commonly refers to chemical treatment of noninfectious diseases, such as cancer.) Chemotherapeutic agents prepared from chemicals in the laboratory are called **synthetic drugs.** Chemicals produced naturally by bacteria and fungi to act against other microorganisms are called **antibiotics.** The success of chemotherapy is based on the fact that some chemicals are more poisonous to microorganisms than to the hosts infected by the microbes. Antimicrobial therapy will be discussed in further detail in Chapter 20.

The First Synthetic Drugs

Paul Ehrlich, a German physician, was the imaginative thinker who fired the first shot in the chemotherapy revolution. As a medical student, Ehrlich speculated about a "magic bullet" that could hunt down and destroy a pathogen without harming the infected host. Ehrlich launched a search for such a bullet. In 1910, after testing hundreds of substances, he found a chemotherapeutic agent called *salvarsan*, an arsenic derivative effective against syphilis. The agent was named salvarsan because it was considered to offer salvation from syphilis and it contained arsenic. Before this discovery, the only known chemical in Europe's medical arsenal was an extract from the bark of a South American tree, *quinine*, which had been used by Spanish conquistadors to treat malaria.

By the late 1930s, researchers had developed several other synthetic drugs that could destroy microorganisms. Most of these drugs were derivatives of dyes. This came about because the dyes synthesized and manufactured for fabrics

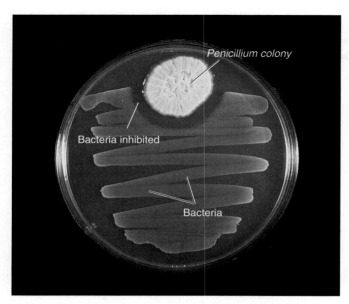

FIGURE 1.5 **Discovering antibiotics.** The antibiotic secreted by the *Penicillium* fungus during its growth inhibited growth of *Staphylococcus aureus* bacteria.

■ New antibiotic-producing organisms are discovered by looking for areas of inhibition around colonies.

were routinely tested for antimicrobial qualities by microbiologists looking for a "magic bullet." In addition, *sulfonamides* (sulfa drugs) were synthesized at about the same time.

A Fortunate Accident—Antibiotics

In contrast to the sulfa drugs, which were deliberately developed from a series of industrial chemicals, the first antibiotic was discovered by accident. Alexander Fleming, a Scottish physician and bacteriologist, almost tossed out some culture plates that had been contaminated by mold. Fortunately, he took a second look at the curious pattern of growth on the contaminated plates. Around the mold was a clear area where bacterial growth had been inhibited (Figure 1.5). Fleming was looking at a mold that could inhibit the growth of a bacterium. The mold was later identified as *Penicillium notatum* (pen-i-sil′lē-um nō-tä′tum), and in 1928 Fleming named the mold's active inhibitor *penicillin*. Thus, penicillin is an antibiotic produced by a fungus. The enormous usefulness of penicillin was not apparent until the 1940s, when it was finally tested clinically and mass-produced.

Since the early discoveries of antibiotics, thousands of others have been discovered. Unfortunately, antibiotics and other chemotherapeutic drugs are not without problems. Many antimicrobial chemicals are too toxic to humans for practical use; they kill the pathogenic microbes, but they also damage the infected host. For reasons we will

discuss later, toxicity to humans is a particular problem in the development of drugs for treating viral diseases. Viral growth is dependent on life processes of normal host cells. Thus, there are very few successful antiviral drugs because a drug that would interfere with viral reproduction would also likely affect uninfected cells of the body.

Another major problem associated with antimicrobial drugs is the emergence and spread of new varieties of microorganisms that are resistant to antibiotics. Over the years, more and more microbes have developed resistance to antibiotics that at one time were very effective against them. Drug resistance results from genetic changes in microbes that enables them to tolerate a certain amount of an antibiotic that would normally inhibit them. These changes might include the production by microbes of chemicals (enzymes) that inactivate antibiotics, changes in the surface of a microbe that prevent an antibiotic from attaching to it, and prevention of an antibiotic from entering the microbe.

The recent appearance of vancomycin-resistant *Staphylococcus aureus* and *Enterococcus faecalis* has alarmed health care professionals because it indicates that some previously treatable bacterial infections may soon be impossible to treat with antibiotics.

Modern Developments in Microbiology

Learning Objectives

- *Define bacteriology, mycology, parasitology, immunology, and virology.*
- *Explain the importance of recombinant DNA technology.*

The quest to solve drug resistance, identify viruses, and develop vaccines requires sophisticated research techniques and correlated studies that were never dreamed of in the days of Koch and Pasteur.

The groundwork laid during the Golden Age of Microbiology provided the basis for several monumental achievements during the twentieth century (Table 1.1). New branches of microbiology were developed, including immunology and virology. Most recently, the development of a set of new methods called recombinant DNA technology has revolutionized research and practical applications in all areas of microbiology.

Bacteriology, Mycology, and Parasitology

Bacteriology, the study of bacteria, began with van Leeuwenhoek's first examination of tooth scrapings. New pathogenic bacteria are still discovered regularly. Many bacteriologists, like their predecessor Pasteur, look at the roles of bacteria in food and the environment. One intriguing discovery came in 1997, when Heide Schulz dis-

TABLE 1.1	**Selected Nobel Prizes Awarded for Research in Microbiology**

Nobel Laureates	Year of Presentation	Country of Birth	Contribution
Emil A. von Behring	1901	Germany	Developed a diphtheria antitoxin.
Ronald Ross	1902	England	Discovered how malaria is transmitted.
Robert Koch	1905	Germany	Cultured tuberculosis bacteria.
Paul Ehrlich	1908	Germany	Developed theories on immunity.
Elie Metchnikoff	1908	Russia	Described phagocytosis, the intake of solid materials by cells.
Alexander Fleming, Ernst Chain, and Howard Florey	1945	Scotland England	Discovered penicillin.
Selman A. Waksman	1952	Ukraine	Discovered streptomycin.
Hans A. Krebs	1953	Germany	Discovered chemical steps of the Krebs cycle in carbohydrate metabolism.
John F. Enders, Thomas H. Weller, and Frederick C. Robbins	1954	United States	Cultured poliovirus in cell cultures.
Joshua Lederberg, George Beadle, and Edward Tatum	1958	United States	Described genetic control of biochemical reactions.
James D. Watson, Frances H. C. Crick, and Maurice A. F. Wilkins	1962	United States England New Zealand	Identified the physical structure of DNA.
François Jacob, Jacques Monod, and André Lwoff	1965	France	Described how protein synthesis is regulated in bacteria.
Robert Holley, Har Gobind Khorana, and Marshall W. Nirenberg	1968	United States India United States	Discovered the genetic code for amino acids.
Max Delbrück, Alfred D. Hershey, and Salvador E. Luria	1969	Germany United States Italy	Described the mechanism of viral infection of bacterial cells.
Gerald M. Edelman and Rodney R. Porter	1972	United States England	Described the nature and structure of antibodies.
Renato Dulbecco, Howard Temin, and David Baltimore	1975	United States	Discovered reverse transcriptase and described how RNA viruses could cause cancer.
Daniel Nathans, Hamilton Smith, and Werner Arber	1978	United States United States Switzerland	Described the action of restriction enzymes (now used in genetic engineering).

covered a bacterium large enough to be seen with the un-aided eye (0.2 mm wide). This bacterium, which she named *Thiomargarita namibiensis* (thī′o-mä-gär-e-tä na′mib-e-ėn-sis), lives in the mud on the African coast. *Thiomargarita* is unusual because of its size and its ecological niche. The bac-terium consumes hydrogen sulfide, which would be toxic to mud-dwelling animals (Figure 11.26 on page 331).

Mycology, the study of fungi, includes medical, agri-cultural, and ecological branches. Recall that Bassi's work leading up to the germ theory of disease was on a fungal

TABLE 1.1	Selected Nobel Prizes Awarded for Research in Microbiology *(continued)*		
Nobel Laureates	**Year of Presentation**	**Country of Birth**	**Contribution**
Peter Mitchell	1978	England	Described the chemiosmotic mechanism for ATP synthesis.
Paul Berg	1980	United States	Performed experiments in gene splicing (genetic engineering).
Walter Gilbert	1980	United States	Discovered a method of DNA sequencing.
Aaron Klug	1982	South Africa	Described the structure of tobacco mosaic virus (TMV).
Barbara McClintock	1983	United States	Discovered transposons (small segments of DNA that can move from one region of a DNA molecule to another).
César Milstein, Georges J.F. Köhler, and Niels Kai Jerne	1984	Argentina Germany Denmark	Developed a technique for producing monoclonal antibodies (single pure antibodies).
Susumu Tonegawa	1987	Japan	Described the genetics of antibody production.
Johann Deisenhofer, Robert Huber, and Hartmut Michel	1988	Germany	Described the structure of bacterial photosynthetic pigments.
J. Michael Bishop and Harold E. Varmus	1989	United States	Discovered cancer-causing genes called oncogenes.
Joseph E. Murray and E. Donnall Thomas	1990	United States	Performed the first successful organ transplants by using immunosuppressive agents.
Edmond H. Fisher and Edwin G. Krebs	1992	United States	Discovered protein kinases, enzymes that regulate cell growth.
Richard J. Roberts and Phillip A. Sharp	1993	Great Britain United States	Discovered that genes can be separated on different segments of DNA.
Kary B. Mullis	1993	United States	Discovered the polymerase chain reaction to amplify (copy) DNA.
Michael Smith	1993	Canada	Discovered a procedure to modify DNA to make new proteins.
Alfred Gilman and Martin Rodbell	1994	United States	Identified G proteins, which translate and integrate external signals that affect a wide range of biological activities.
Peter C. Doherty and Rolf M. Zinkernagel	1996	Australia Switzerland	Discovered how cytotoxic T cells recognize virus-infected cells prior to destroying them.
Stanley B. Prusiner	1997	United States	Discovered and named proteinaceous infectious particles (prions) and demonstrated a relationship between prions and deadly neurological diseases in humans and animals.
Leland H. Hartwell, Paul M. Nurse, and R. Timothy Hunt	2001	United States and Great Britain	Discovered genes that encode proteins regulating cell division

pathogen. Fungal infection rates have been rising during the past decade, accounting for 10% of hospital-acquired infections. Climatic and environmental changes (severe drought) are thought to account for the tenfold increase in *Coccidioides immitis* (kok-sid-ē-oi′dēz im′mi-tis) infections in California. New techniques for diagnosing and treating fungal infections are currently being investigated.

Parasitology is the study of protozoa and parasitic worms. Because many parasitic worms are large enough to be seen with the unaided eye, they have been known by people for thousands of years. One hypothesis is that the

(a)　　　　　　　　　　　**(b)**

FIGURE 1.6 **Parasitology: the study of protozoa and parasitic worms. (a)** The caduceus, the symbol of the medical profession, may have been designed after the procedure for removing parasitic guinea worms. **(b)** A doctor removes a guinea worm *(Dracunculus medinensis)* from the subcutaneous tissue of a patient.

■ How do bacteriology, mycology, and parasitology differ?

medical symbol the caduceus represents the removal of parasitic guinea worms (Figure 1.6).

New parasitic diseases in humans are being discovered as laborers become exposed while clearing rain forests. Previously unknown parasitic diseases are also being found in patients whose immune systems have been suppressed by organ transplants, cancer chemotherapy, and AIDS.

Bacteriology, mycology, and parasitology are currently going through a "golden age of classification." Recent advances in **genomics,** the study of all of an organism's genes, have allowed scientists to classify bacteria and fungi according to their genetic relationships with other bacteria, fungi, and protozoa. Previously these microorganisms were classified according to a limited number of visible characteristics.

Immunology

Immunology, the study of immunity, actually dates back in Western culture to Jenner's first vaccine in 1796. Since then, knowledge about the immune system has accumulated steadily and expanded rapidly during the twentieth century. Vaccines are now available for numerous diseases, including measles, rubella (German measles), mumps, chickenpox, pneumonia, tetanus, tuberculosis, influenza, whooping cough, polio, hepatitis B, and Lyme disease. The smallpox vaccine was so effective that the disease has been eliminated. Public health officials estimate that polio will be eradicated within a few years because of the polio vaccine. In 1960, interferons, substances generated by the

body's own immune system, were discovered. Interferons inhibit replication of viruses and have triggered considerable research related to the treatment of viral diseases and cancer. One of today's biggest challenges for immunologists is learning how the immune system might be stimulated to ward off the virus responsible for AIDS, a disease that destroys the immune system.

A major advance in immunology occurred in 1933, when Rebecca Lancefield proposed that streptococci be classified according to serotypes (variants within a species) based on certain components in the cell walls of the bacteria. Streptococci are responsible for a variety of diseases, such as sore throat (strep throat), streptococcal toxic shock, and septicemia (blood poisoning). Her research permits the rapid identification of specific pathogenic streptococci based on immunological techniques.

Virology

The study of viruses, **virology,** actually originated during the Golden Age of Microbiology. In 1892, Dmitri Iwanowski reported that the organism that caused mosaic disease of tobacco was so small that it passed through filters fine enough to stop all known bacteria. At the time, Iwanowski was not aware that the organism in question was a virus in the sense that we now understand the term. In 1935, Wendell Stanley demonstrated that the organism, called tobacco mosaic virus (TMV), was fundamentally different from other microbes and so simple and homogeneous that it could be crystallized like a chemical compound. Stanley's work facilitated the study of viral structure and chemistry. Since the development of the electron microscope in the 1940s, microbiologists have been able to observe the structure of viruses in detail, and today much is known about their structure and activity.

Recombinant DNA Technology

Microorganisms can now be genetically engineered to manufacture large amounts of human hormones and other urgently needed medical substances. In the late 1960s, Paul Berg showed that fragments of human or animal DNA that code for important proteins (genes) can be attached to bacterial DNA. The resulting hybrid was the first example of **recombinant DNA.** When recombinant DNA is inserted into bacteria (and other microbes), it can be used to make large quantities of the desired protein. The technology that developed from this technique is called **recombinant DNA technology,** or **genetic engineering,** and it had its origins in two related fields. The first, **microbial genetics,** studies the mechanisms by which microorganisms inherit traits. The second, **molecular biology,** specifically studies how genetic information is carried in molecules of DNA and how DNA directs the synthesis of proteins.

Although molecular biology encompasses all organisms, much of our knowledge of how genes determine specific traits has been revealed through experiments with bacteria. Until the 1930s, all genetic research was based on the study of plant and animal cells. But in the 1940s, scientists turned to unicellular organisms, primarily bacteria, which have several advantages for genetic and biochemical research. For one thing, bacteria are less complex than plants and animals. For another, the life cycles of many bacteria require less than an hour, so scientists can cultivate very large numbers of individuals for study in a relatively short time.

Once science turned to the study of unicellular life, progress in genetics began to occur rapidly. In 1941, George W. Beadle and Edward L. Tatum demonstrated the relationship between genes and enzymes. DNA was established as the hereditary material in 1944 by Oswald Avery, Colin MacLeod, and Maclyn McCarty. In 1946, Joshua Lederberg and Edward L. Tatum discovered that genetic material could be transferred from one bacterium to another by a process called conjugation. Then, in 1953, James Watson and Francis Crick proposed a model for the structure and replication of DNA. The early 1960s witnessed a further explosion of discoveries relating to the way DNA controls protein synthesis. In 1961, François Jacob and Jacques Monod discovered messenger RNA (ribonucleic acid), a chemical involved in protein synthesis, and later they made the first major discoveries about the regulation of gene function in bacteria. During the same period, scientists were able to break the genetic code and

thus understand how the information for protein synthesis in messenger RNA is translated into the amino acid sequence for making proteins.

Microbes and Human Welfare

Learning Objective

- List at least four beneficial activities of microorganisms.

As mentioned earlier, only a minority of all microorganisms are pathogenic. Microbes that cause food spoilage, such as soft spots on fruits and vegetables, decomposition of meats, and rancidity of fats and oils, are also a minority. The vast majority of microbes benefit humans, other animals, and plants in many ways. The following sections outline some of these beneficial activities. In later chapters, we will discuss these activities in greater detail.

Recycling Vital Elements

Discoveries made by two microbiologists in the 1880s have formed the basis for today's understanding of the biochemical cycles that support life on Earth. Martinus Beijerinck and Sergei Winogradsky were the first to show how bacteria help recycle vital elements between the soil and the atmosphere. **Microbial ecology,** the study of the relationship between microorganisms and their environment, originated with the work of Beijerinck and Winogradsky. Today, microbial ecology has branched out and includes the study of how microbial populations interact with plants and animals in various environments. Among the concerns of microbial ecologists are water pollution and toxic chemicals in the environment.

The chemical elements carbon, nitrogen, oxygen, sulfur, and phosphorus are essential for life and abundant, but not necessarily in forms that organisms can use. Microorganisms are mostly responsible for converting these elements into forms that can be used by plants and animals. Microorganisms, mostly bacteria and fungi, play a key role in returning carbon dioxide to the atmosphere when they decompose organic wastes and dead plants and animals. Algae, cyanobacteria, and higher plants use the carbon dioxide during photosynthesis to produce carbohydrates for animals, fungi, and bacteria. Nitrogen is abundant in the atmosphere but must be made into a usable form by bacteria so that it be available for plants and animals. Only bacteria can make this conversion naturally.

Sewage Treatment: Using Microbes to Recycle Water

With our society's growing awareness of the need to preserve the environment, people have become more conscious of the responsibility to recycle precious water and prevent the pollution of rivers and oceans. One major pollutant is sewage, which consists of human excrement, waste water, industrial wastes, and surface runoff. Sewage is about 99.9% water, with a few hundredths of 1% suspended solids. The remainder is a variety of dissolved materials.

Sewage treatment plants remove the undesirable materials and harmful microorganisms. Treatments combine various physical and chemical processes with the action of beneficial microbes. Large solids such as paper, wood, glass, gravel, and plastic are removed from sewage; left behind are liquid and organic materials that bacteria convert into such by-products as carbon dioxide, nitrates, phosphates, sulfates, ammonia, hydrogen sulfide, and methane.

Bioremediation: Using Microbes to Clean Up Pollutants

In 1988, scientists began using microbes to clean up pollutants and toxic wastes produced by various industrial processes. For example, some bacteria can actually use pollutants they consume as energy sources; others produce enzymes that break down toxins into less harmful substances. By using bacteria in these ways—a process known as **bioremediation**—toxins can be removed from underground wells, chemical spills, toxic waste sites, and oil spills, such as the *Exxon Valdez* disaster of 1989 (see the box in Chapter 2, page 34). In addition, bacterial enzymes are used in drain cleaners to remove clogs without adding harmful chemicals to the environment. In some cases, microorganisms indigenous to the environment are used; in others, genetically engineered microbes are used. Among the most commonly used bioremedial microbes are certain species of bacteria of the genera *Pseudomonas* (sū-dō-mō′nas) and *Bacillus. Bacillus* enzymes are also used in household detergents to remove spots from clothing.

Insect Pest Control by Microorganisms

Besides spreading diseases, insects can cause devastating crop damage. Insect pest control is therefore important for both agriculture and the prevention of human disease.

The bacterium *Bacillus thuringiensis* (bä-sil′lus thür-in-jē-en′sis) has been used extensively in the United States to control such pests as alfalfa caterpillars, bollworms, corn borers, cabbageworms, tobacco budworms, and fruit tree leaf rollers. It is incorporated into a dusting powder that is applied to the crops these insects eat, or is commercially available in a form that is genetically engineered into plants. The bacteria produce protein crystals that are toxic to the digestive systems of the insects.

By using microbial rather than chemical insect control, farmers can avoid harming the environment. Many chemical insecticides, such as DDT, remain in the soil as toxic pollutants and are eventually incorporated into the food chain.

Modern Biotechnology and Genetic Engineering

Learning Objective

- List two examples of biotechnology that use genetic engineering and two examples that do not.

Earlier, we touched on the commercial use of microorganisms to produce some common foods and chemicals. Such practical applications of microbiology are called **biotechnology.** Although biotechnology has been used in some form for centuries, techniques have become much more sophisticated in the past few decades. In the last several years biotechnology has undergone a revolution through the advent of genetic engineering, the use of recombinant DNA technology to expand the potential of bacteria, viruses, and yeast cells and other fungi as miniature biochemical factories. Cultured plant and animal cells, as well as intact plants and animals, are also used as recombinant cells and organisms.

The applications of genetic engineering are increasing with each passing year. Recombinant DNA techniques have been used thus far to produce a number of natural proteins, vaccines, and enzymes. Such substances have great potential for medical use; some of them are described in Table 9.1 on page 264.

A very exciting and important outcome of recombinant DNA techniques is **gene therapy**—inserting a missing gene or replacing a defective one in human cells. This technique uses a harmless virus to carry the missing or new gene into certain host cells, where the gene is picked up and inserted into the appropriate chromosome. Since 1990, gene therapy has been used to treat patients with adenosine deaminase (ADA) deficiency, a cause of severe combined immunodeficiency disease (SCID), in which cells of the immune system are inactive or missing; Duchenne's muscular dystrophy, a muscle-destroying disease; cystic fibrosis, a disease of the secreting portions of the respiratory passages, pancreas, salivary glands, and sweat glands; and LDL-receptor deficiency, a condition in which low-density lipoprotein (LDL) receptors are defective and LDL cannot enter cells. The LDL remains in the blood in high concentrations and increases the risk of atherosclerosis and coronary artery disease because it leads to fatty plaque formation in blood vessels. Results are still being evaluated. Certain genetic diseases may also be treatable by gene therapy in the future, including hemophilia, an inability of the blood to clot normally; diabetes, elevated blood sugar levels; sickle-cell disease, an abnormal kind of hemoglobin; and one type of hypercholesterolemia, high blood cholesterol.

Beyond medical applications, recombinant DNA techniques have also been applied to agriculture. For example, genetically altered strains of bacteria have been developed to protect fruit against frost damage, and bacteria are being engineered to control insects that damage crops. Bacteria have also been used to improve the appearance, flavor, and shelf life of fruits and vegetables. Potential agricultural uses of genetic engineering include drought resistance, resistance to insects and microbial diseases, and increased temperature tolerance in crops.

Microbes and Human Disease

Learning Objectives
- *Define normal microbiota and resistance.*
- *Define and describe several infectious diseases.*
- *Describe emerging infectious diseases.*

Normal Microbiota

We all live from birth until death in a world filled with microbes, and we all have a variety of microorganisms on and inside our bodies. These microorganisms make up our **normal microbiota,** or *flora** (Figure 1.7). The normal microbiota not only do us no harm, but also in some cases can actually benefit us. For example, some normal microbiota protect us against disease by preventing the overgrowth of harmful microbes, while others produce useful substances such as vitamin K and B vitamins. Unfortunately, under some circumstances normal microbiota can make us sick or infect people we contact. For instance, when some normal microbiota leave their habitat, they can cause disease.

When is a microbe a welcome part of a healthy human, and when is it a harbinger of disease? The distinction between health and disease is in large part a balance between the natural defenses of the body and the disease-producing properties of microorganisms. Whether our bodies overcome the offensive tactics of a particular microbe depends on our **resistance**—the ability to ward off diseases. Important resistance is provided by the barrier of the skin, mucous membranes, cilia, stomach acid, and antimicrobial chemicals such as interferons. Microbes can be destroyed by white blood cells, the inflammatory response, fever, and by specific responses of our immune systems. Sometimes, when our nat-

---

*At one time, bacteria and fungi were thought to be plants and thus the term *flora* was used.

FIGURE 1.7 **Several types of bacteria found as part of the normal microbiota inside the human mouth.**

- Normal microbiota generally do not harm us and, in some cases, actually benefit us.

ural defenses are not strong enough to overcome an invader, they have to be supplemented by antibiotics or other drugs.

Infectious Diseases

An **infectious disease** is one in which pathogens invade a susceptible host, such as a human or an animal. In the process, the pathogen carries out at least part of its life cycle inside the host, and as a result, disease frequently occurs. By the end of World War II, many people believed that infectious diseases were under control. They thought malaria would be eradicated through the use of the insecticide DDT to kill mosquitoes, that a vaccine would prevent diphtheria, and that improved sanitation measures would help prevent cholera transmission. Malaria is far from eliminated. Since 1986, local outbreaks have been identified in New Jersey, California, Florida, New York, and Texas, and the disease infects 300 million people worldwide. In 1994, diphtheria appeared in the United States, brought by travelers from the newly independent states of the former Soviet Union, which were experiencing a massive diphtheria epidemic. The epidemic was brought under control in 1998.

Emerging Infectious Diseases

These recent outbreaks point to the fact that infectious diseases not only are not disappearing, but seem to be reemerging and increasing. In addition, a number of new diseases—**emerging infectious diseases (EIDs)**—have cropped up in recent years. These are diseases that are new or changing and are increasing or have the potential to increase in incidence in the near future. Some of the factors that have contributed to the emergence of EIDs are evolutionary changes in existing organisms; the spread of known diseases to new geographic regions or populations by modern transportation; and increased human exposure to new, unusual infectious agents in areas that are undergoing ecologic changes such as deforestation and construction. An increasing number of incidents in recent years highlights the extent of the problem.

West Nile encephalitis (WNE) is a disease caused by West Nile virus, which can produce encephalitis (inflammation of the brain). WNE was first diagnosed in the West Nile region of Uganda in 1937. In 1999 the virus made its first North American appearance in humans in New York City. By 2002, 3559 people had been infected in 37 states. West Nile virus is now established in nonmigratory birds in 42 states, and WNE has occurred in horses in 15 states. The virus, which is carried by birds, is transmitted between birds, and to horses and humans, by mosquitoes. West Nile virus may have arrived in the United States in an infected traveler or in a mosquito or bird onboard a ship.

In 1996, countries worldwide were refusing to import beef from the United Kingdom, and the U. K. had to kill hundreds of thousands of cattle born after 1988 because of an epidemic of **bovine spongiform encephalopathy** (en-sef-a-lop′a-thē), also called **BSE** or **mad cow disease.** BSE first came to the attention of microbiologists in 1986 as one of a handful of diseases caused by an infectious protein called a *prion.* Studies suggest that the source of disease was cattle feed prepared from sheep infected with their own version of the disease. Cattle are herbivores (plant-eaters), but their growth and health are improved by adding protein to their feed. **Creutzfeldt-Jakob disease** (kroits′felt yä′kôb) or **CJD** is a human disease also caused by a prion. The incidence of CJD in the U. K. is similar to the incidence in other countries. However, by 2002 the U. K. reported 138 human cases of CJD caused by a new variant related to the bovine disease.

Escherichia coli is a normal inhabitant of the large intestine of vertebrates, including humans, and its presence is beneficial because it helps produce certain vitamins and breaks down otherwise undigestible foodstuffs. However, a strain called *E. coli* O157:H7 causes bloody diarrhea when it grows in the intestines. This strain was first recognized in 1982 and since then has emerged as a public health problem. It is now one of the leading causes of diarrhea worldwide. In 1996, some 9000 people in Japan became ill, and 7 died, as a result of infection by *E. coli* O157:H7. The recent outbreaks of *E. coli* O157:H7 in the United States,

associated with contamination of undercooked meat and unpasteurized beverages, have made public health officials aware that new methods of testing for bacteria in food must be developed.

In 1995, infections of so-called flesh-eating bacteria were reported on the front pages of major newspapers. The bacteria are more correctly named invasive group A *Streptococcus* (strep-tō-kok′kus) or IGAS. There has been a trend toward increasing rates of IGAS in the United States, Scandinavia, England, and Wales.

In 1995, a hospital laboratory technician in Congo (Zaire) who had fever and bloody diarrhea underwent surgery for a suspected perforated bowel. Subsequent to surgery, he started hemorrhaging, and his blood began clotting in his blood vessels. A few days later, health care workers in the hospital where he was staying developed similar symptoms. One of them was transferred to a hospital in a different city; personnel in the second hospital who cared for this patient also developed symptoms. By the time the epidemic was over, 315 people had contracted **Ebola hemorrhagic fever** (hem-o-raj′ik), or **EHF,** and over 75% of them died. The epidemic was controlled when microbiologists instituted training on the use of protective equipment and educational measures in the community. Human-to-human transmission occurs when there is close personal contact with infectious blood or other body fluids or tissue.

Microbiologists first isolated Ebola viruses from humans during earlier outbreaks in Congo in 1976. (The virus is named after Congo's Ebola River.) In 1994, a single case of infection from a newly described Ebola virus occurred in Côte d'Ivoire. In 1989 and 1996, outbreaks among monkeys imported into the United States from the Philippines were caused by another Ebola virus but were not associated with human disease. Microbiologists have been studying many animals but have not yet discovered the natural reservoir (source) of EHF virus.

Hantavirus **pulmonary syndrome** caught the attention of the public in 1993 when two people in the same household became ill and died within 5 days of each other. Their illnesses began with a fever and cough and rapidly progressed to respiratory failure. Within a month, 23 additional cases, including 10 deaths, were reported in the Four Corners area of the American Southwest. Worried tourists began canceling vacations to the area, and residents wondered who would get the disease next. Using techniques that only became available in the 1990s, microbiologists determined that the cause was a newly recognized *Hantavirus* called the Sin Nombre virus, which is carried by deer mice. (The virus is named after the Hantaan River in Korea, where the virus was first discovered.) Researchers developed a test to rapidly identify the virus and made recommendations to help people reduce their risk of exposure to potentially infected rodents. It is probably not a new virus. Comparisons of *Hantavirus* genes suggest that the virus probably came to North America with the first rats from the Old World.

Also in 1993, an outbreak of **cryptosporidiosis** (krip-tō-spô-rid-ē-ō′sis) transmitted through the public water supply in Milwaukee, Wisconsin, resulted in diarrheal illness in an estimated 403,000 persons. The microorganism responsible for this outbreak was the protozoan *Cryptosporidium* (krip-tō-spô-ri′dē-um). First reported as a cause of human disease in 1976, it is responsible for up to 30% of the diarrheal illness in developing countries. In the United States, transmission has occurred via drinking water, swimming pools, and contaminated hospital supplies.

AIDS (acquired immunodeficiency syndrome) first came to public attention in 1981 with reports from Los Angeles that a few young homosexual men had died of a previously rare type of pneumonia known as *Pneumocystis* (nü-mō-sis′tis) pneumonia. These men had experienced a severe weakening of the immune system, which normally fights infectious diseases. Soon these cases were correlated with an unusual number of occurrences of a rare form of cancer, Kaposi's sarcoma, among young homosexual men. Similar increases in such rare diseases were found among hemophiliacs and intravenous drug users.

By the end of 2002, nearly 800,000 people in the United States had been diagnosed as having AIDS, and 50% of them had died as a result of the disease. A great many more people had tested positive for the presence of the AIDS virus in their blood. As of 2003, health officials estimated that 800,000–900,000 Americans have HIV infection. In 2002, the World Health Organization (WHO) estimated that over 40 million people worldwide are living with HIV/AIDS and that 14,000 new infections occur every day.

Researchers quickly discovered that the cause of AIDS was a previously unknown virus (see Figure 1.1e). The virus, now called **human immunodeficiency virus (HIV),** destroys certain white blood cells of the immune system called CD4 lymphocytes, one of the cells of the body's defense. Sickness and death result from microorganisms or cancerous cells that might otherwise have been defeated by the body's natural defenses. So far, the disease has been inevitably fatal once symptoms develop.

By studying disease patterns, medical researchers found that HIV could be spread through sexual intercourse, by contaminated needles, from infected mothers to their fetuses before birth, and by blood transfusions—in short, by the transmission of body fluids from one person to another. Since 1985, blood used for transfusions has been carefully checked for the presence of HIV, and it is now quite unlikely that the virus can be spread by this means.

Since 1994, new treatments have extended the life span of people with AIDS; however, approximately 40,000 new cases occur annually in the United States. The majority of individuals with AIDS are in the sexually active age group, and because heterosexual partners of AIDS sufferers are at high risk of infection, public health officials are concerned that even more women and minorities will contract AIDS. In 1997, AIDS diagnoses began increasing among women and minorities. Among the AIDS cases reported for individuals 13 to 24 years of age, 44% were females and 63% were African American.

In the months and years to come, microbiological techniques will continue to be applied to help scientists learn more about the structure of the deadly HIV, how it is transmitted, how it grows in cells and causes disease, how drugs can be directed against it, and whether an effective vaccine can be developed. Public health officials have also focused on prevention through education.

AIDS poses one of this century's most formidable health threats, but it is not the first serious epidemic of a sexually transmitted disease. Syphilis was also once a fatal epidemic disease. As recently as 1941, syphilis caused an estimated 14,000 deaths per year in the United States. With few drugs available for treatment and no vaccines to prevent it, efforts to control the disease focused mainly on restraining sexual behavior and on the use of condoms. The eventual development of drugs to treat syphilis contributed significantly to preventing the spread of the disease. According to the Centers for Disease Control and Prevention (CDC), reported cases of syphilis dropped from a record high of 575,000 in 1943 to 6378 cases in 2002.

Just as microbiological techniques helped researchers in the fight against syphilis and smallpox, they will help scientists discover the causes of new emerging infectious diseases in the twenty-first century. Undoubtedly there will be new diseases. Ebola virus and *Hantavirus* are examples of viruses that may be changing their abilities to infect different host species. Emerging infectious diseases will be discussed further in Chapter 14 on page 425.

Infectious diseases may reemerge because of the development of resistance to antibiotics and through the use of microorganisms as weapons. The breakdown of public health measures for previously controlled infections has resulted in unexpected cases of tuberculosis, whooping cough, and diphtheria.

On October 4, 2001, a 63-year-old man was hospitalized in Florida and later died as a result of a disease called **inhalation anthrax.** This was the first reported case of inhalation anthrax since 1978. Veterinarians and agricultural workers are at risk of cutaneous anthrax. In 2000, a rancher contracted anthrax while disposing of five cows that died from anthrax. Anthrax is not transmitted from person to person. In 2001, as more people developed anthrax, it became apparent that these infections were due to intentional release of a bacterium through the mail. In all, 22 people were infected, half of them with inhalation anthrax. This was the first instance of transmission of the disease as a result of using this bacterium as a biological weapon. Anthrax is a disease caused by the bacterium *Bacillus anthracis* (bä-sil′lus an-thrā′sis). This microbe has another important place in the history of microbiology because, as mentioned earlier, in 1877 the German physician Robert Koch established a sequence of experimental steps relating *B. anthracis* to anthrax, the first instance in which a specific microbe was related to a specific disease. See Figure 14.3.

* * *

The diseases we have mentioned are caused by viruses, bacteria, protozoa, and prions—types of microorganisms. This book introduces you to the enormous variety of microscopic organisms. It shows you how microbiologists use specific techniques and procedures to study the microbes that cause such diseases as AIDS and diarrhea—and diseases that have yet to be discovered. You will also learn about the body's responses to microbial infection and the ways certain drugs combat microbial diseases. Finally, you will learn about the many beneficial roles that microbes play in the world around us.

STUDY OUTLINE

Microbes in Our Lives (pp. 1–2)

1. Living things too small to be seen with the unaided eye are called microorganisms.
2. Microorganisms are important in the maintenance of an ecological balance on Earth.
3. Some microorganisms live in humans and other animals and are needed to maintain the animal's health.
4. Some microorganisms are used to produce foods and chemicals.
5. Some microorganisms cause disease.

Naming and Classifying Microorganisms (pp. 2–6)

1. In a nomenclature system designed by Carolus Linnaeus (1735), each living organism is assigned two names.
2. The two names consist of a genus and a specific epithet, both of which are underlined or italicized.

Types of Microorganisms (pp. 2–5)

Bacteria (pp. 2–5)

1. Bacteria are unicellular organisms. Because they have no nucleus, the cells are described as prokaryotic.
2. The three major basic shapes of bacteria are bacillus, coccus, and spiral.
3. Most bacteria have a peptidoglycan cell wall; they divide by binary fission, and they may possess flagella.
4. Bacteria can use a wide range of chemical substances for their nutrition.

Archaea (p. 5)

1. Archaea consist of prokaryotic cells; they lack peptidoglycan in their cell walls.

2. Archaea include methanogens, extreme halophiles, and extreme thermophiles.

Fungi (p. 5)

1. Fungi (mushrooms, molds, and yeasts) have eukaryotic cells (with a true nucleus). Most fungi are multicellular.
2. Fungi obtain nutrients by absorbing organic material from their environment.

Protozoa (p. 5)

1. Protozoa are unicellular eukaryotes.
2. Protozoa obtain nourishment by absorption or ingestion through specialized structures.

Algae (p. 5)

1. Algae are unicellular or multicellular eukaryotes that obtain nourishment by photosynthesis.
2. Algae produce oxygen and carbohydrates that are used by other organisms.

Viruses (p. 5)

1. Viruses are noncellular entities that are parasites of cells.
2. Viruses consist of a nucleic acid core (DNA or RNA) surrounded by a protein coat. An envelope may surround the coat.

Multicellular Animal Parasites (p. 5)

1. The principal groups of multicellular animal parasites are flatworms and roundworms, collectively called helminths.
2. The microscopic stages in the life cycle of helminths are identified by traditional microbiological procedures.

Classification of Microorganisms (pp. 5–6)

1. All organisms are classified into Bacteria, Archaea, and Eukarya. Eukarya include protists, fungi, plants, and animals.

A Brief History of Microbiology (pp. 6–16)

The First Observations (p. 6)

1. Robert Hooke observed that cork was composed of "little boxes"; he introduced the term *cell* (1665).

2. Hooke's observations laid the groundwork for development of the cell theory, the concept that all living things are composed of cells.

3. Antoni van Leeuwenhoek, using a simple microscope, was the first to observe microorganisms (1673).

The Debate Over Spontaneous Generation (pp. 6–8)

1. Until the mid-1880s, many people believed in spontaneous generation, the idea that living organisms could arise from nonliving matter.

2. Francesco Redi demonstrated that maggots appear on decaying meat only when flies are able to lay eggs on the meat (1668).

3. John Needham claimed that microorganisms could arise spontaneously from heated nutrient broth (1745).

4. Lazzaro Spallanzani repeated Needham's experiments and suggested that Needham's results were due to microorganisms in the air entering his broth (1765).

5. Rudolf Virchow introduced the concept of biogenesis: Living cells can arise only from preexisting cells (1858).

6. Louis Pasteur demonstrated that microorganisms are in the air everywhere and offered proof of biogenesis (1861).

7. Pasteur's discoveries led to the development of aseptic techniques used in laboratory and medical procedures to prevent contamination by microorganisms in the air.

The Golden Age of Microbiology (pp. 8–11)

1. Rapid advances in the science of microbiology were made between 1857 and 1914.

Fermentation and Pasteurization (p. 9)

1. Pasteur found that yeast ferment sugars to alcohol and that bacteria can oxidize the alcohol to acetic acid.

2. A heating process called pasteurization is used to kill bacteria in some alcoholic beverages and milk.

The Germ Theory of Disease (pp. 9–11)

1. Agostino Bassi (1835) and Pasteur (1865) showed a causal relationship between microorganisms and disease.

2. Joseph Lister introduced the use of a disinfectant to clean surgical wounds in order to control infections in humans (1860s).

3. Robert Koch proved that microorganisms cause disease. He used a sequence of procedures called Koch's postulates (1876), which are used today to prove that a particular microorganism causes a particular disease.

Vaccination (p. 11)

1. In a vaccination, immunity (resistance to a particular disease) is conferred by inoculation with a vaccine.

2. In 1798, Edward Jenner demonstrated that inoculation with cowpox material provides humans with immunity to smallpox.

3. About 1880, Pasteur discovered that avirulent bacteria could be used as a vaccine for fowl cholera; he coined the word *vaccine*.

4. Modern vaccines are prepared from living avirulent microorganisms or killed pathogens, from isolated components of pathogens, and by recombinant DNA techniques.

The Birth of Modern Chemotherapy: Dreams of a "Magic Bullet" (pp. 11–12)

1. Chemotherapy is the chemical treatment of a disease.

2. Two types of chemotherapeutic agents are synthetic drugs (chemically prepared in the laboratory) and antibiotics (substances produced naturally by bacteria and fungi to inhibit the growth of other microorganisms).

3. Paul Ehrlich introduced an arsenic-containing chemical called salvarsan to treat syphilis (1910).

4. Alexander Fleming observed that the mold (fungus) *Penicillium* inhibited the growth of a bacterial culture. He named the active ingredient penicillin (1928).

5. Penicillin has been used clinically as an antibiotic since the 1940s.

6. Researchers are tackling the problem of drug-resistant microbes.

Modern Developments in Microbiology (pp. 12–16)

1. Bacteriology is the study of bacteria, mycology is the study of fungi, and parasitology is the study of parasitic protozoa and worms.

2. Microbiologists are using genomics, the study of all of an organism's genes, to classify bacteria, fungi, and protozoa.

3. The study of AIDS, analysis of the action of interferons, and the development of new vaccines are among the current research interests in immunology.

4. New techniques in molecular biology and electron microscopy have provided tools for advancement of our knowledge of virology.

5. The development of recombinant DNA technology has helped advance all areas of microbiology.

Microbes and Human Welfare (pp. 16–18)

1. Microorganisms degrade dead plants and animals and recycle chemical elements to be used by living plants and animals.

2. Bacteria are used to decompose organic matter in sewage.

3. Bioremediation processes use bacteria to clean up toxic wastes.

4. Bacteria that cause diseases in insects are being used as biological controls of insect pests. Biological controls are specific for the pest and do not harm the environment.

5. Using microbes to make products such as foods and chemicals is called biotechnology.

6. Using recombinant DNA, bacteria can produce important substances such as proteins, vaccines, and enzymes.

7. In gene therapy, viruses are used to carry replacements for defective or missing genes into human cells.

8. Genetically engineered bacteria are used in agriculture to protect plants from frost and insects and to improve the shelf life of produce.

Microbes and Human Disease (pp. 18–21)

1. Everyone has microorganisms in and on the body; these make up the normal microbiota or flora.

2. The disease-producing properties of a species of microbe and the host's resistance are important factors in determining whether a person will contract a disease.

3. An infectious disease is one in which pathogens invade a susceptible host.

4. An emerging infectious disease (EID) is a new or changing disease showing an increase in incidence in the recent past or a potential to increase in the near future.

STUDY QUESTIONS

Review

1. How did the idea of spontaneous generation come about?

2. Some proponents of spontaneous generation believed that air is necessary for life. They thought that Spallanzani did not really disprove spontaneous generation because he hermetically sealed his flasks to keep air out. How did Pasteur's experiments address the air question without allowing the microbes in the air to ruin his experiment?

3. Briefly state the role played by microorganisms in each of the following:
   a. Biological control of pests
   b. Recycling of elements
   c. Normal microbiota
   d. Sewage treatment
   e. Human insulin production
   f. Vaccine production

4. Into which field of microbiology would the following scientists best fit?

Researcher Who	Field
___ Studies biodegradation of toxic wastes	(a) Biotechnology
___ Studies the causative agent of *Hantavirus* pulmonary syndrome	(b) Immunology
___ Studies the production of human proteins by bacteria	(c) Microbial ecology
___ Studies the symptoms of AIDS	(d) Microbial genetics
___ Studies the production of toxin by *E. coli*	(e) Microbial physiology
___ Studies the life cycle of *Cryptosporidium*	(f) Molecular biology
___ Develops gene therapy for a disease	(g) Mycology
___ Studies the fungus *Candida albicans*	(h) Virology

5. Match the following people to their contribution toward the advancement of microbiology.

___ Avery, MacLeod, and McCarty	(a) Developed vaccine against smallpox
___ Beadle and Tatum	(b) Discovered how DNA controls protein synthesis in a cell
___ Berg	(c) Discovered penicillin
___ Ehrlich	(d) Discovered that DNA can be transferred from one bacterium to another
___ Fleming	(e) Disproved spontaneous generation
___ Hooke	(f) First to characterize a virus
___ Iwanowski	(g) First to use disinfectants in surgical procedures
___ Jacob and Monod	(h) First to observe bacteria
___ Jenner	(i) First to observe cells in plant material and name them
___ Koch	(j) Observed that viruses are filterable
___ Lancefield	(k) Proved that DNA is the hereditary material
___ Lederberg and Tatum	(l) Proved that microorganisms can cause disease
___ Lister	(m) Said living cells arise from preexisting living cells
___ Pasteur	(n) Showed that genes code for enzymes
___ Stanley	(o) Spliced animal DNA to bacterial DNA
___ van Leeuwenhoek	(p) Used bacteria to produce acetone
___ Virchow	(q) Used the first synthetic chemotherapeutic agent
___ Weizmann	(r) Proposed a classification system for streptococci based on antigens in their cell walls

6. The genus name of a bacterium is "erwinia" and the specific epithet is "carotovora." Write the scientific name of this organism correctly. Using this name as an example, explain how scientific names are chosen.

7. Match the following microorganisms to their descriptions.

__ Archaea	(a) Not composed of cells
__ Algae	(b) Cell wall made of chitin
__ Bacteria	(c) Cell wall made of peptidoglycan
__ Fungi	(d) Cell wall made of cellulose; photosynthetic
__ Helminths	(e) Complex cell structure lacking a cell wall
__ Protozoa	(f) Multicellular animals
__ Viruses	(g) Prokaryote without peptidoglycan cell wall

8. It is possible to purchase the following microorganisms in a retail store. Provide a reason for buying each.
   a. *Bacillus thuringiensis*
   b. *Saccharomyces*

Multiple Choice

1. Which of the following is a scientific name?
   a. *Mycobacterium tuberculosis*
   b. Tubercle bacillus

2. Which of the following is *not* a characteristic of bacteria?
   a. are prokaryotic
   b. have peptidoglycan cell walls
   c. have the same shape
   d. grow by binary fission
   e. have the ability to move

3. Which of the following is the most important element of Koch's germ theory of disease? The animal shows disease symptoms when
   a. the animal has been in contact with a sick animal.
   b. the animal has a lowered resistance.
   c. a microorganism is observed in the animal.
   d. a microorganism is inoculated into the animal.
   e. microorganisms can be cultured from the animal.

4. Recombinant DNA is
   a. DNA in bacteria.
   b. the study of how genes work.
   c. the DNA resulting when genes of two different organisms are mixed.
   d. the use of bacteria in the production of foods.
   e. the production of proteins by genes.

5. Which of the following statements is the best definition of biogenesis?
   a. Nonliving matter gives rise to living organisms.
   b. Living cells can only arise from preexisting cells.
   c. A vital force is necessary for life.
   d. Air is necessary for living organisms.
   e. Microorganisms can be generated from nonliving matter.

6. Which of the following is a beneficial activity of microorganisms?
   a. Some microorganisms are used as food for humans.
   b. Some microorganisms use carbon dioxide.
   c. Some microorganisms provide nitrogen for plant growth.
   d. Some microorganisms are used in sewage treatment processes.
   e. all of the above

7. It has been said that bacteria are essential for the existence of life on Earth. Which of the following would be the essential function performed by bacteria?
   a. control insect populations
   b. directly provide food for humans
   c. decompose organic material and recycle elements
   d. cause disease
   e. produce human growth hormones such as insulin

8. Which of the following is an example of bioremediation?
   a. application of oil-degrading bacteria to an oil spill
   b. application of bacteria to a crop to prevent frost damage
   c. fixation of gaseous nitrogen into usable nitrogen
   d. production by bacteria of a human protein such as interferon
   e. all of the above

9. Spallanzani's conclusion about spontaneous generation was challenged because Lavoisier had just shown that oxygen was the vital component of air. Which of the following statements is true?
   a. All life requires air.
   b. Only disease-causing organisms require air.
   c. Some microbes do not require air.
   d. Pasteur kept air out of his biogenesis experiments.
   e. Lavoisier was mistaken.

10. Which of the following statements about *E. coli* is *not* true?
   a. *E. coli* was the first disease-causing bacterium identified by Koch.
   b. *E. coli* is part of the normal microbiota of humans.
   c. *E. coli* is beneficial in human intestines.
   d. A disease-causing strain of *E. coli* causes bloody diarrhea.
   e. none of the above

Critical Thinking

1. How did the theory of biogenesis lead the way for the germ theory of disease?

2. Even though the germ theory of disease was not demonstrated until 1876, why did Semmelweis (1840) and Lister (1867) argue for the use of aseptic techniques?

3. Find at least three supermarket products made by microorganisms. (*Hint:* The label will state the scientific name of the organism or include the word *culture, fermented,* or *brewed.*)

4. People believed all microbial diseases would be controlled during the twentieth century. List three reasons why we are identifying new diseases now.

Clinical Applications

1. The prevalence of arthritis in the United States is 1 in 100,000 children. However, 1 in 10 children in Lyme, Connecticut, developed arthritis between June and September in 1973. Allen Steere, a rheumatologist at Yale University, investigated the cases in Lyme and found that 25% of the patients remembered having a skin rash during their arthritic episode and that the disease was treatable with penicillin. Steere concluded that this was a new infectious disease and did not have an environmental, genetic, or immunologic cause.
   a. What was the factor that caused Steere to reach his conclusion?
   b. What is the disease?
   c. Why was the disease more prevalent between June and September?

2. In 1864, Lister observed that patients recovered completely from simple fractures, but compound fractures had "disas-trous consequences." He knew that the application of phenol (carbolic acid) to fields in the town of Carlisle prevented cattle disease. In 1864, Lister treated compound fractures with phenol, and his patients recovered without complications. How was Lister influenced by Pasteur's work? Why was Koch's work still needed?

LEARNING WITH TECHNOLOGY

The Microbiology Place CD-ROM/website includes many resources to help you study and succeed in this course. In addition to self-quizzes, web links, and flashcards, you'll find the following activities for this chapter in the "Activities" section of The Microbiology Place:

   Meet the Microbes

   Case Study: Are Ulcers an Infectious Disease?

Chapter 2

Chemical Principles

CYANOBACTERIA. *Light energy drives the oxidation reactions in these photosynthetic bacteria. All organisms use oxidation reactions to get energy.*

We can see a tree rot and smell milk going sour, but we might not realize what is happening on a microscopic level. In both cases, microbes are conducting chemical operations. The tree rots when microorganisms decompose the wood. Milk turns sour from the production of lactic acid by bacteria. Most of the activities of microorganisms are the result of a series of chemical reactions.

Like all organisms, microorganisms must use nutrients to make chemical building blocks that are used for growth and for all the other functions essential to life. For most microorganisms, synthesizing these building blocks requires them to break down nutrient substances and use the energy released to assemble the resulting molecular fragments into new substances. These chemical reactions take place minute by minute in countless microenvironments.

All matter—whether air, rock, or a living organism—is made up of small units called **atoms.** Atoms interact with each other in certain combinations to form **molecules.** Living cells are made up of molecules, some of which are very complex. The science of the interaction between atoms and molecules is called **chemistry.**

The chemistry of microbes is one of the most important concerns of microbiologists. Knowledge of chemistry is essential to understanding the roles of microorganisms in nature, how they cause disease, how methods for diagnosing disease are developed, how the body's defenses combat infection, and how antibiotics and vaccines are produced to combat the harmful effects of microbes. To understand the changes that occur in microorganisms and the changes microbes make in the world around us, we need to know how molecules are formed and how they interact.

The Structure of Atoms

Learning Objective

- *Describe the structure of an atom and its relation to the chemical properties of elements.*

Atoms are the smallest units of matter that enter into chemical reactions. Every atom has a centrally located **nucleus** and particles called **electrons** that move around the nucleus in patterns known as electronic configurations (Figure 2.1). The nuclei of most atoms are stable—that is, they do not change spontaneously—and nuclei do not participate in chemical reactions. The nucleus is made up of positively (+) charged particles called **protons** and uncharged (neutral) particles called **neutrons.** The nucleus, therefore, bears a net positive charge. Neutrons and protons have approximately the same weight, which is about 1840 times that of an electron. The charge on electrons is negative (−), and in all atoms the number of electrons is equal to the number of protons. Because the total positive charge of the nucleus equals the total negative charge of the electrons, each atom is electrically neutral.

The number of protons in an atomic nucleus ranges from one (in a hydrogen atom) to more than 100 (in the largest atoms known). Atoms are often listed by their **atomic number,** the number of protons in the nucleus. The total number of protons and neutrons in an atom is its approximate **atomic weight.**

Chemical Elements

All atoms with the same number of protons behave the same way chemically and are classified as the same **chemical element.** Each element has its own name and a one- or two-letter symbol, usually derived from the English or Latin

TABLE 2.1	**The Elements of Life***		
Element	Symbol	Atomic Number	Approximate Atomic Weight
Hydrogen	H	1	1
Carbon	C	6	12
Nitrogen	N	7	14
Oxygen	O	8	16
Sodium	Na	11	23
Magnesium	Mg	12	24
Phosphorus	P	15	31
Sulfur	S	16	32
Chlorine	Cl	17	35
Potassium	K	19	39
Calcium	Ca	20	40
Iron	Fe	26	56
Iodine	I	53	127

*Hydrogen, carbon, nitrogen, and oxygen are the most abundant chemical elements in living organisms.

name for the element. For example, the symbol for the element hydrogen is H, and the symbol for carbon is C. The symbol for sodium is Na—the first two letters of its Latin name, *natrium*—to distinguish it from nitrogen, N, and from sulfur, S. There are 92 naturally occurring elements. However, only about 26 elements are commonly found in living things. Table 2.1 lists some of the chemical elements found in living organisms, including their atomic numbers and weights. The elements most abundant in living matter are hydrogen, carbon, nitrogen, and oxygen.

Most elements have several **isotopes**—atoms with different numbers of neutrons in their nuclei. All isotopes of an element have the same number of protons in their nuclei, but their atomic weights differ because of the difference in the number of neutrons. For example, in a natural sample of oxygen, all the atoms will contain eight protons. However, 99.76% of the atoms will have eight neutrons, 0.04% will contain nine neutrons, and the remaining 0.2% will contain ten neutrons. Therefore, the three isotopes composing a natural sample of oxygen will have atomic weights of 16, 17, and 18, although all will have the atomic number 8. Atomic numbers are written as a subscript to the left of an element's chemical symbol. Atomic weights are written as a superscript above the atomic number. Thus, natural oxygen isotopes are represented as $^{16}_{8}O$, $^{17}_{8}O$, and $^{18}_{8}O$. Isotopes of certain elements are extremely useful in

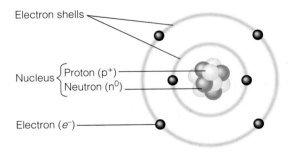

Electron shells

Nucleus { Proton (p⁺)
Neutron (n⁰) }

Electron (e⁻)

FIGURE 2.1 **The structure of an atom.** In this simplified diagram of a carbon atom, note the central location of the nucleus. The nucleus contains six neutrons and six protons, although not all are visible in this view. The six electrons move about the nucleus in regions called electron shells, shown here as circles.

■ What is the atomic number of this atom?

biological research, medical diagnosis, the treatment of some disorders, and in some forms of sterilization.

Electronic Configurations

In an atom, electrons are arranged in **electron shells,** which are regions corresponding to different **energy levels.** The arrangement is called an **electronic configuration.** Shells are layered outward from the nucleus, and each shell can hold a characteristic maximum number of electrons—two electrons in the innermost shell (lowest energy level), eight electrons in the second shell, and eight electrons in the third shell, if it is the atom's outermost (valence) shell. The fourth, fifth, and sixth electron shells can each accommodate 18 electrons, although there are some exceptions to this generalization. Table 2.2 shows the electronic configurations for atoms of some elements found in living organisms.

There is a tendency for the outermost shell to be filled with the maximum number of electrons. An atom can give up, accept, or share electrons with other atoms to fill this shell. The chemical properties of atoms are largely a function of the number of electrons in the outermost electron shell. When its outer shell is filled, the atom is chemically stable, or inert: It does not tend to react with other atoms. Helium (atomic number 2) and neon (atomic number 10) are examples of atoms of inert gases that have filled outer shells.

When an atom's outer electron shell is only partially filled, the atom is chemically unstable. Such an atom reacts with other atoms, and this reaction depends, in part, on the degree to which the outer energy levels are filled. Notice the number of electrons in the outer energy levels of the atoms in Table 2.2. We will see later how the number correlates with the chemical reactivity of the elements.

How Atoms Form Molecules: Chemical Bonds

Learning Objective
- *Define ionic bond, covalent bond, hydrogen bond, molecular weight, and mole.*

When the outermost energy level of an atom is not completely filled by electrons, you can think of it as having either unfilled spaces or extra electrons in that energy level, depending on whether it is easier for the atom to gain or lose electrons. For example, an atom of oxygen, with two electrons in the first energy level and six in the second, has two unfilled spaces in the second electron shell; an atom of magnesium has two extra electrons in its outermost shell. The most chemically stable configuration for any atom is

to have its outermost shell filled, as do the inert gases. Therefore, for these two atoms to attain that state, oxygen must gain two electrons, and magnesium must lose two electrons. All atoms tend to combine so that the extra electrons in the outermost shell of one atom fill the spaces of the outermost shell of the other atom; for example, oxygen and magnesium combine so that the outermost shell of each atom has the full complement of eight electrons.

The **valence,** or combining capacity, of an atom is the number of extra or missing electrons in its outermost electron shell. For example, hydrogen has a valence of 1 (one unfilled space, or one extra electron), oxygen has a valence of 2 (two unfilled spaces), carbon has a valence of 4 (four unfilled spaces, or four extra electrons), and magnesium has a valence of 2 (two extra electrons).

Basically, atoms achieve the full complement of electrons in their outermost energy shells by combining to form molecules, which are made up of atoms of one or more elements. A molecule that contains at least two different kinds of atoms, such as H_2O, the water molecule, is called a **compound.** In H_2O, the subscript 2 indicates that there are two atoms of hydrogen; the absence of a subscript indicates that there is only one atom of oxygen. Molecules hold together because the valence electrons of the combining atoms form attractive forces, called **chemical bonds,** between the atomic nuclei. Therefore, valence may also be viewed as the bonding capacity of an element. Because energy is required for chemical bond formation, each chemical bond possesses a certain amount of potential chemical energy.

In general, atoms form bonds in one of two ways: by either gaining or losing electrons from their outer electron shell, or by sharing outer electrons. When atoms have gained or lost outer electrons, the chemical bond is called an ionic bond. When outer electrons are shared, the bond is called a covalent bond. Although we will discuss ionic and covalent bonds separately, the kinds of bonds actually found in molecules do not belong entirely to either category. Instead, bonds range from the highly ionic to the highly covalent.

Ionic Bonds

Atoms are electrically neutral when the number of positive charges (protons) equals the number of negative charges (electrons). But when an isolated atom gains or loses electrons, this balance is upset. If the atom gains electrons, it acquires an overall negative charge; if the atom loses electrons, it acquires an overall positive charge. Such a negatively or positively charged atom (or group of atoms) is called an **ion.**

Consider the following examples. Sodium (Na) has 11 protons and 11 electrons, with one electron in its outer electron shell. Sodium tends to lose the single outer

TABLE 2.2	**Electronic Configurations for the Atoms of Some Elements Found in Living Organisms**

Element	First Electron Shell	Second Electron Shell	Third Electron Shell	Diagram	Number of Valence (Outermost) Shell Electrons	Number of Unfilled Spaces	Maximum Number of Bonds Formed
Hydrogen	1	—	—		1	1	1
Carbon	2	4	—		4	4	4
Nitrogen	2	5	—		5	3	3
Oxygen	2	6	—		6	2	2
Magnesium	2	8	2		2	6	2
Phosphorus	2	8	5		5	3	5
Sulfur	2	8	6		6	2	2

Key:
- electron
- unfilled space
- atomic nucleus

FIGURE 2.2 **Ionic bond formation.** **(a)** A sodium atom (Na), left, loses one electron to an electron acceptor and forms a sodium ion (Na⁺). A chlorine atom (Cl), right, accepts one electron from an electron donor to become a chloride ion (Cl⁻). **(b)** The sodium and chloride ions are attracted because of their opposite charges and are held together by an ionic bond to form a molecule of sodium chloride.

■ An ionic bond is a chemical attraction that holds together ions with different charges.

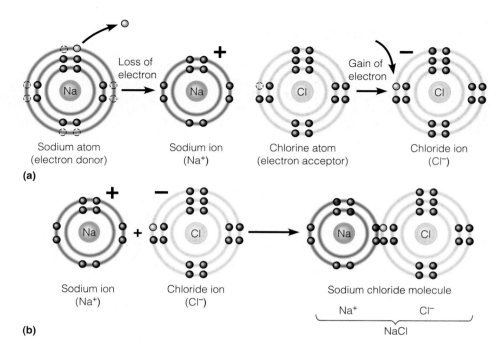

Sodium atom
(electron donor)

Sodium ion
(Na⁺)

Chlorine atom
(electron acceptor)

Chloride ion
(Cl⁻)

(a)

Sodium ion
(Na⁺)

Chloride ion
(Cl⁻)

Sodium chloride molecule

Na⁺   Cl⁻

NaCl

(b)

electron; it is an *electron donor* (Figure 2.2a). When sodium donates an electron to another atom, it is left with 11 protons and only 10 electrons and so has an overall charge of +1. This positively charged sodium atom is called a sodium ion and is written as Na⁺. Chlorine (Cl) has a total of 17 electrons, seven of them in the outer electron shell. Because this outer shell can hold eight electrons, chlorine tends to pick up an electron that has been lost by another atom; it is an *electron acceptor* (Figure 2.2a). By accepting an electron, chlorine totals 18 electrons. However, it still has only 17 protons in its nucleus. The chloride ion therefore has a charge of −1 and is written as Cl⁻.

The opposite charges of the sodium ion (Na⁺) and chloride ion (Cl⁻) attract each other. The attraction, an ionic bond, holds the two atoms together, and a molecule is formed (Figure 2.2b). The formation of this molecule, called sodium chloride (NaCl) or table salt, is a common example of ionic bonding. Thus, an **ionic bond** is an attraction between ions of opposite charge that holds them together to form a stable molecule. Put another way, an ionic bond is an attraction between atoms in which one atom loses electrons and another atom gains electrons. Strong ionic bonds, such as those that hold Na⁺ and Cl⁻ together in salt crystals, have limited importance in living cells. But the weaker ionic bonds formed in aqueous (water) solutions are important in biochemical reactions in microbes and other organisms. For example, weaker ionic bonds assume a role in certain antigen-antibody reactions—that is, reactions in which molecules produced by the immune system (antibodies) combine with foreign substances (antigens) to combat infection.

In general, an atom whose outer electron shell is less than half-filled will lose electrons and form positively charged ions, called **cations.** Examples of cations are the potassium ion (K⁺), calcium ion (Ca²⁺), and sodium ion (Na⁺). When an atom's outer electron shell is more than half-filled, the atom will gain electrons and form negatively charged ions, called **anions.** Examples are the iodide ion (I⁻), chloride ion (Cl⁻), and sulfide ion (S²⁻).

Covalent Bonds

A **covalent bond** is a chemical bond formed by two atoms sharing one or more pairs of electrons. Covalent bonds are stronger and far more common in organisms than are true ionic bonds. In the hydrogen molecule, H₂, two hydrogen atoms share a pair of electrons. Each hydrogen atom has its own electron plus one electron from the other atom (Figure 2.3a). The shared pair of electrons actually orbits the nuclei of both atoms. Therefore, the outer electron shells of both atoms are filled. When only one pair of electrons is shared between atoms, a *single covalent bond* is formed. For simplicity, a single covalent bond is expressed as a single line between the atoms (H—H). When two pairs of electrons are shared between atoms, a *double covalent bond* is formed, expressed as two single lines (=). A *triple covalent bond*, expressed as three single lines (≡), occurs when three pairs of electrons are shared.

The principles of covalent bonding that apply to atoms of the same element also apply to atoms of different elements. Methane (CH₄) is an example of covalent bonding between atoms of different elements (Figure 2.3b). The

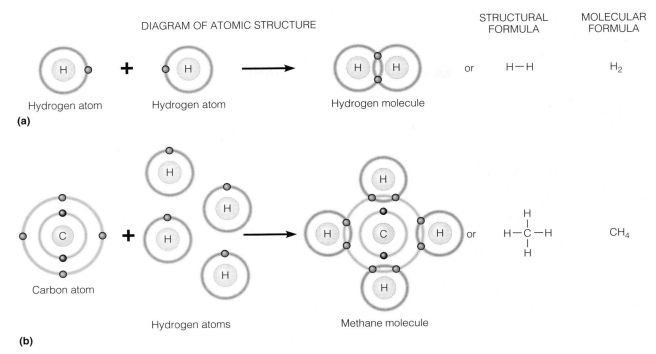

FIGURE 2.3 **Covalent bond formation. (a)** A single covalent bond between two hydrogen atoms. **(b)** Single covalent bonds between four hydrogen atoms and a carbon atom, forming a methane molecule. On the right are simpler ways to represent molecules. In structural formulas, each covalent bond is written as a straight line between the symbols for two atoms. In molecular formulas, the number of atoms in each molecule is noted by subscripts.

■ In a covalent bond, two atoms share one, two, or three pairs of valance electrons.

outer electron shell of the carbon atom can hold eight electrons but has only four; each hydrogen atom can hold two electrons but has only one. Consequently, in the methane molecule the carbon atom gains four hydrogen electrons to complete its outer shell, and each hydrogen atom completes its pair by sharing one electron from the carbon atom. Each outer electron of the carbon atom orbits both the carbon nucleus and a hydrogen nucleus. Each hydrogen electron orbits both its own nucleus and the carbon nucleus.

Elements such as hydrogen and carbon, whose outer electron shells are half-filled, form covalent bonds quite easily. In fact, in living organisms, carbon almost always forms covalent bonds; it almost never becomes an ion. *Remember:* Covalent bonds are formed by the *sharing* of electrons between atoms. Ionic bonds are formed by *attraction* between atoms that have lost or gained electrons and are therefore positively or negatively charged.

Hydrogen Bonds

Another chemical bond of special importance to all organisms is the **hydrogen bond,** in which a hydrogen atom that is covalently bonded to one oxygen or nitrogen atom is attracted to another oxygen or nitrogen atom. Such bonds are weak and do not bind atoms into molecules. However, they do serve as bridges between different molecules or between various portions of the same molecule.

When hydrogen combines with atoms of oxygen or nitrogen, the relatively large nucleus of these larger atoms attracts the hydrogen electron more strongly than does the small hydrogen nucleus. Thus, in a molecule of water (H_2O), all the electrons tend to be closer to the oxygen nucleus than to the hydrogen nuclei. The oxygen portion of the molecule thus has a slightly negative charge, and the hydrogen portion of the molecule has a slightly positive charge (Figure 2.4a). When the positively charged end of one molecule is attracted to the negatively charged end of another molecule, a hydrogen bond is formed (Figure 2.4b). This attraction can also occur between hydrogen and other atoms of the same molecule, especially in large molecules. Oxygen and nitrogen are the elements most frequently involved in hydrogen bonding.

Hydrogen bonds are considerably weaker than either ionic or covalent bonds; they have only about 5% of the

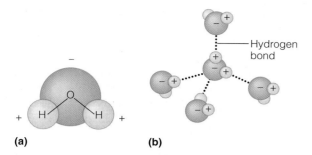

(a)   **(b)**

FIGURE 2.4 **Hydrogen bond formation in water.**
(a) In a water molecule, the electrons of the hydrogen
atoms are strongly attracted to the oxygen atom. Therefore,
the part of the water molecule containing the oxygen atom
has a slightly negative charge, and the part containing
hydrogen atoms has a slightly positive charge. **(b)** In a
hydrogen bond between water molecules, the hydrogen of
one water molecule is attracted to the oxygen of another
water molecule. Many water molecules may be attracted to
each other by hydrogen bonds (black dots).

■ How do hydrogen bonds affect the temperature of water?

strength of covalent bonds. Consequently, hydrogen bonds
are formed and broken relatively easily. This property accounts for the temporary bonding that occurs between certain atoms of large and complex molecules, such as proteins
and nucleic acids. Even though hydrogen bonds are relatively weak, large molecules containing several hundred of
these bonds have considerable strength and stability.

Molecular Weight and Moles

You have seen that bond formation results in the creation
of molecules. Molecules are often discussed in terms of
units of measure called molecular weight and moles. The
molecular weight of a molecule is the sum of the atomic
weights of all its atoms. To relate the molecular level to the
laboratory level, we use a unit called the mole. One **mole**
of a substance is its molecular weight expressed in grams.
For example, 1 mole of water weighs 18 grams because the
molecular weight of H_2O is 18 [$(2 \times 1) + 16$].

Chemical Reactions

Learning Objectives
* *Diagram three basic types of chemical reactions.*

As we said earlier, **chemical reactions** involve the making or breaking of bonds between atoms. After a chemical
reaction, the total number of atoms remains the same, but
there are new molecules with new properties because the
atoms have been rearranged.

Energy in Chemical Reactions

Some change of energy occurs whenever bonds between
atoms are formed or broken during chemical reactions.
This energy is called **chemical energy.** When a chemical
bond is formed, energy is required. Such a chemical reaction that absorbs more energy than it releases is called an
endergonic reaction (*end* = within), meaning that energy is directed inward. When a bond is broken, energy is
released. A chemical reaction that releases more energy
than it absorbs is called an **exergonic reaction** (*ex* =
out), meaning that energy is directed outward.

In this section we will look at three basic types of
chemical reactions common to all living cells. By becoming familiar with these reactions, you will be able to understand the specific chemical reactions we will discuss
later, particularly in Chapter 5.

Synthesis Reactions

When two or more atoms, ions, or molecules combine to
form new and larger molecules, the reaction is called a
synthesis reaction. To synthesize means to put together,
and a synthesis reaction *forms new bonds.* Synthesis reactions can be expressed in the following way:

		Combine to form	
A	+ B	$\longrightarrow$	AB
Atom, ion, or molecule A	Atom, ion, or molecule B		New molecule AB

The combining substances, A and B, are called the *reactants*; the substance formed by the combination, AB, is
the *product.* The arrow indicates the direction in which
the reaction proceeds.

Pathways of synthesis reactions in living organisms are
collectively called anabolic reactions, or simply **anabolism** (an-ab'ō-lizm). The combining of sugar molecules to form starch, and of amino acids to form proteins,
are two examples of anabolism.

Decomposition Reactions

The reverse of a synthesis reaction is a **decomposition reaction.** To decompose means to break down into smaller
parts, and in a decomposition reaction *bonds are broken.*
Typically, decomposition reactions split large molecules
into smaller molecules, ions, or atoms. A decomposition
reaction occurs in the following way:

	Breaks down into		
AB	$\longrightarrow$	A	+ B
Molecule AB		Atom, ion, or molecule A	Atom, ion, or molecule B

Decomposition reactions that occur in living organisms are collectively called catabolic reactions, or simply **catabolism** (ka-tab'ō-lizm). An example of catabolism is the breakdown of sucrose (table sugar) into simpler sugars, glucose and fructose, during digestion. Bacterial decomposition of petroleum is discussed in the box on page 34.

Exchange Reactions

All chemical reactions are based on synthesis and decomposition. Many reactions, such as **exchange reactions,** are actually part synthesis and part decomposition. An exchange reaction works in the following way:

$$\text{AB + CD} \xrightarrow{\substack{\text{Recombine} \\ \text{to form}}} \text{AD + BC}$$

First, the bonds between A and B and between C and D are broken in a decomposition process. New bonds are then formed between A and D and between B and C in a synthesis process. For example, an exchange reaction occurs when sodium hydroxide (NaOH) and hydrochloric acid (HCl) react to form table salt (NaCl) and water (H_2O), as follows:

$$\text{NaOH + HCl} \longrightarrow \text{NaCl + } H_2O$$

The Reversibility of Chemical Reactions

All chemical reactions are, in theory, reversible; that is, they can occur in either direction. In practice, however, some reactions do this more easily than others. A chemical reaction that is readily reversible (when the end product can revert to the original molecules) is termed a **reversible reaction** and is indicated by two arrows, as shown here:

$$\text{A + B} \underset{\text{Breaks down into}}{\overset{\text{Combines to form}}{\rlap{\longrightarrow}\longleftarrow}} \text{AB}$$

Some reversible reactions occur because neither the reactants nor the end products are very stable. Other reactions will reverse only under special conditions:

$$\text{A + B} \underset{\text{Water}}{\overset{\text{Heat}}{\rlap{\longrightarrow}\longleftarrow}} \text{AB}$$

Whatever is written above or below the arrows indicates the special condition under which the reaction in that direction occurs. In this case, A and B react to produce AB only when heat is applied, and AB breaks down into A and B only in the presence of water. See Figure 2.8 on page 39 as another example.

In Chapter 5 we will examine the various factors that affect chemical reactions.

IMPORTANT BIOLOGICAL MOLECULES

Biologists and chemists divide compounds into two principal classes: inorganic and organic. **Inorganic compounds** are defined as molecules, usually small and structurally simple, that typically lack carbon and in which ionic bonds may play an important role. Inorganic compounds include water, oxygen, carbon dioxide, and many salts, acids, and bases.

Organic compounds always contain carbon and hydrogen and are typically structurally complex. Carbon is a unique element because it has four electrons in its outer shell and four unfilled spaces. It can combine with a variety of atoms, including other carbon atoms, to form straight or branched chains and rings. Carbon chains form the basis of many organic compounds in living cells, including sugars, amino acids, and vitamins. Organic compounds are held together mostly or entirely by covalent bonds. Some organic molecules, such as polysaccharides, proteins, and nucleic acids, are very large and usually contain thousands of atoms. Such giant molecules are called *macromolecules.* In the following section we will discuss inorganic and organic compounds that are essential for cells.

Inorganic Compounds

Water

> #### Learning Objective
>
> - List several properties of water that are important to living systems.

All living organisms require a wide variety of inorganic compounds for growth, repair, maintenance, and reproduction. Water is one of the most important, as well as one of the most abundant, of these compounds, and it is particularly vital to microorganisms. Outside the cell, nutrients are dissolved in water, which facilitates their passage through cell membranes. And inside the cell, water is the medium for most chemical reactions. In fact, water is by far the most abundant component of almost all living cells. Water makes up at least 5–95% of every cell, the average being between 65% and 75%. Simply stated, no organism can survive without water.

APPLICATIONS OF MICROBIOLOGY

Bioremediation—Bacteria Clean Up Pollution

Readers of science fiction have long realized that beings from another planet might have quite a different chemical makeup from Earthlings and might be able to eat, drink, and breathe the substances we find deadly. Such aliens could be invaluable in helping clean up pollutants such as crude oil, gasoline, and mercury, which harm plants, animals, and humans alike. Fortunately, however, we need not wait for a visit from outer space to find creatures whose unusual chemistry can be harnessed for environmental cleanup. Although many bacteria have dietary requirements similar to ours—that's why they cause food spoilage—others metabolize (or process chemically) the substances we might expect at a banquet of extraterrestrials: heavy metals, sulfur, nitrogen gas, petroleum, and even polychlorinated biphenyls (PCBs) and mercury.

Bacteria have several advantages as pollution fighters. They can extract pollutants that have combined with soil and water and hence cannot be simply shoveled away. In addition, they may chemically alter a harmful substance so that it becomes harmless or even beneficial. Bacteria that can degrade many pollutants are naturally present in soil and water; using them to degrade pollutants is called *bioremediation*. However, their small numbers make them inefficient in dealing with large-scale contamination. Scientists are now working to improve the efficiency of natural pollution fighters and, in some cases, are altering organisms by genetic engineering to give them exactly the right chemical appetites.

One of the most promising successes for bioremediation occurred on an Alaskan beach following the *Exxon Valdez* oil spill. Several naturally occurring bacteria in the genus *Pseudomonas* are able to degrade oil for their carbon and energy requirements. In the presence of air, they remove two carbons at a time from a large petroleum molecule (see the figures).

The bacteria degrade the oil too slowly to be helpful in cleaning up an oil spill. However, scientists hit upon a very simple way to speed up the process—with no need for genetic engineering. They simply dumped ordinary nitrogen and phosphorus plant fertilizers (bioenhancers) onto the beach. The number of oil-degrading bacteria increased compared with that on unfertilized control beaches, and oil was quickly cleared from the test beach.

Another group of bacteria is being investigated for its ability to clean up mercury contamination. Mercury is present in such common substances as discarded leftover paint and can leak into soil and water from garbage dumps. One species of bacteria that is common in the environment, *Desulfovibrio desulfuricans*, actually makes the mercury more dangerous by adding a methyl group, converting it into the highly toxic substance methyl mercury. Methyl mercury in ponds or marshes sticks to small organisms such as plankton, which are eaten by larger organisms, which in turn are eaten by fish. Fish and human poisonings have been attributed to the ingestion of methyl mercury.

However, other bacteria, such as species of *Pseudomonas*, may offer the solution. To avoid mercury poisoning, these bacteria first convert methyl mercury to mercuric ion:

$$CH_3Hg \rightarrow CH_4 + Hg^{2+}$$

Methyl   Methane   Mercuric
mercury   ion

Many bacteria can then convert the positively charged mercuric ion to the relatively harmless elemental form by adding electrons, which they take from hydrogen atoms:

$$Hg^{2+} + 2H \xrightarrow{2e^-} Hg + 2H^+$$

Mercuric Hydrogen Elemental Hydrogen
ion   atoms   mercury   ions

These bacteria work too slowly in nature for cleaning up human-made toxic spills, but scientists are experimenting with bioenhancers and other techniques to increase their effectiveness. Unlike some forms of environmental cleanup, in which dangerous substances are removed from one place only to be dumped in another, bacterial cleanup eliminates the toxic substance and often returns a harmless or useful substance to the environment.

Typical saturated hydrocarbon found in petroleum

Two-carbon unit can be metabolized in cell

Water has structural and chemical properties that make it particularly suitable for its role in living cells. As we discussed, the total charge on the water molecule is neutral, but the oxygen region of the molecule has a slightly negative charge and the hydrogen region has a slightly positive charge (see Figure 2.4a on page 32). Any molecule having such an unequal distribution of charges is called a **polar molecule.** The polar nature of water gives it four characteristics that make it a useful medium for living cells.

First, every water molecule is capable of forming four hydrogen bonds with nearby water molecules (see Figure 2.4b). This property results in a strong attraction between water molecules. Because of this strong attraction, a great deal of heat is required to separate water molecules from each other to form water vapor; thus, water has a relatively high boiling point (100°C). Because water has such a high boiling point, it exists in the liquid state on most of the Earth's surface. Furthermore, the hydrogen bonding between water molecules affects the density of water, depending on whether it occurs as ice or a liquid. For example, the hydrogen bonds in the crystalline structure of water (ice) make ice take up more space. As a result, ice has fewer molecules than an equal volume of liquid water. This makes its crystalline structure less dense than liquid water. For this reason, ice floats and can serve as an insulating layer on the surfaces of lakes and streams that harbor living organisms.

Second, the polarity of water makes it an excellent dissolving medium or **solvent.** Many polar substances undergo **dissociation,** or separation, into individual molecules in water—that is, they dissolve—because the negative part of the water molecules is attracted to the positive part of the molecules in the **solute,** or dissolving substance, and the positive part of the water molecules is attracted to the negative part of the solute molecules. Substances (such as salts) that are composed of atoms (or groups of atoms) held together by ionic bonds tend to dissociate into separate cations and anions in water. Thus, the polarity of water allows molecules of many different substances to separate and become surrounded by water molecules (Figure 2.5).

Third, polarity accounts for water's characteristic role as a reactant or product in many chemical reactions. Its polarity facilitates the splitting and rejoining of hydrogen ions (H^+) and hydroxide ions (OH^-). Water is a key reactant in the digestive processes of organisms, whereby larger molecules are broken down into smaller ones. Water molecules are also involved in synthetic reactions; water is an important source of the hydrogen and oxygen that are incorporated into numerous organic compounds in living cells.

Finally, the relatively strong hydrogen bonding between water molecules (see Figure 2.4b) makes water an

Key

- Chloride ion   ⊖ Oxygen
+ Sodium ion   ⊕ Hydrogen

Sodium chloride crystal

(a) Sodium ion dissolved in water

(b) Chloride ion dissolved in water

FIGURE 2.5 **How water acts as a solvent for sodium chloride (NaCl). (a)** The positively charged sodium ion (Na^+) is attracted to the negative part of the water molecule. **(b)** The negatively charged chloride ion (Cl^-) is attracted to the positive part of the water molecule. In the presence of water molecules, the bonds between the Na^+ and Cl^- are disrupted, and the NaCl dissolves in the water.

■ Ionization is the separation of molecules into ions.

excellent temperature buffer. A given quantity of water requires a great gain of heat to increase its temperature and a great loss of heat to decrease its temperature, compared with many other substances. Normally, heat absorption by molecules increases their kinetic energy and thus increases their rate of motion and their reactivity. In water, however, heat absorption first breaks hydrogen bonds rather than increasing the rate of motion. Therefore, much more heat must be applied to raise the temperature of water than to raise the temperature of a non–hydrogen-bonded liquid. The reverse is true as water cools. Thus, water more easily maintains a constant temperature than other solvents and tends to protect a cell from fluctuations in environmental temperatures.

Acids, Bases, and Salts

Learning Objective

- *Define acid, base, salt, and pH.*

As we saw in Figure 2.5, when inorganic salts such as sodium chloride (NaCl) are dissolved in water, they undergo **ionization** or *dissociation;* that is, they break apart into ions. Substances called acids and bases show similar behavior.

An **acid** can be defined as a substance that dissociates into one or more hydrogen ions (H^+) and one or more

FIGURE 2.6 **Acids, bases, and salts. (a)** In water, hydrochloric acid (HCl) dissociates into H$^+$ and Cl$^-$. **(b)** Sodium hydroxide (NaOH), a base, dissociates into OH$^-$ and Na$^+$ in water. **(c)** In water, table salt (NaCl) dissociates into positive ions (Na$^+$) and negative ions (Cl$^-$), neither of which are H$^+$ or OH$^-$.

■ How do acids and bases differ?

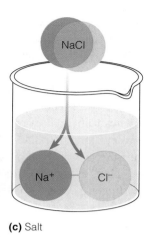

(a) Acid   **(b)** Base   **(c)** Salt

negative ions (anions). Thus, an acid can also be defined as a proton (H$^+$) donor. A **base** dissociates into one or more positive ions (cations) plus one or more negatively charged hydroxide ions (OH$^-$) that can accept, or combine with, protons. Thus, sodium hydroxide (NaOH) is a base because it dissociates to release OH$^-$, which have a strong attraction for protons and are among the most important proton acceptors. A **salt** is a substance that dissociates in water into cations and anions, neither of which is H$^+$ or OH$^-$. Figure 2.6 shows common examples of each type of compound and how they dissociate in water.

Acid–Base Balance

An organism must maintain a fairly constant balance of acids and bases to remain healthy. For example, if a particular acid or base concentration is too high or too low, enzymes change in shape and no longer function effectively in promoting chemical reactions in a cell. In the aqueous environment within organisms, acids dissociate into hydrogen ions (H$^+$) and anions. Bases, in contrast, dissociate into hydroxide ions (OH$^-$) and cations. The more hydrogen ions that are free in a solution, the more acidic the solution is. Conversely, the more hydroxide ions that are free in a solution, the more basic, or alkaline, it is.

Biochemical reactions—that is, chemical reactions in living systems—are extremely sensitive to even small changes in the acidity or alkalinity of the environments in which they occur. In fact, H$^+$ and OH$^-$ are involved in almost all biochemical processes, and the functions of a cell are modified greatly by any deviation from its narrow band of normal H$^+$ and OH$^-$ concentrations. For this reason, the acids and bases that are continually formed in an organism must be kept in balance.

It is convenient to express the amount of H$^+$ in a solution by a logarithmic **pH** scale, which ranges from 0 to 14 (Figure 2.7). The term *pH* means potential of hydrogen.

On a logarithmic scale, a change of one whole number represents a tenfold change from the previous concentration. Thus, a solution of pH 1 has ten times more hydrogen ions than a solution of pH 2 and has 100 times more hydrogen ions than a solution of pH 3.

A solution's pH is calculated as $-\log_{10}[H^+]$, the negative logarithm to the base 10 of the hydrogen ion concentration (denoted by brackets), determined in moles per liter [H$^+$]. For example, if the H$^+$ concentration of a solution is 1.0×10^{-4} moles/liter, or 10^{-4}, its pH equals $-\log_{10} 10^{-4} = -(-4) = 4$; this is about the pH value of wine (see Appendix D). The pH values of some human body fluids and other common substances are also shown in Figure 2.7. In the laboratory, you will usually measure the pH of a solution with a pH meter or with chemical test papers.

Acidic solutions contain more H$^+$ than OH$^-$ and have a pH lower than 7. If a solution has more OH$^-$ than H$^+$, it is a basic, or alkaline, solution. In pure water, a small percentage of the molecules are dissociated into H$^+$ and OH$^-$, so it has a pH of 7. Because the concentrations of H$^+$ and OH$^-$ are equal, this pH is said to be the pH of a neutral solution.

Keep in mind that the pH of a solution can be changed. We can increase its acidity by adding substances that will increase the concentration of hydrogen ions. As a living organism takes up nutrients, carries out chemical reactions, and excretes wastes, its balance of acids and bases tends to change, and the pH fluctuates. Fortunately, organisms possess natural pH **buffers,** compounds that help keep the pH from changing drastically. But the pH in our environment's water and soil can be altered by waste products from organisms, pollutants from industry, or fertilizers used in agricultural fields or gardens. When bacteria are grown in a laboratory medium, they excrete waste products such as acids that can alter the pH of the medium. If this effect were to continue, the medium would become acidic

pH scale

pH	
0	
1	
	Stomach acid
2	Lemon juice
3	Grapefruit juice
4	Wine Tomato juice
5	
6	Urine
	Milk
7	**Pure water**
	Human blood
8	
	Seawater
9	
10	
	Milk of magnesia
11	
	Household ammonia
12	
	Household bleach
13	
	Oven cleaner
14	Limewater

H^+
OH^-
Acidic solution

Neutral solution

Basic solution

FIGURE 2.7 **The pH scale.** As pH values decrease from 14 to 0, the H^+ concentration increases. Thus, the lower the pH, the more acidic the solution; the higher the pH, the more basic the solution. If the pH value of a solution is below 7, the solution is acidic; if the pH is above 7, the solution is basic (alkaline). The approximate pH values of some human body fluids and common substances are shown next to the pH scale.

■ At what pH is the concentration of H^+ and OH^- equal?

enough to inhibit bacterial enzymes and cause the death of the bacteria. To prevent this problem, pH buffers are added to the culture medium. One very effective pH buffer for some culture media uses a mixture of K_2HPO_4 and KH_2PO_4 (see Table 6.3 on page 165).

Different microbes function best within different pH ranges, but most organisms grow best in environments with a pH value between 6.5 and 8.5. Among microbes, fungi are best able to tolerate acidic conditions, whereas the prokaryotes called cyanobacteria tend to do well in alkaline habitats. *Propionibacterium acnes* (prō-pē-on-ē-bak-ti′rē-um ak′nēz), a bacterium that contributes to acne, has as its natural environment human skin, which tends to be slightly acidic, with a pH of about 4. *Thiobacillus ferrooxidans* (thī-ō-bä-sil′lus fer-rō-oks′i-danz) is a bacterium that grows on elemental sulfur and produces sulfuric acid (H_2SO_4). Its pH range for optimum growth is from 1 to 3.5. The sulfuric acid produced by this bacterium in mine water is important in dissolving uranium and copper from low-grade ore (see the box on page 34).

Organic Compounds

Learning Objectives

- *Distinguish between organic and inorganic compounds.*
- *Identify the building blocks of carbohydrates, simple lipids, phospholipids, proteins, and nucleic acids.*

Inorganic compounds, excluding water, constitute about 1–1.5% of living cells. These relatively simple components, whose molecules have only a few atoms, cannot be used by cells to perform complicated biological functions. Organic molecules, whose carbon atoms can combine in an enormous variety of ways with other carbon atoms and with atoms of other elements, are relatively complex and thus are capable of more complicated biological functions.

Structure and Chemistry

In the formation of organic molecules, carbon's four outer electrons can participate in up to four covalent bonds, and carbon atoms can bond to each other to form straight-chain, branched-chain, or ring structures.

In addition to carbon, the most common elements in organic compounds are hydrogen (which can form one bond), oxygen (two bonds), and nitrogen (three bonds). Sulfur (two bonds) and phosphorus (five bonds) appear less often. Other elements are found, but only in a relatively few organic compounds. The elements that are most abundant in living organisms are the same as those that are most abundant in organic compounds (see Table 2.1, page 27).

The chain of carbon atoms in an organic molecule is called the **carbon skeleton;** a huge number of combinations is possible for carbon skeletons. Most of these carbons are bonded to hydrogen atoms. The bonding of other elements with carbon and hydrogen forms characteristic **functional groups,** specific groups of atoms that are most commonly involved in chemical reactions and are responsible for most of the characteristic chemical properties and many of the physical properties of a particular organic compound (Table 2.3).

Different functional groups confer different properties on organic molecules. For example, the hydroxyl group of alcohols is hydrophilic (water-loving) and thus attracts water molecules to it. This attraction helps dissolve organic molecules containing hydroxyl groups. Because the carboxyl group is a source of hydrogen ions, molecules containing it have acidic properties. Amino groups, by contrast, function as bases because they readily accept hydrogen ions. The sulfhydryl group helps stabilize the intricate structure of many proteins.

Functional groups help us classify organic compounds. For example, the —OH group is present in each of the following molecules:

Methanol   Ethanol

Isopropanol

Because the characteristic reactivity of the molecules is based on the —OH group, they are grouped together in a class called alcohols. The —OH group is called the *hydroxyl group* and is not to be confused with the *hydroxide ion* (OH⁻) of bases. The hydroxyl group of alcohols does not ionize at neutral pH; it is covalently bonded to a carbon atom.

When a class of compounds is characterized by a certain functional group, the letter **R** can be used to stand for the remainder of the molecule. For example, alcohols in general may be written R—OH.

TABLE 2.3	**Representative Functional Groups and the Compounds in Which They Are Found**	
Structure	**Name of Group**	**Biological Importance**
R—O—H	Alcohol	Lipids, carbohydates
R—C(=O)—H	Aldehyde*	Reducing sugars such as glucose; polysaccharides
R—C(=O)—R	Ketone*	Metabolic intermediates
R—C(H)(H)—H	Methyl	DNA; energy metabolism
R—C(H)(H)—NH₂	Amino	Proteins
R—C(=O)—O—R'	Ester	Bacteria and eukaryotic plasma membranes
R—C(H)(H)—O—C(H)(H)—R'	Ether	Archaea plasma membranes
R—C(H)(H)—SH	Sulfhydryl	Energy metabolism; protein structure
R—C(=O)—OH	Carboxyl	Organic acids, lipids, proteins
R=O—P(O⁻)(O⁻)—O	Phosphate	ATP, DNA

*In an aldehyde, a C=O is at the end of a molecule, in contrast to the internal C=O in a ketone.

Frequently, more than one functional group is found in a single molecule. For example, an amino acid molecule contains both amino and carboxyl groups. The amino acid glycine has the following structure:

Most of the organic compounds found in living organisms are quite complex; a large number of carbon atoms form the skeleton, and many functional groups are attached. In organic molecules, it is important that each of the four bonds of carbon is satisfied (attached to another atom) and that each of the attaching atoms has its characteristic number of bonds satisfied. Because of this, such molecules are chemically stable.

Small organic molecules can be combined into very large molecules called **macromolecules** (*macro* = *large*). Macromolecules are usually **polymers** (*poly* = many; *mers* = parts), large molecules formed by covalent bonding of many repeating small molecules called **monomers** (*mono* = one). When two monomers join together, the reaction usually involves the elimination of a hydrogen atom from one monomer and a hydroxyl group from the other; the hydrogen atom and the hydroxyl group combine to produce water:

$$R - \boxed{OH + H} - R' \longrightarrow R - R' + \boxed{H_2O}$$

This type of exchange reaction is called **dehydration synthesis** (*de* = from; *hydra* = water), or a **condensation reaction,** because a molecule of water is released (Figure 2.8a). Such macromolecules as carbohydrates, lipids, proteins, and nucleic acids are assembled in the cell, essentially by dehydration synthesis. However, other molecules must also participate to provide energy for bond formation.

ATP, the cell's chief energy provider, is discussed at the end of this chapter.

Carbohydrates

The **carbohydrates** are a large and diverse group of organic compounds that includes sugars and starches. Carbohydrates perform a number of major functions in living systems. For instance, one type of sugar (deoxyribose) is a building block of deoxyribonucleic acid (DNA), the molecule that carries hereditary information. Other sugars help form the cell walls of bacterial cells. Simple carbohydrates are used in the synthesis of amino acids and fats or fatlike substances, which are used to build structures and provide an emergency source of energy. Macromolecular carbohydrates function as food reserves. The principal function of carbohydrates, however, is to fuel cell activities with a ready source of energy.

Carbohydrates are made up of carbon, hydrogen, and oxygen atoms. The ratio of hydrogen to oxygen atoms is always 2:1 in simple carbohydrates. This ratio can be seen in the formulas for the carbohydrates ribose ($C_5H_{10}O_5$), glucose ($C_6H_{12}O_6$), and sucrose ($C_{12}H_{22}O_{11}$). Although there are exceptions, the general formula for carbohydrates is $(CH_2O)_n$, where *n* indicates that there are three or more CH_2O units. Carbohydrates can be classified into three major groups on the basis of size: monosaccharides, disaccharides, and polysaccharides.

Monosaccharides

Simple sugars are called **monosaccharides** (*sacchar* = sugar); each molecule contains from three to seven carbon atoms. The number of carbon atoms in the molecule of a simple sugar is indicated by the prefix in its name. For example, simple sugars with three carbons are called trioses. There are also tetroses (four-carbon sugars), pentoses (five-carbon

FIGURE 2.8 **Dehydration synthesis and hydrolysis. (a)** In dehydration synthesis (left to right), the monosaccharides glucose and fructose combine to form a molecule of the disaccharide sucrose. A molecule of water is released in the reaction. **(b)** In hydrolysis (right to left), the sucrose molecule breaks down into the smaller molecules glucose and fructose. For the hydrolysis reaction to proceed, water must be added to the sucrose.

■ What is the difference between a polymer and a monomer?

sugars), hexoses (six-carbon sugars), and heptoses (seven-carbon sugars). Pentoses and hexoses are extremely important to living organisms. Deoxyribose is a pentose found in DNA. Glucose, a very common hexose, is the main energy-supplying molecule of living cells.

Disaccharides

Disaccharides (*di* = two) are formed when two monosaccharides bond in a dehydration synthesis reaction.* For example, molecules of two monosaccharides, glucose and fructose, combine to form a molecule of the disaccharide sucrose (table sugar) and a molecule of water (Figure 2.8a). Similarly, the dehydration synthesis of the monosaccharides glucose and galactose forms the disaccharide lactose (milk sugar).

It may seem odd that glucose and fructose have the same chemical formula (Figure 2.8), even though they are different monosaccharides. The positions of the oxygens and carbons differ in the two different molecules, and consequently the molecules have different physical and chemical properties. Two molecules with the same chemical formula but different structures and properties are called **isomers** (*iso* = same).

Disaccharides can be broken down into smaller, simpler molecules when water is added. This chemical reaction, the reverse of dehydration synthesis, is called **hydrolysis** (*hydro* = water; *lysis* = to loosen) (Figure 2.8b). A molecule of sucrose, for example, may be hydrolyzed (digested) into its components of glucose and fructose by reacting with the H^+ and OH^- of water.

As you will see in Chapter 4, the cell walls of bacterial cells are composed of disaccharides and proteins (together called peptidoglycan).

Polysaccharides

Carbohydrates in the third major group, the **polysaccharides,** consist of tens or hundreds of monosaccharides joined through dehydration synthesis. Polysaccharides often have side chains branching off the main structure and are classified as macromolecules. Like disaccharides, polysaccharides can be split apart into their constituent sugars through hydrolysis. Unlike monosaccharides and disaccharides, however, they usually lack the characteristic sweetness of sugars such as fructose and sucrose and usually are not soluble in water.

One important polysaccharide is *glycogen,* which is composed of glucose subunits and is synthesized as a stor-

age material by animals and some bacteria. *Cellulose,* another important glucose polymer, is the main component of the cell walls of plants and most algae. Although cellulose is the most abundant carbohydrate on Earth, it can be digested by only a few organisms that have the appropriate enzyme. The polysaccharide *dextran,* which is produced as a sugary slime by certain bacteria, is used in a blood plasma substitute. *Chitin* is a polysaccharide that makes up part of the cell wall of most fungi and the exoskeletons of lobsters, crabs, and insects. *Starch* is a polymer of glucose produced by plants and used as food by humans.

Many animals, including humans, produce enzymes called *amylases* that can break the bonds between the glucose molecules in glycogen. However, this enzyme cannot break the bonds in cellulose. Bacteria and fungi that produce enzymes called *cellulases* can digest cellulose. Cellulases from the fungus *Trichoderma* (trik′ō-dėr-mä) are used for a variety of industrial purposes. One of the more unusual uses is producing stone-washed denim. Because washing the fabric with rocks would damage washing machines, cellulase is used to digest, and therefore soften, the cotton. (See the box in Chapter 9 on page 251.)

Lipids

If lipids were suddenly to disappear from the Earth, all living cells would collapse in a pool of fluid, because lipids are essential to the structure and function of membranes that separate living cells from their environment. **Lipids** (*lip* = fat) are a second major group of organic compounds found in living matter. Like carbohydrates, they are composed of atoms of carbon, hydrogen, and oxygen, but lipids lack the 2:1 ratio between hydrogen and oxygen atoms. Even though lipids are a very diverse group of compounds, they share one common characteristic: They are *nonpolar* molecules; unlike water, they do not have a positive and a negative end (pole). Therefore, most lipids are insoluble in water but dissolve readily in nonpolar solvents, such as ether and chloroform. Lipids function in energy storage and provide part of the structure of membranes and cell walls.

Simple Lipids

Simple lipids, called *fats* or *triglycerides,* contain an alcohol called *glycerol* and a group of compounds known as *fatty acids.* Glycerol molecules have three carbon atoms to which are attached three hydroxyl (—OH) groups (Figure 2.9a). Fatty acids consist of long hydrocarbon chains (composed only of carbon and hydrogen atoms) ending in a carboxyl (—COOH, organic acid) group (Figure 2.9b). Most common fatty acids contain an even number of carbon atoms.

---

*Carbohydrates composed of 2 to about 20 monosaccharides are called **oligosaccharides** (*oligo* = few). Disaccharides are the most common oligosaccharides.

(a) Glycerol

(b) Fatty acid (palmitic acid), $C_{15}H_{31}COOH$

(c) Molecule of fat (triglyceride)

Palmitic acid ($C_{15}H_{31}COOH$) + H_2O
(Saturated)

Stearic acid ($C_{17}H_{35}COOH$) + H_2O
(Saturated)

Oleic acid ($C_{17}H_{33}COOH$) + H_2O
(Unsaturated)

FIGURE 2.9 **Structural formulas of simple lipids. (a)** Glycerol. **(b)** Palmitic acid, a fatty acid. **(c)** The chemical combination of a molecule of glycerol and three fatty acid molecules (palmitic, stearic, and oleic in this example) forms one molecule of fat (triglyceride) and three molecules of water in a dehydration synthesis reaction. The bond between glycerol and each fatty acid is called an ester linkage. The addition of three water molecules to a fat forms glycerol and three fatty acid molecules in a hydrolysis reaction.

■ How do saturated and unsaturated fatty acids differ?

A fat molecule is formed when a molecule of glycerol combines with one to three fatty acid molecules to form a monoglyceride, diglyceride, or triglyceride (Figure 2.9c). In the reaction, one to three molecules of water are formed (dehydration), depending on the number of fatty acid molecules reacting. The chemical bond formed where the water molecule is removed is called an *ester linkage*. In the reverse reaction, hydrolysis, a fat molecule is broken down into its component fatty acid and glycerol molecules.

Because the fatty acids that form lipids have different structures, there is a wide variety of lipids. For example, three molecules of fatty acid A might combine with a glycerol molecule. Or one molecule each of fatty acids A, B, and C might unite with a glycerol molecule (see Figure 2.9c).

The primary function of lipids is the formation of plasma membranes that enclose cells. A plasma membrane supports the cell and allows nutrients and wastes to pass in and out; therefore, the lipids must maintain the same viscosity, regardless of the temperature of the surroundings. The membrane must be about as viscous as olive oil, without getting too fluid when warmed or too thick when cooled. As everyone who has ever cooked a meal knows,

animal fats (such as butter) are usually solid at room temperature, whereas vegetable oils are usually liquid at room temperature. The difference in their respective melting points is caused by the degrees of saturation of the fatty acid chains. A fatty acid is said to be *saturated* when it has no double bonds; then the carbon skeleton contains the maximum number of hydrogen atoms (see Figures 2.9c and 2.10a). Saturated chains become solid more easily because they are relatively straight and are thus able to pack together more closely than unsaturated chains. The double bonds of *unsaturated* chains create kinks in the chain, which keep the chains apart from one another (Figure 2.10b).

Complex Lipids

Complex lipids contain such elements as phosphorus, nitrogen, and sulfur, in addition to the carbon, hydrogen, and oxygen found in simple lipids. The complex lipids called *phospholipids* are made up of glycerol, two fatty acids, and, in place of a third fatty acid, a phosphate group bonded to one of several organic groups (Figure 2.10a). Phospholipids are the lipids that build membranes; they are essential to a cell's survival. Phospholipids have polar as well as nonpolar

Saturated fatty acids (closely packed)

Unsaturated fatty acids (loosely packed)

(b)

(c) Sterol molecules separate fatty acid chains

FIGURE 2.10 **Phospholipid structure and orientation, showing saturated and unsaturated fatty acids and the molecules' polarity. (a)** Phospholipid structure. The fatty acids and the R group (at top) may vary with the particular phospholipid. **(b)** This structure (greatly reduced) is the symbol used to represent phospholipids throughout the text. **(c)** Orientation of phospholipids in a plasma membrane, where they form a bilayer, with the hydrophilic heads in contact with the water and the hydrophobic tails oriented away from the water.

■ Where are phospholipids found in cells?

regions (Figure 2.10a and b; see also Figure 4.13 on page 84). When placed in water, phospholipid molecules twist themselves in such a way that all polar (hydrophilic) portions will orient themselves toward the polar water molecules, with which they then form hydrogen bonds. (Recall that *hydrophilic* means water-loving.) This forms the basic structure of a plasma membrane (Figure 2.10c). Polar portions consist of a phosphate group and glycerol. In contrast to the polar regions, all nonpolar (hydrophobic) parts of the phospholipid make contact only with the nonpolar portions of neighboring molecules. (*Hydrophobic* means water-fearing.) Nonpolar portions consist of fatty acids. This characteristic behavior makes phospholipids particularly suitable to being a major component of the membranes that enclose cells. Phospholipids enable the membrane to act as a barrier that separates the contents of the cell from the water-based environment in which it lives.

Some complex lipids are useful in identifying certain bacteria. For example, the cell wall of *Mycobacterium tuberculosis* (mī-kō-bak-ti′rē-um tü-bėr-kū-lō′sis), the bacterium that causes tuberculosis, is distinguished by its lipid-rich content. The cell wall contains complex lipids such as waxes and glycolipids (lipids with carbohydrates attached) that give the bacterium distinctive staining characteristics. Cell walls rich in such complex lipids are characteristic of all members of the genus *Mycobacterium*.

Steroids

Steroids are structurally very different from lipids. Figure 2.11 shows the structure of the steroid cholesterol, with the four interconnected carbon rings that are characteristic of steroids. When an —OH group is attached to one of the rings, the steroid is called a *sterol* (an alcohol). Sterols are important constituents of the plasma membranes of animal cells and of one group of bacteria (mycoplasmas), and

FIGURE 2.11 Cholesterol, a steroid. Note the four "fused" carbon rings (labeled A–D), which are characteristic of steroid molecules. The hydrogen atoms attached to the carbons at the corners of the rings have been omitted. The —OH group (colored red) makes this molecule a sterol.

■ Where are sterols found in cells?

they are also found in fungi and plants. The sterols separate the fatty acid chains and thus prevent the packing that would harden the plasma membrane at low temperatures (see Figure 2.10c).

Proteins

Proteins are organic molecules that contain carbon, hydrogen, oxygen, and nitrogen. Some also contain sulfur. If you were to separate and weigh all the groups of organic compounds in a living cell, the proteins would tip the scale. Hundreds of different proteins can be found in any single cell, and together they make up 50% or more of a cell's dry weight.

Proteins are essential ingredients in all aspects of cell structure and function. *Enzymes* are the proteins that speed up biochemical reactions. But proteins have other functions as well. *Transporter proteins* help transport certain chemicals into and out of cells. Other proteins, such as the *bacteriocins* produced by many bacteria, kill other bacteria. Certain *toxins,* called exotoxins, produced by some disease-causing microorganisms are also proteins. Some proteins play a role in the *contraction* of animal muscle cells and the *movement* of microbial and other types of cells. Other proteins are integral parts of *cell structures* such as walls, membranes, and cytoplasmic components. Still others, such as the *hormones* of certain organisms, have regulatory functions. As we will see in Chapter 17, proteins called *antibodies* play a role in vertebrate immune systems.

Amino Acids

Just as monosaccharides are the building blocks of larger carbohydrate molecules, and fatty acids and glycerol are the building blocks of fats, **amino acids** are the building blocks of proteins. Amino acids contain at least one carboxyl (—COOH) group and one amino (—NH₂) group attached to the same carbon atom, called an alpha-carbon (written C_α) (Figure 2.12a). Such amino acids are called *alpha-amino*

acids. Also attached to the alpha-carbon is a side group (R group), which is the amino acid's distinguishing feature. The side group can be a hydrogen atom, an unbranched or branched chain of atoms, or a ring structure that is cyclic (all carbon) or heterocyclic (when an atom other than carbon is included in the ring). Figure 2.12b shows the structural formula of tyrosine, an amino acid that has a cyclic side group. The side group can contain functional groups, such as the sulfhydryl group (—SH), the hydroxyl group (—OH), or additional carboxyl or amino groups. These side groups and the carboxyl and alpha-amino groups affect the total structure of a protein, described later. The structures and standard abbreviations of the 20 amino acids found in proteins are shown in Table 2.4.

Most amino acids exist in either of two configurations called **stereoisomers,** designated by D and L. These configurations are mirror images, corresponding to "right-handed" (D) and "left-handed" (L) three-dimensional shapes (Figure 2.13). The amino acids found in proteins are always the L-isomers (except for glycine, the simplest amino acid, which does not have stereoisomers). However, D-amino acids occasionally occur in nature—for example, in certain bacterial cell walls and antibiotics. (Many other kinds of organic molecules also can exist in D and L forms. One example is the sugar glucose, which occurs in nature as D-glucose.)

Although only 20 different amino acids occur naturally in proteins, a single protein molecule can contain from 50 to hundreds of amino acid molecules, which can be arranged in an almost infinite number of ways to make proteins of different lengths, compositions, and structures. The number of proteins is practically endless, and every living cell produces many different proteins.

Peptide Bonds

Amino acids bond between the carbon atom of the carboxyl (—COOH) group of one amino acid and the nitrogen

(a) Generalized amino acid   **(b)** Tyrosine

FIGURE 2.12 Amino acid structure. (a) The general structural formula for an amino acid. The alpha-carbon (C_α) is shown in the center. Different amino acids have different R groups, also called side groups. **(b)** Structural formula for the amino acid tyrosine, which has a cyclic side group.

■ Each amino acid has a unique R group.

TABLE 2.4	The 20 Amino Acids Found in Proteins*

Glycine (Gly)

Hydrogen atom

Alanine (Ala)

Unbranched chain

Valine (Val)

Branched chain

Leucine (Leu)

Branched chain

Isoleucine (Ile)

Branched chain

Serine (Ser)

Hydroxyl (—OH) group

Threonine (Thr)

Hydroxyl (—OH) group

Cysteine (Cys)

Sulphur containing (—SH) group

Methionine (Met)

Thioether (SC) group

Glutamic acid (Glu)

Additional carboxyl (—COOH) group, acidic

Aspartic acid (Asp)

Additional carboxyl (—COOH) group, acidic

Lysine (Lys)

Additional amino (—NH₂) group, basic

Arginine (Arg)

Additional amino (—NH₂) group, basic

Asparagine (Asn)

Additional amino (—NH₂) group, basic

Glutamine (Gln)

Additional amino (—NH₂) group, basic

Phenylalanine (Phe)

Cyclic

Tyrosine (Tyr)

Cyclic

Histidine (His)

Heterocyclic

Tryptophan (Trp)

Heterocyclic

Proline (Pro)

Heterocyclic

*Shown are the amino acid names, including the three-letter abbreviation in parentheses (above), their structural formulas (center), and characteristic R group (below). Note that cysteine and methionine are the only amino acids that contain sulfur.

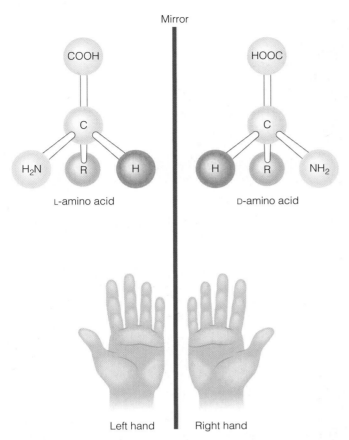

Mirror

L-amino acid

D-amino acid

Left hand   Right hand

FIGURE 2.13 **The L- and D-isomers of an amino acid, shown with ball-and-stick models.** The two isomers, like left and right hands, are mirror images of each other and cannot be superimposed on one another. (Try it!)

■ The L-isomer is always found in proteins.

atom of the amino ($-NH_2$) group of another (Figure 2.14). The bonds between amino acids are called **peptide bonds.** For the formation of every peptide bond between two amino acids, one water molecule is released; thus, peptide bond formation occurs by dehydration synthesis. In Figure 2.14 the resulting compound is called a *dipeptide* because it consists of two amino acids joined by a peptide bond. Adding another amino acid to a dipeptide would form a *tripeptide*. Further additions of amino acids would produce a long, chainlike molecule called a *peptide* (4–9 amino acids) or *polypeptide* (10–2000 or more amino acids).

Levels of Protein Structure

Proteins vary tremendously in structure. Different proteins have different architectures and different three-dimensional shapes. This variation in structure is directly related to their diverse functions.

When a cell makes a protein, the polypeptide chain folds spontaneously to assume a certain shape. One reason for folding of the polypeptide is that some parts of a protein are attracted to water and other parts are repelled by it. In practically every case, the function of a protein depends on its ability to recognize and bind to some other molecule. For example, an enzyme binds specifically with its substrate. A hormonal protein binds to a receptor on a cell whose function it will alter. An antibody binds to an antigen (foreign substance) that has invaded the body. The unique shape of each protein permits it to interact with specific other molecules in order to carry out specific functions.

Proteins are described in terms of four levels of organization: primary, secondary, tertiary, and quaternary. The *primary structure* is the unique sequence in which the amino acids are linked together to form a polypeptide chain (Figure 2.15a). This sequence is genetically determined. Alterations in sequence can have profound metabolic effects. For example, a single incorrect amino acid in a blood protein can produce the deformed hemoglobin molecule characteristic of sickle-cell disease. But proteins do not exist as

FIGURE 2.14 **Peptide bond formation by dehydration synthesis.** The amino acids glycine and alanine combine to form a dipeptide. The newly formed bond between the carbon atom of glycine and the nitrogen atom of alanine is called a peptide bond.

■ Amino acids are the building blocks of proteins.

Glycine   Alanine H_2O Dehydration synthesis (minus H_2O)   Peptide bond   Glycylalanine (a dipeptide)   Water

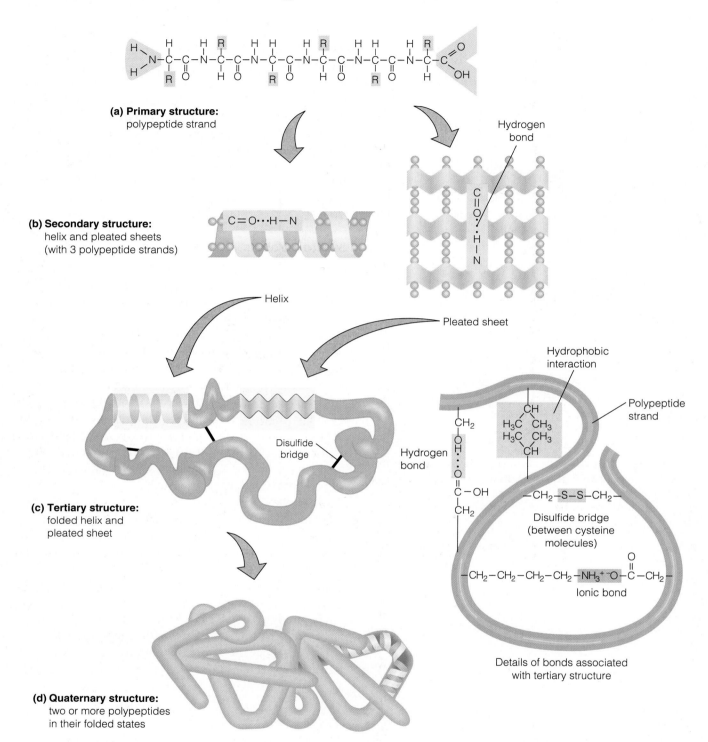

(a) Primary structure:
polypeptide strand

Hydrogen bond

(b) Secondary structure:
helix and pleated sheets
(with 3 polypeptide strands)

$C=O\cdots H-N$

Helix

Pleated sheet

Hydrophobic interaction

Polypeptide strand

Disulfide bridge

Hydrogen bond

(c) Tertiary structure:
folded helix and
pleated sheet

Disulfide bridge
(between cysteine molecules)

Ionic bond

Details of bonds associated
with tertiary structure

(d) Quaternary structure:
two or more polypeptides
in their folded states

FIGURE 2.15 **Protein structure. (a)** Primary structure, the amino acid sequence. **(b)** Secondary structures: helix and pleated sheet. **(c)** Tertiary structure, the overall three-dimensional folding of a polypeptide chain. **(d)** Quaternary structure, the relationship between several polypeptide chains that make up a protein. Shown here is the quaternary structure of a hypothetical protein composed of two polypeptide chains.

■ The unique shape of each protein enables it to carry out specific functions.

long, straight chains. Each polypeptide chain folds and coils in specific ways into a relatively compact structure with a characteristic three-dimensional shape.

A protein's *secondary structure* is the localized, repetitious twisting or folding of the polypeptide chain. This aspect of a protein's shape results from hydrogen bonds joining the atoms of peptide bonds at different locations along the polypeptide chain. The two types of secondary protein structures are clockwise spirals called helices (singular: *helix*) and pleated sheets, which form from roughly parallel portions of the chain (Figure 2.15b). Both structures are held together by hydrogen bonds between oxygen or nitrogen atoms that are part of the polypeptide's backbone.

Tertiary structure refers to the overall three-dimensional structure of a polypeptide chain (Figure 2.15c). The folding is not repetitive or predictable, as in secondary structure. Whereas secondary structure involves hydrogen bonding between atoms of the amino and carboxyl groups involved in the peptide bonds, tertiary structure involves several interactions between various amino acid side groups in the polypeptide chain. For example, amino acids with nonpolar (hydrophobic) side groups usually interact at the core of the protein, out of contact with water. This *hydrophobic interaction* helps contribute to tertiary structure. Hydrogen bonds between side groups, and ionic bonds between oppositely charged side groups, also contribute to tertiary structure. Proteins that contain the amino acid cysteine form strong covalent bonds called *disulfide bridges*. These bridges form when two cysteine molecules are brought close together by the folding of the protein. Cysteine molecules contain sulfhydryl groups (—SH), and the sulfur of one cysteine molecule bonds to the sulfur on another, forming (by the removal of hydrogen atoms) a disulfide bridge (S—S) that holds parts of the protein together.

Some proteins have a *quaternary structure*, which consists of an aggregation of two or more individual polypeptide chains (subunits) that operate as a single functional unit. Figure 2.15d shows a hypothetical protein consisting of two polypeptide chains. More commonly, proteins have two or more kinds of polypeptide subunits. The bonds that hold a quaternary structure together are basically the same as those that maintain tertiary structure. The overall shape of a protein may be globular (compact and roughly spherical) or fibrous (threadlike).

If a protein encounters a hostile environment in terms of temperature, pH, or salt concentrations, it may unravel and lose its characteristic shape. This process is called **denaturation** (see Figure 5.6 on page 118). As a result of denaturation, the protein is no longer functional. This process will be discussed in more detail in Chapter 5 with regard to denaturation of enzymes.

The proteins we have been discussing are *simple proteins*, which contain only amino acids. *Conjugated proteins* are combinations of amino acids with other organic or inorganic components. Conjugated proteins are named by their non–amino acid component. Thus, glycoproteins contain sugars, nucleoproteins contain nucleic acids, metalloproteins contain metal atoms, lipoproteins contain lipids, and phosphoproteins contain phosphate groups. Phosphoproteins are important regulators of activity in eukaryotic cells. Bacterial synthesis of phosphoproteins may be important for the survival of bacteria such as *Legionella pneumophila* that grow inside host cells.

Nucleic Acids

In 1944, three American microbiologists—Oswald Avery, Colin MacLeod, and Maclyn McCarty—discovered that a substance called **deoxyribonucleic acid (DNA)** is the substance of which genes are made. Nine years later, James Watson and Francis Crick, working with molecular models and X-ray information supplied by Maurice Wilkins and Rosalind Franklin, identified the physical structure of DNA. In addition, Crick suggested a mechanism for DNA replication and how it works as the hereditary material. DNA, and another substance called **ribonucleic acid (RNA),** are together referred to as **nucleic acids** because they were first discovered in the nuclei of cells. Just as amino acids are the structural units of proteins, nucleotides are the structural units of nucleic acids.

Each **nucleotide** has three parts: a nitrogen-containing base, a pentose (five-carbon) sugar called **deoxyribose** or **ribose,** and a phosphate group (phosphoric acid). The nitrogen-containing bases are cyclic compounds made up of carbon, hydrogen, oxygen, and nitrogen atoms. The bases are named adenine (A), thymine (T), cytosine (C), guanine (G), and uracil (U). A and G are double-ring structures called **purines,** whereas T, C, and U are single-ring structures referred to as **pyrimidines.**

Nucleotides are named according to their nitrogenous base. Thus, a nucleotide containing thymine is a *thymine nucleotide*, one containing adenine is an *adenine nucleotide*, and so on. The term **nucleoside** refers to the combination of a purine or pyrimidine plus a pentose sugar; it does not contain a phosphate group.

DNA

According to the model proposed by Watson and Crick, a DNA molecule consists of two long strands wrapped around each other to form a **double helix** (Figure 2.16). The double helix looks like a twisted ladder, and each strand is composed of many nucleotides.

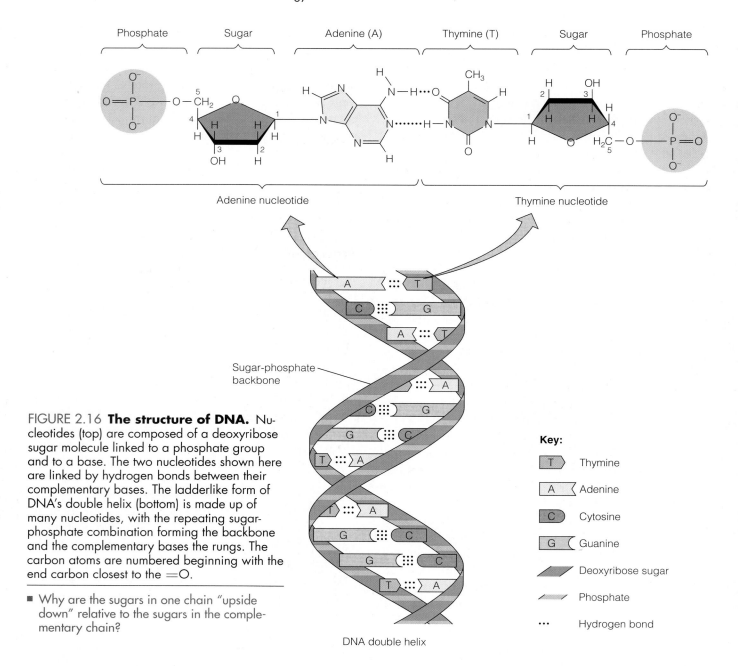

FIGURE 2.16 **The structure of DNA.** Nucleotides (top) are composed of a deoxyribose sugar molecule linked to a phosphate group and to a base. The two nucleotides shown here are linked by hydrogen bonds between their complementary bases. The ladderlike form of DNA's double helix (bottom) is made up of many nucleotides, with the repeating sugar-phosphate combination forming the backbone and the complementary bases the rungs. The carbon atoms are numbered beginning with the end carbon closest to the =O.

■ Why are the sugars in one chain "upside down" relative to the sugars in the complementary chain?

DNA double helix

Key:

T Thymine
A Adenine
C Cytosine
G Guanine
▱ Deoxyribose sugar
▱ Phosphate
... Hydrogen bond

Every strand of DNA composing the double helix has a "backbone" consisting of alternating deoxyribose sugar and phosphate groups. The deoxyribose of one nucleotide is joined to the phosphate group of the next. (Refer to Figure 8.3 to see how nucleotides are bonded.) The nitrogen-containing bases make up the rungs of the ladder. Note that the purine A is always paired with the pyrimidine T and that the purine G is always paired with the pyrimidine C. The bases are held together by hydrogen bonds; A and T are held by two hydrogen bonds, and G and C are held by three.

The order in which the nitrogen base pairs occur along the backbone is extremely specific and in fact contains the genetic instructions for the organism. Nucleotides form genes, and a single DNA molecule may contain thousands of genes. Genes determine all hereditary traits, and they control all the activities that take place within cells.

One very important consequence of nitrogen-containing base pairing is that if the sequence of bases of one strand is known, then the sequence of the other strand is also known. For example, if one strand has the sequence . . . ATGC . . . , then the other strand has the sequence . . . TACG. . . . Because the sequence of bases of one strand is determined by the sequence of bases of the other, the bases are said to be *complementary*. The actual transfer of information becomes possible because of DNA's unique structure and will be discussed further in Chapter 8.

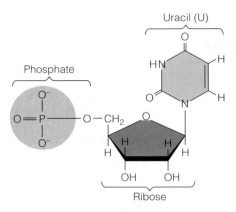

FIGURE 2.17 **A uracil nucleotide of RNA.**

■ How do DNA and RNA differ in structure?

RNA

RNA, the second principal kind of nucleic acid, differs from DNA in several respects. Whereas DNA is double-stranded, RNA is usually single-stranded. The five-carbon sugar in the RNA nucleotide is ribose, which has one more oxygen atom than deoxyribose. Also, one of RNA's bases is uracil (U) instead of thymine (Figure 2.17). The other three bases (A, G, C) are the same as DNA. Three major kinds of RNA have been identified in cells. These are referred to as **messenger RNA (mRNA), ribosomal RNA (rRNA),** and **transfer RNA (tRNA).** As we will see in Chapter 8, each type of RNA has a specific role in protein synthesis.

Adenosine Triphosphate (ATP)

Learning Objective

• *Describe the role of ATP in cellular activities.*

Adenosine triphosphate (ATP) is the principal energy-carrying molecule of all cells and is indispensable to the life of the cell. It stores the chemical energy released by some chemical reactions, and it provides the energy for reactions that require energy. ATP consists of an adenosine unit, composed of adenine and ribose, with three phosphate groups (abbreviated Ⓟ) attached (Figure 2.18). In other words, it

FIGURE 2.18 **The structure of ATP.** High-energy phosphate bonds are indicated by wavy lines. When ATP breaks down to ADP and inorganic phosphate, a large amount of chemical energy is released for use in other chemical reactions.

■ How is ATP similar to a nucleotide in RNA? In DNA?

is an adenine nucleotide (also called adenosine monophosphate, or AMP) with two extra phosphate groups. ATP is called a high-energy molecule because it releases a large amount of usable energy when the third phosphate group is hydrolyzed to become **adenosine diphosphate (ADP).** This reaction can be represented as follows:

$$\text{Adenosine} - Ⓟ - Ⓟ - Ⓟ + H_2O \rightleftharpoons$$
Adenosine   Water
triphosphate

$$\text{Adenosone} - Ⓟ - Ⓟ + Ⓟ_i + \text{Energy}$$
Adenosine   Inorganic
diphosphate   phosphate

A cell's supply of ATP at any particular time is limited. Whenever the supply needs replenishing, the reaction goes in the reverse direction; the addition of a phosphate group to ADP and the input of energy produces more ATP. The energy required to attach the terminal phosphate group to ADP is supplied by the cell's various oxidation reactions, particularly the oxidation of glucose. ATP can be stored in every cell, where its potential energy is not released until needed.

STUDY OUTLINE

Introduction (p. 26)

1. The science of the interaction between atoms and molecules is called chemistry.

2. The metabolic activities of microorganisms involve complex chemical reactions.

3. Nutrients are broken down by microbes to obtain energy and to make new cells.

The Structure of Atoms (pp. 27–28)

1. Atoms are the smallest units of chemical elements that enter into chemical reactions.
2. Atoms consist of a nucleus, which contains protons and neutrons, and electrons that move around the nucleus.
3. The atomic number is the number of protons in the nucleus; the total number of protons and neutrons is the atomic weight.

Chemical Elements (pp. 27–28)

1. Atoms with the same number of protons and the same chemical behavior are classified as the same chemical element.
2. Chemical elements are designated by abbreviations called chemical symbols.
3. About 26 elements are commonly found in living cells.
4. Atoms that have the same atomic number (are of the same element) but different atomic weights are called isotopes.

Electronic Configurations (p. 28)

1. In an atom, electrons are arranged around the nucleus in electron shells.
2. Each shell can hold a characteristic maximum number of electrons.
3. The chemical properties of an atom are due largely to the number of electrons in its outermost shell.

How Atoms Form Molecules: Chemical Bonds (pp. 28–32)

1. Molecules are made up of two or more atoms; molecules consisting of at least two different kinds of atoms are called compounds.
2. Atoms form molecules in order to fill their outermost electron shells.
3. Attractive forces that bind the atomic nuclei of two atoms together are called chemical bonds.
4. The combining capacity of an atom—the number of chemical bonds the atom can form with other atoms—is its valence.

Ionic Bonds (pp. 28–30)

1. A positively or negatively charged atom or group of atoms is called an ion.
2. A chemical attraction between ions of opposite charge is called an ionic bond.
3. To form an ionic bond, one ion is an electron donor, and the other ion is an electron acceptor.

Covalent Bonds (pp. 30–31)

1. In a covalent bond, atoms share pairs of electrons.
2. Covalent bonds are stronger than ionic bonds and are far more common in organisms.

Hydrogen Bonds (pp. 31–32)

1. A hydrogen bond exists when a hydrogen atom covalently bonded to one oxygen or nitrogen atom is attracted to another oxygen or nitrogen atom.
2. Hydrogen bonds form weak links between different molecules or between parts of the same large molecule.

Molecular Weight and Moles (p. 32)

1. The molecular weight is the sum of the atomic weights of all the atoms in a molecule.
2. A mole of an atom, ion, or molecule is equal to its atomic or molecular weight expressed in grams.

Chemical Reactions (pp. 32–33)

1. Chemical reactions are the making or breaking of chemical bonds between atoms.
2. A change of energy occurs during chemical reactions.
3. Endergonic reactions require energy; exergonic reactions release energy.
4. In a synthesis reaction, atoms, ions, or molecules are combined to form a larger molecule.
5. In a decomposition reaction, a larger molecule is broken down into its component molecules, ions, or atoms.
6. In an exchange reaction, two molecules are decomposed, and their subunits are used to synthesize two new molecules.
7. The products of reversible reactions can readily revert to form the original reactants.

Important Biological Molecules (pp. 33–49)

Inorganic Compounds (pp. 33–37)

1. Inorganic compounds are usually small, ionically bonded molecules.
2. Water and many common acids, bases, and salts are examples of inorganic compounds.

Water (pp. 33–35)

1. Water is the most abundant substance in cells.
2. Because water is a polar molecule, it is an excellent solvent.
3. Water is a reactant in many of the decomposition reactions of digestion.
4. Water is an excellent temperature buffer.

Acids, Bases, and Salts (pp. 35–36)

1. An acid dissociates into H^+ and anions.
2. A base dissociates into OH^- and cations.
3. A salt dissociates into negative and positive ions, neither of which is H^+ or OH^-.

Acid–Base Balance (pp. 36–37)

1. The term pH refers to the concentration of H^+ in a solution.
2. A solution of pH 7 is neutral; a pH value below 7 indicates acidity; pH above 7 indicates alkalinity.
3. A pH buffer, which stabilizes the pH inside a cell, can be used in culture media.

Organic Compounds (pp. 37–49)

1. Organic compounds always contain carbon and hydrogen.
2. Carbon atoms form up to four bonds with other atoms.
3. Organic compounds are mostly or entirely covalently bonded, and many of them are large molecules.

Structure and Chemistry (pp. 37–39)

1. A chain of carbon atoms forms a carbon skeleton.
2. Functional groups of atoms are responsible for most of the properties of organic molecules.
3. The letter R may be used to denote the remainder of an organic molecule.
4. Frequently encountered classes of molecules are R—OH (alcohols) and R—COOH (organic acids).
5. Small organic molecules may combine into very large molecules called macromolecules.
6. Monomers usually bond together by dehydration synthesis, or condensation reactions, that form water and a polymer.
7. Organic molecules may be broken down by hydrolysis, a reaction involving the splitting of water molecules.

Carbohydrates (pp. 39–40)

1. Carbohydrates are compounds consisting of atoms of carbon, hydrogen, and oxygen, with hydrogen and oxygen in a 2:1 ratio.
2. Carbohydrates include sugars and starches.
3. Carbohydrates can be classified as monosaccharides, disaccharides, and polysaccharides.
4. Monosaccharides contain from three to seven carbon atoms.
5. Isomers are two molecules with the same chemical formula but different structures and properties—for example, glucose ($C_6H_{12}O_6$) and fructose ($C_6H_{12}O_6$).
6. Monosaccharides may form disaccharides and polysaccharides by dehydration synthesis.

Lipids (pp. 40–43)

1. Lipids are a diverse group of compounds distinguished by their insolubility in water.
2. Simple lipids (fats) consist of a molecule of glycerol and three molecules of fatty acids.

3. A saturated lipid has no double bonds between carbon atoms in the fatty acids; an unsaturated lipid has one or more double bonds. Saturated lipids have higher melting points than unsaturated lipids.
4. Phospholipids are complex lipids consisting of glycerol, two fatty acids, and a phosphate group.
5. Steroids have carbon ring structures; sterols have a functional hydroxyl group.

Proteins (pp. 43–47)

1. Amino acids are the building blocks of proteins.
2. Amino acids consist of carbon, hydrogen, oxygen, nitrogen, and sometimes sulfur.
3. Twenty amino acids occur naturally.
4. By linking amino acids, peptide bonds (formed by dehydration synthesis) allow the formation of polypeptide chains.
5. Proteins have four levels of structure: primary (sequence of amino acids), secondary (helices or pleats), tertiary (overall three-dimensional structure of a polypeptide), and quaternary (two or more polypeptide chains).
6. Conjugated proteins consist of amino acids combined with other organic or inorganic compounds.

Nucleic Acids (pp. 47–49)

1. Nucleic acids—DNA and RNA—are macromolecules consisting of repeating nucleotides.
2. A nucleotide is composed of a pentose, a phosphate group, and a nitrogen-containing base. A nucleoside is composed of a pentose and a nitrogen-containing base.
3. A DNA nucleotide consists of deoxyribose (a pentose) and one of the following nitrogen-containing bases: thymine or cytosine (pyrimidines) or adenine or guanine (purines).
4. DNA consists of two strands of nucleotides wound in a double helix. The strands are held together by hydrogen bonds between purine and pyrimidine nucleotides: AT and GC.
5. Genes consist of sequences of nucleotides.
6. An RNA nucleotide consists of ribose (a pentose) and one of the following nitrogenous bases: cytosine, guanine, adenine, or uracil.

Adenosine Triphosphate (ATP) (p. 49)

1. ATP stores chemical energy for various cellular activities.
2. When the bond to ATP's terminal phosphate group is hydrolyzed, energy is released.
3. The energy from oxidation reactions is used to regenerate ATP from ADP and inorganic phosphate.

STUDY QUESTIONS

Review

1. What is a chemical element?

2. Diagram the electronic configuration of a carbon atom.

3. How does $^{14}_{6}C$ differ from $^{12}_{6}C$?

4. What type of bonding exists between water molecules?

5. What type of bonds holds the following atoms together?
   a. Li^+ and Cl^- in $LiCl$
   b. Carbon and oxygen atoms in CO_2
   c. Oxygen atoms in O_2
   d. A hydrogen atom of one nucleotide to a nitrogen or oxygen atom of another nucleotide in:

Deoxyribose-phosphate   Deoxyribose-phosphate

Guanine   Cytosine

6. Vinegar, pH 3, is how many times more acidic than pure water, pH 7?

7. Calculate the molecular weight of 1 mole of $C_6H_{12}O_6$.

8. Classify the following types of chemical reactions.
   a. Glucose + fructose $\rightarrow$ sucrose + H_2O
   b. Lactose $\rightarrow$ glucose + galactose
   c. $NH_4Cl + H_2O \rightarrow NH_4OH + HCl$
   d. $ATP \rightarrow ADP + \textcircled{P}_i$

9. Bacteria use the enzyme urease to obtain nitrogen in a form they can use from urea in the following reaction:

$$CO(NH_2)_2 + H_2O \longrightarrow 2NH_3 + CO_2$$
   Urea   Ammonia   Carbon
                                     dioxide

   What purpose does the enzyme serve in this reaction? What type of reaction is this?

10. Classify the following as subunits of either a carbohydrate, lipid, protein, or nucleic acid.
    a. $CH_3-(CH_2)_7-CH=CH-(CH_2)_7-COOH$
       Oleic acid
    b.

       $H-\overset{\overset{\displaystyle NH_2}{|}}{\underset{\underset{\displaystyle OH}{|}}{\underset{\displaystyle CH_2}{|}}{C}}-COOH$
       Serine

    c. $C_6H_{12}O_6$
    d. Thymine nucleotide

11. Add the appropriate functional group(s) to this ethyl group to produce each of the following compounds: ethanol, acetic acid, acetaldehyde, ethanolamine, diethyl ether.

12. The artificial sweetener aspartame, or NutraSweet, is made by joining aspartic acid to methylated phenylalanine, as shown below.

    a. What types of molecules are aspartic acid and phenylalanine?
    b. What direction is the hydrolysis reaction (left to right or right to left)?
    c. What direction is the dehydration synthesis reaction?
    d. Circle the atoms involved in the formation of water.
    e. Identify the peptide bond.

13. The energy-carrying property of the ATP molecule is due to energy dynamics that favor the breaking of bonds between _____. What type of bonds are these?

14. The following diagram shows a protein molecule. Indicate the regions of primary, secondary, and tertiary structure. Does this protein have quaternary structure?

15. Draw a simple lipid, and show how it could be modified to a phospholipid.

Multiple Choice

Radioisotopes are frequently used to label molecules in a cell. The fate of atoms and molecules in a cell can then be followed. This process is the basis for questions 1–3.

1. Assume *E. coli* are grown in a nutrient medium containing the radioisotope ^{16}N. After a 48-hour incubation period, the ^{16}N would most likely be found in the *E. coli's*
   a. carbohydrates.   c. proteins.   e. none of the above
   b. lipids.   d. water.

2. If *Pseudomonas* bacteria are supplied with radioactively labeled cytosine, after a 24-hour incubation period this cytosine would most likely be found in the cells'
   a. carbohydrates.   c. lipids.   e. proteins.
   b. DNA.   d. water.

3. If *E. coli* were grown in a medium containing the radioactive isotope ^{32}P, the ^{32}P would be found in all of the following molecules of the cell *except*
   a. ATP.   c. DNA.   e. none of the above
   b. carbohydrates.   d. plasma membrane.

4. A carbonated drink, pH 3, is _____ times more acid than distilled water.
   a. 4   c. 100   e. 10,000
   b. 10   d. 1000

5. The best definition of ATP is that it is:
   a. a molecule stored for food use.
   b. a molecule that supplies energy to do work.
   c. a molecule stored for an energy reserve.
   d. a molecule used as a source of phosphate.

6. Which of the following is an organic molecule?
   a. H_2O (water)   d. FeO (iron oxide)
   b. O_2 (oxygen)   e. $F_2C=CF_2$ (Teflon)
   c. $C_{18}H_{29}SO_3$ (Styrofoam)

Classify the molecules shown in questions 7–10 using the following choices. The dissociation products of the molecules are shown to help you.
   a. acid   b. base   c. salt

7. $HNO_3 \rightarrow H^+ + NO_3^-$ 9. $NaOH \rightarrow Na^+ + OH^-$
8. $H_2SO_4 \rightarrow 2H^+ + SO_4^{2-}$ 10. $MgSO4 \rightarrow Mg^{2+} + SO_4^{2-}$

Critical Thinking

1. When you blow bubbles into a glass of water, the following reactions take place:

$$H_2O + CO_2 \xrightarrow{A} H_2CO_3 \xrightarrow{B} H^+ + HCO_3^-$$

   a. What type of reaction is *A*?
   b. What does reaction *B* tell you about the type of molecule H_2CO_3 is?

2. What are the common structural characteristics of ATP and DNA molecules?

3. What happens to the relative amount of unsaturated lipids in its plasma membrane when *E. coli* grown at 25°C are then grown at 37°C?

4. Giraffes, termites, and koalas eat only plant matter. Because animals cannot digest cellulose, how do you suppose these animals get nutrition from the leaves and wood they eat?

Applications

1. Laundry detergents and drain cleaners contain enzymes from *Bacillus subtilis*. What is the value of these enzymes in clothes washing and drain cleaning?

2. *Thiobacillus ferrooxidans* was responsible for destroying buildings in the Midwest by causing changes in the earth. The original rock, which contained lime ($CaCO_3$) and pyrite (FeS_2), expanded as bacterial metabolism caused gypsum ($CaSO_4$) crystals to form. How did *T. ferrooxidans* bring about the change from lime to gypsum?

3. When growing in an animal, *Bacillus anthracis* produces a capsule that is resistant to phagocytosis. The capsule is composed of D-glutamic acid. Why is this capsule resistant to digestion by the host's phagocytes? (Phagocytes are white blood cells that engulf bacteria.)

4. The antibiotic amphotericin B causes leaks in cells by combining with sterols in the plasma membrane. Would you expect to use amphotericin B against a bacterial infection? A fungal infection? Offer a reason why amphotericin B has severe side effects in humans.

LEARNING WITH TECHNOLOGY

The Microbiology Place CD-ROM/website includes many resources to help you study and succeed in this course. In addition to self-quizzes, web links, and flashcards, you'll find the following activities for this chapter in the "Activities" section of The Microbiology Place:

Atomic Structure

Ionic and Covalent Bonding

Hydrogen Bonds and Water

Building Biomolecules

Case Study: Kesterson National Wildlife Refuge

Chapter 3

Observing Microorganisms Through a Microscope

PARAMECIUM CAUDATUM. *Surface structures including cilia and the oral groove of this Paramecium can be seen with a scanning electron microscope.*

Microorganisms are much too small to be seen with the unaided eye; they must be observed with a microscope. The word microscope is derived from the Latin word micro, which means small, and the Greek word skopos, to look at. Modern microbiologists use microscopes that produce, with great clarity, magnifications that range from ten to thousands of times greater than those of van Leeuwenhoek's single lens (see Figure 1.2b on page 7). This chapter describes how different types of microscopes function and why one type might be used in preference to another.

Some microbes are more readily visible than others because of their larger size or more easily observable features. Many microbes, however, must undergo several staining procedures before their cell walls, membranes, and other structures lose their colorless natural state. The last part of this chapter will explain some of the more commonly used methods of preparing specimens for examination through a light microscope.

You may wonder how we are going to sort, count, and measure the specimens we will study. To answer these questions, this chapter opens with a discussion of how to use the metric system for measuring microbes.

Units of Measurement

Learning Objective

- List the metric units of measurement used for microorganisms, and their metric equivalents.

Because microorganisms and their component parts are so very small, they are measured in units that are unfamiliar to many of us in everyday life. When measuring microorganisms, we use the metric system. The standard unit of length in the metric system is the meter (m). A major advantage of the metric system is that the units are related to each other by factors of 10. Thus, 1 m equals 10 decimeters (dm) or 100 centimeters (cm) or 1000 millimeters (mm). Units in the U.S. system of measure do not have the advantage of easy conversion by a single factor of 10. For example, we use 3 feet or 36 inches to equal 1 yard.

Microorganisms and their structural components are measured in even smaller units, such as micrometers and nanometers. A **micrometer (μm)** is equal to 0.000001 m (10^{-6} m). The prefix *micro* indicates that the unit following it should be divided by 1 million, or 10^6 (see the Exponential Notation section in Appendix E). A **nanometer (nm)** is equal to 0.000000001 m (10^{-9} m). Angstrom (Å) was previously used for 10^{-10} m or 0.1 nm.

Table 3.1 presents the basic metric units of length and some of their U.S. equivalents. In Table 3.1, you can compare the microscopic units of measurement with the commonly known macroscopic units of measurement, such as centimeters, meters, and kilometers. If you look ahead to Figure 3.2 (page 57), you will see the relative sizes of various organisms on the metric scale.

TABLE 3.1	**Metric Units of Length and U.S. Equivalents**			
Metric Unit	**Meaning of Prefix**	**Metric Equivalent**		**U.S. Equivalent**
1 kilometer (km)	*kilo* = 1000	1000 m = 10^3 m		3280.84 ft or 0.62 mi; 1 mi = 1.61 km
1 meter (m)		Standard unit of length		39.37 in or 3.28 ft or 1.09 yd
1 decimeter (dm)	*deci* = 1/10	0.1 m = 10^{-1} m		3.94 in
1 centimeter (cm)	*centi* = 1/100	0.01 m = 10^{-2} m		0.394 in; 1 in = 2.54 cm
1 millimeter (mm)	*milli* = 1/1000	0.001 m = 10^{-3} m		
1 micrometer (μm)	*micro* = 1/1,000,000	0.000001 m = 10^{-6} m		
1 nanometer (nm)	*nano* = 1/1,000,000,000	0.000000001 m = 10^{-9} m		

Microscopy: The Instruments

The simple microscope used by van Leeuwenhoek in the seventeenth century had only one lens and was similar to a magnifying glass. However, van Leeuwenhoek was the best lens grinder in the world in his day. His lenses were ground with such precision that a single lens could magnify a microbe 300×. His simple microscopes enabled him to be the first person to see bacteria (see Figure 1.2).

Contemporaries of van Leeuwenhoek, such as Robert Hooke, built compound microscopes, which have multiple lenses. In fact, a Dutch spectacles maker, Zaccharias Janssen, is credited with making the first compound microscope around 1600. However, these early compound microscopes were of poor quality and could not be used to see bacteria. It was not until about 1830 that a significantly better microscope was developed by Joseph Jackson Lister (the father of Joseph Lister). Various improvements to Lister's microscope resulted in the development of the modern compound microscope, the kind used in microbiology laboratories today. Microscopic studies of live specimens have revealed dramatic interactions between microbes (see the box on page 58).

Light Microscopy

Learning Objectives

- *Diagram the path of light through a compound microscope.*
- *Define total magnification and resolution.*
- *Identify a use for darkfield, phase-contrast, differential interference contrast (DIC), fluorescence, and confocal microscopy, and compare each with brightfield illumination.*

Light microscopy refers to the use of any kind of microscope that uses visible light to observe specimens. Here we examine several types of light microscopy.

Compound Light Microscopy

A modern **compound light microscope (LM)** has a series of lenses and uses visible light as its source of illumination (Figure 3.1a). With a compound light microscope, we can examine very small specimens as well as some of their fine detail. A series of finely ground lenses (Figure 3.1b) forms a clearly focused image that is many times larger than the specimen itself. This magnification is achieved when light rays from an **illuminator,** the light source, pass through a **condenser,** which has lenses that direct the light rays through the specimen. From here, light rays pass into the **objective lenses,** the lenses closest to the specimen. The image of the specimen is magnified again by the **ocular lens,** or *eyepiece.*

We can calculate the **total magnification** of a specimen by multiplying the objective lens magnification (power) by the ocular lens magnification (power). Most microscopes used in microbiology have several objective lenses, including 10× (low power), 40× (high power), and 100× (oil immersion, which is described shortly). Most oculars magnify specimens by a factor of 10. Multiplying the magnification of a specific objective lens with that of the ocular, we see that the total magnifications would be 100× for low power, 400× for high power, and 1000× for oil immersion. Some compound light microscopes can achieve a magnification of 2000× with the oil immersion lens.

Resolution (also called *resolving power*) is the ability of the lenses to distinguish fine detail and structure. Specifically, it refers to the ability of the lenses to distinguish between two points a specified distance apart. For example, if a microscope has a resolving power of 0.4 nm, it can distinguish between two points if they are at least 0.4 nm apart. A general principle of microscopy is that the shorter the wavelength of light used in the instrument, the greater the resolution. The white light

Ocular lens (eyepiece) Remagnifies the image formed by the objective lens

Body tube Transmits the image from the objective lens to the ocular lens

Arm

Objective lenses Primary lenses that magnify the specimen

Stage Holds the microscope slide in position

Condenser Focuses light through specimen

Diaphragm Controls the amount of light entering the condenser

Coarse focusing knob

Illuminator Light source

Base

Fine focusing knob

(a) Principal parts and functions

Ocular lens

Line of vision

Path of light

Prism

Body tube

Objective lenses

Specimen

Condenser lenses

Illuminator

Base with source of illumination

(b) The path of light (bottom to top)

FIGURE 3.1 **The compound light microscope.**

■ How is the total magnification of a compound light microscope calculated?

used in a compound light microscope has a relatively long wavelength and cannot resolve structures smaller than about 0.2 μm. This fact and other practical considerations limit the magnification achieved by even the best compound light microscopes to about 2000×. By comparison, van Leeuwenhoek's microscopes had a resolution of 1 μm.

Figure 3.2 shows various specimens that can be resolved by the human eye, light microscope, and electron microscope.

To obtain a clear, finely detailed image under a compound light microscope, specimens must be made to contrast sharply with their *medium* (substance through which light passes). To attain such contrast, we must change the refractive index of specimens from that of their medium. The **refractive index** is a measure of the light-bending ability of a medium. We change the refractive index of specimens by staining them, a procedure we will discuss shortly. Light rays move in a straight line through a single medium. After staining, when light rays pass through the

two materials (the specimen and its medium) with different refractive indexes, the rays change direction (refract) from a straight path by bending or changing an angle at the boundary between the materials and increase the image's contrast between the specimen and the medium. As the light rays travel away from the specimen, they spread out and enter the objective lens, and the image is thereby magnified.

To achieve high magnification (100×) with good resolution, the objective lens must be small. Although we want light traveling through the specimen and medium to refract differently, we do not want to lose light rays after they have passed through the stained specimen. To preserve the direction of light rays at the highest magnification, immersion oil is placed between the glass slide and the oil immersion objective lens (Figure 3.3). The immersion oil has the same refractive index as glass, so the oil becomes part of the optics of the glass of the microscope. Unless immersion oil is used, light rays are refracted as they enter the air from the slide, and the objective lens

FIGURE 3.2 **Relationships between the sizes of various specimens and the resolution of the human eye, light microscope, and electron microscope.** The area in yellow shows the size range of most of the organisms we will be studying in this book.

■ Resolution is the ability of lenses to distinguish between two points a specified distance apart.

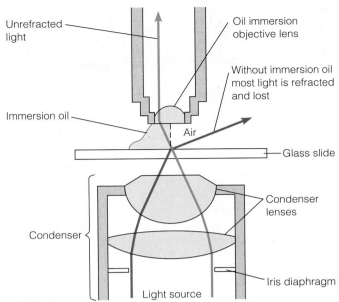

FIGURE 3.3 **Refraction in the compound microscope using an oil immersion objective lens.** Because the refractive indexes of the glass microscope slide and immersion oil are the same, the light rays do not refract when passing from one to the other when an oil immersion objective lens is used. This method produces images with better resolution at magnifications greater than 900×.

sion objective lens, the image becomes fuzzy, with poor resolution.

Under usual operating conditions, the field of vision in a compound light microscope is brightly illuminated. By focusing the light, the condenser produces a **brightfield illumination** (Figure 3.4a).

It is not always desirable to stain a specimen. However, an unstained cell has little contrast with its surroundings and is therefore difficult to see. Unstained cells are more easily observed with the modified compound microscopes described in the next section.

Darkfield Microscopy

A **darkfield microscope** is used for examining live microorganisms that either are invisible in the ordinary light microscope, cannot be stained by standard methods, or are so distorted by staining that their characteristics then cannot be identified. Instead of the normal condenser, a darkfield microscope uses a darkfield condenser that contains an opaque disc. The disc blocks light that would enter the objective lens directly. Only light that is reflected off (turned away from) the specimen enters the objective lens. Because there is no direct background light, the specimen appears light against a

would have to be increased in diameter to capture most of them. The oil has the same effect as increasing the objective lens diameter; therefore, it improves the resolving power of the lenses. If oil is not used with an oil immer-

Can Bacteria Make Food Safe?

Even through a microscope, appearances can be deceptive, and first impressions can be quite misleading. For example, at first glance, the bacterium *Bdellovibrio* looks like a perfectly ordinary 1-μm-long curved, gram-negative rod. *Bdellovibrio* cells exhibit the normal runs and tumbles of flagellated bacteria, although they move about ten times faster than other bacteria of the same size. However, more patient observation, or use of a microscope equipped with a video camera, reveals that *Bdellovibrio*'s behavior is unlike that of any other known bacterium.

A unique microbial drama occurs when *Bdellovibrio* cells are placed in a suspension with other bacteria, such as *E. coli*. Apparently drawn by chemical attractants, *Bdellovibrio* speeds toward and rams the larger, slower *E. coli*. The two cells become attached and rotate rapidly. As the two bacteria whirl about, the *Bdellovibrio* paralyzes the *E. coli* by an unknown mechanism that probably involves one of *Bdellovibrio*'s protein products. *Bdellovibrio* kills *E. coli* without breaking down its contents.

Next, in preparation for moving into its newly acquired home, *Bdellovibrio* adds structural components to the outer membrane of *E. coli*. This activity, which stabilizes the outer membrane, is another ability unique to *Bdellovibrio*. After it penetrates *E. coli*'s cell wall and detaches its flagellum, *Bdellovibrio* snuggles into the periplasm (the space between the plasma membrane and the cell wall) and prepares for a feast.

Macromolecules leak from *E. coli*'s cytoplasm into the periplasm. *Bdellovibrio* digests the macromolecules and grows into a long, coiled filament. In 2–3 hours, the cytoplasmic contents of the host (including its DNA) are digested, and the *Bdellovibrio* filament fragments into as many as 20 individual curved

rods. Each rod develops a flagellum and is released when the host cell is broken open by a *Bdellovibrio* enzyme. Within 4 hours, new *Bdellovibrio* cells scurry away to feed on other bacteria.

Bdellovibrio was once regarded as simply a parasite. However, thorough studies have shown that *Bdellovibrio* does not merely siphon off its host's resources and energy; it actually digests the cell's contents as food, so *Bdellovibrio* is now considered to be a bacterial predator. The ecological importance of *Bdellovibrio* is that it, along with protozoa and bacteriophages (viruses that infect bacteria), helps control the population growth of soil bacteria.

In addition, now that its habits are known, *Bdellovibrio* may have a new career as a bacterial exterminator. Researchers led by Richard Whiting at the U.S. Department of Agriculture (USDA)

are investigating the use of *Bdellovibrio* to control gram-negative foodborne pathogens such as *E. coli* O157:H7 and *Salmonella*. Although these bacteria are destroyed by cooking, foods that contain raw eggs or unpasteurized apple juice are a problem. It isn't easy to kill *E. coli* on meat and fruit without heating. Even washing in organic acids (pH 3.6) does not kill *E. coli* O157:H7. However, *Bdellovibrio* can kill even high concentrations of *E. coli* at 20°C and pH 5.6–8.6. In addition, researchers hope that *Bdellovibrio* can be introduced into chickens via food or water to kill salmonellae.

As harmful as they are to other bacteria, *Bdellovibrio* cells are safe for humans. Because *Bdellovibrio* cells prey on gram-negative bacteria, the USDA researchers hope to use *Bdellovibrio* for the biological control of pathogenic and spoilage organisms in food.

E. coli plasma membrane

Bdellovibrio

E. coli cell wall

TEM 0.5 μm

A *Bdellovibrio* bacterium has penetrated the cell wall of an *E. coli* cell. The *Bdellovibrio*, having killed the prey cell, uses the *E. coli* cell's contents for its elongation and multiplication.

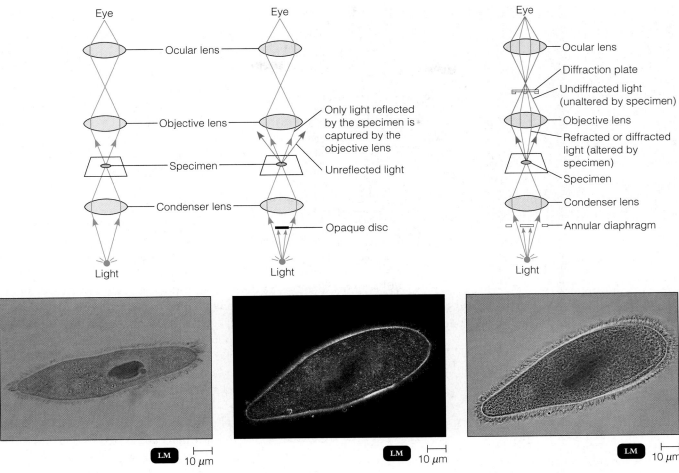

(a) Brightfield. (Top) The path of light in brightfield microscopy, the type of illumination produced by regular compound light microscopes. (Bottom) Brightfield illumination shows internal structures and the outline of the transparent pellicle (external covering).

(b) Darkfield. (Top) The darkfield microscope uses a special condenser with an opaque disc that eliminates all light in the center of the beam. The only light that reaches the specimen comes in at an angle; thus, only light reflected by the specimen (blue rays) reaches the objective lens. (Bottom) Against the black background seen with darkfield microscopy, edges of the cell are bright, some internal structures seem to sparkle, and the pellicle is almost visible.

(c) Phase-contrast. (Top) In phase-contrast microscopy, the specimen is illuminated by light passing through an annular diaphragm. Direct light rays (unaltered by the specimen) travel a different path than light rays that are reflected or diffracted as they pass through the specimen. These two sets of rays are combined at the eye. Reflected or diffracted light rays are indicated in blue; direct rays are red. (Bottom) Phase-contrast microscopy shows greater differentiation of internal structures and also shows the pellicle.

FIGURE 3.4 **Brightfield, darkfield, and phase-contrast microscopy.** The illustrations show the contrasting light pathways of each of these types of microscopy. The photographs compare the same *Paramecium* specimen using these different microscopy techniques.

■ What are the advantages of brightfield, darkfield, and phase-contrast microscopy?

black background—the dark field (Figure 3.4b). This technique is frequently used to examine unstained microorganisms suspended in liquid. One use for darkfield microscopy is the examination of very thin spirochetes, such as *Treponema pallidum* (tre-pō-nē′mä pal′li-dum), the causative agent of syphilis.

Phase-Contrast Microscopy

Another way to observe microorganisms is with a **phase-contrast microscope.** Phase-contrast microscopy is especially useful because it permits detailed examination of internal structures in *living* microorganisms. In addition, it is not necessary to fix (attach the microbes to the microscope

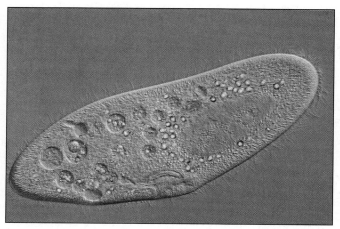

LM   10 µm

FIGURE 3.5 **Differential interference contrast (DIC) microscopy.** Like phase-contrast, DIC uses differences in refractive indexes to produce an image, in this case of the protozoan *Paramecium*. The colors in the image are produced by prisms that split the two light beams used in this process.

■ Why is the resolution of a DIC microscope higher than that of a phase-contrast microscope?

slide) or stain the specimen—procedures that could distort or kill the microorganisms.

The principle of phase-contrast microscopy is based on the wave nature of light rays, and the fact that light rays can be *in phase* (their peaks and valleys match) or *out of phase*. If the wave peak of light rays from one source coincides with the wave peak of light rays from another source, the rays interact to produce *reinforcement* (relative brightness). However, if the wave peak from one light source coincides with the wave trough from another light source, the rays interact to produce *interference* (relative darkness). In a phase-contrast microscope, one set of light rays comes directly from the light source. The other set comes from light that is reflected or diffracted from a particular structure in the specimen. (*Diffraction* is the scattering of light rays as they "touch" a specimen's edge. The diffracted rays are bent away from the parallel light rays that pass farther from the specimen.) When the two sets of light rays—direct rays and reflected or diffracted rays—are brought together, they form an image of the specimen on the ocular lens, containing areas that are relatively light (in phase), through shades of gray, to black (out of phase; Figure 3.4c). In phase-contrast microscopy, the internal structures of a cell become more sharply defined.

Differential Interference Contrast (DIC) Microscopy

Differential interference contrast (DIC) microscopy is similar to phase-contrast microscopy in that it uses differences in refractive indexes. However, a DIC microscope uses two beams of light instead of one. In addition, prisms split each light beam, adding contrasting colors to the specimen. Therefore, the resolution of a DIC microscope is higher than that of a standard phase-contrast microscope. Also, the image is brightly colored and appears nearly three-dimensional (Figure 3.5).

Fluorescence Microscopy

Fluorescence microscopy takes advantage of **fluorescence,** the ability of substances to absorb short wavelengths of light (ultraviolet) and give off light at a longer wavelength (visible). Some organisms fluoresce naturally under ultraviolet light; if the specimen to be viewed does not naturally fluoresce, it is stained with one of a group of fluorescent dyes called *fluorochromes*. When microorganisms stained with a fluorochrome are examined under a fluorescence microscope with an ultraviolet or near-ultraviolet light source, they appear as luminescent, bright objects against a dark background.

Fluorochromes have special attractions for different microorganisms. For example, the fluorochrome auramine O, which glows yellow when exposed to ultraviolet light, is strongly absorbed by *Mycobacterium tuberculosis*, the bacterium that causes tuberculosis. When the dye is applied to a sample of material suspected of containing the bacterium, the bacterium can be detected by the appearance of bright yellow organisms against a dark background (Table 3.2). *Bacillus anthracis*, the causative agent of anthrax, appears apple green when stained with another fluorochrome, fluorescein isothiocyanate (FITC).

The principal use of fluorescence microscopy is a diagnostic technique called the **fluorescent-antibody (FA) technique,** or **immunofluorescence. Antibodies** are natural defense molecules that are produced by humans and many animals in reaction to a foreign substance, or *antigen*. Fluorescent antibodies for a particular antigen are obtained as follows: An animal is injected with a specific antigen, such as a bacterium; the animal then begins to produce specific antibodies against that antigen. After a sufficient time, the antibodies are removed from the serum of the animal. Next, as shown in Figure 3.6a on page 63, a fluorochrome is chemically combined with the antibodies. These fluorescent antibodies are then added to a microscope slide containing an unknown bacterium. If this unknown bacterium is the same bacterium that was injected into the animal, the fluorescent antibodies bind to antigens on the surface of the bacterium, causing it to fluoresce.

TABLE 3.2	**A Summary of Various Types of Microscopes**

Microscope Type	Distinguishing Features	Typical Image	Principal Uses
Light Brightfield	Uses visible light as a source of illumination; cannot resolve structures smaller than about 0.2 μm; specimen appears against a bright background. Inexpensive and easy to use.	 LM ⊢⊣ 10 μm	To observe various stained specimens and to count microbes; does not resolve very small specimens, such as viruses.
Darkfield	Uses a special condenser with an opaque disc that blocks light from entering the objective lens directly; light reflected by specimen enters the objective lens, and the specimen appears light against a black background.	 LM ⊢⊣ 10 μm	To examine living microorganisms that are invisible in brightfield microscopy, do not stain easily, or are distorted by staining; frequently used to detect *Treponema pallidum* in the diagnosis of syphilis.
Phase-contrast	Uses a special condenser containing an annular (ring-shaped) diaphragm. The diaphragm allows direct light to pass through the condenser, focusing light on the specimen and a diffraction plate in the objective lens. Direct and reflected or diffracted light rays are brought together to produce the image. No staining required.	 LM ⊢⊣ 10 μm	To facilitate detailed examination of the internal structures of living specimens.
Differential interference contrast (DIC)	Like phase-contrast, uses differences in refractive indexes to produce images. Uses two beams of light separated by prisms; the specimen appears colored as a result of the prism effect. No staining required.	 LM ⊢⊣ 10 μm	To provide three-dimensional images.
Fluorescence	Uses an ultraviolet or near-ultraviolet source of illumination that causes fluorescent microbes (green-colored) in a specimen to emit light.	 LM ⊢⊣ 10 μm	For fluorescent-antibody techniques (immunofluorescence) to rapidly detect and identify microbes in tissues or clinical specimens.

TABLE 3.2	**A Summary of Various Types of Microscopes** (continued)		
Microscope Type	**Distinguishing Features**	**Typical Image**	**Principal Uses**
Confocal	Uses laser light to illuminate one plane of a specimen at a time.	LM |——| 10 μm	To obtain two- and three-dimensional images of cells for biomedical applications.
Electron Transmission	Uses a beam of electrons instead of light; electrons pass through the specimen; because of the shorter wavelength of electrons, structures smaller than 0.2 μm can be resolved. The image produced is two-dimensional.	TEM |——| 10 μm	To examine viruses or the internal ultrastructure in thin sections of cells (usually magnified 10,000–100,000×).
Scanning	Uses a beam of electrons instead of light; electrons are reflected from the specimen; because of the shorter wavelength of electrons, structures smaller than 0.2 μm can be resolved. The image produced appears three-dimensional.	SEM |——| 10 μm	To study the surface features of cells and viruses (usually magnified 1000–10,000×).
Scanned-probe Scanning tunneling	Uses a thin metal probe that scans a specimen and produces an image revealing the bumps and depressions of the atoms on the surface of the specimen. Resolving power is much greater than that of an electron microscope. No special preparation required.	STM |——| 1 nm	Provides very detailed views of molecules inside cells.
Atomic force	Uses a metal-and-diamond probe gently forced down along the surface of the specimen. Produces a three-dimensional image. No special preparation required.	AFM |——| 2 nm	Provides images of biological molecules in nearly atomic detail and molecular processes.

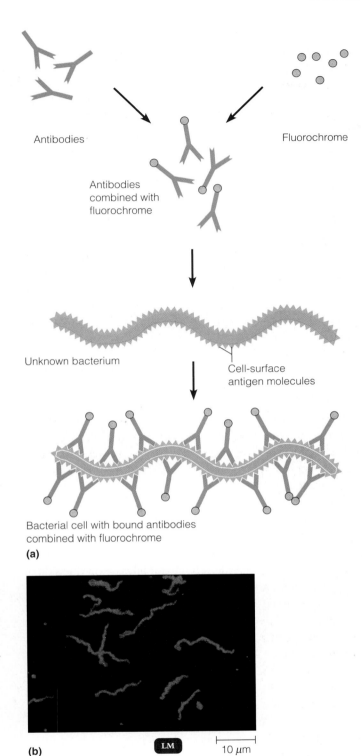

(a)

Antibodies

Fluorochrome

Antibodies combined with fluorochrome

Unknown bacterium

Cell-surface antigen molecules

Bacterial cell with bound antibodies combined with fluorochrome

(b)   LM   10 μm

FIGURE 3.6 The principle of immunofluorescence. (a) A type of fluorochrome is combined with antibodies against a specific type of bacterium. When the preparation is added to bacterial cells on a microscope slide, the antibodies attach to the bacterial cells, and the cells fluoresce when illuminated with ultraviolet light. **(b)** In the fluorescent treponemal antibody absorption (FTA-ABS) test for syphilis shown here, *Treponema pallidum* shows up as light cells against a darker background.

■ Why won't other bacteria fluoresce in the FTA-ABS test?

We will say more about antigen-antibody reactions and immunofluorescence in Chapter 18.

Confocal Microscopy

A fairly recent development in light microscopy is known as **confocal microscopy.** Like fluorescent microscopy, specimens are stained with fluorochromes so they will emit, or return, light. In confocal microscopy, one plane of a small region of a specimen is illuminated with a laser, which passes the returned light through an aperture aligned with the illuminated region. Each plane corresponds to an image of a fine slice that has been physically cut from a specimen. Successive planes and regions are illuminated until the entire specimen has been scanned. Because confocal microscopy uses a pinhole aperture, it eliminates the blurring that occurs with other microscopes. As a result, exceptionally clear two-dimensional images can be obtained, with improved resolution of up to 40% over that of other microscopes.

Most confocal microscopes are used in conjunction with computers to construct three-dimensional images. The scanned planes of a specimen, which resemble a stack of images, are converted to a digital form that can be used by a computer to construct a three-dimensional representation. The reconstructed images can be rotated and viewed in any orientation. This technique has been used to obtain three-dimensional images of entire cells and cellular components (Figure 3.7). In addition, confocal microscopy can be used to evaluate cellular physiology by monitoring the distributions and concentrations of substances such as ATP and calcium ions.

Electron Microscopy

Learning Objectives

- *Explain how electron microscopy differs from light microscopy.*
- *Identify one use for the TEM, SEM, and scanned-probe microscopes.*

This technique can detect bacteria or other pathogenic microorganisms, even within cells, tissues, or other clinical specimens (Figure 3.6b). Of paramount importance, it can be used to identify a microbe in minutes. Immunofluorescence is especially useful in diagnosing syphilis and rabies.

CF   ⊢————⊣ 10 μm

FIGURE 3.7 Confocal microscopy. Confocal microscopy produces three-dimensional images and can be used to look inside cells. Shown here are contractile vacuoles in *Paramecium multimicronucleatum.*

Objects smaller than about 0.2 μm, such as viruses or the internal structures of cells, must be examined with an **electron microscope.** In electron microscopy, a beam of electrons is used instead of light. Free electrons travel in waves. The resolving power of the electron microscope is far greater than that of the other microscopes described here so far. The better resolution of electron microscopes is due to the shorter wavelengths of electrons; the wavelengths of electrons are about 100,000 times smaller than the wavelengths of visible light. Thus, electron microscopes are used to examine structures too small to be resolved with light microscopes. Images produced by electron microscopes are always black and white, but they may be colored artificially to accentuate certain details.

Instead of using glass lenses, an electron microscope uses electromagnetic lenses to focus a beam of electrons onto a specimen. There are two types of electron microscopes: the transmission electron microscope and the scanning electron microscope.

Transmission Electron Microscopy

In the **transmission electron microscope (TEM),** a finely focused beam of electrons from an electron gun passes through a specially prepared, ultrathin section of the specimen (Figure 3.8a). The beam is focused on a small area of the specimen by an electromagnetic condenser lens that performs roughly the same function as the condenser of a light microscope—directing the beam of electrons in a straight line to illuminate the specimen.

Electron microscopes use electromagnetic lenses to control illumination, focus, and magnification. Instead of being placed on a glass slide, as in light microscopes, the specimen is usually placed on a copper mesh grid. The beam of electrons passes through the specimen and then through an electromagnetic objective lens, which magnifies the image. Finally, the electrons are focused by an electromagnetic projector lens, rather than by an ocular lens as in a light microscope, onto a fluorescent screen or photographic plate. The final image, called a *transmission electron micrograph,* appears as many light and dark areas, depending on the number of electrons absorbed by different areas of the specimen.

In practice, the transmission electron microscope can resolve objects as close together as 2.5 nm, and objects are generally magnified 10,000–100,000×. Because most microscopic specimens are so thin, the contrast between their ultrastructures and the background is weak. Contrast can be greatly enhanced by using a "stain" that absorbs electrons and produces a darker image in the stained region. Salts of various heavy metals, such as lead, osmium, tungsten, and uranium, are commonly used as stains. These metals can be fixed onto the specimen (*positive staining*) or used to increase the electron opacity of the surrounding field (*negative staining*). Negative staining is useful for the study of the very smallest specimens, such as virus particles, bacterial flagella, and protein molecules.

In addition to positive and negative staining, a microbe can be viewed by a technique called *shadow casting.* In this procedure, a heavy metal such as platinum or gold is sprayed at an angle of about 45° so that it strikes the microbe from only one side. The metal piles up on one side of the specimen, and the uncoated area on the opposite side of the specimen leaves a clear area behind it as a shadow. This gives a three-dimensional effect to the specimen and provides a general idea of the size and shape of the specimen (see Figure 4.6a on page 78).

Transmission electron microscopy has high resolution and is extremely valuable for examining different layers of specimens. However, it does have certain disadvantages. Because electrons have limited penetrating power, only a very thin section of a specimen (about 100 nm) can be studied effectively. Thus, the specimen has no three-dimensional aspect. In addition, specimens must be fixed, dehydrated, and viewed under a high vacuum to prevent electron scattering. These treatments not only kill the specimen but also cause some shrinkage and distortion, sometimes to the extent that there may appear to be additional structures in a prepared cell. Structures that appear as a result of the method of preparation are called *artifacts.*

(a) Transmission. (Top) In a transmission electron microscope, electrons pass through the specimen and are scattered. Magnetic lenses focus the image onto a fluorescent screen or photographic plate. (Bottom) This colorized transmission electron micrograph (TEM) shows a thin slice of a *Paramecium*. In this type of microscopy, the internal structures present in the slice can be seen.

(b) Scanning. (Top) In a scanning electron microscope, primary electrons sweep across the specimen and knock electrons from its surface. These secondary electrons are picked up by a collector, amplified, and transmitted onto a viewing screen or photographic plate. (Bottom) In this colorized scanning electron micrograph (SEM), the surface structures of a *Paramecium* can be seen. Note the three-dimensional appearance of this cell, in contrast

FIGURE 3.8 **Transmission and scanning electron microscopy.** The illustrations show the pathways of electron beams used to create images of the specimens. The photographs show a *Paramecium* viewed with both of these types of electron microscopes. Although electron micrographs are normally black and white, these and other electron micrographs in this book have been artifically colorized for emphasis.

■ How do TEM and SEM images of the same organism differ?

Scanning Electron Microscopy

The **scanning electron microscope (SEM)** overcomes the problem of sectioning associated with a transmission electron microscope. A scanning electron microscope provides striking three-dimensional views of specimens (Figure 3.8b). In scanning electron microscopy, an electron gun produces a finely focused beam of electrons called the primary electron beam. These electrons pass through electromagnetic lenses and are directed over the surface of the specimen. The primary electron beam knocks electrons out

(a)   STM |— 1 nm —|

(b)   AFM |— 5 nm —|

FIGURE 3.9 **Scanned-probe microscopy. (a)** Scanning tunneling microscopy (STM) of the DNA double helix. **(b)** Atomic force microscopy (AFM) of a bacterial surface layer protein.

of the surface of the specimen, and the secondary electrons thus produced are transmitted to an electron collector, amplified, and used to produce an image on a viewing screen or photographic plate. The image is called a *scanning electron micrograph*. This microscope is especially useful in studying the surface structures of intact cells and viruses. In practice, it can resolve objects as close together as 20 nm, and objects are generally magnified 1000–10,000×.

Scanned-Probe Microscopy

Since the early 1980s, several new types of microscopes, called **scanned-probe microscopes,** have been developed. They use various kinds of probes to examine the surface of a specimen at very close range, and they do so without modifying the specimen or exposing it to damaging, high-energy radiation. Such microscopes can be used to map atomic and molecular shapes, to characterize magnetic and chemical properties, and to determine temperature variations inside cells. Among the new scanned-probe microscopes are the scanning tunneling microscope and the atomic force microscope, discussed next.

Scanning Tunneling Microscopy

Scanning tunneling microscopy (STM) uses a thin metal (tungsten) probe that scans a specimen and produces an image revealing the bumps and depressions of the atoms on the surface of the specimen (Figure 3.9a). The resolving power of an STM is much greater than that of an electron microscope; it can resolve features that are only about 1/100 the size of an atom. Moreover, special preparation of the specimen for observation is not needed. STMs are used to provide incredibly detailed views of molecules such as DNA.

Atomic Force Microscopy

In **atomic force microscopy (AFM),** a metal-and-diamond probe is gently forced down onto a specimen. As the probe moves along the surface of the specimen, its movements are recorded and a three-dimensional image is produced (Figure 3.9b). As with STM, AFM does not require special specimen preparation. AFM is used to image both biological substances (in nearly atomic detail) and molecular processes (such as the assembly of fibrin, a component of a blood clot).

The various types of microscopy described above are summarized in Table 3.2 (pages 61–62).

Preparation of Specimens for Light Microscopy

Learning Objectives

- *Differentiate between an acidic dye and a basic dye.*
- *Compare simple, differential, and special stains.*
- *List the steps in preparing a Gram stain, and describe the appearance of gram-positive and gram-negative cells after each step.*
- *Compare and contrast the Gram stain and the acid-fast stain.*

Because most microorganisms appear almost colorless when viewed through a standard light microscope, we often must prepare them for observation. One of the ways this can be done is by staining (coloring). Next we will discuss several different staining procedures.

Preparing Smears for Staining

Most initial observations of microorganisms are made with stained preparations. **Staining** simply means coloring the microorganisms with a dye that emphasizes certain structures. Before the microorganisms can be stained, however, they must be **fixed** (attached) to the microscope slide. Fixing simultaneously kills the microorganisms and attaches them to the slide. It also preserves various parts of microbes in their natural state with only minimal distortion.

When a specimen is to be fixed, a thin film of material containing the microorganisms is spread over the surface of the slide. This film, called a **smear,** is allowed to air dry. In most staining procedures the slide is then fixed by passing it through the flame of a Bunsen burner several times, smear side up, or by covering the slide with methyl alcohol for 1 minute. Air drying and flaming fix the microorganisms to the slide. Stain is applied and then washed off with water; then the slide is blotted with absorbent paper. Without fixing, the stain might wash the microbes off the slide. The stained microorganisms are now ready for microscopic examination.

Stains are salts composed of a positive and a negative ion, one of which is colored and is known as the *chromophore.* The color of so-called **basic dyes** is in the positive ion; in **acidic dyes,** it is in the negative ion. Bacteria are slightly negatively charged at pH 7. Thus, the colored positive ion in a basic dye is attracted to the negatively charged bacterial cell. Basic dyes, which include crystal violet, methylene blue, malachite green, and safranin, are more commonly used than acidic dyes. Acidic dyes are not attracted to most types of bacteria because the dye's negative ions are repelled by the negatively charged bacterial surface, so the stain colors the background instead. Preparing colorless bacteria against a colored background is called **negative staining.** It is valuable in the observation of overall cell shapes, sizes, and capsules because the cells are made highly visible against a contrasting dark background (see Figure 3.12a on page 70). Distortions of cell size and shape are minimized because heat fixing is not necessary and the cells do not pick up the stain. Examples of acidic dyes are eosin, acid fuchsin, and nigrosin.

To apply acidic or basic dyes, microbiologists use three kinds of staining techniques: simple, differential, and special.

Simple Stains

A **simple stain** is an aqueous or alcohol solution of a single basic dye. Although different dyes bind specifically to different parts of cells, the primary purpose of a simple stain is to highlight the entire microorganism so that cellular shapes and basic structures are visible. The stain is applied to the fixed smear for a certain length of time and then washed off, and the slide is dried and examined. Occasionally, a chemical is added to the solution to intensify the stain; such an additive is called a **mordant.** One function of a mordant is to increase the affinity of a stain for a biological specimen; another is to coat a structure (such as a flagellum) to make it thicker and easier to see after it is stained with a dye. Some of the simple stains commonly used in the laboratory are methylene blue, carbolfuchsin, crystal violet, and safranin.

Differential Stains

Unlike simple stains, **differential stains** react differently with different kinds of bacteria and thus can be used to distinguish among them. The differential stains most frequently used for bacteria are the Gram stain and the acid-fast stain.

Gram Stain

The **Gram stain** was developed in 1884 by the Danish bacteriologist Hans Christian Gram. It is one of the most useful staining procedures because it classifies bacteria into two large groups: gram-positive and gram-negative.

In this procedure (Figure 3.10a),

❶ A heat-fixed smear is covered with a basic purple dye, usually crystal violet. Because the purple stain imparts its color to all cells, it is referred to as a **primary stain.**

❷ After a short time, the purple dye is washed off, and the smear is covered with iodine, a mordant. When the iodine is washed off, both gram-positive and gram-negative bacteria appear dark violet or purple.

❸ Next, the slide is washed with alcohol or an alcohol-acetone solution. This solution is a **decolorizing agent,** which removes the purple from the cells of some species but not from others.

❹ The alcohol is rinsed off, and the slide is then stained with safranin, a basic red dye. The smear is washed again, blotted dry, and examined microscopically.

The purple dye and the iodine combine in the cytoplasm of each bacterium and color it dark violet or purple. Bacteria that retain this color after the alcohol has

(a)

① Application of crystal violet (purple dye)

② Application of iodine (mordant)

③ Alcohol wash (decolorization)

④ Application of safranin (counterstain)

- Crystal violet
- Iodine
- Alcohol
- Safranin

(b)

Rod

Coccus

Vibrio

LM 5 μm

FIGURE 3.10 **Gram staining. (a)** Procedure. **(b)** Micrograph of gram-stained bacteria. The rods and cocci (purple) are gram-positive, and the vibrios (pink) are gram-negative.

■ How can the Gram reaction be useful in prescribing antibiotic treatment?

attempted to decolorize them are classified as **gram-positive;** bacteria that lose the dark violet or purple color after decolorization are classified as **gram-negative** (Figure 3.10b). Because gram-negative bacteria are colorless after the alcohol wash, they are no longer visible. This is why the basic dye safranin is applied; it turns the gram-negative bacteria pink. Stains such as safranin that have a contrasting color to the primary stain are called **counterstains.** Because gram-positive bacteria retain the original purple stain, they are not affected by the safranin counterstain.

As you will see in Chapter 4, different kinds of bacteria react differently to the Gram stain, because structural differences in their cell walls affect the retention or escape of a combination of crystal violet and iodine, called the crystal violet–iodine (CV–I) complex. Among other differences, gram-positive bacteria have a thicker peptidoglycan (disaccharides and amino acids) cell wall than gram-

negative bacteria. In addition, gram-negative bacteria contain a layer of lipopolysaccharide (lipids and polysaccharides) as part of their cell wall (see Figure 4.13 on page 84). When applied to both gram-positive and gram-negative cells, crystal violet and then iodine readily enter the cells. Inside the cells, the crystal violet and iodine combine to form CV–I. This complex is larger than the crystal violet molecule that entered the cells, and, because of its size, it cannot be washed out of the intact peptidoglycan layer of gram-positive cells by alcohol. Consequently, gram-positive cells retain the color of the crystal violet dye. In gram-negative cells, however, the alcohol wash disrupts the outer lipopolysaccharide layer, and the CV–I complex is washed out through the thin layer of peptidoglycan. As a result, gram-negative cells are colorless until counterstained with safranin, after which they are pink.

In summary, gram-positive cells retain the dye and remain purple. Gram-negative cells do not retain the dye; they are colorless until counterstained with a red dye, after which they appear pink.

The Gram method is one of the most important staining techniques in medical microbiology. But Gram staining results are not universally applicable because some bacterial cells stain poorly or not at all. The Gram reaction is most consistent when it is used on young, growing bacteria.

The Gram reaction of a bacterium can provide valuable information for the treatment of disease. Gram-positive bacteria tend to be killed easily by penicillins and cephalosporins. Gram-negative bacteria are generally more resistant because the antibiotics cannot penetrate the lipopolysaccharide layer. Some resistance to these antibiotics among both gram-positive and gram-negative bacteria is due to bacterial inactivation of the antibiotics.

Acid-Fast Stain

Another important differential stain (one that differentiates bacteria into distinctive groups) is the **acid-fast stain,** which binds strongly only to bacteria that have a waxy material in their cell walls. Microbiologists use this stain to identify all bacteria in the genus *Mycobacterium*, including the two important pathogens *Mycobacterium tuberculosis*, the causative agent of tuberculosis, and *Mycobacterium leprae* (lep′rī), the causative agent of leprosy. This stain is also used to identify the pathogenic strains of the genus *Nocardia* (nō-kär′dē-ä).

In the acid-fast staining procedure, the red dye carbolfuchsin is applied to a fixed smear, and the slide is gently heated for several minutes. (Heating enhances penetration and retention of the dye.) Then the slide is cooled and washed with water. The smear is next treated with acid-alcohol, a decolorizer, which removes the red stain from bacteria that are not acid-fast. The acid-fast microorganisms retain the red color because the carbolfuchsin is more soluble in the cell wall lipids than in the acid-alcohol (Figure 3.11). In non–acid-fast bacteria, whose cell walls lack the lipid components, the carbolfuchsin is rapidly removed during decolorization, leaving the cells colorless. The smear is then stained with a methylene blue counterstain. Non–acid-fast cells appear blue after application of the counterstain.

Special Stains

Learning Objective

- *Explain why each of the following is used: capsule stain, endospore stain, flagella stain.*

Special stains are used to color and isolate specific parts of microorganisms, such as endospores and flagella, and to reveal the presence of capsules.

LM |———| 10 μm

FIGURE 3.11 Acid-fast bacteria. The *Mycobacterium leprae* bacteria that have infected this tissue have been stained red with an acid-fast stain.

■ What two diseases can be diagnosed using the acid-fast stain?

Negative Staining for Capsules

Many microorganisms contain a gelatinous covering called a **capsule,** which we will discuss in our examination of the prokaryotic cell in Chapter 4. In medical microbiology, demonstrating the presence of a capsule is a means of determining the organism's **virulence,** the degree to which a pathogen can cause disease.

Capsule staining is more difficult than other types of staining procedures because capsular materials are soluble in water and may be dislodged or removed during rigorous washing. To demonstrate the presence of capsules, a microbiologist can mix the bacteria in a solution containing a fine colloidal suspension of colored particles (usually India ink or nigrosin) to provide a dark background and then stain the bacteria with a simple stain, such as safranin (Figure 3.12a). Because of their chemical composition, capsules do not accept most biological dyes, such as safranin, and thus appear as halos surrounding each stained bacterial cell.

Endospore (Spore) Staining

An **endospore** is a special resistant, dormant structure formed within a cell that protects a bacterium from adverse environmental conditions. Although endospores are relatively uncommon in bacterial cells, they can be formed by a few genera of bacteria. Endospores cannot be stained by ordinary methods, such as simple staining and Gram staining, because the dyes do not penetrate the wall of the endospore.

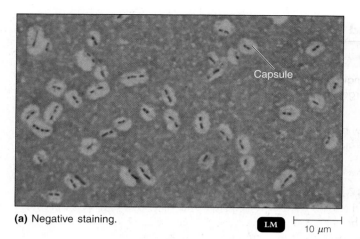

(a) Negative staining. LM ⊢——⊣ 10 μm

(b) Endospore staining. LM ⊢——⊣ 10 μm

FIGURE 3.12 **Special staining. (a)** Capsule staining provides a contrasting background, so the capsules of these bacteria, *Klebsiella pneumoniae,* show up as light areas surrounding the stained cells. **(b)** Endospores are seen as green ovals in these rod-shaped cells of the bacterium *Bacillus cereus,* using the Schaeffer-Fulton endospore stain. **(c)** Flagella are shown as wavy extensions from the ends of these cells of the bacterium *Spirillum volutans.* In relation to the body of the cell, the flagella are much thicker than normal because layers of the stain have accumulated from treatment of the specimen with a mordant.

■ Of what value are capsules, endospores, and flagella to bacteria?

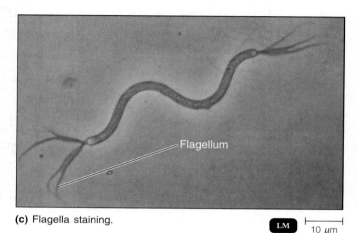

(c) Flagella staining. LM ⊢——⊣ 10 μm

The most commonly used endospore stain is the *Schaeffer-Fulton endospore stain* (Figure 3.12b). Malachite green, the primary stain, is applied to a heat-fixed smear and heated to steaming for about 5 minutes. The heat helps the stain penetrate the endospore wall. Then the preparation is washed for about 30 seconds with water to remove the malachite green from all of the cells' parts except the endospores. Next, safranin, a counterstain, is applied to the smear to stain portions of the cell other than endospores. In a properly prepared smear, the endospores appear green within red or pink cells. Because endospores are highly refractive, they can be detected under the light microscope when unstained, but they cannot be differentiated from inclusions of stored material without a special stain.

Flagella Staining

Bacterial **flagella** (singular: **flagellum**) are structures of locomotion too small to be seen with a light microscope without staining. A tedious and delicate staining procedure uses a mordant and the stain carbolfuchsin to build up the diameters of the flagella until they become visible under the light microscope (Figure 3.12c). Microbiologists use the number and arrangement of flagella as diagnostic aids.

* * *

A summary of stains is presented in Table 3.3. In the next chapter we will take a closer look at the structure of microbes and how they protect, nourish, and reproduce themselves.

TABLE 3.3	**A Summary of Various Stains and Their Uses**
Stain	**Principal Uses**
Simple (methylene blue, carbolfuchsin, crystal violet, safranin)	Aqueous or alcohol solution of a single basic dye. (Sometimes a mordant is added to intensify the stain.) Used to highlight microorganisms to determine cellular shapes and arrangements.
Differential Gram	React differently with different kinds of bacteria in order to distinguish among them. Classifies bacteria into two large groups: gram-positive and gram-negative. Gram-positive bacteria retain the crystal violet stain and appear purple. Gram-negative bacteria do not retain the crystal violet stain and remain colorless until counterstained with safranin and then appear pink.
Acid-fast	Used to distinguish *Mycobacterium* species and some species of *Nocardia*. Acid-fast bacteria, once stained with carbolfuchsin and treated with acid-alcohol, remain red because they retain the carbolfuchsin stain. Non–acid-fast bacteria, when stained and treated the same way and then stained with methylene blue, appear blue because they lose the carbolfuchsin stain and are then able to accept the methylene blue stain.
Special	Used to color and isolate various structures, such as capsules, endospores, and flagella; sometimes used as a diagnostic aid.
Negative	Used to demonstrate the presence of capsules. Because capsules do not accept most stains, the capsules appear as unstained halos around bacterial cells and stand out against a dark background.
Endospore	Used to detect the presence of endospores in bacteria. When malachite green is applied to a heat-fixed smear of bacterial cells, the stain penetrates the endospores and stains them green. When safranin (red) is then applied, it stains the remainder of the cells red or pink.
Flagella	Used to demonstrate the presence of flagella. A mordant is used to build up the diameters of flagella until they become visible microscopically when stained with carbolfuchsin.

STUDY OUTLINE

Units of Measurement (pp. 54–55)

1. The standard unit of length is the meter (m).
2. Microorganisms are measured in micrometers, μm (10^{-6} m), and in nanometers, nm (10^{-9} m).

Microscopy: The Instruments (pp. 55–66)

1. A simple microscope consists of one lens; a compound microscope has multiple lenses.

Light Microscopy (pp. 55–63)

Compound Light Microscopy (pp. 55–57)

1. The most common microscope used in microbiology is the compound light microscope (**LM**).
2. The total magnification of an object is calculated by multiplying the magnification of the objective lens by the magnification of the ocular lens.
3. The compound light microscope uses visible light.
4. The maximum resolution, or resolving power (the ability to distinguish between two points) of a compound light microscope is 0.2 μm; maximum magnification is 2000×.

5. Specimens are stained to increase the difference between the refractive indexes of the specimen and the medium.
6. Immersion oil is used with the oil immersion lens to reduce light loss between the slide and the lens.
7. Brightfield illumination is used for stained smears.
8. Unstained cells are more productively observed using darkfield, phase-contrast, or DIC microscopy.

Darkfield Microscopy (pp. 57–59)

1. The darkfield microscope shows a light silhouette of an organism against a dark background.
2. It is most useful for detecting the presence of extremely small organisms.

Phase-Contrast Microscopy (pp. 59–60)

1. A phase-contrast microscope brings direct and reflected or diffracted light rays together (in phase) to form an image of the specimen on the ocular lens.
2. It allows the detailed observation of living organisms.

Differential Interference Contrast (DIC) Microscopy (p. 60)

1. The DIC microscope provides a colored, three-dimensional image of the object being observed.

2. It allows detailed observations of living cells.

Fluorescence Microscopy (pp. 60–63)

1. In fluorescence microscopy, specimens are first stained with fluorochromes and then viewed through a compound microscope by using an ultraviolet light source.

2. The microorganisms appear as bright objects against a dark background.

3. Fluorescence microscopy is used primarily in a diagnostic procedure called fluorescent-antibody (FA) technique, or immunofluorescence.

Confocal Microscopy (p. 63)

1. In confocal microscopy, a specimen is stained with a fluorescent dye and illuminated one plane at a time.

2. Using a computer to process the images, two-dimensional and three-dimensional images of cells can be produced.

Electron Microscopy (pp. 63–66)

1. A beam of electrons, instead of light, is used with an electron microscope.

2. Electromagnets, instead of glass lenses, control focus, illumination, and magnification.

3. Thin sections of organisms can be seen in an electron micrograph produced using a transmission electron microscope (TEM). Magnification: 10,000–100,000×. Resolving power: 2.5 nm.

4. Three-dimensional views of the surfaces of whole microorganisms can be obtained with a scanning electron microscope (SEM). Magnification: 1000–10,000×. Resolving power: 20 nm.

Scanned-Probe Microscopy (p. 66)

1. Scanning tunneling microscopy (STM) and atomic force microscopy (AFM) produce three-dimensional images of the surface of a molecule.

Preparation of Specimens for Light Microscopy (p. 66–71)

Preparing Smears for Staining (p. 67)

1. Staining means coloring a microorganism with a dye to make some structures more visible.

2. Fixing uses heat or alcohol to kill and attach microorganisms to a slide.

3. A smear is a thin film of material used for microscopic examination.

4. Bacteria are negatively charged, and the colored positive ion of a basic dye will stain bacterial cells.

5. The colored negative ion of an acidic dye will stain the background of a bacterial smear; a negative stain is produced.

Simple Stains (p. 67)

1. A simple stain is an aqueous or alcohol solution of a single basic dye.

2. It is used to make cellular shapes and arrangements visible.

3. A mordant may be used to improve bonding between the stain and the specimen.

Differential Stains (pp. 67–69)

1. Differential stains, such as the Gram stain and acid-fast stain, differentiate bacteria according to their reactions to the stains.

2. The Gram stain procedure uses a purple stain (crystal violet), iodine as a mordant, an alcohol decolorizer, and a red counterstain.

3. Gram-positive bacteria retain the purple stain after the decolorization step; gram-negative bacteria do not and thus appear pink from the counterstain.

4. Acid-fast microbes, such as members of the genera *Mycobacterium* and *Nocardia*, retain carbolfuchsin after acid-alcohol decolorization and appear red; non–acid-fast microbes take up the methylene blue counterstain and appear blue.

Special Stains (pp. 69–70)

1. The endospore stain and flagella stain are special stains that color only certain parts of bacteria.

2. Negative staining is used to make microbial capsules visible.

STUDY QUESTIONS

Review

1. Fill in the following blanks:

   $1 \mu m = $_____ m

   1 _____ $= 10^{-9}$ m

   $1 \mu m = $_____ nm

2. Label the parts of the compound light microscope in the following figure:

(a) _____

(b) _____

(c) _____

(d) _____

(e) _____

a. _____   d. _____

b. _____   e. _____

c. _____

3. Calculate the total magnification of the nucleus of a cell being observed through a compound light microscope with a 10× ocular lens and an oil immersion lens.

4. Which type of microscope would be best to use to observe each of the following?
   a. a stained bacterial smear
   b. unstained bacterial cells when the cells are small and no detail is needed
   c. unstained live tissue when it is desirable to see some intracellular detail
   d. a sample that emits light when illuminated with ultraviolet light
   e. intracellular detail of a cell that is 1 μm long
   f. unstained live cells in which intracellular structures are shown in color

5. An electron microscope differs from a light microscope in that _____ focused by _____ is used instead of light, and the image is viewed not through the ocular lenses but on _____.

6. The maximum magnification of a compound microscope is _____; that of an electron microscope, _____. The maximum resolution of a compound microscope is _____; that of an electron microscope, _____. One advantage of a scanning electron microscope over a transmission electron microscope is _____.

7. Why do basic dyes stain bacterial cells? Why don't acidic dyes stain bacterial cells?

8. When is it most appropriate to use each of the following?
   a. a simple stain   c. a negative stain
   b. a differential stain   d. a flagella stain

9. Why is a mordant used in the Gram stain? In the flagella stain?

10. What is the purpose of a counterstain in the acid-fast stain?

11. What is the purpose of a decolorizer in the Gram stain? In the acid-fast stain?

12. Choose from the following terms to fill in the blanks: counterstain, decolorizer, mordant, primary stain. In the endospore stain, safranin is the _____. In the Gram stain, safranin is the _____.

13. Fill in the following table regarding the Gram stain:

Steps	Appearance After This Step of Gram-Positive Cells	Gram-Negative Cells
Crystal violet		
Iodine		
Alcohol-acetone		
Safranin		

Multiple Choice

a. You are trying to identify an unknown gram-negative bacterium. Which of the following stains is *not* necessary?
   a. negative stain   d. endospore stain
   b. acid-fast stain   e. both b and d
   c. flagella stain

1. Assume you stain *Bacillus* by applying malachite green with heat and then counterstain with safranin. Through the microscope, the green structures are
   a. cell walls.   d. flagella.
   b. capsules.   e. impossible to identify.
   c. endospores.

3. Carbolfuchsin can be used as a simple stain and a negative stain. As a simple stain, the pH is
   a. 2.
   b. higher than the negative stain.
   c. lower than the negative stain.
   d. the same as the negative stain.

4. Looking at the cell of a photosynthetic microorganism, you observe that the chloroplasts are green in brightfield microscopy and red in fluorescence microscopy. You conclude that
   a. chlorophyll is fluorescent.
   b. the magnification has distorted the image.
   c. you're not looking at the same structure in both microscopes.

d. the stain masked the green color.
e. none of the above

5. Which of the following is *not* a functionally analogous pair of stains?
   a. nigrosin and malachite green
   b. crystal violet and carbolfuchsin
   c. safranin and methylene blue
   d. ethanol-acetone and acid-alcohol
   e. none of the above

6. Which of the following pairs is mismatched?
   a. capsule—negative stain
   b. cell arrangement—simple stain
   c. cell size—negative stain
   d. Gram stain—bacterial identification
   e. none of the above

7. Assume you stain *Clostridium* by applying a basic stain, carbolfuchsin, with heat, decolorizing with acid-alcohol, and counterstaining with an acidic stain, nigrosin. Through the microscope, the endospores are _____1_____ and the cells are stained _____2_____.
   a. 1—red; 2—black
   b. 1—black; 2—colorless
   c. 1—colorless; 2—black
   d. 1—red; 2—colorless
   e. 1—black; 2—red

8. Assume that you are viewing a Gram-stained field of red cocci and blue bacilli through the microscope. You can safely conclude that you have
   a. made a mistake in staining.
   b. two different species.
   c. old bacterial cells.
   d. young bacterial cells.
   e. none of the above

9. In 1996, scientists described a new tapeworm parasite that had killed at least one person. The initial examination of the patient's abdominal mass was most likely made using
   a. brightfield microscopy.
   b. darkfield microscopy.
   c. electron microscopy.
   d. phase-contrast microscopy.
   e. fluorescence microscopy.

10. Which of the following is *not* a modification of a compound light microscope?
    a. brightfield microscopy
    b. darkfield microscopy
    c. electron microscopy
    d. phase-contrast microscopy
    e. fluorescence microscopy

Critical Thinking

1. In a Gram stain, one step could be omitted and still allow differentiation between gram-positive and gram-negative cells. What is that one step?

2. Using a good compound light microscope with a resolving power of 0.3 μm, a 10$\times$ ocular lens, and a 100$\times$ oil immersion lens, would you be able to discern two objects separated by 3 μm? 0.3 μm? 300 nm?

3. Why isn't the Gram stain used on acid-fast bacteria? If you did Gram stain acid-fast bacteria, what would their Gram reaction be? What is the Gram reaction of non–acid-fast bacteria?

4. Endospores can be seen as refractile structures in unstained cells and as colorless areas in Gram-stained cells. Why is it necessary to do an endospore stain to verify the presence of endospores?

Clinical Applications

1. In 1882, German bacteriologist Paul Erhlich described a method for staining *Mycobacterium* and noted, "It may be that all disinfecting agents which are acidic will be without effect on this [tubercle] bacillus, and one will have to be limited to alkaline agents." How did he reach this conclusion without testing disinfectants?

2. Laboratory diagnosis of *Neisseria gonorrhoeae* infection is based on microscopic examination of Gram-stained pus. Identify the bacteria in this light micrograph. What is the disease?

LM | 10 μm

3. Assume that you are viewing a Gram-stained sample of vaginal discharge. Large (10 μm) red cells are coated with small (0.5 μm $\times$ 1.5 μm) blue cells on their surfaces. What is the most likely explanation for the red and blue cells?

4. A sputum sample from Calle, a 30-year-old Asian elephant, was smeared onto a slide and air dried. The smear was fixed, covered with carbolfuchsin, and heated for 5 minutes. After washing with water, acid-alcohol was placed on the smear for 30 seconds. Finally, the smear was stained with methylene blue for 30 seconds, washed with water, and dried. On examination at 1000$\times$, the zoo veterinarian saw red rods on the slide. What infections do the results suggest? (Calle was treated and recovered.)

LEARNING WITH TECHNOLOGY

The Microbiology Place CD-ROM/website includes many resources to help you study and succeed in this course. In addition to self-quizzes, web links, and flashcards, you'll find the following activities for this chapter in the "Activities" section of The Microbiology Place:

The Metric System

Exploring Size and Scale

Case Study: Electron Microscopy

Chapter 4

Functional Anatomy of Prokaryotic and Eukaryotic Cells

BACILLUS. *Cells of the genus Bacillus frequently form long chains.*

Despite their complexity and variety, all living cells can be classified into two groups, prokaryotes and eukaryotes, based on their ultrastructure as seen with the electron microscope. Plants and animals are entirely composed of eukaryotic cells. In the microbial world, Bacteria and Archaea are prokaryotes. Other cellular microbes—fungi (yeasts and molds), protozoa, and algae—are eukaryotes.

Viruses, as noncellular elements, do not fit into any organizational scheme of living cells. They are genetic particles that replicate but are unable to perform the usual chemical activities of living cells. Viral structure and activity will be discussed in Chapter 13. In this chapter we will concentrate on describing prokaryotic and eukaryotic cells.

Comparing Prokaryotic and Eukaryotic Cells: An Overview

Learning Objective
- *Compare and contrast the overall cell structure of prokaryotes and eukaryotes.*

Prokaryotes and eukaryotes are chemically similar, in the sense that they both contain nucleic acids, proteins, lipids, and carbohydrates. They use the same kinds of chemical reactions to metabolize food, build proteins, and store energy. It is primarily the structure of cell walls and membranes, and the absence of *organelles* (specialized cellular structures that have specific functions), that distinguish prokaryotes from eukaryotes.

The chief distinguishing characteristics of **prokaryotes** (from the Greek words meaning prenucleus) are as follows:

1. Their DNA (genetic material) is not enclosed within a membrane and is one circular chromosome.
2. Their DNA is not associated with histones (special chromosomal proteins found in eukaryotes); other proteins are associated with the DNA.
3. They lack membrane-enclosed organelles.
4. Their cell walls almost always contain the complex polysaccharide peptidoglycan.
5. They usually divide by **binary fission.** During this process, the DNA is copied and the cell splits into two cells. Binary fission involves fewer structures and processes than eukaryotic cell division.

Eukaryotes (from the Greek words meaning true nucleus) have the following distinguishing characteristics:

1. Their DNA is found in the cell's nucleus, which is separated from the cytoplasm by a nuclear membrane, and the DNA is found in multiple chromosomes.

2. Their DNA is consistently associated with chromosomal proteins called histones and with nonhistones.

3. They have a number of membrane-enclosed organelles, including mitochondria, endoplasmic reticulum, Golgi complex, lysosomes, and sometimes chloroplasts.

4. Their cell walls, when present, are chemically simple.

5. They usually divide by mitosis, in which chromosomes replicate and an identical set is distributed into each of two nuclei. This process is guided by the mitotic spindle, a football-shaped assembly of microtubules (described later). Division of the cytoplasm and other organelles follows so that the two cells produced are identical to each other.

Additional differences between prokaryotic and eukaryotic cells are listed in Table 4.2 on page 96. Next we describe, in detail, the parts of the prokaryotic cell.

THE PROKARYOTIC CELL

The members of the prokaryotic world make up a vast heterogeneous group of very small unicellular organisms. Prokaryotes include bacteria and archaea. The majority of prokaryotes, including the photosynthesizing cyanobacteria, are included in the bacteria. Although bacteria and archaea look similar, they are different in chemical composition, as will be described later. The thousands of species of bacteria are differentiated by many factors, including morphology (shape), chemical composition (often detected by staining reactions), nutritional requirements, biochemical activities, and source of energy (sunlight or chemicals).

The Size, Shape, and Arrangement of Bacterial Cells

Learning Objective

• Identify the three basic shapes of bacteria.

Bacteria come in a great many sizes and shapes. Most bacteria range from 0.2–2.0 μm in diameter and from 2–8 μm in length. They have a few basic shapes: spherical **coccus** (plural: *cocci*, meaning berries), rod-shaped **bacillus** (plural: *bacilli*, meaning little staffs), and **spiral.**

Cocci are usually round but can be oval, elongated, or flattened on one side. When cocci divide to reproduce, the cells can remain attached to one another. Cocci that remain in pairs after dividing are called **diplococci;** those that divide and remain attached in chainlike patterns are called **streptococci** (Figure 4.1a). Those that divide in two planes and remain in groups of four are known as **tetrads** (Figure 4.1b). Those that divide in three planes and remain attached in cubelike groups of eight are called **sarcinae** (Figure 4.1c). Those that divide in multiple planes and form grapelike clusters or broad sheets are called **staphylococci** (Figure 4.1d). These group characteristics are frequently helpful in the identification of certain cocci.

FIGURE 4.1 **Arrangements of cocci. (a)** Division in one plane produces diplococci and streptococci. **(b)** Division in two planes produces tetrads. **(c)** Division in three planes produces sarcinae, and **(d)** division in multiple planes produces staphylococci.

■ The planes in which the cell divides determine the arrangement of cells.

(a)   Single bacillus

(b)   Diplobacilli

(c)   Streptobacilli

(d)   Coccobacillus

SEM   2 μm

SEM   5 μm

SEM   1 μm

FIGURE 4.2 **Bacilli. (a)** Single bacilli. **(b)** Diplobacilli. In the top micrograph, a few joined pairs of bacilli serve as examples of diplobacilli. **(c)** Streptobacilli. **(d)** Coccobacilli.

■ Why don't bacilli form tetrads or clusters?

Bacilli divide only across their short axis, so there are fewer groupings of bacilli than of cocci. Most bacilli appear as single rods (Figure 4.2a). **Diplobacilli** appear in pairs after division (Figure 4.2b), and **streptobacilli** occur in chains (Figure 4.2c). Some bacilli look like straws. Others have tapered ends, like cigars. Still others are oval and look so much like cocci that they are called **coccobacilli** (Figure 4.2d).

"Bacillus" has two meanings in microbiology. As we have just used it, bacillus refers to a bacterial shape. When capitalized and italicized, it refers to a specific genus. For example, the bacterium *Bacillus anthracis* is the causative agent of anthrax. *Bacillus* cells often form long, twisted chains of cells (Figure 4.3).

Spiral bacteria have one or more twists; they are never straight. Bacteria that look like curved rods are called **vibrios** (Figure 4.4a). Others, called **spirilla,** have a helical shape, like a corkscrew, and fairly rigid bodies (Figure 4.4b). Yet another group of spirals are helical and flexible; they are called **spirochetes** (Figure 4.4c). Unlike the spirilla, which use

LM   10 μm

FIGURE 4.3 **A double-stranded helix formed by *Bacillus subtilis*.**

■ *Bacillus* cells often remain attached to each other, forming extended chains.

(a)   Vibrio

SEM   4 μm

(b)   Spirillum

SEM   2 μm

(c)   Spirochete

SEM   1.5 μm

FIGURE 4.4 **Spiral bacteria. (a)** Vibrios. **(b)** Spirilla. **(c)** Spirochete.

■ Spiral bacteria have one or more twists.

(a) Star-shaped bacteria

LM 0.5 μm

(b) Rectangular bacteria

LM 2 μm

FIGURE 4.5 **Star-shaped and rectangular prokaryotes.**
(a) *Stella* (star-shaped).
(b) *Haloarcula,* a genus of halophilic archaea (rectangular cells).

whiplike external appendages called flagella to move, spirochetes move by means of axial filaments, which resemble flagella but are contained within a flexible external sheath.

In addition to the three basic shapes, there are star-shaped cells (genus *Stella*); rectangular, flat cells (halophilic archaea) of the genus *Haloarcula* (Figure 4.5); and triangular cells.

The shape of a bacterium is determined by heredity. Genetically, most bacteria are **monomorphic;** that is, they maintain a single shape. However, a number of environ-

mental conditions can alter that shape. If the shape is altered, identification becomes difficult. Moreover, some bacteria, such as *Rhizobium* (rī-zō′bē-um) and *Corynebacterium* (kô-rī-nē-bak-ti′rē-um), are genetically **pleomorphic,** which means they can have many shapes, not just one.

The structure of a typical prokaryotic cell is shown in Figure 4.6. We will discuss its components according to the following organization: (1) structures external to the cell wall, (2) the cell wall itself, and (3) structures internal to the cell wall.

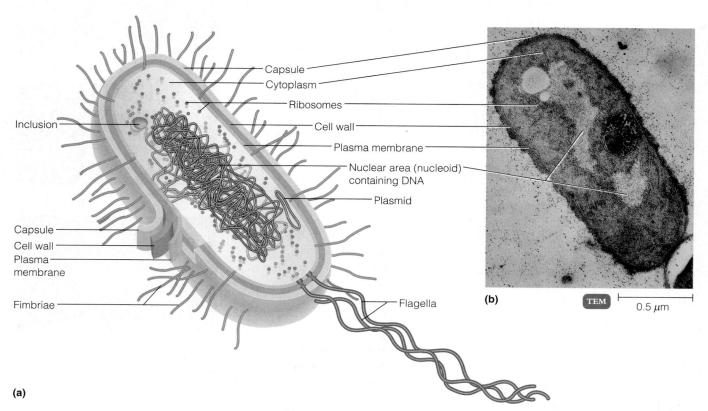

Inclusion

Capsule
Cell wall
Plasma membrane

Fimbriae

Capsule
Cytoplasm
Ribosomes
Cell wall
Plasma membrane
Nuclear area (nucleoid) containing DNA
Plasmid

Flagella

(b)

TEM 0.5 μm

(a)

FIGURE 4.6 **A prokaryotic cell showing typical structures.** Both the drawing **(a)** and micrograph **(b)** show a bacterium lengthwise to reveal the internal composition.

■ Prokaryotic cells are distinguished from eukaryotic cells primarily by the structure of the cell wall and membranes and the absence of organelles.

Structures External to the Cell Wall

Learning Objective

- Describe the structure and function of the glycocalyx, flagella, axial filaments, fimbriae, and pili.

Among the possible structures external to the prokaryotic cell wall are the glycocalyx, flagella, axial filaments, fimbriae, and pili.

Glycocalyx

Many prokaryotes secrete on their surface a substance called glycocalyx. **Glycocalyx** (meaning sugar coat) is the general term used for substances that surround cells. The bacterial glycocalyx is a viscous (sticky), gelatinous polymer that is external to the cell wall and composed of polysaccharide, polypeptide, or both. Its chemical composition varies widely with the species. For the most part, it is made inside the cell and secreted to the cell surface. If the substance is organized and is firmly attached to the cell wall, the glycocalyx is described as a **capsule.** The presence of a capsule can be determined by using negative staining, described in Chapter 3 (see Figure 3.12a on page 70). If the substance is unorganized and only loosely attached to the cell wall, the glycocalyx is described as a **slime layer.**

In certain species, capsules are important in contributing to bacterial virulence (the degree to which a pathogen causes disease). Capsules often protect pathogenic bacteria from phagocytosis by the cells of the host. (As you will see later, phagocytosis is the ingestion and digestion of microorganisms and other solid particles.) For example, *Bacillus anthracis* produces a capsule of D-glutamic acid. (Recall from Chapter 2 that the D forms of amino acids are unusual.) Because only encapsulated *B. anthracis* causes anthrax, it is speculated that the capsule may prevent its being destroyed by phagocytosis.

Another example involves *Streptococcus pneumoniae* (strep-tō-kok'kus nü-mō'nē-ī), which causes pneumonia only when the cells are protected by a polysaccharide capsule. Unencapsulated *S. pneumoniae* cells cannot cause pneumonia and are readily phagocytized. The polysaccharide capsule of *Klebsiella* (kleb-sē-el'lä) also prevents phagocytosis and allows the bacterium to adhere to and colonize the respiratory tract. A glycocalyx made of sugars is called an **extracellular polysaccharide (EPS).** The EPS enables a bacterium to attach to various surfaces in its natural environment in order to survive. Through attachment, bacteria can grow on diverse surfaces such as rocks in fast-moving streams, plant roots, human teeth, medical implants, water pipes, and even other bacteria. *Streptococcus mutans* (mū'tans), an important cause of dental caries, attaches itself to the surface of teeth by a glycocalyx. *S. mutans* may use its capsule as a source of nutrition by breaking it down and utilizing the sugars when energy stores are low. A glycocalyx can protect a cell against dehydration. Also, its viscosity may inhibit the movement of nutrients out of the cell.

Flagella

Some prokaryotic cells have **flagella** (singular: **flagellum,** meaning whip), which are long filamentous appendages that propel bacteria (Figure 4.7).

Bacterial cells have four arrangements of flagella (Figure 4.7): **monotrichous** (a single polar flagellum), **amphitrichous** (a tuft of flagella at each end of the cell), **lophotrichous** (two or more flagella at one pole of the cell), and **peritrichous** (flagella distributed over the entire cell).

A flagellum has three basic parts (Figure 4.8). The long outermost region, the *filament*, is constant in diameter and contains the globular (roughly spherical) protein *flagellin* arranged in several chains that intertwine and form a helix around a hollow core. Certain pathogenic bacteria can be identified by their flagellar proteins. In most bacteria, filaments are not covered by a membrane or sheath, as in eukaryotic cells. The filament is attached to a slightly wider *hook*, consisting of a different protein. The third portion of a flagellum is the *basal body*, which anchors the flagellum to the cell wall and plasma membrane.

The basal body is composed of a small central rod inserted into a series of rings. Gram-negative bacteria contain two pairs of rings; the outer pair of rings is anchored to various portions of the cell wall, and the inner pair of rings is anchored to the plasma membrane. In gram-positive bacteria, only the inner pair is present. As you will see later, the flagella (and cilia) of eukaryotic cells are more complex than those of prokaryotic cells.

Bacteria with flagella are motile; that is, they have the ability to move on their own. Each prokaryotic flagellum is a semirigid, helical structure that moves the cell by rotating from the basal body. The rotation of a flagellum is either clockwise or counterclockwise around its long axis. (Eukaryotic flagella, by contrast, undulate in a wavelike motion.) The movement of a prokaryotic flagellum results from rotation of its basal body and is similar to the movement of the shaft of an electric motor. As the flagella rotate, they form a bundle that pushes against the surrounding liquid and propels the bacterium. Flagellar rotation depends on the cell's continuous generation of energy.

Bacterial cells can alter the speed and direction of rotation of flagella and thus are capable of various patterns of **motility,** the ability of an organism to move by itself. When a bacterium moves in one direction for a length of time, the

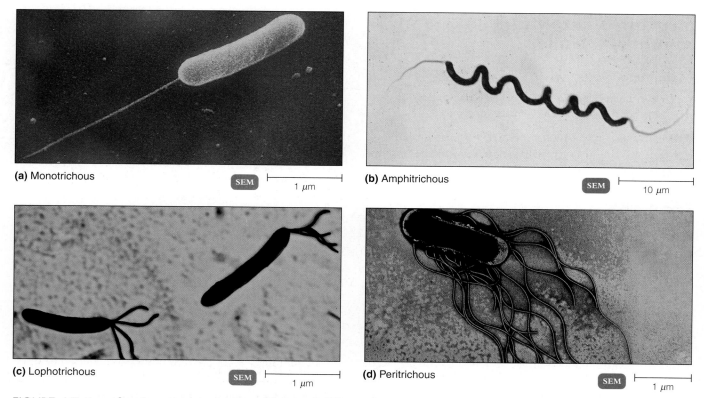

(a) Monotrichous · SEM · 1 μm

(b) Amphitrichous · SEM · 10 μm

(c) Lophotrichous · SEM · 1 μm

(d) Peritrichous · SEM · 1 μm

FIGURE 4.7 **Four basic arrangements of bacterial flagella.**

movement is called a "run" or "swim." "Runs" are interrupted by periodic, abrupt, random changes in direction called "tumbles." Then, a "run" resumes. "Tumbles" are caused by a reversal of flagellar rotation (Figure 4.9a). Some species of bacteria endowed with many flagella—*Proteus* (prō′tē-us), for example (Figure 4.9b)—can "swarm," or show rapid wavelike movement across a solid culture medium.

One advantage of motility is that it enables a bacterium to move toward a favorable environment or away from an adverse one. The movement of a bacterium toward or away from a particular stimulus is called **taxis.** Such stimuli include chemicals **(chemotaxis)** and light **(phototaxis).** Motile bacteria contain receptors in various locations, such as in or just under the cell wall. These receptors pick up chemical stimuli, such as oxygen, ribose, and galactose. In response to the stimuli, information is passed to the flagella. If the chemotactic signal is positive, called an *attractant,* the bacteria move toward the stimulus with many runs and few tumbles. If the chemotactic signal is negative, called a *repellent,* the frequency of tumbles increases as the bacteria move away from the stimulus.

The flagellar protein called H antigen is useful for distinguishing among **serovars,** or variations within a species, of gram-negative bacteria. For example, there are at least 50 different H antigens for *E. coli.* Those serovars

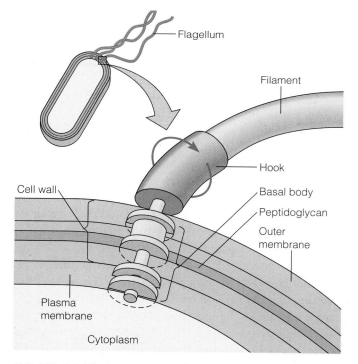

FIGURE 4.8 **The structure of a prokaryotic flagellum.** The parts and attachment of a flagellum of a gram-negative bacterium are shown in this highly schematic diagram.

Labels: Flagellum, Filament, Hook, Basal body, Peptidoglycan, Outer membrane, Cell wall, Plasma membrane, Cytoplasm

■ What are the three basic parts of a flagellum?

(b)

TEM | 2 μm

(a)

FIGURE 4.9 **Flagella and bacterial motility. (a)** A bacterium running and tumbling. Notice that the direction of flagellar rotation determines which of these movements occurs. **(b)** A *Proteus* cell in the swarming stage may have more than 1000 peritrichous flagella.

■ Motility is the ability of microbes to move on their own.

identified as *E. coli* O157:H7 are associated with foodborne epidemics (see Chapter 1, page 19).

Axial Filaments

Spirochetes are a group of bacteria that have unique structure and motility. One of the best-known spirochetes is *Treponema pallidum,* the causative agent of syphilis. Another spirochete is *Borrelia burgdorferi,* the causative agent of Lyme disease. Spirochetes move by means of **axial filaments,** or **endoflagella,** bundles of fibrils that arise at the ends of the cell beneath an outer sheath and spiral around the cell (Figure 4.10).

Axial filaments, which are anchored at one end of the spirochete, have a structure similar to that of flagella. The rotation of the filaments produces a movement of the outer sheath that propels the spirochetes in a spiral motion. This type of movement is similar to the way a corkscrew moves through a cork. This corkscrew motion probably enables a bacterium such as *T. pallidum* to move effectively through body fluids.

Fimbriae and Pili

Many gram-negative bacteria contain hairlike appendages that are shorter, straighter, and thinner than flagella and are used for attachment rather than for motility. These structures, which consist of a protein called *pilin* arranged helically around a central core, are divided into two types, fimbriae and pili, having very different functions. (Some microbiologists use the two terms interchangeably to refer to all such structures, but we distinguish between them.)

Fimbriae (singular: **fimbria**) can occur at the poles of the bacterial cell, or they can be evenly distributed over the entire surface of the cell. They can number anywhere from a few to several hundred per cell (Figure 4.11). Like the glycocalyx, fimbriae enable a cell to adhere to surfaces, including the surfaces of other cells. For example, fimbriae attached to the bacterium *Neisseria gonorrhoeae* (nī-se′rē-ä go-nôr-rē′ī), the causative agent of gonorrhea, help the microbe colonize mucous membranes. Once colonization occurs, the bacteria can cause disease. When fimbriae are absent (because of genetic mutation), colonization cannot happen, and no disease ensues.

Pili (singular: **pilus**) are usually longer than fimbriae and number only one or two per cell. Pili join bacterial cells in preparation for the transfer of DNA from one cell to another. For this reason, they are sometimes also called **sex pili** (see Chapter 8).

(a)

SEM ⊢——⊣ 1 μm

(b)

Outer sheath

Cell wall

Axial filament

Outer sheath
Cell wall
Axial filaments

(c)

FIGURE 4.10 **Axial filaments.** **(a)** A photomicrograph of the spirochete *Leptospira*, showing an axial filament. **(b)** A diagram of axial filaments wrapping around part of a spirochete. **(c)** A cross-sectional diagram of the spirochete, showing the position of axial filaments.

■ How do spirochetes and spirilla differ?

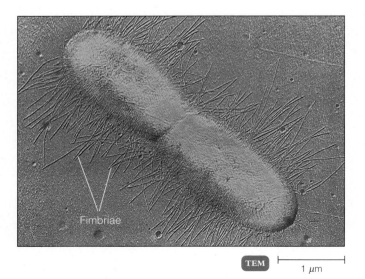

Fimbriae

TEM ⊢——⊣ 1 μm

FIGURE 4.11 **Fimbriae.** The fimbriae seem to bristle from this *E. coli* cell, which is beginning to divide.

■ Fimbriae enable bacterial cells to adhere to surfaces.

The Cell Wall

Learning Objectives

- *Compare and contrast the cell walls of gram-positive bacteria, gram-negative bacteria, archaea, and mycoplasmas.*

- *Differentiate between protoplast, spheroplast, and L form.*

The **cell wall** of the bacterial cell is a complex, semirigid structure responsible for the shape of the cell. The cell wall surrounds the underlying, fragile plasma (cytoplasmic) membrane and protects it and the interior of the cell from adverse changes in the outside environment (see Figure 4.6). Almost all prokaryotes have cell walls.

The major function of the cell wall is to prevent bacterial cells from rupturing when the water pressure inside the cell is greater than that outside the cell. It also helps maintain the shape of a bacterium and serves as a point of anchorage for flagella. As the volume of a bacterial cell increases, its plasma membrane and cell wall extend as needed. Clinically, the cell wall is important because it contributes to the ability of some species to cause disease and is the site of action of some antibiotics. In addition, the chemical composition of the cell wall is used to differentiate major types of bacteria.

Although the cells of some eukaryotes, including plants, algae, and fungi, have cell walls, their walls differ chemically from those of prokaryotes, are simpler in structure, and are less rigid.

Composition and Characteristics

The bacterial cell wall is composed of a macromolecular network called **peptidoglycan** (also known as *murein*), which is present either alone or in combination with other substances. Peptidoglycan consists of a repeating disaccharide attached by polypeptides to form a lattice that surrounds and protects the entire cell. The disaccharide portion is made up of monosaccharides called N-acetylglucosamine (NAG) and N-acetylmuramic acid (NAM) (from *murus*, meaning wall), which are related to glucose. The structural formulas for NAG and NAM are shown in Figure 4.12.

The various components of peptidoglycan are assembled in the cell wall (Figure 4.13a). Alternating NAM and NAG molecules are linked in rows of 10 to 65 sugars to form a carbohydrate "backbone" (the glycan portion of peptidoglycan). Adjacent rows are linked by **polypeptides** (the peptide portion of peptidoglycan). Although the structure of the polypeptide link varies, it always includes *tetrapeptide side chains*, which consist of four amino acids attached to NAMs in the backbone. The amino acids occur in an alternating pattern of D and L forms (see Figure 2.13 on page 45). This is unique because the amino acids found in other proteins are L forms. Parallel tetrapeptide side chains may be directly bonded to each other or linked by a *peptide cross-bridge*, consisting of a short chain of amino acids.

Penicillin interferes with the final linking of the peptidoglycan rows by peptide cross-bridges (Figure 4.13a). As a result, the cell wall is greatly weakened and the cell undergoes **lysis,** destruction caused by rupture of the plasma membrane and the loss of cytoplasm.

Gram-Positive Cell Walls

In most gram-positive bacteria, the cell wall consists of many layers of peptidoglycan, forming a thick, rigid structure (Figure 4.13b). By contrast, gram-negative cell walls contain only a thin layer of peptidoglycan.

In addition, the cell walls of gram-positive bacteria contain *teichoic acids*, which consist primarily of an alcohol (such as glycerol or ribitol) and phosphate. There are two classes of teichoic acids: *lipoteichoic acid,* which spans the

FIGURE 4.12 **N-acetylglucosamine (NAG) and N-acetylmuramic acid (NAM) joined as in peptidoglycan.** The light purple areas show the differences between the two molecules. The linkage between them is called a β-1,4 linkage.

peptidoglycan layer and is linked to the plasma membrane, and *wall teichoic acid*, which is linked to the peptidoglycan layer. Because of their negative charge (from the phosphate groups), teichoic acids may bind and regulate the movement of cations (positive ions) into and out of the cell. They may also assume a role in cell growth, preventing extensive wall breakdown and possible cell lysis. Finally, teichoic acids provide much of the wall's antigenic specificity and thus make it possible to identify bacteria by certain laboratory tests (see Chapter 10).

The cell walls of gram-positive streptococci are covered with various polysaccharides that allow them to be grouped into medically significant types. The cell walls of acid-fast bacteria, such as *Mycobacterium,* consist of as much as 60% mycolic acid, a waxy lipid, whereas the rest is peptidoglycan. These bacteria can be stained with the Gram stain and are considered gram-positive.

Gram-Negative Cell Walls

The cell walls of gram-negative bacteria consist of one or a very few layers of peptidoglycan and an outer membrane (Figure 4.13c). The peptidoglycan is bonded to lipoproteins (lipids covalently linked to proteins) in the outer membrane and is in the *periplasm*, a fluid-filled space between the outer membrane and the plasma membrane. The periplasm contains a high concentration of degradative enzymes and transport proteins. Gram-negative cell walls do not contain teichoic acids. Because the cell walls of gram-negative bacteria contain only a small amount of peptidoglycan, they are more susceptible to mechanical breakage.

N-acetylglucosamine (NAG)

N-acetylmuramic acid (NAM)

Side chain amino acid

Cross-bridge amino acid

Tetrapeptide side chain

Peptide cross-bridge

NAM

NAG

Peptide bond

NAG

Carbohydrate "backbone"

(a) Structure of peptidoglycan in gram-positive bacteria

Wall teichoic acid

Peptidoglycan

Lipoteichoic acid

Cell wall

Plasma membrane

Protein

(b) Gram-positive cell wall

Porin protein

Lipoprotein

Phospholipid

Enzymes and other active substances

Lipopolysaccharide

O polysaccharide

Lipid A

Outer membrane

Cell wall

Peptidoglycan

Plasma membrane

Periplasm

Protein

(c) Gram-negative cell wall

FIGURE 4.13 **Bacterial cell walls. (a)** The structure of peptidoglycan in gram-positive bacteria. Together the carbohydrate backbone (glycan portion) and tetrapeptide side chains (peptide portion) make up peptidoglycan. The frequency of peptide cross-bridges and the number of amino acids in these bridges vary with the species of bacterium. The small arrows indicate where penicillin interferes with the linkage of peptidoglycan rows by peptide cross-bridges. **(b)** A gram-positive cell wall. **(c)** A gram-negative cell wall.

■ What are the major structural differences between gram-positive and gram-negative cell walls?

The *outer membrane* of the gram-negative cell consists of lipopolysaccharides (LPS), lipoproteins, and phospholipids (Figure 4.13c). The outer membrane has several specialized functions. Its strong negative charge is an important factor in evading phagocytosis and the actions of complement (lyses cells and promotes phagocytosis), two components of the defenses of the host (discussed in detail in Chapter 16). The outer membrane also provides a barrier to certain antibiotics (for example, penicillin), digestive enzymes such as lysozyme, detergents, heavy metals, bile salts, and certain dyes. The box in Chapter 7 (page 200) illustrates the importance of the outer membrane in health care.

However, the outer membrane does not provide a barrier to all substances in the environment because nutrients must pass through to sustain the metabolism of the cell. Part of the permeability of the outer membrane is due to proteins in the membrane, called **porins,** that form channels. Porins permit the passage of molecules such as nucleotides, disaccharides, peptides, amino acids, vitamin B_{12}, and iron.

The LPS component of the outer membrane provides two important characteristics of gram-negative bacteria. First, the polysaccharide portion is composed of sugars, called O *polysaccharides*, that function as antigens and are useful for distinguishing species of gram-negative bacteria. For example, the foodborne pathogen *E. coli* O157:H7 is distinguished from other serovars by certain laboratory tests that test for the specific antigens. This role is comparable to that of teichoic acids in gram-positive cells. Second, the lipid portion of the lipopolysaccharide, called *lipid A*, is referred to as *endotoxin* and is toxic when in the host's bloodstream or gastrointestinal tract. It causes fever and shock. The nature and importance of endotoxins and other bacterial toxins will be discussed in Chapter 15.

Cell Walls and the Gram Stain Mechanism

Now that you have studied the Gram stain (in Chapter 3, page 67) and the chemistry of the bacterial cell wall (in the previous section), it is easier to understand the mechanism of the Gram stain. The mechanism is based on differences in the structure of the cell walls of gram-positive and gram-negative bacteria and how each reacts to the various reagents (substances used for producing a chemical reaction). Crystal violet, the primary stain, stains both gram-positive and gram-negative cells purple because the dye enters the cytoplasm of both types of cells. When iodine (the mordant) is applied, it forms large crystals with the dye that are too large to escape through the cell wall. The application of alcohol dehydrates the peptidoglycan

of gram-positive cells to make it more impermeable to the crystal violet-iodine. The effect on gram-negative cells is quite different; alcohol dissolves the outer membrane of gram-negative cells and even leaves small holes in the thin peptidoglycan layer through which crystal violet-iodine diffuse. Because gram-negative bacteria are colorless after the alcohol wash, the addition of safranin (the counterstain) turns the cells pink.

In any population of cells, some gram-positive cells will give a gram-negative response. These cells are usually dead. However, there are a few gram-positive genera that show an increasing number of gram-negative cells as the culture ages. *Bacillus, Clostridium,* and *Mycobacterium* are the most notable of these, and are often described as *gram-variable*.

A comparison of some of the characteristics of gram-positive and gram-negative bacteria is presented in Table 4.1.

Atypical Cell Walls

Among prokaryotes, certain types of cells have no walls or have very little wall material. These include members of the genus *Mycoplasma* (mī-kō-plaz′mä) and related organisms (see Figure 11.19 on page 325). Mycoplasmas are the smallest known bacteria that can grow and reproduce outside living host cells. Because of their size and because they have no cell walls, they pass through most bacterial filters and were first mistaken for viruses. Their plasma membranes are unique among bacteria in having lipids called *sterols*, which are thought to help protect them from lysis (rupture).

Archaea may lack walls or may have unusual walls composed of polysaccharides and proteins but not peptidoglycan. These walls do, however, contain a substance similar to peptidoglycan called *pseudomurein*. Pseudomurein contains N-acetyltalosaminuronic acid instead of NAM and lacks the D amino acids found in bacterial cell walls. A representative of the archaea is described in the box in Chapter 5 (page 143).

Damage to the Cell Wall

Chemicals that damage bacterial cell walls, or interfere with their synthesis, often do not harm the cells of an animal host because the bacterial cell wall is made of chemicals unlike those in eukaryotic cells. Thus, cell wall synthesis is the target for some antimicrobial drugs. One way the cell wall can be damaged is by exposure to the digestive enzyme *lysozyme*. This enzyme occurs naturally in some eukaryotic cells and is a constituent of tears, mucus, and saliva. Lysozyme is particularly active on the major cell wall components of most gram-positive bacteria, making them vulnerable to lysis. Lysozyme catalyzes hydrolysis of the bonds between the

TABLE 4.1	Some Comparative Characteristics of Gram-Positive and Gram-Negative Bacteria	
Characteristic	**Gram-Positive**	**Gram-Negative**
	LM ⊢ 4 µm	LM ⊢ 4 µm
Gram reaction	Retain crystal violet dye and stain dark violet or purple	Can be decolorized to accept counterstain (safranin) and stain red
Peptidoglycan layer	Thick (multilayered)	Thin (single-layered)
Teichoic acids	Present in many	Absent
Periplasmic space	Absent	Present
Outer membrane	Absent	Present
Lipopolysaccharide (LPS) content	Virtually none	High
Lipid and lipoprotein content	Low (acid-fast bacteria have lipids linked to peptidoglycan)	High (due to presence of outer membrane)
Flagellar structure	2 rings in basal body	4 rings in basal body
Toxins produced	Primarily exotoxins	Primarily endotoxins
Resistance to physical disruption	High	Low
Cell wall disruption by lysozyme	High	Low (requires pretreatment to destabilize outer membrane)
Susceptibility to penicillin and sulfonamide	High	Low
Susceptibility to streptomycin, chloramphenicol, and tetracycline	Low	High
Inhibition by basic dyes	High	Low
Susceptibility to anionic detergents	High	Low
Resistance to sodium azide	High	Low
Resistance to drying	High	Low

sugars in the repeating disaccharide "backbone" of peptidoglycan. This act is analogous to cutting the steel supports of a bridge with a cutting torch: The gram-positive cell wall is almost completely destroyed by lysozyme. The cellular contents that remain surrounded by the plasma membrane may remain intact if lysis does not occur; this wall-less cell is termed a **protoplast.** Typically, a protoplast is spherical and is still capable of carrying on metabolism.

Some members of the genus *Proteus,* as well as other genera, can lose their cell walls and swell into irregularly shaped cells called **L forms,** named for the Lister Institute, where they were discovered. They may form spontaneously or develop in response to penicillin, which inhibits cell wall formation, or lysozyme, which removes the cell wall. L forms can live and divide repeatedly or return to the walled state.

When lysozyme is applied to gram-negative cells, usually the wall is not destroyed to the same extent as in gram-positive cells; some of the outer membrane also remains. In this case, the cellular contents, plasma membrane, and remaining outer wall layer are called a **spheroplast,** also a spherical structure. For lysozyme to exert its effect on gram-negative cells, the cells are first treated with ethylenediaminetetraacetic acid (EDTA). EDTA weakens ionic bonds in the outer membrane and thereby damages it, giving the lysozyme access to the peptidoglycan layer.

Protoplasts and spheroplasts burst in pure water or very dilute salt or sugar solutions because the water molecules from the surrounding fluid rapidly move into and enlarge the cell, which has a much lower internal concentration of water. This rupturing is called **osmotic lysis** and will be discussed in detail shortly.

As noted earlier, certain antibiotics, such as penicillin, destroy bacteria by interfering with the formation of the peptide cross-bridges of peptidoglycan, thus preventing the formation of a functional cell wall. Most gram-negative bacteria are not as susceptible to penicillin as gram-positive bacteria are because the outer membrane of gram-negative bacteria forms a barrier that inhibits the entry of this and other substances, and gram-negative bacteria have fewer peptide cross-bridges. However, gram-negative bacteria are quite susceptible to some β-lactam antibiotics that penetrate the outer membrane better than penicillin. Antibiotics will be discussed in more detail in Chapter 20.

Structures Internal to the Cell Wall

Thus far, we have discussed the prokaryotic cell wall and structures external to it. We will now look inside the prokaryotic cell and discuss the structures and functions of the plasma membrane and components within the cytoplasm of the cell.

The Plasma (Cytoplasmic) Membrane

Learning Objectives

- *Describe the structure, chemistry, and functions of the prokaryotic plasma membrane.*
- *Define simple diffusion, facilitated diffusion, osmosis, active transport, and group translocation.*

The **plasma (cytoplasmic) membrane** (or *inner membrane*) is a thin structure lying inside the cell wall and enclosing the cytoplasm of the cell (see Figure 4.6). The plasma membrane of prokaryotes consists primarily of phospholipids (see Figure 2.10 on page 42), which are the most abundant chemicals in the membrane, and proteins. Eukaryotic plasma membranes also contain carbohydrates and sterols, such as cholesterol. Because they lack sterols, prokaryotic plasma membranes are less rigid than eukaryotic membranes. One exception is the wall-less prokaryote *Mycoplasma*, which contains membrane sterols.

Structure

In electron micrographs, prokaryotic and eukaryotic plasma membranes (and the outer membranes of gram-negative bacteria) look like two-layered structures; there are two dark lines with a light space between the lines (Figure 4.14a). The phospholipid molecules are arranged in two parallel rows, called a *phospholipid bilayer* (Figure 4.14b). As introduced in Chapter 2, each phospholipid molecule contains a polar head, composed of a phosphate group and glycerol that is hydrophilic (water-loving) and soluble in water, and nonpolar tails, composed of fatty acids that are hydrophobic (water-fearing) and insoluble in water (Figure 4.14c). The polar heads are on the two surfaces of the phospholipid bilayer, and the nonpolar tails are in the interior of the bilayer.

The protein molecules in the membrane can be arranged in a variety of ways. Some, called *peripheral proteins*, are easily removed from the membrane by mild treatments and lie at the inner or outer surface of the membrane. They may function as enzymes that catalyze chemical reactions, as a "scaffold" for support, and as mediators of changes in membrane shape during movement. Other proteins, called *integral proteins*, can be removed from the membrane only after disrupting the bilayer (by using detergents, for example). Some integral proteins are believed to penetrate the membrane completely and are called transmembrane proteins. Some integral proteins are channels that have a pore or hole through which substances enter and exit the cell.

Studies have demonstrated that the phospholipid and protein molecules in membranes are not static but move quite freely within the membrane surface. This movement is most probably associated with the many functions performed by the plasma membrane. Because the fatty acid tails cling together, phospholipids in the presence of water form a self-sealing bilayer, with the result that breaks and tears in the membrane will heal themselves. The membrane must be about as viscous as olive oil to allow membrane proteins to move freely enough to perform their functions without destroying the structure of the membrane. This dynamic arrangement of phospholipids and proteins is referred to as the **fluid mosaic model.**

Functions

The most important function of the plasma membrane is to serve as a selective barrier through which materials enter and exit the cell. In this function, plasma membranes have **selective permeability** (sometimes called *semipermeability*). This term indicates that certain molecules and

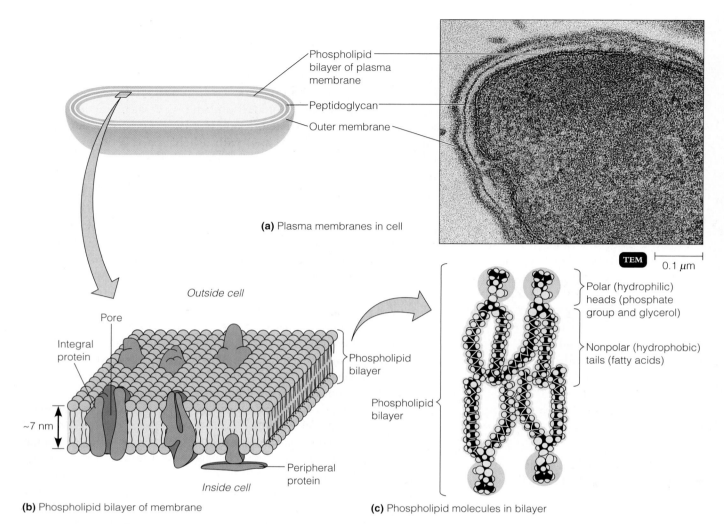

(a) Plasma membranes in cell

TEM   0.1 µm

(b) Phospholipid bilayer of membrane

(c) Phospholipid molecules in bilayer

FIGURE 4.14 **Plasma membrane. (a)** A diagram and micrograph showing the phospholipid bilayer forming the inner plasma membrane of the gram-negative bacterium *Aquaspirillum serpens*. Layers of the cell wall, including the outer membrane, can be seen outside the inner membrane. **(b)** A portion of the inner membrane showing the phospholipid bilayer and proteins. The outer membrane of gram-negative bacteria is also a phospholipid bilayer. **(c)** Space-filling models of several molecules as they are arranged in the phospholipid bilayer.

■ What is the action of polymyxins on plasma membranes?

ions pass through the membrane, but others are prevented from passing through it. The permeability of the membrane depends on several factors. Large molecules (such as proteins) cannot pass through the plasma membrane, possibly because these molecules are larger than the pores in integral proteins that function as channels. But smaller molecules (such as water, oxygen, carbon dioxide, and some simple sugars) usually pass through easily. Ions penetrate the membrane very slowly. Substances that dissolve easily in lipids (such as oxygen, carbon dioxide, and nonpolar organic molecules) enter and exit more easily than

other substances because the membrane consists mostly of phospholipids. The movement of materials across plasma membranes also depends on transporter molecules, which will be described shortly.

Plasma membranes are also important to the breakdown of nutrients and the production of energy. The plasma membranes of bacteria contain enzymes capable of catalyzing the chemical reactions that break down nutrients and produce ATP. In some bacteria, pigments and enzymes involved in photosynthesis are found in infoldings of the plasma membrane that extend into the cytoplasm.

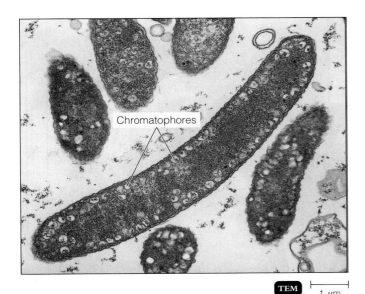

FIGURE 4.15 **Chromatophores.** In this micrograph of *Rhodospirillum rubrum,* a purple (nonsulfur) bacterium, the chromatophores are visible.

■ Chromatophores are sites of photosynthesis.

These membranous structures are called **chromatophores** or **thylakoids** (Figure 4.15).

When viewed with an electron microscope, bacterial plasma membranes often appear to contain one or more large, irregular folds called **mesosomes.** Many functions have been proposed for mesosomes. However, it is now known that they are artifacts, not true cell structures. Mesosomes are believed to be folds in the plasma membrane that develop by the process used for preparing specimens for electron microscopy.

Destruction of the Plasma Membrane by Antimicrobial Agents

Because the plasma membrane is vital to the bacterial cell, it is not surprising that several antimicrobial agents exert their effects at this site. In addition to the chemicals that damage the cell wall and thereby indirectly expose the membrane to injury, many compounds specifically damage plasma membranes. These compounds include certain alcohols and quaternary ammonium compounds, which are used as disinfectants. By disrupting the membrane's phospholipids, a group of antibiotics known as the *polymyxins* cause leakage of intracellular contents and subsequent cell death. This mechanism will be discussed in Chapter 20.

The Movement of Materials Across Membranes

Materials move across plasma membranes of both prokaryotic and eukaryotic cells by two kinds of processes: passive and active. In *passive processes,* substances cross the membrane from an area of high concentration to an area of low concentration (move with the concentration gradient, or difference), without any expenditure of energy (ATP) by the cell. In *active processes,* the cell must use energy (ATP) to move substances from areas of low concentration to areas of high concentration (against the concentration gradient).

Passive Processes Passive processes include simple diffusion, facilitated diffusion, and osmosis.

Simple diffusion is the net (overall) movement of molecules or ions from an area of high concentration to an area of low concentration (Figure 4.16). The movement continues until the molecules or ions are evenly distributed. The point of even distribution is called *equilibrium.* Cells rely on simple diffusion to transport certain small molecules, such as oxygen and carbon dioxide, across their cell membranes.

In **facilitated diffusion,** the substance (glucose, for example) to be transported combines with a plasma membrane protein called a *transporter* (sometimes called a *permease*). In one proposed mechanism for facilitated diffusion, transporters bind a substance on one side of the membrane and, by changing shape, move it to the other

(a) **(b)**

FIGURE 4.16 **The principle of simple diffusion. (a)** After a dye pellet is put into a beaker of water, the molecules of dye in the pellet diffuse into the water from an area of high dye concentration to areas of low dye concentration. **(b)** The dye potassium permanganate in the process of diffusing.

■ In passive processes such as simple diffusion, no energy (ATP) is expended by a cell.

FIGURE 4.17 **Facilitated diffusion.** Transporter proteins in the plasma membrane transport molecules across the membrane from an area of high concentration to one of low concentration (with the concentration gradient). The transporter undergoes a change in shape to transport the substance. The process does not require ATP.

■ How does simple diffusion differ from facilitated diffusion?

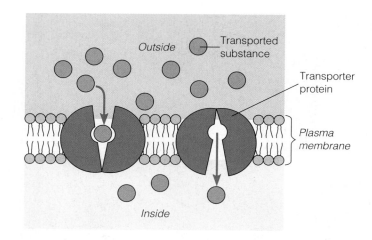

side of the membrane, where it is released (Figure 4.17). Facilitated diffusion is similar to simple diffusion in that the cell does not need to expend energy because the substance moves from a high to a low concentration. The process differs from simple diffusion in its use of transporters.

In some cases, molecules that bacteria need are too large to be transported into the cells by these methods. Most bacteria, however, produce enzymes that can break down large molecules into simpler ones (such as proteins into amino acids, or polysaccharides into simple sugars). Such enzymes, which are released by the bacteria into the surrounding medium, are appropriately called *extracellular enzymes*. Once the enzymes degrade the large molecules, the subunits move into the cell with the help of transporters. For example, specific carriers retrieve DNA bases, such as the purine guanine, from extracellular media and bring them into the cell's cytoplasm.

Osmosis is the net movement of solvent molecules across a selectively permeable membrane from an area with a high concentration of solvent molecules (low concentration of solute molecules) to an area of low concentration of solvent molecules (high concentration of solute molecules). In living systems, the chief solvent is water.

Osmosis may be demonstrated with the apparatus shown in Figure 4.18. A sack constructed from cellophane, which is a selectively permeable membrane, is filled with a solution of 20% sucrose (table sugar). The cellophane sack is placed into a beaker containing distilled water. Initially, the concentrations of water on either side of the membrane are different. Because of the sucrose molecules, the concentration of water is lower inside the cellophane sack. Therefore, water moves from the beaker (where its concentration is higher) into the cellophane sack (where its concentration is lower).

There is no movement of sugar out of the cellophane sack into the beaker, however, because the cellophane is impermeable to molecules of sugar—the sugar molecules are too large to go through the pores of the membrane. As water moves into the cellophane sack, the sugar solution becomes increasingly dilute, and, because the cellophane sack has expanded to its limit as a result of an increased volume of water, water begins to move up the glass tube. In

time, the water that has accumulated in the cellophane sack and the glass tube exerts a downward pressure that forces water molecules out of the cellophane sack and back into the beaker. This movement of water through a selectively permeable membrane produces a pressure called osmotic pressure. **Osmotic pressure** is the pressure required to prevent the movement of pure water (water with no solutes) into a solution containing some solutes. In other words, osmotic pressure is the pressure needed to stop the flow of water across the selectively permeable membrane (cellophane). When water molecules leave and enter the cellophane sack at the same rate, equilibrium is reached.

A bacterial cell may be subjected to any of three kinds of osmotic solutions: isotonic, hypotonic, or hypertonic. An **isotonic (isosmotic) solution** is a medium in which the overall concentration of solutes equals that found inside a cell (*iso* means equal). Water leaves and enters the cell at the same rate (no net change); the cell's contents are in equilibrium with the solution outside the cell wall (Figure 4.18c).

Earlier we mentioned that lysozyme and certain antibiotics damage bacterial cell walls, causing the cells to rupture, or lyse. Such rupturing occurs because bacterial cytoplasm usually contains such a high concentration of solutes that, when the wall is weakened or removed, additional water enters the cell by osmosis. The damaged (or removed) cell wall cannot constrain the swelling of the cytoplasmic membrane, and the membrane bursts. This is an example of osmotic lysis caused by immersion in a hypotonic solution. A **hypotonic (hypoosmotic) solution** outside the cell is a medium whose concentration of solutes is lower than that inside the cell (*hypo* means under or less). Most bacteria live in hypotonic solutions, and swelling is contained by the cell wall. Cells with weak cell walls, such as gram-negative bacteria, may burst or undergo osmotic lysis as a result of excessive water intake (Figure 4.18d).

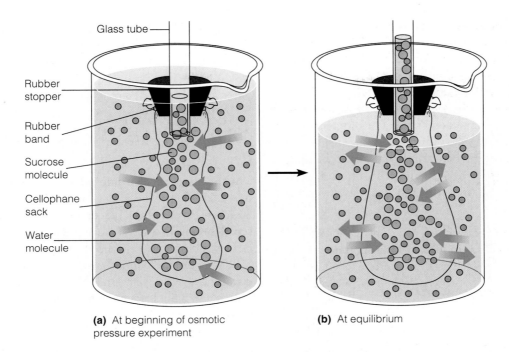

Glass tube

Rubber stopper

Rubber band

Sucrose molecule

Cellophane sack

Water molecule

(a) At beginning of osmotic pressure experiment

(b) At equilibrium

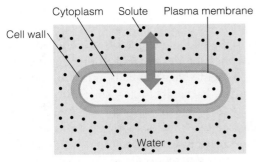

Cytoplasm Solute Plasma membrane

Cell wall

Water

(c) Isotonic (isosmotic) solution— no net movement of water

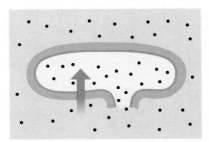

(d) Hypotonic (hypoosmotic) solution— water moves into the cell and may cause the cell to burst if the wall is weak or damaged (osmotic lysis)

(e) Hypertonic (hyperosmotic) solution— water moves out of the cell, causing its cytoplasm to shrink (plasmolysis)

FIGURE 4.18 **The principle of osmosis. (a)** Setup at the beginning of an osmotic pressure experiment. Water molecules start to move from the beaker into the sack along the concentration gradient. **(b)** Setup at equilibrium. The osmotic pressure exerted by the solution in the sack pushes water molecules from the sack back into the beaker to balance the rate of water entry into the sack. The height of solution in the glass tube at equilibrium is a measure of the osmotic pressure. **(c)–(e)** The effects of various solutions on bacterial cells.

■ Osmosis is the movement of water across a selectively permeable membrane from an area of higher water concentration to an area of lower water concentration.

A **hypertonic (hyperosmotic) solution** is a medium having a higher concentration of solutes than inside the cell has (*hyper* means above or more). Most bacterial cells placed in a hypertonic solution shrink and collapse or plasmolyze because water leaves the cells by osmosis (Figure 4.18e). Keep in mind that the terms *isotonic, hypotonic,* and *hypertonic* describe the concentration of solutions outside the cell *relative to* the concentration inside the cell.

Active Processes Simple diffusion and facilitated diffusion are useful mechanisms for transporting substances into cells when the concentrations of the substances are greater outside the cell. However, when a bacterial cell is in an environment in which nutrients are in low concentration, the cell must use active processes, such as active transport and group translocation, to accumulate the needed substances.

In performing **active transport,** the cell *uses energy* in the form of adenosine triphosphate (ATP) to move substances across the plasma membrane. The movement of a substance in active transport is usually from outside to inside, even though the concentration might be much higher inside the cell. Like facilitated diffusion, active transport depends on transporter proteins in the plasma membrane (see Figure 4.17). There appears to be a different transporter for each transported substance or group of closely related transported substances.

In active transport, the substance that crosses the membrane is not altered by transport across the membrane. In **group translocation,** a special form of active transport that occurs exclusively in prokaryotes, the substance is chemically altered during transport across the membrane. Once the substance is altered and inside the cell, the plasma membrane is impermeable to it, so it remains inside the cell. This important mechanism enables a cell to accumulate various substances even though they may be in low concentrations outside the cell. Group translocation requires energy supplied by high-energy phosphate compounds, such as phosphoenolpyruvic acid (PEP).

One example of group translocation is the transport of the sugar glucose, which is often used in growth media for bacteria. While a specific carrier protein is transporting the glucose molecule across the membrane, a phosphate group is added to the sugar. This phosphorylated form of glucose, which cannot be transported out, can then be used in the cell's metabolic pathways.

Some eukaryotic cells (those without cell walls) can use two additional active transport processes called phagocytosis and pinocytosis. These processes, which do not occur in bacteria, are described in later sections.

Cytoplasm

For a prokaryotic cell, the term **cytoplasm** refers to the substance of the cell inside the plasma membrane (see Figure 4.6 on page 78). Cytoplasm is about 80% water and contains primarily proteins (enzymes), carbohydrates, lipids, inorganic ions, and many low-molecular-weight compounds. Inorganic ions are present in much higher concentrations in cytoplasm than in most media. Cytoplasm is thick, aqueous, semitransparent, and elastic. The major structures in the cytoplasm of prokaryotes are a nuclear area (containing DNA), particles called ribosomes, and reserve deposits called inclusions.

Prokaryotic cytoplasm lacks certain features of eukaryotic cytoplasm, such as a cytoskeleton and cytoplasmic streaming. These features will be described later.

The Nuclear Area

Learning Objective
- Identify the functions of the nuclear area, ribosomes, and inclusions.

The nuclear area, or **nucleoid,** of a bacterial cell (see Figure 4.6) contains a single long, continuous, circularly arranged thread of double-stranded DNA called the **bacterial chromosome.** This is the cell's genetic information, which carries all the information required for the cell's structures and functions. Unlike the chromosomes of eukaryotic cells, bacterial chromosomes are not surrounded by a nuclear envelope (membrane) and do not include histones. The nuclear area can be spherical, elongated, or dumbbell-shaped. In actively growing bacteria, as much as 20% of the cell volume is occupied by DNA because such cells presynthesize nuclear material for future cells. The chromosome is attached to the plasma membrane. Proteins in the plasma membrane are believed to be responsible for replication of the DNA and segregation of the new chromosomes to daughter cells in cell division.

In addition to the bacterial chromosome, bacteria often contain small circular, double-stranded DNA molecules called **plasmids** (see the F factor in Figure 8.27a on page 239). These molecules are extrachromosomal genetic elements; that is, they are not connected to the main bacterial chromosome, and they replicate independently of chromosomal DNA. Research indicates that plasmids are associated with plasma membrane proteins. Plasmids usually contain from 5 to 100 genes that are generally not crucial for the survival of the bacterium under normal environmental conditions; plasmids may be gained or lost without harming the cell. Under certain conditions, however, plasmids are an advantage to cells. Plasmids may carry genes for such activities as antibiotic resistance, tolerance to toxic metals, the production of toxins, and the synthesis of enzymes. Plasmids can be transferred from one bacterium to another. In fact, plasmid DNA is used for gene manipulation in biotechnology.

Ribosomes

All eukaryotic and prokaryotic cells contain **ribosomes,** which function as the sites of protein synthesis. Cells that have high rates of protein synthesis, such as those that are actively growing, have a large number of ribosomes. The cytoplasm of a prokaryotic cell contains tens of thousands of these very small structures, which give the cytoplasm a granular appearance (see Figure 4.6).

Ribosomes are composed of two subunits, each of which consists of *protein* and a type of RNA called *ribosomal RNA (rRNA)*. Prokaryotic ribosomes differ from eukaryotic ribosomes in the number of proteins and rRNA molecules they contain; they are also somewhat smaller and less dense than ribosomes of eukaryotic cells. Accordingly, prokaryotic ribosomes are called 70S ribosomes (Figure 4.19), and those of eukaryotic cells are known as 80S ribosomes. The letter S refers to Svedberg units, which indicate the relative rate of sedimentation during ultra–high-speed centrifugation. Sedimentation rate is a function of the size, weight, and shape of a particle. The subunits of a 70S ribosome are a small 30S subunit containing one molecule of rRNA and a larger 50S subunit containing two molecules of rRNA.

Several antibiotics work by inhibiting protein synthesis on prokaryotic ribosomes. Antibiotics such as streptomycin and gentamicin attach to the 30S subunit and interfere with protein synthesis. Other antibiotics, such as erythromycin and chloramphenicol, interfere with protein synthesis by attaching to the 50S subunit. Because of differences in prokaryotic and eukaryotic ribosomes, the microbial cell can be killed by the antibiotic while the eukaryotic host cell remains unaffected.

Inclusions

Within the cytoplasm of prokaryotic cells are several kinds of reserve deposits, known as **inclusions.** Cells may accumulate certain nutrients when they are plentiful and use them when the environment is deficient. Evidence suggests that macromolecules concentrated in inclusions avoid the increase in osmotic pressure that would result if the molecules were dispersed in the cytoplasm. Some inclusions are common to a wide variety of bacteria, whereas others are limited to a small number of species and therefore serve as a basis for identification.

Metachromatic Granules

Metachromatic granules are large inclusions that take their name from the fact that they sometimes stain red with certain blue dyes such as methylene blue. Collectively they are known as **volutin.** Volutin represents a reserve of inorganic phosphate (polyphosphate) that can be used in the synthesis of ATP. It is generally formed by cells that grow in phosphate-rich environments. Metachromatic granules are found in algae, fungi, and protozoa, as well as in bacteria. These granules are characteristic of *Corynebacterium diphtheriae* (kô-rī-nē-bak-ti′rē-um dif-thi′rē-ī), the causative agent of diphtheria; thus, they have diagnostic significance.

(a) Small subunit   **(b)** Large subunit   **(c)** Complete 70S ribosome

FIGURE 4.19 The prokaryotic ribosome. (a) A small 30S subunit and **(b)** a large 50S subunit make up **(c)** the complete 70S prokaryotic ribosome.

■ What is the importance of the differences between prokaryotic and eukaryotic ribosomes with regard to antibiotic therapy?

Polysaccharide Granules

Inclusions known as **polysaccharide granules** typically consist of glycogen and starch, and their presence can be demonstrated when iodine is applied to the cells. In the presence of iodine, glycogen granules appear reddish brown and starch granules appear blue.

Lipid Inclusions

Lipid inclusions appear in various species of *Mycobacterium, Bacillus, Azotobacter* (ä-zō-tō-bak′tėr), *Spirillum* (spī-ril′lum), and other genera. A common lipid-storage material, one unique to bacteria, is the polymer *poly-β-hydroxybutyric acid.* Lipid inclusions are revealed by staining cells with fat-soluble dyes, such as Sudan dyes.

Sulfur Granules

Certain bacteria—for example, the "sulfur bacteria" that belong to the genus *Thiobacillus*—derive energy by oxidizing sulfur and sulfur-containing compounds. These bacteria may deposit **sulfur granules** in the cell, where they serve as an energy reserve.

Carboxysomes

Carboxysomes are inclusions that contain the enzyme ribulose 1,5-diphosphate carboxylase. Bacteria that use carbon dioxide as their sole source of carbon require this enzyme for carbon dioxide fixation. Among the bacteria containing carboxysomes are nitrifying bacteria, cyanobacteria, and thiobacilli.

Gas Vacuoles

Hollow cavities found in many aquatic prokaryotes, including cyanobacteria, anoxygenic photosynthetic bacteria, and halobacteria are called **gas vacuoles.** Each vacuole consists of rows of several individual *gas vesicles,* which are hollow cylinders covered by protein. Gas

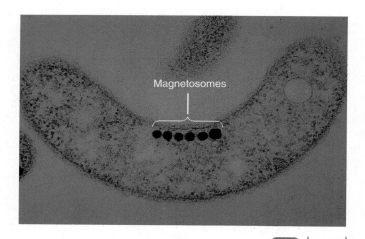

TEM   ⊢————⊣
1 μm

FIGURE 4.20 **Magnetosomes.** This micrograph of *Aquaspirillum magnetotacticum* shows a chain of magnetosomes. The outer membrane of the gram-negative wall is also visible.

■ Magnetosomes are iron-oxide inclusions, formed by some gram-negative bacteria, that act like magnets.

vacuoles maintain buoyancy so that the cells can remain at the depth in the water appropriate for them to receive sufficient amounts of oxygen, light, and nutrients.

Magnetosomes

Magnetosomes are inclusions of iron oxide (Fe_3O_4), formed by several gram-negative bacteria such as *Aquaspirillum magnetotacticum*, that act like magnets (Figure 4.20). Bacteria may use magnetosomes to move downward until they reach a suitable attachment site. In vitro, magnetosomes can decompose hydrogen peroxide, which forms in cells in the presence of oxygen. Researchers speculate that magnetosomes may protect the cell against hydrogen peroxide accumulation. Industrial microbiologists are developing culture methods to obtain large quantities of magnetite from bacteria to use in the production of magnetic tapes for sound and data recording.

Endospores

Learning Objective

- Describe the functions of endospores, sporulation, and endospore germination.

When essential nutrients are depleted, certain gram-positive bacteria, such as those of the genera *Clostridium* and *Bacillus*, form specialized "resting" cells called **endospores** (Figure 4.21). As you will see later, some members of the genus *Clostridium* cause diseases such as gangrene, tetanus, botulism, and food poisoning. Some members of the genus

Bacillus cause anthrax and food poisoning. Unique to bacteria, endospores are highly durable dehydrated cells with thick walls and additional layers. They are formed internal to the bacterial cell membrane.

When released into the environment, they can survive extreme heat, lack of water, and exposure to many toxic chemicals and radiation. For example, 7500-year-old endospores of *Thermoactinomyces vulgaris* (thĕr-mō-ak-tin-ō-mī′sēs vul-ga′ris) from the freezing muds of Elk Lake in Minnesota have germinated when rewarmed and placed in a nutrient medium, and 25- to 40-million-year-old endospores found in the gut of a stingless bee entombed in amber (hardened tree resin) in the Dominican Republic are reported to have germinated when placed in nutrient media. Although true endospores are found in gram-positive bacteria, one gram-negative species, *Coxiella burnetii* (käks-ē-el′lä bĕr-ne′tē-ē), the cause of Q fever, forms endospore-like structures that resist heat and chemicals and can be stained with endospore stains.

The process of endospore formation within a vegetative (parent) cell takes several hours and is known as **sporulation** or **sporogenesis** (Figure 4.21a). Vegetative cells of endospore-forming bacteria begin sporulation when a key nutrient, such as the carbon or nitrogen source, becomes scarce or unavailable. In the first observable stage of sporulation, a newly replicated bacterial chromosome and a small portion of cytoplasm are isolated by an ingrowth of the plasma membrane called a *spore septum*. The spore septum becomes a double-layered membrane that surrounds the chromosome and cytoplasm. This structure, entirely enclosed within the original cell, is called a *forespore*. Thick layers of peptidoglycan are laid down between the two membrane layers. Then a thick *spore coat* of protein forms around the outside membrane. This coat is responsible for the resistance of endospores to many harsh chemicals.

The diameter of the endospore may be the same as, smaller than, or larger than the diameter of the vegetative cell. Depending on the species, the endospore might be located *terminally* (at one end), *subterminally* (near one end; Figure 4.21b), or *centrally* inside the vegetative cell. When the endospore matures, the vegetative cell wall ruptures (lyses), killing the cell, and the endospore is freed.

Most of the water present in the forespore cytoplasm is eliminated by the time sporulation is complete, and endospores do not carry out metabolic reactions. The highly dehydrated endospore core contains only DNA, small amounts of RNA, ribosomes, enzymes, and a few important small molecules. The latter include a strikingly large amount of an organic acid called *dipicolinic acid* (found in the cytoplasm), which is accompanied by a large number of calcium ions. These cellular components are essential for resuming metabolism later.

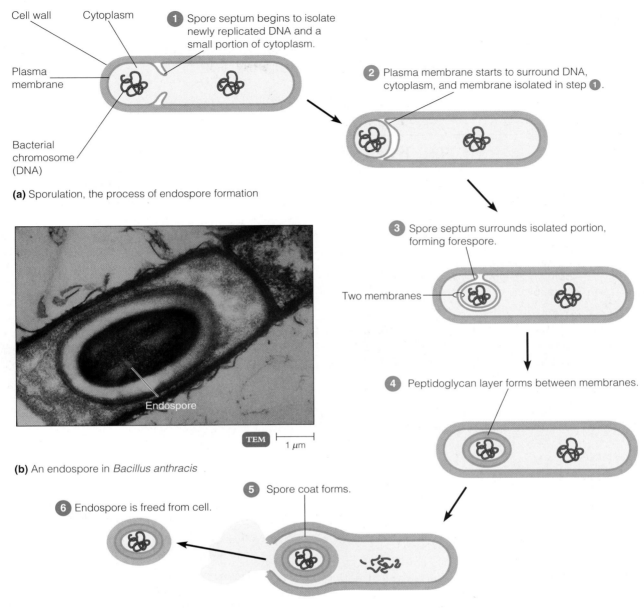

Cell wall   Cytoplasm

1 Spore septum begins to isolate newly replicated DNA and a small portion of cytoplasm.

Plasma membrane

Bacterial chromosome (DNA)

(a) Sporulation, the process of endospore formation

2 Plasma membrane starts to surround DNA, cytoplasm, and membrane isolated in step **1**.

3 Spore septum surrounds isolated portion, forming forespore.

Two membranes

4 Peptidoglycan layer forms between membranes.

Endospore

TEM | 1 μm

(b) An endospore in *Bacillus anthracis*

5 Spore coat forms.

6 Endospore is freed from cell.

FIGURE 4.21 **Formation of endospores by sporulation.**

■ Under what conditions are endospores formed by bacteria?

Endospores can remain dormant for thousands of years. An endospore returns to its vegetative state by a process called **germination.** Germination is triggered by physical or chemical damage to the endospore's coat. The endospore's enzymes then break down the extra layers surrounding the endospore, water enters, and metabolism resumes. Because one vegetative cell forms a single endospore, which, after germination, remains one cell, sporulation in bacteria is *not* a means of reproduction. This process does not increase the number of cells.

Endospores are important from a clinical viewpoint and in the food industry because they are resistant to processes that normally kill vegetative cells. Such processes include heating, freezing, desiccation, use of chemicals, and radiation. Whereas most vegetative cells are killed by temperatures above 70°C, endospores can survive in boiling water for several hours or more. Endospores of thermophilic (heat-loving) bacteria can survive in boiling water for 19 hours. Endospore-forming bacteria are a problem in the food industry because they are likely to survive underprocessing, and if conditions for growth occur, some species produce toxins and disease. Special methods for controlling organisms that produce endospores are discussed in Chapter 7.

As noted in Chapter 3, endospores are difficult to stain for detection. Thus, a specially prepared stain must be used along with heat. (The Schaeffer-Fulton endospore stain is commonly used.)

* * *

Having examined the functional anatomy of the prokaryotic cell, we will now look at the functional anatomy of the eukaryotic cell.

THE EUKARYOTIC CELL

As mentioned earlier, eukaryotic organisms include algae, protozoa, fungi, higher plants, and animals. The eukaryotic cell is typically larger and structurally more complex than the prokaryotic cell (Figure 4.22). By comparing the structure of the prokaryotic cell in Figure 4.6 with that of the eukaryotic cell, the differences between the two types of cells become apparent. The principal differences between prokaryotic and eukaryotic cells are summarized in Table 4.2.

The following discussion of eukaryotic cells will parallel our discussion of prokaryotic cells by starting with structures that extend to the outside of the cell.

TABLE 4.2	**Principal Differences Between Prokaryotic and Eukaryotic Cells**	
Characteristic	**Prokaryotic**	**Eukaryotic**
Size of cell	Typically 0.2–2.0 μm in diameter	Typically 10–100 μm in diameter
Nucleus	No nuclear membrane or nucleoli	True nucleus, consisting of nuclear membrane and nucleoli
Membrane-enclosed organelles	Absent	Present; examples include lysosomes, Golgi complex, endoplasmic reticulum, mitochondria, and chloroplasts
Flagella	Consist of two protein building blocks	Complex; consist of multiple microtubules
Glycocalyx	Present as a capsule or slime layer	Present in some cells that lack a cell wall
Cell wall	Usually present; chemically complex (typical bacterial cell wall includes peptidoglycan)	When present, chemically simple
Plasma membrane	No carbohydrates and generally lacks sterols	Sterols and carbohydrates that serve as receptors present
Cytoplasm	No cytoskeleton or cytoplasmic streaming	Cytoskeleton; cytoplasmic streaming
Ribosomes	Smaller size (70S)	Larger size (80S); smaller size (70S) in organelles
Chromosome (DNA)	Single circular chromosome; lacks histones	Multiple linear chromosomes with histones arrangement
Cell division	Binary fission	Mitosis
Sexual recombination	No meiosis; transfer of DNA fragments only	Involves meiosis

(a) Highly schematic diagram of a composite eukaryotic cell, half plant and half animal

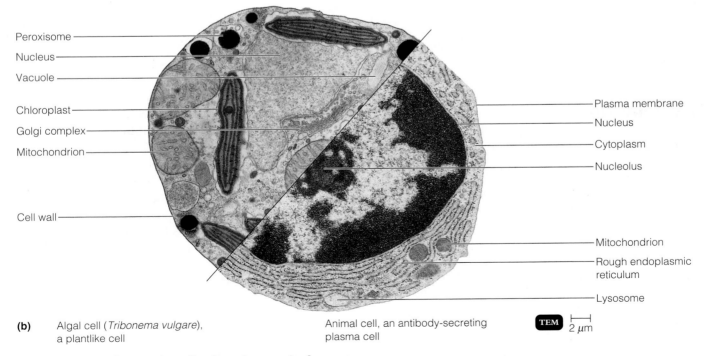

(b) Algal cell (*Tribonema vulgare*), a plantlike cell

Animal cell, an antibody-secreting plasma cell

TEM |—| 2 μm

FIGURE 4.22 **Eukaryotic cells showing typical structures.**

■ What kingdoms contain eukaryotic organisms?

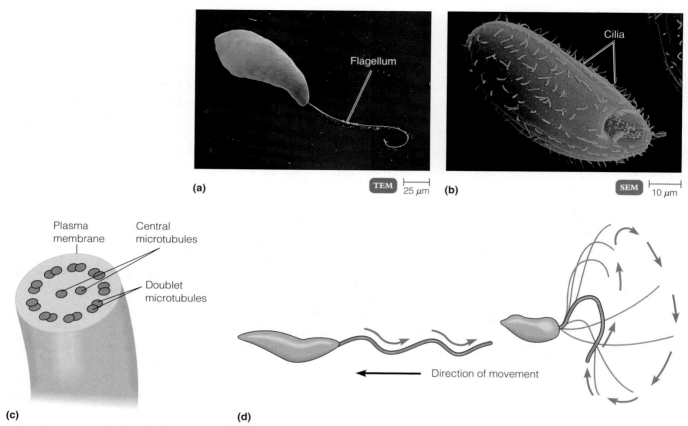

FIGURE 4.23 **Eukaryotic flagella and cilia. (a)** A micrograph of *Euglena*, a chlorophyll-containing alga, with its flagellum. **(b)** A micrograph of *Tetrahymena*, a common freshwater protozoan, with cilia. **(c)** The internal structure of a flagellum (or cilium), showing the 9 + 2 arrangement of microtubules. **(d)** The pattern of movement of a eukaryotic flagellum.

■ How do prokaryotic and eukaryotic flagella differ?

Flagella and Cilia

Learning Objective

• *Differentiate between prokaryotic and eukaryotic flagella.*

Many types of eukaryotic cells have projections that are used for cellular locomotion or for moving substances along the surface of the cell. These projections contain cytoplasm and are enclosed by the plasma membrane. If the projections are few and are long in relation to the size of the cell, they are called **flagella.** If the projections are numerous and short, resembling hairs, they are called **cilia** (singular: **cilium**).

Algae of the genus *Euglena* use a flagellum for locomotion, whereas protozoa, such as *Tetrahymena* (tet-rä-hī′me-nä), use cilia for locomotion (Figure 4.23a and b). Both flagella and cilia are anchored to the plasma membrane by a basal body, and both consist of nine pairs of microtubules (doublets) arranged in a ring, plus another two micro-

tubules in the center of the ring, an arrangement called a *9 + 2 array* (Figure 4.23c). **Microtubules** are long, hollow tubes made up of a protein called tubulin. A prokaryotic flagellum rotates, but a eukaryotic flagellum moves in a wavelike manner (Figure 4.23d). To help keep foreign material out of the lungs, ciliated cells of the human respiratory system move the material along the surface of the cells in the bronchial tubes and trachea toward the throat and mouth (see Figure 16.4 on page 461).

The Cell Wall and Glycocalyx

Learning Objective

• *Compare and contrast prokaryotic and eukaryotic cell walls and glycocalyxes.*

Most eukaryotic cells have cell walls, although they are generally much simpler than those of prokaryotic cells.

Many algae have cell walls consisting of the polysaccharide *cellulose* (as do all plants); other chemicals may be present as well. Cell walls of some fungi also contain cellulose, but in most fungi the principal structural component of the cell wall is the polysaccharide *chitin*, a polymer of N-acetylglucosamine (NAG) units. (Chitin is also the main structural component of the exoskeleton of crustaceans and insects.) The cell walls of yeasts contain the polysaccharides *glucan* and *mannan*. In eukaryotes that lack a cell wall, the plasma membrane may be the outer covering; however, cells that have direct contact with the environment may have coatings outside the plasma membrane. Protozoa do not have a typical cell wall; instead, they have a flexible outer covering called a *pellicle*.

In other eukaryotic cells, including animal cells, the plasma membrane is covered by a **glycocalyx,** a layer of material containing substantial amounts of sticky carbohydrates. Some of these carbohydrates are covalently bonded to proteins and lipids in the plasma membrane, forming glycoproteins and glycolipids that anchor the glycocalyx to the cell. The glycocalyx strengthens the cell surface, helps attach cells together, and may contribute to cell–cell recognition.

Eukaryotic cells do not contain peptidoglycan, the framework of the prokaryotic cell wall. This is significant medically because antibiotics, such as penicillins and cephalosporins, act against peptidoglycan and therefore do not affect human eukaryotic cells.

The Plasma (Cytoplasmic) Membrane

Learning Objective

* Compare and contrast prokaryotic and eukaryotic plasma membranes.

The **plasma membrane** of eukaryotic and prokaryotic cells is very similar in function and basic structure. There are, however, differences in the types of proteins found in the membranes. Eukaryotic membranes also contain carbohydrates, which serve as receptor sites that assume a role in such functions as cell–cell recognition. These carbohydrates also provide attachment sites for bacteria. Eukaryotic plasma membranes also contain *sterols*, complex lipids not found in prokaryotic plasma membranes (with the exception of *Mycoplasma* cells). Sterols seem to be associated with the ability of the membranes to resist lysis resulting from increased osmotic pressure.

Substances can cross eukaryotic and prokaryotic plasma membranes by simple diffusion, facilitated diffusion, osmosis, or active transport. Group translocation does not occur in eukaryotic cells. However, eukaryotic cells can use a mechanism called **endocytosis.** This occurs when a segment of the plasma membrane surrounds a particle or large molecule, encloses it, and brings it into the cell. Endocytosis is one of the ways viruses can enter animal cells.

Two very important types of endocytosis are phagocytosis and pinocytosis. During *phagocytosis*, cellular projections called pseudopods engulf particles and bring them into the cell. Phagocytosis is used by white blood cells to destroy bacteria and foreign substances (see Figure 16.8 and further discussion in Chapter 16). In *pinocytosis*, the plasma membrane folds inward, bringing extracellular fluid into the cell, along with whatever substances are dissolved in the fluid.

Cytoplasm

Learning Objective

* Compare and contrast prokaryotic and eukaryotic cytoplasms.

The **cytoplasm** of eukaryotic cells encompasses the substance inside the plasma membrane and outside the nucleus (see Figure 4.22). The cytoplasm is the substance in which various cellular components are found. (The term **cytosol** refers to the fluid portion of cytoplasm.) A major difference between eukaryotic and prokaryotic cytoplasm is that eukaryotic cytoplasm has a complex internal structure, consisting of exceedingly small rods called *microfilaments* and *intermediate filaments* and cylinders called *microtubules*. Together, they form the **cytoskeleton.** The cytoskeleton provides support and shape and assists in transporting substances through the cell (and even in moving the entire cell, as in phagocytosis). The movement of eukaryotic cytoplasm from one part of the cell to another, which helps distribute nutrients and move the cell over a surface, is called **cytoplasmic streaming.** Another difference between prokaryotic and eukaryotic cytoplasm is that many of the important enzymes found in the cytoplasmic fluid of prokaryotes are sequestered in the organelles of eukaryotes.

Organelles

Learning Objectives

* Define organelle.
* Describe the functions of the nucleus, endoplasmic reticulum, ribosomes, Golgi complex, lysosomes, vacuoles, mitochondria, chloroplasts, peroxisomes, and centrosomes.

Organelles are structures with specific shapes and specialized functions and are characteristic of eukaryotic cells. They include the nucleus, endoplasmic reticulum, ribosome, Golgi complex, lysosomes, vacuoles, mitochondria, chloroplasts, peroxisomes, and centrosomes. In addition, eukaryotic cells

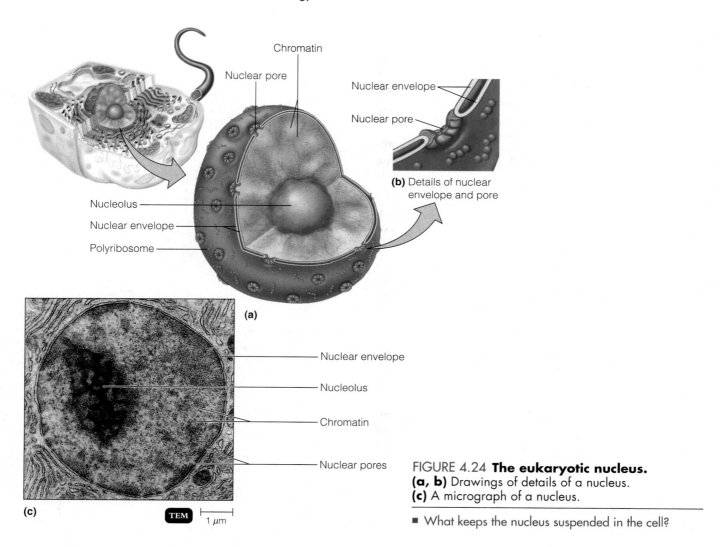

(a)

(b) Details of nuclear envelope and pore

Nuclear envelope

Nucleolus

Chromatin

Nuclear pores

(c)   TEM ⊢——⊣ 1 μm

FIGURE 4.24 The eukaryotic nucleus.
(a, b) Drawings of details of a nucleus.
(c) A micrograph of a nucleus.

■ What keeps the nucleus suspended in the cell?

contain ribosomes that are larger and denser than prokaryotic ribosomes. Not all of the organelles described are found in all cells. Certain cells have their own type and distribution of organelles based on specialization, age, and level of activity.

The Nucleus

The most characteristic eukaryotic organelle is the nucleus (see Figure 4.22). The **nucleus** (Figure 4.24) is usually spherical or oval, is frequently the largest structure in the cell, and contains almost all of the cell's hereditary information (DNA). Some DNA is also found in mitochondria and in the chloroplasts of photosynthetic organisms.

The nucleus is surrounded by a double membrane called the **nuclear envelope.** Both membranes resemble the plasma membrane in structure. Tiny channels in the membrane called **nuclear pores** allow the nucleus to communicate with the cytoplasm (see Figure 4.24). Nuclear pores control the movement of substances between the nucleus and cytoplasm. Within the nuclear envelope

are one or more spherical bodies called **nucleoli** (singular: **nucleolus**). Nucleoli are actually condensed regions of chromosomes where ribosomal RNA is being synthesized. Ribosomal RNA is an essential component of ribosomes (described shortly).

The nucleus also contains most of the cell's DNA, which is combined with several proteins, including some basic proteins called **histones** and nonhistones. The combination of about 165 base pairs of DNA and 9 molecules of histones is referred to as a *nucleosome*. When the cell is not reproducing, the DNA and its associated proteins appear as a threadlike mass called **chromatin.** During nuclear division, the chromatin coils into shorter and thicker rodlike bodies called **chromosomes.** Prokaryotic chromosomes do not undergo this process, do not have histones, and are not enclosed in a nuclear envelope.

Eukaryotic cells require two elaborate mechanisms: mitosis and meiosis to segregate chromosomes prior to cell division. Neither process occurs in prokaryotic cells.

Endoplasmic Reticulum

Within the cytoplasm of eukaryotic cells is the **endoplasmic reticulum** or **ER,** an extensive network of flattened membranous sacs or tubules called **cisterns** (Figure 4.25). The ER network is continuous with the nuclear envelope (see Figure 4.22a).

Most eukaryotic cells contain two distinct, but interrelated, forms of ER that differ in structure and function. The membrane of **rough ER** is continuous with the nuclear membrane and usually unfolds into a series of flattened sacs. The outer surface of rough ER is studded with ribosomes, the sites of protein synthesis. Proteins synthesized by ribosomes attached to rough ER enter cisterns within the ER for processing and sorting. In some cases, enzymes within the cisterns attach the proteins to carbohydrates to form glycoproteins. In other cases, enzymes attach the proteins to phospholipids, also synthesized by rough ER. These molecules may be incorporated into organelle membranes or the plasma membrane. Thus, rough ER is a factory for synthesizing secretory proteins and membrane molecules.

Smooth ER extends from the rough ER to form a network of membrane tubules (Figure 4.25). Unlike rough ER, smooth ER does not have ribosomes on the outer surface of its membrane. However, smooth ER contains unique enzymes that make it functionally more diverse than rough ER. Although it does not synthesize proteins, smooth ER does synthesize phospholipids, as does rough ER. Smooth ER also synthesizes fats and steroids, such as estrogens and testosterone. In liver cells, enzymes of the smooth ER help release glucose into the bloodstream and inactivate or detoxify drugs and other potentially harmful substances (for example, alcohol). In muscle cells, calcium ions released from the sarcoplasmic reticulum, a form of smooth ER, trigger the contraction process.

Ribosomes

Attached to the outer surface of rough ER are **ribosomes** (see Figure 4.25), which are also found free in the cytoplasm. As in prokaryotes, ribosomes are the sites of protein synthesis in the cell.

The ribosomes of eukaryotic ER and cytoplasm are somewhat larger and denser than those of prokaryotic cells. These eukaryotic ribosomes are 80S ribosomes, each of which consists of a large 60S subunit containing three molecules of rRNA and a smaller 40S subunit with one molecule of rRNA. The subunits are made separately in the nucleolus and, once produced, exit the nucleus and join together in the cytosol. Chloroplasts and mitochondria contain 70S ribosomes, which may indicate their evolution from prokaryotes. This theory is discussed later in

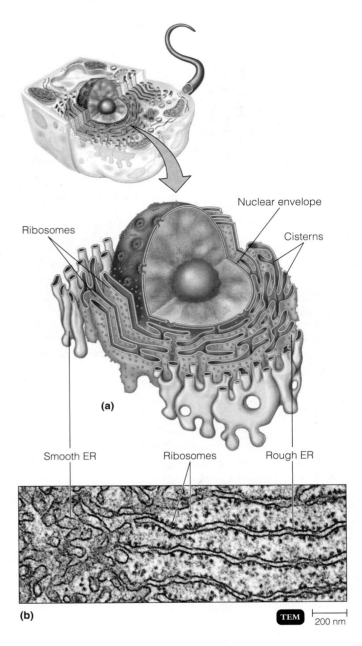

(a)

Nuclear envelope

Ribosomes

Cisterns

Smooth ER   Ribosomes   Rough ER

(b)   TEM |—— 200 nm

FIGURE 4.25 **Rough endoplasmic reticulum and ribosomes. (a)** A drawing of details of endoplasmic reticulum. **(b)** A micrograph of endoplasmic reticulum and ribosomes.

■ What is the difference between rough ER and smooth ER?

this chapter. The role of ribosomes in protein synthesis will be discussed in more detail in Chapter 8.

Some ribosomes, called *free ribosomes*, are unattached to any structure in the cytoplasm. Primarily, free ribosomes synthesize proteins used *inside* the cell. Other ribosomes, called *membrane-bound ribosomes*, attach to the nuclear membrane and the endoplasmic reticulum. These ribosomes

FIGURE 4.26 Golgi complex. (a) A drawing of details of a Golgi complex. **(b)** A micrograph of a Golgi complex.

- The Golgi complex modifies, sorts, and packages proteins received from the ER; discharges proteins via exocytosis; replaces portions of the plasma membrane; and forms lysosomes.

synthesize proteins destined for insertion in the plasma membrane or for export from the cell. Ribosomes are also located within mitochondria, where they synthesize mitochondrial proteins. Sometimes 10–20 ribosomes join together in a stringlike arrangement called a *polyribosome*.

Golgi Complex

Most of the proteins synthesized by ribosomes attached to rough ER are ultimately transported to other regions of the cell. The first step in the transport pathway is through an organelle called the **Golgi complex.** It consists of 3–20 cisterns that resemble a stack of pita bread (Figure 4.26). The cisterns are often curved, giving the Golgi complex a cuplike shape.

Proteins synthesized by ribosomes on the rough ER are surrounded by a portion of the ER membrane, which eventually buds from the membrane surface to form a **transport vesicle.** The transport vesicle fuses with a cistern of the Golgi complex, releasing proteins into the cistern. The proteins are modified and move from one cistern to another via **transfer vesicles** that bud from the cisterns' edges. Enzymes in the cisterns modify the proteins to form glycoproteins, glycolipids, and lipoproteins. Some of the processed proteins leave the cisterns in **secretory vesicles,** which detach from the cistern and deliver the proteins to the plasma membrane, where they are discharged by exocytosis. Other processed proteins leave the cisterns in vesicles that deliver their contents to the plasma membrane for incorporation into the membrane. Finally, some processed proteins leave the cisterns in vesicles

that are called **storage vesicles.** The major storage vesicle is a lysosome, whose structure and functions are discussed next.

Lysosomes

Lysosomes are formed from Golgi complexes and look like membrane-enclosed spheres. Unlike mitochondria, lysosomes have only a single membrane and lack internal structure (see Figure 4.22). But they contain as many as 40 different kinds of powerful digestive enzymes capable of breaking down various molecules. Moreover, these enzymes can also digest bacteria that enter the cell. Human white blood cells, which use phagocytosis to ingest bacteria, contain large numbers of lysosomes.

Vacuoles

A **vacuole** (see Figure 4.22) is a space or cavity in the cytoplasm of a cell that is enclosed by a membrane called a *tonoplast.* In plant cells, vacuoles may occupy 5–90% of the cell volume, depending on the type of cell. Vacuoles are derived from the Golgi complex and have several diverse functions. Some vacuoles serve as temporary storage organelles for substances such as proteins, sugars, organic acids, and inorganic ions. Other vacuoles form during endocytosis to help bring food into the cell. Many plant cells also store metabolic wastes and poisons that would otherwise be injurious if they accumulated in the cytoplasm. Finally, vacuoles may take up water, enabling plant cells to increase in size and also providing rigidity to leaves and stems.

Matrix   Cristae   Inner membrane   Outer membrane

FIGURE 4.27 **Mitochondria. (a)** A drawing of details of a mitochondrion. **(b)** A micrograph of a mitochondrion from a rat pancreas cell.

■ How are mitochondria similar to prokaryotic cells?

TEM   0.5 µm

Mitochondria

Spherical or rod-shaped organelles called **mitochondria** (singular: **mitochondrion**) appear throughout the cytoplasm of most eukaryotic cells (see Figure 4.22). The number of mitochondria per cell varies greatly among different types of cells. For example, the protozoan *Giardia* has no mitochondria, whereas liver cells contain 1000–2000 per cell. A mitochondrion consists of a double membrane similar in structure to the plasma membrane (Figure 4.27). The outer mitochondrial membrane is smooth, but the inner mitochondrial membrane is arranged in a series of folds called **cristae** (singular: **crista**). The center of the mitochondrion is a semifluid substance called the **matrix.** Because of the nature and arrangement of the cristae, the inner membrane provides an enormous surface area on which chemical reactions can occur. Some proteins that function in cellular respiration, including the enzyme that makes ATP, are located on the cristae of the inner mitochondrial membrane, and many of the metabolic steps involved in cellular respiration are concentrated in the matrix (see Chapter 5). Mitochondria are often called the "powerhouses of the cell" because of their central role in ATP production.

Mitochondria contain 70S ribosomes and some DNA of their own, as well as the machinery necessary to replicate, transcribe, and translate the information encoded by their DNA. In addition, mitochondria can reproduce more or less on their own by growing and dividing in two.

Chloroplasts

Algae and green plants contain a unique organelle called a **chloroplast** (Figure 4.28), a membrane-enclosed structure that contains both the pigment chlorophyll and the enzymes required for the light-gathering phases of photosynthesis (see Chapter 5). The chlorophyll is contained in flattened membrane sacs called **thylakoids;** stacks of thylakoids are called *grana* (singular: *granum*) (Figure 4.28).

Like mitochondria, chloroplasts contain 70S ribosomes, DNA, and enzymes involved in protein synthesis. They are capable of multiplying on their own within the cell. The way both chloroplasts and mitochondria multiply—by increasing in size and then dividing in two—is strikingly reminiscent of bacterial multiplication.

Granum

(a)

Chloroplast   Thylakoids

FIGURE 4.28 **Chloroplasts.**
Photosynthesis occurs in chloroplasts; the light-trapping pigments
are located on the thylakoids.
(a) A drawing of details of a
chloroplast, showing grana.
(b) A micrograph of chloroplasts
in a plant cell.

■ What are the similarities between
chloroplasts and prokaryotic
cells?

(b)

TEM |— 0.5 μm

Peroxisomes

Organelles similar in structure to lysosomes, but smaller,
are called **peroxisomes** (see Figure 4.22). Although peroxisomes were once thought to form by budding off the
ER, it is now generally agreed that they form by the division of preexisting peroxisomes.

Peroxisomes contain one or more enzymes that can oxidize various organic substances. For example, substances
such as amino acids and fatty acids are oxidized in peroxisomes as part of normal metabolism. In addition, enzymes
in peroxisomes oxidize toxic substances, such as alcohol. A
by-product of the oxidation reactions is hydrogen peroxide
(H_2O_2), a potentially toxic compound. However, peroxisomes also contain the enzyme *catalase*, which decomposes
H_2O_2 (see Chapter 6, page 163). Because the generation
and degradation of H_2O_2 occurs within the same organelle, peroxisomes protect other parts of the cell from
the toxic effects of H_2O_2.

Centrosome

The **centrosome,** located near the nucleus, consists of
two components: the pericentriolar area and centrioles
(see Figure 4.22). The *pericentriolar material* is a region of
the cytosol composed of a dense network of small protein
fibers. This area is the organizing center for the mitotic
spindle, which plays a critical role in cell division, and for
microtubule formation in nondividing cells. Within the
pericentriolar material is a pair of cylindrical structures
called *centrioles*, each of which is composed of nine clusters
of three microtubules (triplets) arranged in a circular pattern, an arrangement called a 9 + 0 *array*. The 9 refers to
the nine clusters of microtubules, and the 0 refers to the
absence of microtubules in the center. The long axis of one
centriole is at a right angle to the long axis of the other.

Why Microbiologists Study Termites

Although termites are famous for their ability to eat wood, causing damage to wooden structures and recycling cellulose in the soil, they are unable to digest the wood that they eat. To break down the cellulose, termites enlist the help of a variety of microorganisms. Some termites, for example, dig tunnels in the wood, then inoculate the tunnels with fungi that grow on the wood. These termites then eat the fungi, not the wood itself.

What microbiologists find more interesting are the termites that contain, within their digestive tracts, symbiotic microorganisms that digest the cellulose that the termites chew and swallow. Even more fascinating to microbiologists is the fact that these microorganisms themselves can survive only because of even smaller symbionts that live on and within them, without which they would not even be able to move. By studying how a single termite survives, microbiologists have begun to gain an entirely new understanding of symbiosis.

The termite's dependence on nitrogen-fixing bacteria to supply its nitrogen and on protozoans such as *Trichonympha sphaerica* to digest cellulose is an example of endosymbiosis, a symbiotic relationship with an organism that lives inside the body of the host organism (in this case, within the hindgut of the termite).

Theoretically, the protozoan produces the cellulolytic enzymes that digest the cellulose. The picture is more complicated than this, however, for *T. sphaerica* itself is unable to digest cellulose without the aid of bacteria that live within its body: in other words, the protozoan has its own endosymbionts.

Certain hindgut ciliates such as *T. sphaerica* also demonstrate another form of symbiosis—ectosymbiosis, a symbiotic relationship with organisms that live outside its body. Recent advances in microscopy have shown that these ciliates are covered by precise rows consisting of thousands of bacteria, either rods or spirochetes. If these bacteria are killed, the protozoan is unable to move. Evidently, the protozoan does not use its cilia to move; instead, the rows of bacteria move the protozoan like rows of oarsmen in a boat.

The protozoan *Mixotricha*, for example, has rows of spirochetes on its surface. As shown in part (a) of the figure, the end of each spirochete abuts against a swelling known as a bracket. The spirochetes undulate in unison, thereby creating waves of motion along *Mixotricha*'s surface.

Rod-shaped bacteria aligned in grooves that cover the surface of devescovinids, another group of termite-hindgut protozoans. Each rod has twelve flagella that overlap the flagella of the adjacent bacteria to form a continuous filament along the groove (see part (b) of the figure). The bacteria beat their flagella to create coordinated waves along all these rows of filaments, thus propelling the protozoan.

An important question remains: Do the bacteria detect a food source and move the protozoan toward it, or does the protozoan force the bacteria to respond to its commands?

(a)

(b)

Arrangements of bacteria on the surfaces of two protozoans. **(a)** Spirochetes attached to brackets over the surface of a *Mixotricha* protozoan align themselves and move in unison. **(b)** On this devescovinid protozoan, the flagella from one rod-shaped bacterium overlap the next to form a continuous filament.

The Evolution of Eukaryotes

Learning Objective

- *Discuss evidence that supports the endosymbiotic theory of eukaryotic evolution.*

Biologists generally believe that life arose on Earth in the form of very simple organisms, similar to prokaryotic cells, about 3.5 to 4 billion years ago. About 2.5 billion years ago, the first eukaryotic cells evolved from prokaryotic cells. Recall that prokaryotes and eukaryotes differ mainly in that eukaryotes contain highly specialized organelles. The theory explaining the origin of eukaryotes from prokaryotes, pioneered by Lynn Margulis, is the **endosymbiotic theory.** According to this theory, larger bacterial cells lost their cell walls and engulfed smaller bacterial cells. This relationship, in which one organism lives within another, is called *endosymbiosis* (*symbiosis* = living together).

According to the endosymbiotic theory, the ancestral eukaryote developed a rudimentary nucleus when the plasma membrane folded around the chromosome (see Figure 10.2, page 280). This cell, called a nucleoplasm, may have ingested aerobic bacteria. Some ingested bacteria lived inside the host nucleoplasm. This arrangement evolved into a symbiotic relationship in which the host nucleoplasm supplied nutrients and the endosymbiotic bacterium produced energy that could be used by the nucleoplasm. Similarly, chloroplasts may be descendants of photosynthetic prokaryotes ingested by this early nucleoplasm. Eukaryotic flagella and cilia are believed to have originated from symbiotic associations between the plasma membrane of early eukaryotes and motile spiral bacteria called spirochetes. A living example that suggests how flagella developed is described in the box on page 105.

Studies comparing prokaryotic and eukaryotic cells provide evidence for the endosymbiotic theory. For example, both mitochondria and chloroplasts resemble bacteria in size and shape. Further, these organelles contain circular DNA, which is typical of prokaryotes, and the organelles can reproduce independently of their host cell. Moreover, mitochondrial and chloroplast ribosomes resemble those of prokaryotes, and their mechanism of protein synthesis is more similar to that found in bacteria than eukaryotes. Also, the same antibiotics that inhibit protein synthesis on ribosomes in bacteria also inhibit protein synthesis on ribosomes in mitochondria and chloroplasts.

Our next concern is to examine microbial metabolism. In Chapter 5, you will learn about the importance of enzymes to microorganisms, and the ways microbes produce and use energy.

STUDY OUTLINE

Comparing Prokaryotic and Eukaryotic Cells: An Overview (pp. 75–76)

1. Prokaryotic and eukaryotic cells are similar in their chemical composition and chemical reactions.
2. Prokaryotic cells lack membrane-enclosed organelles (including a nucleus).
3. Peptidoglycan is found in prokaryotic cell walls but not in eukaryotic cell walls.
4. Eukaryotic cells have a membrane-bound nucleus and other organelles.

THE PROKARYOTIC CELL (pp. 76–96)

1. Bacteria are unicellular, and most of them multiply by binary fission.
2. Bacterial species are differentiated by morphology, chemical composition, nutritional requirements, biochemical activities, and source of energy.

The Size, Shape, and Arrangement of Bacterial Cells (pp. 76–78)

1. Most bacteria are 0.2–2.0 μm in diameter and 2–8 μm in length.

2. The three basic bacterial shapes are coccus (spherical), bacillus (rod-shaped), and spiral (twisted).

3. Pleomorphic bacteria can assume several shapes.

Structures External to the Cell Wall (pp. 79–81)

Glycocalyx (p. 79)

1. The glycocalyx (capsule, slime layer, or extracellular polysaccharide) is a gelatinous polysaccharide and/or polypeptide covering.
2. Capsules may protect pathogens from phagocytosis.
3. Capsules enable adherence to surfaces, prevent desiccation, and may provide nutrients.

Flagella (pp. 79–81)

1. Flagella are relatively long filamentous appendages consisting of a filament, hook, and basal body.
2. Prokaryotic flagella rotate to push the cell.
3. Motile bacteria exhibit taxis; positive taxis is movement toward an attractant, and negative taxis is movement away from a repellent.
4. Flagellar (H) protein functions as an antigen.

Axial Filaments (p. 81)

1. Spiral cells that move by means of an axial filament (endoflagellum) are called spirochetes.
2. Axial filaments are similar to flagella, except that they wrap around the cell.

Fimbriae and Pili (p. 81)

1. Fimbriae and pili are short, thin appendages.
2. Fimbriae help cells adhere to surfaces.
3. Pili join cells for the transfer of DNA from one cell to another.

The Cell Wall (pp. 82–87)

Composition and Characteristics (pp. 83–85)

1. The cell wall surrounds the plasma membrane and protects the cell from changes in water pressure.
2. The bacterial cell wall consists of peptidoglycan, a polymer consisting of NAG and NAM and short chains of amino acids.
3. Penicillin interferes with peptidoglycan synthesis.
4. Gram-positive cell walls consist of many layers of peptidoglycan and also contain teichoic acids.
5. Gram-negative bacteria have a lipopolysaccharide-lipoprotein-phospholipid outer membrane surrounding a thin peptidoglycan layer.
6. The outer membrane protects the cell from phagocytosis and from penicillin, lysozyme, and other chemicals.
7. Porins are proteins that permit small molecules to pass through the outer membrane; specific channel proteins allow other molecules to move through the outer membrane.
8. The lipopolysaccharide component of the outer membrane consists of sugars (O polysaccharides) that function as antigens and lipid A, which is an endotoxin.

Cell Walls and the Gram Stain Mechanism (p. 85)

1. The crystal violet–iodine complex combines with peptidoglycan.
2. The decolorizer removes the lipid outer membrane of gram-negative bacteria and washes out the crystal violet.

Atypical Cell Walls (p. 85)

1. *Mycoplasma* is a bacterial genus that naturally lacks cell walls.
2. Archaea have pseudomurein; they lack peptidoglycan.

Damage to the Cell Wall (pp. 85–87)

1. In the presence of lysozyme, gram-positive cell walls are destroyed, and the remaining cellular contents are referred to as a protoplast.
2. In the presence of lysozyme, gram-negative cell walls are not completely destroyed, and the remaining cellular contents are referred to as a spheroplast.
3. L forms are gram-positive or gram-negative bacteria that do not make a cell wall.
4. Antibiotics such as penicillin interfere with cell wall synthesis.

Structures Internal to the Cell Wall (pp. 87–96)

The Plasma (Cytoplasmic) Membrane (pp. 87–92)

1. The plasma membrane encloses the cytoplasm and is a phospholipid bilayer with peripheral and integral proteins (the fluid mosaic model).
2. The plasma membrane is selectively permeable.
3. Plasma membranes carry enzymes for metabolic reactions, such as nutrient breakdown, energy production, and photosynthesis.
4. Mesosomes, irregular infoldings of the plasma membrane, are artifacts, not true cell structures.
5. Plasma membranes can be destroyed by alcohols and polymyxins.

The Movement of Materials Across Membranes (pp. 89–92)

1. Movement across the membrane may be by passive processes, in which materials move from areas of higher to lower concentration, and no energy is expended by the cell.
2. In simple diffusion, molecules and ions move until equilibrium is reached.
3. In facilitated diffusion, substances are transported by transporter proteins across membranes from areas of high to low concentration.
4. Osmosis is the movement of water from areas of high to low concentration across a selectively semipermeable membrane until equilibrium is reached.
5. In active transport, materials move from areas of low to high concentration by transporter proteins, and the cell must expend energy.
6. In group translocation, energy is expended to modify chemicals and transport them across the membrane.

Cytoplasm (p. 92)

1. Cytoplasm is the fluid component inside the plasma membrane.
2. The cytoplasm is mostly water, with inorganic and organic molecules, DNA, ribosomes, and inclusions.

The Nuclear Area (p. 92)

1. The nuclear area contains the DNA of the bacterial chromosome.
2. Bacteria can also contain plasmids, which are circular, extrachromosomal DNA molecules.

Ribosomes (pp. 92–93)

1. The cytoplasm of a prokaryote contains numerous 70S ribosomes; ribosomes consist of rRNA and protein.

2. Protein synthesis occurs at ribosomes; it can be inhibited by certain antibiotics.

Inclusions (pp. 93–94)

1. Inclusions are reserve deposits found in prokaryotic and eukaryotic cells.

2. Among the inclusions found in bacteria are metachromatic granules (inorganic phosphate), polysaccharide granules (usually glycogen or starch), lipid inclusions, sulfur granules, carboxysomes (ribulose 1,5-diphosphate carboxylase), magnetosomes (Fe_3O_4), and gas vacuoles.

Endospores (pp. 94–96)

1. Endospores are resting structures formed by some bacteria for survival during adverse environmental conditions.

2. The process of endospore formation is called sporulation; the return of an endospore to its vegetative state is called germination.

THE EUKARYOTIC CELL (pp. 96–106)

Flagella and Cilia (p. 98)

1. Flagella are few and long in relation to cell size; cilia are numerous and short.

2. Flagella and cilia are used for motility, and cilia also move substances along the surface of the cells.

3. Both flagella and cilia consist of an arrangement of nine pairs and two single microtubules.

The Cell Wall and Glycocalyx (pp. 98–99)

1. The cell walls of many algae and some fungi contain cellulose.

2. The main material of fungal cell walls is chitin.

3. Yeast cell walls consist of glucan and mannan.

4. Animal cells are surrounded by a glycocalyx, which strengthens the cell and provides a means of attachment to other cells.

The Plasma (Cytoplasmic) Membrane (p. 99)

1. Like the prokaryotic plasma membrane, the eukaryotic plasma membrane is a phospholipid bilayer containing proteins.

2. Eukaryotic plasma membranes contain carbohydrates attached to the proteins and sterols not found in prokaryotic cells (except Mycoplasma bacteria).

3. Eukaryotic cells can move materials across the plasma membrane by the passive processes used by prokaryotes, in addition to active transport and endocytosis (phagocytosis and pinocytosis).

Cytoplasm (p. 99)

1. The cytoplasm of eukaryotic cells includes everything inside the plasma membrane and external to the nucleus.

2. The chemical characteristics of the cytoplasm of eukaryotic cells resemble those of the cytoplasm of prokaryotic cells.

3. Eukaryotic cytoplasm has a cytoskeleton and exhibits cytoplasmic streaming.

Organelles (pp. 99–104)

1. Organelles are specialized membrane-enclosed structures in the cytoplasm of eukaryotic cells.

2. The nucleus, which contains DNA in the form of chromosomes, is the most characteristic eukaryotic organelle.

3. The nuclear envelope is connected to a system of membranes in the cytoplasm called the endoplasmic reticulum (ER).

4. The ER provides a surface for chemical reactions, serves as a transporting network, and stores synthesized molecules. Protein synthesis and transport occur on rough ER; lipid synthesis occurs on smooth ER.

5. 80S ribosomes are found in the cytoplasm or attached to the rough ER.

6. The Golgi complex consists of flattened sacs called cisterns. It functions in membrane formation and protein secretion.

7. Lysosomes are formed from Golgi complexes. They store digestive enzymes.

8. Vacuoles are membrane-enclosed cavities derived from the Golgi complex or endocytosis. They are usually found in plant cells that store various substances, help bring food into the cell, increase cell size, and provide rigidity to leaves and stems.

9. Mitochondria are the primary sites of ATP production. They contain 70S ribosomes and DNA, and they multiply by binary fission.

10. Chloroplasts contain chlorophyll and enzymes for photosynthesis. Like mitochondria, they contain 70S ribosomes and DNA and multiply by binary fission.

11. A variety of organic compounds are oxidized in peroxisomes. Catalase in peroxisomes destroys H_2O_2.

12. The centrosome consists of the pericentriolar material and centrioles. Centrioles are 9 triplet microtubules involved in formation of the mitotic spindle and microtubules.

The Evolution of Eukaryotes (p. 106)

1. According to the endosymbiotic theory, eukaryotic cells evolved from symbiotic prokaryotes living inside other prokaryotic cells.

STUDY QUESTIONS

Review

1. Diagram each of the following flagellar arrangements:
   a. lophotrichous
   b. monotrichous
   c. peritrichous

2. Endospore formation is called _____. It is initiated by _____. Formation of a new cell from an endospore is called _____. This process is triggered by _____.

3. Draw the following bacterial shapes (listed in a, b, and c). Show how d, e, and f are special conditions of a, b, and c, respectively.
   a. spiral
   b. bacillus
   c. coccus
   d. spirochetes
   e. streptobacilli
   f. staphylococci

4. Match the structures to their functions.
   ____ cell wall
   ____ endospore
   ____ fimbriae
   ____ flagella
   ____ glycocalyx
   ____ pili
   ____ plasma membrane
   ____ ribosomes

   a. attachment to surfaces
   b. cell wall formation
   c. motility
   d. protection from osmotic lysis
   e. protection from phagocytes
   f. resting
   g. protein synthesis
   h. selective permeability
   i. transfer of genetic material

5. Why is an endospore called a resting structure? Of what advantage is an endospore to a bacterial cell?

6. Compare and contrast the following:
   a. simple diffusion and facilitated diffusion
   b. active transport and facilitated diffusion
   c. active transport and group translocation

7. Why are mycoplasmas resistant to antibiotics that interfere with cell wall synthesis?

8. Compare and contrast the following:
   a. Spheroplast and L form
   b. *Mycoplasma* and L form

9. Answer the following questions using the diagrams provided, which represent cross sections of bacterial cell walls.
   a. Which diagram represents a gram-positive bacterium? How can you tell?

b. Explain how the Gram stain works to distinguish between these two types of cell walls.
   c. Why does penicillin have no effect on most gram-negative cells?
   d. How do essential molecules enter cells through each wall?
   e. Which cell wall is toxic to humans?

10. Starch is readily metabolized by many cells, but a starch molecule is too large to cross the plasma membrane. How does a cell obtain the glucose molecules from a starch polymer? How does the cell transport these glucose molecules across the plasma membrane?

11. Match the following characteristics of eukaryotic cells with their functions.
    ____ pericentriolar material
    ____ chloroplasts
    ____ Golgi complex
    ____ lysosomes
    ____ mitochondria
    ____ peroxisomes
    ____ rough ER

    a. digestive enzyme storage
    b. oxidation of fatty acids
    c. microtubule formation
    d. photosynthesis
    e. protein synthesis
    f. respiration
    g. secretion

12. Eukaryotic cells might have evolved from early prokaryotic cells living in close association. What do you know about eukaryotic organelles that would support this theory?

13. What process would a eukaryotic cell use to ingest a prokaryotic cell? To ingest a virus?

14. The antibiotic erythromycin binds with the 50S portion of a ribosome. What effect does this have on a prokaryotic cell? A eukaryotic cell?

Multiple Choice

1. Which of the following is *not* a distinguishing characteristic of prokaryotic cells?
   a. They have a single, circular chromosome.
   b. They lack membrane-enclosed organelles.
   c. They have cell walls containing peptidoglycan.
   d. Their DNA is not associated with histones.
   e. They lack a plasma membrane.

Use the following choices to answer questions 2–4.
   a. No change will result; the solution is isotonic.
   b. Water will move into the cell.
   c. Water will move out of the cell.
   d. The cell will undergo osmotic lysis.
   e. Sucrose will move into the cell from an area of higher concentration to one of lower concentration.

2. Which statement best describes what happens when a gram-positive bacterium is placed in distilled water and penicillin?

3. Which statement best describes what happens when a gram-negative bacterium is placed in distilled water and penicillin?

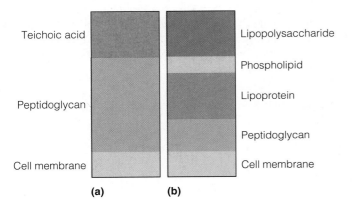

Teichoic acid

Peptidoglycan

Cell membrane

Lipopolysaccharide

Phospholipid

Lipoprotein

Peptidoglycan

Cell membrane

(a)   **(b)**

4. Which statement best describes what happens when a gram-positive bacterium is placed in an aqueous solution of lysozyme and 10% sucrose?

5. Which of the following statements best describes what happens to a cell exposed to polymyxins that destroy phospholipids?
   a. In an isotonic solution, nothing will happen.
   b. In a hypotonic solution, the cell will lyse.
   c. Water will move into the cell.
   d. Intracellular contents will leak from the cell.
   e. Any of the above might happen.

6. Which of the following is *not* true about fimbriae?
   a. They are composed of protein.
   b. They may be used for attachment.
   c. They are found on gram-negative cells.
   d. They are composed of pilin.
   e. They may be used for motility.

7. Which of the following pairs is mismatched?
   a. glycocalyx—adherence
   b. pili—reproduction
   c. membrane—DNA synthesis
   d. cell wall—protection
   e. plasma membrane—transport

8. Which of the following pairs is mismatched?
   a. metachromatic granules—stored phosphates
   b. polysaccharide granules—stored starch
   c. lipid inclusions—poly-β-hydroxybutyric acid
   d. sulfur granules—energy reserve
   e. ribosomes—protein storage

9. You have isolated a motile, gram-positive cell with no visible nucleus. You can assume this cell has
   a. ribosomes.
   b. mitochondria.
   c. an endoplasmic reticulum.
   d. a Golgi complex.
   e. all of the above

10. The antibiotic amphothericin B disrupts plasma membranes by combining with sterols; it will affect all of the following cells *except*
   a. animal cells.
   b. bacterial cells.
   c. fungal cells.
   d. *Mycoplasma* cells
   e. plant cells.

Critical Thinking

1. Why can prokaryotic cells be smaller than eukaryotic cells and still carry on all the functions of life?

2. The smallest eukaryotic cell is the motile alga *Micromonas*. What is the minimum number of organelles this alga must have?

3. Two types of prokaryotic cells have been distinguished: bacteria and archaea. How do these cells differ from each other? How are they similar?

4. In 1985, a 0.5-mm cell was discovered in surgeonfish and named *Epulopiscium fishelsoni* (see Figure 11.15 on page 321). It was presumed to be a protozoan. In 1993, researchers determined that *Epulopiscium* was actually a gram-positive bacterium. What do you suppose caused the initial identification of this organism as a protozoan? What evidence would change the classification to bacterium?

5. When *E. coli* cells are exposed to a hypertonic solution, the bacteria produce a transporter protein that can move K+ (potassium ions) into the cell. Of what value is the active transport of K+, which requires ATP?

Clinical Applications

1. A child with a bloodborne *Neisseria* infection was treated with gentamicin. After treatment, *Neisseria* could not be cultured from her blood, indicating that the bacteria were killed. However, her symptoms became worse. Annually, nearly half of similar patients die. Explain why antibiotic treatment made her symptoms increase.

2. *Clostridium botulinum* is a strict anaerobe; that is, it is killed by the molecular oxygen (O_2) present in air. Humans can die of botulism from eating foods in which *C. botulinum* is growing. How does this bacterium survive on plants picked for human consumption? Why are home-canned foods most often the source of botulism?

3. Within a 3-day period at a large hospital, five patients undergoing hemodialysis developed fever and chills. *Pseudomonas aeruginosa* and *Klebsiella pneumoniae* were isolated from three of the patients. *P. aeruginosa*, *K. pneumoniae*, and *Enterobacter agglomerans* were isolated from the dialysis system. Why do all three bacteria cause similar symptoms?

4. A South San Francisco child enjoyed bath time at his home because of the colorful orange and red water. The water did not have this rusty color at its source, and the water department could not culture the *Thiobacillus* bacteria responsible for the rusty color from the source. How were the bacteria getting into the household water? What bacterial structures make this possible?

5. Live cultures of *Bacillus thuringiensis* (Dipel) and *B. subtilis* (Kodiak) are sold for agricultural use. What bacterial structures make it possible to package and sell these bacteria? For what purpose is each product used? (*Hint:* Refer to Chapter 11.)

LEARNING WITH TECHNOLOGY

The Microbiology Place CD-ROM/website includes many resources to help you study and succeed in this course. In addition to self-quizzes, web links, and flashcards, you'll find the following activities for this chapter in the "Activities" section of The Microbiology Place:

Cell Structure and Function

Case Study: *Coxiella burnetti*

Chapter 5

Microbial Metabolism

FERMENTATION TEST. *Biochemical tests such as fermentation are used to study bacterial metabolism and to understand their role in the environment.*

Catabolic and Anabolic Reactions

Learning Objectives
- *Define metabolism, and describe the fundamental differences between anabolism and catabolism.*
- *Identify the role of ATP as an intermediate between catabolism and anabolism.*

Now that you are familiar with the structure of prokaryotic cells, we can discuss the activities that enable these microbes to thrive. The life-support processes of even the most structurally simple organism involve a large number of complex biochemical reactions. Most, although not all, of the biochemical processes of bacteria also occur in eukaryotic microbes and in the cells of multicellular organisms, including humans. However, the reactions that are unique to bacteria are fascinating because they allow microorganisms to do things we cannot do. For example,

some bacteria can live on cellulose, while others can live on petroleum. Through their metabolism, bacteria recycle elements after other organisms have used them. Chemoautotrophs can live on diets of such inorganic substances as carbon dioxide, iron, sulfur, hydrogen gas, and ammonia.

This chapter examines some representative chemical reactions that either produce energy (the catabolic reactions) or use energy (the anabolic reactions) in microorganisms. We will also look at how these various reactions are integrated within the cell.

We use the term **metabolism** to refer to the sum of all chemical reactions within a living organism. Because chemical reactions either release or require energy, metabolism can be viewed as an energy-balancing act. Accordingly, metabolism can be divided into two classes of chemical reactions: those that release energy and those that require energy.

In living cells, the enzyme-regulated chemical reactions that release energy are generally the ones involved in **catabolism,** the breakdown of complex organic compounds into simpler ones. These reactions are called *catabolic,* or *degradative,* reactions. Catabolic reactions are generally hydrolytic reactions (reactions that use water and in which chemical bonds are broken), and they are exergonic (produce more energy than they consume). An example of catabolism occurs when cells break down sugars into carbon dioxide and water.

The enzyme-regulated energy-requiring reactions are mostly involved in **anabolism,** the building of complex organic molecules from simpler ones. These reactions are called *anabolic,* or *biosynthetic,* reactions. Anabolic processes often involve dehydration synthesis reactions (reactions that release water), and they are endergonic (consume more energy than they produce). Examples of anabolic processes are the formation of proteins from amino

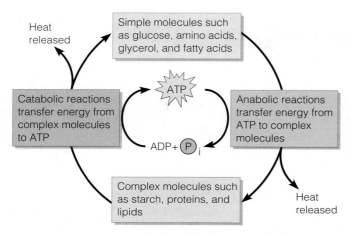

FIGURE 5.1 **The role of ATP in coupling anabolic and catabolic reactions.** When complex molecules are split apart (catabolism), some of the energy is transferred to and trapped in ATP, and the rest is given off as heat. When simple molecules are combined to form complex molecules (anabolism), ATP provides the energy for synthesis, and again some energy is given off as heat.

■ The coupling of anabolic and catabolic reactions is achieved through ATP.

acids, nucleic acids from nucleotides, and polysaccharides from simple sugars. These biosynthetic reactions generate the materials for cell growth.

Catabolic reactions provide building blocks for anabolic reactions and furnish the energy needed to drive anabolic reactions. This coupling of energy-requiring and energy-releasing reactions is made possible through the molecule adenosine triphosphate (ATP). (You can review its structure in Figure 2.18 on page 49.) ATP stores energy derived from catabolic reactions and releases it later to drive anabolic reactions and perform other cellular work. Recall from Chapter 2 that a molecule of ATP consists of an adenine, a ribose, and three phosphate groups. When the terminal phosphate group is split from ATP, adenosine diphosphate (ADP) is formed, and energy is released to drive anabolic reactions. Using ⓟ to represent a phosphate group, we write this reaction as:

$$ATP \longrightarrow ADP + ⓟ_i + energy$$

Then, the energy from catabolic reactions is used to combine ADP and a ⓟ to resynthesize ATP:

$$ADP + ⓟ_i + energy \longrightarrow ATP$$

Thus, anabolic reactions are coupled to ATP breakdown, and catabolic reactions are coupled to ATP synthesis. This concept of coupled reactions is very important; you will see why by the end of this chapter. For now, you should know that the chemical composition of a living cell is constantly

changing: Some molecules are broken down while others are being synthesized. This balanced flow of chemicals and energy maintains the life of a cell.

The role of ATP in coupling anabolic and catabolic reactions is shown in Figure 5.1. Only part of the energy released in catabolism is actually available for cellular functions, for part of the energy is lost to the environment as heat. Because the cell must use energy to maintain life, it has a continuous need for new external sources of energy.

Before we discuss how cells produce energy, let's first consider the principal properties of a group of proteins involved in almost all biologically important chemical reactions. These proteins, the enzymes, were described briefly in Chapter 2. It is important to understand that a cell's **metabolic pathways** (sequences of chemical reactions) are determined by its enzymes, which are in turn determined by the cell's genetic makeup.

Enzymes

Learning Objectives
- *Identify the components of an enzyme.*
- *Describe the mechanism of enzymatic action.*
- *List the factors that influence enzymatic activity.*
- *Define ribozyme.*

Collision Theory

We indicated in Chapter 2 that chemical reactions occur when chemical bonds are formed or broken. In order for reactions to take place, atoms, ions, or molecules must collide. The **collision theory** explains how chemical reactions occur and how certain factors affect the rates of those reactions. The basis of the collision theory is that all atoms, ions, and molecules are continuously moving and are thus continuously colliding with one another. The energy transferred by the particles in the collision can disrupt their electron structures enough so that chemical bonds are broken or new bonds are formed.

Several factors determine whether a collision will cause a chemical reaction: the velocities of the colliding particles, their energy, and their specific chemical configurations. Up to a point, the higher the particles' velocities, the greater the probability that their collision will cause a reaction. Also, each chemical reaction requires a specific level of energy. But even if colliding particles possess the minimum energy needed for reaction, no reaction will take place unless the particles are properly oriented toward each other.

Let's assume that molecules of substance AB (the reactant) are to be converted to molecules of substances A and B (the products). In a given population of molecules of

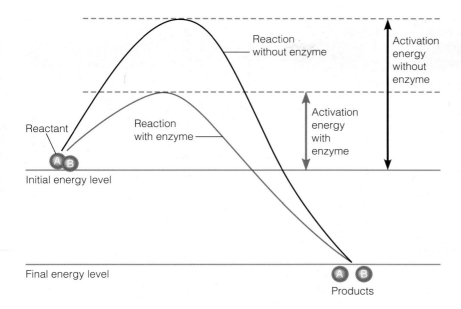

FIGURE 5.2 **Energy requirements of a chemical reaction.** This graph shows the progress of the reaction AB → A + B both without (black line) and with (magenta line) an enzyme. The presence of an enzyme lowers the activation energy of the reaction (see arrows). Thus, more molecules of reactant AB are converted to products A and B because more molecules of reactant AB possess the activation energy needed for the reaction.

■ How do enzymes speed up chemical reactions?

substance AB, at a specific temperature, some molecules will possess relatively little energy; the majority of the population will possess an average amount of energy; and a small portion of the population will have high energy. If only the high-energy AB molecules are able to react and be converted to A and B molecules, then only relatively few molecules at any one time possess enough energy to react in a collision. The collision energy required for a chemical reaction is its **activation energy,** which is the amount of energy needed to disrupt the stable electronic configuration of any specific molecule so that the electrons can be rearranged.

The **reaction rate**—the frequency of collisions containing sufficient energy to bring about a reaction—depends on the number of reactant molecules at or above the activation energy level. One way to increase the reaction rate of a substance is to raise its temperature. By causing the molecules to move faster, heat increases both the frequency of collisions and the number of molecules that attain activation energy. The number of collisions also increases when pressure is increased or when the reactants are more concentrated (because the distance between molecules is thereby decreased). In living systems, enzymes increase the reaction rate without raising the temperature.

Enzymes and Chemical Reactions

Substances that can speed up a chemical reaction without being permanently altered themselves are called **catalysts.** In living cells, **enzymes** serve as biological catalysts. As catalysts, enzymes are specific. Each acts on a specific substance, called the enzyme's **substrate** (or substrates, when there are two or more reactants), and each catalyzes only one reaction. For example, sucrose (table sugar) is the substrate of the enzyme sucrase, which catalyzes the hydrolysis of sucrose to glucose and fructose.

As catalysts, enzymes typically accelerate chemical reactions. The three-dimensional enzyme molecule has an *active site*, a region that will interact with a specific chemical substance (see Figure 5.4 on page 116). The chemical substance (reactant) an enzyme acts on is referred to as the enzyme's substrate.

The enzyme orients the substrate into a position that increases the probability of a reaction. The **enzyme–substrate complex** formed by the temporary binding of enzyme and reactants enables the collisions to be more effective and lowers the activation energy of the reaction (Figure 5.2). The enzyme therefore speeds up the reaction by increasing the number of AB molecules that attain sufficient activation energy to react.

An enzyme's ability to accelerate a reaction without the need for an increase in temperature is crucial to living systems because a significant temperature increase would destroy cellular proteins. The crucial function of enzymes, therefore, is to speed up biochemical reactions at a temperature that is compatible with the normal functioning of the cell.

Enzyme Specificity and Efficiency

The specificity of enzymes is made possible by their structures. Enzymes are generally large globular proteins that range in molecular weight from about 10,000 to several million. Each of the thousands of known enzymes has a characteristic three-dimensional shape with a specific surface configuration as a result of its primary, secondary, and tertiary structures (see Figure 2.15 on page 46). The

TABLE 5.1	**Enzyme Classification Based on Type of Chemical Reaction Catalyzed**	
Class	**Type of Chemical Reaction Catalyzed**	**Examples**
Oxidoreductase	Oxidation-reduction in which oxygen and hydrogen are gained or lost	Cytochrome oxidase, lactate dehydrogenase
Transferase	Transfer of functional groups, such as an amino group, acetyl group, or phosphate group	Acetate kinase, alanine deaminase
Hydrolase	Hydrolysis (addition of water)	Lipase, sucrase
Lyase	Removal of groups of atoms without hydrolysis	Oxalate decarboxylase, isocitrate lyase
Isomerase	Rearrangement of atoms within a molecule	Glucose-phosphate isomerase, alanine racemase
Ligase	Joining of two molecules (using energy usually derived from the breakdown of ATP)	Acetyl-CoA synthetase, DNA ligase

unique configuration of each enzyme enables it to "find" the correct substrate from among the large number of diverse molecules in the cell.

Enzymes are extremely efficient. Under optimum conditions, they can catalyze reactions at rates 10^8–10^{10} times (up to 10 billion times) higher than those of comparable reactions without enzymes. The **turnover number** (maximum number of substrate molecules an enzyme molecule converts to product each second) is generally between 1 and 10,000 and can be as high as 500,000. For example, the enzyme DNA polymerase I, which participates in the synthesis of DNA, has a turnover number of 15, whereas the enzyme lactate dehydrogenase, which removes hydrogen atoms from lactic acid, has a turnover number of 1000.

Many enzymes exist in the cell in both active and inactive forms. The rate at which enzymes switch between these two forms is determined by the cellular environment.

Naming Enzymes

The names of enzymes usually end in -ase. All enzymes can be grouped into six classes, according to the type of chemical reaction they catalyze (Table 5.1). Enzymes within each of the major classes are named according to the more specific types of reactions they assist. For example, the class called oxidoreductases is involved with oxidation-reduction reactions (described shortly). Enzymes in the oxidoreductase class that remove hydrogen from a substrate are called *dehydrogenases*; those that add molecular oxygen (O_2) are called *oxidases*. As you will see later, dehydrogenase and oxidase enzymes have even more specific names, such as lactate dehydrogenase and cytochrome oxidase, depending on the specific substrates on which they act.

Enzyme Components

Although some enzymes consist entirely of proteins, most consist of both a protein portion called an **apoenzyme** and a nonprotein component called a **cofactor.** Ions of iron, zinc, magnesium, or calcium are examples of cofactors. If the cofactor is an organic molecule, it is called a **coenzyme.** Apoenzymes are inactive by themselves; they must be activated by cofactors. Together, the apoenzyme and cofactor form a **holoenzyme,** or whole, active enzyme (Figure 5.3). If the cofactor is removed, the apoenzyme will not function.

Coenzymes may assist the enzyme by accepting atoms removed from the substrate or by donating atoms required by the substrate. Some coenzymes act as electron carriers, removing electrons from the substrate and donating them to other molecules in subsequent reactions. Many coenzymes are derived from vitamins (Table 5.2). Two of the most important coenzymes in cellular metabolism are **nicotinamide adenine dinucleotide (NAD^+)** and **nicotinamide adenine dinucleotide phosphate ($NADP^+$).** Both compounds contain derivatives of the B vitamin nicotinic acid (niacin), and both function as electron carriers. Whereas NAD^+ is primarily involved in catabolic (energy-yielding) reactions, $NADP^+$ is primarily involved in anabolic (energy-requiring) reactions. The flavin coenzymes, such as **flavin mononucleotide (FMN)** and **flavin adenine dinucleotide (FAD),** contain derivatives of the B vitamin riboflavin and are also electron carriers. Another important coenzyme, **coenzyme A (CoA),** contains a derivative of pantothenic acid, another B vitamin. This coenzyme plays an important role in the synthesis and breakdown of fats and in a series of oxidizing reactions called the Krebs cycle. We will come across all of these coenzymes in our discussion of metabolism later in the chapter.

Substrate

Coenzyme

Apoenzyme
(protein portion),
inactive

+

Cofactor
(nonprotein portion),
activator

Holoenzyme
(whole enzyme),
active

FIGURE 5.3 **Components of a holoenzyme.** Many enzymes require both an apoenzyme (protein portion) and a cofactor (nonprotein portion) to become active. The cofactor can be a metal ion, or if it is an organic molecule, it is called a coenzyme (as shown here). The apoenzyme and cofactor together make up the holoenzyme, or whole enzyme. The substrate is the reactant acted upon by the enzyme.

- What is the function of enzymes in living organisms?

As noted earlier, some cofactors are metal ions, including iron, copper, magnesium, manganese, zinc, calcium, and cobalt. Such cofactors may help catalyze a reaction by forming a bridge between the enzyme and a substrate. For example, magnesium (Mg^{2+}) is required by many phosphorylating enzymes (enzymes that transfer a phosphate group from ATP to another substrate). The Mg^{2+} can form a link between the enzyme and the ATP molecule. Most trace elements required by living cells are probably used in some such way to activate cellular enzymes.

The Mechanism of Enzymatic Action

Enzymes lower the activation energy of chemical reactions. The general sequence of events in enzyme action is as follows (Figure 5.4a):

1. The surface of the substrate contacts a specific region of the surface of the enzyme molecule called the **active site.**

2. A temporary intermediate compound forms, called an **enzyme–substrate complex.**

3. The substrate molecule is transformed by the rearrangement of existing atoms, the breakdown of the substrate molecule, or in combination with another substrate molecule.

4. The transformed substrate molecules—the products of the reaction—are released from the enzyme molecule because they no longer fit in the active site of the enzyme.

5. The unchanged enzyme is now free to react with other substrate molecules.

TABLE 5.2	Selected Vitamins and Their Coenzymatic Functions
Vitamin	**Function**
Vitamin B_1 (thiamine)	Part of coenzyme cocarboxylase; has many functions, including the metabolism of pyruvic acid
Vitamin B_2 (riboflavin)	Coenzyme in flavoproteins; active in electron transfers
Niacin (nicotinic acid)	Part of NAD molecule; active in electron transfers
Vitamin B_6 (pyridoxine)	Coenzyme in amino acid metabolism
Vitamin B_{12} (cyanocobalamin)	Coenzyme (methyl cyanocobalamide) involved in the transfer of methyl groups; active in amino acid metabolism
Pantothenic acid	Part of coenzyme A molecule; involved in the metabolism of pyruvic acid and lipids
Biotin	Involved in carbon dioxide fixation reactions and fatty acid synthesis
Folic acid	Coenzyme used in the synthesis of purines and pyrimidines
Vitamin E	Needed for cellular and macromolecular syntheses
Vitamin K	Coenzyme used in electron transport (naphthoquinones and quinones)

FIGURE 5.4 **The mechanism of enzymatic action. (a)** ❶ The substrate contacts the active site on the enzyme to form ❷ an enzyme–substrate complex. ❸ The substrate is then transformed into products, ❹ the products are released, and ❺ the enzyme is recovered unchanged. In the example shown, the transformation into products involves a breakdown of the substrate into two products. Other transformations, however, may occur. **(b)** Left: A molecular model of the enzyme in step ❶ of part a. The active site of the enzyme can be seen here as a groove on the surface of the protein. Right: As the enzyme and substrate meet in step ❷ of part a, they change shape slightly to fit together more tightly.

As a result of these events, an enzyme speeds up a chemical reaction.

As mentioned earlier, enzymes have *specificity* for particular substrates. For example, a specific enzyme may be able to hydrolyze a peptide bond only between two specific amino acids. Other enzymes can hydrolyze starch but not cellulose; even though both starch and cellulose are polysaccharides composed of glucose subunits, the orientations of the subunits in the two polysaccharides differ. Enzymes have this specificity because the three-dimensional shape of the active site fits the substrate somewhat as a lock fits with its key (Figure 5.4b). However, the active site and substrate are flexible, and they change shape somewhat as they meet to fit together more tightly. The substrate is usually much smaller than the enzyme, and relatively few of the enzyme's amino acids make up the active site.

A certain compound can be a substrate for several different enzymes that catalyze different reactions, so the fate of a compound depends on the enzyme that acts upon it. Glucose 6-phosphate, a molecule important in cell metabolism, can be acted upon by at least four different enzymes, and each reaction will yield a different product.

Factors Influencing Enzymatic Activity

Enzymes are subject to various cellular controls. Two primary types are the control of enzyme *synthesis* (see Chapter 8) and the control of enzyme *activity* (how much enzyme is present versus how active it is).

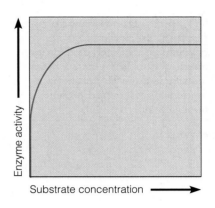

(a) Temperature. The enzymatic activity (rate of reaction catalyzed by the enzyme) increases with increasing temperature until the enzyme, a protein, is denatured by heat and inactivated. At this point, the reaction rate falls steeply.

(b) pH. The enzyme illustrated is most active at about pH 5.0.

(c) Substrate concentration. With increasing concentration of substrate molecules, the rate of reaction increases until the active sites on all the enzyme molecules are filled, at which point the maximum rate of reaction is reached.

FIGURE 5.5 **Factors that influence enzymatic activity, plotted for a hypothetical enzyme.**

■ How will this enzyme act at 25°C? At 45°C? At pH 7?

Several factors influence the activity of an enzyme. Among the more important are temperature, pH, substrate concentration, and the presence or absence of inhibitors.

Temperature

The rate of most chemical reactions increases as the temperature increases. Molecules move more slowly at lower temperatures than at higher temperatures and so may not have enough energy to cause a chemical reaction. For enzymatic reactions, however, elevation beyond a certain temperature (the optimal temperature) drastically reduces the rate of reaction (Figure 5.5a). The optimal temperature for most disease-producing bacteria in the human body is between 35°C and 40°C. The reduced rate of reaction beyond the optimal temperature is due to the enzyme's **denaturation,** the loss of its characteristic three-dimensional structure (tertiary configuration) (Figure 5.6). Denaturation of a protein involves the breakage of hydrogen bonds and other noncovalent bonds; a common example is the transformation of uncooked egg white (a protein called albumin) to a hardened state by heat.

Denaturation of an enzyme changes the arrangement of the amino acids in the active site, altering its shape and causing the enzyme to lose its catalytic ability. In some cases, denaturation is partially or fully reversible. However, if denaturation continues until the enzyme has lost its solubility and coagulates, the enzyme cannot re-

gain its original properties. Enzymes can also be denatured by concentrated acids, bases, heavy-metal ions (such as lead, arsenic, or mercury), alcohol, and ultraviolet radiation.

pH

Most enzymes have an optimum pH at which their activity is characteristically maximal. Above or below this pH value, enzyme activity, and therefore the reaction rate, decline (Figure 5.5b). When the H^+ concentration (pH) in the medium is changed drastically, the protein's three-dimensional structure is altered. Extreme changes in pH can cause denaturation. Acids (and bases) alter a protein's three-dimensional structure because the H^+ (and OH^-) compete with hydrogen and ionic bonds in an enzyme, resulting in the enzyme's denaturation.

Substrate Concentration

There is a maximum rate at which a certain amount of enzyme can catalyze a specific reaction. Only when the concentration of substrate(s) is extremely high can this maximum rate be attained. Under conditions of high substrate concentration, the enzyme is said to be in **saturation;** that is, its active site is always occupied by substrate or product molecules. In this condition, a further increase in substrate concentration will not affect the reaction rate because all active sites are already in use (Figure 5.5c). Under normal cellular conditions, enzymes are not saturated with sub-

FIGURE 5.6 **Denaturation of a protein.** Breakage of the non-covalent bonds (such as hydrogen bonds) that hold the active protein in its three-dimensional shape renders the denatured protein nonfunctional.

▪ What factors may cause denaturation?

Active (functional) protein　　　　　　Denatured protein

strate(s). At any given time, many of the enzyme molecules are inactive for lack of substrate; thus, the rate of reaction is likely to be influenced by the substrate concentration.

Inhibitors

An effective way to control the growth of bacteria is to control their enzymes. Certain poisons, such as cyanide, arsenic, and mercury, combine with enzymes and prevent them from functioning. As a result, the cells stop functioning and die.

Enzyme inhibitors are classified as either competitive or noncompetitive inhibitors (Figure 5.7). **Competitive inhibitors** fill the active site of an enzyme and compete with the normal substrate for the active site. A competitive inhibitor can do this because its shape and chemical structure are similar to those of the normal substrate (Figure 5.7b). However, unlike the substrate, it does not undergo any reaction to form products. Some competitive inhibitors bind irreversibly to amino acids in the active site, preventing any further interactions with the substrate. Others bind reversibly, alternately occupying and leaving the active site; these slow the enzyme's interaction with the substrate. Reversible competitive inhibition can be overcome by increasing the substrate concentration. As active sites become available, more substrate molecules than competitive inhibitor molecules are available to attach to the active sites of enzymes.

One good example of a competitive inhibitor is sulfanilamide (a sulfa drug), which inhibits the enzyme whose normal substrate is para-aminobenzoic acid (PABA):

Sulfanilamide　　　　PABA

PABA is an essential nutrient used by many bacteria in the synthesis of folic acid, a vitamin that functions as a coenzyme. When sulfanilamide is administered to bacteria, the enzyme that normally converts PABA to folic acid combines instead with the sulfanilamide. Folic acid is not synthesized, and the bacteria cannot grow. Because human cells do not use PABA to make their folic acid, sulfanilamide kills bacteria but does not harm human cells.

Noncompetitive inhibitors do not compete with the substrate for the enzyme's active site; instead, they interact with another part of the enzyme (Figure 5.7c). In this process, called **allosteric** ("other space") **inhibition,** the inhibitor binds to a site on the enzyme other than the substrate's binding site, called the **allosteric site.** This binding causes the active site to change its shape, making it

FIGURE 5.7 **Enzyme inhibitors. (a)** An uninhibited enzyme and its normal substrate. **(b)** A competitive inhibitor. **(c)** One type of noncompetitive inhibitor, causing allosteric inhibition.

▪ Competitive inhibitors fill the active site of an enzyme and compete with the normal substrate for the active site.

NORMAL BINDING OF SUBSTRATE　　　　ACTION OF ENZYME INHIBITORS

(a)　　　　　(b)　　　　　(c)

nonfunctional. As a result, the enzyme's activity is reduced. This effect can be either reversible or irreversible, depending on whether or not the active site can return to its original shape. In some cases, allosteric interactions can activate an enzyme rather than inhibit it. Another type of noncompetitive inhibition can operate on enzymes that require metal ions for their activity. Certain chemicals can bind or tie up the metal ion activators and thus prevent an enzymatic reaction. Cyanide can bind the iron in iron-containing enzymes, and fluoride can bind calcium or magnesium. Substances such as cyanide and fluoride are sometimes called *enzyme poisons* because they permanently inactivate enzymes. In low concentrations, fluoride kills bacteria in the mouth that can contribute to tooth decay.

Feedback Inhibition

Allosteric inhibitors play a role in a kind of biochemical control called **feedback inhibition,** or **end-product inhibition.** This control mechanism stops the cell from wasting chemical resources by making more of a substance than it needs. In some metabolic reactions, several steps are required for the synthesis of a particular chemical compound, called the *end-product.* The process is similar to an assembly line, with each step catalyzed by a separate enzyme (Figure 5.8). In many anabolic pathways, the final product can allosterically inhibit the activity of one of the enzymes earlier in the pathway. This phenomenon is feedback inhibition.

Feedback inhibition generally acts on the first enzyme in a metabolic pathway (similar to shutting down an assembly line by stopping the first worker). Because the enzyme is inhibited, the product of the first enzymatic reaction in the pathway is not synthesized. Because that unsynthesized product would normally be the substrate for the second enzyme in the pathway, the second reaction stops immediately as well. Thus, even though only the first enzyme in the pathway is inhibited, the entire pathway shuts down and no new end-product is formed. By inhibiting the first enzyme in the pathway, the cell also keeps metabolic intermediates from accumulating. As the existing end-product is used up by the cell, the first enzyme's allosteric site will more often remain unbound, and the pathway will resume activity.

The bacterium *E. coli* can be used to demonstrate feedback inhibition in the synthesis of the amino acid isoleucine, which is required for the cell's growth. In this metabolic pathway, five steps are taken to enzymatically convert the amino acid threonine to isoleucine. If isoleucine is added to the growth medium for *E. coli*, it inhibits the first enzyme in the pathway, and the bacteria stop synthesizing isoleucine. This condition is maintained

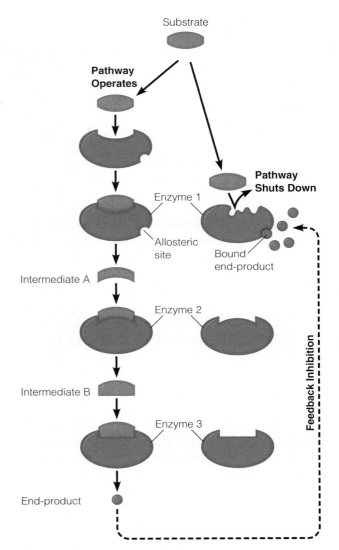

FIGURE 5.8 Feedback inhibition.

■ In feedback inhibition, a series of enzymes makes an end-product that inhibits the first enzyme in the series, thus shutting down the entire pathway when sufficient end-product has been made.

until the supply of isoleucine is depleted. This type of feedback inhibition is also involved in regulating the cells' production of other amino acids, as well as vitamins, purines, and pyrimidines.

Ribozymes

Prior to 1982, it was believed that only protein molecules had enzymatic activity. Researchers working on microbes discovered a unique type of RNA called a **ribozyme.** Like protein enzymes, ribozymes function as catalysts, have active sites that bind to substrates, and are not used up in a chemical reaction. Ribozymes specifically act on strands of

RNA by removing sections and splicing together the remaining pieces. In this respect, ribozymes are more restricted than protein enzymes in terms of the diversity of substrates with which they interact.

Energy Production

Learning Objectives

- *Explain what is meant by oxidation-reduction.*
- *List and provide examples of three types of phosphorylation reactions that generate ATP.*

Nutrient molecules, like all molecules, have energy associated with the electrons that form bonds between their atoms. When it is spread throughout the molecule, this energy is difficult for the cell to use. Various reactions in catabolic pathways, however, concentrate the energy into the bonds of ATP, which serves as a convenient energy carrier. ATP is generally referred to as having "high-energy" bonds. Actually, a better term is probably unstable bonds. Although the amount of energy in these bonds is not exceptionally large, it can be released quickly and easily. In a sense, ATP is similar to a highly flammable liquid such as kerosene. Although a large log might eventually burn to produce more heat than a cup of kerosene, the kerosene is easier to ignite and provides heat more quickly and conveniently. In a similar way, the "high-energy" unstable bonds of ATP provide the cell with readily available energy for anabolic reactions.

Before discussing the catabolic pathways, we will consider two general aspects of energy production: the concept of oxidation-reduction and the mechanisms of ATP generation.

Oxidation-Reduction Reactions

Oxidation is the removal of electrons (e^-) from an atom or molecule, a reaction that often produces energy. Figure 5.9 shows an example of an oxidation in which molecule A loses an electron to molecule B. Molecule A has undergone oxidation (meaning that it has lost one or more electrons), whereas molecule B has undergone **reduction** (meaning that it has gained one or more electrons).* Oxi-

---

*The terms do not seem logical until one considers the history of the discovery of these reactions. When mercury is roasted, it gains weight as mercuric oxide is formed; this was called *oxidation*. Later it was determined that the mercury actually *lost* electrons, and the observed *gain* in oxygen was a direct result of this. Oxidation, therefore, is a *loss* of electrons, and reduction is a *gain* of electrons, but the gain and loss of electrons is not usually apparent as chemical-reaction equations are usually written. For example, in the equations for aerobic respiration on page 130, notice that each carbon in glucose had only one oxygen originally, and later, as carbon dioxide, each carbon now has two oxygens. However, the gain or loss of electrons actually responsible for this is not apparent.

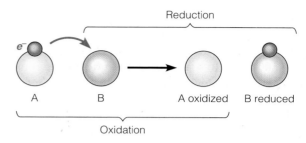

FIGURE 5.9 **Oxidation-reduction.** An electron is transferred from molecule A to molecule B. In the process, molecule A is oxidized and molecule B is reduced.

■ Oxidation is the removal of electrons, and reduction is the addition of electrons.

dation and reduction reactions are always coupled; in other words, each time one substance is oxidized, another is simultaneously reduced. The pairing of these reactions is called **oxidation-reduction** or a **redox reaction.**

In many cellular oxidations, electrons and protons (hydrogen ions, H^+) are removed at the same time; this is equivalent to the removal of hydrogen atoms, because a hydrogen atom is made up of one proton and one electron (see Table 2.2, page 29). Because most biological oxidations involve the loss of hydrogen atoms, they are also called **dehydrogenation** reactions. Figure 5.9 shows an example of a biological oxidation. An organic molecule is oxidized by the loss of two hydrogen atoms, and a molecule of NAD^+ is reduced. Recall from our earlier discussion of coenzymes that NAD^+ assists enzymes by accepting hydrogen atoms removed from the substrate, in this case the organic molecule. As shown in Figure 5.10, NAD^+ accepts two electrons and one proton. One proton (H^+) is left over and is released into the surrounding medium. The reduced coenzyme, NADH, contains more energy than NAD^+. This energy can be used to generate ATP in later reactions.

An important point to remember about biological oxidation-reduction reactions is that cells use them in catabolism to extract energy from nutrient molecules. Cells take nutrients, some of which serve as energy sources, and degrade them from highly reduced compounds (with many hydrogen atoms) to highly oxidized compounds. For example, when a cell oxidizes a molecule of glucose $(C_6H_{12}O_6)$ to CO_2 and H_2O, the energy in the glucose molecule is removed in a stepwise manner and ultimately is trapped by ATP, which can then serve as an energy source for energy-requiring reactions. Compounds such as glucose that have many hydrogen atoms are highly reduced compounds, containing a large amount of potential energy. Thus, glucose is a valuable nutrient for organisms.

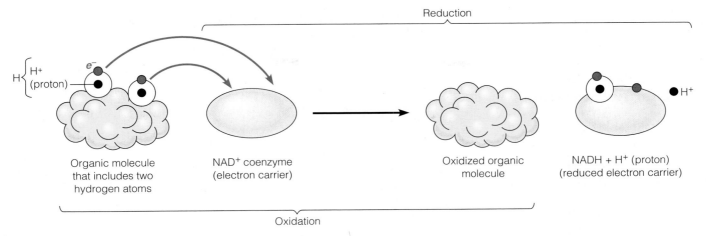

FIGURE 5.10 **Representative biological oxidation.** Two electrons and two protons (altogether equivalent to two hydrogen atoms) are transferred from an organic substrate molecule to a coenzyme, NAD$^+$. NAD$^+$ actually receives one hydrogen atom and one electron, and one proton is released into the medium. NAD$^+$ is reduced to NADH, which is a more energy-rich molecule.

■ Organisms use oxidation-reduction reactions in catabolism to extract energy from nutrient molecules, such as glucose.

The Generation of ATP

Much of the energy released during oxidation-reduction reactions is trapped within the cell by the formation of ATP. Specifically, a phosphate group, Ⓟ, is added to ADP with the input of energy to form ATP:

$$\overbrace{\text{Adenosine}—Ⓟ\sim Ⓟ}^{\text{ADP}} + \text{Energy} + Ⓟ \rightarrow$$
$$\underbrace{\text{Adenosine}—Ⓟ\sim Ⓟ\sim Ⓟ}_{\text{ATP}}$$

The symbol ~ designates a "high-energy" bond—that is, one that can readily be broken to release usable energy. The high-energy bond that attaches the third Ⓟ in a sense contains the energy stored in this reaction. When this Ⓟ is removed, usable energy is released. The addition of Ⓟ to a chemical compound is called **phosphorylation.** Organisms use three mechanisms of phosphorylation to generate ATP from ADP.

Substrate-Level Phosphorylation

In **substrate-level phosphorylation,** ATP is usually generated when a high-energy Ⓟ is directly transferred from a phosphorylated compound (a substrate) to ADP. Generally, the Ⓟ has acquired its energy during an earlier reaction in which the substrate itself was oxidized. The following example shows only the carbon skeleton and the Ⓟ of a typical substrate:

$$\text{C}—\text{C}—\text{C} \sim Ⓟ + \text{ADP} \longrightarrow \text{C}—\text{C}—\text{C} + \text{ATP}$$

Oxidative Phosphorylation

In **oxidative phosphorylation,** electrons are transferred from organic compounds to one group of electron carriers (usually to NAD$^+$ and FAD). Then, the electrons are passed through a series of different electron carriers to molecules of oxygen (O_2) or other oxidized inorganic and organic molecules. This process occurs in the plasma membrane of prokaryotes and in the inner mitochondrial membrane of eukaryotes. The sequence of electron carriers used in oxidative phosphorylation is called an **electron transport chain (system)** (see Figure 5.14 on page 128). The transfer of electrons from one electron carrier to the next releases energy, some of which is used to generate ATP from ADP through a process called *chemiosmosis,* to be described shortly.

Photophosphorylation

The third mechanism of phosphorylation, **photophosphorylation,** occurs only in photosynthetic cells, which contain light-trapping pigments such as chlorophylls. In photosynthesis, organic molecules, especially sugars, are synthesized with the energy of light from the energy-poor building blocks carbon dioxide and water. Photophosphorylation starts this process by converting light energy to the chemical energy of ATP and NADPH, which, in turn, are used to synthesize organic molecules. As in oxidative phosphorylation, an electron transport chain is involved.

Metabolic Pathways of Energy Production

Learning Objective
- *Explain the overall function of metabolic pathways.*

Organisms release and store energy from organic molecules by a series of controlled reactions rather than in a single burst. If the energy were released all at once, as a large amount of heat, it could not be readily used to drive chemical reactions and would, in fact, damage the cell. To extract energy from organic compounds and store it in chemical form, organisms pass electrons from one compound to another through a series of oxidation-reduction reactions.

As noted earlier, a sequence of enzymatically catalyzed chemical reactions occurring in a cell is called a metabolic pathway. Below is a hypothetical pathway that converts starting material A to end-product F in a series of five steps:

$$NAD^+ \qquad NADH + H^+$$

$$A \xrightarrow{\;\;\;①\;\;\;} B \xrightarrow{②}$$

Starting material

$$ADP + ⓟ \quad ATP \qquad\qquad O_2$$

$$C \xrightarrow{\;\;\;③\;\;\;} D \rightleftharpoons_{④} E \xrightarrow{⑤} F$$

$$CO_2 \; H_2O \quad \text{End-product}$$

The first step is the conversion of molecule A to molecule B. The curved arrow indicates that the reduction of coenzyme NAD^+ to NADH is coupled to that reaction; the electrons and protons come from molecule A. Similarly, the two arrows in ❸ show a coupling of two reactions. As C is converted to D, ADP is converted to ATP; the energy needed comes from C as it transforms into D. The reaction converting D to E is readily reversible, as indicated by the double arrow. In the fifth step, the curved arrow leading from O_2 indicates that O_2 is a reactant in the reaction. The curved arrows leading to CO_2 and H_2O indicate that these substances are secondary products produced in the reaction, in addition to F, the end-product that (presumably) interests us the most. Secondary products such as CO_2 and H_2O shown here are sometimes called "by-products" or "waste products." Keep in mind that almost every reaction in a metabolic pathway is catalyzed by a specific enzyme; sometimes the name of the enzyme is printed near the arrow.

Carbohydrate Catabolism

Learning Objectives
- *Describe the chemical reactions of glycolysis.*
- *Explain the products of the Krebs cycle.*
- *Describe the chemiosmotic model for ATP generation.*

Most microorganisms oxidize carbohydrates as their primary source of cellular energy. **Carbohydrate catabolism,** the breakdown of carbohydrate molecules to produce energy, is therefore of great importance in cell metabolism. Glucose is the most common carbohydrate energy source used by cells. Microorganisms can also catabolize various lipids and proteins for energy production, as you will see later.

To produce energy from glucose, microorganisms use two general processes: *cellular respiration* and *fermentation*. (In discussing cellular respiration, we frequently refer to the process simply as respiration, but it should not be confused with breathing.) Both processes usually start with the same first step, glycolysis, but follow different subsequent pathways (Figure 5.11). Before examining the details of glycolysis, respiration, and fermentation, we will first look at a general overview of the processes.

As shown in Figure 5.11, the respiration of glucose typically occurs in three principal stages: glycolysis, the Krebs cycle, and the electron transport chain (system).

❶ Glycolysis is the oxidation of glucose to pyruvic acid with the production of some ATP and energy-containing NADH.

❷ The Krebs cycle is the oxidation of acetyl (a derivative of pyruvic acid) to carbon dioxide, with the production of some ATP, energy-containing NADH, and another reduced electron carrier, $FADH_2$.

❸ In the electron transport chain (system), NADH and $FADH_2$ are oxidized, contributing the electrons they have carried from the substrates to a "cascade" of oxidation-reduction reactions involving a series of additional electron carriers. Energy from these reactions is used to generate a considerable amount of ATP. In respiration, most of the ATP is generated in the third step.

Because respiration involves a long series of oxidation-reduction reactions, the entire process can be thought of as involving a flow of electrons from the energy-rich glucose molecule to the relatively energy-poor CO_2 and H_2O molecules. The coupling of ATP production to this flow is somewhat analogous to the production of electrical power by using energy from a flowing stream. Carrying the analogy further, you could imagine a stream flowing down a gentle slope during glycolysis and the Krebs cycle, supply-

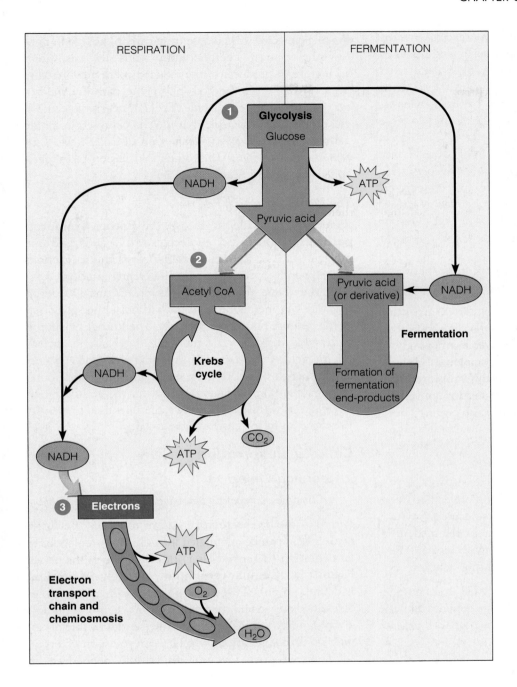

FIGURE 5.11 **An overview of respiration and fermentation.** ❶ Glycolysis produces ATP and reduces NAD⁺ to NADH while oxidizing glucose to pyruvic acid. In respiration, the pyruvic acid is converted into the first reactant in ❷ the Krebs cycle, which produces ATP and reduces NAD⁺ (and another electron carrier called FADH₂) while giving off CO_2. The NADH from both processes carries electrons to ❸ the electron transport chain, in which their energy is used to produce a great deal of ATP. In fermentation, the pyruvic acid and the electrons carried by NADH from glycolysis are incorporated into fermentation end-products. A small version of this figure will be included in figures throughout the chapter to indicate the relationships of different reactions to the overall processes.

■ What is the basic difference between respiration and fermentation?

ing energy to turn two old-fashioned waterwheels. Then the stream rushes down a steep slope in the electron transport chain, supplying energy for a large modern power plant. In a similar way, glycolysis and the Krebs cycle generate a small amount of ATP and also supply the electrons that generate a great deal of ATP at the electron transport chain stage.

Typically, the initial stage of fermentation is also glycolysis (Figure 5.11). However, once glycolysis has taken place, the pyruvic acid is converted into one or more different products, depending on the type of cell. These products might include alcohol (ethanol) and lactic acid. Unlike respiration, there is no Krebs cycle or electron transport chain in fermentation. Accordingly, the ATP yield, which comes only from glycolysis, is much lower.

Glycolysis

Glycolysis, the oxidation of glucose to pyruvic acid, is usually the first stage in carbohydrate catabolism. Most microorganisms use this pathway; in fact, it occurs in most living cells.

Glycolysis is also called the *Embden-Meyerhof pathway.* The word *glycolysis* means splitting of sugar, and this is exactly what happens. The enzymes of glycolysis catalyze the splitting of glucose, a six-carbon sugar, into two three-carbon sugars. These sugars are then oxidized, releasing energy, and their atoms are rearranged to form two molecules of pyruvic acid. During glycolysis NAD^+ is reduced to NADH, and there is a net production of two ATP molecules by substrate-level phosphorylation. Glycolysis does not require oxygen; it can occur whether oxygen is present or not. This pathway is a series of ten chemical reactions, each catalyzed by a different enzyme. The steps are outlined in Figure 5.12; see also Appendix C for a more detailed representation of glycolysis.

To summarize the process, glycolysis consists of two basic stages, a preparatory stage and an energy-conserving stage:

1. First, in the preparatory stage (steps ❶–❹ in Figure 5.12), two molecules of ATP are used as a six-carbon glucose molecule is phosphorylated, restructured, and split into two three-carbon compounds: glyceraldehyde 3-phosphate (GP) and dihydroxyacetone phosphate (DHAP). ❺ DHAP is readily converted to GP. (The reverse reaction may also occur.) The conversion of DHAP into GP means that from this point on in glycolysis, two molecules of GP are fed into the remaining chemical reactions.

2. In the energy-conserving stage (steps ❻–❿ in Figure 5.12), the two three-carbon molecules are oxidized in several steps to two molecules of pyruvic acid. In these reactions, two molecules of NAD^+ are reduced to NADH, and four molecules of ATP are formed by substrate-level phosphorylation.

Because two molecules of ATP were needed to get glycolysis started and four molecules of ATP are generated by the process, *there is a net gain of two molecules of ATP for each molecule of glucose that is oxidized.*

Alternatives to Glycolysis

Many bacteria have another pathway in addition to glycolysis for the oxidation of glucose. The most common alternative is the pentose phosphate pathway; another alternative is the Entner-Doudoroff pathway.

The Pentose Phosphate Pathway

The **pentose phosphate pathway** (or *hexose monophosphate shunt*) operates simultaneously with glycolysis and provides a means for the breakdown of five-carbon sugars (pentoses) as well as glucose (see the detailed figure in Appendix C). A key feature of this pathway is that it produces important intermediate pentoses used in the synthesis of (1) nucleic acids, (2) glucose from carbon dioxide in photosynthesis, and (3) certain amino acids. The pathway is an important producer of the reduced coenzyme NADPH from $NADP^+$. The pentose phosphate pathway yields a net gain of only one molecule of ATP for each molecule of glucose oxidized. Bacteria that use the pentose phosphate pathway include *Bacillus subtilis* (sub'til-us), *E. coli*, *Leuconostoc mesenteroides* (lü-kō-nos'tok mes-en-ter-oi'dēz), and *Enterococcus faecalis* (fē-kāl'is).

The Entner-Doudoroff Pathway

From each molecule of glucose, the **Entner-Doudoroff pathway** produces two molecules of NADPH and one molecule of ATP for use in cellular biosynthetic reactions (see Appendix C for a more detailed representation). Bacteria that have the enzymes for the Entner-Doudoroff pathway can metabolize glucose without either glycolysis or the pentose phosphate pathway. The Entner-Doudoroff pathway is found in some gram-negative bacteria, including *Rhizobium, Pseudomonas* (sū-dō-mō'nas), and *Agrobacterium* (ag-rō-bak-ti'rē-um); it is generally not found among gram-positive bacteria. Tests for the ability to oxidize glucose by this pathway are sometimes used to identify *Pseudomonas* in the clinical laboratory.

Cellular Respiration

Learning Objective

- *Compare and contrast aerobic and anaerobic respiration.*

After glucose has been broken down to pyruvic acid, the pyruvic acid can be channeled into the next step of either fermentation (described later) or cellular respiration (see Figure 5.11). **Cellular respiration,** or simply **respiration,** is defined as an ATP-generating process in which molecules are oxidized and the final electron acceptor is (almost always) an inorganic molecule. An essential feature of respiration is the operation of an electron transport chain.

There are two types of respiration, depending on whether an organism is an **aerobe,** which uses oxygen, or an **anaerobe,** which does not use oxygen and may even be killed by it. In **aerobic respiration,** the final electron acceptor is O_2; in **anaerobic respiration,** the final electron acceptor is an inorganic molecule other than molecular oxygen or, rarely, an organic molecule. First we will describe respiration as it typically occurs in an aerobic cell.

Aerobic Respiration

The Krebs Cycle The **Krebs cycle,** also called the *tricarboxylic acid (TCA) cycle* or *citric acid cycle,* is a series of biochemical reactions in which the large amount of potential chemical energy stored in acetyl CoA is released

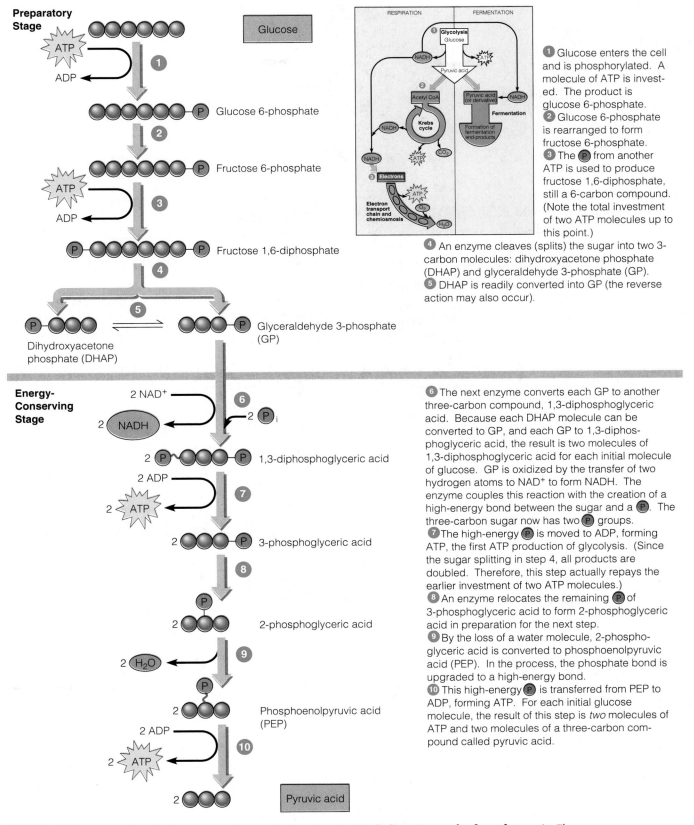

Preparatory Stage

ATP → ADP ①

Glucose

Glucose 6-phosphate

②

Fructose 6-phosphate

ATP → ADP ③

Fructose 1,6-diphosphate

④

⑤

Dihydroxyacetone phosphate (DHAP)

Glyceraldehyde 3-phosphate (GP)

Energy-Conserving Stage

2 NAD⁺

2 NADH

⑥

2 P$_i$

2 1,3-diphosphoglyceric acid

2 ADP

2 ATP ⑦

2 3-phosphoglyceric acid

⑧

2 2-phosphoglyceric acid

2 H₂O ⑨

2 Phosphoenolpyruvic acid (PEP)

2 ADP

2 ATP ⑩

2 Pyruvic acid

RESPIRATION   FERMENTATION

① Glycolysis Glucose

NADH

Pyruvic acid

② Acetyl CoA

Krebs cycle

NADH

③ Electrons

Electron transport chain and chemiosmosis

Pyruvic acid (or derivative) NADH

Fermentation

Formation of fermentation end-products

CO₂

O₂

H₂O

ATP

① Glucose enters the cell and is phosphorylated. A molecule of ATP is invested. The product is glucose 6-phosphate.
② Glucose 6-phosphate is rearranged to form fructose 6-phosphate.
③ The ℗ from another ATP is used to produce fructose 1,6-diphosphate, still a 6-carbon compound. (Note the total investment of two ATP molecules up to this point.)
④ An enzyme cleaves (splits) the sugar into two 3-carbon molecules: dihydroxyacetone phosphate (DHAP) and glyceraldehyde 3-phosphate (GP).
⑤ DHAP is readily converted into GP (the reverse action may also occur).

⑥ The next enzyme converts each GP to another three-carbon compound, 1,3-diphosphoglyceric acid. Because each DHAP molecule can be converted to GP, and each GP to 1,3-diphosphoglyceric acid, the result is two molecules of 1,3-diphosphoglyceric acid for each initial molecule of glucose. GP is oxidized by the transfer of two hydrogen atoms to NAD⁺ to form NADH. The enzyme couples this reaction with the creation of a high-energy bond between the sugar and a ℗. The three-carbon sugar now has two ℗ groups.
⑦ The high-energy ℗ is moved to ADP, forming ATP, the first ATP production of glycolysis. (Since the sugar splitting in step 4, all products are doubled. Therefore, this step actually repays the earlier investment of two ATP molecules.)
⑧ An enzyme relocates the remaining ℗ of 3-phosphoglyceric acid to form 2-phosphoglyceric acid in preparation for the next step.
⑨ By the loss of a water molecule, 2-phosphoglyceric acid is converted to phosphoenolpyruvic acid (PEP). In the process, the phosphate bond is upgraded to a high-energy bond.
⑩ This high-energy ℗ is transferred from PEP to ADP, forming ATP. For each initial glucose molecule, the result of this step is *two* molecules of ATP and two molecules of a three-carbon compound called pyruvic acid.

FIGURE 5.12 An outline of the reactions of glycolysis (Embden-Meyerhof pathway). The inset indicates the relationship of glycolysis to the overall processes of respiration and fermentation. A more detailed version of glycolysis is presented in Appendix C.

■ Glycolysis is the oxidation of glucose to pyruvic acid with the production of ATP and NADH.

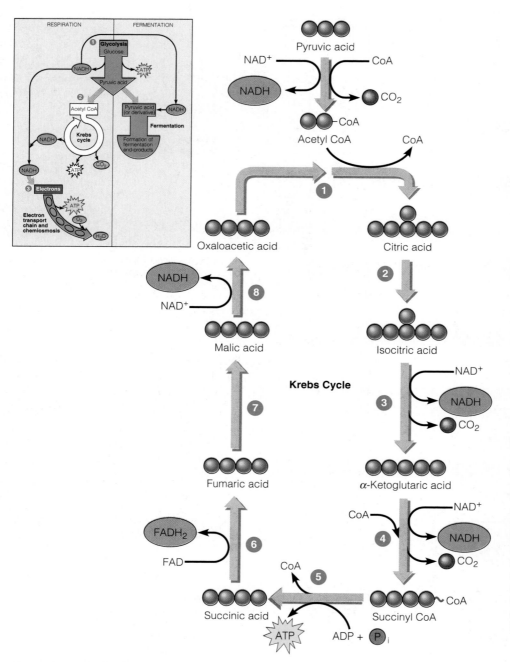

FIGURE 5.13 The Krebs cycle. The inset indicates the relationship of the Krebs cycle to the overall process of respiration.

- The Krebs cycle is the oxidation of acetyl groups to CO_2 with the production of ATP, NADH, and $FADH_2$.

① A turn of the cycle begins as enzymes strip off the CoA portion from acetyl CoA and combine the remaining two-carbon acetyl group with oxalo-acetic acid. Adding the acetyl group produces the six-carbon molecule citric acid.
② – ④ Oxidations generate NADH. Step 2 is a rearrange-ment. Steps 3 and 4 combine oxidations and decarboxy-lations to dispose of the two carbon atoms that came from oxaloacetic acid. The carbons are released as CO_2, and the oxidations generate NADH from NAD^+. During the second oxidation (step 4), CoA is added into the cycle, forming the com-pound succinyl CoA.
⑤ ATP is produced by sub-strate-level phosphorylation. CoA is removed from succinyl CoA, leaving succinic acid.
⑥ – ⑧ Enzymes rearrange chemical bonds, producing three different molecules before regenerating oxaloacetic acid. In step 6, an oxidation pro-duces $FADH_2$. In step 8, a final oxidation generates NADH and converts malic acid to oxalo-acetic acid, which is ready to enter another round of the Krebs cycle. See Appendix C.

step by step (see Figure 5.11). In this cycle, a series of oxi-dations and reductions transfer that potential energy, in the form of electrons, to electron carrier coenzymes, chiefly NAD^+. The pyruvic acid derivatives are oxidized; the coenzymes are reduced.

Pyruvic acid, the product of glycolysis, cannot enter the Krebs cycle directly. In a preparatory step, it must lose

one molecule of CO_2 and become a two-carbon compound (Figure 5.13, at top). This process is called **decarboxyla-tion.** The two-carbon compound, called an *acetyl group,* attaches to coenzyme A through a high-energy bond; the resulting complex is known as *acetyl coenzyme A (acetyl CoA).* During this reaction, pyruvic acid is also oxidized and NAD^+ is reduced to NADH.

Remember that the oxidation of one glucose molecule produces two molecules of pyruvic acid, so for each molecule of glucose, two molecules of CO_2 are released in this preparatory step, two molecules of NADH are produced, and two molecules of acetyl CoA are formed. Once the pyruvic acid has undergone decarboxylation and its derivative (the acetyl group) has attached to CoA, the resulting acetyl CoA is ready to enter the Krebs cycle.

As acetyl CoA enters the Krebs cycle, CoA detaches from the acetyl group. The two-carbon acetyl group combines with a four-carbon compound called oxaloacetic acid to form the six-carbon citric acid. This synthesis reaction requires energy, which is provided by the cleavage of the high-energy bond between the acetyl group and CoA. The formation of citric acid is thus the first step in the Krebs cycle. The major chemical reactions of this cycle are outlined in Figure 5.13; a more detailed representation of the Krebs cycle is provided in Appendix C. Keep in mind that each reaction is catalyzed by a specific enzyme.

The chemical reactions of the Krebs cycle fall into several general categories; one of these is decarboxylation. For example, in step ❸ isocitric acid, a six-carbon compound, is decarboxylated to the five-carbon compound called α-ketoglutaric acid. Another decarboxylation takes place in step ❹. Because one decarboxylation has taken place in the preparatory step and two in the Krebs cycle, all three carbon atoms in pyruvic acid are eventually released as CO_2 by the Krebs cycle. This represents the conversion to CO_2 of all six carbon atoms contained in the original glucose molecule.

Another general category of Krebs cycle chemical reactions is oxidation-reduction. For example, in step ❸, two hydrogen atoms are lost during the conversion of the six-carbon isocitric acid to a five-carbon compound. In other words, the six-carbon compound is oxidized. Hydrogen atoms are also released in the Krebs cycle in steps ❹, ❻, and ❽ and are picked up by the coenzymes NAD^+ and FAD. Because NAD^+ picks up two electrons but only one additional proton, its reduced form is represented as NADH; however, FAD picks up two complete hydrogen atoms and is reduced to $FADH_2$.

If we look at the Krebs cycle as a whole, we see that for every two molecules of acetyl CoA that enter the cycle, four molecules of CO_2 are liberated by decarboxylation, six molecules of NADH and two molecules of $FADH_2$ are produced by oxidation-reduction reactions, and two molecules of ATP are generated by substrate-level phosphorylation. Many of the intermediates in the Krebs cycle also play a role in other pathways, especially in amino acid biosynthesis (discussed later in the chapter).

The CO_2 produced in the Krebs cycle is ultimately liberated into the atmosphere as a gaseous by-product of aerobic respiration. (Humans produce CO_2 from the Krebs cycle in most cells of the body and discharge it through the lungs during exhalation.) The reduced coenzymes NADH and $FADH_2$ are the most important products of the Krebs cycle because they contain most of the energy originally stored in glucose. During the next phase of respiration, a series of reductions indirectly transfers the energy stored in those coenzymes to ATP. These reactions are collectively called the electron transport chain.

The Electron Transport Chain (System) An **electron transport chain (system)** consists of a sequence of carrier molecules that are capable of oxidation and reduction. As electrons are passed through the chain, there is a stepwise release of energy, which is used to drive the chemiosmotic generation of ATP, to be described shortly. The final oxidation is irreversible. In eukaryotic cells, the electron transport chain is contained in the inner membrane of mitochondria; in prokaryotic cells, it is found in the plasma membrane.

There are three classes of carrier molecules in electron transport chains. The first are **flavoproteins.** These proteins contain flavin, a coenzyme derived from riboflavin (vitamin B_2), and are capable of performing alternating oxidations and reductions. One important flavin coenzyme is flavin mononucleotide (FMN). The second class of carrier molecules are **cytochromes,** proteins with an iron-containing group (heme) capable of existing alternately as a reduced form (Fe^{2+}) and an oxidized form (Fe^{3+}). The cytochromes involved in electron transport chains include cytochrome b (cyt b), cytochrome c_1 (cyt c_1), cytochrome c (cyt c), cytochrome a (cyt a), and cytochrome a_3 (cyt a_3). The third class is known as **ubiquinones,** or **coenzyme Q,** symbolized Q; these are small nonprotein carriers.

The electron transport chains of bacteria are somewhat diverse, in that the particular carriers used by a bacterium and the order in which they function may differ from those of other bacteria and from those of eukaryotic mitochondrial systems. Even a single bacterium may have several types of electron transport chains. However, keep in mind that all electron transport chains achieve the same basic goal, that of releasing energy as electrons are transferred from higher-energy compounds to lower-energy compounds. Much is known about the electron transport chain in the mitochondria of eukaryotic cells, so this is the chain we will describe.

The first step in the mitochondrial electron transport chain involves the transfer of high-energy electrons from NADH to FMN, the first carrier in the chain (Figure 5.14). This transfer actually involves the passage of a hydrogen atom with two electrons to FMN, which then picks

FIGURE 5.14 An electron transport chain (system). The inset indicates the relationship of the electron transport chain to the overall process of respiration. In the mitochondrial electron transport chain shown, the electrons pass along the chain in a gradual and stepwise fashion, so energy is released in manageable quantities. To learn where ATP is formed, see Figure 5.16.

■ In the electron transport chain, NADH and FADH$_2$ are oxidized, and electron carriers are oxidized and reduced to produce ATP.

up an additional H$^+$ from the surrounding aqueous medium. As a result of the first transfer, NADH is oxidized to NAD$^+$, and FMN is reduced to FMNH$_2$. In the second step in the electron transport chain, FMNH$_2$ passes 2H$^+$ to the other side of the mitochondrial membrane (see Figure 5.16) and passes two electrons to Q. As a result, FMNH$_2$ is oxidized to FMN. Q also picks up an additional 2H$^+$ from the surrounding aqueous medium and releases it on the other side of the membrane.

The next part of the electron transport chain involves the cytochromes. Electrons are passed successively from Q to cyt b, cyt c_1, cyt c, cyt a, and cyt a_3. Each cytochrome in the chain is reduced as it picks up electrons and is oxidized as it gives up electrons. The last cytochrome, cyt a_3, passes its electrons to molecular oxygen (O$_2$), which becomes negatively charged and then picks up protons from the surrounding medium to form H$_2$O.

Notice that Figure 5.14 shows FADH$_2$, which is derived from the Krebs cycle, as another source of electrons. However, FADH$_2$ adds its electrons to the electron transport chain at a lower level than NADH. Because of this, the electron transport chain produces about one-third less energy for ATP generation when FADH$_2$ donates electrons than when NADH is involved.

An important feature of the electron transport chain is the presence of some carriers, such as FMN and Q, that ac-

cept and release protons as well as electrons, and other carriers, such as cytochromes, that transfer electrons only. Electron flow down the chain is accompanied at several points by the active transport (pumping) of protons from the matrix side of the inner mitochondrial membrane to the opposite side of the membrane. The result is a buildup of protons on one side of the membrane. Just as water behind a dam stores energy that can be used to generate electricity, this buildup of protons provides energy for the generation of ATP by the chemiosmotic mechanism.

The Chemiosmotic Mechanism of ATP Generation The mechanism of ATP synthesis using the electron transport chain is called **chemiosmosis.** To understand chemiosmosis, we need to recall several concepts that were introduced in Chapter 4 as part of the section on the movement of materials across membranes (page 89). Recall that substances diffuse passively across membranes from areas of high concentration to areas of low concentration; this diffusion yields energy. Recall also that the movement of substances *against* such a concentration gradient *requires* energy, and that in such an active transport of molecules or ions across biological membranes, the required energy is usually provided by ATP. In chemiosmosis, the energy released when a substance moves along a gradient is used to *synthesize* ATP. The "substance" in this case refers to

FIGURE 5.15 **Chemiosmosis.** An overview of the mechanism of chemiosmosis. The membrane shown could be a prokaryotic plasma membrane, a eukaryotic mitochondrial membrane, or a photosynthetic thylakoid. The numbered steps are described in the text.

■ What is the proton motive force?

protons. In respiration, chemiosmosis is responsible for most of the ATP that is generated. The steps of chemiosmosis are as follows (Figures 5.15 and 5.16):

❶ As energetic electrons from NADH (or chlorophyll) pass down the electron transport chain, some of the carriers in the chain pump—actively transport—protons across the membrane. Such carrier molecules are called *proton pumps*.

❷ The phospholipid membrane is normally impermeable to protons, so this one-directional pumping establishes a proton gradient (a difference in the concentrations of protons on the two sides of the membrane). In addition to a concentration gradient, there is an electrical charge gradient. The excess H^+ on one side of the membrane makes that side positively charged compared with the other side. The resulting electrochemical gradient has potential energy, called the *proton motive force*.

❸ The protons on the side of the membrane with the higher proton concentration can diffuse across the membrane only through special protein channels that contain an enzyme called *adenosine triphosphatase (ATP synthase)*. When this flow occurs, energy is released and is used by the enzyme to synthesize ATP from ADP and ℗ᵢ.

Figure 5.16 shows in detail how the electron transport chain operates in eukaryotes to drive the chemiosmotic mechanism. ❶ Energetic electrons from NADH pass

down the electron transport chains. Within the inner mitochondrial membrane, the carriers of the electron transport chain are organized into three complexes, with Q transporting electrons between the first and second complexes, and cyt *c* transporting them between the second and third complexes. ❷ Three components of the system pump protons: the first and second complexes and Q. At the end of the chain, electrons join with protons and oxygen (O_2) in the matrix fluid to form water (H_2O). Thus, O_2 is the final electron acceptor.

Both prokaryotic and eukaryotic cells use the chemiosmotic mechanism to generate energy for ATP production. However, in eukaryotic cells, ❸ the inner mitochondrial membrane contains the electron transport carriers and ATP synthase, whereas in most prokaryotic cells, the plasma membrane does so. An electron transport chain also operates in photophosphorylation and is located in the thylakoid membrane of cyanobacteria and eukaryotic chloroplasts.

A Summary of Aerobic Respiration The electron transport chain regenerates NAD^+ and FAD^+, which can be used again in glycolysis and the Krebs cycle. The various electron transfers in the electron transport chain generate about 34 molecules of ATP from each molecule of glucose oxidized: approximately three from each of the ten molecules of NADH (a total of 30), and approximately two from each of the two molecules of $FADH_2$ (a total of four). To arrive at the total number of ATP molecules generated

FIGURE 5.16 Electron transport and the chemiosmotic generation of ATP.
In the mitochondrial membrane, electron carriers are organized into three complexes, and protons (H^+) are pumped across the membrane at three points. In a eukaryotic cell, they are pumped from the matrix side of the mitochondrial membrane to the opposite side. In a prokaryotic cell, protons are pumped across the plasma membrane from the cytoplasmic side. The flow of electrons is shown in magenta.

■ How does chemiosmosis differ in eukaryotes and prokaryotes?

for each molecule of glucose, the 34 from chemiosmosis are added to those generated by oxidation in glycolysis and the Krebs cycle. In aerobic respiration among prokaryotes, a total of 38 molecules of ATP can be generated from one molecule of glucose. Note that four of those ATPs come from substrate-level phosphorylation in glycolysis and the Krebs cycle. Table 5.3 provides a detailed accounting of the ATP yield during prokaryotic aerobic respiration.

Aerobic respiration among eukaryotes produces a total of only 36 molecules of ATP. There are fewer ATPs than in prokaryotes because some energy is lost when electrons are shuttled across the mitochondrial membranes that separate glycolysis (in the cytoplasm) from the electron transport chain. No such separation exists in prokaryotes. We

can now summarize the overall reaction for aerobic respiration in prokaryotes as follows:

$$C_6H_{12}O_6 + 6O_2 + 38\ ADP + 38\ \textcircled{P}_i \longrightarrow$$
Glucose   Oxygen

$$6\ CO_2 + 6\ H_2O + 38\ ATP$$
Carbon   Water
dioxide

A summary of the various stages of aerobic respiration in prokaryotes is presented in Figure 5.17.

Anaerobic Respiration

In anaerobic respiration, the final electron acceptor is an inorganic substance other than oxygen (O_2). Some bacteria,

TABLE 5.3	ATP Yield During Prokaryotic Aerobic Respiration of One Glucose Molecule
Source	**ATP Yield (Method)**

Glycolysis
1. Oxidation of glucose to pyruvic acid   2 ATP (substrate-level phosphorylation)

2. Production of 2 NADH   6 ATP (oxidative phosphorylation in electron transport chain)

Preparatory Step
1. Formation of acetyl CoA produces 2 NADH   6 ATP (oxidative phosphorylation in electron transport chain)

Krebs Cycle
1. Oxidation of succinyl CoA to succinic acid   2 GTP (equivalent of ATP; substrate-level phosphorylation)

2. Production of 6 NADH   18 ATP (oxidative phosphorylation in electron transport chain)

3. Production of 2 FADH   <u>4</u> ATP (oxidative phosphorylation in electron transport chain)

Total: 38 ATP

Electron transport chain and chemiosmosis

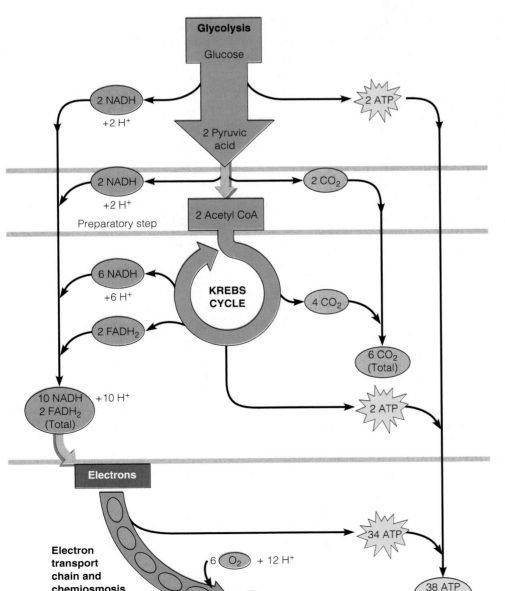

FIGURE 5.17 **A summary of aerobic respiration in prokaryotes.** Glucose is broken down completely to carbon dioxide and water, and ATP is generated. This process has three major phases: glycolysis, the Krebs cycle, and the electron transport chain. The preparatory step is between glycolysis and the Krebs cycle. The key event in aerobic respiration is that electrons are picked up from intermediates of glycolysis and the Krebs cycle by NAD^+ or FAD and are carried by NADH or $FADH_2$ to the electron transport chain. NADH is also produced in the conversion of pyruvic acid to acetyl CoA. Most of the ATP generated by aerobic respiration is made by the chemiosmotic mechanism during the electron transport chain phase; this is called oxidative phosphorylation.

■ How do aerobic and anaerobic respiration differ?

What Is Fermentation?

To many people, fermentation simply means the production of alcohol: Grains and fruits are fermented to produce beer and wine. If a food soured, you might say it was "off" or fermented. Here are some definitions of fermentation. They range from informal, general usage to more scientific definitions. Fermentation is

1. Any spoilage of food by microorganisms (general use).

2. Any process that produces alcoholic beverages or acidic dairy products (general use).

3. Any large-scale microbial process occurring with or without air (common definition used in industry).

4. Any energy-releasing metabolic process that takes place only under anaerobic conditions (becoming more scientific).

5. Any metabolic process that releases energy from a sugar or other organic molecule, does not require oxygen or an electron transport system, and uses an organic molecule as the final electron acceptor. (This is the definition we use in this book.)

such as *Pseudomonas* and *Bacillus*, can use a nitrate ion (NO_3^-) as a final electron acceptor; the nitrate ion is reduced to a nitrite ion (NO_2^-), nitrous oxide (N_2O), or nitrogen gas (N_2). Other bacteria, such as *Desulfovibrio* (dē-sul-fō-vib′rē-ō), use sulfate (SO_4^{2-}) as the final electron acceptor to form hydrogen sulfide (H_2S). Still other bacteria use carbonate (CO_3^{2-}) to form methane (CH_4). Anaerobic respiration by bacteria using nitrate and sulfate as final acceptors is essential for the nitrogen and sulfur cycles that occur in nature. The amount of ATP generated in anaerobic respiration varies with the organism and the pathway. Because only part of the Krebs cycle operates under anaerobic conditions, and since not all the carriers in the electron transport chain participate in anaerobic respiration, the ATP yield is never as high as in aerobic respiration. Accordingly, anaerobes tend to grow more slowly than aerobes.

Fermentation

Learning Objective

* Describe the chemical reactions of, and list some products of, fermentation.

After glucose has been broken down into pyruvic acid, the pyruvic acid can be completely broken down in respiration, as previously described, or it can be converted to an organic product in fermentation, whereupon NAD^+ and $NADP^+$ are regenerated and can enter another round of glycolysis (see Figure 5.11). **Fermentation** can be defined in several ways (see the box), but we define it here as a process that

1. releases energy from sugars or other organic molecules, such as amino acids, organic acids, purines, and pyrimidines;

2. does not require oxygen (but sometimes can occur in its presence);

3. does not require use of the Krebs cycle or an electron transport chain;

4. uses an organic molecule as the final electron acceptor;

5. produces only small amounts of ATP (only one or two ATP molecules for each molecule of starting material) because much of the original energy in glucose remains in the chemical bonds of the organic end-products, such as lactic acid or ethanol.

During fermentation, electrons are transferred (along with protons) from reduced coenzymes (NADH, NADPH) to pyruvic acid or its derivatives (Figure 5.18a). Those final electron acceptors are reduced to the end-products shown in Figure 5.18b. An essential function of the second stage of fermentation is to ensure a steady supply of NAD^+ and $NADP^+$ so that glycolysis can continue. In fermentation, ATP is generated only during glycolysis.

Various microorganisms can ferment various substrates; the end-products depend on the particular microorganism, the substrate, and the enzymes that are present and active. Chemical analyses of these end-products are useful in identifying microorganisms. We next consider two of the more important processes: lactic acid fermentation and alcohol fermentation.

Lactic Acid Fermentation

During glycolysis, which is the first phase of **lactic acid fermentation,** a molecule of glucose is oxidized to two molecules of pyruvic acid (Figure 5.19 on page 134; see also Figure 5.10). This oxidation generates the energy that

(a)

(b)

FIGURE 5.18 **Fermentation.** The inset indicates the relationship of fermentation to the overall energy-producing processes. **(a)** An overview of fermentation. The first step is glycolysis, the conversion of glucose to pyruvic acid. In the second step, the reduced coenzymes from glycolysis or its alternatives (NADH, NADPH) donate their electrons and hydrogen ions to pyruvic acid or a derivative to form a fermentation end-product. **(b)** End-products of various microbial fermentations.

■ In fermentation, ATP is generated only during glycolysis.

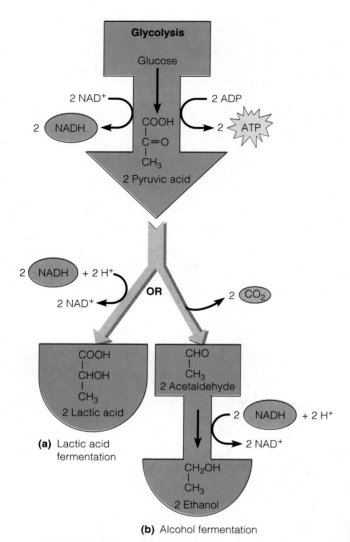

FIGURE 5.19 Types of fermentation.

■ What is the difference between homolactic and heterolactic fermentation?

Alcohol Fermentation

Alcohol fermentation also begins with the glycolysis of a molecule of glucose to yield two molecules of pyruvic acid and two molecules of ATP. In the next reaction, the two molecules of pyruvic acid are converted to two molecules of acetaldehyde and two molecules of CO_2 (Figure 5.19b). The two molecules of acetaldehyde are next reduced by two molecules of NADH to form two molecules of ethanol. Again, alcohol fermentation is a low-energy-yield process because most of the energy contained in the initial glucose molecule remains in the ethanol, the end-product.

Alcohol fermentation is carried out by a number of bacteria and yeasts. The ethanol and carbon dioxide produced by the yeast *Saccharomyces* (sak-ä-rō-mī′sēs) are waste products for yeast cells but are useful to humans. Ethanol made by yeasts is the alcohol in alcoholic beverages, and carbon dioxide made by yeasts causes bread dough to rise (see the box in Chapter 1, page 3).

Organisms that produce lactic acid as well as other acids or alcohols are known as **heterolactic** (or *heterofermentative*) and often use the pentose phosphate pathway.

Table 5.4 lists some of the various microbial fermentations used by industry to convert inexpensive raw materials into useful end-products. A summary comparison of aerobic respiration, anaerobic respiration, and fermentation is given in Table 5.5.

Lipid and Protein Catabolism

Learning Objective

- *Describe how lipids and proteins undergo catabolism.*

Our discussion of energy production has emphasized the oxidation of glucose, the main energy-supplying carbohydrate. However, microbes also oxidize lipids and proteins, and the oxidations of all these nutrients are related.

Recall that fats are lipids consisting of fatty acids and glycerol. Microbes produce extracellular enzymes called *lipases* that break fats down into their fatty acid and glycerol components. Each component is then metabolized separately (Figure 5.20). The Krebs cycle functions in the oxidation of glycerol and fatty acids. Many bacteria that hydrolyze fatty acids can use the same enzymes to degrade petroleum products. Although these bacteria are a nuisance when they grow in a fuel storage tank, they are beneficial when they grow in oil spills. Beta oxidation (the oxidation of fatty acids) of petroleum is discussed in the box in Chapter 2 (page 34).

Proteins are too large to pass unaided through plasma membranes. Microbes produce extracellular *proteases* and *peptidases*, enzymes that break down proteins into their

---

is used to form the two molecules of ATP. In the next step, the two molecules of pyruvic acid are reduced by two molecules of NADH to form two molecules of lactic acid (Figure 5.19a). Because lactic acid is the end-product of the reaction, it undergoes no further oxidation, and most of the energy produced by the reaction remains stored in the lactic acid. Thus, this fermentation yields only a small amount of energy.

Two important genera of lactic acid bacteria are *Streptococcus* and *Lactobacillus* (lak-tō-bä-sil′lus). Because these microbes produce only lactic acid, they are referred to as **homolactic** (or *homofermentative*). Lactic acid fermentation can result in food spoilage. However, the process can also produce yogurt from milk, sauerkraut from fresh cabbage, and pickles from cucumbers.

TABLE 5.4	Some Industrial Uses for Different Types of Fermentations		
Fermentation End-Product(s)	**Industrial or Commercial Use**	**Starting Material**	**Microorganism**
Ethanol	Beer	Malt extract	*Saccharomyces cerevisiae* (yeast, a fungus)
	Wine	Grape or other fruit juices	*Saccharomyces cerevisiae* var. *ellipsoideus*
	Fuel	Agricultural wastes	*Saccharomyces cerevisiae*
Acetic acid	Vinegar	Ethanol	*Acetobacter* (bacterium)
Lactic acid	Cheese, yogurt	Milk	*Lactobacillus, Streptococcus* (bacteria)
	Rye bread	Grain, sugar	*Lactobacillus bulgaricus* (bacterium)
	Sauerkraut	Cabbage	*Lactobacillus plantarum* (bacterium)
	Summer sausage	Meat	*Pediococcus* (bacterium)
Propionic acid and carbon dioxide	Swiss cheese	Lactic acid	*Propionibacterium freudenreichii* (bacterium)
Acetone and butanol	Pharmaceutical, industrial uses	Molasses	*Clostridium acetobutylicum* (bacterium)
Glycerol	Pharmaceutical, industrial uses	Molasses	*Saccharomyces cerevisiae*
Citric acid	Flavoring	Molasses	*Aspergillus* (fungus)
Methane	Fuel	Acetic acid	*Methanosarcina* (bacterium)
Sorbose	Vitamin C (ascorbic acid)	Sorbitol	*Gluconobacter*

TABLE 5.5	Aerobic Respiration, Anaerobic Respiration, and Fermentation Compared			
Energy-Producing Process	**Growth Conditions**	**Final Hydrogen (Electron) Acceptor**	**Type of Phosphorylation Used to Generate ATP**	**ATP Molecules Produced per Glucose Molecule**
Aerobic respiration	Aerobic	Molecular oxygen (O_2)	Substrate-level and oxidative	36 or 38*
Anaerobic respiration	Anaerobic	Usually an inorganic substance (such as NO_3^-, SO_4^{2-}, or CO_3^{2-}), but not molecular oxygen (O_2)	Substrate-level and oxidative	Variable (fewer than 38 but more than 2)
Fermentation	Aerobic or anaerobic	An organic molecule	Substrate-level	2

*In prokaryotic aerobic respiration, 38 ATP molecules are produced; in eukaryotic aerobic respiration, 36 ATP molecules are produced.

component amino acids, which can cross the membranes. However, before amino acids can be catabolized, they must be enzymatically converted to other substances that can enter the Krebs cycle. In one such conversion, called **deamination,** the amino group of an amino acid is removed and converted to an ammonium ion (NH_4^+),

which can be excreted from the cell. The remaining organic acid can enter the Krebs cycle. Other conversions involve **decarboxylation** (the removal of —COO) and **dehydrogenation.**

A summary of the interrelationships of carbohydrate, lipid, and protein catabolism is shown in Figure 5.21.

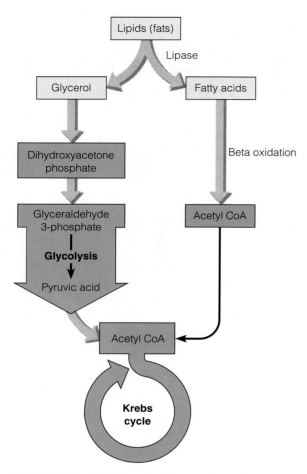

FIGURE 5.20 Lipid catabolism. Glycerol is converted into dihydroxyacetone phosphate (DHAP) and catabolized via glycolysis and the Krebs cycle. Fatty acids undergo beta oxidation, in which carbon fragments are split off two at a time to form acetyl CoA, which is catabolized via the Krebs cycle.

- Glycerol and fatty acids are metabolized along different metabolic pathways.

Biochemical Tests and Bacterial Identification

Learning Objective

- *Describe two examples of the use of biochemical tests to identify bacteria in the laboratory.*

Biochemical testing is frequently used to identify bacteria and yeasts because different species produce different enzymes. Such biochemical tests are designed to detect the presence of enzymes. One type of biochemical test is the detection of amino acid catabolizing enzymes involved in decarboxylation and dehydrogenation (discussed on page 135; Figure 5.22 on page 138). Another is a **fermentation test.** The test medium contains protein, a single carbohydrate, a pH indicator, and an inverted Durham tube to capture gas

(Figure 5.23). Bacteria inoculated into the tube can use the protein or carbohydrate as a carbon and energy source. If they catabolize the carbohydrate and produce acid, the pH indicator changes color. Some organisms produce gas as well as acid from carbohydrate catabolism. The presence of a bubble in the Durham tube indicates gas formation (Figure 5.23b–d on page 138). Another example of the use of biochemical tests is shown in Figure 10.8 on page 289.

Note that in some instances, the waste products of one microorganism can be used as a carbon and energy source by another species. *Acetobacter* (ä-sē-tō-bak′tėr) bacteria oxidize ethanol made by yeast. *Propionibacterium* (prō-pē-on-ē-bak-ti′re-um) can use lactic acid produced by other bacteria. Propionibacteria convert lactic acid to pyruvic acid in preparation for the Krebs cycle. During the Krebs cycle, propionic acid and CO_2 are made. The holes in Swiss cheese are formed by the accumulation of the CO_2 gas.

Photosynthesis

Learning Objectives

- *Compare and contrast cyclic and noncyclic photophosphorylation.*

- *Compare and contrast the light-dependent and light-independent reactions of photosynthesis.*

- *Compare and contrast oxidative phosphorylation and photophosphorylation.*

In all of the metabolic pathways just discussed, organisms obtain energy for cellular work by oxidizing organic compounds. But where do organisms obtain these organic compounds? Some, including animals and many microbes, feed on matter produced by other organisms. For example, bacteria may catabolize compounds from dead plants and animals or may obtain nourishment from a living host.

Other organisms synthesize complex organic compounds from simple inorganic substances. The major mechanism for such synthesis is a process called **photosynthesis,** which is used by plants and many microbes. Essentially, photosynthesis is the conversion of light energy from the sun into chemical energy. The chemical energy is then used to convert CO_2 from the atmosphere to more reduced carbon compounds, primarily sugars. The word *photosynthesis* summarizes the process: *photo* means light, and *synthesis* refers to the assembly of organic compounds. This synthesis of sugars by using carbon atoms from CO_2 gas is also called **carbon fixation.** Continuation of life as we know it on Earth depends on the recycling of carbon in this way. Cyanobacteria, algae, and green plants all contribute to this vital recycling with photosynthesis.

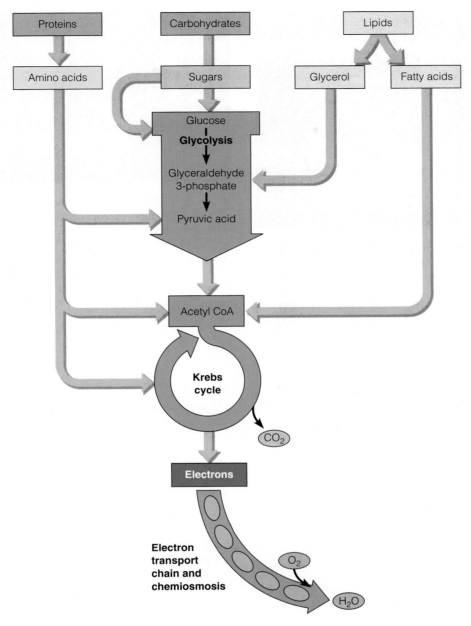

FIGURE 5.21 **Catabolism of various organic food molecules.** Proteins, carbohydrates, and lipids can all be sources of electrons and protons for respiration. These food molecules enter glycolysis or the Krebs cycle at various points.

■ Glycolysis and the Krebs cycle are catabolic funnels through which high-energy electrons from all kinds of organic molecules flow on their energy-releasing pathways.

Photosynthesis can be summarized as follows:

$$6 \, CO_2 + 12 \, H_2O + \text{Light energy} \longrightarrow$$
$$C_6H_{12}O_6 + 6 \, O_2 + 6 \, H_2O$$

In the course of photosynthesis, electrons are taken from the hydrogen atoms of water, an energy-poor molecule, and

incorporated into sugar, an energy-rich molecule. The energy boost is supplied by light energy, although indirectly.

Photosynthesis takes place in two stages. In the first stage, called the **light-dependent (light) reactions,** light energy is used to convert ADP and Ⓟ to ATP. In addition, in the predominant form of the light-dependent reactions,

FIGURE 5.22 Detecting amino acid catabolizing enzymes in the lab. Bacteria are inoculated in tubes containing glucose, a pH indicator, and a specific amino acid. **(a)** The pH indicator turns to yellow when bacteria produce acid from glucose. **(b)** Alkaline products from decarboxylation turn the indicator to purple.

■ What is decarboxylation?

FIGURE 5.23 A fermentation test. (a) An uninoculated fermentation tube containing the carbohydrate mannitol. **(b)** *Staphylococcus epidermidis* grew on the protein but did not use the carbohydrate. This organism is described as mannitol −. **(c)** *Staphylococcus aureus* produced acid but not gas. This species is mannitol +. **(d)** *Escherichia coli* is also mannitol + and produced acid and gas from mannitol. The gas is trapped in the inverted Durham tube.

■ What is the importance of biochemical testing?

the electron carrier $NADP^+$ is reduced to NADPH. The coenzyme NADPH, like NADH, is an energy-rich carrier of electrons. In the second stage, the **light-independent (dark) reactions,** these electrons are used along with energy from ATP to reduce CO_2 to sugar.

The Light-Dependent Reactions: Photophosphorylation

Photophosphorylation is one of the three ways ATP is formed, and it occurs only in photosynthetic cells. In this mechanism, light energy is absorbed by chlorophyll molecules in the photosynthetic cell, exciting some of the molecules' electrons. The chlorophyll principally used by green plants, algae, and cyanobacteria is *chlorophyll a*. It is located in the membranous thylakoids of chloroplasts in algae and green plants (see Figure 4.28 on page 104) and in the thylakoids found in the photosynthetic structures of cyanobacteria. Other bacteria use *bacteriochlorophylls*.

The excited electrons jump from the chlorophyll to the first of a series of carrier molecules, an electron transport chain similar to that used in respiration. As electrons are passed along the series of carriers, protons are pumped across the membrane, and ADP is converted to ATP by chemiosmosis. In **cyclic photophosphorylation** the electron eventually returns to chlorophyll (Figure 5.24a). In **noncyclic photophosphorylation,** which is the more common process, the electrons released from chlorophyll do not return to chlorophyll but become incorporated into NADPH (Figure 5.24b). The electrons lost from chloro-

phyll are replaced by electrons from a reducing substance other than H_2O or another oxidizable compound, such as hydrogen sulfide (H_2S). To summarize: The products of noncyclic photophosphorylation are ATP (formed by chemiosmosis using energy released in an electron transport chain), O_2 (from water molecules), and NADPH (in which the hydrogen electrons and protons were derived ultimately from water).

The Light-Independent Reactions: The Calvin-Benson Cycle

The light-independent (dark) reactions are so named because no light is directly required for them to occur. They include a complex cyclic pathway called the **Calvin-Benson cycle,** in which CO_2 is "fixed"—that is, used to synthesize sugars (Figure 5.25 on page 140; see also Appendix C).

A Summary of Energy Production Mechanisms

Learning Objective

• Write a sentence to summarize energy production in cells.

In the living world, energy passes from one organism to another in the form of the potential energy contained in the bonds of chemical compounds. Organisms obtain the energy from oxidation reactions. To obtain energy in a usable form, a cell must have an electron (or hydrogen) donor,

(a) Cyclic photophosphorylation

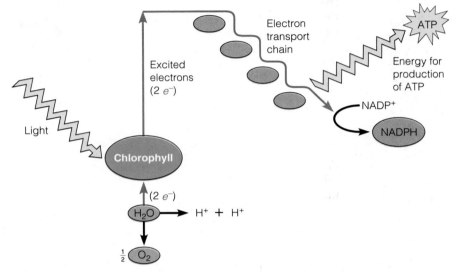

(b) Noncyclic photophosphorylation

FIGURE 5.24 **Photophosphorylation. (a)** In cyclic photophosphorylation, electrons released from chlorophyll by light return to chlorophyll after passage along the electron transport chain. The energy from electron transfer is converted to ATP. **(b)** In noncyclic photophosphorylation, electrons released from chlorophyll are replaced by electrons from water. The chlorophyll electrons are passed along the electron transport chain to the electron acceptor $NADP^-$. $NADP^+$ combines with electrons and with hydrogen ions from water, forming NADPH.

■ How are oxidative phosphorylation and photophosphorylation similar?

which serves as an initial energy source within the cell. Electron donors can be as diverse as photosynthetic pigments, glucose or other organic compounds, elemental sulfur, ammonia, or hydrogen gas (Figure 5.26 on page 141). Next, electrons removed from the chemical energy sources are transferred to electron carriers, such as the coenzymes NAD^+, $NADP^+$, and FAD (see Figure 5.26). This transfer is an oxidation-reduction reaction; the initial energy source is oxidized as this first electron carrier is reduced. During this phase, some ATP is produced. In the third stage, electrons are transferred from electron carriers to their final electron acceptors in further oxidation-reduction reactions (see Figure 5.26), producing more ATP.

In aerobic respiration, oxygen (O_2) serves as the final electron acceptor. In anaerobic respiration, inorganic substances other than oxygen, such as nitrate ions (NO_3^-) or sulfate ions (SO_4^{2-}), serve as the final electron acceptors. In fermentation, organic compounds serve as the final electron acceptors. In aerobic and anaerobic respiration, a series of electron carriers called an electron transport chain releases energy that is used by the mechanism of chemiosmosis to synthesize ATP. Regardless of their energy sources, all organisms use similar oxidation-reduction reactions to transfer electrons, and similar mechanisms to use the energy released to produce ATP.

Metabolic Diversity Among Organisms

Learning Objective

- Categorize the various nutritional patterns among organisms according to carbon source and mechanisms of carbohydrate catabolism and ATP generation.

We have looked in detail at some of the energy-generating metabolic pathways that are used by animals and plants, as well as by many microbes. Microbes are distinguished by their great metabolic diversity, however, and some can sustain themselves on inorganic substances by using pathways that are unavailable to either plants or animals. All

FIGURE 5.25 **A simplified version of the Calvin-Benson cycle.** This diagram shows three turns of the cycle, in which three molecules of CO_2 are fixed and one molecule of glyceraldehyde 3-phosphate is produced and leaves the cycle. Two molecules of glyceraldehyde 3-phosphate are needed to make one molecule of glucose. Therefore, the cycle must turn six times for each glucose molecule produced, requiring a total investment of 6 molecules of CO_2, 18 molecules of ATP, and 12 molecules of NADPH. A more detailed version of this cycle is presented in Appendix C.

■ In the Calvin-Benson cycle, CO_2 is used to synthesize sugars.

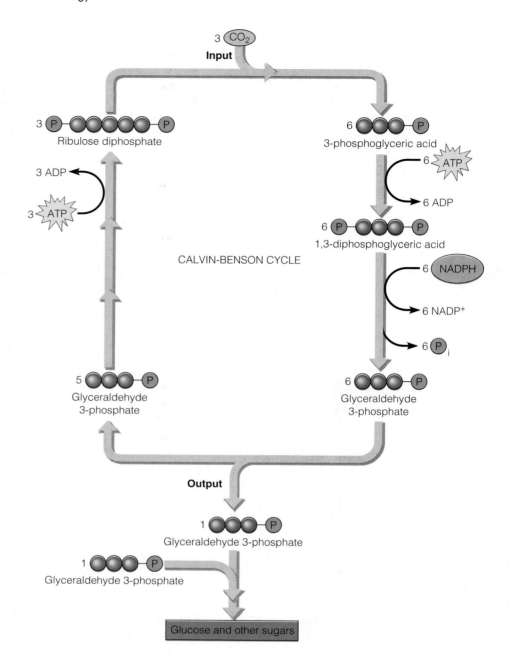

organisms, including microbes, can be classified metabolically according to their *nutritional pattern*—their source of energy and their source of carbon.

First considering the energy source, we can generally classify organisms as phototrophs or chemotrophs. **Phototrophs** use light as their primary energy source, whereas **chemotrophs** depend on oxidation-reduction reactions of inorganic or organic compounds for energy. For their principal carbon source, **autotrophs** (self-feeders) use carbon dioxide, and **heterotrophs** (feeders on others) require an organic carbon source. Autotrophs are also referred to as *lithotrophs* (rock eating), and heterotrophs are also referred to as *organotrophs*.

If we combine the energy and carbon sources, we derive the following nutritional classifications for organisms: *photoautotrophs*, *photoheterotrophs*, *chemoautotrophs*, and *chemoheterotrophs* (Figure 5.27 on page 142). Almost all of the medically important microorganisms discussed in this book are chemoheterotrophs. Typically, infectious organisms catabolize substances obtained from the host.

Photoautotrophs

Photoautotrophs use light as a source of energy and carbon dioxide as their chief source of carbon. They include photosynthetic bacteria (green and purple bacteria and cyanobacteria), algae, and green plants. In the photosynthetic reactions of cyanobacteria, algae, and green plants, the hydrogen atoms of water are used to reduce carbon dioxide, and oxygen gas is given off. Because this photosynthetic process produces O_2, it is sometimes called **oxygenic.**

In addition to the cyanobacteria (see Figure 11.16 on page 322), there are several other families of photosynthetic prokaryotes. Each is classified according to the way it reduces CO_2. These bacteria cannot use H_2O to reduce CO_2 and cannot carry on photosynthesis when oxygen is present (they must have an anaerobic environment). Consequently, their photosynthetic process does not produce O_2 and is called **anoxygenic.** The anoxygenic photoautotrophs are the green and purple bacteria. The **green bacteria,** such as *Chlorobium* (klô-rō′bē-um), use sulfur (S), sulfur compounds (such as hydrogen sulfide, H_2S), or hydrogen gas (H_2) to reduce carbon dioxide and form organic compounds. Applying the energy from light and the appropriate enzymes, these bacteria oxidize sulfide (S^{2-}) or sulfur (S) to sulfate (SO_4^{2-}), or hydrogen gas to water (H_2O). The **purple bacteria,** such as *Chromatium* (krō-mā′tē-um), also use sulfur, sulfur compounds, or hydrogen gas to reduce carbon dioxide. They are distinguished from the green bacteria by their type of chlorophyll, location of stored sulfur, and ribosomal RNA.

The chlorophylls used by these photosynthetic bacteria are called *bacteriochlorophylls,* and they absorb light at longer wavelengths than that absorbed by chlorophyll *a.* Bacteriochlorophylls of green sulfur bacteria are found in vesicles called *chlorosomes* (or *chlorobium vesicles*) underlying and attached to the plasma membrane. In the purple sulfur bacteria, the bacteriochlorophylls are located in invaginations of the plasma membrane (*intracytoplasmic membranes*).

Several characteristics that distinguish eukaryotic photosynthesis from prokaryotic photosynthesis are presented in Table 5.6. See the box on page 143 for a discussion of an exceptional photosynthetic system that exists in *Halobacterium.* The system does not use chlorophyll.

Photoheterotrophs

Photoheterotrophs use light as a source of energy but cannot convert carbon dioxide to sugar; rather, they use organic compounds, such as alcohols, fatty acids, other organic acids, and carbohydrates, as sources of carbon. They are anoxygenic. The **green nonsulfur bacteria,** such as *Chloroflexus* (klô-rō-flex′us), and **purple nonsulfur bacteria,** such as *Rhodopseudomonas* (rō-dō-sū-dō-mō′nas), are photoheterotrophs.

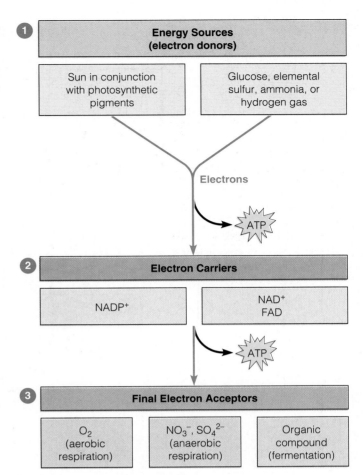

FIGURE 5.26 **Requirements of ATP production.** The production of ATP requires ❶ an energy source (electron donor), ❷ the transfer of electrons to an electron carrier during an oxidation-reduction reaction, and ❸ the transfer of electrons to a final electron acceptor.

Chemoautotrophs

Chemoautotrophs use the electrons from reduced inorganic compounds as a source of energy, and they use CO_2 as their principal source of carbon. Inorganic sources of energy for these organisms include hydrogen sulfide (H_2S) for *Beggiatoa* (bej-jē-ä-tō′ä); elemental sulfur (S) for *Thiobacillus thiooxidans;* ammonia (NH_3) for *Nitrosomonas* (nī-trō-sō-mō′näs); nitrite ions (NO_2^-) for *Nitrobacter* (nī-trō-bak′tėr); hydrogen gas (H_2) for *Hydrogenomonas* (hī-drō-je-nō-mō′näs); ferrous iron (Fe^{2+}) for *Thiobacillus ferrooxidans;* and carbon monoxide (CO) for *Pseudomonas carboxydohydrogena.* The energy derived from the oxidation of these inorganic compounds is eventually stored in ATP, which is produced by oxidative phosphorylation.

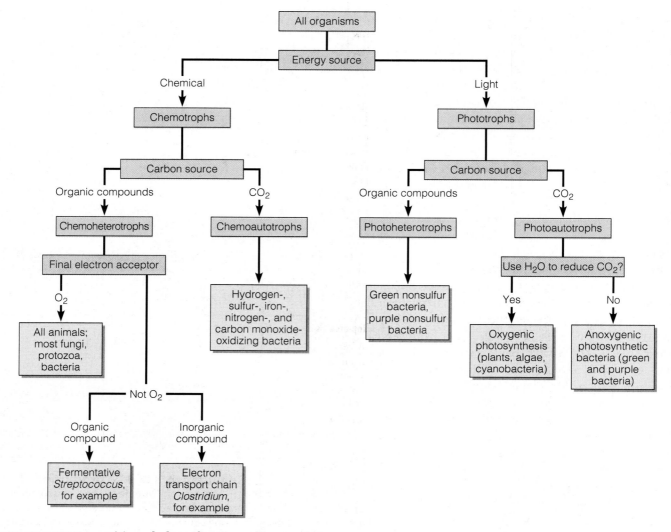

FIGURE 5.27 A nutritional classification of organisms.

■ What is the basic difference between chemotrophs and phototrophs?

TABLE 5.6	**Photosynthesis Compared in Selected Eukaryotes and Prokaryotes**			
Characteristic	**Eukaryotes**		**Prokaryotes**	
	Algae, Plants	**Cyanobacteria**	**Green Bacteria**	**Purple Bacteria**
Substance that reduces CO_2	H atoms of H_2O	H atoms of H_2O	Sulfur, sulfur compounds, H_2 gas	Sulfur, sulfur compounds, H_2 gas
Oxygen production	Oxygenic	Oxygenic (and anoxygenic)	Anoxygenic	Anoxygenic
Type of chlorophyll	Chlorophyll *a*	Chlorophyll *a*	Bacteriochlorophyll *a*	Bacteriochlorophyll *a* or *b*
Site of photosynthesis	Chloroplasts with thylakoids	Thylakoids	Chlorosomes	Intracytoplasmic membrane
Environment	Aerobic	Aerobic (and anaerobic)	Anaerobic	Anaerobic

Bacteria Make a Faster, Smarter Computer

An interesting member of the Archaea, *Halobacterium*, lives where very little else can grow. This bacterium is found in salt lakes, salt licks on ranches, salt flats, or any other environment with a concentration of salt that is five to seven times that of the ocean. Halobacteria are easy to detect because they turn their environment purple (see the photograph). These bacteria cannot ferment carbohydrates and do not contain chlorophyll, so it was assumed that all their energy comes from oxidative phosphorylation. The exciting discovery of a new system of photophosphorylation arose through the study of the plasma membrane of *Halobacterium halobium*.

Researchers found that the plasma membrane of *H. halobium* fragments into two fractions (red and purple) when the cell is broken down and its components are sorted. The red fraction, which constitutes most of the membrane, contains cytochromes, flavoproteins, and other parts of the electron transport chain, which carries out oxidative phosphorylation. The purple fraction is more interesting. This purple membrane occurs in distinct patches of hexagonal lattices within the plasma membrane. The purple color comes from a protein that makes up 75% of the purple membrane. This protein is similar to the retinal pigment in the rod cells of the human eye, rhodopsin, so the protein was named bacteriorhodopsin. At the time it was discovered, its function was not known.

Further studies showed that *H. halobium* can grow in the presence of either light or oxygen but cannot grow when neither is present.

This unexpected result suggested that *Halobacterium* can obtain energy by using either of two systems, one that operates in the presence of oxygen (oxidative phosphorylation) and one that operates in the presence of light (some kind of photophosphorylation). The rate of ATP synthesis by *H. halobium* was found to be highest when the cells receive light that is between 550 and 600 nm in wavelength; this range exactly corresponds to the absorption spectrum of bacteriorhodopsin.

Researchers hypothesized that bacteriorhodopsin, like chlorophyll-containing systems, acts as a proton pump to create a proton gradient across a cell membrane; in this case, the gradient is created across the purple membrane. The proton gradient can do cellular work—can drive the synthesis of ATP or transport solutes.

Halobacterium Replaces the Silicon Chip

At Syracuse University's Center of Molecular Electronics, Robert Birge grows *Halobacterium* for 5 days in 5-liter batches and then extracts bacteriorhodopsin from the cells for a novel use. Birge has developed a computer chip made of a thin layer of bacteriorhodopsin.

Conventional computers store information on thin wafers of silicon. Computers process information by "reading" a series of zeros and ones produced as electrons flow through switches etched in the silicon. Electrons passing through a switch represent a one; a switch that stops the electron flow represents a zero. However, silicon can't hold enough information or process information fast enough for such applications as artificial intelligence or robot vision.

In contrast, the bacteriorhodopsin chip will be able to store more information than a silicon chip and process the information faster, more like a human brain. The bacteriorhodopsin chip works with light, which of course moves at the speed of light, much faster than the flow of electrons. Green light causes the protein to fold; a folded protein is read as a one, whereas an unfolded protein represents a zero. Laser light is used to "see" the configuration of the protein.

At present, the protein chip needs to be stored at −4°C to maintain its structure, but Birge and his coworkers are hopeful they will solve this problem. Russian scientists have made a protein processor for military radar, and the U.S. military is apparently using the protein chips in their combat planes. If such a plane crashes, the cooling system will go off, thus destroying the chip and keeping classified information from being stolen. Eventually, these smaller, faster, and higher-capacity chips will probably make it possible to develop computers that perform functions closer to human intelligence, such as acting as eyes for blind people.

Halobacterium (purple color) grows in the high-salt concentration of solar evaporation ponds used for manufacturing salt around San Francisco Bay.

Chemoheterotrophs

When we discuss photoautotrophs, photoheterotrophs, and chemoautotrophs, it is easy to categorize the energy source and carbon source because they occur as separate entities. However, in chemoheterotrophs, the distinction is not as clear because the energy source and carbon source are usually the same organic compound—glucose, for example. **Chemoheterotrophs** specifically use the electrons from hydrogen atoms in organic compounds as their energy source.

Heterotrophs are further classified according to their source of organic molecules. **Saprophytes** live on dead organic matter, and **parasites** derive nutrients from a living host. Most bacteria, and all fungi, protozoa, and animals, are chemoheterotrophs.

Bacteria and fungi can use a wide variety of organic compounds for carbon and energy sources. This is why they can live in diverse environments. Understanding microbial diversity is scientifically interesting and economically important. In some situations microbial growth is undesirable, such as when rubber-degrading bacteria destroy a gasket or shoe sole. However, these same bacteria might be beneficial if they decomposed discarded rubber products such as tires. *Rhodococcus erythropolis* (rō-dō-kok′kus er-i-throp′ō-lis) is widely distributed in soil and can cause disease in humans and other animals. This same species is able to replace sulfur atoms in petroleum with atoms of oxygen. A Texas company is currently using *R. erythropolis* to produce desulfurized oil.

*   *   *

We will next consider how cells use ATP pathways for the synthesis of organic compounds such as carbohydrates, lipids, proteins, and nucleic acids.

Metabolic Pathways of Energy Use

Learning Objective

*   Describe the major types of anabolism and their relationship to catabolism.

Up to now we have been considering energy production. Through the oxidation of organic molecules, organisms produce energy by aerobic respiration, anaerobic respiration, and fermentation. Much of this energy is given off as heat. The complete metabolic oxidation of glucose to carbon dioxide and water is considered a very efficient process, but about 45% of the energy of glucose is lost as heat. Cells use the remaining energy, which is trapped in the bonds of ATP, in a variety of ways. Microbes use ATP to provide energy for the transport of substances across plasma membranes—the process called active transport that we discussed in Chapter 4. Microbes also use some of their energy for flagellar motion (also discussed in Chapter 4). Most of the ATP, however, is used in the production of new cellular components. This production is a continuous process in cells, and, in general, is faster in prokaryotic cells than in eukaryotic cells.

Autotrophs build their organic compounds by fixing carbon dioxide in the Calvin-Benson cycle (see Figure 5.25). This requires both energy (ATP) and electrons (from the oxidation of NADPH). Heterotrophs, by contrast, must have a ready source of organic compounds for biosynthesis—the production of needed cellular components, usually from simpler molecules. The cells use these compounds as both the carbon source and the energy source. We will next consider the biosynthesis of a few representative classes of biological molecules: carbohydrates, lipids, amino acids, purines, and pyrimidines. As we do so, keep in mind that synthesis reactions require a net input of energy.

Polysaccharide Biosynthesis

Microorganisms synthesize sugars and polysaccharides. The carbon atoms required to synthesize glucose are derived from the intermediates produced during processes such as glycolysis and the Krebs cycle and from lipids or amino acids. After synthesizing glucose (or other simple sugars), bacteria may assemble it into more complex polysaccharides such as glycogen. For bacteria to build glucose into glycogen, glucose units must be phosphorylated and linked. The product of glucose phosphorylation is glucose 6-phosphate. Such a process involves the expenditure of energy, usually in the form of ATP. In order for bacteria to synthesize glycogen, a molecule of ATP is added to glucose 6-phosphate to form *adenosine diphosphoglucose (ADPG)* (Figure 5.28). Once ADPG is synthesized, it is linked with similar units to form glycogen.

Using a nucleotide called uridine triphosphate (UTP) as a source of energy and glucose 6-phosphate, animals synthesize glycogen (and many other carbohydrates) from *uridine diphosphoglucose, UDPG* (see Figure 5.28). A compound related to UDPG, called *UDP-N-acetylglucosamine (UDPNAc)*, is a key starting material in the biosynthesis of peptidoglycan, the substance that forms bacterial cell walls. UDPNAc is formed from fructose 6-phosphate, and the reaction also uses UTP.

Lipid Biosynthesis

Because lipids vary considerably in chemical composition, they are synthesized by a variety of routes. Cells synthesize fats by joining glycerol and fatty acids. The glycerol

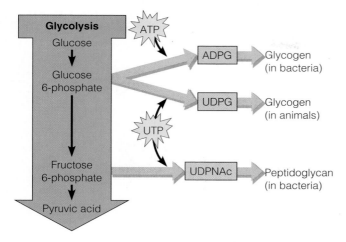

FIGURE 5.28 The biosynthesis of polysaccharides.

■ How are polysaccharides used in cells?

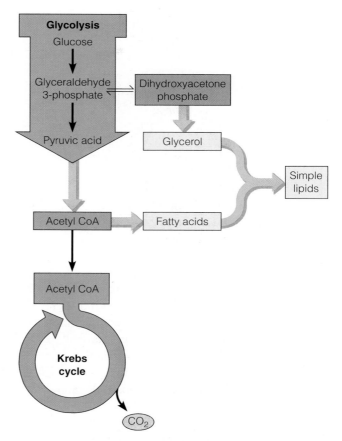

FIGURE 5.29 The biosynthesis of simple lipids.

■ What is the primary use of lipids in cells?

portion of the fat is derived from dihydroxyacetone phosphate, an intermediate formed during glycolysis. Fatty acids, which are long-chain hydrocarbons (hydrogen linked to carbon), are built up when two-carbon fragments of acetyl CoA are successively added to each other (Figure 5.29). As with polysaccharide synthesis, the building units of fats and other lipids are linked via dehydration synthesis reactions that require energy, not always in the form of ATP.

The most important role of lipids is as structural components of biological membranes, and most membrane lipids are phospholipids. A lipid of a very different structure, cholesterol, is also found in plasma membranes of eukaryotic cells. Waxes are lipids that are important components of the cell wall of acid-fast bacteria. Other lipids, such as carotenoids, provide the red, orange, and yellow pigments of some microorganisms. Some lipids form portions of chlorophyll molecules. Lipids also function in energy storage. Recall that the breakdown products of lipids after biological oxidation feed into the Krebs cycle.

Amino Acid and Protein Biosynthesis

Amino acids are required for protein biosynthesis. Some microbes, such as *E. coli,* contain the enzymes necessary to use starting materials, such as glucose and inorganic salts, for the synthesis of all the amino acids they need. Organisms with the necessary enzymes can synthesize all amino acids directly or indirectly from intermediates of carbohydrate metabolism (Figure 5.30a). Other microbes require that the environment provide some preformed amino acids.

One important source of the *precursors* (intermediates) used in amino acid synthesis is the Krebs cycle. Adding an amine group to pyruvic acid or to an appropriate organic acid of the Krebs cycle converts the acid into an amino acid. This process is called **amination.** If the amine group comes from a preexisting amino acid, the process is called **transamination** (Figure 5.30b).

Most amino acids within cells are destined to be building blocks for protein synthesis. Proteins play major roles in the cell as enzymes, structural components, and toxins, to name just a few uses. The joining of amino acids to form proteins involves dehydration synthesis and requires energy in the form of ATP. The mechanism of protein synthesis involves genes and is discussed in Chapter 8.

Purine and Pyrimidine Biosynthesis

Recall from Chapter 2 that the informational molecules DNA and RNA consist of repeating units called *nucleotides,* each of which consists of a purine or pyrimidine, a pentose (five-carbon sugar), and a phosphate group. The

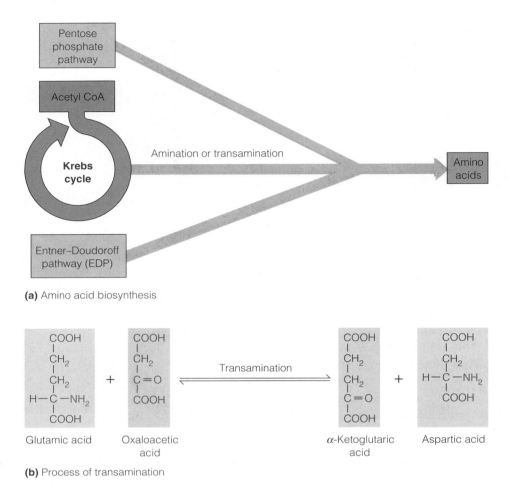

(a) Amino acid biosynthesis

(b) Process of transamination

FIGURE 5.30 **The biosynthesis of amino acids. (a)** Pathways of amino acid biosynthesis through amination or transamination of intermediates of carbohydrate metabolism from the Krebs cycle, pentose phosphate pathway, and Entner-Doudoroff pathway. **(b)** Transamination, a process by which new amino acids are made with the amine groups from old amino acids. Glutamic acid and aspartic acid are both amino acids; the other two compounds are intermediates in the Krebs cycle.

■ What is the function of amino acids in cells?

five-carbon sugars of nucleotides are derived from either the pentose phosphate pathway or the Entner-Doudoroff pathway. Certain amino acids—aspartic acid, glycine, and glutamine—made from intermediates produced during glycolysis and in the Krebs cycle participate in the biosyntheses of purines and pyrimidines (Figure 5.31). The carbon and nitrogen atoms derived from these amino acids form the purine and pyrimidine rings, and the energy for synthesis is provided by ATP. DNA contains all the information necessary to determine the specific structures and functions of cells. Both RNA and DNA are required for protein synthesis. In addition, such nucleotides as ATP, NAD^+, and $NADP^+$ assume roles in stimulating and inhibiting the rate of cellular metabolism. The synthesis of DNA and RNA from nucleotides will be discussed in Chapter 8.

The Integration of Metabolism

Learning Objective

• Define amphibolic pathways.

We have seen thus far that the metabolic processes of microbes produce energy from light, inorganic compounds, and organic compounds. Reactions also occur in which energy is used for biosynthesis. With such a variety of activity, you might imagine that anabolic and catabolic reactions occur independently of each other in space and time. Actually, anabolic and catabolic reactions are joined through a group of common intermediates (identified as key intermediates in Figure 5.32). Both anabolic and catabolic reactions also share some metabolic pathways, such

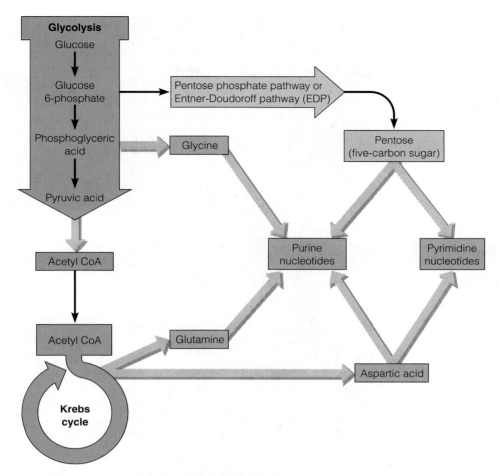

FIGURE 5.31 **The biosynthesis of purine and pyrimidine nucleotides.**

■ What are the functions of nucleotides in a cell?

as the Krebs cycle. For example, reactions in the Krebs cycle not only participate in the oxidation of glucose but also produce intermediates that can be converted to amino acids. Metabolic pathways that function in both anabolism and catabolism are called **amphibolic pathways,** meaning that they are dual-purpose.

Amphibolic pathways bridge the reactions that lead to the breakdown and synthesis of carbohydrates, lipids, proteins, and nucleotides. Such pathways enable simultaneous reactions to occur in which the breakdown product formed in one reaction is used in another reaction to synthesize a different compound, and vice versa. Because various intermediates are common to both anabolic and catabolic reactions, mechanisms exist that regulate synthesis and breakdown pathways and allow these reactions to occur simultaneously. One such mechanism involves the use of different coenzymes for opposite pathways. For example, NAD^+ is involved in catabolic reactions, whereas $NADP^+$ is involved in anabolic reactions. Enzymes can also coordinate anabolic and catabolic reactions by accelerating or inhibiting the rates of biochemical reactions.

The energy stores of a cell can also affect the rates of biochemical reactions. For example, if ATP begins to accumulate, an enzyme shuts down glycolysis; this control helps to synchronize the rates of glycolysis and the Krebs cycle. Thus, if citric acid consumption increases, either because of a demand for more ATP or because anabolic pathways are draining off intermediates of the citric acid cycle, glycolysis accelerates and meets the demand.

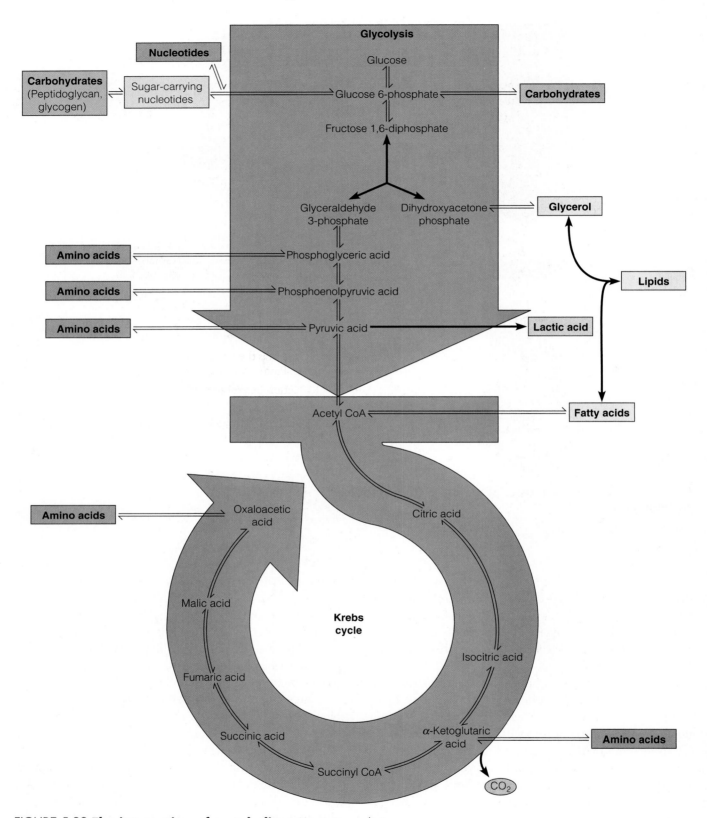

FIGURE 5.32 **The integration of metabolism.** Key intermediates are shown. Although not indicated in the figure, amino acids and ribose are used in the synthesis of purine and pyrimidine nucleotides (see Figure 5.31). The double arrows indicate amphibolic pathways.

■ What is an amphibolic pathway?

STUDY OUTLINE

Catabolic and Anabolic Reactions (pp. 111–112)

1. The sum of all chemical reactions within a living organism is known as metabolism.

2. Catabolism refers to chemical reactions that result in the breakdown of more complex organic molecules into simpler substances. Catabolic reactions usually release energy.

3. Anabolism refers to chemical reactions in which simpler substances are combined to form more complex molecules. Anabolic reactions usually require energy.

4. The energy of catabolic reactions is used to drive anabolic reactions.

5. The energy for chemical reactions is stored in ATP.

Enzymes (pp. 112–120)

1. Enzymes are proteins, produced by living cells, that catalyze chemical reactions by lowering the activation energy.

2. Enzymes are generally globular proteins with characteristic three-dimensional shapes.

3. Enzymes are efficient, can operate at relatively low temperatures, and are subject to various cellular controls.

Naming Enzymes (p. 114)

1. Enzyme names usually end in *-ase*.

2. The six classes of enzymes are defined on the basis of the types of reactions they catalyze.

Enzyme Components (pp. 114–115)

1. Most enzymes are holoenzymes, consisting of a protein portion (apoenzyme) and a nonprotein portion (cofactor).

2. The cofactor can be a metal ion (iron, copper, magnesium, manganese, zinc, calcium, or cobalt) or a complex organic molecule known as a coenzyme (NAD^+, $NADP^+$, FMN, FAD, or coenzyme A).

The Mechanism of Enzymatic Action (pp. 115–116)

1. When an enzyme and substrate combine, the substrate is transformed, and the enzyme is recovered.

2. Enzymes are characterized by specificity, which is a function of their active sites.

Factors Influencing Enzymatic Activity (pp. 116–119)

1. At high temperatures, enzymes undergo denaturation and lose their catalytic properties; at low temperatures, the reaction rate decreases.

2. The pH at which enzymatic activity is maximal is known as the optimum pH.

3. Within limits, enzymatic activity increases as substrate concentration increases.

4. Competitive inhibitors compete with the normal substrate for the active site of the enzyme. Noncompetitive inhibitors act on other parts of the apoenzyme or on the cofactor and decrease the enzyme's ability to combine with the normal substrate.

Feedback Inhibition (p. 119)

1. Feedback inhibition occurs when the end-product of a metabolic pathway inhibits an enzyme's activity near the start of the pathway.

Ribozymes (pp. 119–120)

1. Ribozymes are enzymatic RNA molecules that cut and splice RNA in eukaryotic cells.

Energy Production (pp. 120–122)

Oxidation-Reduction Reactions (p. 120)

1. Oxidation is the removal of one or more electrons from a substrate. Protons (H^+) are often removed with the electrons.

2. Reduction of a substrate refers to its gain of one or more electrons.

3. Each time a substance is oxidized, another is simultaneously reduced.

4. NAD^+ is the oxidized form; NADH is the reduced form.

5. Glucose is a reduced molecule; energy is released during a cell's oxidation of glucose.

The Generation of ATP (pp. 121–122)

1. Energy released during certain metabolic reactions can be trapped to form ATP from ADP and Ⓟ (phosphate). Addition of a Ⓟ to a molecule is called phosphorylation.

2. During substrate-level phosphorylation, a high-energy Ⓟ from an intermediate in catabolism is added to ADP.

3. During oxidative phosphorylation, energy is released as electrons are passed to a series of electron acceptors (an electron transport chain) and finally to O_2 or another inorganic compound.

4. During photophosphorylation, energy from light is trapped by chlorophyll, and electrons are passed through a series of electron acceptors. The electron transfer releases energy used for the synthesis of ATP.

Metabolic Pathways of Energy Production (p. 122)

1. A series of enzymatically catalyzed chemical reactions called metabolic pathways store energy in and release energy from organic molecules.

Carbohydrate Catabolism (pp. 122–134)

1. Most of a cell's energy is produced from the oxidation of carbohydrates.

2. Glucose is the most commonly used carbohydrate.

3. The two major types of glucose catabolism are respiration, in which glucose is completely broken down, and fermentation, in which it is partially broken down.

Glycolysis (pp. 123–124)

1. The most common pathway for the oxidation of glucose is glycolysis. Pyruvic acid is the end-product.

2. Two ATP and two NADH molecules are produced from one glucose molecule.

Alternatives to Glycolysis (p. 124)

1. The pentose phosphate pathway is used to metabolize five-carbon sugars; one ATP and 12 NADPH molecules are produced from one glucose molecule.

2. The Entner-Doudoroff pathway yields one ATP and two NADPH molecules from one glucose molecule.

Cellular Respiration (pp. 124–132)

1. During respiration, organic molecules are oxidized. Energy is generated from the electron transport chain.

2. In aerobic respiration, O_2 functions as the final electron acceptor.

3. In anaerobic respiration, the final electron acceptor is an inorganic molecule other than O_2.

Aerobic Respiration (pp. 124–130)

The Krebs Cycle (pp. 124–127)

1. Decarboxylation of pyruvic acid produces one CO_2 molecule and one acetyl group.

2. Two-carbon acetyl groups are oxidized in the Krebs cycle. Electrons are picked up by NAD^+ and FAD for the electron transport chain.

3. From one molecule of glucose, oxidation produces six molecules of NADH, two molecules of $FADH_2$, and two molecules of ATP.

4. Decarboxylation produces six molecules of CO_2.

The Electron Transport Chain (System) (pp. 127–128)

1. Electrons are brought to the electron transport chain by NADH.

2. The electron transport chain consists of carriers, including flavoproteins, cytochromes, and ubiquinones.

The Chemiosmotic Mechanism of ATP Generation (pp. 128–129)

1. Protons being pumped across the membrane generate a proton motive force as electrons move through a series of acceptors or carriers.

2. Energy produced from movement of the protons back across the membrane is used by ATP synthase to make ATP from ADP and Ⓟ.

3. In eukaryotes, electron carriers are located in the inner mitochondrial membrane; in prokaryotes, electron carriers are in the plasma membrane.

A Summary of Aerobic Respiration (pp. 129–130)

1. In aerobic prokaryotes, 38 ATP molecules can be produced from complete oxidation of a glucose molecule in glycolysis, the Krebs cycle, and the electron transport chain.

2. In eukaryotes, 36 ATP molecules are produced from complete oxidation of a glucose molecule.

Anaerobic Respiration (pp. 130–132)

1. The final electron acceptors in anaerobic respiration include NO_3^-, SO_4^{2-}, and CO_3^{2-}.

2. The total ATP yield is less than in aerobic respiration because only part of the Krebs cycle operates under anaerobic conditions.

Fermentation (pp. 132–134)

1. Fermentation releases energy from sugars or other organic molecules by oxidation.

2. O_2 is not required in fermentation.

3. Two ATP molecules are produced by substrate-level phosphorylation.

4. Electrons removed from the substrate reduce NAD^+.

5. The final electron acceptor is an organic molecule.

6. In lactic acid fermentation, pyruvic acid is reduced by NADH to lactic acid.

7. In alcohol fermentation, acetaldehyde is reduced by NADH to produce ethanol.

8. Heterolactic fermenters can use the pentose phosphate pathway to produce lactic acid and ethanol.

Lipid and Protein Catabolism (pp. 134–136)

1. Lipases hydrolyze lipids into glycerol and fatty acids.

2. Fatty acids and other hydrocarbons are catabolized by beta oxidation.

3. Catabolic products can be further broken down in glycolysis and the Krebs cycle.

4. Before amino acids can be catabolized, they must be converted to various substances that enter the Krebs cycle.

5. Transamination, decarboxylation, and dehydrogenation reactions convert the amino acids to be catabolized.

Biochemical Tests and Bacterial Identification (p. 136)

1. Bacteria and yeast can be identified by detecting action of their enzymes.

2. Fermentation tests are used to determine whether an organism can ferment a carbohydrate to produce acid and gas.

Photosynthesis (pp. 136–138)

1. Photosynthesis is the conversion of light energy from the sun into chemical energy; the chemical energy is used for carbon fixation.

The Light-Dependent Reactions: Photophosphorylation (p. 138)

1. Chlorophyll *a* is used by green plants, algae, and cyanobacteria; it is found in thylakoid membranes.

2. Electrons from chlorophyll pass through an electron transport chain, from which ATP is produced by chemiosmosis.

3. In cyclic photophosphorylation, the electrons return to the chlorophyll.

4. In noncyclic photophosphorylation, the electrons are used to reduce $NADP^+$, and electrons are returned to chlorophyll from H_2O or H_2S.

5. When H_2O is oxidized by green plants, algae, and cyanobacteria, O_2 is produced.

The Light-Independent Reactions: The Calvin-Benson Cycle (p. 138)

1. CO_2 is used to synthesize sugars in the Calvin-Benson cycle.

A Summary of Energy Production Mechanisms (pp. 138–139)

1. Sunlight is converted to chemical energy in oxidation-reduction reactions carried on by phototrophs. Chemotrophs can use this chemical energy.

2. In oxidation-reduction reactions, energy is derived from the transfer of electrons.

3. To produce energy, a cell needs an electron donor (organic or inorganic), a system of electron carriers, and a final electron acceptor (organic or inorganic).

Metabolic Diversity Among Organisms (pp. 139–144)

1. Photoautotrophs obtain energy by photophosphorylation and fix carbon from CO_2 via the Calvin-Benson cycle to synthesize organic compounds.

2. Cyanobacteria are oxygenic phototrophs. Green bacteria and purple bacteria are anoxygenic phototrophs.

3. Photoheterotrophs use light as an energy source and an organic compound for their carbon source and electron donor.

4. Chemoautotrophs use inorganic compounds as their energy source and carbon dioxide as their carbon source.

5. Chemoheterotrophs use complex organic molecules as their carbon and energy sources.

Metabolic Pathways of Energy Use (pp. 144–146)

Polysaccharide Biosynthesis (p. 144)

1. Glycogen is formed from ADPG.

2. UDPNAc is the starting material for the biosynthesis of peptidoglycan.

Lipid Biosynthesis (pp. 144–145)

1. Lipids are synthesized from fatty acids and glycerol.

2. Glycerol is derived from dihydroxyacetone phosphate, and fatty acids are built from acetyl CoA.

Amino Acid and Protein Biosynthesis (p. 145)

1. Amino acids are required for protein biosynthesis.

2. All amino acids can be synthesized either directly or indirectly from intermediates of carbohydrate metabolism, particularly from the Krebs cycle.

Purine and Pyrimidine Biosynthesis (pp. 145–146)

1. The sugars composing nucleotides are derived from either the pentose phosphate pathway or the Entner-Doudoroff pathway.

2. Carbon and nitrogen atoms from certain amino acids form the backbones of the purines and pyrimidines.

The Integration of Metabolism (pp. 146–148)

1. Anabolic and catabolic reactions are integrated through a group of common intermediates.

2. Such integrated metabolic pathways are referred to as amphibolic pathways.

STUDY QUESTIONS

Review

1. Define metabolism.

2. Distinguish between catabolism and anabolism. How are these processes related?

3. Using the diagrams to the right, show:
   a. where the substrate will bind.
   b. where the competitive inhibitor will bind.
   c. where the noncompetitive inhibitor will bind.
   d. which of the four elements could be the inhibitor in feedback inhibition.

Enzyme   Substrate   Competitive inhibitor   Noncompetitive inhibitor

4. What will the effect of the reactions in question 3 be?

5. Name the pathways diagrammed in a, b, and c of the figure to the right.

6. Show where glycerol is catabolized and where fatty acids are catabolized.

7. Show where the amino acid glutamic acid is catabolized:

$$\text{HOOC}-\text{CH}_2-\text{CH}_2-\overset{\overset{\displaystyle H}{|}}{\underset{\underset{\displaystyle NH_2}{|}}{C}}-\text{COOH}$$

8. Show how these pathways are related.

9. Where is ATP required in pathways a and b?

10. Where is CO_2 released in pathways b and c?

11. Show where a long-chain hydrocarbon such as petroleum is catabolized.

12. Where is NADH (or $FADH_2$ or NADPH) used and produced in these pathways?

13. Identify four places where anabolic and catabolic pathways are integrated.

14. Why are most enzymes active at one particular temperature? Why are enzymes less active below this temperature? What happens above this temperature?

15. List four compounds that can be made from pyruvic acid by an organism that uses fermentation only.

16. Fill in the following table with the carbon source and energy source of each type of organism.

Organism	Carbon Source	Energy Source
Photoautotroph		
Photoheterotroph		
Chemoautotroph		
Chemoheterotroph		

17. There are three mechanisms for the phosphorylation of ADP to produce ATP. Write the name of the mechanism that describes each of the reactions in the following table.

ATP Generated by	Reaction
_____	An electron, liberated from chlorophyll by light, is passed down an electron transport chain.
_____	Cytochrome *c* passes two electrons to cytochrome *a*.
_____	$\underset{\substack{\text{Phosphoenolpyruvic}\\\text{acid}}}{\overset{\displaystyle CH_2}{\underset{\underset{\displaystyle COOH}{|}}{\overset{\|}{C}-O\sim Ⓟ}}} \rightarrow \underset{\text{Pyruvic acid}}{\overset{\displaystyle CH_3}{\underset{\underset{\displaystyle COOH}{|}}{C=O}}}$

Use the following diagrams for questions 5–13.

(a)

(b)

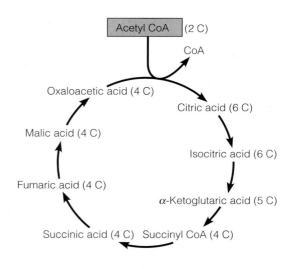

(c)

18. Define oxidation-reduction, and differentiate between the following terms:
    a. aerobic and anaerobic respiration
    b. respiration and fermentation
    c. cyclic and noncyclic photophosphorylation

19. The pentose phosphate pathway produces only one ATP. List four advantages of this pathway for the cell.

20. All of the energy-producing biochemical reactions that occur in cells, such as photophosphorylation and glycolysis, are _____ reactions.

21. Explain how ATP is a key intermediate in metabolism.

22. An enzyme and substrate are combined. The rate of reaction begins as shown in the following graph. To complete the graph, show the effect of increasing substrate concentration on a constant enzyme concentration. Show the effect of increasing temperature.

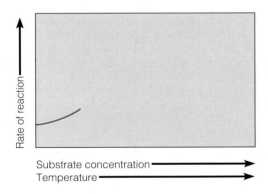

Multiple Choice

1. Which substance in the following reaction is being reduced?

$$\begin{array}{ccc}
\overset{\displaystyle H}{\underset{\displaystyle CH_3}{\overset{|}{\underset{|}{C}}}}{=}O + NADH+H^+ & \rightarrow & H{-}\overset{\displaystyle H}{\underset{\displaystyle CH_3}{\overset{|}{\underset{|}{C}}}}{-}OH + NAD^+
\end{array}$$

   Acetaldehyde   Ethanol

   a. Acetaldehyde
   b. NADH
   c. Ethanol
   d. NAD$^+$

2. Which of the following reactions produces the most molecules of ATP during aerobic metabolism?
   a. glucose → glucose-6-P
   b. phosphoenolpyruvic acid → pyruvic acid
   c. glucose → pyruvic acid
   d. acetyl CoA → CO_2 + H_2O
   e. succinic acid → fumaric acid

3. Which of the following processes does not generate ATP?
   a. photophosphorylation
   b. the Calvin-Benson cycle
   c. oxidative phosphorylation
   d. substrate-level phosphorylation
   e. none of the above

4. Which of the following compounds has the greatest amount of energy for a cell?
   a. CO_2
   b. ATP
   c. glucose
   d. O_2
   e. lactic acid

5. Which of the following is the best definition of the Krebs cycle?
   a. the oxidation of pyruvic acid
   b. the way cells produce CO_2
   c. a series of chemical reactions in which NADH is produced from the oxidation of pyruvic acid
   d. a method of producing ATP by phosphorylating ADP
   e. a series of chemical reactions in which ATP is produced from the oxidation of pyruvic acid.

6. Which of the following is the best definition of respiration?
   a. a sequence of carrier molecules with O_2 as the final electron acceptor
   b. a sequence of carrier molecules with an inorganic molecule as the final electron acceptor
   c. a method of generating ATP
   d. the complete oxidation of glucose to CO_2 and H_2O
   e. a series of reactions in which pyruvic acid is oxidized to CO_2 and H_2O

Use the following choices to answer questions 7–10.
   a. *E. coli* growing in glucose broth at 35°C with O_2 for 5 days
   b. *E. coli* growing in glucose broth at 35°C without O_2 for 5 days
   c. both a and b
   d. neither a nor b

7. Which culture produces the most lactic acid?

8. Which culture produces the most ATP?

9. Which culture uses NAD$^+$?

10. Which culture uses the most glucose?

Critical Thinking

1. Write your own definition of the chemiosmotic mechanism of ATP generation. On Figure 5.16, mark the following using the appropriate letter:
   a. the acidic side of the membrane
   b. the side with a positive electrical charge
   c. potential energy
   d. kinetic energy

2. Explain why, even under ideal conditions, *Streptococcus* grows slowly.

3. Why must NADH be reoxidized? How does this happen in an organism that uses respiration? Fermentation?

4. The following graph shows the normal rate of reaction of an enzyme and its substrate (black) and the rate when an excess of competitive inhibitor is present (red). Explain why the graph appears as it does.

5. Compare and contrast carbohydrate catabolism and energy production in the following bacteria:
   a. *Pseudomonas*, an aerobic chemoheterotroph
   b. *Spirulina*, an oxygenic photoautotroph
   c. *Ectothiorhodospira*, an anoxygenic photoautotroph

6. How much ATP could be obtained from the complete oxidation of one molecule of glucose? From one molecule of butterfat containing one glycerol and three 12-carbon chains?

7. The chemoautotroph *Thiobacillus* can obtain energy from the oxidation of arsenic ($As^{3+} \rightarrow As^{5+}$). How does this reaction provide energy? How can this bacterium be put to use by humans?

Clinical Applications

1. *Haemophilus influenzae* requires hemin (X factor) to synthesize cytochromes and NAD^+ (V factor) from other cells. For what does it use these two growth factors? What diseases does *H. influenzae* cause?

2. The drug HIVID®, also called ddC, inhibits DNA synthesis. It is used to treat HIV infection and AIDS. Compare the following illustration to Figure 2.16 on page 48. How does this drug work?

3. The bacterial enzyme streptokinase is used to digest fibrin (blood clots) in patients with atherosclerosis. Why doesn't injection of streptokinase cause a streptococcal infection? How do we know the streptokinase will digest fibrin only and not good tissues?

LEARNING WITH TECHNOLOGY

The Microbiology Place CD-ROM/website includes many resources to help you study and succeed in this course. In addition to self-quizzes, web links, and flashcards, you'll find the following activities for this chapter in the "Activities" section of The Microbiology Place:

Mechanism of Enzymatic Action

Factors that Affect Enzymatic Activity

Enzyme Inhibitors

Oxidation-Reduction

Glycolysis: Preparatory Stage

Glycolysis: Energy-Conserving Stage

Fermentation

Krebs Cycle

Cellular Respiration: Oxidation of Glucose

Cellular Respiration: Chemiosmotic Model

Cellular Respiration: Aerobic vs. Anaerobic

Photosynthesis: Photophosphorylation

Photosynthesis: Calvin-Benson Cycle

Case Study: The Pasteur Effect

Chapter 6

Microbial Growth

STAPHYLOCOCCUS AUREUS. *Colonies of coagulase-positive* S. aureus *can be identified by their black colonies on tellurite-glycine medium.*

When we talk about microbial growth, we are really referring to the number of cells, not the size of the cells. Microbes that are "growing" are increasing in number, accumulating into colonies (groups of cells large enough to be seen without a microscope) of hundreds of thousands of cells, or populations of billions of cells. For the most part, we are not concerned with the growth of individual cells. Although individual cells approximately double in size during their lifetime, this change is not very significant compared with the size increases observed during the lifetime of plants and animals.

Microbial populations can become incredibly large in a very short time, as we will see later in this chapter. By understanding the conditions necessary for microbial growth, we can determine how to control the growth of microbes that cause diseases and food spoilage. We can also learn how to encourage the growth of helpful microbes and those we wish to study. The manager of a sewage treatment plant and the manager of a brewery may not seem to have much in common, but both would be interested in promoting rapid microbial activity.

In this chapter we will examine the physical and chemical requirements for microbial growth, the various kinds of culture media, bacterial division, the phases of microbial growth, and the methods of measuring microbial growth. In later chapters we will see how unwanted microbial growth can be controlled.

The Requirements for Growth

The requirements for microbial growth can be divided into two main categories: physical and chemical. Physical aspects include temperature, pH, and osmotic pressure. Chemical requirements include sources of carbon, nitrogen, sulfur, phosphorus, trace elements, oxygen, and organic growth factors.

Physical Requirements

Learning Objectives

- *Classify microbes into five groups on the basis of preferred temperature range.*
- *Identify how and why the pH of culture media is controlled.*
- *Explain the importance of osmotic pressure to microbial growth.*

Temperature

Most microorganisms grow well at the temperatures favored by humans. However, certain bacteria are capable of growing at extremes of temperature that would certainly hinder the survival of almost all eukaryotic organisms.

Microorganisms are classified into three primary groups on the basis of their preferred range of temperature: **psychrophiles** (cold-loving microbes), **mesophiles** (moderate-temperature–loving microbes), and **thermophiles** (heat-loving microbes). Most bacteria grow only within a limited range of temperatures, and their maximum and minimum growth temperatures are only about 30°C apart. They grow poorly at the high and low temperature extremes within their range.

Each bacterial species grows at particular minimum, optimum, and maximum temperatures. The **minimum growth temperature** is the lowest temperature at which the species will grow. The **optimum growth temperature** is the temperature at which the species grows best. The **maximum growth temperature** is the highest temperature at which growth is possible. By graphing the growth response over a temperature range, we can see that the optimum growth temperature is usually near the top of the range; above that temperature the rate of growth drops off rapidly (Figure 6.1). This happens presumably because the high temperature has inactivated necessary enzymatic systems of the cell.

The ranges and maximum growth temperatures that define bacteria as psychrophiles, mesophiles, or thermophiles are not rigidly defined. Psychrophiles, for example, were originally considered simply to be organisms capable of growing at 0°C. However, there seem to be two fairly distinct groups capable of growth at that temperature. One group, composed of psychrophiles in the strictest sense, can grow at 0°C but has an optimum growth temperature of about 15°C. Most of these organisms are so sensitive to higher temperatures that they will not even grow in a reasonably warm room (25°C). Found mostly in the oceans' depths or in certain polar regions, such organisms seldom cause problems in food preservation. The other group that can grow at 0°C has higher optimum temperatures, usually 20–30°C and cannot grow above about 40°C. Organisms of this type are much more common than psychrophiles and are the most likely to be encountered in low-temperature food spoilage because they grow fairly well at refrigerator temperatures. We will use the term **psychrotrophs,** which is favored by food microbiologists, for

FIGURE 6.1 **Typical growth rates of different types of microorganisms in response to temperature.** Optimum growth (fastest reproduction) is represented by the peak of the curve. Notice that the reproductive rate drops off very quickly at temperatures only a little above the optimum. At either extreme of the temperature range, the reproductive rate is much lower than the rate at the optimum temperature.

■ Why is it difficult to define psychrophile, mesophile, and thermophile?

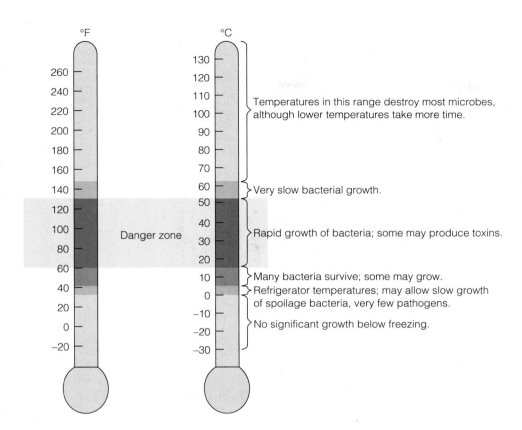

FIGURE 6.2 **Food spoilage temperatures.** Low temperatures decrease microbial reproduction rates, which is the basic principle of refrigeration. There are always some exceptions to the temperature responses shown here; for example, certain bacteria grow well at temperatures that would kill most bacteria, and a few bacteria can actually grow at temperatures well below freezing.

■ Which bacterium would theoretically be the more likely to grow at refrigerator temperatures, a human intestinal pathogen or a soil-borne plant pathogen?

Text within figure:

Temperatures in this range destroy most microbes, although lower temperatures take more time.

Very slow bacterial growth.

Danger zone

Rapid growth of bacteria; some may produce toxins.

Many bacteria survive; some may grow.
Refrigerator temperatures; may allow slow growth of spoilage bacteria, very few pathogens.
No significant growth below freezing.

this group of spoilage microorganisms. Many environmental microbiologists prefer to call them *moderate psychrophiles* or *facultative psychrophiles*.

Refrigeration is the most common method of preserving household food supplies. It is based on the principle that microbial reproductive rates decrease at low temperatures. Although microbes usually survive even subfreezing temperatures (they might become entirely dormant), they gradually decline in number. Some species decline faster than others. Psychrotrophs actually do not grow well at low temperatures, except in comparison with other organisms, but given time they are able to slowly degrade food. Such spoilage might take the form of mold mycelium, slime on food surfaces, or off-tastes or off-colors in foods. The temperature inside a properly set refrigerator will greatly slow the growth of most spoilage organisms and will entirely prevent the growth of all but a few pathogenic bacteria. Figure 6.2 illustrates the importance of low temperatures for preventing the growth of spoilage and disease organisms. When large amounts of food must be refrigerated, it is important to keep in mind the slow cooling rate of a large quantity of warm food (Figure 6.3).

Mesophiles, with an optimum growth temperature of 25–40°C, are the most common type of microbe. Organisms that have adapted to live in the bodies of animals usually have an optimum temperature close to that of their hosts. The optimum temperature for many pathogenic bacteria is about 37°C, and incubators for clinical cultures are usually set at about this temperature. The mesophiles include most of the common spoilage and disease organisms.

Thermophiles are microorganisms capable of growth at high temperatures. Many of these organisms have an optimum growth temperature of 50–60°C, about the temperature of water from a hot water tap. Such temperatures can also be reached in sunlit soil and in thermal waters such as hot springs. Remarkably, many thermophiles cannot grow at temperatures below about 45°C. Endospores formed by thermophilic bacteria are unusually heat resistant and may survive the usual heat treatment given canned goods. Although elevated storage temperatures may cause surviving endospores to germinate and grow, thereby spoiling the food, these thermophilic bacteria are not considered a public health problem. Thermophiles are important in organic compost piles, in which the temperature can rise rapidly to 50–60°C.

Some microbes, members of the Archaea (page 6), have an optimum growth temperature of 80°C or higher. These organisms are called **hyperthermophiles,** or sometimes **extreme thermophiles.** Most of these organisms live in hot springs associated with volcanic activity; sulfur is usually important in their metabolic activity. The known record for bacterial growth at high temperatures is

FIGURE 6.3 **The effect of the amount of food on its cooling rate in a refrigerator and its chance of spoilage.** Notice that in this example the pan of rice with a depth of 5 cm (2 in) cooled through the incubation temperature range of the *Bacillus cereus* in about 1 hour, whereas the pan of rice with a depth of 15 cm (6 in) remained in this temperature range for about 5 hours.

- The fact that large amounts of food cool slowly is an essential consideration for people who prepare food for large groups.

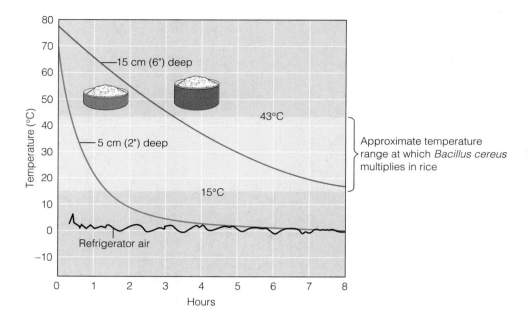

pH

Recall from Chapter 2 (page 36) that pH refers to the acidity or alkalinity of a solution. Most bacteria grow best in a narrow pH range near neutrality, between pH 6.5 and 7.5. Very few bacteria grow at an acidic pH below about pH 4. This is why a number of foods, such as sauerkraut, pickles, and many cheeses, are preserved from spoilage by acids produced by bacterial fermentation. Nonetheless, some bacteria, called **acidophiles,** are remarkably tolerant of acidity. One type of chemoautotrophic bacteria, which is found in the drainage water from coal mines and oxidizes sulfur to form sulfuric acid, can survive at a pH value of 1. Molds and yeasts will grow over a greater pH range than bacteria will, but the optimum pH of molds and yeasts is generally below that of bacteria, usually about pH 5–6. Alkalinity also inhibits microbial growth but is rarely used to preserve foods.

When bacteria are cultured in the laboratory, they often produce acids that eventually interfere with their own growth. To neutralize the acids and maintain the proper pH, chemical buffers are included in the growth medium. The peptones and amino acids in some media act as buffers, and many media also contain phosphate salts. Phosphate salts have the advantage of exhibiting their buffering effect in the pH growth range of most bacteria. They are also nontoxic; in fact, they provide phosphorus, an essential nutrient.

about 110°C near deep-sea hydrothermal vents. The immense pressure in the ocean depths prevents water from boiling even at temperatures well above 100°C.

Osmotic Pressure

Microorganisms obtain almost all their nutrients in solution from the surrounding water. Thus, they require water for growth and are made up of 80–90% water. High osmotic pressures have the effect of removing necessary water from a cell. When a microbial cell is in a solution that has a higher concentration of solutes than in the cell (*hypertonic*), the cellular water passes out through the plasma membrane to the high solute concentration. (See the discussion of osmosis in Chapter 4, page 90, and review Figure 4.18 for the three types of solution environments a cell may encounter.) This osmotic loss of water causes **plasmolysis,** or shrinkage of the cell's cytoplasm (Figure 6.4).

The importance of this phenomenon is that the growth of the cell is inhibited as the plasma membrane pulls away from the cell wall. Thus, the addition of salts (or other solutes) to a solution, and the resulting increase in osmotic pressure, can be used to preserve foods. Salted fish, honey, and sweetened condensed milk are preserved largely by this mechanism; the high salt or sugar concentrations draw water out of any microbial cells that are present and thus prevent their growth. These effects of osmotic pressure are roughly related to the number of dissolved molecules and ions in a volume of solution.

Some organisms, called **extreme halophiles,** have adapted so well to high salt concentrations that they actually require them for growth. In this case, they may be termed **obligate halophiles.** Organisms from such saline waters as the Dead Sea often require nearly 30% salt, and the inoculating loop (a device for handling bacteria in the laboratory) used to transfer them must first be dipped into a saturated

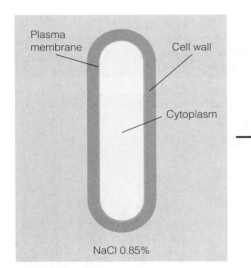

(a) Normal cell in isotonic solution. Under these conditions, the osmotic pressure in the cell is equivalent to a solute concentration of 0.85% sodium chloride (NaCl).

(b) Plasmolyzed cell in hypertonic solution. If the concentration of solutes such as NaCl is higher in the surrounding medium than in the cell (hypertonic), water tends to leave the cell. Growth of the cell is inhibited.

FIGURE 6.4 **Plasmolysis.**

■ The addition of salts to a solution increases osmotic pressure and is used to preserve food.

salt solution. More common are **facultative halophiles,** which do not require high salt concentrations but are able to grow at salt concentrations up to 2%, a concentration that inhibits the growth of many other organisms. A few species of facultative halophiles can even tolerate 15% salt.

Most microorganisms, however, must be grown in a medium that is nearly all water. For example, the concentration of agar (a complex polysaccharide isolated from marine algae) used to solidify microbial growth media is usually about 1.5%. If markedly higher concentrations are used, the increased osmotic pressure can inhibit the growth of some bacteria.

If the osmotic pressure is unusually low (*hypotonic*)— such as in distilled water, for example—water tends to enter the cell rather than leave it. Some microbes that have a relatively weak cell wall may be lysed by such treatment.

Chemical Requirements

Learning Objectives

- *Provide a use for each of the four elements (carbon, nitrogen, sulfur, and phosphorus) needed in large amounts for microbial growth.*

- *Explain how microbes are classified on the basis of oxygen requirements.*

- *Identify ways in which aerobes avoid damage by toxic forms of oxygen.*

Carbon

Besides water, one of the most important requirements for microbial growth is carbon. Carbon is the structural back-bone of living matter; it is needed for all the organic compounds that make up a living cell. Half the dry weight of a typical bacterial cell is carbon. Chemoheterotrophs get most of their carbon from the source of their energy— organic materials such as proteins, carbohydrates, and lipids. Chemoautotrophs and photoautotrophs derive their carbon from carbon dioxide.

Nitrogen, Sulfur, and Phosphorus

In addition to carbon, other elements are needed by microorganisms for the synthesis of cellular material. For example, protein synthesis requires considerable amounts of nitrogen as well as some sulfur. The syntheses of DNA and RNA also require nitrogen and some phosphorus, as does the synthesis of ATP, the molecule so important for the storage and transfer of chemical energy within the cell. Nitrogen makes up about 14% of the dry weight of a bacterial cell, and sulfur and phosphorus together constitute about another 4%.

Organisms use nitrogen primarily to form the amino group of the amino acids of proteins. Many bacteria meet this requirement by decomposing protein-containing material and reincorporating the amino acids into newly synthesized proteins and other nitrogen-containing compounds. Other bacteria use nitrogen from ammonium ions (NH_4^+), which are already in the reduced form and are usually found in organic cellular material. Still other bacteria are able to derive nitrogen from nitrates (compounds that dissociate to give the nitrate ion, NO_3^-, in solution).

Some important bacteria, including many of the photo-synthesizing cyanobacteria (page 317), use gaseous nitrogen

APPLICATIONS OF MICROBIOLOGY

Studying Hydrothermal Bacteria

Until humans explored the deep ocean floor, scientists believed that only a few forms of life could survive in that high-pressure, completely dark, oxygen-poor environment. Then, in 1977, the first manned vehicle capable of penetrating to the bottom of the deepest oceans carried two scientists 2600 meters below the surface at the Galápagos Rift, about 350 km northeast of the Galápagos Islands. There, amid the vast expanse of barren basalt rocks, the scientists found unexpectedly rich oases of life, including mollusks, crustaceans, and worms (see the photograph). How do such creatures survive in these harsh conditions? Many bacterial samples have been taken to the surface and are being studied.

Ecosystem of the Hydrothermal Vents

Life at the surface of the world's oceans depends on photosynthetic organisms, such as bacteria and algae, which harness the sun's energy to fix carbon dioxide (CO_2) to make carbohydrates. At the deep ocean floor, where no light penetrates, photosynthesis is not possible. The scientists found that the primary producers at the ocean floor are chemoautotrophic bacteria. Using chemical energy from hydrogen sulfide (H_2S) as a source of energy to fix CO_2, the chemoautotrophs create an environment that supports higher life forms.

1 m

O$_2$
CO$_2$
HS$^-$

Host cell

Bacterial cell

$HS^- + 2O_2 \longrightarrow SO_4^{-2}$

Energy

$CO_2 + R \longrightarrow$ Organic molecules

Tubeworms and mussels at the Galápagos vent.

(N_2) directly from the atmosphere. This process is called **nitrogen fixation.** Some organisms that can use this method are free-living, mostly in the soil, but others live cooperatively in **symbiosis** with the roots of legumes such as clover, soybeans, alfalfa, beans, and peas. The nitrogen fixed in the symbiosis is used by both the plant and the bacterium.

Sulfur is used to synthesize sulfur-containing amino acids and vitamins such as thiamine and biotin. The box above describes an unusual ecosystem based on a supply of hydrogen sulfide (H_2S) that would be toxic to most organisms. Important natural sources of sulfur include the sulfate ion (SO_4^{2-}), hydrogen sulfide, and the sulfur-containing amino acids.

Phosphorus is essential for the synthesis of nucleic acids and the phospholipids of cell membranes. Among other places, it is also found in the energy bonds of ATP. An important source of phosphorus is the phosphate ion (PO_4^{3-}). Potassium, magnesium, and calcium are also elements that microorganisms require, often as cofactors for enzymes (see Chapter 5, page 114).

Studying Hydrothermal Bacteria (*continued*)

Hydrothermal vents in the seafloor supply the H_2S and CO_2. As superheated water from within the Earth rises through fractures in the Earth's crust called vents, it reacts with surrounding rock and dissolves metal ions, sulfides, and CO_2. The ecosystems of these vents depend on an abundance of sulfur compounds in the hot water. The concentration of sulfide (S^{2-}) is three times greater than the concentration of molecular oxygen in and around the vents. Such high sulfide concentrations are toxic to many organisms, but not to the creatures that inhabit this exotic environment.

Mats of bacteria grow along the sides of the vents, where temperatures exceed 100°C. These are the highest temperatures that any organism is known to tolerate. Above the vent, where temperatures are about 30°C, the concentration of bacteria is about four times greater than that in water farther from the vents, and the growth rate of bacteria is equal to that found in productive, sunlit coastal waters. The bacteria in and around the vents create an environment in which bacteria are the producers supporting consumers (such as clams, mussels, and worms) and decomposers.

Biotechnological Benefits from Hydrothermal Vents

Researchers at Oak Ridge National Laboratory in Tennessee have identified two archaea living near deep-sea vents that hold promise for a renewable energy source. *Thermoplasma acidophilus* and *Pyrococcus furiosus* can produce the fuel, hydrogen gas, and an extracellular polysaccharide from glucose. Since glucose can be produced by bacteria and plants, it is a renewable resource. An additional benefit is that the polysaccharide is pro-

duced in large quantities and may have industrial applications—for example, as a thickener in foods or other products.

DNA polymerases (enzymes that synthesize specific polymers) isolated from two archaea living near deep-sea vents are being used in the polymerase chain reaction (PCR), a technique for making many copies of DNA. In PCR, single-stranded DNA is made by heating a chromosome fragment to 98°C and cooling it so that DNA polymerase can copy each strand. DNA polymerase from *Thermococcus litoralis*, called Vent$_R$, and from *Pyrococcus*, called Deep Vent$_R$, are not denatured at 98°C. These enzymes can be used in automatic thermalcyclers to repeat the heating and cooling cycles, allowing many copies of DNA to be made easily and quickly. Vent$_R$ and Deep Vent$_R$ add bases to DNA at a rate of about 1000 bases per minute.

Trace Elements

Microbes require very small amounts of other mineral elements, such as iron, copper, molybdenum, and zinc; these are referred to as **trace elements.** Most are essential for the functions of certain enzymes, usually as cofactors. Although these elements are sometimes added to a laboratory medium, they are usually assumed to be naturally present in tap water and other components of media. Even most distilled waters contain adequate amounts, but tap water is sometimes specified to ensure that these trace minerals will be present in culture media.

Oxygen

We are accustomed to thinking of molecular oxygen (O_2) as a necessity of life, but it is actually in a sense a poisonous gas. Very little molecular oxygen existed in the atmosphere during most of Earth's history—in fact, it is possible that life could not have arisen had oxygen been present. However, many current forms of life have metabolic systems that require oxygen for aerobic respiration. As we have seen, hydrogen atoms that have been stripped from organic compounds combine with oxygen to form water, as shown in Figure 5.14. This process yields a great deal of energy while neutralizing a potentially toxic gas—a very neat solution, all in all.

Microbes that use molecular oxygen (aerobes) produce more energy from nutrients than microbes that do not use oxygen (anaerobes). Organisms that require oxygen to live are called **obligate aerobes** (Table 6.1a).

Obligate aerobes are at a disadvantage because oxygen is poorly soluble in the water of their environment. Therefore, many of the aerobic bacteria have developed, or retained, the ability to continue growing in the absence of oxygen. Such organisms are called **facultative anaerobes** (Table 6.1b). In other words, facultative anaerobes can use oxygen when it is present but are able to continue growth by using fermentation or anaerobic respiration when oxygen is not available. However, their efficiency in producing energy decreases in the absence of oxygen. Examples of facultative anaerobes are the familiar *Escherichia coli* that is found in the human intestinal tract, and many yeasts. Recall from the discussion of anaerobic respiration in Chapter 5, page 130, that many microbes are able to substitute other electron acceptors, such as nitrate ions, for oxygen.

Obligate anaerobes (Table 6.1c) are bacteria that are unable to use molecular oxygen for energy-yielding reactions. In fact, most are harmed by it. The genus *Clostridium* (klôs-tri′dē-um), which contains the species that cause tetanus and botulism, is the most familiar example. These

TABLE 6.1	**The Effect of Oxygen on the Growth of Various Types of Bacteria**				
	a. Obligate Aerobes	**b. Facultative Anaerobes**	**c. Obligate Anaerobes**	**d. Aerotolerant Anaerobes**	**e. Micro-aerophiles**
Effect of Oxygen on Growth	Only aerobic growth; oxygen required.	Both aerobic and anaerobic growth; greater growth in presence of oxygen.	Only anaerobic growth; ceases in presence of oxygen.	Only anaerobic growth; but continues in presence of oxygen.	Only aerobic growth; oxygen required in low concentration.
Bacterial Growth in Tube of Solid Growth Medium					
Explanation of Growth Patterns	Growth occurs only where high concentrations of oxygen have diffused into the medium.	Growth is best where most oxygen is present, but occurs throughout tube.	Growth occurs only where there is no oxygen.	Growth occurs evenly; oxygen has no effect.	Growth occurs only where a low concentration of oxygen has diffused into medium.
Explanation of Oxygen's Effects	Presence of enzymes catalase and superoxide dismutase (SOD) allows toxic forms of oxygen to be neutralized; can use oxygen.	Presence of enzymes catalase and SOD allows toxic forms of oxygen to be neutralized; can use oxygen.	Lacks enzymes to neutralize harmful forms of oxygen; cannot tolerate oxygen.	Presence of one enzyme, SOD, allows harmful forms of oxygen to be partially neutralized; tolerates oxygen.	Produce lethal amounts of toxic forms of oxygen if exposed to normal atmospheric oxygen.

bacteria do use oxygen atoms present in cellular materials; the atoms are usually obtained from water.

Understanding how organisms can be harmed by oxygen requires a brief discussion of the toxic forms of oxygen:

1. **Singlet oxygen** is normal molecular oxygen (O_2) that has been boosted into a higher-energy state and is extremely reactive. It is present in phagocytic cells (see Figure 16.8), which play an important part in the body's defenses against pathogens such as bacteria. Once the phagocytic cell ingests bacteria, they are killed (with a few resistant exceptions) by exposure to singlet oxygen.

2. **Superoxide free radicals** (O_2^-) are formed in small amounts during the normal respiration of organisms that use oxygen as a final electron acceptor, forming water. In the presence of oxygen, obligate anaerobes also appear to form some superoxide free radicals, which are so toxic to cellular components that all organisms attempting to grow in

atmospheric oxygen must produce an enzyme, **superoxide dismutase (SOD),** to neutralize them. Their toxicity is caused by their great instability, which leads them to steal an electron from a neighboring molecule, which in turn becomes a radical and steals an electron, and so on. Aerobic bacteria, facultative anaerobes growing aerobically, and aerotolerant anaerobes (discussed shortly) produce SOD, with which they convert the superoxide free radical into molecular oxygen (O_2) and hydrogen peroxide (H_2O_2):

$$O_2^- + O_2^- + 2\,H^+ \longrightarrow H_2O_2 + O_2$$

3. The hydrogen peroxide produced in this reaction contains the **peroxide anion** O_2^{2-} and is also toxic. In Chapter 7, page 201 we will encounter it as the active principle in the antimicrobial agents hydrogen peroxide and benzoyl peroxide. Because the hydrogen peroxide produced during normal aerobic respiration is toxic, microbes have developed

enzymes to neutralize it. The most familiar of these is **catalase,** which converts it into water and oxygen:

$$2 H_2O_2 \longrightarrow 2 H_2O + O_2$$

Catalase is easily detected by its action on hydrogen peroxide. When a drop of hydrogen peroxide is added to a colony of bacterial cells producing catalase, oxygen bubbles are released. Anyone who has put hydrogen peroxide on a wound will recognize that cells in human tissue also contain catalase. The other enzyme that breaks down hydrogen peroxide is **peroxidase,** which differs from catalase in that its reaction does not produce oxygen:

$$H_2O_2 + 2 H^+ \longrightarrow 2 H_2O$$

4. The **hydroxyl radical** (OH·) is another intermediate form of oxygen and probably the most reactive. It is formed in the cellular cytoplasm by ionizing radiation. Most aerobic respiration produces traces of hydroxyl radicals but they are transient.

Obligate anaerobes usually produce neither superoxide dismutase nor catalase. Because aerobic conditions probably lead to an accumulation of superoxide free radicals in their cytoplasm, obligate anaerobes are extremely sensitive to oxygen.

Aerotolerant anaerobes (Table 6.1d) cannot use oxygen for growth, but they tolerate it fairly well. On the surface of a solid medium, they will grow without the special techniques (discussed later) required for obligate anaerobes. Many of the aerotolerant bacteria characteristically ferment carbohydrates to lactic acid. As lactic acid accumulates, it inhibits the growth of aerobic competitors and establishes a favorable ecological niche for lactic acid producers. A common example of lactic acid–producing aerotolerant anaerobes is the lactobacilli used in the production of many acidic fermented foods, such as pickles and cheese. In the laboratory, they are handled and grown much like any other bacteria, but they make no use of the oxygen in the air. These bacteria can tolerate oxygen because they possess SOD or an equivalent system that neutralizes the toxic forms of oxygen previously discussed.

A few bacteria are **microaerophiles** (Table 6.1e). They are aerobic; they do require oxygen. However, they grow only in oxygen concentrations lower than those in air. In a test tube of solid nutrient medium, they grow only at a depth where small amounts of oxygen have diffused into the medium; they do not grow near the oxygen-rich surface or below the narrow zone of adequate oxygen. This limited tolerance is probably due to their sensitivity to superoxide free radicals and peroxides, which they produce in lethal concentrations under oxygen-rich conditions.

Organic Growth Factors

Essential organic compounds an organism is unable to synthesize are known as **organic growth factors;** they must be directly obtained from the environment. One group of organic growth factors for humans is vitamins. Most vitamins function as coenzymes, the organic cofactors required by certain enzymes in order to function. Many bacteria can synthesize all their own vitamins and are not dependent on outside sources. However, some bacteria lack the enzymes needed for the synthesis of certain vitamins, and for them those vitamins are organic growth factors. Other organic growth factors required by some bacteria are amino acids, purines, and pyrimidines.

Culture Media

Learning Objectives
- Distinguishing between chemically defined and complex media.
- Justify the use of each of the following: anaerobic techniques, living host cells, candle jars, selective and differential media, enrichment medium.

A nutrient material prepared for the growth of microorganisms in a laboratory is called a **culture medium.** Some bacteria can grow well on just about any culture medium; others require special media, and still others cannot grow on any nonliving medium yet developed. When microbes are introduced into a culture medium to initiate growth, they are called an **inoculum.** The microbes that grow and multiply in or on a culture medium are referred to as a **culture.**

Suppose we want to grow a culture of a certain microorganism, perhaps the microbes from a particular clinical specimen. What criteria must the culture medium meet? First, it must contain the right nutrients for the specific microorganism we want to grow. It should also contain sufficient moisture, a properly adjusted pH, and a suitable level of oxygen, perhaps none at all. The medium must initially be **sterile**—that is, it must initially contain no living microorganisms—so that the culture will contain only the microbes (and their offspring) we add to the medium. Finally, the growing culture should be incubated at the proper temperature.

A wide variety of media are available for the growth of microorganisms in the laboratory. Most of these media, which are available from commercial sources, have premixed components and require only the addition of water and then sterilization. Media are constantly being developed or revised for use in the isolation and identification of bacteria that are of interest to researchers in such fields as food, water, and clinical microbiology.

TABLE 6.2	A Chemically Defined Medium for Growing a Typical Chemoheterotroph, Such as *E. coli*
Constituent	**Amount**
Glucose	5.0 g
Ammonium phosphate, monobasic ($NH_4H_2PO_4$)	1.0 g
Sodium chloride (NaCl)	5.0 g
Magnesium sulfate ($MgSO_4 \cdot 7H_2O$)	0.2 g
Potassium phosphate, dibasic (K_2HPO_4)	1.0 g
Water	1 liter

When it is desirable to grow bacteria on a solid medium, a solidifying agent such as agar is added to the medium. A complex polysaccharide derived from a marine alga, **agar** has long been used as a thickener in foods such as jellies and ice cream.

Agar has some very important properties that make it valuable to microbiology, and no satisfactory substitute has ever been found. Few microbes can degrade agar, so it remains solid. Also, agar liquefies at about 100°C (the boiling point of water) and at sea level remains liquid until the temperature drops to about 40°C. For laboratory use, agar is held in water baths at about 50°C. At this temperature, it does not injure most bacteria when it is poured over them (as shown in Figure 6.16a). Once the agar has solidified, it can be incubated at temperatures approaching 100°C before it again liquefies; this property is particularly useful when thermophilic bacteria are being grown.

Agar media are usually contained in test tubes or *Petri dishes*. The test tubes are called *slants* when they are allowed to solidify with the tube held at an angle so that a large surface area for growth is available. When the agar solidifies in a vertical tube, it is called a *deep*. Petri dishes, named for their inventor, are shallow dishes with a lid that nests over the bottom to prevent contamination; when filled, they are called *Petri* (or culture) *plates*.

Chemically Defined Media

To support microbial growth, a medium must provide an energy source, as well as sources of carbon, nitrogen, sulfur, phosphorus, and any organic growth factors the organism is unable to synthesize. A **chemically defined medium** is one whose exact chemical composition is known. For a chemoheterotroph, the chemically defined medium must contain organic growth factors that serve as a source of carbon and energy. For example, as shown in Table 6.2, glu-cose is included in the medium for growing the chemoheterotroph *E. coli*.

As Table 6.3 shows, many organic growth factors must be provided in the chemically defined medium used to cultivate a species of *Neisseria* (page 311). Organisms that require many growth factors are described as "fastidious." Organisms of this type, such as *Lactobacillus* (page 322), are sometimes used in tests that determine the concentration of a particular vitamin in a substance. To perform such a *microbiological assay*, a growth medium is prepared that contains all the growth requirements of the bacterium except the vitamin being assayed. Then the medium, test substance, and bacterium are combined, and the growth of bacteria is measured. This bacterial growth, which is reflected by the amount of lactic acid produced, will be proportional to the amount of vitamin in the test substance. The more lactic acid, the more the *Lactobacillus* cells have been able to grow, so the more vitamin is present.

Complex Media

Chemically defined media are usually reserved for laboratory experimental work or for the growth of autotrophic bacteria. Most heterotrophic bacteria and fungi, such as you would work with in an introductory lab course, are routinely grown on **complex media,** made up of nutrients such as extracts from yeasts, meat, or plants, or digests of proteins from these and other sources. The exact chemical composition varies slightly from batch to batch. Table 6.4 shows one widely used recipe.

In complex media, the energy, carbon, nitrogen, and sulfur requirements of the growing microorganisms are primarily provided by protein. Protein is a large, relatively insoluble molecule that a minority of microorganisms can utilize directly, but a partial digestion by acids or enzymes reduces protein to shorter chains of amino acids called *peptones*. These small, soluble fragments can be digested by most bacteria.

Vitamins and other organic growth factors are provided by meat extracts or yeast extracts. The soluble vitamins and minerals from the meats or yeasts are dissolved in the extracting water, which is then evaporated so that these factors are concentrated. (These extracts also supplement the organic nitrogen and carbon compounds.) Yeast extracts are particularly rich in the B vitamins. If a complex medium is in liquid form, it is called **nutrient broth.** When agar is added, it is called **nutrient agar.** (This terminology can be confusing; just remember that agar itself is not a nutrient.)

Anaerobic Growth Media and Methods

The cultivation of anaerobic bacteria poses a special problem. Because anaerobes might be killed by exposure to oxygen, special media called **reducing media** must be used. These media contain ingredients, such as sodium

TABLE 6.3	**A Chemically Defined Medium for Growing a Fastidious Chemoheterotrophic Bacterium, Such as *Neisseria gonorrhoeae***			
Constituent	**Amount**	**Constituent**	**Amount**
Carbon and energy sources		Amino acids	
Glucose	9.1 g	Cysteine	1.5 g
Starch	9.1 g	Arginine, proline (each)	0.3 g
Sodium acetate	1.8 g	Glutamic acid, methionine (each)	0.2 g
Sodium citrate	1.4 g	Asparagine, isoleucine, serine (each)	0.2 g
Oxaloacetate	0.3 g	Cystine	0.06 g
Salts		Organic growth factors	
Potassium phosphate, dibasic (K_2HPO_4)	12.7 g	Calcium pantothenate	0.02 g
Sodium chloride (NaCl)	6.4 g	Thiamine	0.02 g
Potassium phosphate, monobasic (KH_2PO_4)	5.5 g	Nicotinamide adenine dinucleotide	0.01 g
Sodium bicarbonate ($NaHCO_3$)	1.2 g	Uracil	0.006 g
Potassium sulfate (K_2SO_4)	1.1 g	Biotin	0.005 g
Sodium sulfate (Na_2SO_4)	0.9 g	Hypoxanthine	0.003 g
Magnesium chloride ($MgCl_2$)	0.5 g	Reducing agent	
Ammonium chloride (NH_4Cl)	0.4 g	Sodium thioglycolate	0.00003 g
Potassium chloride (KCl)	0.4 g	Water	1 liter
Calcium chloride ($CaCl_2$)	0.006 g		
Ferric nitrate [$Fe(NO_3)_3$]	0.006 g		

SOURCE: R. M. Atlas, *Handbook of Microbiological Media*, Ann Arbor, MI: CRC Press, 1993.

TABLE 6.4	**Composition of Nutrient Agar, a Complex Medium for the Growth of Heterotrophic Bacteria**
Constituent	**Amount**
Peptone (partially digested protein)	5.0 g
Beef extract	3.0 g
Sodium chloride	8.0 g
Agar	15.0 g
Water	1 liter

thioglycolate, that chemically combine with dissolved oxygen and deplete the oxygen in the culture medium. To routinely grow and maintain pure cultures of obligate anaerobes, microbiologists use reducing media stored in ordinary, tightly capped test tubes. These media are heated shortly before use, to drive off absorbed oxygen.

When the culture must be grown in Petri plates to observe individual colonies, special anaerobic jars are used (Figure 6.5). The culture plates are placed in the jar, and oxygen is removed by the following process: A packet of chemicals (sodium bicarbonate and sodium borohydride) in the jar is moistened with a few milliliters of water, and the jar is sealed. Hydrogen and carbon dioxide are produced by the reaction of the chemicals with the water. A palladium catalyst in the jar combines the oxygen in the jar with the hydrogen produced by the chemical reaction, and water is formed. As a result, the oxygen quickly disappears. Moreover, the carbon dioxide that is produced aids the growth of many anaerobic bacteria.

A relatively new technique to provide an anaerobic environment makes use of an enzyme, oxyrase, that reduces oxygen to water. Oxyrase is a respiratory enzyme derived from the plasma membranes of certain bacteria. When it is added to growth media, it transforms the Petri plate, OxyPlate, into a self-contained anaerobic chamber. This method, which can often avoid the need for more cumbersome apparatus, is now increasingly used in clinical laboratories.

FIGURE 6.5 **A jar for cultivating anaerobic bacteria on Petri plates.** When water is mixed with the chemical packet containing sodium bicarbonate and sodium borohydride, hydrogen and carbon dioxide are generated. Reacting on the surface of a palladium catalyst in a screened reaction chamber, which may also be incorporated into the chemical packet, the hydrogen and atmospheric oxygen in the jar combine to form water. The oxygen is thus removed. Also in the jar is an anaerobic indicator, containing methylene blue, which is blue when oxidized (as shown here) and turns colorless when the oxygen is removed.

■ The CO_2 produced in the jar aids the growth of many anaerobic bacteria.

Researchers regularly working with anaerobes use transparent anaerobic chambers equipped with air locks and filled with inert gases (Figure 6.6). Technicians can manipulate the equipment by inserting their hands into airtight rubber gloves called glove ports, which are fitted to the wall of the chamber.

Special Culture Techniques

Many bacteria have never been successfully grown on artificial laboratory media. *Mycobacterium leprae*, the leprosy bacillus, is now usually grown in armadillos, which have a relatively low body temperature that matches the requirements of the microbe. Another example is the syphilis spirochete, although certain nonpathogenic strains of this microbe have been grown on laboratory media. With few

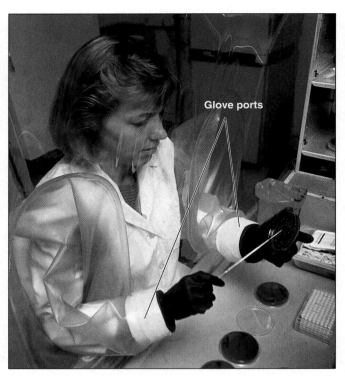

FIGURE 6.6 **An anaerobic chamber.** The technician is holding a Petri plate and an inoculating loop that are inside an anaerobic chamber filled with an inert, oxygen-free gas. Organisms and materials enter and leave through an air lock, and equipment is manipulated by means of glove ports.

■ The air lock used to allow entry and exit of samples is similar to those used to enter or leave the Space Laboratory in the vacuum of outer space.

exceptions, the obligate intracellular bacteria, such as the rickettsias and the chlamydias, do not grow on artificial media. Like viruses, they can reproduce only in a living host cell. See the discussion of cell culture, page 385.

Many clinical laboratories have special *carbon dioxide incubators* in which to grow aerobic bacteria that require concentrations of CO_2 higher or lower than that found in the atmosphere. Desired CO_2 levels are maintained by electronic controls. High CO_2 levels are also obtained with simple *candle jars* (Figure 6.7a). Cultures are placed in a large sealed jar containing a lighted candle, which consumes oxygen. The candle stops burning when the air in the jar has a lowered concentration of oxygen (but one still adequate for the growth of aerobic bacteria). An elevated concentration of CO_2 is also present. Microbes that grow better at high CO_2 concentrations are called **capnophiles.** The low-oxygen, high-CO_2 conditions resemble those found in the intestinal tract, respiratory tract, and other body tissues where pathogenic bacteria grow.

Candle jars are still used occasionally, but more often commercially available chemical packets are used to

(a) Candle jar
Plates and tubes inoculated with, for example, *Neisseria meningitidis* are placed in a jar with a lighted candle, and the jar is sealed. This will provide a CO_2 atmosphere of approximately 3%.

(b) CO₂-generating packet
The packet consists of a bag containing a Petri plate and a CO_2 gas generator. The gas generator is crushed to mix the chemicals it contains and start the reaction that produces CO_2. This gas reduces the oxygen concentration in the bag to about 5% and provides a CO_2 concentration of about 10%.

FIGURE 6.7 Equipment for producing CO₂-rich environments.

■ Microbes that grow better at high CO_2 concentrations are called capnophiles.

generate carbon dioxide atmospheres in containers (Figure 6.7b). When only one or two Petri plates of cultures are to be incubated, clinical laboratory investigators often use small plastic bags with self-contained chemical gas generators that are activated by crushing the packet or moistening it with a few milliliters of water. These packets are sometimes specially designed to provide precise concentrations of carbon dioxide (usually higher than can be obtained in candle jars) and oxygen, for culturing organisms such as the microaerophilic *Campylobacter* bacteria (page 317).

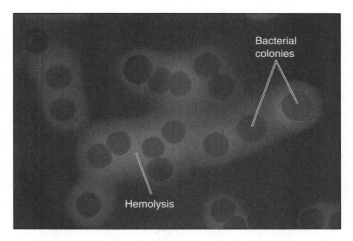

FIGURE 6.8 Blood agar, a differential medium containing red blood cells. The bacteria have lysed the red blood cells (beta-hemolysis), causing the clear areas around the colonies.

■ Differential media are designed to distinguish colonies of the desired organism from other colonies growing on the same plate.

Selective and Differential Media

In clinical and public health microbiology, it is frequently necessary to detect the presence of specific microorganisms associated with disease or poor sanitation. For this task, selective and differential media are used. **Selective media** are designed to suppress the growth of unwanted bacteria and encourage the growth of the desired microbes. For example, bismuth sulfite agar is one medium used to isolate the typhoid bacterium, the gram-negative *Salmonella typhi* (tī′fē), from feces. Bismuth sulfite inhibits gram-positive bacteria and most gram-negative intestinal bacteria (other than *S. typhi,*) as well. Dyes such as brilliant green selectively inhibit gram-positive bacteria, and this dye is the basis of a medium called brilliant green agar that is used to isolate the gram-negative *Salmonella.*

Sabouraud's dextrose agar, which has a pH of 5.6, is used to isolate fungi that outgrow most bacteria at this pH. **Differential media** make it easier to distinguish colonies of the desired organism from other colonies growing on the same plate. Similarly, pure cultures of microorganisms have identifiable reactions with differential media in tubes or plates. Blood agar (which contains red blood cells) is a medium that microbiologists often use to identify bacterial species that destroy red blood cells. These species, such as *Streptococcus pyogenes* (pī-äj′en-ēz), the bacterium that causes strep throat, show a clear ring around their colonies (beta-hemolysis, page 323) where they have lysed the surrounding blood cells (Figure 6.8).

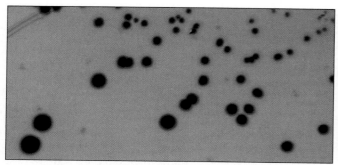

(a) *Staphylococcus aureus* on tellurite-glycine medium.

(b) *Escherichia coli* on eosin methylene blue (EMB) medium. The black-centered colonies are surrounded by a characteristic metallic green sheen.

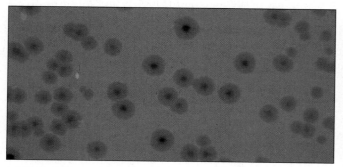

(c) *Enterobacter aerogenes* on EMB medium showing characteristic dark-centered colonies.

(d) On pseudomonas agar P (PSP) medium, *Pseudomonas aeruginosa* produces a blue-green water-soluble pigment.

FIGURE 6.9 **Bacterial colonies on several differential media.**

■ Which media pictured here are both selective and differential?

Sometimes, selective and differential characteristics are combined in a single medium. Suppose we want to isolate the common bacterium *Staphylococcus aureus*, found in the nasal passages. This organism has a tolerance for high concentrations of sodium chloride; it can also ferment the carbohydrate mannitol to form acid. Mannitol salt agar contains 7.5% sodium chloride, which will discourage the growth of competing organisms and thus *select for* (favor the growth of) *S. aureus*. This salty medium also contains a pH indicator that changes color if the mannitol in the medium is fermented to acid; the mannitol-fermenting colonies of *S. aureus* are thus *differentiated from* colonies of bacteria that do not ferment mannitol. Bacteria that grow at the high salt concentration *and* ferment mannitol to acid can be readily identified by the color change. These are probably colonies of *S. aureus*, and their identification can be confirmed by additional tests.

Another medium that is both selective and differential is MacConkey agar. This medium contains bile salts and crystal violet, which inhibit the growth of gram-positive bacteria. Because this medium also contains lactose, gram-negative bacteria that produce acid from the metabolism of lactose can be differentiated from similar bacteria that cannot. The ability to distinguish between lactose fermenters (red or pink colonies) and nonfermenters (colorless colonies) is useful in distinguishing between the pathogenic *Salmonella* bacteria and other related bacteria. Figure 6.9 shows the appearance of bacterial colonies on several differential media.

Enrichment Culture

Because bacteria present in small numbers can be missed, especially if other bacteria are present in much larger numbers, it is sometimes necessary to use an **enrichment culture.** This is often the case for soil or fecal samples. The medium (enrichment medium) for an enrichment culture is usually liquid and provides nutrients and environmental conditions that favor the growth of a particular microbe but not others. In this sense, it is also a selective medium, but it is designed to increase very small numbers of the desired type of organism to detectable levels.

Suppose we want to isolate from a soil sample a microbe that can grow on phenol and is present in much

TABLE 6.5	**Culture Media**
Type	**Purpose**
Chemically defined	Growth of chemoautotrophs and photoautotrophs, and microbiological assays.
Complex	Growth of most chemoheterotrophic organisms.
Reducing	Growth of obligate anaerobes.
Selective	Suppression of unwanted microbes; encouraging desired microbes.
Differential	Differentiation of colonies of desired microbes from others.
Enrichment	Similar to selective media but designed to increase numbers of desired microbes to detectable levels.

smaller numbers than other species. If the soil sample is placed in a liquid enrichment medium in which phenol is the only source of carbon and energy, microbes unable to metabolize phenol will not grow. The culture medium is allowed to incubate for a few days, and then a small amount of it is transferred into another flask of the same medium. After a series of such transfers, the surviving population will consist of bacteria capable of metabolizing phenol. The bacteria are given time to grow in the medium between transfers; this is the enrichment stage. Any nutrients in the original inoculum are rapidly diluted out with the successive transfers. When the last dilution is streaked onto a solid medium of the same composition, only those colonies of organisms capable of using phenol should grow. A remarkable aspect of this particular technique is that phenol is normally lethal to most bacteria.

* * *

Table 6.5 summarizes the purposes of the main types of culture media.

Obtaining Pure Cultures

Learning Objectives
- *Define colony.*
- *Describe how pure cultures can be isolated by using the streak plate method.*

Most infectious materials, such as pus, sputum, and urine, contain several different kinds of bacteria; so do samples of soil, water, or food. If these materials are plated out onto the surface of a solid medium, colonies will form that are exact copies of the original organism. A visible **colony** theoretically arises from a single spore or vegetative cell or from a group of the same microorganisms attached to one another in clumps or chains. Microbial colonies often have a distinctive appearance that distinguishes one microbe from another (see Figure 6.9). The bacteria must be distributed widely enough so that the colonies are visibly separated from each other.

Most bacteriological work requires pure cultures, or clones, of bacteria. The isolation method most commonly used to get pure cultures is the **streak plate method** (Figure 6.10). A sterile inoculating loop is dipped into a mixed culture that contains more than one type of microbe and is streaked in a pattern over the surface of the nutrient medium. As the pattern is traced, bacteria are rubbed off the loop onto the medium. The last cells to be rubbed off the loop are far enough apart to grow into isolated colonies. These colonies can be picked up with an inoculating loop and transferred to a test tube of nutrient medium to form a pure culture containing only one type of bacterium.

The streak plate method works well when the organism to be isolated is present in large numbers relative to the total population. However, when the microbe to be isolated is present only in very small numbers, its numbers must be greatly increased by selective enrichment before it can be isolated with the streak plate method.

Preserving Bacterial Cultures

Learning Objective
- *Explain how microorganisms are preserved by deep-freezing and lyophilization (freeze-drying).*

Refrigeration can be used for the short-term storage of bacterial cultures. Two common methods of preserving microbial cultures for long periods are deep-freezing and lyophilization. **Deep-freezing** is a process in which a pure culture of microbes is placed in a suspending liquid and quick-frozen at temperatures ranging from −50° to −95°C. The culture can usually be thawed and cultured even several years later. During **lyophilization (freeze-drying),** a suspension of microbes is quickly frozen at temperatures ranging from −54° to −72°C, and the water is removed by a high vacuum (sublimation). While under vacuum, the container is sealed by melting the glass with a high-temperature torch. The remaining powderlike residue that contains the surviving microbes can be stored for years. The organisms can be revived at any time by hydration with a suitable liquid nutrient medium.

(a) The direction of streaking is indicated by arrows. Streak series 1 is made from the original bacterial culture. The inoculating loop is sterilized following each streak series. In series 2 and 3, the loop picks up bacteria from the previous series, diluting the number of cells each time. There are numerous variants of such patterns.

(b) In series 3 of this example, notice that well-isolated colonies of bacteria of two different types, red and white, have been obtained.

FIGURE 6.10 **The streak plate method for isolating pure bacterial cultures.**

■ Streak plates are the most common technique used to obtain pure cultures of microorganisms.

The Growth of Bacterial Cultures

Learning Objective

• *Define bacterial growth, including binary fission.*

Being able to represent graphically the enormous populations resulting from the growth of bacterial cultures is an essential part of microbiology. It is also necessary to be able to determine microbial numbers, either directly, by counting, or indirectly, by measuring their metabolic activity.

Bacterial Division

As we mentioned at the beginning of the chapter, bacterial growth refers to an increase in bacterial numbers, not an increase in the size of the individual cells. Bacteria normally reproduce by **binary fission** (Figure 6.11).

A few bacterial species reproduce by **budding;** they form a small initial outgrowth (a bud) that enlarges until its size approaches that of the parent cell, and then it separates. Some filamentous bacteria (certain actinomycetes) reproduce by producing chains of conidiospores carried externally at the tips of the filaments. A few filamentous species simply fragment, and the fragments initiate the growth of new cells.

Generation Time

For purposes of calculating the generation time of bacteria, we will consider only reproduction by binary fission, which is by far the most common method. As you can see in Figure 6.12, one cell's division produces two cells, two cells' divisions produce four cells, and so on. When the number of cells in each generation is expressed as a power of 2, the exponent tells the number of doublings (generations) that have occurred.

The time required for a cell to divide (and its population to double) is called the **generation time.** It varies considerably among organisms and with environmental conditions, such as temperature. Most bacteria have a generation time of 1–3 hours; others require more than 24 hours per generation. (The math required to calculate generation times is presented in Appendix D.) If binary fission continues unchecked, an enormous number of cells will be produced. If a doubling occurred every 20 minutes—which is the case for *E. coli* under favorable conditions—after 20 generations a single initial cell would increase to over 1 million cells. This would require a little less than 7 hours. In 30 generations, or 10 hours, the population would be 1 billion, and in 24 hours it would be a number trailed by 21 zeros. It is difficult to graph population

(1) Cell elongates and DNA is replicated

(2) Cell wall and plasma membrane begin to divide

(3) Cross-wall forms completely around divided DNA

(4) Cells separate

Cell wall   Plasma membrane

DNA (nuclear area)

DNA (nuclear area)

Partially formed cross-wall

Cell wall

TEM   0.5 μm

(a) A diagram of the sequence of cell division.

(b) A thin section of a cell of *Bacillus licheniformis* starting to divide.

FIGURE 6.11 **Binary fission in bacteria.**

■ As a result of binary fission, two individual cells are formed, each identical to the parent cell.

Numbers of Cells	Numbers Expressed as a Power of 2	Visual Representation of Numbers
1	2^0	
2	2^1	
4	2^2	
8	2^3	
16	2^4	
32	2^5	

(a)

Generation Number	Number of Cells	Log$_{10}$ of Number of Cells
0	1	0
5 (2^5) =	32	1.51
10 (2^{10}) =	1,024	3.01
15 (2^{15}) =	32,768	4.52
16 (2^{16}) =	65,536	4.82
17 (2^{17}) =	131,072	5.12
18 (2^{18}) =	262,144	5.42
19 (2^{19}) =	524,288	5.72
20 (2^{20}) =	1,048,576	6.02

(b)

FIGURE 6.12 **Cell division. (a)** Visual representation of increase in bacterial number over five generations. The number of bacteria doubles in each generation. The superscript indicates the generation, that is, $2^5 = 5$ generations. **(b)** Conversion of the number of cells in a population into the logarithmic expression of this number. To arrive at the numbers in the center column, use the y^x key on your calculator. Enter 2 on the calculator; then press the key; then enter 5; then the = sign. The calculator will show the number 32. Thus, the fifth generation population of bacteria will total 32 cells. To arrive at the numbers in the right-hand column, use the log key on your calculator. Enter the number 32, then press the log key. The calculator will show, rounded off, that the log$_{10}$ of 32 is 1.51.

■ If a single bacterium reproduced every 20 minutes, how many would there be in 2 hours?

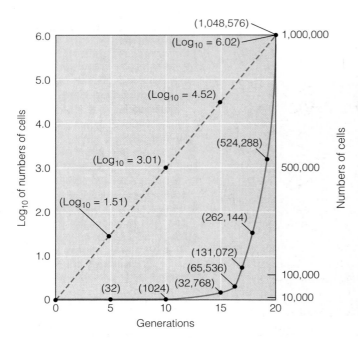

FIGURE 6.13 **A growth curve for an exponentially increasing population, plotted logarithmically (dashed line) and arithmetically (solid line).**

■ If the arithmetic numbers (solid line) were plotted for two more generations, would it still be on the page?

changes of such enormous magnitude by using arithmetic numbers. This is why logarithmic scales are generally used to graph bacterial growth. Understanding logarithmic representations of bacterial populations requires some use of mathematics and is necessary for anyone studying microbiology. (See Appendix D.)

Logarithmic Representation of Bacterial Populations

To illustrate the difference between logarithmic and arithmetic graphing of bacterial populations, let's express 20 bacterial generations both logarithmically and arithmetically. In five generations (2^5), there would be 32 cells; in ten generations (2^{10}), there would be 1024 cells, and so on. (If your calculator has a y^x key and a log key, you can duplicate the numbers in the third column of Figure 6.12.)

In Figure 6.13, notice that the arithmetically plotted line (solid) does not clearly show the population changes in the early stages of the growth curve at this scale. In fact, the first ten generations do not appear to leave the baseline. Furthermore, another one or two arithmetic generations graphed to the same scale would greatly increase the height of the graph and take the line off the page.

The dashed line in Figure 6.13 shows how these plotting problems can be avoided by graphing the $\log_{10}$ of the

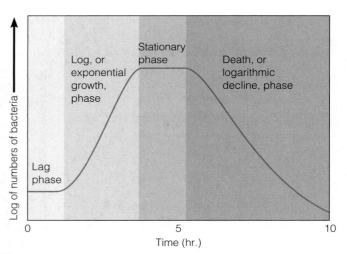

FIGURE 6.14 **A bacterial growth curve, showing the four typical phases of growth.**

■ A bacterial growth curve shows the growth of a population over time.

population numbers. The $\log_{10}$ of the population is plotted at 5, 10, 15, and 20 generations. Notice that a straight line is formed and that a thousand times this population (1,000,000,000, or $\log_{10}$ 9.0) could be accommodated in relatively little extra space. However, this advantage is obtained at the cost of distorting our "common sense" perception of the actual situation. We are not accustomed to thinking in logarithmic relationships, but it is necessary for a proper understanding of graphs of microbial populations.

Phases of Growth

Learning Objective

• Compare the phases of microbial growth, and describe their relation to generation time.

When a few bacteria are inoculated into a liquid growth medium and the population is counted at intervals, it is possible to plot a **bacterial growth curve** that shows the growth of cells over time (Figure 6.14). There are four basic phases of growth: the lag, log, stationary, and death phases.

The Lag Phase

For a while, the number of cells changes very little because the cells do not immediately reproduce in a new medium. This period of little or no cell division is called the **lag phase,** and it can last for 1 hour or several days. During this time, however, the cells are not dormant. The microbial population is undergoing a period of intense metabolic activity involving, in particular, synthesis of enzymes and various molecules. (The situation is analogous to a factory being equipped to produce automobiles; there is considerable tooling-up activity but no immediate increase in the automobile population.)

The Log Phase

Eventually, the cells begin to divide and enter a period of growth, or logarithmic increase, called the **log phase,** or **exponential growth phase.** Cellular reproduction is most active during this period, and generation time reaches a constant minimum. Because the generation time is constant, a logarithmic plot of growth during the log phase is a straight line. The log phase is the time when cells are most active metabolically and is preferred for industrial purposes where, for example, a product needs to be produced efficiently.

However, during their log phase of growth, microorganisms are particularly sensitive to adverse conditions. Radiation and many antimicrobial drugs—for example, the antibiotic penicillin—exert their effect by interfering with some important step in the growth process and are therefore most harmful to cells during this phase.

The Stationary Phase

If exponential growth continues unchecked, startlingly large numbers of cells could arise. For example, a single bacterium (at a weight of 9.5×10^{-13} g per cell) dividing every 20 minutes for only 25.5 hours can theoretically produce a population equivalent in weight to that of an 80,000-ton aircraft carrier. In reality, this does not happen. Eventually, the growth rate slows, the number of microbial deaths balances the number of new cells, and the population stabilizes. The metabolic activities of individual surviving cells also slow at this stage. This period of equilibrium is called the **stationary phase.**

What causes exponential growth to stop is not always clear. The exhaustion of nutrients, accumulation of waste products, and harmful changes in pH may all play a role. In a specialized apparatus called a *chemostat,* a population can be kept in the exponential growth phase indefinitely by draining off spent medium and adding fresh medium. This type of *continuous* culture is used in industrial fermentations.

The Death Phase

The number of deaths eventually exceeds the number of new cells formed, and the population enters the **death phase,** or **logarithmic decline phase.** This phase continues until the population is diminished to a tiny fraction of the number of cells in the previous phase, or the population dies out entirely. Some species pass through the entire series of phases in only a few days; others retain some surviving cells almost indefinitely. Microbial death will be discussed further in Chapter 7.

Direct Measurement of Microbial Growth

Learning Objective
- Explain four direct methods of measuring cell growth.

The growth of microbial populations can be measured in a number of ways. Some methods measure cell numbers; other methods measure the population's total mass, which is often directly proportional to cell numbers. Population numbers are usually recorded as the number of cells in a milliliter of liquid or in a gram of solid material. Because bacterial populations are usually very large, most methods of counting them are based on direct or indirect counts of very small samples; calculations then determine the size of the total population. Assume, for example, that a millionth of a milliliter (10^{-6} ml) of sour milk is found to contain 70 bacterial cells. Then there must be 70 times 1 million, or 70 million, cells per milliliter.

However, it is not practical to measure out a millionth of a milliliter of liquid or a millionth of a gram of food. Therefore, the procedure is done indirectly, in a series of dilutions. For example, if we add 1 ml of milk to 99 ml of water, each milliliter of this dilution now has one-hundredth as many bacteria as each milliliter of the original sample had. By making a series of such dilutions, we can readily estimate the number of bacteria in our original sample. To count microbial populations in foods (such as hamburger), an homogenate of one part food to nine parts water is finely ground in a food blender. Samples of this initial one-tenth dilution can then be transferred with a pipette for further dilutions or cell counts.

Plate Counts

The most frequently used method of measuring bacterial populations is the **plate count.** An important advantage of this method is that it measures the number of viable cells. One disadvantage may be that it takes some time, usually 24 hours or more, for visible colonies to form. This can be a serious problem in some applications, such as quality control of milk, when it is not possible to hold a particular lot for this amount of time.

Plate counts assume that each live bacterium grows and divides to produce a single colony. This is not always true because bacteria frequently grow linked in chains or as clumps (see Figure 4.1 on page 76). Therefore, a colony often results, not from a single bacterium, but from short segments of a chain or from a bacterial clump. To reflect this reality, plate counts are often reported as **colony-forming units (CFU).**

When a plate count is performed, it is important that only a limited number of colonies develop in the plate. When too many colonies are present, some cells are

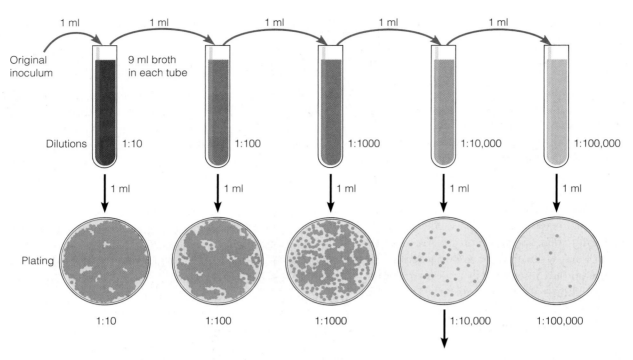

Calculation: Number of colonies on plate × reciprocal of dilution of sample = number of bacteria/ml
(For example, if 32 colonies are on a plate of $^1/_{10,000}$ dilution, then the count is 32 × 10,000 = 320,000/ml in sample.)

FIGURE 6.15 Plate counts and serial dilutions. In serial dilutions, the original inoculum is diluted in a series of dilution tubes. In our example, each succeeding dilution tube will have only one-tenth the number of microbial cells as the preceding tube. Then samples of the dilution are used to inoculate Petri plates, on which colonies grow and can be counted. This count is then used to estimate the number of bacteria in the original sample.

■ Why were the dilutions of 1:1000 and 1:100,000 not counted? Theoretically, how many colonies should appear on the 1:1000 plate?

overcrowded and do not develop; these conditions cause inaccuracies in the count. The Food and Drug Administration convention is to count only plates with 25–250 colonies, but many microbiologists prefer plates with 30–300 colonies. To ensure that some colony counts will be within this range, the original inoculum is diluted several times in a process called **serial dilution** (Figure 6.15).

Serial Dilutions Let's say, for example, that a milk sample has 10,000 bacteria per milliliter. If 1 ml of this sample were plated out, there would theoretically be 10,000 colonies formed in the Petri plate of medium. Obviously, this would not produce a countable plate. If 1 ml of this sample were transferred to a tube containing 9 ml of sterile water, each milliliter of fluid in this tube would now contain 1000 bacteria. If 1 ml of this sample were inoculated into a Petri plate, there would still be too many potential colonies to count on a plate. Therefore, another serial dilution could be made. One milliliter containing 1000 bacteria would be transferred to a second tube of 9 ml of water.

Each milliliter of this tube would now contain only 100 bacteria, and if 1 ml of the contents of this tube were plated out, potentially 100 colonies would be formed—an easily countable number. Learning how to do serial dilutions is an important part of certain experiments in microbiology laboratory classes.

Pour Plates and Spread Plates A plate count is done by either the pour plate method or the spread plate method. The **pour plate method** follows the procedure shown in Figure 6.16a. Either 1.0 ml or 0.1 ml of dilutions of the bacterial suspension is introduced into a Petri dish. The nutrient medium, in which the agar is kept liquid by holding it in a water bath at about 50°C, is poured over the sample, which is then mixed into the medium by gentle agitation of the plate. When the agar solidifies, the plate is incubated. With the pour plate technique, colonies will grow within the nutrient agar (from cells suspended in the nutrient medium as the agar solidifies) as well as on the surface of the agar plate.

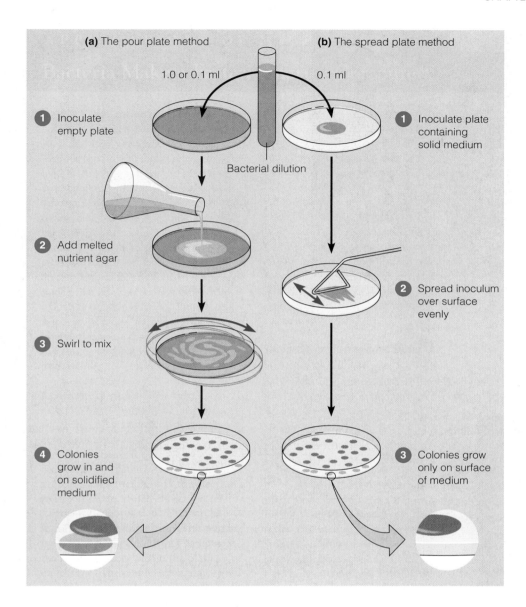

1.0 or 0.1 ml

(b) The spread plate method

0.1 ml

Bacterial dilution

(a) The pour plate method

1 Inoculate empty plate

2 Add melted nutrient agar

3 Swirl to mix

4 Colonies grow in and on solidified medium

(b) The spread plate method

1 Inoculate plate containing solid medium

2 Spread inoculum over surface evenly

3 Colonies grow only on surface of medium

FIGURE 6.16 **Methods of preparing plates for plate counts. (a)** The pour plate method. **(b)** The spread plate method.

- What are the advantages of each method?

This technique has some drawbacks because some relatively heat-sensitive microorganisms may be damaged by the melted agar and will therefore be unable to form colonies. Also, when certain differential media are used, the distinctive appearance of the colony on the surface is essential for diagnostic purposes. Colonies that form beneath the surface of a pour plate are not satisfactory for such tests. To avoid these problems, the **spread plate method** is frequently used instead (Figure 6.16b). A 0.1-ml inoculum is added to the surface of a prepoured, solidified agar medium. The inoculum is then spread uniformly over the surface of the medium with a specially shaped, sterilized glass rod. This method positions all the colonies on the surface and avoids contact of the cells with melted agar.

Filtration

When the quantity of bacteria is very small, as in lakes or relatively pure streams, bacteria can be counted by **filtration** methods (Figure 6.17). In this technique, at least 100 ml of water are passed through a thin membrane filter whose pores are too small to allow bacteria to pass. Thus, the bacteria are filtered out and retained on the surface of the filter. This filter is then transferred to a Petri dish containing a pad soaked in liquid nutrient medium, where colonies arise from the bacteria on the filter's surface. This method is applied frequently to detection and enumeration of coliform bacteria, which are indicators of fecal pollution of food or water. The colonies formed by these bacteria are distinctive when a differential nutrient medium is used. (The colonies shown in Figures 6.9b and c are examples of coliforms.)

FIGURE 6.17 **Counting bacteria by filtration.**

■ Bacteria can be counted by filtration when their concentration is very small.

(a) The bacteria in 100 ml of water were sieved out onto the surface of a membrane filter.

SEM ⊢————⊣ 1 μm

(b) A filter such as shown in photo (a), with the bacteria much more widely spaced, was placed on a pad saturated with liquid Endo medium, which is selective for gram-negative bacteria. The individual bacteria grew into visible colonies. One hundred twenty-four colonies are visible, so we would record 124 bacteria per 100 ml of water sample.

The Most Probable Number (MPN) Method

Another method for determining the number of bacteria in a sample is the **most probable number (MPN) method,** illustrated in Figure 6.18. This statistical estimating technique is based on the fact that the greater the number of bacteria in a sample, the more dilution is needed to reduce the density to the point at which no bacteria are left to grow in the tubes in a dilution series. The MPN method is most useful when the microbes being counted will not grow on solid media (such as the chemoautotrophic nitrifying bacteria). It is also useful when the growth of bacteria in a liquid differential medium is used to identify the microbes (such as coliform bacteria, which selectively ferment lactose to acid, in water testing). The MPN is only a statement that there is a 95% chance that the bacterial population falls within a certain range and that the MPN is statistically the most probable number.

Direct Microscopic Count

In the method known as the **direct microscopic count,** a measured volume of a bacterial suspension is placed within a defined area on a microscope slide. In the *Breed count method,* which is used to count the number of bacteria in milk, for example, a 0.01-ml sample is spread over a marked square centimeter of slide, stain is added so that the bacteria can be seen, and the sample is viewed under the oil immersion objective lens. The area of the viewing field of this objective can be determined. Once the number of bacteria

has been counted in several different fields, the average number of bacteria per viewing field can be calculated. From these data, the number of bacteria in the square centimeter over which the sample was spread can also be calculated. Because this area on the slide contained 0.01 ml of sample, the number of bacteria in each milliliter of the suspension is the number of bacteria in the sample times 100.

A specially designed slide called a *Petroff-Hausser cell counter* is also used in direct microscopic counts (Figure 6.19).

❶ In the center of the microscope slide the surface is inscribed with a grid of squares of known area. Above this grid is a space of known volume formed because it is slightly lower than the portions of the slide that support the cover glass.

❷ The volume over the grid is filled with the microbial suspension.

❸–❹ The average number of bacteria in each of a series of these squares is calculated and then multiplied by the factor that produces the count per milliliter.

Motile bacteria are difficult to count by this method, and, as happens with other microscopic methods, dead cells are about as likely to be counted as live ones. In addition to these disadvantages, a rather high concentration of cells is required to be countable—about 10 million bacteria per milliliter. The chief advantage of microscopic counts is that no incubation time is required, and they are usually reserved for applications in which time is the primary consideration. This

Volume of Inoculum for Each Set of Five Tubes	Tubes of Nutrient Medium (Sets of Five Tubes)	Number of Positive Tubes in Set
10 ml		5
1 ml		3
0.1 ml		1

(a) Most probable number (MPN) dilution series. In this example, there are three sets of tubes and five tubes in each set. Each tube in the first set of five tubes receives 10 ml of the inoculum, such as a sample of water. Each tube in the second set of five tubes receives 1 ml of the sample, and the third set, 0.1 ml each. There were enough bacteria in the sample so that all five tubes in the first set showed bacterial growth and were recorded as positive. In the second set, which received only one-tenth as much inoculum, only three tubes were positive. In the third set, which received one-hundredth as much inoculum, only one tube was positive.

Combination of Positives	MPN Index/ 100 ml	95% Confidence Limits	
		Lower	Upper
4-2-0	22	9	56
4-2-1	26	12	65
4-3-0	27	12	67
4-3-1	33	15	77
4-4-0	34	16	80
5-0-0	23	9	86
5-0-1	30	10	110
5-0-2	40	20	140
5-1-0	30	10	120
5-1-1	50	20	150
5-1-2	60	30	180
5-2-0	50	20	170
5-2-1	70	30	210
5-2-2	90	40	250
5-3-0	80	30	250
5-3-1	110	40	300
5-3-2	140	60	360

(b) MPN table. MPN tables enable us to calculate for a sample the microbial numbers that are statistically likely to lead to such a result. The number of positive tubes is recorded for each set: In the shaded example, 5, 3, and 1. If we look up this combination in an MPN table, we find that the MPN index per 100 ml is 110. Statistically, this means that 95% of the water samples that give this result contain 40–300 bacteria, with 110 being the most probable number.

FIGURE 6.18 **The most probable number (MPN) method.**

- Under what circumstances is the MPN method used to determine the number of bacteria in a sample?

Grid with 25 large squares

Cover glass

Slide

1 Bacterial suspension is added here and fills the shallow volume over the squares by capillary action.

Bacterial suspension

Cover glass

Slide

Location of squares

2 Cross section of a cell counter. The depth under the cover glass is known, and the area of the squares is known, so the volume of the bacterial suspension over the squares can be calculated (depth × area).

3 Microscopic count: All cells in several large squares are counted, and the numbers are averaged. The large square shown here has 14 bacterial cells.

4 The volume of fluid over the large square is 1/1,250,000 of a milliliter. If it contains 14 cells, as shown here, then there are 14 times 1,250,000 (17,500,000) cells in a milliliter.

FIGURE 6.19 **Direct microscopic count of bacteria with a Petroff-Hausser cell counter.** The average number of cells within a large square multiplied by a factor of 1,250,000 gives the number of bacteria per milliliter.

- Direct microscopic counts are useful for certain applications but have a disadvantage of requiring rather high microbial populations for them to be countable.

advantage also holds for *electronic cell counters*, sometimes known as *Coulter counters*, which automatically count the number of cells in a measured volume of liquid. These instruments are used in some research laboratories and hospitals.

Estimating Bacterial Numbers by Indirect Methods

Learning Objectives

- *Differentiate between direct and indirect methods of measuring cell growth.*
- *Explain three indirect methods of measuring cell growth.*

It is not always necessary to count microbial cells to estimate their numbers. In science and industry, microbial numbers and activity are determined by some of the following indirect means as well.

Turbidity

For some types of experimental work, estimating *turbidity* is a practical way of monitoring bacterial growth. As bacteria multiply in a liquid medium, the medium becomes turbid, or cloudy with cells.

The instrument used to measure turbidity is a *spectrophotometer* (or colorimeter). In the spectrophotometer, a beam of light is transmitted through a bacterial suspension to a light-sensitive detector (Figure 6.20). As bacterial numbers increase, less light will reach the detector. This change of light will register on the instrument's scale as the *percentage of transmission*. Also printed on the instrument's scale is a logarithmic expression called the *absorbance* (sometimes called *optical density*, or *OD*); a value derived from the percentage of transmission may also be reported. The absorbance is used to plot bacterial growth. When the bacteria are in logarithmic growth or decline, a graph of absorbance versus time will form an approximately straight line. If absorbance readings are matched with plate counts of the same culture, this correlation can be used in future estimations of bacterial numbers obtained by measuring turbidity.

More than a million cells per milliliter must be present for the first traces of turbidity to be visible. About 10 million to 100 million cells per milliliter are needed to make a suspension turbid enough to be read on a spectrophotometer. Therefore, turbidity is not a useful measure of contamination of liquids by relatively small numbers of bacteria.

Metabolic Activity

Another indirect way to estimate bacterial numbers is to measure a population's *metabolic activity*. This method assumes that the amount of a certain metabolic product, such as acid or CO_2, is in direct proportion to the number of bacteria present. An example of a practical application of a metabolic test is the microbiological assay in which acid production is used to determine amounts of vitamins.

Dry Weight

For filamentous bacteria and molds, the usual measuring methods are less satisfactory. A plate count would not measure this increase in filamentous mass. In plate counts

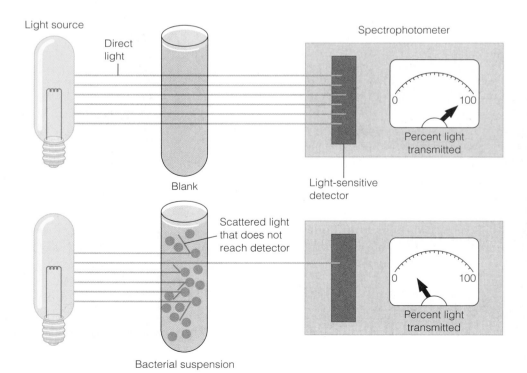

FIGURE 6.20 Turbidity estimation of bacterial numbers. The amount of light striking the light-sensitive detector on the spectrophotometer is inversely proportional to the number of bacteria under standardized conditions. The less light transmitted, the more bacteria in the sample.

■ Is turbidity a direct or an indirect method of measuring bacterial growth?

of actinomycetes (Figure 11.20) and molds, it is mostly the number of asexual spores that is counted instead. This is not a good measure of growth. One of the better ways to measure the growth of filamentous organisms is by *dry weight*. In this procedure, the fungus is removed from the growth medium, filtered to remove extraneous material, and dried in a desiccator. It is then weighed. For bacteria, the same basic procedure is followed.

* * *

You now have a basic understanding of the requirements for, and measurements of, microbial growth. In Chapter 7, we will look at how this growth is controlled in laboratories, hospitals, industry, and our homes.

STUDY OUTLINE

The Requirements for Growth (pp. 156–163)

1. The growth of a population is an increase in the number of cells.

2. The requirements for microbial growth are both physical and chemical.

Physical Requirements (pp. 156–159)

1. On the basis of preferred temperature ranges, microbes are classified as psychrophiles (cold-loving), mesophiles (moderate-temperature–loving), and thermophiles (heat-loving).

2. The minimum growth temperature is the lowest temperature at which a species will grow, the optimum growth temperature is the temperature at which it grows best, and the maximum growth temperature is the highest temperature at which growth is possible.

3. Most bacteria grow best at a pH value between 6.5 and 7.5.

4. In a hypertonic solution, most microbes undergo plasmolysis; halophiles can tolerate high salt concentrations.

Chemical Requirements (pp. 159–163)

1. All organisms require a carbon source; chemoheterotrophs use an organic molecule, and autotrophs typically use carbon dioxide.

2. Nitrogen is needed for protein and nucleic acid synthesis. Nitrogen can be obtained from the decomposition of proteins or from NH_4^+ or NO_3^-; a few bacteria are capable of nitrogen (N_2) fixation.

3. On the basis of oxygen requirements, organisms are classified as obligate aerobes, facultative anaerobes, obligate anaerobes, aerotolerant anaerobes, and microaerophiles.

4. Aerobes, facultative anaerobes, and aerotolerant anaerobes must have the enzymes superoxide dismutase ($2 O_2^- + 2 H^+ \longrightarrow O_2 + H_2O_2$) and either catalase ($2 H_2O_2 \longrightarrow 2 H_2O + O_2$) or peroxidase ($H_2O_2 + 2 H^+ \longrightarrow 2 H_2O$).

5. Other chemicals required for microbial growth include sulfur, phosphorus, trace elements, and, for some microorganisms, organic growth factors.

Culture Media (pp. 163–169)

1. A culture medium is any material prepared for the growth of bacteria in a laboratory.

2. Microbes that grow and multiply in or on a culture medium are known as a culture.

3. Agar is a common solidifying agent for a culture medium.

Chemically Defined Media (p. 164)

1. A chemically defined medium is one in which the exact chemical composition is known.

Complex Media (p. 164)

1. A complex medium is one in which the exact chemical composition varies slightly from batch to batch.

Anaerobic Growth Media and Methods (pp. 164–166)

1. Reducing media chemically remove molecular oxygen (O_2) that might interfere with the growth of anaerobes.

2. Petri plates can be incubated in an anaerobic jar, anaerobic chamber, or OxyPlate.

Special Culture Techniques (pp. 166–167)

1. Some parasitic and fastidious bacteria must be cultured in living animals or in cell cultures.

2. CO_2 incubators or candle jars are used to grow bacteria requiring an increased CO_2 concentration.

Selective and Differential Media (pp. 167–168)

1. By inhibiting unwanted organisms with salts, dyes, or other chemicals, selective media allow growth of only the desired microbes.

2. Differential media are used to distinguish among different organisms.

Enrichment Culture (pp. 168–169)

1. An enrichment culture is used to encourage the growth of a particular microorganism in a mixed culture.

Obtaining Pure Cultures (p. 169)

1. A colony is a visible mass of microbial cells that theoretically arose from one cell.

2. Pure cultures are usually obtained by the streak plate method.

Preserving Bacterial Cultures (p. 169)

1. Microbes can be preserved for long periods of time by deep-freezing or lyophilization (freeze-drying).

The Growth of Bacterial Cultures (pp. 170–179)

Bacterial Division (p. 170)

1. The normal reproductive method of bacteria is binary fission, in which a single cell divides into two identical cells.

2. Some bacteria reproduce by budding, aerial spore formation, or fragmentation.

Generation Time (pp. 170–172)

1. The time required for a cell to divide or a population to double is known as the generation time.

Logarithmic Representation of Bacterial Populations (p. 172)

1. Bacterial division occurs according to a logarithmic progression (two cells, four cells, eight cells, etc.).

Phases of Growth (pp. 172–173)

1. During the lag phase, there is little or no change in the number of cells, but metabolic activity is high.

2. During the log phase, the bacteria multiply at the fastest rate possible under the conditions provided.

3. During the stationary phase, there is an equilibrium between cell division and death.

4. During the death phase, the number of deaths exceeds the number of new cells formed.

Direct Measurement of Microbial Growth (pp. 173–178)

1. A standard plate count reflects the number of viable microbes and assumes that each bacterium grows into a single colony; plate counts are reported as number of colony-forming units (CFU).

2. A plate count may be done by either the pour plate method or the spread plate method.

3. In filtration, bacteria are retained on the surface of a membrane filter and then transferred to a culture medium to grow and subsequently be counted.

4. The most probable number (MPN) method can be used for microbes that will grow in a liquid medium; it is a statistical estimation.

5. In a direct microscopic count, the microbes in a measured volume of a bacterial suspension are counted with the use of a specially designed slide.

Estimating Bacterial Numbers by Indirect Methods (pp. 178–179)

1. A spectrophotometer is used to determine turbidity by measuring the amount of light that passes through a suspension of cells.

2. An indirect way of estimating bacterial numbers is measuring the metabolic activity of the population (for example, acid production or oxygen consumption).

3. For filamentous organisms such as fungi, measuring dry weight is a convenient method of growth measurement.

STUDY QUESTIONS

Review

1. Describe binary fission.

2. Draw a typical bacterial growth curve. Label and define each of the four phases.

3. Macronutrients (needed in relatively large amounts) are often listed as CHONPS. What does each of these letters indicate, and why are they needed by the cell?

4. Most bacteria grow best at pH _____.

5. Why can high concentrations of salt or sugar be used to preserve food?

6. Define and explain the importance of each of the following:
   a. catalase
   b. hydrogen peroxide
   c. peroxidase
   d. superoxide free radical
   e. superoxide dismutase

7. *Clostridium* can be cultured in an anaerobic incubator or in the presence of atmospheric oxygen if thioglycolate is added to the nutrient broth. Compare these two techniques. Using terms from question 6, explain why elaborate culture techniques are used for *Clostridium*.

8. Eight methods of measuring microbial growth were explained in this chapter. Categorize each as either a direct or an indirect method.

9. By deep-freezing, bacteria can be stored without harm for extended periods. Why do refrigeration and freezing preserve foods?

10. A pastry chef accidentally inoculated a cream pie with six *S. aureus* cells. If *S. aureus* has a generation time of 60 minutes, how many cells would be in the cream pie after 7 hours?

11. Nitrogen and phosphorus added to beaches following an oil spill encourage the growth of natural oil-degrading bacteria.

Explain why the bacteria do not grow if nitrogen and phosphorus are not added.

12. Differentiate between complex and chemically defined media.

13. Draw the following growth curves for *E. coli*, starting with 100 cells with a generation time of 30 minutes at 35°C.
    a. The cells are incubated for 5 hours at 35°C.
    b. After 5 hours, the temperature is changed to 20°C for 2 hours.
    c. After 5 hours at 35°C, the temperature is changed to 5°C for 2 hours followed by 35°C for 5 hours.

Multiple Choice

Use the following information to answer questions 1 and 2. Two culture media were inoculated with four different bacteria. After incubation, the following results were obtained:

Organism	Medium 1	Medium 2
Escherichia coli	Red colonies	No growth
Staphylococcus aureus	No growth	Growth
S. epidermidis	No growth	Growth
Salmonella enterica	Colorless colonies	No growth

1. Medium 1 is
   a. selective.
   b. differential.
   c. both selective and differential.

2. Medium 2 is
   a. selective.
   b. differential.
   c. both selective and differential.

Use the following graph to answer questions 3 and 4.

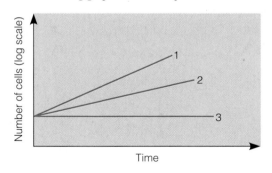

3. Which of the lines best depicts the log phase of a thermophile incubated at room temperature?

4. Which of the lines best depicts the log phase of *Listeria monocytogenes* growing in a human?

5. Assume you inoculated 100 facultatively anaerobic cells onto nutrient agar and incubated the plate aerobically. You then inoculated 100 cells of the same species onto nutrient agar and incubated the second plate anaerobically. After incubation for 24 hours, you should have
   a. more colonies on the aerobic plate.
   b. more colonies on the anaerobic plate.
   c. the same number of colonies on both plates.

6. The term *trace elements* refers to
   a. the elements CHONPS.
   b. vitamins.
   c. nitrogen, phosphorus, and sulfur.
   d. small mineral requirements.
   e. toxic substances.

7. Which one of the following temperatures would most likely kill a mesophile?
   a. −50°C   **c.** 9°C   **e.** 60°C
   b. 0°C   **d.** 37°C

8. All of the following are true about agar *except:*
   a. It is a source of nutrients in culture media.
   b. It is a polysaccharide.
   c. It liquefies at 100°C.
   d. It solidifies at approximately 40°C.
   e. It is metabolized by few bacteria.

9. Which of the following types of media would *not* be used to culture aerobes?
   a. selective media
   b. reducing media
   c. enrichment media
   d. differential media
   e. complex media

10. An organism that has peroxidase and superoxide dismutase but lacks catalase is most likely an
    a. aerobe.
    b. aerotolerant anaerobe.
    c. obligate anaerobe.

Critical Thinking

1. *E. coli* was incubated with aeration in a nutrient medium containing two carbon sources, and the following growth curve was made from this culture.
   a. Explain what happened at the time marked *x*.
   b. Which substrate provided "better" growth conditions for the bacteria? How can you tell?

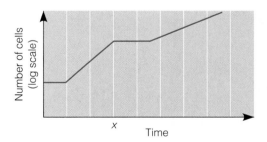

2. *Clostridium* and *Streptococcus* are both catalase-negative. *Streptococcus* grows by fermentation. Why is *Clostridium* killed by oxygen, whereas *Streptococcus* is not?

3. Most laboratory media contain a fermentable carbohydrate and peptone because the majority of bacteria require carbon, nitrogen, and energy sources in these forms. How are these three needs met by glucose–minimal salts medium? (*Hint:* See Table 6.2.)

4. Flask A contains yeast cells in glucose–minimal salts broth incubated at 30°C with aeration. Flask B contains yeast cells in glucose–minimal salt broth incubated at 30°C in an anaerobic jar. The yeasts are facultative anaerobes.
   a. Which culture produced more ATP?
   b. Which culture produced more alcohol?
   c. Which culture had the shorter generation time?
   d. Which culture had the greater cell mass?
   e. Which culture had the higher absorbance?

Clinical Applications

1. Assume that after washing your hands, you leave ten bacterial cells on a new bar of soap. You then decide to do a plate count of the soap after it was left in the soap dish for 24 hours. You dilute 1 g of the soap 1:10^6 and plate it on standard plate count agar. After 24 hours of incubation, there are 168 colonies. How many bacteria were on the soap? How did they get there?

2. Heat lamps are commonly used to maintain foods at about 50°C for as long as 12 hours in cafeteria serving lines. The following experiment was conducted to determine whether this practice poses a potential health hazard.

   Beef cubes were surface-inoculated with 500,000 bacterial cells and incubated at 43–53°C to establish temperature limits for bacterial growth. The following results were obtained from standard plate counts performed on beef cubes at 6 and 12 hours after inoculation:

	Temp. (°C)	Bacteria/Gram of Beef After 6 hr	12 hr
S. aureus	43	140,000,000	740,000,000
	51	810,000	59,000
	53	650	300
S. typhimurium	43	3,200,000	10,000,000
	51	950,000	83,000
	53	1,200	300
C. perfringens	43	1,200,000	3,600,000
	51	120,000	3,800
	53	300	300

Draw the growth curves for each organism. What holding temperature would you recommend? Assuming that cooking kills bacteria in foods, how could these bacteria contaminate the cooked foods? What disease does each organism cause?

3. The number of bacteria in saliva samples was determined by collecting the saliva, making serial dilutions, and inoculating nutrient agar by the pour plate method. The plates were incubated aerobically for 48 hours at 37°C.

	Bacteria/ml Saliva Before Using Mouthwash	After Using Mouthwash
Product 1	13.1 × 10^6	10.9 × 10^6
Product 2	11.7 × 10^6	14.2 × 10^5
Product 3	9.3 × 10^5	7.7 × 10^5

What can you conclude from these data? Did all the bacteria present in each saliva sample grow?

LEARNING WITH TECHNOLOGY

The Microbiology Place CD-ROM/website includes many resources to help you study and succeed in this course. In addition to self-quizzes, web links, and flashcards, you'll find the following activities for this chapter in the "Activities" section of The Microbiology Place:

Binary Fission

Arithmetic and Logarithmic Growth Curves

Interpreting Growth Curves

Case Study: Determining the Effectiveness of a Food Preservative

Chapter 7

The Control of Microbial Growth

BACTERIA TRAPPED IN A MEMBRANE FILTER. *Filtration can be used to remove microorganisms from water and solutions.*

The scientific control of microbial growth began only about 100 years ago. Recall from Chapter 1 that Pasteur's work on microorganisms led scientists to believe that microbes were a possible cause of disease. In the mid-1800s, the Hungarian physician Ignatz Semmelweis and English physician Joseph Lister used this thinking to develop some of the first microbial control practices for medical procedures. These practices included hand washing with microbe-killing chloride of lime and **aseptic surgery** techniques to prevent microbial contamination of surgical wounds. Until that time, hospital-acquired

infections, or nosocomial infections, were the cause of death in at least 10% of surgical cases, and deaths of delivering mothers were as high as 25%. Ignorance of microbes was such that, during the American Civil War, a surgeon might have cleaned his scalpel on his bootsole between incisions.

Over the last century, scientists have continued to develop a variety of physical methods and chemical agents to control microbial growth. In Chapter 20 we will discuss methods for the control of microbes once infection has occurred, mainly antibiotic chemotherapy.

The Terminology of Microbial Control

Learning Objective
- Define the following key terms related to microbial control: sterilization, disinfection, antisepsis, degerming, sanitization, biocide, germicide, bacteriostasis, and asepsis.

A word frequently used, and misused, in discussing the control of microbial growth is *sterilization*. **Sterilization** is the removal or destruction of *all forms* of microbial life. Heating is the most common method used for killing microbes, including the most resistant forms such as endospores. Sterilization by removal of microbes from liquids or gases can be done by filtration.

One would think that canned food in the supermarket is completely sterile. In reality, the heat treatment required to ensure absolute sterility would unnecessarily degrade the quality of the food. Instead, food is subjected only to enough heat to destroy the endospores of *Clostridium botulinum*, which can produce a deadly toxin. This limited heat treatment is termed **commercial sterilization.** The endospores of a number of thermophilic bacteria, capable of causing food spoilage but not human disease, are considerably more resistant to heat than *C. botulinum*. If present, they will survive, but their survival is usually of no practical consequence; they will not grow at normal food storage temperatures. If canned foods in a supermarket were incubated at temperatures in the growth range of these thermophiles (above about 45°C), significant food spoilage would occur.

Complete sterilization is often not required in other settings. For example, the body's normal defenses can cope

TABLE 7.1	**Terminology Relating to the Control of Microbial Growth**	
	Definition	**Comments**
Sterilization	Destruction or removal of all forms of microbial life, including endospores.	Usually done by steam under pressure or a sterilizing gas such as ethylene oxide.
Commercial Sterilization	Sufficient heat treatment to kill endospores of *Clostridium botulinum* in canned food.	More-resistant endospores of thermophilic bacteria may survive, but they will not germinate and grow under normal storage conditions.
Disinfection	Destruction of vegetative pathogens.	May make use of physical or chemical methods.
Antisepsis	Destruction of vegetative pathogens on living tissue.	Treatment is almost always by chemical antimicrobials.
Degerming	Removal of microbes from a limited area, such as the skin around an injection site.	Mostly a mechanical removal by an alcohol-soaked swab.
Sanitization	Treatment intended to lower microbial counts on eating and drinking utensils to safe public health levels.	May be done with high-temperature washing or by dipping into a chemical disinfectant.

with a few microbes entering a surgical wound. A drinking glass or a fork in a restaurant requires only enough microbial control to prevent the transmission of possibly pathogenic microbes from one person to another.

Control directed at destroying harmful microorganisms is called **disinfection.** It usually refers to the destruction of vegetative (non–endospore-forming) pathogens, which is not the same thing as complete sterility. Disinfection might make use of chemicals, ultraviolet radiation, boiling water, or steam. In practice, the term is most commonly applied to the use of a chemical (a *disinfectant*) to treat an inert surface or substance. When this treatment is directed at living tissue, it is called **antisepsis,** and the chemical is then called an *antiseptic*. Therefore, in practice the same chemical might be called a disinfectant for one use and an antiseptic for another. Of course, many chemicals suitable for swabbing a tabletop would be too harsh to use on living tissue.

There are modifications of disinfection and antisepsis. For example, when someone is about to receive an injection, the skin is swabbed with alcohol—the process of **degerming** (or *degermation*), which mostly results in the mechanical removal, rather than the killing, of most of the microbes in a limited area. Restaurant glassware, china, and tableware are subjected to **sanitization,** which is intended to lower microbial counts to safe public health levels and minimize the chances of disease transmission from one user to another. This is usually accomplished by high-temperature washing or, in the case of glassware in a bar, washing in a sink followed by a dip in a chemical disinfectant.

Table 7.1 summarizes the terminology relating to the control of microbial growth.

Names of treatments that cause the outright death of microbes have the suffix -*cide*, meaning kill. A **biocide,** or **germicide,** kills microorganisms (usually with certain exceptions, such as endospores); a *fungicide* kills fungi; a *virucide* inactivates viruses; and so on. Other treatments only inhibit the growth and multiplication of bacteria; their names have the suffix -*stat* or -*stasis*, meaning to stop or to steady, as in **bacteriostasis.** Once a bacteriostatic agent is removed, growth might resume.

Sepsis, from the Greek for decay or putrid, indicates bacterial contamination, as in septic tanks for sewage treatment. *Aseptic* means that an object or area is free of pathogens. Recall from Chapter 1 that **asepsis** is the absence of significant contamination. Aseptic techniques are important in surgery to minimize contamination from the instruments, operating personnel, and the patient.

The Rate of Microbial Death

Learning Objective
- *Describe the patterns of microbial death caused by treatments with microbial control agents.*

When bacterial populations are heated or treated with antimicrobial chemicals, they usually die at a constant rate.

TABLE 7.2	Microbial Death Rate: An Example	
Time (mm)	**Deaths per Minute**	**Number of Survivors**
0	0	1,000,000
1	900,000	100,000
2	90,000	10,000
3	9000	1000
4	900	100
5	90	10
6	9	1

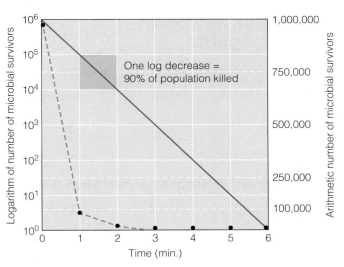

(a) The curve is plotted logarithmically (solid line) and arithmetically (broken line). In this case, the cells are dying at a rate of 90% each minute.

For example, suppose a population of 1 million microbes has been treated for 1 minute, and 90% of the population has died. We are now left with 100,000 microbes. If the population is treated for another minute, 90% of *those* microbes die, and we are left with 10,000 survivors. In other words, for each minute the treatment is applied, 90% of the remaining population is killed (Table 7.2). If the death curve is plotted logarithmically, the death rate is constant, as shown by the straight line in Figure 7.1a.

Several factors influence the effectiveness of antimicrobial treatments:

- *The number of microbes.* The more microbes there are to begin with, the longer it takes to eliminate the entire population (Figure 7.1b).

- *Environmental influences.* The presence of organic matter often inhibits the action of chemical antimicrobials. In hospitals, the presence of organic matter in blood, vomit, or feces influences the selection of disinfectants. Microbes in surface biofilms are difficult for biocides to reach effectively. Because their activity is due to temperature-dependent chemical reactions, disinfectants work somewhat better under warm conditions. Directions on disinfectant containers frequently specify the use of a warm solution.

  The nature of the suspending medium is also a factor in heat treatment. Fats and proteins are especially protective, and a medium rich in these substances protects microbes, which will then have a higher survival rate. Heat is also measurably more effective under acidic conditions.

- *Time of exposure.* Chemical antimicrobials often require extended exposure for more-resistant microbes or endospores to be affected. In heat treatments, a

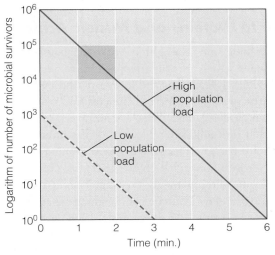

(b) The effect of high or low initial load of microbes. If the rate of killing is the same, it will take longer to kill all members of a larger population than a smaller one. This is true for both heat and chemical treatments.

FIGURE 7.1 **A microbial death curve.**

■ If the graph in part (a) reflects the experience with a vegetative bacterium, what would the logarithmic curve look like for the endospores of the same bacterium? Less steep? Steeper?

longer exposure can compensate for a lower temperature, a phenomenon of particular importance to pasteurization of dairy products.

- *Microbial characteristics.* The concluding section of this chapter discusses how microbial characteristics affect chemical and physical control methods.

Actions of Microbial Control Agents

Learning Objective

- *Describe the effects of microbial control agents on cellular structures.*

In this section, we examine the ways various agents actually kill or inhibit microbes.

Alteration of Membrane Permeability

A microorganism's plasma membrane (see Figure 4.13), located just inside the cell wall, is the target of many microbial control agents. This membrane actively regulates the passage of nutrients into the cell and the elimination of wastes from the cell. Damage to the lipids or proteins of the plasma membrane by antimicrobial agents causes cellular contents to leak into the surrounding medium and interferes with the growth of the cell.

Damage to Proteins and Nucleic Acids

Bacteria are sometimes thought of as "little bags of enzymes." Enzymes, which are primarily protein, are vital to all cellular activities. Recall that the functional properties of proteins are the result of their three-dimensional shape (see Figure 2.15 on page 46). This shape is maintained by chemical bonds that link adjoining portions of the amino acid chain as it folds back and forth upon itself. Some of those bonds are hydrogen bonds, which are susceptible to breakage by heat or certain chemicals; breakage results in denaturation of the protein. Covalent bonds, which are stronger, are also subject to attack. For example, disulfide bridges, which play an important role in protein structure by joining amino acids with exposed sulfhydryl (—SH) groups, can be broken by certain chemicals or sufficient heat.

The nucleic acids DNA and RNA are the carriers of the cell's genetic information. Damage to these nucleic acids by heat, radiation, or chemicals is frequently lethal to the cell; the cell can no longer replicate, nor can it carry out normal metabolic functions such as the synthesis of enzymes.

Physical Methods of Microbial Control

Learning Objectives

- *Compare the effectiveness of moist heat (boiling, autoclaving, pasteurization) and dry heat.*
- *Describe how filtration, low temperatures, high pressure, desiccation, and osmotic pressure suppress microbial growth.*
- *Explain how radiation kills cells.*

As early as the Stone Age, it is likely that humans were already using some physical methods of microbial control to preserve foods. Drying (desiccation) and salting (osmotic pressure) were probably among the earliest techniques.

When selecting methods of microbial control, consideration must be given to effects on things besides the microbes. For example, certain vitamins or antibiotics in a solution might be inactivated by heat. Many laboratory or hospital materials, such as rubber and latex tubing, are damaged by repeated heating. There are also economic considerations; for example, it may be less expensive to use presterilized, disposable plasticware than to repeatedly wash and resterilize glassware.

Heat

A visit to any supermarket will demonstrate that heat-preserved canned goods represent one of the most common methods of food preservation. Laboratory media and glassware, and hospital instruments, are also usually sterilized by heat. Heat appears to kill microorganisms by denaturing their enzymes; the resultant changes to the three-dimensional shapes of these proteins inactivate them (see Figure 5.6 on page 118).

Heat resistance varies among different microbes; these differences can be expressed through the concept of thermal death point. **Thermal death point (TDP)** is the lowest temperature at which all the microorganisms in a particular liquid suspension will be killed in 10 minutes.

Another factor to be considered in sterilization is the length of time required. This is expressed as **thermal death time (TDT),** the minimal length of time for all bacteria in a particular liquid culture to be killed at a given temperature. Both TDP and TDT are useful guidelines that indicate the severity of treatment required to kill a given population of bacteria.

Decimal reduction time (DRT, or *D value)* is a third concept related to bacterial heat resistance. DRT is the time, in minutes, in which 90% of a population of bacteria at a given temperature will be killed (in Table 7.2 and Figure 7.1a, DRT is 1 minute).

Moist Heat

Moist heat kills microorganisms primarily by the coagulation of proteins (denaturation), which is caused by breakage of the hydrogen bonds that hold the proteins in their three-dimensional structure. This coagulation process is familiar to anyone who has watched an egg white frying.

One type of moist heat sterilization is boiling, which kills vegetative forms of bacterial pathogens, almost all viruses, and fungi and their spores within about 10 minutes, usually much faster. Free-flowing (unpressurized) steam is equivalent in temperature to boiling water. Endospores and

Steam to chamber Safety valve Pressure gauge Operating valve (controls steam from jacket to chamber)

FIGURE 7.2 **An autoclave.** The entering steam forces the air out of the bottom (blue arrows). The automatic ejector valve remains open as long as an air-steam mixture is passing out of the waste line. When all the air has been ejected, the higher temperature of the pure steam closes the valve, and the pressure in the chamber increases.

■ Autoclaving is the preferred method of sterilization, provided the material to be sterilized will not be damaged by heat or moisture.

Steam

Steam chamber

Door

Air

Perforated shelf

Sediment screen

Steam jacket

Thermometer

To waste line

Automatic ejector valve is thermostatically controlled and closes on contact with pure steam when air is exhausted.

Pressure regulator for steam supply

Steam supply

some viruses, however, are not destroyed this quickly. Some hepatitis viruses, for example, can survive up to 30 minutes of boiling, and some bacterial endospores can resist boiling for more than 20 hours. Boiling is therefore not always a reliable sterilization procedure. However, brief boiling, even at high altitudes, will kill most pathogens. The use of boiling to sanitize baby bottles is a familiar example.

Reliable sterilization with moist heat requires temperatures above that of boiling water. These high temperatures are most commonly achieved by steam under pressure in an **autoclave** (Figure 7.2). Autoclaving is the preferred method of sterilization, unless the material to be sterilized can be damaged by heat or moisture.

The higher the pressure in the autoclave, the higher the temperature. For example, when free-flowing steam at a temperature of 100°C is placed under a pressure of 1 atmosphere above sea level pressure—that is, about 15 pounds of pressure per square inch (psi)—the temperature rises to 121°C. Increasing the pressure to 20 psi raises the temperature to 126°C. The relationship between temperature and pressure is shown in Table 7.3.

TABLE 7.3	**The Relationship Between the Pressure and Temperature of Steam at Sea Level***
Pressure (psi in excess of atmospheric pressure)	**Temperature (°C)**
0 psi	100
5 psi	110
10 psi	116
15 psi	121
20 psi	126
30 psi	135

*At higher altitudes the atmospheric pressure is less, which must be taken into account in operation of an autoclave. For example, in order to reach sterilizing temperatures (121°C) in Denver, Colorado, whose altitude is 5280 feet (1600 meters), the pressure shown on the autoclave gauge would need to be higher than the 15 psi shown in the table.

TABLE 7.4	**The Effect of Container Size on Autoclave Sterilization Times for Liquid Solutions***	
Container Size	**Liquid Volume**	**Sterilization Time (min)**
Test tube: 18 × 150 mm	10 ml	15
Erlenmeyer flask: 125 ml	95 ml	15
Erlenmeyer flask: 2000 ml	1500 ml	30
Fermentation bottle: 9000 ml	6750 ml	70

*Sterilization times in the autoclave include the time for the contents of the containers to reach sterilization temperatures. For smaller containers, this is only 5 min or less; but for a 9000-ml bottle, it might be as much as 70 min. A container is usually not filled past 75% of its capacity.

Sterilization in an autoclave is most effective when the organisms are either contacted by the steam directly or are contained in a small volume of aqueous (primarily water) liquid. Under these conditions, steam at a pressure of about 15 psi (121°C) will kill *all* organisms and their endospores in about 15 minutes.

Autoclaving is used to sterilize culture media, instruments, dressings, intravenous equipment, applicators, solutions, syringes, transfusion equipment, and numerous other items that can withstand high temperatures and pressures. Large industrial autoclaves are called *retorts*, but the same principle applies for the common household pressure cooker used in the home canning of foods.

Heat requires extra time to reach the center of solid materials, such as canned meats, because such materials do not develop the efficient heat-distributing convection currents that occur in liquids. Heating large containers also requires extra time. Table 7.4 shows the different time requirements for sterilizing liquids in various container sizes. Unlike sterilizing aqueous solutions, sterilizing the surface of a solid requires that steam actually contact it. To sterilize dry glassware, bandages, and the like, care must be taken to ensure that steam contacts all surfaces.

For example, aluminum foil is impervious to steam and should not be used to wrap dry materials that are to be sterilized; paper should be used instead. Care should also be taken to avoid trapping air in the bottom of a dry container because trapped air will not be replaced by steam, which is lighter than air. The trapped air is the equivalent of a small hot-air oven, which, as we will see shortly, requires a higher temperature and longer time to sterilize materials. Containers that can trap air should be placed in a tipped position so that the steam will force out the air. Products that do not permit penetration by moisture, such as mineral oil or petroleum jelly, are not sterilized by the same methods that would sterilize aqueous solutions.

Several commercially available methods can indicate whether sterilization has been achieved by heat treatment. Some of these are chemical reactions in which an indicator changes color when the proper times and temperatures have been reached (Figure 7.3). In some designs, the word "sterile"or "autoclaved" appears on wrappings or tapes. In another method, a pellet contained within a glass vial melts. A widely used test consists of preparations of specified species of bacterial endospores impregnated into paper strips. After autoclaving, these can then be aseptically inoculated into culture media. Growth in the culture media indicates survival of the endospores and therefore inadequate processing. Other designs use endospore suspensions that can be released, after heating, into a surrounding culture medium within the same vial.

Steam under pressure fails to sterilize when the air is not completely exhausted. This can happen with the premature closing of the autoclave's automatic ejector valve (see Figure 7.2). The principles of heat sterilization have a direct bearing on home canning. As anyone familiar with home canning knows, the steam must flow vigorously out of the valve in the lid for several minutes to carry with it all the air before the pressure cooker is sealed. If the air is not completely exhausted, the container will not reach the temperature expected for a given pressure. Because of the possibility of botulism, a kind of food poisoning resulting from improper canning methods, people involved in home canning should obtain reliable directions and follow them exactly.

Pasteurization

Recall from Chapter 1 that in the early days of microbiology, Louis Pasteur found a practical method of preventing the spoilage of beer and wine. Pasteur used mild heating, which was sufficient to kill the organisms that caused the particular spoilage problem without seriously damaging the taste of the product. The same principle was later applied to milk to produce what we now call pasteurized milk. The intent of pasteurization of milk was to eliminate pathogenic microbes. It also lowers microbial numbers, which prolongs milk's good quality under refrigeration. Many relatively heat-resistant (**thermoduric**) bacteria survive pasteurization, but these are unlikely to cause disease or cause refrigerated milk to spoil.

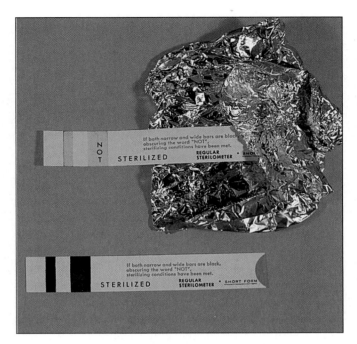

FIGURE 7.3 Examples of sterilization indicators.
The strips indicate if the item has been properly sterilized;
the word *NOT* appears if heating has been inadequate.
In the illustration, the indicator that was wrapped with
aluminum foil was not sterilized because steam couldn't
penetrate the foil.

■ What should have been used instead of aluminum foil to
wrap the items?

Products other than milk, such as ice cream, yogurt,
and beer, all have their own pasteurization times and temperatures, which often differ considerably. There are several reasons for these variations. For example, heating is
less efficient in foods that are more viscous, and fats in food
can have a protective effect on microorganisms. The dairy
industry routinely uses a test to determine whether products have been pasteurized: the *phosphatase test* (phosphatase is an enzyme naturally present in milk). If the
product has been pasteurized, phosphatase will have been
inactivated.

In the classic pasteurization treatment of milk, the
milk was exposed to a temperature of about 63°C for 30
minutes. Most milk pasteurization today uses higher temperatures, at least 72°C, but for only 15 seconds. This
treatment, known as **high-temperature short-time
(HTST) pasteurization,** is applied as the milk flows continuously past a heat exchanger. In addition to killing
pathogens, HTST pasteurization lowers total bacterial
counts, so the milk keeps well under refrigeration.

Milk can also be sterilized—something quite different
from pasteurization—by **ultra-high-temperature (UHT)
treatments** so that it can be stored without refrigeration.

This is more useful in parts of the world where refrigeration
facilities are not always available. In the United States,
sterilization is sometimes used on the small containers of
coffee creamers found in restaurants. To avoid giving the
milk a cooked taste, a UHT system is used in which the liquid milk never touches a surface hotter than the milk itself
while being heated by steam. The milk falls in a thin film
through a chamber of superheated steam and reaches
140°C in less than a second. It is held for 3 seconds in a
holding tube and then cooled in a vacuum chamber, where
the steam flashes off. With this process, in less than 5 seconds the milk temperature rises from 74°C to 140°C and
drops back to 74°C.

The heat treatments we have just discussed illustrate
the concept of **equivalent treatments:** As the temperature is increased, much less time is needed to kill the same
number of microbes. For example, the destruction of
highly resistant endospores might take 70 minutes at
115°C, whereas only 7 minutes might be needed at 125°C.
Both treatments yield the same result. The concept of
equivalent treatments also explains why classic pasteurization at 63°C for 30 minutes, HTST treatment at 72°C for
15 seconds, and UHT treatment at 140°C for less than a
second can have similar effects.

Dry Heat Sterilization

Dry heat kills by oxidation effects. A simple analogy is the
slow charring of paper in a heated oven, even when the
temperature remains below the ignition point of paper.
One of the simplest methods of dry heat sterilization is direct **flaming.** You will use this procedure many times in
the microbiology laboratory when you sterilize inoculating
loops. To effectively sterilize the inoculating loop, you heat
the wire to a red glow. A similar principle is used in *incineration,* an effective way to sterilize and dispose of contaminated paper cups, bags, and dressings.

Another form of dry heat sterilization is **hot-air sterilization.** Items to be sterilized by this procedure are placed
in an oven. Generally, a temperature of about 170°C maintained for nearly 2 hours ensures sterilization. The longer
period and higher temperature (relative to moist heat) are
required because the heat in water is more readily transferred to a cool body than is the heat in air. For example,
imagine the different effects of immersing your hand in
boiling water at 100°C (212°F) and of holding it in a hot-
air oven at the same temperature for the same amount of
time.

Filtration

Recall from Chapter 6 that *filtration* is the passage of a liquid or gas through a screenlike material with pores small

FIGURE 7.4 **Filter sterilization with a disposable, presterilized plastic unit.** The sample is placed into the upper chamber and forced through the membrane filter by a vacuum in the lower chamber. Pores in the membrane filter are smaller than the bacteria, so bacteria are retained on the filter. The sterilized sample can then be decanted from the lower chamber. Similar equipment with removable filter disks is used to count bacteria in samples (see Figure 6.17a).

- Filters are available with pores small enough to retain viruses.

enough to retain microorganisms (often the same apparatus used for counting; see Figure 6.17 on page 176). A vacuum that is created in the receiving flask helps gravity pull the liquid through the filter. Filtration is used to sterilize heat-sensitive materials, such as some culture media, enzymes, vaccines, and antibiotic solutions.

Some operating theaters and rooms occupied by burn patients receive filtered air to lower the numbers of airborne microbes. **High-efficiency particulate air (HEPA) filters** remove almost all microorganisms larger than about 0.3 μm in diameter.

In the early days of microbiology, hollow candle-shaped filters of unglazed porcelain were used to filter liquids. The long and indirect passageways through the walls of the filter adsorbed the bacteria. Unseen pathogens that passed through the filters (causing such diseases as rabies) were called *filterable viruses*.

In recent years, **membrane filters,** composed of such substances as cellulose esters or plastic polymers, have popular for industrial and laboratory use (Figure 7.4). These filters are only 0.1 mm thick. The pores of mem-

brane filters include, for example, 0.22-μm and 0.45-μm sizes, which are intended for bacteria. Some very flexible bacteria, such as spirochetes, or the wall-less mycoplasma, will sometimes pass through such filters, however. Filters are available with pores as small as 0.01 μm, a size that will retain viruses and even some large protein molecules.

Low Temperatures

The effect of low temperatures on microorganisms depends on the particular microbe and the intensity of the application. For example, at temperatures of ordinary refrigerators (0–7°C), the metabolic rate of most microbes is so reduced that they cannot reproduce or synthesize toxins. In other words, ordinary refrigeration has a bacteriostatic effect. Yet psychrotrophs do grow slowly at refrigerator temperatures and will alter the appearance and taste of foods after a time. For example, a single microbe reproducing only three times a day would reach a population of more than 2 million within a week. Pathogenic bacteria generally will not grow at refrigerator temperatures.

Surprisingly, some bacteria can grow at temperatures several degrees below freezing. Most foods remain unfrozen until −2°C or lower. Rapidly attained subfreezing temperatures tend to render microbes dormant but do not necessarily kill them. Slow freezing is more harmful to bacteria; the ice crystals that form and grow disrupt the cellular and molecular structure of the bacteria. Thawing, being inherently slower, is actually the more damaging part of a freeze-thaw cycle. Once frozen, one-third of the population of some vegetative bacteria might survive a year, whereas other species might have very few survivors after this time. Many eukaryotic parasites, such as the roundworms that cause trichinosis, are killed by several days of freezing temperatures. Some important temperatures associated with microorganisms and food spoilage are shown in Figure 6.2 on page 157.

High Pressure

High pressure applied to liquid suspensions is transferred instantly and evenly throughout the sample. If the pressure is high enough, the molecular structures of proteins and carbohydrates are altered, resulting in the rapid inactivation of vegetative bacterial cells. Endospores are relatively resistant to high pressure. They can, however, be killed by other techniques, such as combining high pressure with elevated temperatures or by alternating pressure cycles that cause spore germination, followed by pressure-caused death of the resulting vegetative cells. Fruit juices preserved by high-pressure treatments have been marketed in Japan and the United States. An advantage is that these treatments preserve the flavors, colors, and nutrient values of the products.

Desiccation

In the absence of water, a condition known as **desiccation,** microorganisms cannot grow or reproduce but can remain viable for years. Then, when water is made available to them, they can resume their growth and division. This ability is used in the laboratory when microbes are preserved by lyophilization, or freeze-drying, a process described in Chapter 6 on page 169. Certain foods are also freeze-dried (for example, coffee and some fruit additives for dry cereals).

The resistance of vegetative cells to desiccation varies with the species and the organism's environment. For example, the gonorrhea bacterium can withstand dryness for only about an hour, but the tuberculosis bacterium can remain viable for months. Viruses are generally resistant to desiccation, but they are not as resistant as bacterial endospores, some of which have survived for centuries. This ability of certain dried microbes and endospores to remain viable is important in a hospital setting. Dust, clothing, bedding, and dressings might contain infectious microbes in dried mucus, urine, pus, and feces.

Osmotic Pressure

The use of high concentrations of salts and sugars to preserve food is based on the effects of *osmotic pressure*. High concentrations of these substances create a hypertonic environment that causes water to leave the microbial cell (see Figure 6.4 on page 159). This process resembles preservation by desiccation, in that both methods deny the cell the moisture it needs for growth. The principle of osmotic pressure is used in the preservation of foods. For example, concentrated salt solutions are used to cure meats, and thick sugar solutions are used to preserve fruits.

As a general rule, molds and yeasts are much more capable than bacteria of growing in materials with low moisture or high osmotic pressures. This property of molds, sometimes combined with their ability to grow under acidic conditions, is the reason fruits and grains are spoiled by molds rather than by bacteria. It is also part of the reason molds are able to form mildew on a damp wall or a shower curtain.

Radiation

Radiation has various effects on cells, depending on its wavelength, intensity, and duration. Radiation that kills microorganisms (sterilizing radiation) is of two types: ionizing and nonionizing.

Ionizing radiation—gamma rays, X rays, or high-energy electron beams—has a wavelength shorter than that of nonionizing radiation, less than about 1 nm. Therefore, it carries much more energy (Figure 7.5). *Gamma*

FIGURE 7.5 The radiant energy spectrum. Visible light and other forms of radiant energy radiate through space as waves of various lengths. Ionizing radiation, such as gamma rays and X rays, has a wavelength shorter than 1 nm. Nonionizing radiation, such as ultraviolet (UV) light, has a wavelength between 1 nm and about 380 nm, where the visible spectrum begins.

■ What effect might increased UV radiation (due to decrease in the ozone layer) have on the Earth's ecosystems?

rays are emitted by certain radioactive elements such as cobalt, and electron beams are produced by accelerating electrons to high energies in special machines. *X rays*, which are produced by machines in a manner similar to the production of electron beams, are similar in nature to gamma rays. Gamma rays penetrate deeply but may require hours to sterilize large masses; *high-energy electron beams* have much lower penetrating power but usually require only a few seconds of exposure. The principal effect of ionizing radiation is the ionization of water, which forms highly reactive hydroxyl radicals (see the discussion of toxic forms of oxygen in Chapter 6, pages 161–163). These radicals react with organic cellular components, especially DNA.

The so-called target theory of damage by radiation supposes that ionizing particles, or packets of energy, pass through or close to vital portions of the cell; these constitute "hits." One, or a few, hits may only cause nonlethal mutations, some of them conceivably useful. More hits are likely to cause sufficient mutations to kill the microbe.

The food industry has recently renewed its interest in the use of radiation for food preservation. Low-level ionizing radiation, used for years in many countries, has been approved in the United States for processing spices and certain meats and vegetables. Ionizing radiation, especially high-energy electron beams, is used for the sterilization of pharmaceuticals and disposable dental and medical supplies, such as plastic syringes, surgical gloves, suturing materials, and catheters. As a protection against bioterrorism, the postal service often uses electron beam radiation to sterilize certain classes of mail.

Nonionizing radiation has a wavelength longer than that of ionizing radiation, usually greater than about 1 nm. The best example of nonionizing radiation is ultraviolet (UV) light. UV light damages the DNA of exposed cells by causing bonds to form between adjacent pyrimidine bases (page 230), usually thymines, in DNA chains (see Figure 8.20). These *thymine dimers* inhibit correct replication of the DNA during reproduction of the cell. The UV wavelengths most effective for killing microorganisms are about 260 nm; these wavelengths are specifically absorbed by cellular DNA. UV radiation is also used to control microbes in the air. A UV or "germicidal" lamp is commonly found in hospital rooms, nurseries, operating rooms, and cafeterias. UV light is also used to disinfect vaccines and other medical products. A major disadvantage of UV light as a disinfectant is that the radiation is not very penetrating, so the organisms to be killed must be directly exposed to the rays. Organisms protected by solids and such coverings as paper, glass, and textiles are not affected. Another potential problem is that UV light can damage human

eyes, and prolonged exposure can cause burns and skin cancer in humans.

Sunlight contains some UV radiation, but the shorter wavelengths—those most effective against bacteria—are screened out by the ozone layer of the atmosphere. The antimicrobial effect of sunlight is due almost entirely to the formation of singlet oxygen in the cytoplasm (see Chapter 6 on page 162). Many pigments produced by bacteria provide protection from sunlight.

Microwaves do not have much direct effect on microorganisms, and bacteria can readily be isolated from the interior of recently operated microwave ovens. Moisture-containing foods are heated by microwave action, and the heat will kill most vegetative pathogens. Solid foods heat unevenly because of the uneven distribution of moisture. For this reason, pork cooked in a microwave oven has been responsible for outbreaks of trichinosis.

* * *

Table 7.5 summarizes the physical methods of microbial control.

Chemical Methods of Microbial Control

Chemical agents are used to control the growth of microbes on both living tissue and inanimate objects. Unfortunately, few chemical agents achieve sterility; most of them merely reduce microbial populations to safe levels or remove vegetative forms of pathogens from objects. A common problem in disinfection is the selection of an agent. No single disinfectant is appropriate for all circumstances.

Principles of Effective Disinfection

Learning Objective
- *List the factors related to effective disinfection.*

By reading the label, we can learn a great deal about a disinfectant's properties. The label will usually indicate what groups of organisms the disinfectant will be effective against. Remember that the concentration of a disinfectant affects its action, so it should always be diluted exactly as specified by the manufacturer.

Also consider the nature of the material being disinfected. For example, are organic materials present that might interfere with the action of the disinfectant? Similarly, the pH of the medium often has a great effect on a disinfectant's activity.

Another very important consideration is whether the disinfectant will easily make contact with the microbes. An area might need to be scrubbed and rinsed before the

TABLE 7.5	**Physical Methods Used to Control Microbial Growth**		

Method	Mechanism of Action	Comment	Preferred Use
Heat			
1. Moist heat			
a. Boiling or flowing steam	Protein denaturation	Kills vegetative bacterial and fungal pathogens and almost all viruses within 10 min; less effective on endospores.	Dishes, basins, pitchers, various equipment
b. Autoclaving	Protein denaturation	Very effective method of sterilization; at about 15 psi of pressure (121°C), all vegetative cells and their endospores are killed in about 15 min.	Microbiological media, solutions, linens, utensils, dressings, equipment, and other items that can withstand temperature and pressure
2. Pasteurization	Protein denaturation	Heat treatment for milk (72°C for about 15 sec) that kills all pathogens and most nonpathogens.	Milk, cream, and certain alcoholic beverages (beer and wine)
3. Dry heat			
a. Direct flaming	Burning contaminants to ashes	Very effective method of sterilization.	Inoculating loops
b. Incineration	Burning to ashes	Very effective method of sterilization.	Paper cups, contaminated dressings, animal carcasses, bags, and wipes
c. Hot-air sterilization	Oxidation	Very effective method of sterilization, but requires temperature of 170°C for about 2 hr.	Empty glassware, instruments, needles, and glass syringes
Filtration	Separation of bacteria from suspending liquid	Removes microbes by passage of a liquid or gas through a screen-like material. Most filters in use consist of cellulose acetate or nitrocellulose.	Useful for sterilizing liquids (enzymes, vaccines) that are destroyed by heat
Cold			
1. Refrigeration	Decreased chemical reactions and possible changes in proteins	Has a bacteriostatic effect	Food, drug, and culture preservation
2. Deep-freezing (see Chapter 6, page 169)	Decreased chemical reactions and possible changes in proteins	An effective method for preserving microbial cultures, in which cultures are quick-frozen between −50° and −95°C.	Food, drug, and culture preservation
3. Lyophilization (see Chapter 6, page 169)	Decreased chemical reactions and possible changes in proteins	Most effective method for long-term preservation of microbial cultures; water removed by high vacuum at low temperature.	Food, drug, and culture preservation
High pressure	Alteration of molecular structure of proteins and carbohydrates	Preservation of colors, flavors, nutrient values.	Fruit juices
Desiccation	Disruption of metabolism	Involves removing water from microbes; primarily bacteriostatic.	Food preservation
Osmotic Pressure	Plasmolysis	Results in loss of water from microbial cells.	Food preservation
Radiation			
1. Ionizing	Destruction of DNA	Not widespread in routine sterilization.	Used for sterilizing pharmaceuticals and medical and dental supplies
2. Nonionizing	Damage to DNA	Radiation not very penetrating.	Control of closed environment with UV (germicidal) lamp

disinfectant is applied. In general, disinfection is a gradual process. Thus, to be effective, a disinfectant might need to be left on a surface for several hours.

Evaluating a Disinfectant

Learning Objective

- Interpret the results of use-dilution tests and the disk-diffusion method.

Use-Dilution Tests

There is a need to evaluate the effectiveness of disinfectants and antiseptics. For many years the standard test was the *phenol coefficient test*, which compared the activity of a given disinfectant with that of phenol. However, the current standard is the American Official Analytical Chemist's **use-dilution test.** For most purposes, the three bacteria used in this test are *Salmonella choleraesuis, Staphylococcus aureus*, and *Pseudomonas aeruginosa*. Metal carrier rings are dipped into standardized cultures of the test bacteria grown in liquid media, removed, and dried at 37°C for a short time. The dried cultures are then placed into a solution of the disinfectant at the concentration recommended by the manufacturer and left there for 10 minutes at 20°C. Following this exposure, the carrier rings are transferred to a medium that will permit the growth of any surviving bacteria. The effectiveness of the disinfectant can then be determined by the number of cultures that grow.

Variations of this method are used for testing the effectiveness of antimicrobial agents against endospores, mycobacteria that cause tuberculosis, and fungi, because they are difficult to control with chemicals. Also, tests of antimicrobials intended for special purposes, such as dairy utensil disinfection, may substitute other test bacteria. Virucidal chemicals are usually tested against cultures of Newcastle disease virus (a disease of birds and domestic fowl). After exposure to the chemical, the cultures are injected into embryonated chicken eggs; if any viruses survive, they will kill the embryos.

The Disk-Diffusion Method

The *disk-diffusion method* is used in teaching laboratories to evaluate the efficacy of a chemical agent. A disk of filter paper is soaked with a chemical and placed on an agar plate that has been previously inoculated and incubated with the test organism. After incubation, if the chemical is effective, a clear zone representing inhibition of growth can be seen around the disk (Figure 7.6).

Disks containing antibiotics are commercially available and used to determine microbial susceptibility to antibiotics (see Figure 20.17 on page 579).

Types of Disinfectants

Learning Objectives

- Identify the methods of action and preferred uses of chemical disinfectants.

- Differentiate between halogens used as antiseptics and as disinfectants.

- Identify the appropriate uses for surface-active agents.

- List the advantages of glutaraldehyde over other chemical disinfectants.

- Identify the method of sterilizing plastic labware.

Phenol and Phenolics

Lister was the first to use **phenol** (carbolic acid) to control surgical infections in the operating room. Its use had been suggested by its effectiveness in controlling odor in sewage. It is now rarely used as an antiseptic or disinfectant because it irritates the skin and has a disagreeable odor. It is often used in throat lozenges for its local anesthetic effect but has little antimicrobial effect at the low concentrations used. At concentrations above 1% (such as in some throat sprays), however, phenol has a significant antibacterial effect. The structure of a phenol molecule is shown in Figure 7.7a.

Derivatives of phenol, called **phenolics,** contain a molecule of phenol that has been chemically altered to reduce its irritating qualities or increase its antibacterial activity in combination with a soap or detergent. Phenolics exert antimicrobial activity by injuring lipid-containing plasma membranes, which results in leakage of cellular contents. The cell wall of mycobacteria, the causes of tuberculosis and leprosy, are rich in lipids, which make them susceptible to phenol derivatives. A useful property of phenolics as disinfectants is that they remain active in the presence of organic compounds, they are stable, and they persist for long periods after application. For these reasons, phenolics are suitable agents for disinfecting pus, saliva, and feces.

One of the most frequently used phenolics is derived from coal tar, a group of chemicals called *cresols*. A very important cresol is *O-phenylphenol* (see Figures 7.6 and 7.7b), the main ingredient in most formulations of Lysol. Cresols are very good surface disinfectants.

Bisphenols

Bisphenols are derivatives of phenol that contain two phenolic groups connected by a bridge (*bis* indicates two). One bisphenol, *hexachlorophene* (Figures 7.6 and 7.7c), is an ingredient of a prescription lotion, pHisoHex, used for surgical and hospital microbial control procedures. Gram-positive staphylococci and streptococci, which can cause skin infections in newborns, are particularly susceptible to

Zone of inhibition

Chlorine
Hexachlorophene
O-phenylphenol
Quat

Staphylococcus aureus
(gram-positive)

Chlorine
Hexachlorophene
O-phenylphenol
Quat

Escherichia coli
(gram-negative)

Chlorine
Hexachlorophene
O-phenylphenol
Quat

Pseudomonas aeruginosa
(gram-negative)

FIGURE 7.6 **Evaluation of disinfectants by the disk-diffusion method.** In this experiment, paper disks are soaked in a solution of disinfectant and placed on the surface of a nutrient medium on which a culture of test bacteria has been spread to produce uniform growth.

At the top of each plate, the tests show that chlorine (as sodium hypochlorite) was effective against all the test bacteria but was more effective against gram-positive bacteria.

At the bottom row of each plate, the tests show that the quaternary ammonium compound ("quat") was also more effective against the gram-positive bacteria, but it did not affect the pseudomonads at all.

At the left side of each plate, the tests show that hexachlorophene was effective against gram-positive bacteria only.

At the right sides, O-phenylphenol was ineffective against pseudomonads but was almost equally effective against the gram-positive bacteria and the gram-negative bacteria.

All four chemicals worked against the gram-positive test bacteria, but only one of the four chemicals affected pseudomonads.

■ Which group of bacteria is most resistant to the disinfectants tested?

hexachlorophene, so it is often used to control such infections in nurseries. However, excessive use of this bisphenol, such as bathing infants with it several times a day, can lead to neurological damage.

Another widely used bisphenol is *triclosan* (Figure 7.7d), an ingredient in antibacterial soaps and at least one toothpaste. Triclosan has even been incorporated into kitchen cutting boards and the handles of knives and other plastic kitchenware. Its use is now so widespread that resistant bacteria have been reported, and concerns about its effect on microbes' resistance to certain antibiotics have been raised. Triclosan inhibits an enzyme needed for the biosynthesis of fatty acids (lipids), which affects the integrity of the plasma membrane. It is especially effective against gram-positive bacteria but also works well against fungi and gram-negative bacteria. There are certain exceptions, such as *Pseudomonas aeruginosa*, a gram-negative bacterium that is very resistant to triclosan, as well as to

many other antibiotics and disinfectants (see the discussion on pages 312 and 423).

Biguanides

Chlorhexidine is a member of the **biguanide** group with a broad spectrum of activity. It is frequently used for microbial control on skin and mucous membranes. Combined with a detergent or alcohol, chlorhexidine is also used for surgical hand scrubs and preoperative skin preparation in patients. In such applications, its strong affinity for binding to the skin or mucous membranes is an advantage, as is its low toxicity. However, contact with the eyes can cause damage. Its killing effect is related to the injury it causes to the plasma membrane. It is biocidal against most vegetative bacteria and fungi. Mycobacteria are relatively resistant, and endospores and protozoan cysts are not affected. The only viruses affected are certain enveloped (lipophilic) types (see Chapter 13).

(a) Phenol **(b)** O-phenylphenol

(c) Hexachlorophene (a bisphenol)

(d) Triclosan (a bisphenol)

FIGURE 7.7 The structure of phenolics and bisphenols.

- At concentrations above 1%, phenol has a significant antibacterial effect.

Halogens

The **halogens,** particularly iodine and chlorine, are effective antimicrobial agents, both alone and as constituents of inorganic or organic compounds. *Iodine* (I_2) is one of the oldest and most effective antiseptics. It is effective against all kinds of bacteria, many endospores, various fungi, and some viruses. One proposed mechanism for the activity of iodine is that it combines with certain amino acids of enzymes and other cellular proteins. The exact mode of action is not known.

Iodine is available as a **tincture**—that is, in solution in aqueous alcohol—and as an iodophor. An **iodophor** is a combination of iodine and an organic molecule, from which the iodine is released slowly. Iodophors have the antimicrobial activity of iodine, but they do not stain and are less irritating. The most common commercial preparations are Betadine and Isodine. These are *povidone-iodines;* povidone is a surface-active iodophor that improves the wetting action and serves as a reservoir of free iodine. Iodines are used mainly for skin disinfection and wound treatment. Many campers are familiar with iodine for water treatment. To treat water, iodine tablets are added or the water can be passed through iodine-treated resin filters.

Chlorine (Cl_2), as a gas or in combination with other chemicals, is another widely used disinfectant. Its germicidal action is caused by the hypochlorous acid (HOCl) that forms when chlorine is added to water:

(1)
$$Cl_2 + H_2O \rightleftharpoons H^+ + Cl^- + HOCl$$

Chlorine   Water   Hydrogen ion   Chloride ion   Hypochlorous acid

(2)
$$HOCl \rightleftharpoons H^+ + OCl^-$$

Hypochlorous acid   Hydrogen ion   Hypochlorite ion

Exactly how hypochlorous acid exerts its killing power is not known. It is a strong oxidizing agent that prevents much of the cellular enzyme system from functioning. Hypochlorous acid is the most effective form of chlorine because it is neutral in electrical charge and diffuses as rapidly as water through the cell wall. Because of its negative charge, the hypochlorite ion (OCl^-) cannot enter the cell freely.

A liquid form of compressed chlorine gas is used extensively for disinfecting municipal drinking water, water in swimming pools, and sewage. Several compounds of chlorine are also effective disinfectants. For example, solutions of *calcium hypochlorite* [$Ca(OCl)_2$] are used to disinfect dairy equipment and restaurant eating utensils. This compound, once called chloride of lime, was used as early as 1825, long before the concept of a germ theory for disease, to soak hospital dressings in Paris hospitals. It was also the disinfectant used in the 1840s by Semmelweis to control hospital infections during childbirth, as mentioned in Chapter 1, page 9. Another chlorine compound, *sodium hypochlorite* (NaOCl; see Figure 7.6), is used as a household disinfectant and bleach (Clorox), and as a disinfectant in dairies, food-processing establishments, and hemodialysis systems. When the quality of drinking water is in question, household bleach can provide a rough equivalent of municipal chlorination. After two drops of bleach are added to a liter of water (four drops if the water is cloudy) and the mixture has sat for 30 minutes, the water is considered safe for drinking under emergency conditions. U.S. military forces in the field are issued tablets (Chlor-Floc) that contain *sodium dichloroisocyanurate*, a form of chlorine combined with an agent that flocculates (coagulates) suspended materials in a water sample, causing them to settle out, thus clarifying it.

Chlorine dioxide (ClO_2) is a gaseous form of chlorine, occasionally used for area disinfection, most notably, to kill endospores of the anthrax bacterium.

Another group of chlorine compounds, the *chloramines,* consist of chlorine and ammonia. They are used as disinfectants, antiseptics, or sanitizing agents. Chloramines are very stable compounds that release chlorine over long periods. They are relatively effective in organic matter, but they have the disadvantages of acting more

TABLE 7.6	Biocidal Action of Various Concentrations of Ethanol in Aqueous Solution Against *Streptococcus pyogenes*				
Concentration of Ethanol (%)	**Time (sec)**				
	10	**20**	**30**	**40**	**50**
100	–	–	–	–	–
95	+	+	+	+	+
90	+	+	+	+	+
80	+	+	+	+	+
70	+	+	+	+	+
60	+	+	+	+	+
50	–	–	+	+	+
40	–	–	–	–	–

NOTE: A minus sign indicates no biocidal action (bacterial growth); a plus sign indicates biocidal action (no bacterial growth). The highlighted area represents bacteria killed by biocidal action.

slowly and being less effective purifiers than many other chlorine compounds. Chloramines are used to sanitize glassware and eating utensils and to treat dairy and food-manufacturing equipment. Ammonia is usually mixed with chlorine in municipal water-treatment systems to form chloramines. The chloramines control taste and odor problems caused by the reaction of chlorine with other nitrogenous compounds in the water. Because chloramines are less effective as germicides, sufficient chlorine must be added to ensure a residual of chlorine in the form of HOCl. (Chloramines are toxic to aquarium fish, but pet shops sell chemicals to neutralize them.)

Alcohols

Alcohols effectively kill bacteria and fungi but not endospores and nonenveloped viruses. The mechanism of action of alcohol is usually protein denaturation, but alcohol can also disrupt membranes and dissolve many lipids, including the lipid component of enveloped viruses. Alcohols have the advantage of acting and then evaporating rapidly and leaving no residue. When the skin is swabbed (degermed) before an injection, most of the microbial control activity comes from simply wiping away dirt and microorganisms, along with skin oils. However, alcohols are unsatisfactory antiseptics when applied to wounds. They cause coagulation of a layer of protein under which bacteria continue to grow.

Two of the most commonly used alcohols are ethanol and isopropanol. The recommended optimum concentration of *ethanol* is 70%, but concentrations between 60% and 95% seem to kill as well (Table 7.6). Pure ethanol is less effective than aqueous solutions (ethanol mixed with water) because denaturation requires water. *Isopropanol*, often sold as rubbing alcohol, is slightly superior to ethanol as an antiseptic and disinfectant. Moreover, it is less volatile, less expensive, and more easily obtained than ethanol.

Ethanol and isopropanol are often used to enhance the effectiveness of other chemical agents. For example, an aqueous solution of Zephiran (described on page 198) kills about 40% of the population of a test organism in two minutes, whereas a tincture of Zephiran kills about 85% in the same period. To compare the effectiveness of tinctures and aqueous solutions, see Figure 7.10 on page 199.

Heavy Metals and Their Compounds

Several heavy metals can be biocidal or antiseptic, including silver, mercury, and copper. The ability of very small amounts of heavy metals, especially silver and copper, to exert antimicrobial activity is referred to as **oligodynamic action** (*oligo* means few). This action can be seen when we place a coin or other clean piece of metal containing silver or copper on a culture on an inoculated Petri plate. Extremely small amounts of metal diffuse from the coin and inhibit the growth of bacteria for some distance around the coin (Figure 7.8). This effect is produced by the action of heavy metal ions on microbes. When the metal ions combine with the sulfhydryl groups on cellular proteins, denaturation results.

Silver is used as an antiseptic in a 1% *silver nitrate* solution. At one time, many states required that the eyes of newborns be treated with a few drops of silver nitrate to guard against an infection of the eyes called gonorrheal ophthalmia neonatorum, which the infants might have contracted as they passed through the birth canal. In recent years, antibiotics have replaced silver nitrate for this purpose.

Recently, there has been renewed interest in silver as an antimicrobial agent. Silver-impregnated dressings that slowly release silver ions have proven especially useful when antibiotic-resistant bacteria are a problem. A combination of silver and the drug sulfadiazine, *silver-sulfadiazine*, is the most common formulation. It is available as a topical cream for use on burns. Silver can also be incorporated into indwelling catheters, which are a common source of hospital infections, and in wound dressings. *Surfacine*, a combination of silver iodide and biguanide, is a promising, very persistent antimicrobial agent that can be used to kill a wide variety of microbes

FIGURE 7.8 **Oligodynamic action of heavy metals.**
Clear zones where bacterial growth has been inhibited are
seen around the sombrero charm (pushed aside), the dime,
and the penny. The charm and the dime contain silver; the
penny contains copper.

■ The coins used in this demonstration were minted many
years ago; why were current coins not used?

that contact skin and environmental surfaces. When approved by the FDA, it is expected to find wide use on medical devices and as a hand antiseptic.

Inorganic mercury compounds, such as *mercuric chloride*, have a long history of use as disinfectants. They have a very broad spectrum of activity; their effect is primarily bacteriostatic. However, their use is now limited because of their toxicity, corrosiveness, and ineffectiveness in organic matter. At present, the primary use of mercurials is to control mildew in paints. In addition, mercurochrome antiseptic, an organic mercury compound, is often found in home medicine chests.

Copper in the form of *copper sulfate* or other copper-containing additives is used chiefly to destroy green algae (algicide) that grow in reservoirs, stock ponds, swimming pools, and fish tanks. If the water does not contain excessive organic matter, copper compounds are effective in concentrations of one part per million of water. To prevent mildew, copper compounds such as copper 8-hydroxy-quinoline are sometimes included in paint.

Another metal used as an antimicrobial is zinc. The effect of trace amounts of zinc can be seen on roofs of buildings down-slope from galvanized (zinc-coated) fittings.

The roof is lighter-colored where biological growth is impeded. Copper- and zinc-treated shingles are available. *Zinc chloride* is a common ingredient in mouthwashes, and *zinc oxide* is probably the most widely used antifungal agent in paints, mainly because it is often part of the pigment formulation.

Surface-Active Agents

Surface-active agents, or **surfactants,** can decrease surface tension among molecules of a liquid. Such agents include soaps and detergents. Soap has little value as an antiseptic, but it does have an important function in the mechanical removal of microbes through scrubbing. The skin normally contains dead cells, dust, dried sweat, microbes, and oily secretions from oil glands. Soap breaks the oily film into tiny droplets, a process called *emulsification,* and the water and soap together lift up the emulsified oil and debris and float them away as the lather is washed off. In this sense, soaps are good degerming agents.

Acid-anionic surface-active sanitizers are very important in the cleaning of dairy utensils and equipment. Their sanitizing ability is related to the negatively charged portion (anion) of the molecule, which reacts with the plasma membrane. These sanitizers, which act on a wide spectrum of microbes, including troublesome thermoduric bacteria, are nontoxic, noncorrosive, and fast acting.

Quaternary Ammonium Compounds (Quats)

The most widely used surface-active agents are the cationic detergents, especially the **quaternary ammonium compounds (quats).** Their cleansing ability is related to the positively charged portion—the cation—of the molecule. Their name is derived from the fact that they are modifications of the four-valence ammonium ion, NH_4^+ (Figure 7.9). Quaternary ammonium compounds are strongly bactericidal against gram-positive bacteria and less active against gram-negative bacteria (see Figure 7.6).

Quats are also fungicidal, amoebicidal, and virucidal against enveloped viruses. They do not kill endospores or mycobacteria. (See the box on page 200.) Their chemical mode of action is unknown, but they probably affect the plasma membrane. They change the cell's permeability and cause the loss of essential cytoplasmic constituents, such as potassium.

Two popular quats are Zephiran, a brand name of *benzalkonium chloride* (see Figure 7.9), and Cepacol, a brand name of *cetylpyridinium chloride.* They are strongly antimicrobial, colorless, odorless, tasteless, stable, easily diluted, and nontoxic, except at high concentrations. If your mouthwash bottle fills with foam when shaken, the mouthwash probably contains a quat. However, organic

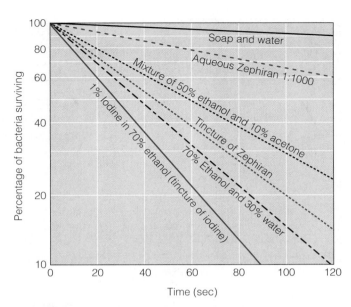

FIGURE 7.9 **The ammonium ion and a quaternary ammonium compound, benzalkonium chloride (Zephiran).** Notice how other groups replace the hydrogens of the ammonium ion.

■ Quaternary ammonium compounds are strongly bactericidal against gram-positive bacteria but less effective against gram-negative bacteria.

matter interferes with their activity, and they are rapidly neutralized by soaps and anionic detergents.

Anyone associated with medical applications of quats should remember that certain bacteria, such as some species of *Pseudomonas,* not only survive in quaternary ammonium compounds but actively grow in them. This resistance occurs not only to the disinfectant solution but also to moistened gauze and bandages, whose fibers tend to neutralize the quats. (See the box on page 200.)

Before we move on to the next group of chemical agents, refer to Figure 7.10, which compares the effectiveness of some of the antiseptics we have discussed so far.

Chemical Food Preservatives

Chemical preservatives are frequently added to foods to retard spoilage. *Sulfur dioxide* (SO_2) has long been used as a disinfectant, especially in wine-making. Homer mentioned its use in *The Odyssey,* written nearly 2800 years ago. Among the more common additives are sodium benzoate, sorbic acid, and calcium propionate. These chemicals are simple organic acids, or salts of organic acids, which the body readily metabolizes and which are generally judged to be safe in foods. *Sorbic acid,* or its more soluble salt *potassium sorbate,* and *sodium benzoate* prevent molds from growing in certain acidic foods, such as cheese and soft drinks. Such foods, usually with a pH of 5.5 or lower, are most susceptible to mold-type spoilage. *Calcium propionate,* an effective fungistat used in bread, prevents the growth of surface molds and the *Bacillus* bacterium that causes ropy bread. These organic acids inhibit mold growth, not by affecting the pH but by interfering with the mold's metabolism or the integrity of the plasma membrane.

Sodium nitrate and *sodium nitrite* are added to many meat products, such as ham, bacon, hot dogs, and sausage. The active ingredient is sodium nitrite, which certain bacteria in the meats can also produce from sodium nitrate. These bacteria use nitrate as a substitute for oxygen under

FIGURE 7.10 **A comparison of the effectiveness of various antiseptics.** The steeper the downward slope of the killing curve of the antiseptic, the more effective it is. A 1% iodine in 70% ethanol solution is the most effective; soap and water are the least effective. Notice that a tincture of Zephiran is more effective than an aqueous solution of the same antiseptic.

■ Why is the tincture of Zephiran more effective than the aqueous solution?

anaerobic conditions. The nitrite has two main functions: to preserve the pleasing red color of the meat by reacting with blood components in the meat, and to prevent the germination and growth of any botulism endospores that might be present. There has been some concern that the reaction of nitrites with amino acids can form certain carcinogenic products known as **nitrosamines,** and the amount of nitrites added to foods has generally been reduced recently for this reason. However, the use of nitrites continues because of their established value in preventing botulism. Because nitrosamines are formed in the body from other sources, the added risk posed by a limited use of nitrates and nitrites in meats is lower than was once thought.

Antibiotics

The antimicrobials discussed in this chapter are not useful for ingestion or injection to treat disease. Antibiotics are used for this purpose. The use of antibiotics is highly restricted; however, at least two have considerable use in food preservation. Neither is of value for clinical purposes. *Nisin,* which is often added to cheese to inhibit the growth of certain endospore-forming spoilage bacteria, is an example of a bacteriocin, a protein that is produced by one

A Hospital-Acquired Infection

As you read through this box you will encounter a series of questions that clinicians ask themselves as they formulate a diagnosis and treatment. Try to answer each question before going on to the next one.

1. During a 17-month period, nine patients in eight hospitals acquired surgical-site infections within 2 months after liposuction. The patients' symptoms included fever, local inflammation, and infected surgical wounds.
What would you do next?

2. Gram stains and acid-fast stains of pus showed gram-positive, acid-fast rods.
What are possible organisms?

3. Slow-growing mycobacteria, including M. *tuberculosis* and M. *leprae*, are human pathogens. These infections, however, were caused by rapidly growing mycobacteria (RGM): *Mycobacterium chelonae*, M. *fortuitum*, and M. *abscessus*.

Where are other species of mycobacteria normally found?

4. RGM are found in soil and water. The liposuction procedure involves making a small surgical wound and using a cannula (needle) for fat suctioning. The cannulae were cleaned with tap water and soap followed by a quat.
What was wrong with this procedure?

5. RGM can be found in tap water and soap used to remove dust and microbes. Quats do have some disinfecting properties but are considered low-level disinfectants because they are ineffective against endospores and mycobacteria.
Why are mycobacteria resistant to some disinfectants?

6. The lipid-rich cell wall prevents entry of most biocides into the cells.
How could you find the source of the RGM?

7. Nosocomial infections associated with contaminated quats that had been used to disinfect patient-care supplies or equipment have been reported. Nosocomial infections have also been associated with RGM growing in a hospital's hot water system. In this case, cultures of the quat and environmental cultures did not yield bacteria or mycobacteria.
What changes in procedures would you recommend?

8. Surgical instruments used in liposuction are intended to enter normally sterile tissue and should be sterilized between patient procedures. The hospitals modified their procedures by replacing the quat with either a high-level disinfectant using 2% glutaraldehyde or ethylene oxide gas sterilization.

SOURCE: Adapted from *MMWR* 47(49): 1065–1067 (12/18/98).

---

bacterium and inhibits another (see Chapter 8, page 241). Nisin is present naturally in small amounts in many dairy products. It is tasteless, readily digested, and nontoxic. *Natamycin* (pimaricin) is an antifungal antibiotic approved for use in foods, mostly cheese.

Aldehydes

Aldehydes are among the most effective antimicrobials. Two examples are formaldehyde and glutaraldehyde. They inactivate proteins by forming covalent cross-links with several organic functional groups on proteins ($-NH_2$, $-OH$, $-COOH$, and $-SH$). *Formaldehyde gas* is an excellent disinfectant. However, it is more commonly available as *formalin*, a 37% aqueous solution of formaldehyde gas. Formalin was once used extensively to preserve biological specimens and inactivate bacteria and viruses in vaccines.

Glutaraldehyde is a chemical relative of formaldehyde that is less irritating and more effective than formaldehyde. Glutaraldehyde is used to disinfect hospital instruments, including respiratory-therapy equipment. When used in a 2% solution (Cidex), it is bactericidal, tubercu-

locidal, and virucidal in 10 minutes and sporicidal in 3–10 hours. Glutaraldehyde is one of the few liquid chemical disinfectants that can be considered a sterilizing agent. However, 30 minutes is often considered the maximum time allowed for a sporicide to act, which is a criterion glutaraldehyde cannot meet. Both glutaraldehyde and formalin are used by morticians for embalming.

A possible replacement for glutaraldehyde for many uses is *ortho-phthalaldehyde* (OPA), which is more effective against many microbes and has fewer irritating properties.

Gaseous Chemosterilizers

Gaseous chemosterilizers are chemicals that sterilize in a closed chamber (similar to an autoclave). A gas suitable for this method is *ethylene oxide*:

$$H_2C - CH_2$$
$$\backslash \quad /$$
$$O$$

Its activity depends on the denaturation of proteins: The proteins' labile hydrogens, such as $-SH$, $-COOH$, or

—OH, are replaced by alkyl groups (*alkylation*), such as —CH_2CH_2OH. Ethylene oxide kills all microbes and endospores but requires a lengthy exposure period of 4–18 hours. It is toxic and explosive in its pure form, so it is usually mixed with a nonflammable gas, such as carbon dioxide or nitrogen. One of its advantages is that it is highly penetrating, so much so that ethylene oxide was chosen to sterilize spacecraft sent to land on the moon and Mars. Using heat to sterilize the electronic gear on these vehicles was not practical.

Because of their ability to sterilize without heat, gases like ethylene oxide are also widely used on medical supplies and equipment. Many large hospitals have ethylene oxide chambers, some large enough to sterilize mattresses, as part of their sterilizing equipment. Propylene oxide and beta-propiolactone are also used for gaseous sterilization:

$$H_3C-CH-CH_2 \qquad\qquad H_2C-CH_2$$

Propylene oxide   Beta-propiolactone

A disadvantage of all these gases is that they are suspected carcinogens, especially beta-propiolactone. For this reason, there has been concern about the exposure of hospital workers to ethylene oxide from such sterilizers. Because of this hazard, these gases may eventually be replaced by *plasma gas sterilization*. This makes use of vapors of hydrogen peroxide (discussed shortly) subjected to radio frequencies or microwave radiation to produce reactive free radicals. No by-products toxic to humans are produced, and it is an effective sterilant.

Peroxygens (Oxidizing Agents)

Peroxygens exert antimicrobial activity by oxidizing cellular components of the treated microbes. Examples are ozone, hydrogen peroxide, and peracetic acid. *Ozone* (O_3) is a highly reactive form of oxygen that is generated by passing oxygen through high-voltage electrical discharges. It is responsible for the air's rather fresh odor after a lightning storm, in the vicinity of electrical sparking, or around an ultraviolet light. Ozone is often used to supplement chlorine in the disinfection of water, because it helps neutralize tastes and odors. Although ozone is a more effective killing agent, its residual activity is difficult to maintain in water, and it is more expensive than chlorine.

Hydrogen peroxide is an antiseptic found in many household medicine cabinets and in hospital supply rooms. It is not a good antiseptic for open wounds; in fact, it may slow wound healing. It is quickly broken down to water and gaseous oxygen by the action of the enzyme catalase, which is present in human cells (see Chapter 6, page 163). However, hydrogen peroxide does effectively disinfect inanimate objects, an application in which it is even sporicidal, especially at elevated temperatures. On a nonliving surface, the normally protective enzymes of aerobic bacteria and facultative anaerobes are overwhelmed by the high concentrations of peroxide used. Because of these factors, the food industry is increasing its use of hydrogen peroxide for aseptic packaging. The packaging materials pass through a hot solution of the chemical before being assembled into a container. In addition, many wearers of contact lenses are familiar with disinfection by hydrogen peroxide. After disinfection, a platinum catalyst in the lens-disinfecting kit destroys residual hydrogen peroxide so that it does not persist on the lens, where it might be an irritant.

Oxidizing agents are useful for irrigating deep wounds, where the oxygen released makes an environment that inhibits the growth of anaerobic bacteria.

Benzoyl peroxide is another compound useful for treating wounds infected by anaerobic pathogens, but it is probably more familiar as the main ingredient in over-the-counter medications for acne, which is caused by a type of anaerobic bacterium infecting hair follicles.

Peracetic acid is one of the most effective liquid chemical sporicides available and is considered a sterilant. It is generally effective on endospores and viruses within 30 minutes, and kills vegetative bacteria and fungi in less than 5 minutes. Peracetic acid has many applications in the disinfection of food-processing and medical equipment because it leaves no toxic residues and is minimally affected by the presence of organic matter.

Microbial Characteristics and Microbial Control

Learning Objective

- Explain how the control of microbial growth is affected by the type of microbe.

Many biocides tend to be more effective against gram-positive bacteria, as a group, than against gram-negative bacteria. This is illustrated in Figure 7.11, which presents a simplified hierarchy of relative resistance of major microbial groups to biocides. A principal factor in this relative resistance to biocides is the external lipopolysaccharide layer of gram-negative bacteria. Within gram-negative bacteria, members of the genera *Pseudomonas* and *Burkholderia* are of special interest. These closely related bacteria are unusually resistant to biocides (see Figure 7.6) and will even grow actively in some disinfectants and antiseptics, most notably the quaternary ammonium compounds. (See the box on page 200). In Chapter 20, you will see that these bacteria are also resistant to many antibiotics. This

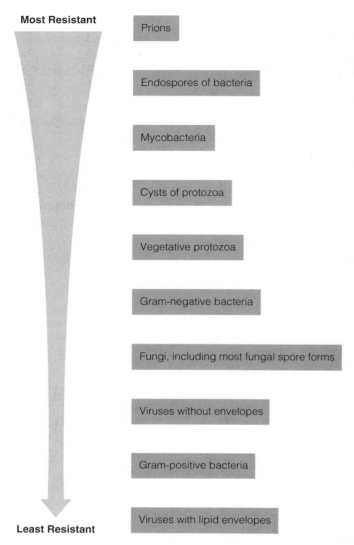

Most Resistant

Prions

Endospores of bacteria

Mycobacteria

Cysts of protozoa

Vegetative protozoa

Gram-negative bacteria

Fungi, including most fungal spore forms

Viruses without envelopes

Gram-positive bacteria

Viruses with lipid envelopes

Least Resistant

FIGURE 7.11 **Decreasing order of resistance of microorganisms to chemical biocides.**

■ Why are viruses with lipid-containing envelopes relatively susceptible to certain biocides?

TABLE 7.7	**The Effectiveness of Chemical Antimicrobials Against Endospores and Mycobaceria**	
Chemical Agent	**Endospores**	**Mycobacteria**
Mercury	No activity	No activity
Phenolics	Poor	Good
Bisphenols	No activity	No activity
Quaternary ammonium compounds	No activity	No activity
Chlorines	Fair	Fair
Iodine	Poor	Good
Alcohols	Poor	Good
Glutaraldehyde	Fair	Good
Chlorhexidine	No activity	Fair

been developed to evaluate the effectiveness of biocides against this bacterial group.

Bacteria endospores are affected by relatively few biocides. (The activity of the major chemical antimicrobial groups against mycobacteria and endospores is summarized in Table 7.7.) The cysts and oocysts of protozoa are also relatively resistant to chemical disinfection.

Viruses are not especially resistant to biocides, but a distinction must be made between those that possess a lipid-containing envelope and those that do not. Antimicrobials that are lipid-soluble are more likely to be effective against enveloped viruses. If so, this will usually be indicated on the label by a statement that they are effective against lipophilic viruses. Nonenveloped viruses, with only a protein coat, are more resistant—fewer biocides are active against them.

A special problem, not yet completely solved, is the reliable killing of prions (see Chapter 13, page 401). Prions are infectious proteins that are the cause of neurological diseases, the spongiform encephalopathies, such as the popularly named mad cow disease. To destroy prions, infected animal carcasses are incinerated. A major problem is the disinfection of surgical instruments exposed to prion contamination. Normal autoclaving has proven to be inadequate. The World Health Organization has recommended the combined use of a solution of sodium hydroxide and autoclaving at 134°C. However, a recent report described another technique, whereby instruments were soaked in a strong solution of sodium hydroxide for an hour followed by an hour of autoclaving at 136°C.

resistance to chemical antimicrobials is mostly related to the characteristics of their *porins* (structural openings in the wall of gram-negative bacteria; see Figure 4.13c on page 84). Porins are highly selective of molecules that they permit to enter the cell.

The mycobacteria are another group of non–endospore-forming bacteria that exhibit greater than normal resistance to chemical biocides. This group includes *Mycobacterium tuberculosis,* the pathogen that causes tuberculosis. The cell wall of this organism and other members of this genus have a waxy, lipid-rich component. Instruction labels on disinfectants often state whether they are tuberculocidal, indicating if they are effective against mycobacteria. Special tuberculocidal tests have

TABLE 7.8	**Chemical Agents Used to Control Microbial Growth**		
Chemical Agent	**Mechanism of Action**	**Preferred Use**	**Comment**
Phenol and Phenolics			
1. Phenol	Disruption of plasma membrane, denaturation of enzymes.	Rarely used, except as a standard of comparison.	Seldom used as a disinfectant or antiseptic because of its irritating qualities and disagreeable odor.
2. Phenolics	Disruption of plasma membrane, denaturation of enzymes.	Environmental surfaces, instruments, skin surfaces, and mucous membranes.	Derivatives of phenol that are reactive even in the presence of organic material; O-phenylphenol is an example.
3. Bisphenols	Probably disruption of plasma membrane.	Disinfectant hand soaps and skin lotions.	Triclosan is an especially common example of a bisphenol. Broad spectrum, but most effective against gram-positives.
Biguanides (Chlorhexidine)	Disruption of plasma membrane.	Skin disinfection, especially for surgical scrubs.	Bactericidal to gram-positives and gram-negatives; nontoxic, persistent.
Halogens	Iodine inhibits protein function and is a strong oxidizing agent; chlorine forms the strong oxidizing agent hypochlorous acid, which alters cellular components.	Iodine is an effective antiseptic available as a tincture and an iodophor; chlorine gas is used to disinfect water; chlorine compounds are used to disinfect dairy equipment, eating utensils, household items, and glassware.	Iodine and chlorine may act alone or as components of inorganic and organic compounds.
Alcohols	Protein denaturation and lipid dissolution.	Thermometers and other instruments; in swabbing the skin with alcohol before an injection, most of the disinfecting action probably comes from a simple wiping away (degerming) of dirt and some microbes.	Bactericidal and fungicidal, but not effective against endospores or nonenveloped viruses; commonly used alcohols are ethanol and isopropanol.

➤

Nevertheless, the report described the treatment as only "fairly effective." Surgeons sometimes resort to the use of disposable instruments.

In summary, it is important to remember that microbial control methods, especially biocides, are not uniformly effective against all microbes.

* * *

Table 7.8 summarizes chemical agents used to control microbial growth.

The compounds discussed in this chapter are not generally useful in the treatment of diseases. Because antibiotics are used in chemotherapy, antibiotics and the pathogens against which they are active will be discussed together, in Chapter 20.

TABLE 7.8	**Chemical Agents Used to Control Microbial Growth** (continued)		
Chemical Agent	**Mechanism of Action**	**Preferred Use**	**Comment**
Heavy Metals and Their Compounds	Denaturation of enzymes and other essential proteins.	Silver nitrate may be used to prevent gonorrheal ophthalmia neonatorum; mercurochrome disinfects skin and mucous membranes; copper sulfate is an algicide.	Heavy metals such as silver and mercury are biocidal.
Surface-Active Agents 1. Soaps and acid-anionic detergents*	Mechanical removal of microbes through scrubbing.	Skin degerming and removal of debris.	Many antibacterial soaps contain antimicrobials.
2. Acid-anionic detergents	Not certain; may involve enzyme inactivation or disruption.	Sanitizers in dairy and food-processing industries.	Wide spectrum of activity; nontoxic, noncorrosive, fast-acting.
3. Cationic detergents (quaternary ammonium compounds)	Enzyme inhibition, protein denaturation, and disruption of plasma membranes.	Antiseptic for skin, instruments, utensils, rubber goods.	Bactericidal, bacteriostatic, fungicidal, and virucidal against enveloped viruses; examples of quats are Zephiran and Cepacol.
Organic Acids	Metabolic inhibition, mostly affecting molds; action not related to their acidity.	Sorbic acid and benzoic acid effective at low pH; parabens much used in cosmetics, shampoos; calcium propionate used in bread.	Widely used to control molds and some bacteria in foods and cosmetics.
Aldehydes	Protein denaturation.	Glutaraldehyde (Cidex) is less irritating than formaldehyde and is used for disinfection of medical equipment.	Very effective antimicrobials.
Gaseous Sterilants	Protein denaturation.	Excellent sterilizing agent, especially for objects that would be damaged by heat.	Ethylene oxide is the most commonly used.
Peroxygens (Oxidizing Agents)	Oxidation.	Contaminated surfaces; some deep wounds, in which they are very effective against oxygen-sensitive anaerobes.	Ozone is widely used as a supplement for chlorination; hydrogen peroxide is a poor antiseptic but a good disinfectant. Peracetic acid is especially effective.

STUDY OUTLINE

The Terminology of Microbial Control (pp. 183–184)

1. The control of microbial growth can prevent infections and food spoilage.

2. Sterilization is the process of removing or destroying all microbial life on an object.

3. Commercial sterilization is heat treatment of canned foods to destroy *C. botulinum* endospores.

4. Disinfection is the process of reducing or inhibiting microbial growth on a nonliving surface.

5. Antisepsis is the process of reducing or inhibiting microorganisms on living tissue.

6. The suffix *-cide* means to kill; the suffix *-stat* means to inhibit.

7. Sepsis is bacterial contamination.

The Rate of Microbial Death (pp. 184–185)

1. Bacterial populations subjected to heat or antimicrobial chemicals usually die at a constant rate.

2. Such a death curve, when plotted logarithmically, shows this constant death rate as a straight line.

3. The time it takes to kill a microbial population is proportional to the number of microbes.

4. Microbial species and life cycle phases (e.g., endospores) have different susceptibilities to physical and chemical controls.

5. Organic matter may interfere with heat treatments and chemical control agents.

6. Longer exposure to lower heat can produce the same effect as shorter time at higher heat.

Actions of Microbial Control Agents (p. 186)

Alteration of Membrane Permeability (p. 186)

1. The susceptibility of the plasma membrane is due to its lipid and protein components.

2. Certain chemical control agents damage the plasma membrane by altering its permeability.

Damage to Proteins and Nucleic Acids (p. 186)

1. Some microbial control agents damage cellular proteins by breaking hydrogen bonds and covalent bonds.

2. Other agents interfere with DNA and RNA replication and protein synthesis.

Physical Methods of Microbial Control (pp. 186–192)

Heat (pp. 186–189)

1. Heat is frequently used to eliminate microorganisms.

2. Moist heat kills microbes by denaturing enzymes.

3. Thermal death point (TDP) is the lowest temperature at which all the microbes in a liquid culture will be killed in 10 minutes.

4. Thermal death time (TDT) is the length of time required to kill all bacteria in a liquid culture at a given temperature.

5. Decimal reduction time (DRT) is the length of time in which 90% of a bacterial population will be killed at a given temperature.

6. Boiling (100°C) kills many vegetative cells and viruses within 10 minutes.

7. Autoclaving (steam under pressure) is the most effective method of moist heat sterilization. The steam must directly contact the material to be sterilized.

8. In HTST pasteurization, a high temperature is used for a short time (72°C for 15 seconds) to destroy pathogens without altering the flavor of the food. Ultra-high-temperature (UHT) treatment (140°C for 3 seconds) is used to sterilize dairy products.

9. Methods of dry heat sterilization include direct flaming, incineration, and hot-air sterilization. Dry heat kills by oxidation.

10. Different methods that produce the same effect (reduction in microbial growth) are called equivalent treatments.

Filtration (pp. 189–190)

1. Filtration is the passage of a liquid or gas through a filter with pores small enough to retain microbes.

2. Microbes can be removed from air by high-efficiency particulate air filters.

3. Membrane filters composed of nitrocellulose or cellulose acetate are commonly used to filter out bacteria, viruses, and even large proteins.

Low Temperatures (p. 190)

1. The effectiveness of low temperatures depends on the particular microorganism and the intensity of the application.

2. Most microorganisms do not reproduce at ordinary refrigerator temperatures (0–7°C).

3. Many microbes survive (but do not grow) at the subzero temperatures used to store foods.

High Pressure (p. 190)

1. High pressure denatures proteins in vegetative cells.

Desiccation (p. 191)

1. In the absence of water, microorganisms cannot grow but can remain viable.

2. Viruses and endospores can resist desiccation.

Osmotic Pressure (p. 191)

1. Microorganisms in high concentrations of salts and sugars undergo plasmolysis.
2. Molds and yeasts are more capable than bacteria of growing in materials with low moisture or high osmotic pressure.

Radiation (pp. 191–192)

1. The effects of radiation depend on its wavelength, intensity, and duration.
2. Ionizing radiation (gamma rays, X rays, and high-energy electron beams) has a high degree of penetration and exerts its effect primarily by ionizing water and forming highly reactive hydroxyl radicals.
3. Ultraviolet (UV) radiation, a form of nonionizing radiation, has a low degree of penetration and causes cell damage by making thymine dimers in DNA that interfere with DNA replication; the most effective germicidal wavelength is 260 nm.
4. Microwaves can kill microbes indirectly as materials get hot.

Chemical Methods of Microbial Control (pp. 192–201)

1. Chemical agents are used on living tissue (as antiseptics) and on inanimate objects (as disinfectants).
2. Few chemical agents achieve sterility.

Principles of Effective Disinfection (pp. 192–194)

1. Careful attention should be paid to the properties and concentration of the disinfectant to be used.
2. The presence of organic matter, degree of contact with microorganisms, and temperature should also be considered.

Evaluating a Disinfectant (p. 194)

1. In the use-dilution test, bacterial (*S. choleraesuis, S. aureus,* and *P. aeruginosa*) survival in the manufacturer's recommended dilution of a disinfectant is determined.
2. Viruses, endospore-forming bacteria, mycobacteria, and fungi can also be used in the use-dilution test.
3. In the disk-diffusion method, a disk of filter paper is soaked with a chemical and placed on an inoculated agar plate; a zone of inhibition indicates effectiveness.

Types of Disinfectants (pp. 194–201)

Phenol and Phenolics (p. 194)

1. Phenolics exert their action by injuring plasma membranes.

Bisphenols (pp. 194–195)

1. Bisphenols such as triclosan (over the counter) and hexachlorophene (prescription) are widely used in household products.

Biguanides (pp. 195–196)

1. Chlorhexidine damages plasma membranes of vegetative cells.

Halogens (pp. 196–197)

1. Some halogens (iodine and chlorine) are used alone or as components of inorganic or organic solutions.
2. Iodine may combine with certain amino acids to inactivate enzymes and other cellular proteins.
3. Iodine is available as a tincture (in solution with alcohol) or as an iodophor (combined with an organic molecule).
4. The germicidal action of chlorine is based on the formation of hypochlorous acid when chlorine is added to water.
5. Chlorine is used as a disinfectant in gaseous form (Cl_2 or ClO_2) or in the form of a compound, such as calcium hypochlorite, sodium hypochlorite, sodium dichloroisocyanurate, and chloramines.

Alcohols (p. 197)

1. Alcohols exert their action by denaturing proteins and dissolving lipids.
2. In tinctures, they enhance the effectiveness of other antimicrobial chemicals.
3. Aqueous ethanol (60–95%) and isopropanol are used as disinfectants.

Heavy Metals and Their Compounds (pp. 197–198)

1. Silver, mercury, copper, and zinc are used as germicidals.
2. They exert their antimicrobial action through oligodynamic action. When heavy metal ions combine with sulfhydryl (—SH) groups, proteins are denatured.

Surface-Active Agents (p. 198)

1. Surface-active agents decrease the surface tension among molecules of a liquid; soaps and detergents are examples.
2. Soaps have limited germicidal action but assist in the removal of microorganisms through scrubbing.
3. Acid-anionic detergents are used to clean dairy equipment.

Quaternary Ammonium Compounds (Quats) (pp. 198–199)

1. Quats are cationic detergents attached to NH_4^+.
2. By disrupting plasma membranes, they allow cytoplasmic constituents to leak out of the cell.
3. Quats are most effective against gram-positive bacteria.

Chemical Food Preservatives (p. 199)

1. SO_2, sorbic acid, benzoic acid, and propionic acid inhibit fungal metabolism and are used as food preservatives.
2. Nitrate and nitrite salts prevent germination of *Clostridium botulinum* endospores in meats.

Antibiotics (pp. 199–200)

1. Nisin and natamycin are anibiotics used to preserve foods, especially cheese.

Aldehydes (p. 200)

1. Aldehydes such as formaldehyde and glutaraldehyde exert their antimicrobial effect by inactivating proteins.

2. They are among the most effective chemical disinfectants.

Gaseous Chemosterilizers (pp. 200–201)

1. Ethylene oxide is the gas most frequently used for sterilization.

2. It penetrates most materials and kills all microorganisms by protein denaturation.

Peroxygens (Oxidizing Agents) (p. 201)

1. Ozone, peroxide, and peracetic acid are used as antimicrobial agents.

2. They exert their effect by oxidizing molecules inside cells.

Microbial Characteristics and Microbial Control (pp. 201–204)

1. Gram-negative bacteria are generally more resistant than gram-positive bacteria to disinfectants and antiseptics.

2. Mycobacteria, endospores, and protozoan cysts and oocysts are very resistant to disinfectants and antiseptics.

3. Nonenveloped viruses are generally more resistant than enveloped viruses to disinfectants and antiseptics.

4. Prions are resistant to disinfection and autoclaving.

STUDY QUESTIONS

Review

1. Name the cause of cell death resulting from damage to each of the following:
   a. cell wall
   b. plasma membrane
   c. proteins
   d. nucleic acids

2. The thermal death time for a suspension of *Bacillus subtilis* endospores is 30 minutes in dry heat and less than 10 minutes in an autoclave. Which type of heat is more effective? Why?

3. If pasteurization does not achieve sterilization, why is food treated by pasteurization?

4. Thermal death point is not considered an accurate measure of the effectiveness of heat sterilization. List three factors that can alter thermal death point.

5. The antimicrobial effect of gamma radiation is due to _____. The antimicrobial effect of ultraviolet radiation is due to _____.

6. A bacterial culture was in log phase in the following figure. At time *x*, an antibacterial compound was added to the culture. Which line indicates addition of a bactericidal compound? A bacteriostatic compound? How can you tell? Explain why the viable count does not immediately drop to zero at *x*.

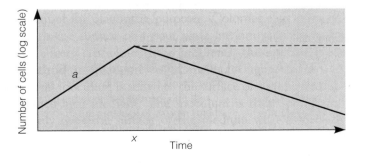

7. Fill in the following table:

Method of Sterilization	Temp.	Time	Type of Heat	Pref'd Use	Types of Action
Autoclaving					
Hot Air					
Pasteurization					

8. How do the examples in question 7 illustrate the concept of equivalent treatments?

9. How do salts and sugars preserve foods? Why are these considered physical rather than chemical methods of microbial control? Name one food that is preserved with sugar and one preserved with salt. How do you account for the occasional growth of *Penicillium* mold in jelly, which is 50% sucrose?

10. List five factors to consider before selecting a disinfectant.

11. Give the method of action and at least one standard use of each of the following types of disinfectants:
   a. phenolics
   b. iodine
   c. chlorine
   d. alcohol
   e. heavy metals
   f. aldehydes
   g. ethylene oxide
   h. oxidizing agents

12. The use-dilution values for two disinfectants tested under the same conditions are: Disinfectant A—1:2; Disinfectant B—1:10,000. If both disinfectants are designed for the same purpose, which would you select?

13. A large hospital washes burn patients in a stainless steel tub. After each patient, the tub is cleaned with a quat. It was noticed that 14 of 20 burn patients acquired *Pseudomonas* infections after being bathed. Provide an explanation for this high rate of infection.

Multiple Choice

1. Which of the following does *not* kill endospores?
   - **a.** autoclaving
   - **b.** incineration
   - **c.** hot-air sterilization
   - **d.** pasteurization
   - **e.** none of the above

2. Which of the following is most effective for sterilizing mattresses and plastic Petri dishes?
   - **a.** chlorine
   - **b.** ethylene oxide
   - **c.** glutaraldehyde
   - **d.** autoclaving
   - **e.** nonionizing radiation

3. Which of these disinfectants does *not* act by disrupting the plasma membrane?
   - **a.** phenolics
   - **b.** phenol
   - **c.** quaternary ammonium compounds
   - **d.** halogens
   - **e.** biguanides

4. Which of the following *cannot* be used to sterilize a heat-labile solution stored in a plastic container?
   - **a.** gamma radiation
   - **b.** ethylene oxide
   - **c.** nonionizing radiation
   - **d.** autoclaving
   - **e.** short-wavelength radiation

5. Which of the following is *not* a characteristic of quaternary ammonium compounds?
   - **a.** bactericidal against gram-positive bacteria
   - **b.** sporicidal
   - **c.** amoebicidal
   - **d.** fungicidal
   - **e.** kills enveloped viruses

6. A classmate is trying to determine how a disinfectant might kill cells. You observed that when he spilled the disinfectant in your reduced litmus milk, the litmus turned blue again. You suggest to your classmate that
   - **a.** the disinfectant might inhibit cell wall synthesis.
   - **b.** the disinfectant might oxidize molecules.
   - **c.** the disinfectant might inhibit protein synthesis.
   - **d.** the disinfectant might denature proteins.
   - **e.** he take his work away from yours.

7. Which of the following is most likely to be bactericidal?
   - **a.** membrane filtration
   - **b.** ionizing radiation
   - **c.** freeze-drying
   - **d.** deep-freezing
   - **e.** all of the above

8. Which of the following is used to control microbial growth in foods?
   - **a.** organic acids
   - **b.** alcohols
   - **c.** aldehydes
   - **d.** heavy metals
   - **e.** all of the above

Use the following information to answer questions 9 and 10. The data were obtained from a use-dilution test comparing four disinfectants against *Salmonella choleraesuis.*

	Bacterial Growth After Exposure to			
Dilution	Disinfectant A	Disinfectant B	Disinfectant C	Disinfectant D
1:2	–	+	–	–
1:4	–	+	–	+
1:8	–	+	+	+
1:16	+	+	+	+

9. Which disinfectant is the most effective?

10. Which disinfectant(s) is (are) bactericidal?
   - **a.** A, B, C, and D
   - **b.** A, C, and D
   - **c.** A only
   - **d.** B only
   - **e.** none of the above

Critical Thinking

1. The disk-diffusion method was used to evaluate three disinfectants. The results were as follows:

Disinfectant	Zone of Inhibition
X	0 mm
Y	5 mm
Z	10 mm

   - **a.** Which disinfectant was the most effective against the organism?
   - **b.** Can you determine whether compound Y was bactericidal or bacteriostatic?

2. Why is each of the following bacteria often resistant to disinfectants?
   - **a.** *Mycobacterium*
   - **b.** *Pseudomonas*
   - **c.** *Bacillus*

3. A use-dilution test was used to evaluate two disinfectants against *Salmonella choleraesuis.* The results were as follows:

	Bacterial Growth After Exposures		
Time of Exposure (min)	Disinfectant A	Disinfectant B Diluted with Distilled Water	Disinfectant B Diluted with Tap Water
10	+	–	+
20	+	–	–
30	–	–	–

   - **a.** Which disinfectant was the most effective?
   - **b.** Which disinfectant should be used against *Staphylococcus*?

4. To determine the lethal action of microwave radiation, two 10^5 suspensions of *E. coli* were prepared. One cell suspension was exposed to microwave radiation while wet,

whereas the other was lyophilized (freeze-dried) and then exposed to radiation. The results are shown in the following figure. Dashed lines indicate the temperature of the samples. What is the most likely method of lethal action of microwave radiation? How do you suppose these data might differ for *Clostridium*?

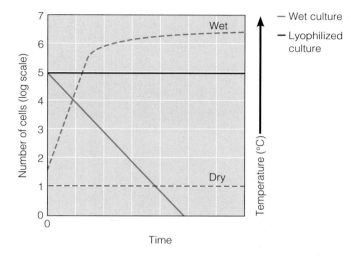

Clinical Applications

1. *Entamoeba histolytica* and *Giardia lamblia* were isolated from the stool sample of a 45-year-old man, and *Shigella sonnei* was isolated from the stool sample of an 18-year-old woman. Both patients experienced diarrhea and severe abdominal cramps, and prior to onset of digestive symptoms both had been treated by the same chiropractor. The chiropractor had administered colonic irrigations (enemas) to these patients. The device used for this treatment was a gravity-dependent apparatus using 12 liters of tap water. There were no check valves to prevent backflow, so all parts of the apparatus could have become contaminated with feces during each colonic treatment. The chiropractor provided colonic treatment to four or five patients per day. Between patients, the adaptor piece that is inserted into the rectum was placed in a "hot-water sterilizer."

   What two errors were made by the chiropractor?

2. Between March 9 and April 12, five chronic peritoneal dialysis patients at one hospital became infected with *Pseudomonas aeruginosa*. Four patients developed peritonitis (inflammation of the abdominal cavity), and one developed a skin infection at the catheter insertion site. All patients with peritonitis had low-grade fever, cloudy peritoneal fluid, and abdominal pain. All patients had permanent indwelling peritoneal catheters, which the nurse wiped with gauze that had been soaked with an iodophor solution each time the catheter was connected to or disconnected from the machine tubing. Aliquots of the iodophor were transferred from stock bottles to small in-use bottles. Cultures from the dialysate concentrate and the internal areas of the dialysis machines were negative; iodophor from a small in-use plastic container yielded a pure culture of *P. aeruginosa*.

   What improper technique led to this infection?

3. Eleven patients received injections of methylprednisolone and lidocaine to relieve the pain and inflammation of arthritis at the same orthopedic surgery office. All of them developed septic arthritis caused by *Serratia marcescens*. Unopened bottles of methylprednisolone from the same lot numbers tested sterile; the methylprednisolone was preserved with a quat. Cotton balls were used to wipe multiple-use injection vials before the medication was drawn into a disposable syringe. The site of injection on each patient was also wiped with a cotton ball. The cotton balls were soaked in benzalkonium chloride, and fresh cotton balls were added as the jar was emptied. Opened methylprednisolone containers and the jar of cotton balls contained *S. marcescens*.

   How was the infection transmitted? What part of the routine procedure caused the contamination?.

LEARNING WITH TECHNOLOGY

The Microbiology Place CD-ROM/website includes many resources to help you study and succeed in this course. In addition to self-quizzes, web links, and flashcards, you'll find the following activities for this chapter in the "Activities" section of The Microbiology Place:

Case Study: Microbial Contamination of Cosmetics

Chapter 8

Microbial Genetics

BACTERIAL CHROMOSOME. *The single chromosome, normally tightly packed inside a bacterial cell, has burst from an E. coli cell after the cell wall and plasma membrane were damaged.*

Virtually all the microbial traits you have read about in earlier chapters are controlled or influenced by heredity. The inherited traits of microbes include their shape and structural features, their metabolism, their ability to move or behave in various ways, and their ability to interact with other organisms—perhaps causing disease. Individual organisms transmit these characteristics to their offspring through genes, the units of the hereditary material that contain the information that determines these characteristics.

An understanding of genetics is key to grasping a number of concepts in microbiology. For example, many antibiotics act by inhibiting steps in protein synthesis. Knowing how biological information flows from genes to proteins can help us understand how some antibiotics work, thereby enabling us to develop new weapons against disease.

Researchers are trying to solve the difficult medical problem of microbes' developing antibiotic resistance. A microorganism can become resistant to antibiotics in any of several ways, all of which depend on genetic information. Resistant microbes have somehow obtained a gene or set of genes that will prevent the action of the antibiotic.

Emerging diseases provide another example of the importance of understanding genetics. New diseases are the result of genetic changes in some existing organism. Currently, biologists are using genetics to discover relatedness among organisms and how life evolved on Earth. It was by looking at genes that Carl Woese discovered that there are two types of prokaryotes, bacteria and archaea.

Structure and Function of the Genetic Material

Learning Objectives

- *Define genetics, genome, chromosome, gene, genetic code, genotype, phenotype, and genomics.*

- *Describe how DNA serves as genetic information.*

Genetics is the science of heredity; it includes the study of what genes are, how they carry information, how they are replicated and passed to subsequent generations of cells or passed between organisms, and how the expression of their information within an organism determines the particular characteristics of that organism. The genetic information in a cell is called the **genome.** A cell's genome is organized into chromosomes. **Chromosomes** are structures containing DNA that physically carry hereditary information; the chromosomes contain the genes. **Genes** are segments of DNA (except in some viruses, in which they are made of RNA) that code for functional products. We saw in Chapter 2, on page 47, that DNA is a macromolecule composed of repeating units called *nucleotides.* Recall that each nucleotide consists of a nitrogenous base (adenine, thymine, cytosine, or guanine), deoxyribose (a pentose sugar), and a phosphate group (see Figure 2.16 on page 48). The DNA within a cell exists as long strands of nucleotides twisted together in pairs to form a double helix. Each strand has a string of alternating sugar and phosphate groups—its *sugar-phosphate backbone*—and a nitrogenous base is attached to each sugar in the backbone. The two strands are held together by hydrogen bonds between their nitrogenous bases. The **base pairs** always occur in a specific way: Adenine always pairs with thymine, and cytosine always pairs with guanine. Because of this specific base pairing, the base sequence of one DNA strand determines the sequence of the other strand. The two strands of DNA are thus *complementary.* You can think of these complementary DNA sequences as being like a positive photograph and its negative.

The structure of DNA helps explain two primary features of biological information storage. First, the linear sequence of bases provides the actual information. Genetic information is encoded by the sequence of bases along a strand of DNA, in much the same way as our written language uses a linear sequence of letters to form words and sentences. The genetic language, however, uses an alphabet with only four letters—the four kinds of nitrogenous bases in DNA (or RNA). But 1000 of these four bases, the number contained in an average-sized gene, can be arranged in 4^{1000} different ways. This astronomically large number explains how genes can be varied enough to provide all the information a cell needs to grow and perform its functions.

The **genetic code,** the set of rules that determines how a nucleotide sequence is converted into the amino acid sequence of a protein, is discussed in more detail later in the chapter.

Second, the complementary structure allows for the precise duplication of DNA during cell division. Again, think of the photograph analogy: If you have a negative, you can always make another copy of the positive print. Likewise with DNA: If you know the sequence of one strand, you also know the sequence of the complementary strand.

Much of cellular metabolism is concerned with translating the genetic message of genes into specific proteins. A gene usually codes for a messenger RNA (mRNA) molecule, which ultimately results in the formation of a protein. Alternatively, the gene product can be a ribosomal RNA (rRNA) or a transfer RNA (tRNA). As we will see, all of these types of RNA are involved in the process of protein synthesis. When the ultimate molecule for which a gene codes (a protein, for example) has been produced, we say that the gene has been *expressed.*

Genotype and Phenotype

The **genotype** of an organism is its genetic makeup, the information that codes for all the particular characteristics of the organism. The genotype represents *potential* properties, but not the properties themselves. **Phenotype** refers to *actual, expressed* properties, such as the organism's ability to perform a particular chemical reaction. Phenotype, then, is the manifestation of genotype.

In molecular terms, an organism's genotype is its collection of genes, its entire DNA. What constitutes the organism's phenotype in molecular terms? In a sense, an organism's phenotype is its collection of proteins. Most of a cell's properties derive from the structures and functions of its proteins. In microbes, most proteins are either enzymatic (catalyze particular reactions) or structural (participate in large functional complexes such as membranes or ribosomes). Even phenotypes that depend on structural macromolecules other than proteins (such as lipids or polysaccharides) rely indirectly on proteins. For instance, the structure of a complex lipid or polysaccharide molecule results from the catalytic activities of enzymes that synthesize, process, and degrade those molecules. Thus, although it is not completely accurate to say that phenotypes are due only to proteins, it is a useful simplification.

DNA and Chromosomes

Bacteria typically have a single circular chromosome consisting of a single circular molecule of DNA with associated proteins. The chromosome is looped and folded (Figure 8.1a) and attached at one or several points to the

FIGURE 8.1 **Chromosomes. (a)** A prokaryotic chromosome. The tangled mass and looping strands of DNA emerging from this disrupted *E. coli* cell are part of its single chromosome. **(b)** A genetic map of the chromosome of *E. coli*. The units are minutes based on the length of time it takes to transfer the genes during mating between two cells.

■ The *E. coli* genome consists of 4.6 million base pairs making about 4300 genes.

plasma membrane. The DNA of *E. coli*, the most-studied bacterial species, has about 4.6 million base pairs and is about 1 mm long—1000 times longer than the entire cell. However, the chromosome takes up only about 10% of the cell's volume because the DNA is twisted or *supercoiled* by an enzyme called **topoisomerase II** or **DNA gyrase.**

The location of genes on a bacterial chromosome is determined by experiments on the transfer of genes from one cell to another. These processes will be discussed later in this chapter. The bacterial chromosome map that results is marked in minutes corresponding to when the genes are transferred from a donor cell to a recipient cell (Figure 8.1b).

In recent years, the complete base sequences of several bacterial chromosomes have been determined. The sequencing and molecular characterization of genomes is called **genomics.** The most important genetic questions being addressed concern how cells use the genetic information available to them.

The Flow of Genetic Information

DNA replication makes possible the flow of genetic information from one generation to the next. As shown in Figure 8.2, the DNA of a cell replicates before cell division, so that each daughter cell receives a chromosome identical to the parent's. Within each metabolizing cell, the genetic information contained in DNA also flows in another way: It is transcribed into mRNA and then translated into protein. We describe the processes of transcription and translation later in this chapter.

DNA Replication

Learning Objective

• *Describe the process of DNA replication.*

In DNA replication, one "parental" double-stranded DNA molecule is converted to two identical "daughter" molecules. The complementary structure of the nitrogenous base sequences in the DNA molecule is the key to understanding DNA replication. Because the bases along the two strands of double-helical DNA are complementary, one strand can act as a template for the production of the other strand (Figure 8.3 on page 214).

DNA replication requires the presence of several cellular proteins that direct a particular sequence of events. When replication begins, the two strands of parental DNA are unwound and separated from each other in one

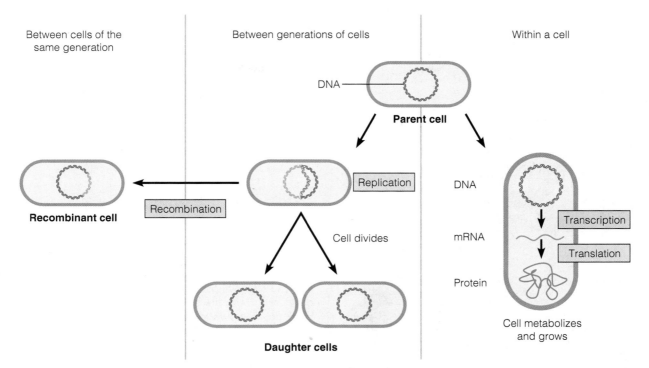

FIGURE 8.2 **An overview of the flow of genetic information.** Genetic information can be transferred between generations of cells, through DNA replication. Occasionally, genetic information can be transferred between cells of the same generation, through recombination. Genetic information is also used within a cell to produce the proteins the cell needs to function, through transcription and translation. The cell represented here is a bacterium with a single circular chromosome. A small version of this figure will be included in figures throughout this chapter to indicate the relationships of different processes.

■ All of these processes can occur at the same time in a bacterial cell. Which process results in reproduction?

small DNA segment after another. Free nucleotides present in the cytoplasm of the cell are matched up to the exposed bases of the single-stranded parental DNA. Where thymine is present on the original strand, only adenine can fit into place on the new strand; where guanine is present on the original strand, only cytosine can fit into place, and so on. Any bases that are improperly base-paired are removed and replaced by replication enzymes. Once aligned, the newly added nucleotide is joined to the growing DNA strand by an enzyme called **DNA polymerase.** Then the parental DNA is unwound a bit further to allow the addition of the next nucleotides. The point at which replication occurs is called the *replication fork*.

As the replication fork moves along the parental DNA, each of the unwound single strands combines with new nucleotides. The original strand and this newly synthesized daughter strand then rewind. Because each new

double-stranded DNA molecule contains one original, conserved strand and one new strand, the process of replication is referred to as **semiconservative replication.**

Before looking at DNA replication in more detail, let's take a closer look at the structure of DNA (see Figure 2.16 on page 48). It is important to understand the concept that the paired DNA strands are oriented in opposite directions relative to each other. Notice in Figure 2.16 that the carbon atoms of the sugar component of each nucleotide are numbered 1′ (pronounced "one prime") to 5′. In order for the paired bases to be next to each other, the sugar components in one strand are upside-down relative to the other. The end with the hydroxyl attached to the 3′ carbon is called the 3′ end of the DNA strand; the end having a phosphate attached to the 5′ carbon is called the 5′ end. The way in which the two strands fit together dictates that the 5′ → 3′ direction of one strand runs counter to the 5′ → 3′ direction of the other strand (Figure 8.4). This structure of DNA

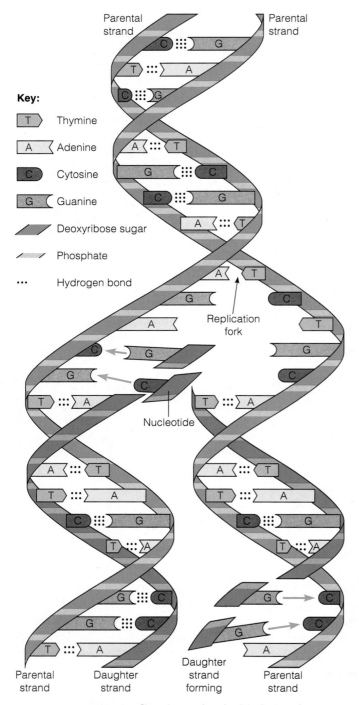

Key:

T	Thymine
A	Adenine
C	Cytosine
G	Guanine
▱	Deoxyribose sugar
▱	Phosphate
⋯	Hydrogen bond

FIGURE 8.3 **DNA replication.** The double helix of the parental DNA separates as weak hydrogen bonds between the nucleotides on opposite strands break in response to the action of replication enzymes. Next, hydrogen bonds form between new complementary nucleotides and each strand of the parental template to form new base pairs. Enzymes catalyze the formation of sugar-phosphate bonds between sequential nucleotides on each resulting daughter strand.

■ What is meant by semiconservative replication?

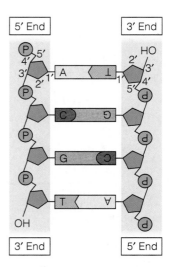

FIGURE 8.4 **The two strands of DNA are antiparallel.** The sugar phosphate backbone of one strand is upside-down relative to the backbone of the other strand. Turn the book upside-down to demonstrate this.

■ Why is one strand "upside-down" relative to the other strand; why can't both strands "face" the same way?

affects the replication process because DNA polymerases can add new nucleotides to the 3′ end only. Therefore, as the replication fork moves along the parental DNA, the two new strands must grow in different directions.

DNA replication requires a great deal of energy. The energy is supplied from the nucleotides, which are actually nucleoside triphosphates. You already know about ATP; the only difference between ATP and the adenine nucleotide in DNA is the sugar component. Deoxyribose is the sugar in the nucleosides used to synthesize DNA, and nucleoside triphosphates with ribose are used to synthesize RNA. Two phosphate groups are removed to add nucleotide to a growing strand of DNA; hydrolysis of the nucleoside is exergonic and provides energy to make the new bonds in the DNA strand (Figure 8.5).

Figure 8.6 provides more detail about the many steps that go into this complex process.

❶–❷ Once the parental DNA is unwound and stabilized, the replication fork forms at a fixed site called the origin of replication.

❸ One new DNA strand, called the **leading strand,** is synthesized continuously as the DNA polymerase moves toward the replication fork making DNA in the 5′ → 3′ direction.

❹ Remember that DNA polymerase can only add new nucleotides to the 3′ end, so a short piece of RNA called an **RNA primer** is needed to start synthesis. DNA polymerase can then add nucleotides to the 3′

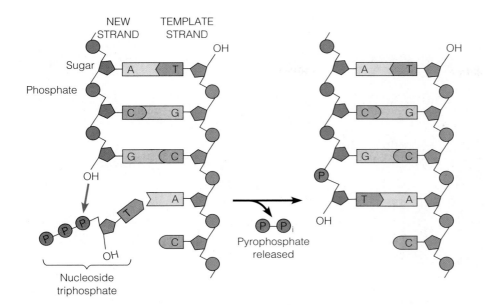

FIGURE 8.5 **Adding a nucleotide to DNA.** When a nucleoside triphosphate bonds to the sugar in a growing DNA strand, it loses two phosphates. Hydrolysis of the phosphate bonds provides the energy for the reaction.

- Synthesis of DNA requires dATP, dTTP, dGTP, dCTP; d stands for deoxyribose.

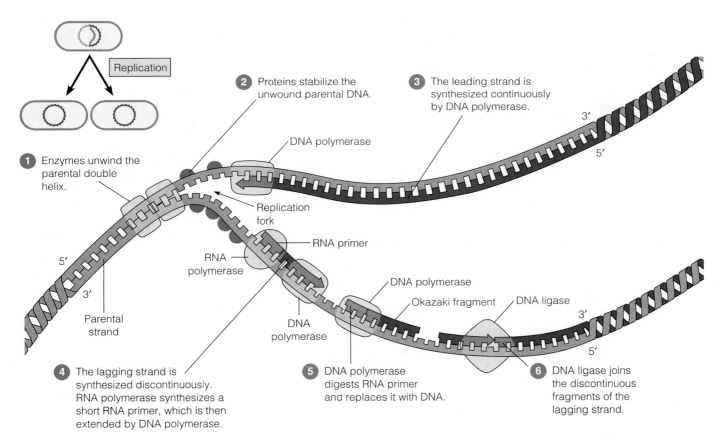

① Enzymes unwind the parental double helix.

② Proteins stabilize the unwound parental DNA.

③ The leading strand is synthesized continuously by DNA polymerase.

④ The lagging strand is synthesized discontinuously. RNA polymerase synthesizes a short RNA primer, which is then extended by DNA polymerase.

⑤ DNA polymerase digests RNA primer and replaces it with DNA.

⑥ DNA ligase joins the discontinuous fragments of the lagging strand.

FIGURE 8.6 **A summary of events at the DNA replication fork.** Enzymes at the replication fork unwind the parental double helix. DNA polymerase synthesizes a continuous strand of new DNA, using one of the parental strands as a template. DNA polymerase also uses the other parental strand as a template, but because the orientation of the sugars is opposite, RNA polymerase starts the synthesis by adding a short stretch of RNA called an RNA primer. The DNA polymerase digests away the RNA as it makes a small piece of DNA. The small units are subsequently joined by DNA ligase.

- Why is one strand of DNA synthesized discontinuously?

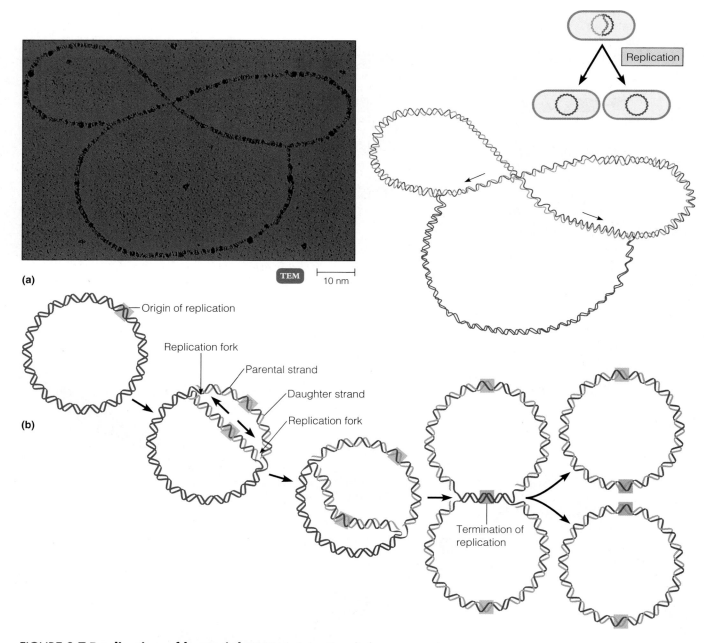

FIGURE 8.7 **Replication of bacterial DNA. (a)** An *E. coli* chromosome in the process of replicating. (In the corresponding diagram at right, the arrows point to the two replication forks.) The chromosome is about one-third replicated. Notice that one of the new helixes is crossed over the other one. **(b)** A diagram of the bidirectional replication of a circular bacterial DNA molecule. The new strands are shown in blue.

■ What is the origin of replication?

end of the RNA. Consequently, the **lagging strand** of new DNA is synthesized in pieces consisting of about 1000 nucleotides, called *Okazaki fragments*, as the DNA polymerase moves away from the replication fork.

❺ DNA polymerase removes the RNA primer, and

❻ the enzyme **DNA ligase** joins the newly made DNA fragments.

DNA replication of some bacteria, such as *E. coli*, goes *bidirectionally* around the chromosome (Figure 8.7). Two replication forks move in opposite directions away from the origin of replication. Because the bacterial chromosome is a closed loop, the replication forks eventually meet when replication is completed. Much evidence shows an association between the bacterial plasma membrane and the origin of replication. After duplication, if each copy of the origin binds to the membrane at opposite poles, then each daughter cell would receive one copy of the DNA molecule—that is, one complete chromosome.

DNA replication is an amazingly accurate process. Typically, mistakes are made at a rate of only 1 in every 10^{10} bases incorporated. Such accuracy is largely due to the *proofreading* capability of DNA polymerase. As each new base is added, the enzyme evaluates whether it forms the proper complementary base-pairing structure. If not, the enzyme excises the improper base and replaces it with the correct one. In this way, DNA replication can be performed very accurately, allowing each daughter chromosome to be virtually identical to the parental DNA.

RNA and Protein Synthesis

Learning Objective

• *Describe protein synthesis, including transcription, RNA processing, and translation.*

How is the information in DNA used to make the proteins that control cell activities? In the process of *transcription*, genetic information in DNA is copied, or transcribed, into a complementary base sequence of RNA. The cell then uses the information encoded in this RNA to synthesize specific proteins through the process of *translation*. We now take a closer look at these two processes as they occur in a bacterial cell.

Transcription

Transcription is the synthesis of a complementary strand of RNA from a DNA template. We will discuss transcription in prokaryotic cells here. Transcription in eukaryotes is discussed on page 221. As mentioned earlier, there are three kinds of RNA in bacterial cells: messenger RNA, ribosomal RNA, and transfer RNA. Ribosomal RNA forms an integral part of ribosomes, the cellular machinery for protein synthesis. Transfer RNA is also involved in protein synthesis, as we will see. **Messenger RNA (mRNA)** carries the coded information for making specific proteins from DNA to ribosomes, where proteins are synthesized.

During transcription, a strand of mRNA is synthesized using a specific gene—a portion of the cell's DNA—as a template. In other words, the genetic information stored in the sequence of nitrogenous bases of DNA is rewritten so that the same information appears in the base sequence of

mRNA. As in DNA replication, a G in the DNA template dictates a C in the mRNA being made, a C in the DNA template dictates a G in the mRNA, and a T in the DNA template dictates an A in the mRNA. However, an A in the DNA template dictates a uracil (U) in the mRNA because RNA contains U instead of T. (U has a chemical structure slightly different from T, but it base-pairs in the same way.) If, for example, the template portion of DNA has the base sequence ATGCAT, the newly synthesized mRNA strand will have the complementary base sequence UACGUA.

The process of transcription requires both an enzyme called *RNA polymerase* and a supply of RNA nucleotides (Figure 8.8 on page 218). Transcription begins when

❶ RNA polymerase binds to the DNA at a site called the **promoter.** Only one of the two DNA strands serves as the template for RNA synthesis for a given gene. Like DNA, RNA is synthesized in the 5′ → 3′ direction.

❷ RNA polymerase assembles free nucleotides into a new chain, using complementary base pairing as a guide.

❸ As the new RNA chain grows, RNA polymerase moves along the DNA.

❹ RNA synthesis continues until RNA polymerase reaches a site on the DNA called the **terminator.**

❺ When this happens, RNA polymerase and the newly formed, single-stranded mRNA are released from the DNA.

The process of transcription allows the cell to produce short-term copies of genes that can be used as the direct source of information for protein synthesis. Messenger RNA acts as an intermediate between the permanent storage form, DNA, and the process that uses the information, translation.

Translation

We have seen how the genetic information in DNA is transferred to mRNA during transcription. Now we will see how mRNA serves as the source of information for the synthesis of proteins. Protein synthesis is called **translation** because it involves decoding the "language" of nucleic acids and converting that information into the "language" of proteins.

The language of mRNA is in the form of **codons,** groups of three nucleotides, such as AUG, GGC, or AAA. The sequence of codons on an mRNA molecule determines the sequence of amino acids that will be in the protein being synthesized. Each codon "codes" for a particular amino acid. This is the genetic code (Figure 8.9 on page 219).

FIGURE 8.8 **The process of transcription.** The orienting diagram indicates the relationship of transcription to the overall flow of genetic information within a cell.

■ Transcription refers to copying or transcribing the genetic information in DNA into a complementary base sequence of RNA.

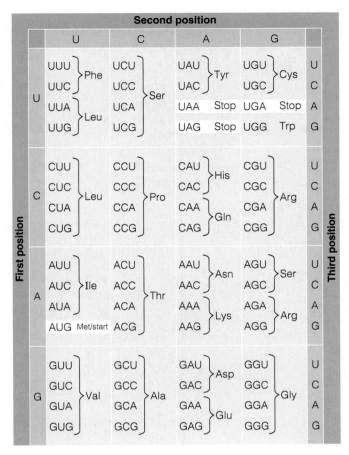

FIGURE 8.9 The genetic code. The three nucleotides in an mRNA codon are designated, respectively, as the first position, second position, and third position of the codon on the mRNA. Each set of three nucleotides specifies a particular amino acid, represented by a three-letter abbreviation (see Table 2.4 on page 44). The codon AUG, which specifies the amino acid methionine, is also the start of protein synthesis. The word *Stop* identifies the nonsense codons that signal the termination of protein synthesis.

■ Why is the genetic code described as degenerate?

Codons are written in terms of their base sequence in mRNA. Notice that there are 64 possible codons but only 20 amino acids. This means that most amino acids are signaled by several alternative codons, a situation referred to as the **degeneracy** of the code. For example, leucine has six codons and alanine has four codons. Degeneracy allows for a certain amount of change, or mutation, in the DNA without affecting the protein ultimately produced.

Of the 64 codons, 61 are sense codons and 3 are nonsense codons. **Sense codons** code for amino acids, and **nonsense codons** (also called *stop codons*) do not. Rather, the nonsense codons—UAA, UAG, and UGA—signal the end of the protein molecule's synthesis. The start codon that initiates the synthesis of the protein molecule is AUG, which is

also the codon for methionine. In bacteria, the start AUG codes for formylmethione rather than the methionine found in other parts of the protein. The initiating methionine is often removed later, so not all proteins begin with methionine.

The codons of mRNA are converted into protein through the process of translation. The codons of an mRNA are "read" sequentially, and in response to each codon, the appropriate amino acid is assembled into a growing chain. The site of translation is the ribosome, and **transfer RNA (tRNA)** molecules both recognize the specific codons and transport the required amino acids.

Each tRNA molecule has an **anticodon,** a sequence of three bases that is complementary to a codon. In this way, a tRNA molecule can base-pair with its associated codon. Each tRNA can also carry on its other end the amino acid coded for by the codon that the tRNA recognizes. The functions of the ribosome are to direct the orderly binding of tRNAs to codons and to assemble the amino acids brought there into a chain, ultimately producing a protein. Figure 8.10 on page 220 shows the details of translation.

❶ The necessary components assemble: the two ribosomal subunits, a tRNA with the anticodon UAC, and the mRNA molecule to be translated, along with several additional protein factors. This sets up the initiator codon (AUG) in the proper position to allow translation to begin.

❷ The first tRNA binds to the start codon, bringing with it the amino acid methionine.

❸ When the tRNA that recognizes the second codon moves into position on the ribosome, the first amino acid is transferred by the ribosome.

❹ After the ribosome joins the two amino acids with a peptide bond, the first tRNA molecule leaves the ribosome.

❺ The ribosome then moves along the mRNA to the next codon.

❻ As the proper amino acids are brought into line one by one, peptide bonds are formed between them, and a polypeptide chain results.

❼ Translation ends when one of the three nonsense codons in the mRNA is reached.

❽ When the ribosome arrives at this codon, it comes apart into its two subunits, and the mRNA and newly synthesized polypeptide chain are released. The ribosome, the mRNA, and the tRNAs are then available to be used again.

The ribosome moves along the mRNA in the 5′ → 3′ direction. As a ribosome moves along the mRNA, it will soon allow the start codon to be exposed. Additional ribosomes

1 Components needed to begin translation come together.

2 On the assembled ribosome, a tRNA carrying the first amino acid is paired with the start codon on the mRNA. A tRNA carrying the second amino acid approaches.

3 The place on the ribosome where the first tRNA sits is called the P site. In the A site next to it, the second codon of the mRNA pairs with a tRNA carrying the second amino acid.

4 The first amino acid joins to the second by a peptide bond, and the first tRNA is released.

5 The ribosome moves along the mRNA until the second tRNA is in the P site, and the process continues.

6 The ribosome continues to move along the mRNA, and new amino acids are added to the polypeptide.

FIGURE 8.10 **The process of translation.** The overall goal of translation is to produce proteins using mRNAs as the source of biological information. The complex cycle of events illustrated here shows the primary role of tRNA and ribosomes in the decoding of this information. The ribosome acts as the site where the mRNA-encoded information is decoded, as well as the site where individual amino acids are connected into polypeptide chains. The tRNA molecules act as the actual "translators"—one end of each tRNA recognizes a specific mRNA codon, while the other end carries the amino acid coded for by that codon.

■ Translation is the process by which the base sequence of mRNA determines the amino acid sequence of a protein.

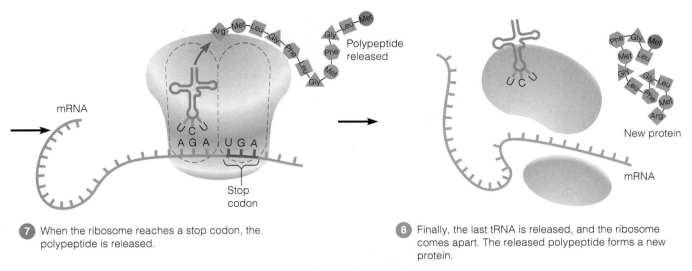

7 When the ribosome reaches a stop codon, the polypeptide is released.

8 Finally, the last tRNA is released, and the ribosome comes apart. The released polypeptide forms a new protein.

FIGURE 8.10 **The process of translation** (continued).

can then assemble and begin synthesizing protein. In this way, there are usually a number of ribosomes attached to a single mRNA, all at various stages of protein synthesis. In prokaryotic cells, the translation of mRNA into protein can begin even before transcription is complete (Figure 8.11). Because mRNA is produced in the cytoplasm, the start codons of an mRNA being transcribed are available to ribosomes before the entire mRNA molecule is even made. Thus, translation can begin before transcription is complete.

In eukaryotic cells, transcription takes place in the nucleus. The mRNA must be completely synthesized and moved through the nuclear membrane to the cytoplasm before translation can begin. In addition, the RNA undergoes processing before it leaves the nucleus. In eukaryotic cells the regions of genes that code for proteins are often interrupted by noncoding DNA. Thus, eukaryotic genes are composed of **exons,** the regions of DNA *expressed*, and **introns,** the *intervening* regions of DNA that do not encode protein (Figure 8.12 on page 222).

1 In the nucleus of a eukaryotic cell, RNA polymerase synthesizes an RNA molecule containing exons and introns called an *RNA transcript*.

2 This long RNA is then processed by ribozymes, which remove the intron-derived RNA and splice together the exon-derived RNA, producing an mRNA.

FIGURE 8.11 **Simultaneous transcription and translation in bacteria.** The micrograph and diagram show these processes in a single bacterial gene. Many molecules of mRNA are being synthesized simultaneously. The longest mRNA molecules were the first to be transcribed at the promoter. Notice the ribosomes attached to the newly forming mRNA. The newly synthesized polypeptides are not shown.

■ Why can translation begin before transcription is complete in prokaryotes but not in eukaryotes?

FIGURE 8.12 **RNA processing in eukaryotic cells.**

■ Why can't the RNA transcript be used for translation?

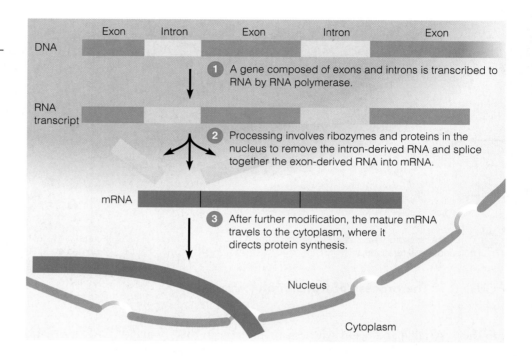

1 A gene composed of exons and introns is transcribed to RNA by RNA polymerase.

2 Processing involves ribozymes and proteins in the nucleus to remove the intron-derived RNA and splice together the exon-derived RNA into mRNA.

3 After further modification, the mature mRNA travels to the cytoplasm, where it directs protein synthesis.

Nucleus

Cytoplasm

❸ The resulting mRNA leaves the nucleus to be used by rRNA and tRNA for protein synthesis.

* * *

To summarize, genes are the units of biological information encoded by the sequence of nucleotide bases in DNA. A gene is expressed, or turned into a product within the cell, through the processes of transcription and translation. The genetic information carried in DNA is transferred to a temporary mRNA molecule by transcription. Then, during translation, the mRNA directs the assembly of amino acids into a polypeptide chain: mRNA attaches to a ribosome, tRNAs deliver the amino acids to the ribosome as directed by the mRNA codon sequence, and the ribosome assembles the amino acids into the chain that will be the newly synthesized protein.

The Regulation of Bacterial Gene Expression

Learning Objective

• *Explain the regulation of gene expression in bacteria by induction, repression, and catabolite repression.*

A cell's genetic machinery and its metabolic machinery are integrated and interdependent. Recall from Chapter 5 that the bacterial cell carries out an enormous number of metabolic reactions. The common feature of all metabolic reactions is that they are catalyzed by enzymes. Also recall

from Chapter 5 (page 119) that feedback inhibition stops a cell from performing unneeded chemical reactions. Feedback inhibition stops enzymes that have already been synthesized. We will now look at mechanisms to prevent synthesis of enzymes that are not needed.

We have seen that genes, through transcription and translation, direct the synthesis of proteins, many of which serve as enzymes—the very enzymes used for cellular metabolism. Because protein synthesis requires a tremendous expenditure of energy, the regulation of protein synthesis is important to the cell's energy economy. The cell conserves energy by making only those proteins needed at a particular time. We will next look at how chemical reactions are regulated by controlling the synthesis of the enzymes.

Many genes, perhaps 60–80%, are not regulated but are instead *constitutive*, meaning that their products are constantly produced at a fixed rate. Usually these genes, which are effectively turned on all the time, code for enzymes that the cell needs in fairly large amounts for its major life processes; the enzymes of glycolysis are examples. The production of other enzymes is regulated so that they are present only when needed. *Trypanosoma*, the protozoan parasite that causes African sleeping sickness, has hundreds of genes coding for surface glycoproteins. Each protozoan cell turns on only one glycoprotein gene at a time. As the host's immune system kills parasites with one type of surface molecule, parasites expressing a different surface glycoprotein can continue to grow.

Repression and Induction

Two genetic control mechanisms known as repression and induction regulate the transcription of mRNA and consequently the synthesis of enzymes from them. These mechanisms control the formation and amounts of enzymes in the cell, not the activities of the enzymes.

Repression

The regulatory mechanism that inhibits gene expression and decreases the synthesis of enzymes is called **repression.** Repression is usually a response to the overabundance of an end-product of a metabolic pathway; it causes a decrease in the rate of synthesis of the enzymes leading to the formation of that product. Repression is mediated by regulatory proteins called **repressors,** which block the ability of RNA polymerase to initiate transcription from the repressed genes (Figure 8.13).

Induction

The process that turns on the transcription of a gene or genes is **induction.** A substance that acts to induce transcription of a gene is called an **inducer,** and enzymes that are synthesized in the presence of inducers are *inducible enzymes.* The genes required for lactose metabolism in *E. coli* are a well-known example of an inducible system. One of these genes codes for the enzyme β-galactosidase, which splits the substrate lactose into two simple sugars, glucose and galactose. (β refers to the type of linkage that joins the glucose and galactose.) If *E. coli* is placed into a medium in which no lactose is present, the organisms contain almost no β-galactosidase; however, when lactose is added to the medium, the bacterial cells produce a large quantity of the enzyme. Lactose is converted in the cell to the related compound allolactose, which is the inducer for these genes; the presence of lactose thus indirectly induces the cells to synthesize more enzyme. This response, which is under genetic control, is termed **enzyme induction.**

The Operon Model of Gene Expression

Details of the control of gene expression by induction and repression are described by the operon model. François Jacob and Jacques Monod formulated this general model in 1961 to account for the regulation of protein synthesis. They based their model on studies of the induction of the enzymes of lactose catabolism in *E. coli*. In addition to β-galactosidase, these enzymes include permease, which is involved in the transport of lactose into the cell, and transacetylase, which metabolizes certain disaccharides other than lactose.

Without repressor:

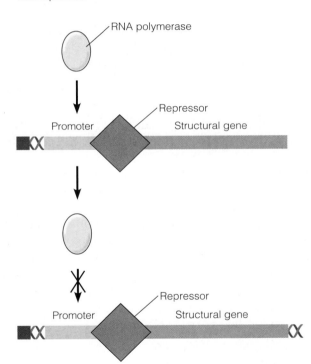

With repressor:

FIGURE 8.13 **A model for repression.** The repressor protein is encoded by the *I* gene. When a repressor protein is present, it either blocks RNA polymerase from binding to the promoter, or blocks its progress along the DNA. In either case, the repressor effectively stops transcription of the gene.

■ Repression is a regulatory mechanism that inhibits gene expression and decreases enzyme synthesis.

The genes for the three enzymes involved in lactose uptake and utilization are next to each other on the bacterial chromosome and are regulated together (Figure 8.14a). These genes, which determine the structures of proteins, are called **structural genes** to distinguish them from an adjoining control region on the DNA. When lactose is introduced into the culture medium, the *lac* structural genes are all transcribed and translated rapidly and simultaneously. We will now see how this regulation occurs.

In the control region of the *lac* operon are two relatively short segments of DNA. One, the *promoter*, is the region of DNA where RNA polymerase initiates transcription. The other is the **operator,** which is like a traffic light that acts as a go or stop signal for transcription of the structural genes. A set of operator and promoter sites, and the structural genes they control, are what define an **operon;** thus, the combination of the three *lac* structural genes and the adjoining control regions is called the *lac* operon.

❶ Near the *lac* operon on the bacterial DNA is a regulatory gene called the *I* gene, which codes for a repressor protein.

❷ When lactose is absent, the repressor protein binds tightly to the operator site. This binding prevents RNA polymerase from transcribing the adjacent structural genes; consequently, no mRNA is made and no enzymes are synthesized.

❸ But when lactose is present, some of it is transported into the cells and converted into the inducer allolactose. The inducer binds to the repressor protein and alters it so it cannot bind to the operator site. In the absence of an operator-bound repressor protein, RNA polymerase can transcribe the structural genes into mRNA, which is then translated into enzymes. This is why, in the presence of lactose, enzymes are produced. Lactose is said to induce enzyme synthesis, and the *lac* operon is called an inducible operon.

In repressible operons, the structural genes are transcribed until they are turned off or *repressed* (Figure 8.14b).

❶ The genes for the enzymes involved in the synthesis of tryptophan are regulated in this manner.

❷ The structural genes are transcribed and translated leading to tryptophan synthesis.

❸ When excess tryptophan is present, the tryptophan acts as a **corepressor** binding to the repressor protein. The repressor protein can now bind to the operator, stopping further tryptophan synthesis.

Regulation of the lactose operon also depends on the level of glucose in the medium, which in turn controls the intracellular level of the small molecule **cyclic AMP (cAMP),** a substance derived from ATP that serves as a cellular alarm signal. Enzymes that metabolize glucose are constitutive, and cells grow at their maximal rate with glucose as their carbon source because they can use it most efficiently (Figure 8.15 on page 226). When glucose is no longer available, cAMP accumulates in the cell. The cAMP binds to the allosteric site of *cAMP receptor protein (CRP)*. CRP then binds to the *lac* promoter which initiates transcription by making it easier for RNA polymerase to bind to the promoter. Thus transcription of the *lac* operon requires both the presence of lactose and the absence of glucose (see Figure 8.14a).

Cyclic AMP is an example of an *alarmone*, a chemical alarm signal the cell uses to respond to environmental or nutritional stress. (In this case, the stress is the lack of glucose.) The same mechanism involving cAMP allows the cell to grow on other sugars. Inhibition of the metabolism of alternative carbon sources by glucose is termed **catabolite repression** (or the *glucose effect*). When glucose is available, the level of cAMP in the cell is low, and consequently CRP is not bound.

Mutation: Change in the Genetic Material

A **mutation** is a change in the base sequence of DNA. Such a change in the base sequence of a gene will sometimes cause a change in the product encoded by that gene. For example, when the gene for an enzyme mutates, the enzyme encoded by the gene may become inactive or less active because its amino acid sequence has changed. Such a change in genotype may be disadvantageous, or even lethal, if the cell loses a phenotypic trait it needs. However, a mutation can be beneficial if, for instance, the altered enzyme encoded by the mutant gene has a new or enhanced activity that benefits the cell.

Many simple mutations are silent (neutral); the change in DNA base sequence causes no change in the activity of the product encoded by the gene. Silent mutations commonly occur when one nucleotide is substituted for another in the DNA, especially at a location corresponding to the third position of the mRNA codon. Because of the degeneracy of the genetic code, the resulting new codon might still code for the same amino acid. Even if the amino acid is changed, the function of the protein may not change if the amino acid is in a nonvital portion of the protein, or is chemically very similar to the original amino acid.

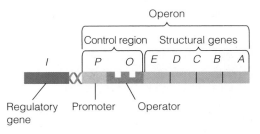

1 **Structure of the operon.** The operon consists of the promoter (*P*), and operator (*O*) sites, and structural genes which code for the protein. The operon is regulated by the product of the regulatory gene (*I*).

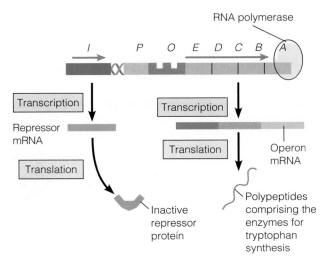

2 **Repressor active, operon off.** The repressor protein binds with the operator, preventing transcription from the operon.

2 **Repressor inactive, operon on.** The repressor is inactive and transcription and translation proceed leading to the synthesis of tryptophan.

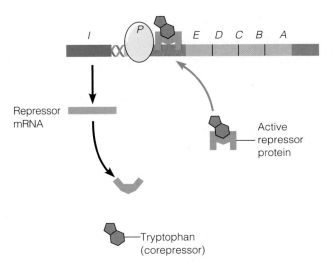

3 **Repressor inactive, operon on.** When the inducer allolactose binds to the repressor protein, the inactivated repressor can no longer block transcription. The structural genes are transcribed, ultimately resulting in the production of the enzymes needed for lactose catabolism.

3 **Repressor active, operon off.** When the corepressor tryptophan binds to the repressor protein, the activated repressor binds with the operator, preventing transcription from the operon.

(a) An inducible operon

(b) A repressible operon

FIGURE 8.14 **The operon: regulation of gene expression. (a)** Lactose is digested by a catabolic pathway catalyzed by inducible enzymes. **(b)** Tryptophan is an amino acid produced by an anabolic pathway catalyzed by repressible enzymes.

■ How does a repressible enzyme differ from an inducible enzyme?

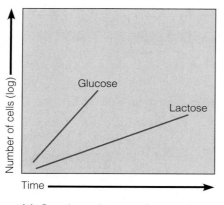

(a) Growth on glucose or lactose alone

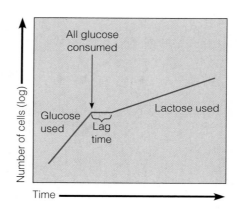

(b) Growth on glucose and lactose combined

FIGURE 8.15 **The growth rate of *E. coli* on glucose and lactose.** The steeper the straight line, the faster the growth. **(a)** Bacteria growing on glucose as the sole carbon source grow faster than on lactose. **(b)** Bacteria growing in a medium containing glucose and lactose first consume the glucose, and then, after a short lag time, the lactose. During the lag time, intracellular cyclic AMP increases, the *lac* operon is transcribed, more lactose is transported into the cell, and β-galactosidase is synthesized to break down lactose.

■ Will transcription of the *lac* operon occur in the presence of lactose and glucose? In the presence of lactose and the absence of glucose? In the presence of glucose and the absence of lactose?

Types of Mutations

Learning Objective

• *Classify mutations by type, and describe how mutations are prevented or repaired.*

The most common type of mutation involving single base pairs is **base substitution** (or *point mutation*), in which a single base at one point in the DNA sequence is replaced with a different base. When the DNA replicates, the result is a substituted base pair (Figure 8.16). For example, AT might be substituted for GC, or CG for GC. If a base substitution occurs within a gene that codes for a protein, the mRNA transcribed from the gene will carry an incorrect base at that position. When the mRNA is translated into protein, the incorrect base may cause the insertion of an incorrect amino acid in the protein. If the base substitution results in an amino acid substitution in the synthesized protein, this change in the DNA is known as a **missense mutation** (Figure 8.17a and b on page 228).

The effects of such mutations can be dramatic. For example, sickle-cell disease is caused by a single change in the gene for globin, the protein component of hemoglobin. Hemoglobin is primarily responsible for transporting oxygen from the lungs to the tissues. A single missense mutation, a change from an A to a T at a specific site, re-

sults in the change from glutamic acid to valine in the protein. The effect of this change is that the shape of the hemoglobin molecule changes under conditions of low oxygen, altering the shape of the red blood cells such that movement of the cells through small capillaries is greatly impeded.

By creating a stop (nonsense) codon in the middle of an mRNA molecule, some base substitutions effectively prevent the synthesis of a complete functional protein; only a fragment is synthesized. A base substitution resulting in a nonsense codon is thus called a **nonsense mutation** (Figure 8.17c on page 228).

Besides base-pair mutations, there are also changes in DNA called **frameshift mutations,** in which one or a few nucleotide pairs are deleted or inserted in the DNA (Figure 8.17d on page 228). This mutation can shift the "translational reading frame"—that is, the three-by-three grouping of nucleotides recognized as codons by the tRNAs during translation. For example, deleting one nucleotide pair in the middle of a gene causes changes in many amino acids downstream from the site of the original mutation. Frameshift mutations almost always result in a long stretch of altered amino acids and the production of an inactive protein from the mutated gene. In most cases, a nonsense codon will eventually be encountered and thereby terminate translation.

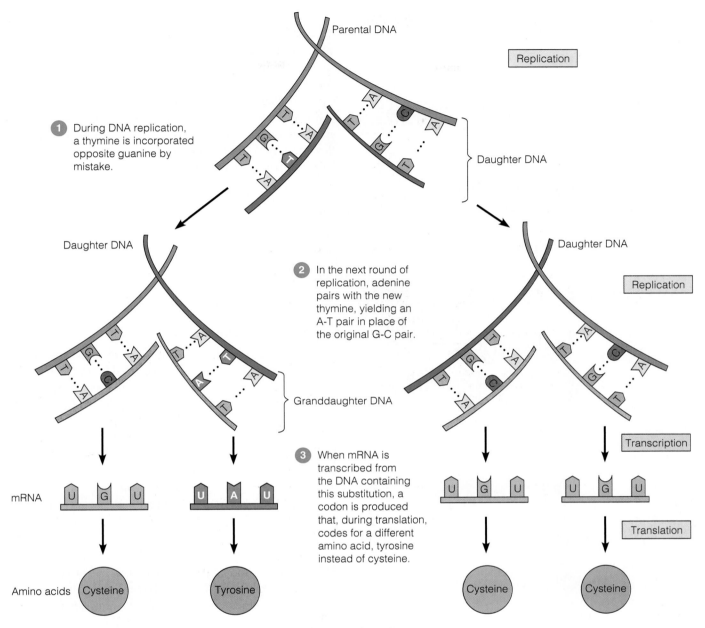

1 During DNA replication, a thymine is incorporated opposite guanine by mistake.

Replication

Parental DNA

Daughter DNA

Daughter DNA

2 In the next round of replication, adenine pairs with the new thymine, yielding an A-T pair in place of the original G-C pair.

Granddaughter DNA

Daughter DNA

Replication

3 When mRNA is transcribed from the DNA containing this substitution, a codon is produced that, during translation, codes for a different amino acid, tyrosine instead of cysteine.

Transcription

Translation

mRNA   U G U   U A U   U G U   U G U

Amino acids   Cysteine   Tyrosine   Cysteine   Cysteine

FIGURE 8.16 **Base substitution.** This mutation leads to an altered protein in a granddaughter cell.

■ In a base substitution, a single base at one point in the DNA sequence is replaced with a different base.

Occasionally, mutations occur where significant numbers of bases are added to (inserted into) a gene. Huntington's disease, for example, is a progressive neurological disorder caused by extra bases inserted into a particular gene. The reason these insertions occur in this particular gene is still being studied.

Base substitutions and frameshift mutations may occur spontaneously because of occasional mistakes made during DNA replication. These **spontaneous mutations** apparently occur in the absence of any mutation-causing agents. Agents in the environment, such as certain chemicals and radiation, that directly or indirectly bring about mutations are called **mutagens.** Almost any agent that can chemically or physically react with DNA can potentially cause mutations. A wide variety of chemicals, many of which are common in nature or in households, are

DNA (coding strand)

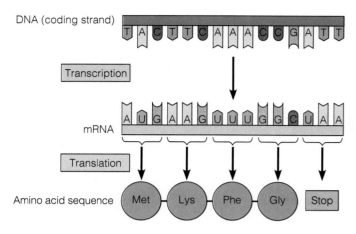

mRNA

Translation

Amino acid sequence

(a) Normal DNA molecule

(b) Missense mutation

(c) Nonsense mutation

(d) Frameshift mutation

known to be mutagens. Many forms of radiation, including X rays and ultraviolet light, are also mutagenic, as discussed shortly.

In the microbial world, certain mutations result in resistance to antibiotics or altered pathogenicity. A mutation in a gene encoding the outer membrane may increase pathogenicity; for example, *Salmonella typhimurium* with an altered outer membrane can survive in phagocytes. A mutation in a capsule-encoding gene may result in decreased pathogenicity because phagocytes can destroy the bacteria, as in the cases of *Streptococcus pneumoniae*, *Haemophilus influenzae*, and *Neisseria meningitidis*.

Mutagens

Learning Objectives
- *Define mutagen.*
- *Describe two ways mutations can be repaired.*

Chemical Mutagens

One of the many chemicals known to be a mutagen is nitrous acid. Figure 8.18 shows how exposure of DNA to nitrous acid can convert the base adenine (A) to a form that no longer pairs with thymine (T) but instead pairs with cytosine (C). When DNA containing such modified adenines replicates, one daughter DNA molecule will have a base-pair sequence different from that of the parent DNA. Eventually, some AT base pairs of the parent will have been changed to GC base pairs in a granddaughter cell. Nitrous acid makes a specific base-pair change in DNA. Like all mutagens, it alters DNA at random locations.

Another type of chemical mutagen is the **nucleoside analog**. These molecules are structurally similar to normal nitrogenous bases, but they have slightly altered base-pairing properties. Examples are 2-aminopurine and 5-bromouracil. The 2-aminopurine molecule is incorporated into DNA in place of adenine but can sometimes pair with cytosine (Figure 8.19a). The 5-bromouracil molecule is incorporated into DNA in place of thymine (T) but often pairs with guanine (Figure 8.19b). When nucleoside analogs are given to growing cells, the analogs

FIGURE 8.17 **Types of mutations and their effects on the amino acid sequences of proteins.**

■ On what basis are missense, nonsense, and frameshift mutations distinguished?

Normal daughter DNA

Mutated granddaughter DNA

Normal parent DNA

Altered parent DNA

Altered daughter DNA

Altered granddaughter DNA

FIGURE 8.18 **Nitrous acid (HNO₂) as a mutagen.** The nitrous acid alters an adenine in such a way that it pairs with cytosine instead of thymine.

■ A mutagen is any environmental agent that can directly or indirectly bring about a mutation.

are randomly incorporated into cellular DNA in place of the normal bases. Then, during DNA replication, the analogs cause mistakes in base pairing. The incorrectly paired bases will be copied during subsequent replication of the DNA, resulting in base-pair substitutions in the progeny cells. Some antiviral and antitumor drugs are nucleoside analogs, including AZT (azidothymidine), one of the primary drugs used to treat HIV infection.

Still other chemical mutagens cause small deletions or insertions, which can result in frameshifts. For instance, under certain conditions, benzpyrene, which is present in smoke and soot, is an effective *frameshift mutagen*. And aflatoxin—produced by *Aspergillus flavus* (a-spér-jil′lus flā′vus), a mold that grows on peanuts and grain—is a frameshift mutagen, as are the acridine dyes used experimentally against herpesvirus infections. Frameshift mutagens usually have the right size and chemical properties to slip between the stacked base pairs of the DNA double helix. They may work by slightly off-setting the two strands of DNA, leaving a gap or bulge in one strand or the other. When the staggered DNA strands are copied during DNA synthesis, one or more

NORMAL NITROGENOUS BASE

ANALOG

(a)   Adenine nucleosides

2-Aminopurine nucleosides

(b)   Thymine nucleosides

5-Bromouracil nucleosides

FIGURE 8.19 **Nucleoside analogs and the nitrogenous bases they replace.**
(a) Adenine and 2-aminopurine nucleosides. **(b)** Thymine and 5-bromouracil nucleosides.

■ What is a nucleoside analog?

base pairs can be inserted or deleted in the new double-stranded DNA. Interestingly, frameshift mutagens are often potent carcinogens.

Radiation

X rays and gamma rays are forms of radiation that are potent mutagens because of their ability to ionize atoms and molecules. The penetrating rays of ionizing radiation cause electrons to pop out of their usual shells (see Chapter 2). These electrons bombard other molecules and cause more damage, and many of the resulting ions and free radicals (molecular fragments with unpaired electrons) are very reactive. Some of these ions can combine with bases in DNA, resulting in errors in DNA replication and repair that produce mutations. An even more serious outcome is the breakage of covalent bonds in the sugar-phosphate backbone of DNA, which causes physical breaks in chromosomes.

Another form of mutagenic radiation is ultraviolet (UV) light, a nonionizing component of ordinary sunlight. However, the most mutagenic component of UV light (wavelength 260 nm) is screened out by the ozone layer of the atmosphere. The most important effect of direct UV light on DNA is the formation of harmful covalent bonds between certain bases. Adjacent thymines in a DNA strand can cross-link to form thymine dimers. Such dimers, unless repaired, may cause serious damage or death to the cell because it cannot properly transcribe or replicate such DNA.

Bacteria and other organisms have enzymes that can repair UV-induced damage. **Light-repair enzymes,** called *photolyases*, use visible light energy to separate the dimer back to the original two thymines. **Nucleotide excision repair,** shown in Figure 8.20, is not restricted to UV-induced damage; it can repair mutations from other causes as well. Enzymes cut out the distorted cross-linked thymines by opening a wide gap. They then fill in the gap with newly synthesized DNA that is complementary to the undamaged strand. By this means, the original base-pair sequence is restored. The last step is the covalent sealing of the DNA backbone by the enzyme DNA ligase. Occasionally, such a repair process makes an error, and the original base-pair sequence is not properly restored. The result of this error is a mutation.

Exposure to UV light in humans, such as by excessive suntanning, causes a large number of thymine dimers in skin cells. Unrepaired dimers may result in skin cancers. Humans with xeroderma pigmentosum, an inherited condition that results in increased sensitivity to UV light, have a defect in nucleotide excision repair; consequently, they have an increased risk of skin cancer.

Ultraviolet light

Thymine dimer

New DNA

1 Exposure to ultraviolet light causes adjacent thymines to become cross-linked, forming a thymine dimer and disrupting their normal base pairing.

2 An enzyme cuts out and removes the damaged DNA.

3 DNA polymerase fills the gap by synthesizing new DNA, using the intact strand as a template.

4 DNA ligase seals the remaining gap by joining the old and new DNA.

FIGURE 8.20 **The creation and repair of a thymine dimer caused by ultraviolet light.** After exposure to UV light, adjacent thymines can become cross-linked, forming a thymine dimer. In the absence of visible light, the nucleotide excision repair mechanism is used in a cell to repair the damage.

■ Bacteria and other organisms have enzymes that can repair radiation damage.

The Frequency of Mutation

Learning Objective

- *Describe the effect of mutagens on the mutation rate.*

The **mutation rate** is the probability that a gene will mutate when a cell divides. The rate is usually stated as a power of 10, and because mutations are very rare, the exponent is always a negative number. For example, if there is one chance in 10,000 that a gene will mutate when the cell divides, the mutation rate is 1/10,000, which is expressed as 10^{-4}. Spontaneous mistakes in DNA replication occur at a very low rate, perhaps only once in 10^9 replicated base pairs (a mutation rate of 10^{-9}). Because the average gene has about 10^3 base pairs, the spontaneous rate of mutation is about once in 10^6 (a million) replicated genes.

Mutations usually occur more or less randomly along a chromosome. The occurrence of random mutations at low frequency is an essential aspect of the adaptation of species to their environment, for evolution requires that genetic diversity be generated randomly and at a low rate. For example, in a bacterial population of significant size—say, greater than 10^7 cells—a few new mutant cells will always be produced in every generation. Most mutations either are harmful and likely to be removed from the gene pool when the individual cell dies, or are neutral. However, a few mutations may be beneficial. For example, a mutation that confers antibiotic resistance is beneficial to a population of bacteria that is regularly exposed to antibiotics. Once such a trait has appeared through mutation, cells carrying the mutated gene are more likely than other cells to survive and reproduce as long as the environment stays the same. Soon most of the cells in the population will have the gene; an evolutionary change will have occurred, although on a small scale.

A mutagen usually increases the spontaneous rate of mutation, which is about once in 10^6 replicated genes, by a factor of 10–1000 times. In other words, in the presence of a mutagen, the normal rate of 10^{-6} mutations per replicated gene becomes a rate of 10^{-5} to 10^{-3} per replicated gene. Mutagens are used experimentally to enhance the production of mutant cells for research on the genetic properties of microorganisms, and for commercial purposes.

Identifying Mutants

Learning Objective

- *Outline the methods of direct and indirect selection of mutants.*

Mutants can be detected by selecting or testing for an altered phenotype. Whether or not a mutagen is used, mutant cells with specific mutations are always rare compared with other cells in the population. The problem is detecting such a rare event.

Experiments are usually performed with bacteria because they reproduce rapidly, so large numbers of organisms (more than 10^9 per milliliter of nutrient broth) can easily be used. Furthermore, because bacteria generally have only one copy of each gene per cell, the effects of a mutated gene are not masked by the presence of a normal version of the gene, as in many eukaryotic organisms.

Positive (direct) selection involves the detection of mutant cells by rejection of the unmutated parent cells. For example, suppose we were trying to find mutant bacteria that are resistant to penicillin. When the bacterial cells are plated on a medium containing penicillin, the mutant can be identified directly. The few cells in the population that are resistant (mutants) will grow and form colonies, whereas the normal, penicillin-sensitive parental cells cannot grow.

To identify mutations in other kinds of genes, **negative (indirect) selection** can be used. This process selects a cell that cannot perform a certain function, using the technique of **replica plating.** For example, suppose we wanted to use replica plating to identify a bacterial cell that has lost the ability to synthesize the amino acid histidine (Figure 8.21 on page 232). First, about 100 bacterial cells are inoculated onto an agar plate. This plate, called the master plate, contains a medium with histidine on which all cells will grow. After 18 to 24 hours of incubation, each cell reproduces to form a colony. Then a pad of sterile material, such as latex, filter paper, or velvet, is pressed over the master plate, and some of the cells from each colony adhere to the velvet. Next, the velvet is pressed down onto two (or more) sterile plates. One plate contains a medium without histidine, and one contains a medium with histidine on which the original, nonmutant bacteria can grow. Any colony that grows on the medium with histidine on the master plate but that cannot synthesize its own histidine will not be able to grow on the medium without histidine. The mutant colony can then be identified on the master plate. Of course, because mutants are so rare (even those induced by mutagens), many plates must be screened with this technique to isolate a specific mutant.

Replica plating is a very effective means of isolating mutants that require one or more new growth factors. Any mutant microorganism having a nutritional requirement that is absent in the parent is known as an **auxotroph.** For example, an auxotroph may lack an enzyme needed to synthesize a particular amino acid and will therefore require that amino acid as a growth factor in its nutrient medium.

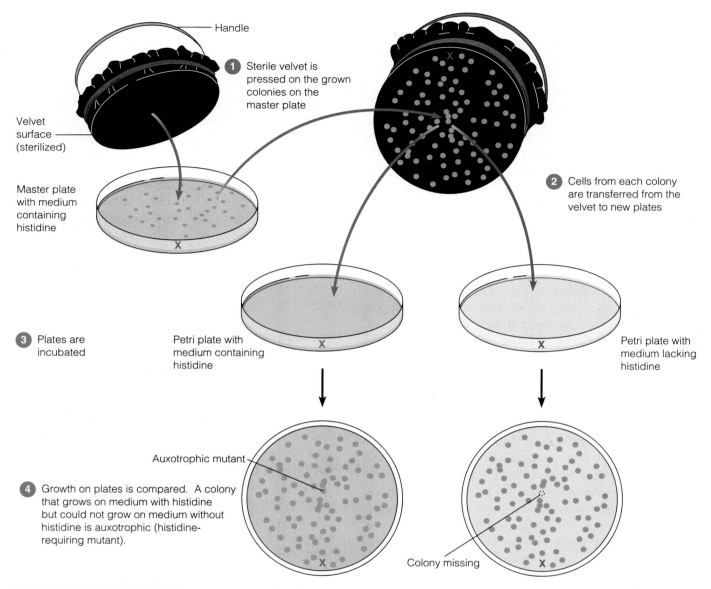

Handle

1 Sterile velvet is pressed on the grown colonies on the master plate

Velvet surface (sterilized)

Master plate with medium containing histidine

2 Cells from each colony are transferred from the velvet to new plates

3 Plates are incubated

Petri plate with medium containing histidine

Petri plate with medium lacking histidine

Auxotrophic mutant

4 Growth on plates is compared. A colony that grows on medium with histidine but could not grow on medium without histidine is auxotrophic (histidine-requiring mutant).

Colony missing

FIGURE 8.21 Replica plating. In this example, the auxotrophic mutant cannot synthesize histidine. The plates must be carefully marked (with an X here) to maintain orientation so that colony positions are known in relation to the original master plate.

■ Replica plating is used to identify auxotrophic mutants, bacteria that have lost the ability to synthesize an essential nutrient.

Identifying Chemical Carcinogens

Learning Objective

- *Identify the purpose of and outline the procedure for the Ames test.*

Many known mutagens have been found to be **carcinogens,** substances that cause cancer in animals, including humans. In recent years, chemicals in the environment,

the workplace, and the diet have been implicated as causes of cancer in humans. The usual subjects of tests to determine potential carcinogens are animals, and the testing procedures are time-consuming and expensive. Now there are faster and less expensive procedures for the preliminary screening of potential carcinogens. One of these, called the **Ames test,** uses bacteria as carcinogen indicators.

The Ames test is based on the observation that exposure of mutant bacteria to mutagenic substances may cause

Experimental Sample

Suspected mutagen

Rat liver extract

Experimental plate

Incubation

Cultures of histidine-dependent *Salmonella*

Media lacking histidine

Colonies of revertant bacteria

Control (No Suspected Mutagen Added)

Rat liver extract

Incubation

Control plate

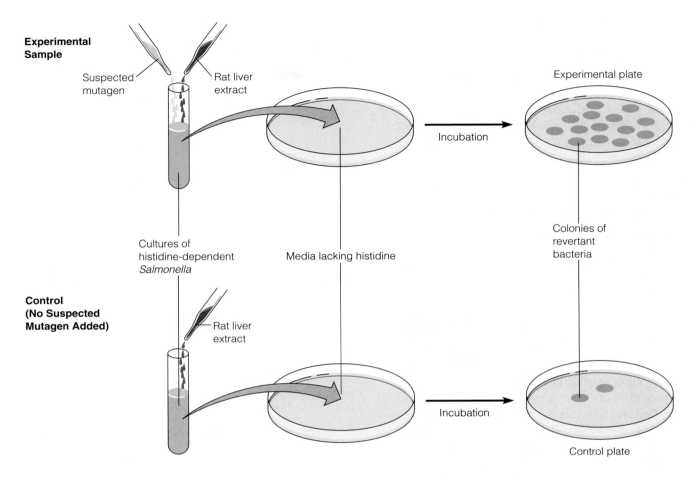

① Two cultures are prepared of *Salmonella* bacteria that have lost the ability to synthesize histidine (histidine-dependent).

② The suspected mutagen is added to the experimental sample only; rat liver extract (an activator) is added to both samples.

③ Each sample is poured onto a plate of medium lacking histidine. The plates are then incubated at 37°C for two days. Only bacteria whose histidine-dependent phenotype has mutated back (reverted) to histidine-synthesizing will grow into colonies.

④ The numbers of colonies on the experimental and control plates are compared. The control plate may show a few spontaneous histidine-synthesizing revertants. The test plates will show an increase in the number of histidine-synthesizing revertants if the test chemical is indeed a mutagen and potential carcinogen. The higher the concentration of mutagen used, the more revertant colonies will result.

FIGURE 8.22 **The Ames test.**

■ The Ames test is used to screen mutagens suspected of causing cancer in humans.

new mutations that reverse the effect (the change in phenotype) of the original mutation. These are called back-mutations, or *reversions*. Specifically, the test measures the reversion of histidine auxotrophs of *Salmonella* (his⁻ cells, mutants that have lost the ability to synthesize histidine) to histidine-synthesizing cells (his⁺) after treatment with a mutagen (Figure 8.22). Bacteria are incubated in both the presence and absence of the substance being tested. Because many chemicals must be activated (transformed chemically into forms that are chemically reactive) by animal enzymes for mutagenic or carcinogenic activity to ap-

pear, the chemical to be tested and the mutant bacteria are incubated together with rat liver extract, a rich source of activation enzymes. If the substance being tested is mutagenic, it will cause the reversion of his⁻ bacteria to his⁺ bacteria at a rate higher than the spontaneous reversion rate. The number of observed revertants provides an indication of the degree to which a substance is mutagenic and therefore possibly carcinogenic.

The test can be used in many ways. Several potential mutagens can be qualitatively tested by spotting the individual chemicals on small paper disks on a single plate inoculated with bacteria. In addition, mixtures such as wine, blood, smoke condensates, and extracts of foods can also be tested to see if they contain mutagenic substances. For an example of the use of the Ames test in exploring the role of bacteria in cancer, see the box on page 235.

About 90% of the substances found by the Ames test to be mutagenic have also been shown to be carcinogenic in animals. By the same token, the more mutagenic substances have generally been found to be more carcinogenic.

Genetic Transfer and Recombination

Learning Objectives

- *Compare the mechanisms of genetic recombination in bacteria.*
- *Differentiate between horizontal and vertical gene transfer.*

Genetic recombination refers to the exchange of genes between two DNA molecules to form new combinations of genes on a chromosome. Figure 8.23 shows one type of genetic recombination occurring between two pieces of DNA, which we will call A and B and regard as chromosomes for the sake of simplicity. If these two chromosomes break and rejoin as shown—a process called **crossing over**—some of the genes carried by these chromosomes are shuffled. The original chromosomes have recombined, so that each now carries a portion of the other's genes.

If A and B represent DNA from different individuals, how are they brought close enough together to recombine? In eukaryotes, genetic recombination is an ordered process that usually occurs as part of the sexual cycle of the organism. Recombination generally takes place during the formation of reproductive cells, such that these cells contain recombinant DNA. In bacteria, genetic recombination can happen in a number of ways, which we will discuss in the following sections.

Like mutation, genetic recombination contributes to a population's genetic diversity, which is the source of variation in evolution. In highly evolved organisms such as present-day microbes, recombination is more likely than mutation to be beneficial because recombination will less likely destroy a gene's function and may bring together combinations of genes that enable the organism to carry out a valuable new function.

The major protein that constitutes the flagella of *Salmonella* is also one of the primary proteins that causes our immune systems to respond. However, these bacteria have the capability of producing two different flagellar proteins. As our immune system mounts a response against those

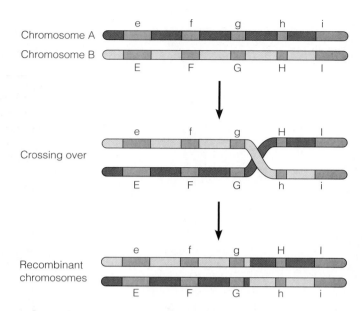

FIGURE 8.23 Genetic recombination by crossing over between two related chromosomes. Chromosomes A and B each carry one version of genes E through I. The chromosomes cross over by breaking and rejoining, sometimes in more than one location. The result is two recombinant chromosomes, each of which carries genes originating from both chromosomes.

■ Genetic recombination is the exchange of genes between two DNA molecules to form new combinations of genes.

cells containing one form of the flagellar protein, those organisms producing the second are not affected. Which flagellar protein is produced is determined by a recombination event that apparently occurs somewhat randomly within the chromosomal DNA. Thus, by altering the flagellar protein produced, *Salmonella* can better avoid the defenses of the host.

Vertical gene transfer occurs when genes are passed from an organism to its offspring. Plants and animals transmit their genes by vertical transmission. Bacteria can pass their genes not only to their offspring, but also laterally, to other microbes of the same generation. This is known as **horizontal gene transfer.** Horizontal gene transfer between bacteria occurs in several ways. In all of the mechanisms, the transfer involves a **donor cell** that gives a portion of its total DNA to a **recipient cell.** Once transferred, part of the donor's DNA is usually incorporated into the recipient's DNA; the remainder is degraded by cellular enzymes. The recipient cell that incorporates donor DNA into its own DNA is called a *recombinant.* The transfer of genetic material between bacteria is by no means a frequent event; it may occur in only 1% or less of an entire population. Let's examine in detail the specific types of genetic transfer.

The Role of Bacteria in Cancer

In 1996, the International Agency for Research on Cancer classified the bacterium *Helicobacter pylori* as a carcinogen. No specific mechanism was proposed to explain the relationship between *H. pylori* and gastric cancer. Researchers have known since the early 1970s that there is a relationship between diet and certain types of cancer. However, it may not be the diet itself that causes cancer but the interaction between diet and normal microbiota, the hundreds of species of microorganisms that flourish in the average person.

The importance of microorganisms in cancer inductions was first noticed when rats were fed cycasin, a substance that occurs naturally in the nut of the cycad plant. The bacterial enzyme β-glucosidase converts cycasin to methylazoxymethanol, which caused cancer in the rats. Germ-free rats were unaffected by the cycasin, but rats with the usual intestinal microbiota developed colon cancer.

To determine the role of intestinal bacteria in converting chemicals to carcinogens, Elena McCoy and her co-workers at New York Medical College used the Ames test to compare the mutagenic abilities of 2-aminofluorene when activated by liver cell enzymes, intestinal cell enzymes alone, *Bacteroides fragilis* enzymes alone, and intestinal cell enzymes plus *B. fragilis* enzymes. They found that liver enzymes produced the highest conversion of this chemical to a mutagenic form. Intestinal enzymes alone and bacterial enzymes alone produced some mutagenic conversion. And a combination of intestinal and bacterial enzymes had an effect almost as great as that produced by liver enzymes.

To determine the role of bacteria in bladder cancer, researchers at the Kyushu Medical Center in Japan used the Ames test to evaluate the mutagenic abilities of human urine activated by bacteria. The bacteria were isolated from patients with urinary tract infections. The bacterial isolates were tested for their ability to reduce nitrate ion (NO_3^-) to nitrite ion (NO_2^-). The mutagenic activity seen with the addition of nitrite ion indicates that nitrite is mutagenic (see Table A). Their results suggest that a chemical reaching the bladder can be activated to a carcinogen by bacterial enzymes.

In 2000, researchers used the Ames test to determine whether *H. pylori* causes cancer by inducing mutations. *H. pylori* cultures were isolated from patients with gastric cancer and patients with gastritis or peptic ulcers and were grown in a culture medium. After incubation, the culture was centrifuged to separate the cells from their supernatant (culture medium). The Ames test was performed on the bacteria and on the supernatant; distilled water provided a negative control, and a carcinogen the positive control. The bacteria isolated from patients were not mutagenic (Table B). There was slightly more *Salmonella* growth after treatment with the supernatant, compared to the sterile culture. This suggests that some substance secreted by *H. pylori* has mutagenic activity.

TABLE A

Test Substance	Relative Amounts of Bacterial Growth in Ames Test
Normal human urine	No growth
Urine + NO_3^--reducing bacteria	+++
Urine + non-NO_3^--reducing bacteria	+
Urine + NO_2^-	+++

TABLE B

	Number of Colonies in Ames Test
H. pylori cells	211
H. pylori supernatant	330
Sterile culture medium	240
Distilled water	246
Methylnitrosourea (a carcinogen)	507

Transformation in Bacteria

During the process of **transformation,** genes are transferred from one bacterium to another as "naked" DNA in solution. This process was first demonstrated over 70 years ago, although it was not understood at the time. Not only did transformation show that genetic material could be transferred from one bacterial cell to another, but study of this phenomenon eventually led to the conclusion that DNA is the genetic material. The initial experiment on transformation was performed by Frederick Griffith in England in 1928 while he was working with two strains of *Streptococcus pneumoniae*. One, a virulent (pathogenic) strain, has a polysaccharide capsule that prevents phagocytosis.

Recombination

1 Living encapsulated bacteria injected into mouse

2 Mouse died

3 Colonies of encapsulated bacteria were isolated from dead mouse

(a)

1 Living nonencapsulated bacteria injected into mouse

2 Mouse remained healthy

3 A few colonies of nonencapsulated bacteria were isolated from mouse; phagocytes destroyed nonencapsulated bacteria

(b)

1 Heat-killed encapsulated bacteria injected into mouse

2 Mouse remained healthy

3 No colonies were isolated from mouse

(c)

1 Living nonencapsulated and heat-killed encapsulated bacteria injected into mouse

2 Mouse died

3 Colonies of encapsulated bacteria were isolated from dead mouse

(d)

FIGURE 8.24 Griffith's experiment demonstrating genetic transformation.
(a) Living encapsulated bacteria caused disease and death when injected into a mouse.
(b) Living nonencapsulated bacteria are readily destroyed by the phagocytic defenses of the host, so the mouse remained healthy after injection. **(c)** After being killed by heat, encapsulated bacteria lost the ability to cause disease. **(d)** However, the combination of living nonencapsulated bacteria and heat-killed encapsulated bacteria (neither of which alone cause disease) did cause disease. Somehow, the live nonencapsulated bacteria were transformed by the dead encapsulated bacteria so that they acquired the ability to form a capsule and therefore cause disease. Subsequent experiments proved the transforming factor to be DNA.

■ Why did encapsulated bacteria kill the mouse while nonencapsulated bacteria did not? What killed the mouse in (d)?

The bacteria grow and cause pneumonia. The other, an avirulent strain, lacks the capsule and does not cause disease.

Griffith was interested in determining whether injections of heat-killed bacteria of the encapsulated strain could be used to vaccinate mice against pneumonia. As he expected, injections of living encapsulated bacteria killed the mouse (Figure 8.24a); injections of live nonen-capsulated bacteria (Figure 8.24b) or dead encapsulated bacteria did not kill the mouse (Figure 8.24c). However, when the dead encapsulated bacteria were mixed with live nonencapsulated bacteria and injected into the mice, many of the mice died. In the blood of the dead mice, Griffith found living, encapsulated bacteria. Hereditary material (genes) from the dead bacteria had entered the live cells and changed them genetically so that their progeny were encapsulated and therefore virulent (Figure 8.24d).

Subsequent investigations based on Griffith's research revealed that bacterial transformation could be carried out without mice. A broth was inoculated with live nonencapsulated bacteria. Dead encapsulated bacteria were then added to the broth. After incubation, the culture was found to contain living bacteria that were encapsulated and virulent. The nonencapsulated bacteria had been transformed; they had acquired a new hereditary trait by incorporating genes from the killed encapsulated bacteria.

The next step was to extract various chemical components from the killed cells to determine which component caused the transformation. These crucial experiments were performed in the United States by Oswald T. Avery and his associates Colin M. MacLeod and Maclyn McCarty. After years of research, they announced in 1944 that the component responsible for transforming harmless *S. pneumoniae* into virulent strains was DNA. Their results provided one of the conclusive indications that DNA was indeed the carrier of genetic information.

Since the time of Griffith's experiment, considerable information has been gathered about transformation. In nature, some bacteria, perhaps after death and cell lysis, release their DNA into the environment. Other bacteria can then encounter the DNA and, depending on the particular species and growth conditions, take up fragments of DNA and integrate them into their own chromosomes by recombination. A recipient cell with this new combination of genes is a kind of hybrid, or recombinant cell (Figure 8.25). All the descendants of such a recombinant cell will be identical to it. Transformation occurs naturally among very few genera of bacteria, including *Bacillus*, *Haemophilus* (hē-mä′fi-lus), *Neisseria*, *Acinetobacter* (a-si-ne′tō-bak-tèr), and certain strains of the genera *Streptococcus* and *Staphylococcus*.

Transformation works best when the donor and recipient cells are very closely related. Even though only a small portion of a cell's DNA is transferred to the recipient, the molecule that must pass through the recipient cell wall and membrane is still very large. When a recipient cell is in a physiological state in which it can take up the donor DNA, it is said to be competent. **Competence** results from alterations in the cell wall that make it permeable to large DNA molecules.

The well-understood and widely used bacterium *E. coli* is not naturally competent for transformation. However, a simple laboratory treatment enables *E. coli* to readily take up DNA. The discovery of this treatment has enabled researchers to use *E. coli* for genetic engineering, discussed in Chapter 9.

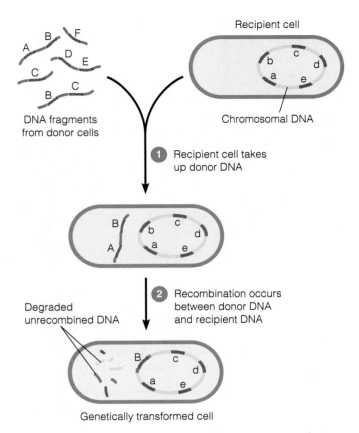

FIGURE 8.25 The mechanism of genetic transformation in bacteria.

■ A recipient cell with a new combination of genes is a hybrid or recombinant cell.

Conjugation in Bacteria

Another mechanism by which genetic material is transferred from one bacterium to another is known as **conjugation.** Conjugation is mediated by one kind of *plasmid*, a circular piece of DNA that replicates independently from the cell's chromosome (discussed on page 240). However, plasmids differ from bacterial chromosomes in that the genes they carry are usually not essential for the growth of the cell under normal conditions. The plasmids responsible for conjugation are transmissible between cells during conjugation.

Conjugation differs from transformation in two major ways. First, conjugation requires direct cell-to-cell contact. Second, the conjugating cells must generally be of opposite mating type; donor cells must carry the plasmid, and recipient cells usually do not. In gram-negative bacteria, the plasmid carries genes that code for the synthesis of *sex pili*, projections from the donor's cell surface that contact the recipient and help bring the two cells into direct contact. Gram-positive bacterial cells produce sticky surface

FIGURE 8.26 **Bacterial conjugation.** The sex pilus connecting these cells undergoing conjugation allows the transfer of genetic information. At the actual time of genetic exchange, the cells' connecting bridge contracts and the cells are much closer together. Notice that one cell has numerous fimbriae.

■ In gram-negative bacteria, the plasmid carries genes that code for the synthesis of sex pili.

molecules that cause cells to come into direct contact with each other. In the process of conjugation, the plasmid is replicated during the transfer of a single-stranded copy of the plasmid DNA to the recipient, where the complementary strand is synthesized.

Because most experimental work on conjugation has been done with *E. coli*, we will describe the process in this organism. In *E. coli*, the **F factor (fertility factor)** was the first plasmid observed to be transferred between cells during conjugation (Figure 8.26). Donors carrying F factors (F⁺ cells) transfer the plasmid to recipients (F⁻ cells), which become F⁺ cells as a result (Figure 8.27a). In some cells carrying F factors, the factor integrates into the chromosome, converting the F⁺ cell to an **Hfr cell** (high frequency of recombination) (Figure 8.27b). When conjugation occurs between an Hfr cell and an F⁻ cell, the Hfr cell's chromosome (with its integrated F factor) replicates, and a parental strand of the chromosome is transferred to the recipient cell (Figure 8.27c). Replication of the Hfr chromosome begins in the middle of the integrated F factor, and a small piece of the F factor leads the chromosomal genes into the F⁻ cell. Usually, the chromosome breaks before it is completely transferred. Once within the recipient cell, donor DNA can recombine with the recipient's DNA. (Donor DNA that is not integrated is degraded.) Therefore, by conjugation with an Hfr cell, an F⁻ cell may acquire new versions of chromosomal genes (just as in transformation). However, it remains an F⁻ cell because it did not receive a complete F factor during conjugation.

Conjugation is used to map the location of genes on a bacterial chromosome (see Figure 8.1b). Notice that genes for the synthesis of threonine (*thr*) and leucine (*leu*) are first, reading clockwise from 0. Their locations were determined by conjugation experiments. Assume that conjugation is allowed for only 1 minute between an Hfr strain that is his⁺, pro⁺, thr⁺ and leu⁺, and an F⁻ strain that is his⁻, pro⁻, thr⁻, and leu⁻. If the F⁻ acquired the ability to synthesize threonine, then the *thr* gene is located early in the chromosome, between 0 and 1 minute. If after 2 minutes the F⁻ cell now becomes thr⁺ and leu⁺, the order of these two genes on the chromosome must be *thr, leu*.

Transduction in Bacteria

A third mechanism of genetic transfer between bacteria is **transduction.** In this process, bacterial DNA is transferred from a donor cell to a recipient cell inside a virus that infects bacteria, called a **bacteriophage,** or **phage.** (Phages will be discussed further in Chapter 13.)

To understand how transduction works, we will consider the life cycle of one type of transducing phage of *E. coli*; this phage carries out **generalized transduction** (Figure 8.28 on page 240).

❶ In the process of infection, the phage attaches to the donor bacterial cell wall and injects its DNA into the bacterium.

❷ The phage DNA acts as a template for the synthesis of new phage DNA and also directs the synthesis of phage protein coats. During phage development inside the infected cell, the bacterial chromosome is broken apart by phage enzymes, and

❸ at least some pieces of bacterial DNA are mistakenly packaged inside phage protein coats. (Even plasmid DNA or DNA of another virus that is inside the cell can be given phage protein coats.) The resulting phage particles then carry bacterial DNA instead of phage DNA.

❹ When the released phage particles later infect a new population of bacteria, bacterial genes will be transferred to the newly infected recipient cells at low frequency.

❺ Transduction of cellular DNA by a virus can lead to recombination between the DNA of the donor host cell and the DNA of the recipient host cell. The process of generalized transduction is typical of bacteriophages such as phage P1 of *E. coli* and phage P22 of *Salmonella*.

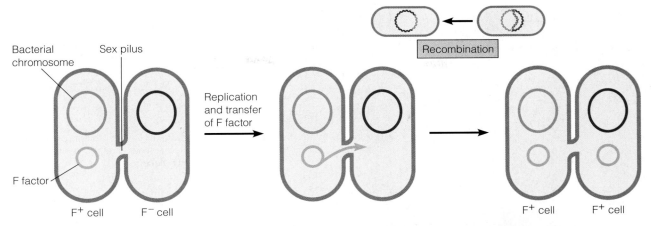

Bacterial chromosome
Sex pilus
F factor
F⁺ cell
F⁻ cell
Replication and transfer of F factor
Recombination
F⁺ cell
F⁺ cell

(a) When an F factor (a plasmid) is transferred from a donor (F⁺) to a recipient (F⁻), the F⁻ cell is converted into an F⁺ cell.

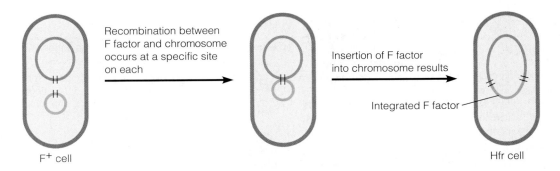

F⁺ cell
Recombination between F factor and chromosome occurs at a specific site on each
Insertion of F factor into chromosome results
Integrated F factor
Hfr cell

(b) When an F factor becomes integrated into the chromosome of an F⁺ cell, it makes the cell a high frequency of recombination (Hfr) cell.

Hfr cell
F⁻ cell
Replication and transfer of part of the chromosome
In the recipient, recombination occurs between the Hfr chromosome fragment and the F⁻ chromosome
Hfr cell
Recombinant F⁻ cell

(c) When an Hfr donor passes a portion of its chromosome into an F⁻ recipient, a recombinant F⁻ cell results.

FIGURE 8.27 **Conjugation in *E. coli.***

■ How does conjugation differ from transformation?

All genes contained within a bacterium infected by a generalized transducing phage are equally likely to be packaged in a phage coat and transferred. In another type of transduction, called **specialized transduction,** only certain bacterial genes are transferred. In one type

of specialized transduction, the phage codes for certain toxins produced by their bacterial hosts, such as *Corynebacterium diphtheriae* (kôr′-i-nē-bak-ti-rē-um dif-thi′-re-ī) for diphtheria toxin, *Streptococcus pyogenes* for erythrogenic toxin, and *E. coli* O157:H7 for Shiga toxin

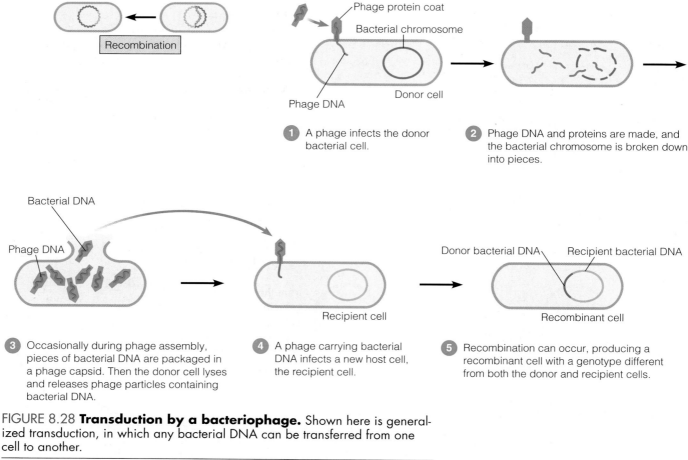

1 A phage infects the donor bacterial cell.

2 Phage DNA and proteins are made, and the bacterial chromosome is broken down into pieces.

3 Occasionally during phage assembly, pieces of bacterial DNA are packaged in a phage capsid. Then the donor cell lyses and releases phage particles containing bacterial DNA.

4 A phage carrying bacterial DNA infects a new host cell, the recipient cell.

5 Recombination can occur, producing a recombinant cell with a genotype different from both the donor and recipient cells.

FIGURE 8.28 **Transduction by a bacteriophage.** Shown here is generalized transduction, in which any bacterial DNA can be transferred from one cell to another.

■ What is transduction?

that causes bloody diarrhea. Specialized transduction will be discussed in Chapter 13 on page 390. In addition to mutation, transformation, and conjugation, transduction is another way bacteria acquire new genotypes.

Plasmids and Transposons

Learning Objective

• Describe the functions of plasmids and transposons.

Plasmids and transposons are genetic elements that provide additional mechanisms for genetic change. They occur in both prokaryotic and eukaryotic organisms, but this discussion focuses on their role in genetic change in prokaryotes.

Plasmids

Recall from Chapter 4 (page 92) that plasmids are self-replicating, gene-containing circular pieces of DNA about 1–5% the size of the bacterial chromosome (Figure 8.29a). They are found mainly in bacteria but also in some eu-

karyotic microorganisms, such as *Saccharomyces cerevisiae*. The F factor is a **conjugative plasmid** that carries genes for sex pili and for the transfer of the plasmid to another cell. Although plasmids are usually dispensable, under certain conditions genes carried by plasmids can be crucial to the survival and growth of the cell. For example, **dissimilation plasmids** code for enzymes that trigger the catabolism of certain unusual sugars and hydrocarbons. Some species of *Pseudomonas* can actually use such exotic substances as toluene, camphor, and hydrocarbons of petroleum as primary carbon and energy sources because they have catabolic enzymes encoded by genes carried on plasmids. Such specialized capabilities permit the survival of those microorganisms in very diverse and challenging environments. Because of their ability to degrade and detoxify a variety of unusual compounds, many of them are being investigated for possible use in the cleanup of environmental wastes. See the box in Chapter 2 on page 34.

Other plasmids code for proteins that enhance the pathogenicity of a bacterium. The strain of *E. coli* that causes infant diarrhea and traveler's diarrhea carries plasmids

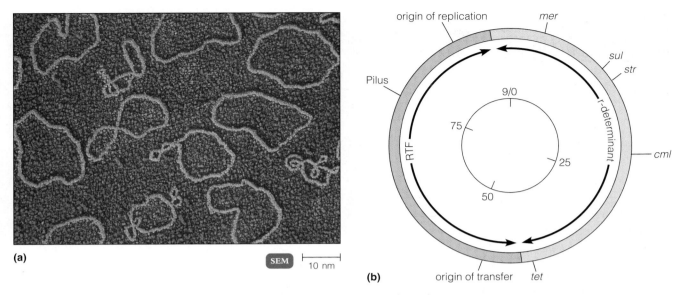

(a)

SEM | 10 nm

(b)

FIGURE 8.29 **R factor, a type of plasmid. (a)** Plasmids isolated from the bacteria *Bacteroides fragilis* that encode resistance to the antibiotic clindamycin. **(b)** A diagram of an R factor, which has two parts: The RTF contains genes needed for plasmid replication and transfer of the plasmid by conjugation; the r-determinant carries genes for resistance to four different antibiotics.

■ Why are R factors important in the treatment of infectious diseases?

that code for toxin production and for bacterial attachment to intestinal cells. Without these plasmids, *E. coli* is a harmless resident of the large intestine; with them, it is pathogenic. Other plasmid-encoded toxins include the exfoliative toxin of *Staphylococcus aureus*, *Clostridium tetani* neurotoxin, and toxins of *Bacillus anthracis*. Still other plasmids contain genes for the synthesis of **bacteriocins,** toxic proteins that kill other bacteria. These plasmids have been found in many bacterial genera, and they are useful markers for the identification of certain bacteria in clinical laboratories.

Resistance factors (R factors) are plasmids that have significant medical importance. They were first discovered in Japan in the late 1950s after several dysentery epidemics. In some of these epidemics, the infectious agent was resistant to the usual antibiotic. Following isolation, the pathogen was also found to be resistant to a number of different antibiotics. In addition, other normal bacteria from the patients (such as *E. coli*) proved to be resistant as well. Researchers soon discovered that these bacteria acquired resistance through the spread of genes from one organism to another. The plasmids that mediated this transfer are R factors.

R factors carry genes that confer upon their host cell resistance to antibiotics, heavy metals, or cellular toxins. Many R factors contain two groups of genes. One group is called the **resistance transfer factor (RTF)** and includes

genes for plasmid replication and conjugation. The other group, the **r-determinant,** has the resistance genes; it codes for the production of enzymes that inactivate certain drugs or toxic substances (Figure 8.29b). Different R factors, when present in the same cell, can recombine to produce R factors with new combinations of genes in their r-determinants.

In some cases, the accumulation of resistance genes within a single plasmid is quite remarkable. For example, Figure 8.29b shows a genetic map of resistance plasmid R100. Carried on this plasmid are resistance genes for sulfonamides, streptomycin, chloramphenicol, and tetracycline, as well as genes for resistance to mercury. This particular plasmid can be transferred between a number of enteric species, including *Escherichia*, *Klebsiella*, and *Salmonella*.

R factors present very serious problems for the treatment of infectious diseases with antibiotics. The widespread use of antibiotics in medicine and agriculture (many types of animal feed contain antibiotics) has led to the preferential survival (selection) of bacteria that have R factors, so populations of resistant bacteria grow larger and larger. The transfer of resistance between bacterial cells of a population, and even between bacteria of different genera, also contributes to the problem. The ability to reproduce sexually with members of its own species defines a eukaryotic species. However, a bacterial species can conjugate and transfer plasmids to other species. *Neisseria* may have acquired its

(a) Insertion sequence "IS1"

(b) Complex transposon "Tn5"

(c) R plasmid

FIGURE 8.30 **Transposons and insertion sequences. (a)** An insertion sequence (IS), the simplest transposon, contains a gene for transposase, the enzyme that catalyzes transposition. The transposase gene is bounded at each end by inverted repeat (IR) sequences that function as recognition sites for the transposon. IS1 is one example of an insertion sequence, shown here with simplified IR sequences. **(b)** Complex transposons carry other genetic material in addition to transposase genes. The example shown here, Tn5, carries the gene for kanamycin resistance and has complete copies of the insertion sequence IS1 at each end. **(c)** Transposition of the transposon Tn5 into R100 plasmid. Notice that a plasmid can acquire IS elements.

- Transposons are small segments of DNA that can move from one region of a DNA molecule to another.

penicillinase-producing plasmid from *Streptococcus*, and *Agrobacterium* can transfer plasmids to plant cells (see Figure 9.18 on page 268). Nonconjugative plasmids may be transferred from one cell to another by inserting themselves into a conjugative plasmid or a chromosome or by transformation when released from a dead cell. Insertion is made possible by an insertion sequence, which will be discussed shortly.

Plasmids are an important tool for genetic engineering, discussed in Chapter 9 on page 256.

Transposons

Transposons are small segments of DNA that can move (be "transposed") from one region of a DNA molecule to another. These pieces of DNA are 700–40,000 base pairs long.

In the 1950s, American geneticist Barbara McClintock discovered transposons in corn, but they occur in all organisms and have been studied most thoroughly in microorganisms. They may move from one site to another site on the same chromosome, or to another chromosome or plasmid. As you might imagine, the frequent movement of transposons could wreak havoc inside a cell. For example, as transposons move about on chromosomes, they may insert themselves *within* genes, inactivating them. Fortunately, transposition occurs relatively rarely. The fre-

quency of transposition is comparable to the spontaneous mutation rate that occurs in bacteria—that is, from 10^{-5} to 10^{-7} per generation.

All transposons contain the information for their own transposition. As shown in Figure 8.30a, the simplest transposons, also called **insertion sequences (IS),** contain only a gene that codes for an enzyme (*transposase*, which catalyzes the cutting and resealing of DNA that occurs in transposition) and recognition sites. *Recognition sites* are short inverted repeat sequences of DNA that the enzyme recognizes as recombination sites between the transposon and the chromosome.

Complex transposons also carry other genes not connected with the transposition process. For example, bacterial transposons may contain genes for enterotoxin or for antibiotic resistance (Figure 8.30b). Plasmids such as R factors are frequently made up of a collection of transposons (Figure 8.30c).

Transposons with antibiotic resistance genes are of practical interest, but there is no limitation on the kinds of genes that transposons can have. Thus, transposons provide a natural mechanism for the movement of genes from one chromosome to another. Furthermore, because they may be carried between cells on plasmids or viruses, they can also spread from one organism—or even species—to another. Transposons are thus a potentially powerful mediator of evolution in organisms.

Genes and Evolution

Learning Objective

• *Discuss how genetic mutation and recombination provide material for natural selection to act on.*

We have now seen how gene activity can be controlled by the cell's internal regulatory mechanisms and how genes themselves can be altered or rearranged by mutation, transposition, and recombination. All these processes provide diversity in the descendants of cells. Diversity provides the raw material for evolution, and natural selection provides its driving force. Natural selection will act on diverse populations to ensure the survival of those fit for that particular environment. The different kinds of microorganisms that exist today are the result of a long history of evolution. Microorganisms have continually changed by alterations in their genetic properties and acquisition of adaptations to many different habitats.

STUDY OUTLINE

Structure and Function of the Genetic Material (pp. 211–222)

1. Genetics is the study of what genes are, how they carry information, how their information is expressed, and how they are replicated and passed to subsequent generations or other organisms.

2. DNA in cells exists as a double-stranded helix; the two strands are held together by hydrogen bonds between specific nitrogenous base pairs: AT and CG.

3. A gene is a segment of DNA, a sequence of nucleotides, that codes for a functional product, usually a protein.

4. When a gene is expressed, DNA is transcribed to produce RNA; mRNA is then translated into proteins.

5. The DNA in a cell is duplicated before the cell divides, so each daughter cell receives the same genetic information.

Genotype and Phenotype (p. 211)

1. Genotype is the genetic composition of an organism, its entire complement of DNA.

2. Phenotype is the expression of the genes: the proteins of the cell and the properties they confer on the organism.

DNA and Chromosomes (pp. 211–212)

1. The DNA in a chromosome exists as one long double helix associated with various proteins that regulate genetic activity.

2. Bacterial DNA is circular; the chromosome of *E. coli*, for example, contains about 4 million base pairs and is approximately 1000 times longer than the cell.

3. Genomics is the molecular characterization of genomes.

4. Information contained in the DNA is transcribed into RNA and translated into proteins.

DNA Replication (pp. 212–217)

1. During DNA replication, the two strands of the double helix separate at the replication fork, and each strand is used as a template by DNA polymerases to synthesize two new strands of DNA according to the rules of nitrogenous base pairing.

2. The result of DNA replication is two new strands of DNA, each having a base sequence complementary to one of the original strands.

3. Because each double-stranded DNA molecule contains one original and one new strand, the replication process is called semiconservative.

4. DNA is synthesized in one direction designated $5' \rightarrow 3'$. At the replication fork, the leading strand is synthesized continuously and the lagging strand discontinuously.

5. DNA polymerase proofreads new molecules of DNA and removes mismatched bases before continuing DNA synthesis.

6. Each daughter bacterium receives a chromosome that is virtually identical to the parent's.

RNA and Protein Synthesis (pp. 217–222)

1. During transcription, the enzyme RNA polymerase synthesizes a strand of RNA from one strand of double-stranded DNA, which serves as a template.

2. RNA is synthesized from nucleotides containing the bases A, C, G, and U, which pair with the bases of the DNA strand being transcribed.

3. The starting point for transcription, where RNA polymerase binds to DNA, is the promoter; the region of DNA that is the end point of transcription is the terminator; RNA is synthesized in the $5' \rightarrow 3'$ direction.

4. Translation is the process in which the information in the nucleotide base sequence of mRNA is used to dictate the amino acid sequence of a protein.

5. The mRNA associates with ribosomes, which consist of rRNA and protein.

6. Three-base segments of mRNA that specify amino acids are called codons.

7. The genetic code refers to the relationship among the nucleotide base sequence of DNA, the corresponding codons of mRNA, and the amino acids for which the codons code.

8. The genetic code is degenerate; that is, most amino acids are coded for by more than one codon.

9. Of the 64 codons, 61 are sense codons (which code for amino acids), and 3 are nonsense codons (which do not code for amino acids and are stop signals for translation).

10. The start codon, AUG, codes for methionine.

11. Specific amino acids are attached to molecules of tRNA. Another portion of the tRNA has a base triplet called an anticodon.

12. The base pairing of codon and anticodon at the ribosome results in specific amino acids being brought to the site of protein synthesis.

13. The ribosome moves along the mRNA strand as amino acids are joined to form a growing polypeptide; mRNA is read in the $5' \rightarrow 3'$ direction.

14. Translation ends when the ribosome reaches a stop codon on the mRNA.

15. In prokaryotes, translation can begin before transcription is complete.

The Regulation of Bacterial Gene Expression (pp. 222–224)

1. Regulating protein synthesis at the gene level is energy-efficient because proteins are synthesized only as they are needed.

2. Constitutive enzymes produce products at a fixed rate. Examples are genes for the enzymes in glycolysis.

3. For these gene regulatory mechanisms, the control is aimed at mRNA synthesis.

Repression and Induction (p. 223)

1. Repression controls the synthesis of one or several (repressible) enzymes.

2. When cells are exposed to a particular end-product, the synthesis of enzymes related to that product decreases.

3. In the presence of certain chemicals (inducers), cells synthesize more enzymes. This process is called induction.

4. An example of induction is the production of β-galactosidase by *E. coli* in the presence of lactose; lactose can then be metabolized.

The Operon Model of Gene Expression (pp. 223–224)

1. The formation of enzymes is determined by structural genes.

2. In bacteria, a group of coordinately regulated structural genes with related metabolic functions, plus the promoter and operator sites that control their transcription, are called an operon.

3. In the operon model for an inducible system, a regulatory gene codes for the repressor protein.

4. When the inducer is absent, the repressor binds to the operator, and no mRNA is synthesized.

5. When the inducer is present, it binds to the repressor so that it cannot bind to the operator; thus, mRNA is made, and enzyme synthesis is induced.

6. In repressible systems, the repressor requires a corepressor in order to bind to the operator site; thus, the corepressor controls enzyme synthesis.

7. Transcription of structural genes for catabolic enzymes (such as β-galactosidase) is induced by the absence of glucose. Cyclic AMP and CRP must bind to a promoter in the presence of an alternative carbohydrate.

8. The presence of glucose inhibits the metabolism of alternative carbon sources by catabolite repression.

Mutation: Change in the Genetic Material (pp. 224–234)

1. A mutation is a change in the nitrogenous base sequence of DNA; that change causes a change in the product coded for by the mutated gene.

2. Many mutations are neutral, some are disadvantageous, and others are beneficial.

Types of Mutations (pp. 226–228)

1. A base substitution occurs when one base pair in DNA is replaced with a different base pair.

2. Alterations in DNA can result in missense mutations (which cause amino acid substitutions) or nonsense mutations (which create stop codons).

3. In a frameshift mutation, one or a few base pairs are deleted or added to DNA.

4. Mutagens are agents in the environment that cause permanent changes in DNA.

5. Spontaneous mutations occur without the presence of any mutagen.

Mutagens (pp. 228–230)

1. Chemical mutagens include base-pair mutagens (for example, nitrous acid), nucleoside analogs (e.g., 2-aminopurine and 5-bromouracil), and frameshift mutagens (e.g., benzpyrene).

2. Ionizing radiation causes the formation of ions and free radicals that react with DNA; base substitutions or breakage of the sugar-phosphate backbone results.

3. Ultraviolet (UV) radiation is nonionizing; it causes bonding between adjacent thymines.

4. Damage to DNA caused by UV radiation can be repaired by enzymes that cut out and replace the damaged portion of DNA.

5. Light-repair enzymes repair thymine dimers in the presence of visible light.

The Frequency of Mutation (p. 231)

1. Mutation rate is the probability that a gene will mutate when a cell divides; the rate is expressed as 10 to a negative power.

2. Mutations usually occur randomly along a chromosome.

3. A low rate of spontaneous mutations is beneficial in providing the genetic diversity needed for evolution.

Identifying Mutants (p. 231)

1. Mutants can be detected by selecting or testing for an altered phenotype.

2. Positive selection involves the selection of mutant cells and the rejection of nonmutated cells.

3. Replica plating is used for negative selection—to detect, for example, auxotrophs that have nutritional requirements not possessed by the parent (nonmutated) cell.

Identifying Chemical Carcinogens (pp. 232–234)

1. The Ames test is a relatively inexpensive and rapid test for identifying possible chemical carcinogens.

2. The test assumes that a mutant cell can revert to a normal cell in the presence of a mutagen, and that many mutagens are carcinogens.

3. Histidine auxotrophs of *Salmonella* are exposed to an enzymatically treated potential carcinogen, and reversions to the nonmutant state are selected.

Genetic Transfer and Recombination (pp. 234–243)

1. Genetic recombination, the rearrangement of genes from separate groups of genes, usually involves DNA from different organisms; it contributes to genetic diversity.

2. In crossing over, genes from two chromosomes are recombined into one chromosome containing some genes from each original chromosome.

3. Vertical gene transfer occurs during reproduction when genes are passed from an organism to its offspring.

4. Horizontal gene transfer in bacteria involves a portion of the cell's DNA being transferred from donor to recipient.

5. When some of the donor's DNA has been integrated into the recipient's DNA, the resultant cell is called a recombinant.

Transformation in Bacteria (pp. 235–237)

1. During this process, genes are transferred from one bacterium to another as "naked" DNA in solution.

2. This process was first demonstrated in *Streptococcus pneumoniae,* and occurs naturally among a few genera of bacteria.

Conjugation in Bacteria (pp. 237–238)

1. This process requires contact between living cells.

2. One type of genetic donor cell is an F^+; recipient cells are F^-. F cells contain plasmids called F factors; these are transferred to the F^- cells during conjugation.

3. When the plasmid becomes incorporated into the chromosome, the cell is called an Hfr (high frequency of recombination) cell.

4. During conjugation, an Hfr cell can transfer chromosomal DNA to an F^- cell. Usually, the Hfr chromosome breaks before it is fully transferred.

Transduction in Bacteria (pp. 238–240)

1. In this process, DNA is passed from one bacterium to another in a bacteriophage and is then incorporated into the recipient's DNA.

2. In generalized transduction, any bacterial genes can be transferred.

Plasmids and Transposons (pp. 240–243)

1. Plasmids are self-replicating circular molecules of DNA carrying genes that are not usually essential for the cell's survival.

2. There are several types of plasmids, including conjugative plasmids, dissimilation plasmids, plasmids carrying genes for toxins or bacteriocins, and resistance factors.

3. Transposons are small segments of DNA that can move from one region to another region of the same chromosome, or to a different chromosome or a plasmid.

4. Transposons are found in the main chromosomes of organisms, in plasmids, and in the genetic material of viruses. They vary from simple (insertion sequences) to complex.

5. Complex transposons can carry any type of gene, including antibiotic-resistance genes, and are thus a natural mechanism for moving genes from one chromosome to another.

Genes and Evolution (p. 243)

1. Diversity is the precondition for evolution.

2. Genetic mutation and recombination provide a diversity of organisms, and the process of natural selection allows the growth of those best adapted to a given environment.

STUDY QUESTIONS

Review

1. Briefly describe the components of DNA, and explain its functional relationship to RNA and protein.

2. Draw a diagram showing a portion of a chromosome undergoing replication.

a. Identify the replication fork.

b. What is the role of DNA polymerase? Of RNA polymerase?

c. How does this process represent semiconservative replication?

3. The following is a code for a strand of DNA.

```
DNA 3'     A T A T _ _ _ T T T _ _ _ _ _ _ _ _ _
           1 2 3 4 5 6 7 8 9 10 11 12 13 14 15 16 17 18 19
mRNA                       C G U       U G A
tRNA                             U G G
Amino acid         Met     ____  ____  ____  ____
ATAT = Promoter sequence
```

a. Using the genetic code provided in Figure 8.9, fill in the blanks to complete the segment of DNA shown.
b. Fill in the blanks to complete the sequence of amino acids coded for by this strand of DNA.
c. Write the code for the complementary strand of DNA completed in part a.
d. What would be the effect if C was substituted for T at base 10?
e. What would be the effect if A was substituted for G at base 11?
f. What would be the effect if G was substituted for T at base 14?
g. What would be the effect if C was inserted between bases 9 and 10?
h. How would UV radiation affect this strand of DNA?
i. Identify a nonsense sequence in this strand of DNA.

4. Describe translation, and be sure to include the following terms: ribosome, rRNA, amino acid activation, tRNA, anticodon, and codon.

5. Explain how you would find an antibiotic-resistant mutant by direct selection and how you would find an antibiotic-sensitive mutant by indirect selection.

6. Match the following examples of mutagens.

_____ A mutagen that is incorporated into DNA in place of a normal base
_____ A mutagen that causes the formation of highly reactive ions
_____ A mutagen that alters adenine so that it base-pairs with cytosine
_____ A mutagen that causes insertions
_____ A mutagen that causes the formation of pyrimidine dimers

a. Frameshift mutagen
b. Nucleoside analog
c. Base-pair mutagen
d. Ionizing radiation
e. Nonionizing radiation

7. Describe the principle of the Ames test for identifying chemical carcinogens.

8. Define plasmids, and explain the relationship between F factors and conjugation.

9. Use the following metabolic pathway to answer the questions that follow it.

    enzyme *a*   enzyme *b*
Substrate A ⟶ Intermediate B ⟶ End-product C
a. If enzyme *a* is inducible and is not being synthesized at present, a _____ protein must be bound tightly to the _____ site. When the inducer is present, it will bind to the _____ so that _____ can occur.
b. If enzyme *a* is repressible, end-product C, called a _____, causes the _____ to bind to the _____. What causes derepression?
c. If enzyme *a* is constitutive, what effect, if any, will the presence of A or C have on it?

10. Identify three ways of preventing mistakes in DNA.

11. Define the following terms:
a. genotype
b. phenotype
c. recombination

12. Which sequence is the best target for damage by UV radiation: AGGCAA, CTTTGA, or GUAAAU? Why aren't all bacteria killed when they are exposed to sunlight?

13. You are provided with cultures with the following characteristics:
Culture 1: F⁺, genotype A⁺ B⁺ C⁺
Culture 2: F⁻, genotype A⁻ B⁻ C⁻
a. Indicate the possible genotypes of a recombinant cell resulting from the conjugation of cultures 1 and 2.
b. Indicate the possible genotypes of a recombinant cell resulting from conjugation of the two cultures after the F⁺ has become an Hfr-cell.

14. Why are semiconservative replication and degeneracy of the genetic code advantageous to the survival of species?

15. Why are mutation and recombination important in the process of natural selection and the evolution of organisms?

Multiple Choice

Match the following terms to the definitions in questions 1 and 2.
a. conjugation
b. transcription
c. transduction
d. transformation
e. translation

1. The transfer of DNA from a donor to a recipient cell by a bacteriophage.
2. The transfer of DNA from a donor to a recipient as naked DNA in solution.
3. Feedback inhibition differs from repression because feedback inhibition
a. is less precise.
b. is slower acting.
c. stops the action of preexisting enzymes.
d. stops the synthesis of new enzymes.
e. all of the above
4. Bacteria can acquire antibiotic resistance by
a. mutation.
b. insertion of transposons.
c. acquiring plasmids.
d. all of the above
e. none of the above
5. Suppose you inoculate three flasks of minimal salts broth with *E. coli*. Flask A contains glucose. Flask B contains glucose and lactose. Flask C contains lactose. After a few

hours of incubation, you test the flasks for the presence of β-galactosidase. Which flask(s) do you predict will have this enzyme?

a. A
b. B
c. C
d. A and B
e. B and C

6. Plasmids differ from transposons because plasmids
   a. become inserted into chromosomes.
   b. are self-replicated outside the chromosome.
   c. move from chromosome to chromosome.
   d. carry genes for antibiotic resistance.
   e. none of the above

Use the following choices to answer questions 7 and 8.

a. catabolite repression
b. DNA polymerase
c. induction
d. repression
e. translation

7. The mechanism by which the presence of glucose inhibits the *lac* operon.

8. The mechanism by which lactose controls the *lac* operon.

9. Two daughter cells are most likely to inherit which one of the following from the parent cell?
   a. a change in a nucleotide in mRNA
   b. a change in a nucleotide in tRNA
   c. a change in a nucleotide in rRNA
   d. a change in a nucleotide in DNA
   e. a change in a protein

10. Which of the following is *not* a method of horizontal gene transfer?
    a. binary fission
    b. conjugation
    c. integration of a transposon
    d. transduction
    e. transformation

Critical Thinking

1. Nucleoside analogs and ionizing radiation are used in the treatment of cancer. These mutagens can cause cancer, so how do you suppose they are used to treat the disease?

2. Replication of the *E. coli* chromosome takes 40–45 minutes, but the organism has a generation time of 26 minutes. How does the cell have time to make complete chromosomes for each daughter cell? For each granddaughter cell?

3. *Pseudomonas* has a plasmid containing the *mer* operon, which includes the gene for mercuric reductase. This enzyme catalyzes the reduction of the mercuric ion Hg^{2+} to the uncharged form of mercury, Hg^0. Hg^{2+} is quite toxic to cells; Hg^0 is not.
   a. What do you suppose is the inducer for this operon?
   b. The protein encoded by one of the *mer* genes binds Hg^{2+} in the periplasm and brings it into the cell. Why would a cell bring in a toxin?
   c. What is the value of the *mer* operon to *Pseudomonas*?

Clinical Applications

1. Ciprofloxacin, erythromycin, and acyclovir are used to treat microbial infections. Ciprofloxacin inhibits DNA gyrase. Erythromycin binds in front of the A site on the 50S subunit of a ribosome. Acyclovir is a guanine analog.
   a. What steps in protein synthesis are inhibited by each drug?
   b. Which drug is more effective against bacteria? Why?
   c. Which drug is more effective against viruses? Why?
   d. Which drugs will have effects on the host's cells? Why?
   e. Use the index to identify the disease for which acyclovir is primarily used. Why is it more effective than erythromycin for treating this disease?

2. HIV, the virus that causes AIDS, was isolated from three individuals, and the amino acid sequences for the viral coat were determined. Of the amino acid sequences shown below, which two of the viruses are most closely related? How can these amino acid sequences be used to identify the source of a virus?

Patient	Viral Amino Acid Sequence
A	Asn Gln Thr Ala Ala Ser Lys Asn Ile Asp Ala Leu
B	Asn Leu His Ser Asp Lys Ile Asn Ile Ile Leu Leu
C	Asn Gln Thr Ala Asp Ser Ile Val Ile Asp Ala Leu

3. Human herpesvirus-8 (HHV-8) is common in certain parts of Africa, the Middle East, and the Mediterranean, but it is rare elsewhere except in AIDS patients. Genetic analyses indicate that the African strain is not changing, whereas the Western strain is accumulating changes. Using the portions of the HHV-8 genomes (shown below) that encode one of the viral proteins, how similar are these two viruses? What mechanism can account for the changes? What disease does HHV-8 cause?

Western   3' ATGGAGTTCTTCTGGACAAGA
African   3' ATAAACTTTTTCTTGACAACG

LEARNING WITH TECHNOLOGY

The Microbiology Place CD-ROM/website includes many resources to help you study and succeed in this course. In addition to self-quizzes, web links, and flashcards, you'll find the following activities for this chapter in the "Activities" section of The Microbiology Place:

DNA Replication

Transcription: Flow of Genetic Information

Transcription: How DNA is Transcribed

Mutation

Recombination

The *lac* Operon

Case Study: Mapping a Bacterial Chromosome

Chapter 9

Biotechnology and Recombinant DNA

ESCHERICHIA COLI. *This bacterium has been genetically engineered to produce a human protein, gamma interferon. Unlike human cells, it doesn't secrete the protein so the cells will be lysed to harvest the protein.*

For thousands of years, people have been consuming foods that are produced by the action of microorganisms. Bread, chocolate, and soy sauce are some of the best known examples. But it was only just over 100 years ago that scientists showed that microorganisms are responsible for these products. This knowledge opened the way for using microorganisms to produce other important products. Since World War I, microbes have been used to produce a variety of chemicals, such as ethanol, acetone, and citric acid. Since World War II, microorganisms have been grown on a large scale to produce antibiotics. More recently, microbes and their enzymes are replacing a variety of chemical processes involved in manufacturing such products as paper, textiles, and fructose. Using microbes or their enzymes instead of chemical syntheses offers several advantages: microbes may use inexpensive, abundant raw materials, such as starch; microbes work at normal temperatures and pressure, thereby avoiding the need for expensive and dangerous pressurized systems; and microbes don't produce toxic, hard-to-treat wastes. Currently, the search for microbes that produce new antibiotics and useful enzymes continues to broaden and expand (see the box on page 251).

In this chapter you will learn the tools and techniques that are used to research and develop a product.

Introduction to Biotechnology

Learning Objective

- Compare and contrast biotechnology, genetic engineering, and recombinant DNA technology.

All the products just discussed are the result of **biotechnology,** the use of microorganisms, cells, or cell components to make a product. Microbes have been used in the commercial production of foods, vaccines, antibiotics, and vitamins for years. Bacteria are also used in mining to extract valuable elements from ore. Additionally, animal cells have been used to produce viral vaccines since the 1950s. Until the 1980s, products made by living cells were all naturally made by the cells; the role of scientists was to find the appropriate cell and develop a method for large-scale cultivation of the cells.

Now, microorganisms as well as entire plants are being used as "factories" to produce chemicals that the organisms don't naturally make. The latter is made possible by inserting genes into cells by a process called **genetic engineering.** The development of genetic engineering is expanding the practical applications of biotechnology almost beyond imagination.

Recombinant DNA Technology

Recall from Chapter 8 that recombination of DNA occurs naturally in microbes. In the 1970s and 1980s, scientists developed artificial techniques for making recombinant DNA. These techniques are called **recombinant DNA technology.**

A gene from a vertebrate animal, including humans, can be inserted into the DNA of a bacterium, or a gene from a virus into a yeast. In many cases, the recipient can then be made to express the gene, which may code for a commercially useful product. Thus, bacteria with genes for human insulin are now being used to produce insulin for treating diabetes, and a vaccine for hepatitis B is being made by yeast carrying a gene for part of the hepatitis virus (the yeast produces a viral coat protein). Scientists hope that such an approach may prove useful in producing vaccines against other infectious agents, eliminating the need to use whole organisms, as in conventional vaccines.

The new recombinant DNA techniques can also be used to make thousands of copies of the same DNA molecule— to *amplify* DNA, thus generating sufficient DNA for various kinds of experimentation and analysis. This technique has practical application for identifying microbes, such as viruses, that can't be cultured.

An Overview of Recombinant DNA Procedures

Learning Objective

- Identify the roles of a clone and a vector in genetic engineering.

Figure 9.1 presents an overview of some of the procedures typically used in genetic engineering, along with some promising applications.

1–**2** After several preparatory steps,

3 the gene of interest is inserted into the vector DNA in vitro. In this example, the vector is a plasmid. The DNA molecule chosen as a vector must be a self-replicating type, such as a plasmid or a viral genome.

4 Next, this recombinant vector DNA is taken up by a cell such as a bacterium, where it can multiply.

5 The cell containing the recombinant vector is then grown in culture to form a **clone** of many genetically identical cells, each of which carries a copy of the vector. This cell clone therefore contains many copies of the gene of interest. This is why DNA vectors are often called *gene-cloning vectors*, or simply *cloning vectors*. (In addition to referring to a culture of identical cells, the word *clone* is also routinely used as a verb, to describe the entire process, as in "to clone a gene.")

The final step varies according to whether the gene itself or the product of the gene is of interest.

6A From the cell clone, the genetic engineer may isolate ("harvest") large quantities of the gene of interest, which may then be used for a variety of purposes. The gene may even be inserted into another vector for introduction into another kind of cell (such as a plant or animal cell). Alternatively,

6B if the gene of interest is expressed (transcribed and translated) in the cell clone,

7 its protein product can be harvested and used for a variety of purposes.

The advantages of genetic engineering for obtaining such proteins is illustrated by one of its early successes, human growth hormone (hGH). Some individuals do not produce adequate amounts of hGH, and their growth is stunted. In the past, hGH needed to correct this deficiency had to be obtained from human pituitary glands at autopsy. (hGH from other animals is not effective in humans.) This practice was not only expensive but also dangerous because on several occasions neurological diseases were

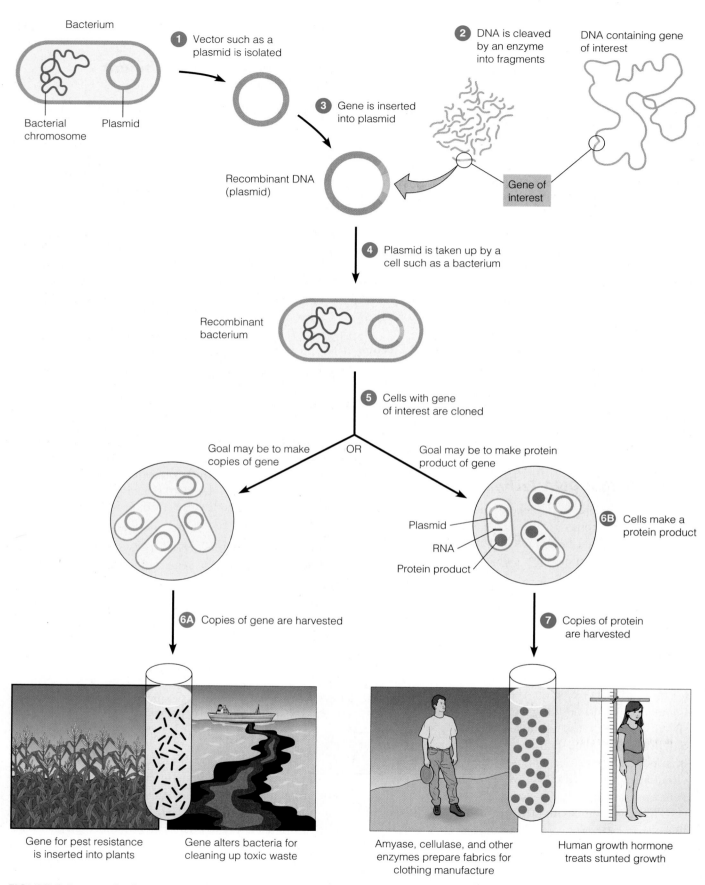

FIGURE 9.1 **A typical genetic engineering procedure, with examples of applications.**

■ What is genetic engineering?

APPLICATIONS OF MICROBIOLOGY

Designer Jeans

Denim blue jeans have become increasingly popular ever since they were first made for California gold miners in 1873 by Levi Strauss and Jacob Davis. Now, many companies manufacture jeans for sale around the world. Over the years, very little has changed in the denim-making process.

The denim for jeans is made from cotton. Cotton plants require water, costly fertilizer, and potentially harmful pesticides. The availability of cotton is dependent on weather and the plants' resistance to disease. After harvesting, cotton is made into yarn that is bleached with chlorine. Gaseous chlorine is quite hazardous but the safer liquid form (hypochlorite) is more expensive. After weaving yarn into cloth, the cloth is treated with starch to hold its shape, and dyed with indigo. The original indigo was made by fermentation of the indigo plant by *Clostridium*. However, this process was replaced over a century ago by a chemical process that requires mutagenic chemicals and a very high pH.

Now, companies that manufacture blue jeans and other products are returning to microbiology to develop environmentally sound production methods that minimize toxic wastes and the costs associated with treating toxic wastes. Moreover, microbiological methods can reduce their production costs and provide abundant, renewable raw materials.

In the 1980s, a softer denim, called "stone-washed," was introduced. The fabric is not really washed with rocks. Enzymes, called cellulases, from *Trichoderma* fungus are used to digest some of the cellulose in the cotton, thereby soft-ening it. Other microbial enzymes may be able to make the entire product safely. Thousands of enzymes have been identified in microbes and because enzymes are proteins they are readily degraded for removal from wastewater. Enzymes catalyze reactions in seconds rather than the days or weeks required of chemical methods and they usually operate at safe temperatures and pHs as opposed to high temperatures and extreme pHs of chemical reactions.

Bleaching

Peroxide is a safer bleaching agent than chlorine and can be easily removed from fabric and wastewater by enzymes. Researchers at Novo Nordisk Biotech cloned a mushroom peroxidase gene in yeast and grew the yeasts in washing machine conditions. The yeast that survived the washing machine were selected as the peroxidase producers.

Fabric

Cotton production requires large tracts of land, and the crop yield depends on the weather. Polyester is a synthetic fabric produced from petroleum. Bacteria can produce both cotton and polyester. *Acetobacter xylinum* bacteria make cellulose by attaching glucose units to simple chains in the outer membrane of the bacterial cell wall. The cellulose microfibrils are extruded through pores in the outer membrane, and bundles of microfibrils then twist into ribbons. Polyester is traditionally derived from a by-product of the oil industry, and its production produces toxic waste products. An environmentally friendly polyester can be made by genetically modified bacteria that convert sugar to glycerol using a bacterial enzyme, then glycerol to trimethylene glycol (3G) using a yeast enzyme. 3G can then be polymerized into polyester. The 3G-producing bacteria can be grown on agricultural waste, and the liquid waste is biodegradable.

Indigo

The current chemical synthesis of indigo requires a high pH and produces waste that explodes in contact with air. This will change if biotechnology companies have their way. A California biotechnology company, Genencor, is giving samples of microbially produced indigo to blue-jean makers. In the Genencor labs, *E. coli* turned blue when researchers put the gene for conversion of indole to indigo from a soil bacterium, *Pseudomonas putida* into the *E. coli*. Then, using site-directed mutagenesis and modifications of the fermentation medium, Genencor scientists increased *E. coli*'s production of indigo. Another company, W. R. Grace, has developed a mutated fungal culture that also produces indigo.

Plastic

Microbes can even make plastic zippers and packaging material for the jeans. Over 25 bacteria make polyhydoxyalkanoate (PHA) inclusion granules as a nutrient reserve. PHAs are similar to common plastics and, because they are made by bacteria, they are also readily degraded by many bacteria. PHAs could provide a biodegradable alternative to conventional plastic, which is made from petroleum.

---

transmitted with the hormone. Human growth hormone produced by genetically engineered *E. coli* is a pure and cost-effective product. Recombinant DNA techniques also result in faster production of the hormone than traditional methods might allow.

Tools of Biotechnology

Learning Objective

• *Compare selection and mutation.*

Research scientists and technicians isolate bacteria and fungi from natural environments such as soil and water to find or *select* the organisms that produce a desired product. The selected organism can be mutated to make more product or to make a better product.

Selection

In nature, organisms with characteristics that enhance survival are more likely to survive and reproduce than are variants that lack the desirable traits. This is called *natural selection*. Humans use **artificial selection** to select desirable breeds of animals or strains of plants to cultivate. As microbiologists learned how to isolate and grow microorganisms in pure culture, they were able to select the ones that could accomplish the desired objective, such as brewing beer more efficiently, for example, or producing a new antibiotic. Over 2000 strains of antibiotic-producing bacteria have been discovered by testing soil bacteria and selecting the strains that produce an antibiotic.

Mutation

As we saw in Chapter 8, mutations are responsible for much of the diversity of life. A bacterium with a mutation that confers resistance to an antibiotic will survive and reproduce in the presence of that antibiotic. Biologists working with antibiotic-producing microbes discovered that they could create new strains by exposing microbes to mutagens. After random mutations were created in penicillin-producing *Penicillium* by exposing fungal cultures to radiation, the highest-yielding variant among the survivors was selected for another exposure to a mutagen. Using mutations, biologists increased the amount of penicillin produced by the fungus over 1000 times.

Screening each mutant for penicillin production is a tedious process. **Site-directed mutagenesis** can be used to make a specific change in a gene. Suppose you determine that changing one amino acid will make a laundry enzyme work better in cold water. Using the genetic code (see Figure 8.9 on page 219), you could produce the sequence of DNA that encodes that amino acid and insert it into the gene for that enzyme using the techniques described next.

The science of genetic engineering has advanced to such a degree that many routine cloning procedures are performed using prepackaged materials and procedures that are very much like cookbook recipes. Genetic engineers have a grab-bag of methods from which to choose, depending on the ultimate application of their experiments. Next we describe some of the most important tools and techniques, and later we will consider some specific applications.

Restriction Enzymes

Learning Objective

- Define restriction enzymes, and outline how they are used to make recombinant DNA.

Genetic engineering has its technical roots in the discovery of **restriction enzymes,** a special class of DNA-cutting enzymes that exist in many bacteria. First isolated in 1970, restriction enzymes in nature had actually been observed earlier, when certain bacteriophages were found to have a restricted host range. If these phages were used to infect bacteria other than their usual hosts, restriction enzymes in the new host destroyed almost all the phage DNA. Restriction enzymes protect a bacterial cell by hydrolyzing phage DNA. The bacterial DNA is protected from digestion because the cell **methylates** (adds methyl groups to) some of the cytosines in its DNA. The purified forms of these bacterial enzymes are used by today's genetic engineers.

What is important for genetic engineering is that a restriction enzyme recognizes and cuts, or digests, only one particular sequence of nucleotide bases in DNA, and it cuts this sequence in the same way each time. Typical restriction enzymes used in cloning experiments recognize four-, six-, or eight-base sequences. Many restriction enzymes make staggered cuts in the two strands of a DNA molecule—cuts that are not directly opposite each other (Figure 9.2).

❶ Notice in this figure that whereas the blue base sequences on the two strands are the same, they run in opposite directions.

❷ Staggered cuts leave stretches of single-stranded DNA at the ends of the DNA fragments. These are called *sticky ends* because they can "stick" to complementary stretches of single-stranded DNA by base pairing. Hundreds of restriction enzymes are known, each producing DNA fragments with characteristic ends.

❸ If two fragments of DNA from different sources have been produced by the action of the same restriction enzyme, the two pieces will have identical sets of sticky ends and can be spliced (recombined) in vitro.

❹ The sticky ends first join spontaneously by hydrogen bonding (base pairing) in either a linear or a circular form.

❺ Then the enzyme DNA ligase is used to covalently link the backbones of the DNA pieces, producing a recombinant DNA molecule.

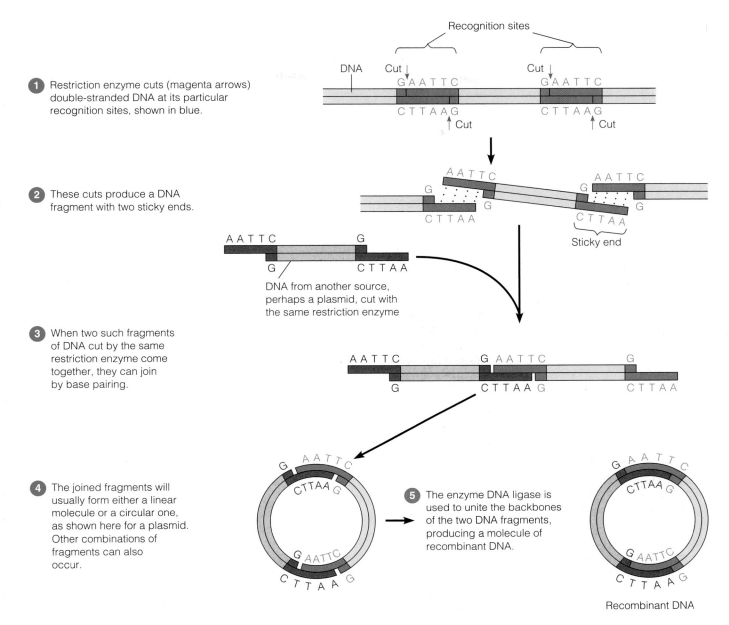

1 Restriction enzyme cuts (magenta arrows) double-stranded DNA at its particular recognition sites, shown in blue.

2 These cuts produce a DNA fragment with two sticky ends.

DNA from another source, perhaps a plasmid, cut with the same restriction enzyme

3 When two such fragments of DNA cut by the same restriction enzyme come together, they can join by base pairing.

4 The joined fragments will usually form either a linear molecule or a circular one, as shown here for a plasmid. Other combinations of fragments can also occur.

5 The enzyme DNA ligase is used to unite the backbones of the two DNA fragments, producing a molecule of recombinant DNA.

Recombinant DNA

FIGURE 9.2 **The role of a restriction enzyme in making recombinant DNA.**

▪ Why are restriction enzymes used in genetic engineering?

Vectors

Learning Objectives

- List the four properties of vectors.
- Describe the use of plasmid and viral vectors.

A great variety of different types of DNA molecules can serve as vectors, provided that they have certain properties. The most important property is self-replication; once in a cell, a vector must be capable of replicating. Any DNA that is cloned in the vector will be replicated in the process. Thus, vectors serve as vehicles for the replication of desired DNA sequences.

Vectors also need to be of a size that allows them to be manipulated outside the cell during recombinant DNA procedures. Smaller vectors are more easily manipulated than larger DNA molecules, which tend to be more fragile. Preservation is another important property of vectors. The circular form of DNA molecules is important in protecting the DNA of the vector from destruction by the recipient of the vector. Notice in Figure 9.3 that the DNA of a plasmid

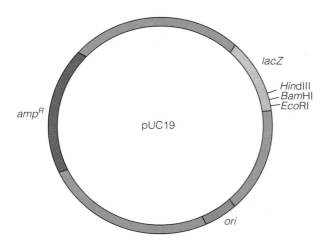

FIGURE 9.3 **A plasmid used for cloning.** pUC19 is a plasmid vector used for cloning in the bacterium *E. coli*. An origin of replication *(ori)* allows the plasmid to be self-replicating. Two genes, one encoding resistance to the antibiotic ampicillin *(amp)* and one encoding the enzyme *β*-galactosidase *(lacZ)*, serve as marker genes. Foreign DNA can be inserted at the restriction enzyme sites.

■ What is a vector in genetic engineering?

is circular. Another preservation mechanism occurs when the DNA of a virus inserts itself quickly into the chromosome of the host (see Chapter 8, page 238).

When it is necessary to retrieve cells containing the vector, a marker gene contained within the vector can often help make selection easy. Common selectable marker genes are for antibiotic resistance or for an enzyme that carries out an easily identified reaction.

Plasmids are one of the primary vectors in use, particularly variants of R factor plasmids. Plasmid DNA can be cut with the same restriction enzymes as the DNA to be cloned, so that all pieces of the DNA will have the same sticky ends. When the pieces are mixed, the DNA to be cloned will become inserted into the plasmid (see Figure 9.3). Note that other possible combinations of fragments can occur as well, including the plasmid re-forming a circle with no DNA inserted.

Some plasmids are capable of existing in several different species. They are called **shuttle vectors** and can be used to move cloned DNA sequences among organisms, such as among bacterial, yeast, and mammalian cells, or among bacterial, fungal, and plant cells. Shuttle vectors can be very useful in the process of genetically modifying multicellular organisms—for example, by trying to insert herbicide resistance genes into plants.

A different kind of vector is viral DNA. This type of vector can usually accept much larger pieces of foreign DNA than plasmids can. After the DNA has been inserted

into the viral vector, it can be cloned in the virus's host cells. The choice of a suitable vector depends on many factors, including the organism that will receive the new gene and the size of the DNA to be cloned. Retroviruses, adenoviruses, and herpesviruses are being used to insert corrective genes into human cells that have defective genes. Gene therapy is discussed on page 263.

Polymerase Chain Reaction

Learning Objective

- *Outline the steps in PCR, and provide an example of its use.*

The **polymerase chain reaction (PCR)** is a technique by which small samples of DNA can be quickly amplified, that is, increased to quantities that are large enough for analysis.

Starting with just one gene-sized piece of DNA, PCR can be used to make literally billions of copies in only a few hours. The PCR process is shown in Figure 9.4.

❶ Each strand of the target DNA will serve as a template for DNA synthesis.

❷ To this DNA is added a supply of the four nucleotides (for assembly into new DNA) and the enzyme for catalyzing the synthesis, DNA polymerase (see Chapter 8, page 213). Short pieces of nucleic acid called primers are also added to help start the reaction. The primers are complementary to the ends of the target DNA and

❸ will hybridize to the fragments to be amplified.

❹ Then, the polymerase synthesizes new complementary strands.

❺ After each cycle of synthesis, the DNA is heated to convert all the new DNA into single strands. Each newly synthesized DNA strand serves in turn as a template for more new DNA.

As a result, the process proceeds exponentially. All of the necessary reagents are added to a tube, which is placed in a *thermalcycler*. The thermalcycler can be set for the desired temperatures, times, and number of cycles. Use of an automated thermalcycler is made possible by the use of DNA polymerase taken from a thermophilic bacterium such as *Thermus aquaticus*; the enzyme from such organisms can survive the heating phase without being destroyed (see the box in Chapter 6, page 160). Thirty cycles, completed in just a few hours, will increase the amount of target DNA by more than a billion times.

Note that PCR can only be used to amplify relatively small, specific sequences of DNA as determined by the choice of primers. It cannot be used to amplify an entire genome.

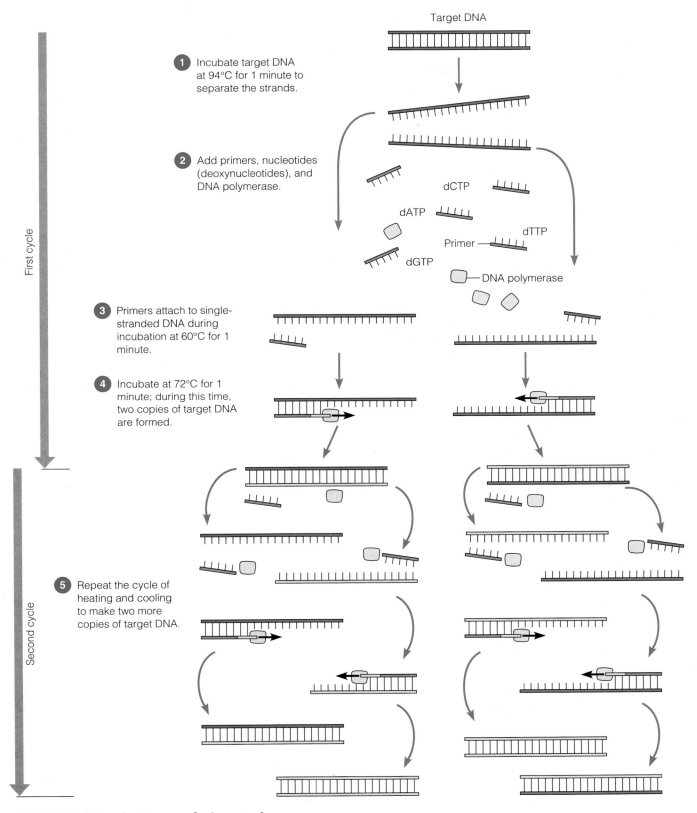

FIGURE 9.4 **The polymerase chain reaction.**

■ The polymerase chain reaction is used to increase samples of DNA to quantities large enough for analysis.

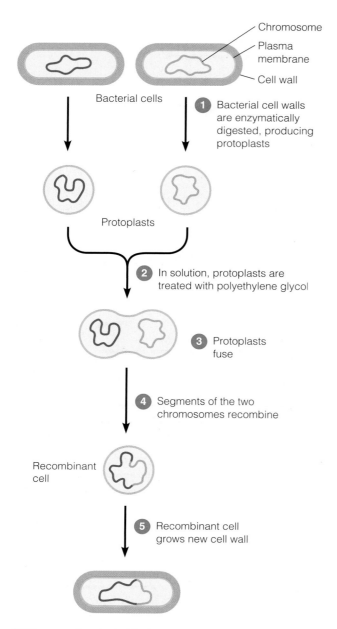

(a) Process of protoplast fusion

FIGURE 9.5 **Protoplast fusion. (a)** A diagram of protoplast fusion with bacterial cells. **(b)** Protoplast algal cells are shown fusing. Removal of the cell wall left only the delicate plasma membrane to bind the cell contents together, allowing the exchange of DNA.

■ Protoplasts are produced by enzymatically removing the cell wall.

PCR can be applied to any situation that requires the amplification of DNA. Especially noteworthy are diagnostic tests that use PCR to detect the presence of infectious agents in situations in which they would otherwise be undetectable.

(b)

LM   ⊢——⊣ 10 μm

Techniques of Genetic Engineering

Inserting Foreign DNA into Cells

Learning Objective

• Describe five ways of getting DNA into a cell.

Recombinant DNA procedures require that DNA molecules be manipulated outside the cell and then returned to living cells. There are several ways to introduce DNA into cells. The choice of method is usually determined by the type of vector and host cell being used.

In nature, plasmids are usually transferred between closely related microbes by cell-to-cell contact, such as in conjugation. In genetic engineering, a plasmid must be inserted into a cell by **transformation,** a procedure during which cells can take up DNA from the surrounding environment (see Chapter 8, page 235). Many cell types, including *E. coli*, yeast, and mammalian cells, do not naturally transform; however, simple chemical treatments can make all of these cell types *competent*, or able to take up external DNA. For *E. coli*, the procedure for making cells competent is to soak them in a solution of calcium chloride for a brief period. Following this treatment, the now-competent cells are mixed with the cloned DNA and given a mild heat shock. Some of these cells will then take up the DNA.

There are other ways to transfer DNA to cells. A process called **electroporation** uses an electrical current to form microscopic pores in the membranes of cells; the DNA then enters the cells through the pores. Electroporation is generally applicable to all cells; those with cell walls often must be converted to protoplasts first (see Chapter 4, page 86). **Protoplasts** are produced by enzymatically removing the cell wall, thereby allowing more direct access to the plasma membrane.

The process of **protoplast fusion** also takes advantage of the properties of protoplasts. Protoplasts in solution will fuse at a low but significant rate; the addition of polyethylene glycol increases the frequency of fusion (Figure 9.5a). In the new hybrid cell, the DNA derived from the two "parent" cells may undergo natural recombination. This method is especially valuable in the genetic manipulation of plant and algal cells (Figure 9.5b).

A remarkable way of introducing foreign DNA into plant cells is to literally shoot it directly through the thick cellulose walls using a gene gun (Figure 9.6). Microscopic particles of tungsten or gold are coated with DNA and propelled by a burst of helium through the plant cell walls. Some of the cells express the introduced DNA as though it were their own.

DNA can be introduced directly into an animal cell by **microinjection.** This technique requires the use of a glass micropipette with a diameter that is much smaller than the cell. The micropipette punctures the plasma membrane, and DNA can be injected through it (Figure 9.7).

Thus, there is a great variety of different restriction enzymes, vectors, and methods of inserting DNA into cells. But foreign DNA will survive only if it is either present on

20 μm

FIGURE 9.7 **The microinjection of foreign DNA into a fertilized mouse egg.** The egg is first immobilized by applying mild suction to the large, blunt, holding pipette (right). Several hundred copies of the gene of interest are then injected into the nucleus of the cell through the tiny end of the micropipette (left).

■ Microinjection is the introduction of DNA into a cell using a glass micropipette that is smaller in diameter than the cell.

a self-replicating vector or incorporated into one of the cell's chromosomes by recombination.

Obtaining DNA

We have seen how genes can be cloned into vectors by using restriction enzymes, and how genes can be transformed or transferred into a variety of cell types. But how do genetic engineers obtain the genes they are interested in? There are two main sources of genes: (1) gene libraries containing either natural copies of genes or cDNA copies of genes made from mRNA, and (2) synthetic DNA.

Gene Libraries

Learning Objective

• *Describe how a gene library is made.*

Isolating specific genes as individual pieces of DNA is seldom practical. Therefore, researchers interested in genes from a particular organism start by extracting the organism's DNA, which can be obtained from cells of any organism, whether plant, animal, or microbe, by lysing the cells and precipitating the DNA. This process results in a DNA mass that includes the organism's entire genome. After the DNA is digested by restriction enzymes, the

FIGURE 9.6 **A gene gun, which can be used to insert DNA-coated "bullets" into a cell.**

■ Name four other methods of inserting DNA into a cell.

Plasmid Library   **Phage Library**

FIGURE 9.8 Gene libraries. Each fragment of DNA, containing about one gene, is carried by a vector, either a plasmid within a bacterial cell or a phage.

■ A gene library is a collection of a large number of DNA fragments from a genome.

restriction fragments are then spliced into plasmid or phage vectors, and the recombinant vectors are introduced into bacterial cells. The goal is to make a collection of clones large enough to ensure that at least one clone exists for every gene in the organism. This collection of clones containing different DNA fragments is called a **gene library;** each "book" is a bacterial or phage strain that contains a fragment of the genome (Figure 9.8). Such libraries are essential for maintaining and retrieving DNA clones; they can even be purchased commercially.

Cloning genes from eukaryotic organisms presents a specific problem. Genes of eukaryotic cells generally contain both **exons,** stretches of DNA that code for protein, and **introns,** intervening stretches of DNA that do not code for protein. When the RNA transcript of such a gene is converted to mRNA, the introns are removed (see Figure 8.12 on page 222). In cloning genes of eukaryotic cells, it is desirable to use a version of the gene that lacks introns because a gene that includes introns may be too large to work with easily. In addition, if such a gene is put into a bacterial cell, the bacterium will not usually be able to remove the introns from the RNA transcript, and therefore it will not be able to make the correct protein product. However, an artificial gene that contains only exons can be produced by

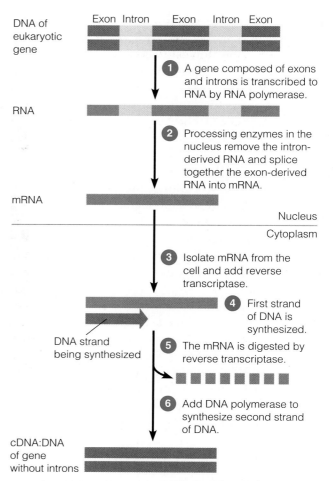

FIGURE 9.9 Making complementary DNA (cDNA) for a eukaryotic gene. Reverse transcriptase catalyzes the synthesis of double-stranded DNA from an RNA template.

■ How does reverse transcriptase differ from DNA polymerase?

using an enzyme called **reverse transcriptase** to synthesize **complementary DNA (cDNA)** from an mRNA template (Figure 9.9). This synthesis is the reverse of the normal DNA-to-RNA transcription process. A DNA copy of mRNA is produced by reverse transcriptase. Following this, the mRNA is enzymatically digested away. DNA polymerase then synthesizes a complementary strand of DNA, creating a double-stranded piece of DNA containing the information from the mRNA. Molecules of cDNA produced from a mixture of all the mRNAs from a tissue or cell type can then be cloned to form a cDNA library.

The cDNA method is the most common method of obtaining eukaryotic genes. A difficulty with this method is that long molecules of mRNA may not be completely reverse-transcribed into DNA; the reverse transcription often aborts, forming only parts of the desired gene.

FIGURE 9.10 **A DNA synthesis machine.** Short sequences of DNA can be synthesized by instruments such as this one.

■ What are some of the disadvantages of a DNA synthesis machine?

Synthetic DNA

Learning Objective

• Differentiate cDNA from synthetic DNA.

Under certain circumstances, genes can be made in vitro with the help of DNA synthesis machines (Figure 9.10). A keyboard on the machine is used to enter the desired sequence of nucleotides, much as letters are entered into a word processor to compose a sentence. A microprocessor controls the synthesis of the DNA from stored supplies of nucleotides and the other necessary reagents. A chain of over 120 nucleotides can be synthesized by this method. Unless the gene is very small, at least several chains must be synthesized separately and linked together to form an entire gene.

The difficulty of this approach, of course, is that the sequence of the gene must be known before it can be synthesized. If the gene has not already been isolated, then the only way to predict the DNA sequence is by knowing the amino acid sequence of the protein product of the gene. If this amino acid sequence is known, in principle one can work backward through the genetic code to obtain the DNA sequence. Unfortunately, the degeneracy of the code prevents an unambiguous determination; thus, if the protein contains a leucine, for example, which of the six codons for leucine is the one in the gene?

For these reasons, it is rare to clone a gene by synthesizing it directly, although some commercial products, such as insulin, interferon, and somatostatin are produced from chemically synthesized genes. Desired restriction sites were added to the synthetic genes so that the genes could be inserted into plasmid vectors for cloning in *E. coli*. Synthetic DNA plays a much more useful role in selection procedures, as we will see.

Selecting a Clone

Learning Objective

• Explain how each of the following is used to locate a clone: antibiotic-resistance genes, DNA probes, gene product.

In cloning, it is necessary to select the particular cell that contains the specific gene of interest. This is difficult because out of millions of cells, only a very few cells might contain the desired gene. Here we will examine a typical screening procedure known as *blue-white screening*, from the color of the bacterial colonies formed at the end of the screening process.

The plasmid vector used contains a gene (amp^R) coding for resistance to the antibiotic ampicillin. The host bacterium will not be able to grow on the test medium, which contains ampicillin, unless the vector has transferred the ampicillin-resistance gene. The plasmid vector also contains a second gene, this one for the enzyme β-galactosidase (*lacZ*). Notice in Figure 9.3 (page 254) that there are several sites in *lacZ* that can be cut by restriction enzymes.

The procedure is shown in Figure 9.11. The two genes, called marker genes, are used so that the insertion of plasmid DNA into the host bacterium can be determined.

❶ The plasmid vector and foreign DNA are digested with the same restriction enzyme.

❷ The foreign DNA will insert into the gene for β-galactosidase. Therefore, the bacterium receiving the plasmid vector will not produce the enzyme β-galactosidase if foreign DNA has been inserted into the plasmid.

❸ The recombinant plasmid is introduced into a culture of ampicillin-sensitive bacteria by transformation.

❹ In the blue-white screening procedure, a library of bacteria is cultured in a medium called X-gal. X-gal contains two essential components other than those necessary to support normal bacterial growth. One is the antibiotic ampicillin, which prevents the growth of any bacterium that has not successfully received the ampicillin-resistance gene from the plasmid. The other, called X-gal, is a substrate for β-galactosidase.

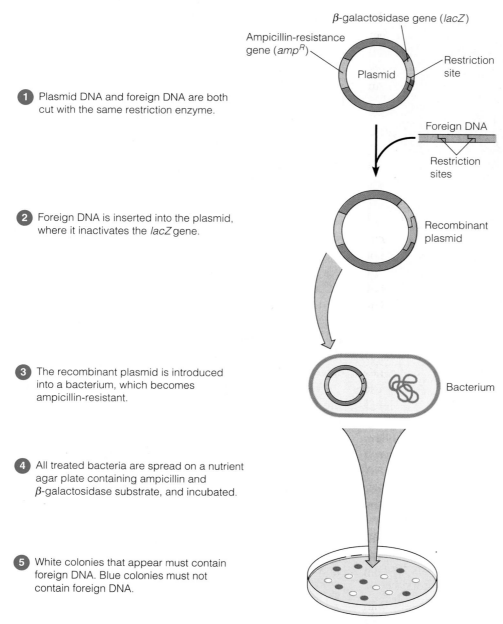

1 Plasmid DNA and foreign DNA are both cut with the same restriction enzyme.

β-galactosidase gene (*lacZ*)

Ampicillin-resistance gene (*amp^R*)

Plasmid

Restriction site

Foreign DNA

Restriction sites

2 Foreign DNA is inserted into the plasmid, where it inactivates the *lacZ* gene.

Recombinant plasmid

3 The recombinant plasmid is introduced into a bacterium, which becomes ampicillin-resistant.

Bacterium

4 All treated bacteria are spread on a nutrient agar plate containing ampicillin and β-galactosidase substrate, and incubated.

5 White colonies that appear must contain foreign DNA. Blue colonies must not contain foreign DNA.

FIGURE 9.11 **One method of selecting recombinant bacteria.**

▪ Why are some colonies blue and others white?

5 Only bacteria that picked up the plasmid will grow—because they are now ampicillin resistant. Bacteria that picked up the recombinant plasmid—in which the new gene was inserted into the *lacZ* gene—will not hydrolyze lactose and will produce white colonies. If a bacterium received the original plasmid containing the intact *lacZ* gene, the cells will hydrolyze X-gal to produce a blue-colored compound; the colony will be blue.

What remains to be done can still be difficult. The above procedure has isolated white colonies known to contain foreign DNA, but it is still not known whether this is the desired fragment of foreign DNA. A second procedure is needed to identify these bacteria. If the foreign DNA in the plasmid codes for the production of an identifiable product, the bacterial isolate only needs to be grown in culture and tested. However, in some cases the gene itself must be identified in the host bacterium.

Colony hybridization is a common method of identifying cells that carry a specific cloned gene. **DNA probes,** short segments of single-stranded DNA that are complementary to the desired gene, are synthesized. If the DNA probe finds a match, it will adhere to the target gene. The DNA probe is labeled with a radioactive element or fluorescent dye so its presence can be determined. A typical colony hybridization experiment is shown in Figure 9.12.

Making a Gene Product

Learning Objective

- List one advantage of engineering each of the following: E. coli, Saccharomyces cerevisiae, *mammalian cells, plant cells.*

We have just seen how to identify cells carrying a particular gene. The gene products are frequently the objective of genetic engineering. Such products are themselves, of course, a frequent objective of genetic engineering. Most of the earliest work in genetic engineering used *E. coli* to synthesize the gene products. *E. coli* is easily grown, and researchers are very familiar with this bacterium and its genetics. For example, some inducible promoters, such as that of the *lac* operon, have been cloned, and cloned genes can be attached to such promoters. The synthesis of great amounts of the cloned gene product can then be directed by the addition of an inducer. Such a method has been used to produce gamma interferon in *E. coli* (Figure 9.13). However, *E. coli* also has several disadvantages. Like other gram-negative bacteria, it produces endotoxins as part of the outer layer of its cell wall. Because endotoxins cause fever and shock in animals, their accidental presence in products intended for human use would be a serious problem.

Another disadvantage of *E. coli* is that it does not usually secrete protein products. To obtain a product, cells must usually be broken open and the product purified from the resulting "soup" of cell components. Recovering the product from such a mixture is expensive when done on an industrial scale. It is more economical to have an organism secrete the product so that it can be recovered continuously from the growth medium. One approach has been to link the product to a natural *E. coli* protein that the bacterium does secrete. However, gram-positive bacteria, such as *Bacillus subtilis*, are more likely to secrete their products and are often preferred industrially for that reason.

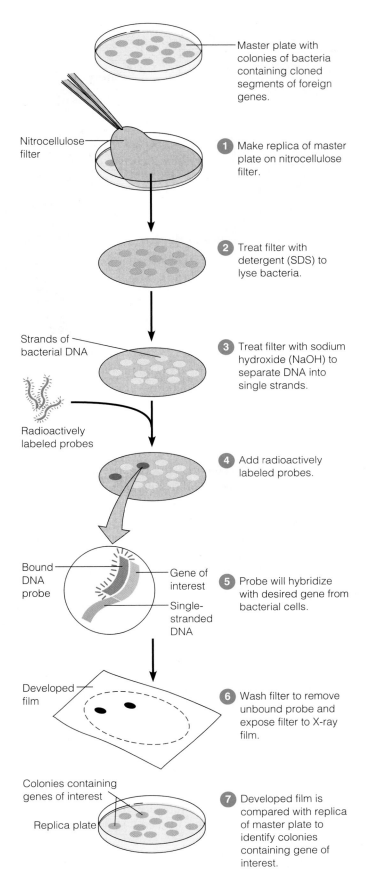

FIGURE 9.12 **Colony hybridization: using a DNA probe to identify a cloned gene of interest.**

- A probe is a single-stranded DNA with a base sequence complementary to that of the gene of interest, that will hybridize with the gene from bacterial cells.

Master plate with colonies of bacteria containing cloned segments of foreign genes.

Nitrocellulose filter

1 Make replica of master plate on nitrocellulose filter.

2 Treat filter with detergent (SDS) to lyse bacteria.

Strands of bacterial DNA

3 Treat filter with sodium hydroxide (NaOH) to separate DNA into single strands.

Radioactively labeled probes

4 Add radioactively labeled probes.

Bound DNA probe

Gene of interest

Single-stranded DNA

5 Probe will hybridize with desired gene from bacterial cells.

Developed film

6 Wash filter to remove unbound probe and expose filter to X-ray film.

Colonies containing genes of interest

Replica plate

7 Developed film is compared with replica of master plate to identify colonies containing gene of interest.

TEM |— 0.25 µm —|

FIGURE 9.13 *E. coli* genetically engineered to produce gamma interferon, a human protein that promotes an immune response. The product, visible here as an orange-colored substance, can be released by lysis of the cell.

- What is one advantage of using *E. coli* for genetic engineering? One disadvantage?

Another microbe being used as a vehicle for expressing genetically engineered genes is baker's yeast, *Saccharomyces cerevisiae*. Its genome is only about four times larger than that of *E. coli* and is probably the best understood eukaryotic genome. Yeasts may carry plasmids, and the plasmids are easily transferred into yeast cells after their cell walls have been removed. As eukaryotic cells, yeasts may be more successful in expressing foreign eukaryotic genes than bacteria. Furthermore, yeasts are likely to continuously secrete the product. Because of all these factors, yeasts have become the eukaryotic workhorse of biotechnology.

Mammalian cells in culture, even human cells, can be used much like bacteria to produce genetically engineered products. Scientists have developed effective methods of growing certain mammalian cells in culture as hosts for growing viruses (see Chapter 13, page 385). In genetic engineering, mammalian cells are often the best suited to making protein products for medical use; these products include hormones, cytokines (which regulate cells of the immune system), and interferon (a natural antiviral substance also used to treat some cancers). Using mammalian cells to make foreign gene products on an industrial scale often requires a preliminary step of cloning the gene in bacteria. Consider the example of colony-stimulating

factor (CSF). A protein produced naturally in tiny amounts by white blood cells, CSF is valuable because it stimulates the growth of certain cells that protect against infection. To produce huge amounts of CSF industrially, the gene is first inserted into a plasmid, and bacteria are used to make multiple copies of the plasmid (see Figure 9.1 on page 250). The recombinant plasmids are inserted into mammalian cells that are grown in bottles.

Plant cells can also be grown in culture, altered by recombinant DNA techniques, and then used to generate genetically engineered plants. Such plants may prove useful as sources of valuable products, such as plant alkaloids (the painkiller codeine, for example), the isoprenoids that are the basis of synthetic rubber, and melanin (the animal skin pigment) for use in sunscreens. (We will return to the topic of genetic engineering of plants later in the chapter, on page 267.)

Applications of Genetic Engineering

Learning Objective

- List at least five applications of genetic engineering.

We have now described the entire sequence of events in cloning a gene. As indicated earlier, such cloned genes can be applied in a variety of ways. One is to produce useful substances more efficiently and less expensively (see the box on page 251). Another is to obtain information from the cloned DNA that is useful for either basic research or medical applications. A third is to use cloned genes to alter the characteristics of cells or organisms.

Therapeutic Applications

An extremely valuable pharmaceutical product is the hormone insulin, a small protein produced by the pancreas that controls the body's uptake of glucose from blood. For many years, people with insulin-dependent diabetes have controlled their disease by injecting insulin obtained from the pancreases of slaughtered animals. Obtaining this insulin is an expensive process, and the insulin from animals is not as effective as human insulin.

Because of the value of human insulin and the small size of the protein, the production of human insulin by recombinant DNA techniques was an early goal for the pharmaceutical industry. To produce the hormone, synthetic genes were first constructed for each of the two short polypeptide chains that make up the insulin molecule. The small size of these chains—only 21 and 30 amino acids long—made it possible to use synthetic genes. Following the procedure described earlier, each of the two synthetic genes was inserted into a plasmid vector and linked to the end of a gene coding for the bacterial enzyme β-galactosidase, so that the insulin polypeptide was coproduced with the enzyme. Two different *E. coli* bacterial cultures were used, one to produce each of the insulin polypeptide chains. The polypeptides were then recovered from the bacteria, separated from the β-galactosidase, and chemically joined to make human insulin. This accomplishment was one of the early commercial successes of genetic engineering, and it illustrates a number of the principles and procedures discussed in this chapter.

Another human hormone that is now being produced commercially by genetic engineering of *E. coli* is somatostatin. At one time 500,000 sheep brains were needed to produce 5 mg of animal somatostatin for experimental purposes. By contrast, only 8 liters of a genetically engineered bacterial culture are now required to obtain the equivalent amount of the human hormone.

Tissue-plasminogen activator (t-PA) and streptokinase dissolve blood clots. Consequently, they accelerate recovery from heart attack and help prevent recurrences, particularly if they are administered very soon after an initial attack. Tissue-plasminogen activator, a product of genetic engineering, is produced by mammalian cell culture; streptokinase, an older product, is a natural product of bacteria.

Subunit vaccines, consisting only of a protein portion of a pathogen, are being made by genetically engineering yeasts. Subunit vaccines have been produced for a number of diseases, notably hepatitis B. One of the advantages of a subunit vaccine is that there is no chance of becoming infected from the vaccine. The protein is harvested from engineered cells and purified for use as a vaccine. Animal viruses such as vaccinia virus can be engineered to carry a gene for another microbe's surface protein. When injected, the virus acts as a vaccine against the other microbe. Table 9.1 lists some other important genetically engineered products used in medical therapy.

The importance of recombinant DNA technology to medical research cannot be emphasized enough. For example, the cloning of HIV genes has formed the basis of current subunit vaccine testing (see the box in Chapter 19, p. 553). Moreover, the *protease inhibitors*, drugs that prevent HIV maturation, resulted from a detailed study of the HIV protease gene.

The rapid advances in genetic engineering of different organisms have resulted in a wide variety of medically related products and procedures. Artificial blood for use in transfusions can now be prepared using human hemoglobin produced in genetically engineered pigs. Sheep have also been genetically engineered to produce a number of drugs in their milk. This procedure has no apparent effect upon the sheep, and they provide a ready source of raw material for the engineered product that does not require sacrificing animals.

Gene therapy may eventually provide cures for some genetic diseases. It is possible to imagine removing some cells from a person and transforming them with a normal gene to replace a defective or mutated gene. When these cells are returned to the person, they should function normally. For example, gene therapy has been used to treat hemophilia B and severe combined immunodeficiency. Adenoviruses and retroviruses are used most often to deliver genes; however, some researchers are working with plasmid vectors. The first gene therapy to treat hemophilia in humans was done in 1999. An attenuated retrovirus was used as the vector. Several gene therapy trials are in progress using genetically engineered adenovirus carrying human gene *p53* to treat a variety of cancers. The *p53* gene, which encodes a tumor-suppressing protein, is the most frequently mutated gene in cancer cells.

The number of gene therapy trials will increase as technical improvements are made and initial attempts are successful. However, there is a great deal of preliminary work to do, and cures may not be possible for all genetic diseases. Antisense DNA (see page 269) introduced into cells is also being explored to treat hepatitis, cancers, and one type of coronary artery disease.

The Human Genome Project

Learning Objective

* Discuss the value of the Human Genome Project.

One monumental project involving much of the new technology is the Human Genome Project. The goal of this project is to map the 35,000–70,000 genes in human DNA and to sequence the entire genome, approximately 3 billion nucleotide pairs. Thousands of people in 18 countries participated in this project. Researchers collected blood (female) or sperm (male) samples from a large number of donors. Only a few samples were processed as DNA resources, and the source names are protected so that neither donors nor scientists know whose samples were used. Development of shotgun sequencing (discussed on page 265 and shown in Figure 9.14) greatly speeded the process, and the genome is nearly complete. Researchers are filling in the gaps and identifying and mapping all of the genes encoded by this DNA. This is a daunting task. One reference points out that if DNA were as wide as railroad tracks, it would be a million miles long. Moreover, only about 10% of the 3 billion nucleotide pairs actually encode a functional product—the other 90% include introns, the chromosome ends (called telomeres), and other nonencoding

TABLE 9.1	**Some Pharmaceutical Products of Genetic Engineering**
Product	**Comments**
Alpha-interferon	Therapy for leukemia, melanoma, and hepatitis; produced by *Escherichia coli* and *Saccharomyces cerevisiae* (yeast).
Antitrypsin	Assists emphysema patients; produced by genetically modified sheep.
Beta-interferon	Treatment for multiple sclerosis; produced by mammalian cell culture.
Bone morphogenic proteins	Induces new bone formation; useful in healing fractures and reconstructive surgery; produced by mammalian cell culture.
Colony-stimulating factor (CSF)	Counteracts effects of chemotherapy; improves resistance to infectious disease such as AIDS; treatment of leukemia; produced by *E. coli* and *S. cerevisiae*.
Epidermal growth factor (EGF)	Heals wounds, burns, ulcers; produced by *E. coli*.
Erythropoietin (EPO)	Treatment of anemia; produced by mammalian cell culture.
Factor VIII	Treatment of hemophilia; improves clotting; produced by mammalian cell culture.
Gamma-interferon	Treatment of chronic granulomatous disease; produced by *E. coli*.
Hepatitis B vaccine	Produced by *S. cerevisiae* that carries hepatitis-virus gene on a plasmid.
Human growth hormone (hGH)	Corrects growth deficiencies in children; produced by *E. coli*.
Human insulin	Therapy for diabetes; better tolerated than insulin extracted from animals; produced by *E. coli*.
Influenza vaccine	Trial vaccine made from *E. coli* or *S. cerevisiae* carrying virus genes.
Interleukins	Regulate the immune system; possible treatment for cancer; produced by *E. coli*.
Monoclonal antibodies	Possible therapy for cancer and transplant rejection; used in diagnostic tests; produced by mammalian cell culture (from fusion of cancer cell and antibody-producing cell).
Orthoclone	Monoclonal antibody used in transplant patients to help suppress the immune system, reducing the chance of tissue rejection; produced by mouse cells.
Prourokinase	Anticoagulant; therapy for heart attacks; produced by *E. coli* and yeast.
Pulmozyme (rhDNase)	Enzyme used to break down mucous secretions in cystic fibrosis patients; produced by mammalian cell culture.
Relaxin	Used to ease childbirth; produced by *E. coli*.
Superoxide dismutase (SOD)	Minimizes damage caused by oxygen free radicals when blood is resupplied to oxygen-deprived tissues; produced by *S. cerevisiae* and *Pichia pastoris* (yeast).
Taxol	Plant product used for treatment for ovarian cancer; produced in *E. coli*.
Tissue plasminogen activator (Activase)	Dissolves the fibrin of blood clots; therapy for heart attacks; produced by mammalian cell culture.
Tumor necrosis factor (TNF)	Causes disintegration of tumor cells; produced by *E. coli*.

sequences. To date, chromosomes 21 and 22 have been completely mapped.

The next goal of researchers is the Human Proteome Project, which will map all the proteins expressed in human cells. Even before it is completed, however, it will yield data that are of immense value to our understanding of biology. It will also eventually be of great medical benefit, especially for the diagnosis and treatment of genetic diseases.

Scientific Applications

Learning Objectives

- *Define the following terms:* random shotgun sequencing, bioinformatics, proteomics.
- *Diagram the Southern blotting procedure, and provide an example of its use.*
- *Diagram DNA fingerprinting, and provide an example of its use.*

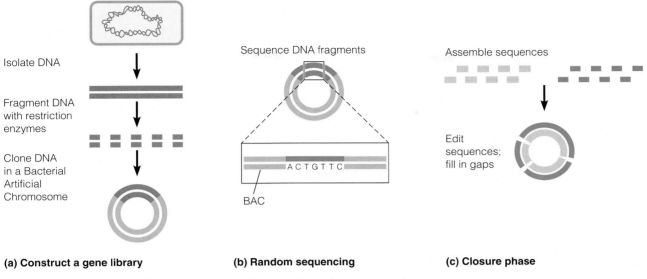

(a) Construct a gene library

Isolate DNA

Fragment DNA
with restriction
enzymes

Clone DNA
in a Bacterial
Artificial
Chromosome

(b) Random sequencing

Sequence DNA fragments

ACTGTTC

BAC

(c) Closure phase

Assemble sequences

Edit
sequences;
fill in gaps

FIGURE 9.14 **Random shotgun sequencing.** In this technique a genome is cut into pieces, and each piece is sequenced. Then the pieces are fit together. There may be gaps if a specific DNA fragment was not sequenced.

■ Does this technique identify genes and their locations?

Recombinant DNA technology can be used to make products, but this is not its only important application. Because of its ability to produce many copies of DNA, it can serve as a sort of DNA "printing press." Once a large amount of a particular piece of DNA is available, various analytic techniques, discussed in this section, can be used to "read" the information contained in the DNA.

What kind of information can be obtained from cloned DNA? One kind is provided by the process of **DNA sequencing**—the determination of the exact sequence of nucleotide bases in DNA.

One technique for genome sequencing is **random shotgun sequencing.** Small pieces of a genome are sequenced, and the sequences are then assembled using a computer. Any gaps between the pieces then have to be found and sequenced (Figure 9.14). The sequences of entire viral genomes are now relatively easy to obtain. The genomes of *Saccharomyces cerevisiae* yeast, *E. coli*, and over 40 other microbes have been mapped; another 100 are in progress. The published maps are between 70% and 99% complete, with some gaps remaining to be filled. The Human Genome Project was a massive undertaking resulting in sequencing the human genome (see page 263). Most of the gaps are repeated sequences that do not encode genes. For example, the *S. cerevisiae* genome is 93% complete, but the gene encoding regions are 100% complete. Computer applications can also be used to look for protein encoding regions, which can then be "translated" by computer software.

DNA sequencing has produced an enormous amount of information that has spawned the new field of **bioinformatics,** the science of understanding the function of genes through computer-assisted analysis. DNA sequences are stored in web-based databases referred to as GenBank. Genomic information can be searched with computer programs to find specific sequences or to look for similar patterns in the genomes of different organisms. Microbial genes are now being searched to identify molecules that are the virulence factors of pathogens. By comparing genomes, researchers discovered that *Chlamydia trachomatis* produces a toxin similar to that of *Clostridium difficile*. The next goal is to identify proteins encoded by these genes. **Proteomics** is the science of determining all of the proteins expressed in a cell. The Institute for Genomic Research (TIGR) is developing new techniques that can be used to quickly analyze gene function.

An example of the use of human DNA sequencing is the identification and cloning of the mutant gene that causes cystic fibrosis (CF). CF is characterized by the oversecretion of mucus, leading to blocked respiratory passageways. The sequence of the mutated gene can be used as a diagnostic tool in a hybridization technique called **Southern blotting** (Figure 9.15), named for Ed Southern, who developed the technique in 1975.

In this technique,

❶ human DNA is first digested with a restriction enzyme, yielding thousands of fragments of various sizes.

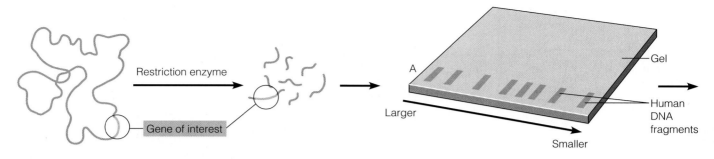

1 DNA containing the gene of interest is extracted from human cells and cut into fragments by restriction enzymes.

2 The fragments are separated according to size by gel electrophoresis. Each band consists of many copies of a particular DNA fragment. The bands are invisible but can be made visible by staining.

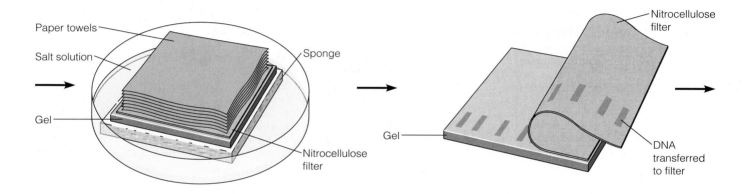

3 The DNA bands are transferred to a nitrocellulose filter by blotting. The solution passes through the gel and filter to the paper towels.

4 This produces a nitrocellulose filter with DNA fragments positioned exactly as on the gel.

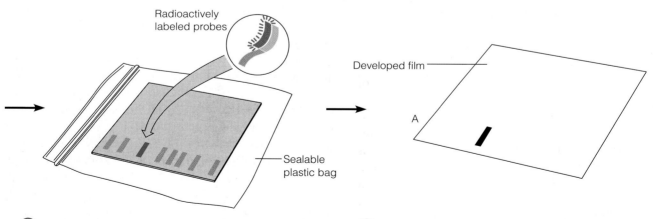

5 The filter is exposed to a radioactively labeled probe for a specific gene. The probe will base-pair (hybridize) with a short sequence present on the gene.

6 The filter is then exposed to X-ray film. The fragment containing the gene of interest is identified by a band on the developed film.

FIGURE 9.15 **Southern blotting.**

■ What is the purpose of Southern blotting?

The different fragments are then separated by **gel electrophoresis.**

❷ The fragments are put in a well at one end of a layer of agarose gel. Then an electrical current is passed through the gel. While the charge is applied, the different-sized pieces of DNA migrate through the gel at different rates. The fragments are called **RFLPs,** for restriction fragment length polymorphisms.

❸–❹ The separated fragments are transferred onto a filter by blotting.

❺ The fragments on the filter are then exposed to a radioactive probe made from the cloned gene of interest, in this case the CF gene. The probe will hybridize to this mutant gene but not to the normal gene.

❻ Fragments to which the probe bind are identified by exposing the filter to X-ray film. With this method, any person's DNA can be tested for the presence of the mutated gene.

This process, called **genetic screening,** can now be used to screen for several hundred genetic diseases. Such screening procedures can be performed on prospective parents and also on fetal tissue. Two of the more commonly screened genes are those associated with inherited forms of breast cancer and the gene responsible for Huntington's disease.

Several important diagnostic tools are now available as a result of genetic engineering, many of which rely on the technique of hybridization. Recall that this procedure enables the identification of a particular DNA sequence among many others. This is exactly what is necessary in most diagnostic situations—the identification of a particular pathogen among many others. Viral infections can often be diagnosed through hybridization: A probe specific for a particular viral gene is synthesized and used in Southern blotting procedures to determine whether that virus is present in the blood or tissue of a patient. Genetic engineering has also been important in the development of ELISA diagnostic tools (see Chapter 18, page 522).

RFLPs are also used in a method of identification known as **DNA fingerprinting,** which can identify bacterial or viral pathogens (Figure 9.16). DNA fingerprinting is also used in forensic medicine to determine paternity or to prove that blood on a murder suspect's clothes came from the murder victim. Southern blotting requires substantial amounts of DNA. As mentioned earlier, small samples of DNA can be quickly amplified, or increased to quantities large enough for analysis, by PCR.

DNA can often be extracted from preserved and fossilized materials, including mummies and extinct plants and animals. Although such material is very rare, and usually partially degraded, PCR enables researchers to study this

FIGURE 9.16 DNA fingerprints used to track an infectious disease. This figure shows the relationship among DNA patterns of bacterial isolates from an outbreak of *Escherichia coli* O157:H7. The isolates from apple juice are identical to the patterns of isolates from patients who drank the contaminated juice but different from those from patients whose infections were not juice-related. (CDC)

■ DNA fingerprints of bacteria that cause foodborne diseases are kept in a database at the Centers for Disease Control and Prevention.

genetic material that no longer exists in its natural form. The study of unusual organisms has also led to advances in basic taxonomy; this will be discussed in Chapter 10.

DNA probes such as the ones used to screen gene libraries are promising tools for the rapid identification of microorganisms. For use in medical diagnosis, these probes are derived from the DNA of a pathogenic microbe and are labeled (with a radioactive tag, for example). The probe then serves a diagnostic function by combining with the DNA of the pathogen to reveal its location in body tissue (or perhaps its presence in food). Probes are also being used in nonmedical aspects of microbiology—for example, for locating and identifying specific microbes in soil. We will discuss PCR and DNA probes further in Chapter 10.

Agricultural Applications

Learning Objective

- *Outline genetic engineering with* Agrobacterium.

The process of selecting for genetically desirable plants has always been a time-consuming one. Performing conventional plant crosses is laborious and involves waiting

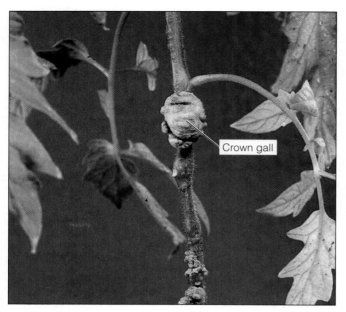

FIGURE 9.17 **Crown gall disease on a tomato plant.** The tumorlike growth is stimulated by a gene on the Ti plasmid carried by the bacterium *Agrobacterium tumefaciens,* which has infected the plant.

■ What are some of the agricultural applications of recombinant DNA technology?

for the planted seed to germinate and for the plant to mature. Plant breeding has been revolutionized by the use of plant cells grown in culture. Clones of plant cells, including cells that have been genetically altered by recombinant DNA techniques, can be grown in large numbers. These cells can then be induced to regenerate whole plants, from which seeds can be harvested.

Recombinant DNA can be introduced into plant cells in several ways. Previously we mentioned protoplast fusion and the use of DNA-coated "bullets." The most elegant method, however, makes use of a plasmid called the **Ti plasmid** (Ti stands for tumor-inducing), which occurs naturally in the bacterium *Agrobacterium tumefaciens*. This bacterium infects certain plants, in which the Ti plasmid causes the formation of a tumorlike growth called a crown gall (Figure 9.17). A part of the Ti plasmid, called T-DNA, integrates into the genome of the infected plant. The T-DNA stimulates local cellular growth (the crown gall) and simultaneously causes the production of certain products used by the bacteria as a source of nutritional carbon and nitrogen.

For plant scientists, the attraction of the Ti plasmid is that it provides a vehicle for introducing genetically engineered DNA into a plant (Figure 9.18). A scientist can insert foreign genes into the T-DNA, put the recombinant

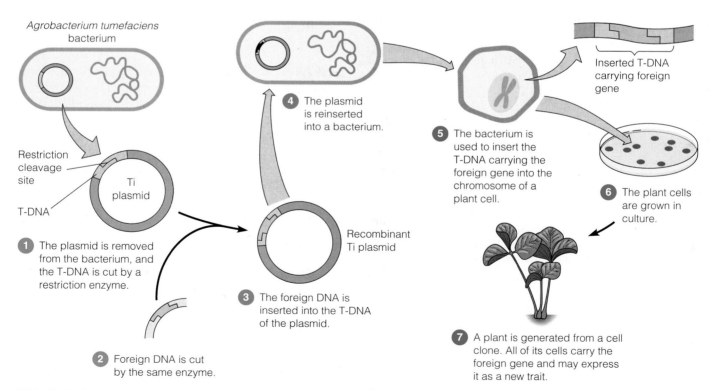

FIGURE 9.18 **Using the Ti plasmid as a vector for genetic engineering in plants.**

■ Why is the Ti plasmid important to biotechnology?

plasmid back into the *Agrobacterium* cell, and use the bacterium to insert the recombinant Ti plasmid into a plant cell. The plant cell with the foreign gene can then be used to generate a new plant. With luck, the new plant will express the foreign gene. Unfortunately, *Agrobacterium* does not naturally infect grasses, so it cannot be used to improve grains such as wheat, rice, or corn.

Noteworthy accomplishments of this approach are the introduction into plants of resistance to the herbicide glyphosate, and a *Bacillus thuringiensis*–derived insecticidal toxin (Bt). Normally, the herbicide kills both weeds and useful plants by inhibiting an enzyme necessary for making certain essential amino acids. *Salmonella* bacteria happen to have this enzyme, and some salmonellae have a mutant enzyme that is resistant to the herbicide. When the DNA for this enzyme is introduced into a crop plant, the crop becomes resistant to the herbicide, which then kills only the weeds. There is now a variety of plants in which different herbicide and pesticide resistances have been engineered. Resistance to drought, viral infection, and several other environmental stresses has also been engineered into crop plants. *Bacillus thuringiensis* bacteria are pathogenic to some insects because they produce a protein called Bt toxin that interferes with the insect digestive tract. The Bt gene has been inserted into a variety of crop plants, including cotton and potatoes, so insects that eat the plants will be killed.

Another example involves MacGregor tomatoes, which stay firm after harvest because the gene for polygalacturonase (PG), the enzyme that breaks down pectin, is suppressed. The suppression was accomplished by **antisense DNA technology.** First, a length of DNA complementary to the PG mRNA is synthesized. This antisense DNA is taken up by the cell and binds to the mRNA to inhibit translation. The DNA-RNA hybrid is broken down by the cell's enzymes, freeing the antisense DNA to disable another mRNA.

Perhaps the most exciting potential use of genetic engineering in plants concerns nitrogen fixation, the ability to convert the nitrogen gas in the air to compounds that living cells can use. The availability of such nitrogen-containing nutrients is usually the main factor limiting crop growth. But in nature, only certain bacteria have genes for carrying out this process. Some plants, such as alfalfa, benefit from a symbiotic relationship with these microbes. Species of the symbiotic bacterium *Rhizobium* have already been genetically engineered for enhanced nitrogen fixation. In the future, *Rhizobium* strains may be designed that can colonize such crop plants as corn and wheat, perhaps eliminating their requirement for nitrogen fertilizer. The ultimate goal would be to introduce functioning nitrogen-fixation genes directly into the plants.

Although this goal cannot be achieved with our current knowledge, work toward it will continue because of its potential for dramatically increasing the world's food supply.

An example of a genetically engineered bacterium now in agricultural use is *Pseudomonas fluorescens* that has been engineered to produce a toxin normally produced by *Bacillus thuringiensis*. This toxin kills certain plant pathogens, such as the European corn borer. The genetically altered *Pseudomonas*, which produces much more toxin than *B. thuringiensis*, can be added to plant seeds and in time will enter the vascular system of the growing plant. Its toxin is ingested by the feeding borer larvae and kills them (but is harmless to humans and other warm-blooded animals).

Animal husbandry has also benefited from genetic engineering. We have seen how one of the early commercial products of genetic engineering was human growth hormone. By similar methods it is possible to manufacture bovine growth hormone (bGH). When bGH is injected into beef cattle, it increases their weight gain; in dairy cows, it also causes a 10% increase in milk production. Such procedures have met with resistance from consumers, especially in Europe, primarily as a result of as-yet unsubstantiated fears that some of the bGH would be present in the milk or meat of these cattle and might be harmful to humans.

Table 9.2 lists these and several other genetically engineered products used in agriculture and animal husbandry.

Safety Issues and the Ethics of Genetic Engineering

Learning Objective

- List the advantages of, and problems associated with, the use of genetic engineering techniques.

There will always be concern about the safety of any new technology, and genetic engineering and biotechnology are certainly no exceptions. One reason for this concern is that it is nearly impossible to prove that something is entirely safe under all conceivable conditions. People worry that the same techniques that can alter a microbe or plant to make them useful to humans could also inadvertently make them pathogenic to humans or otherwise dangerous to living organisms, or could create an ecological nightmare. Therefore, laboratories engaged in recombinant DNA research must meet rigorous standards of control to avoid either accidentally releasing genetically engineered organisms into the environment or exposing humans to any risk of infection. To reduce risk further, microbiologists engaged in genetic engineering often delete from the microbes' genomes certain genes that are essential for

TABLE 9.2	**Some Agriculturally Important Products of Genetic Engineering**
Product	**Comments**
Agricultural Products	
Bt cotton and Bt corn	Plants have toxin-producing gene from *Bacillus thuringiensis;* toxin kills insects that eat plants.
MacGregor tomatoes	Antisense gene blocks pectin degradation so fruits have longer shelf life.
Pseudomonas fluorescens bacterium	Has toxin-producing gene from insect pathogen *B. thuringiensis;* toxin kills root-eating insects that ingest bacteria.
Pseudomonas syringae, ice-minus bacterium	Lacks normal protein product that initiates undesirable ice formation on plants.
Rhizobium meliloti bacterium	Modified for enhanced nitrogen fixation.
Round-Up (glyphosate)-resistant crops	Plants have bacterial gene; allows use of herbicide on weeds without damaging crops.
Animal Husbandry Products	
Bovine growth hormone (bGH)	Improves weight gain and milk production in cattle; produced by *E. coli.*
Porcine growth hormone (pGH)	Improves weight gain in swine; produced by *E. coli.*
Transgenic animals	Genetically alters animals to produce medically useful products in their milk.
Other Food Production Products	
Cellulase	Enzymes that degrade cellulose to make animal feedstocks; produced by *E. coli.*
Rennin	Causes formation of milk curds for dairy products; produced by *Aspergillus niger.*

growth in environments outside the laboratory. Genetically modified organisms intended for use in the environment (in agriculture, for example) may be engineered to contain "suicide genes"—genes that eventually turn on to produce a toxin that kills the microbes, thus ensuring that they will not survive in the environment for very long after they have accomplished their task.

The safety issues in agricultural biotechnology are similar to those concerning chemical pesticides: toxicity to humans and to nonpest species. Although not shown to be harmful, genetically engineered foods have not been popular with consumers. In 1999, researchers in Ohio noticed that humans may develop allergies to *Bacillus thuringiensis* (Bt) toxin after working in fields sprayed with the insecticide. And an Iowa study showed that the caterpillar stage of Monarch butterflies could be killed by ingesting windblown Bt-carrying pollen that landed on milkweed, the caterpillars' normal food. Crop plants can be genetically engineered for herbicide resistance so that fields can be sprayed to eliminate weeds without killing the desired crop. However, if the engineered plants pollinate related weed species, weeds could become resistant to herbicides, making it more diffi-

cult to control unwanted plants. An unanswered question is whether releasing genetically modified organisms will alter evolution as genes move to wild species.

These developing technologies also raise a variety of moral and ethical issues. If genetic screening for diseases becomes routine, who should have access to this information? Should employers or insurance companies have the right to know the results of such tests? Restricting access to such information will be very difficult, which raises questions concerning the right to privacy. How can we be assured that such information will not be used to discriminate against certain groups?

Applications of genetic screening techniques are not limited to adults. The ability to diagnose a genetic disease in a fetus adds even more controversy to the abortion debate. Genetic counseling, which provides advice and counseling to prospective parents with family histories of genetic disease, is becoming more important in considerations about whether or not to have children. As more is learned about genetic causes for various diseases, such as cancer or Huntington's disease, reproductive choices might become more difficult for some families.

What extra burdens does genetic engineering place on our already overtaxed health care system? Genetic screening and gene therapy are expensive procedures, and we need to consider how they will be delivered to the public as the technology develops. Will a sufficient number of genetic counselors be available to work with people? Will expensive medical cures and treatments be available only to those who can afford them?

There are probably just as many harmful applications of a new technology as there are helpful ones. It is particularly easy to imagine genetic engineering being used to develop new and powerful biological weapons. In addition, because such research efforts are performed under top-secret conditions, it is virtually impossible for the general public to learn of them.

Perhaps more than most new technologies, genetic engineering holds the promise of affecting human life in previously unimaginable ways. It is important that society and individuals be given every opportunity to understand the potential impact of these new developments.

Like the invention of the microscope, the development of recombinant DNA techniques is causing profound changes in science, agriculture, and human health care. With this technology only slightly more than 30 years old, it is difficult to predict exactly what changes will occur. However, it is likely that within another 30 years, many of the treatments and diagnostic methods discussed in this book will have been replaced by far more powerful techniques based on the unprecedented ability to manipulate DNA precisely.

STUDY OUTLINE

Introduction to Biotechnology (pp. 249–251)

1. Biotechnology is the use of microorganisms, cells, or cell components to make a product.

Recombinant DNA Technology (p. 249)

1. Closely related organisms can exchange genes in natural recombination.
2. Genes can be transferred among unrelated species via laboratory manipulation, called genetic engineering.
3. Recombinant DNA is DNA that has been artificially manipulated to combine genes from two different sources.

An Overview of Recombinant DNA Technologies (pp. 249–251)

1. A desired gene is inserted into a DNA vector, such as a plasmid or a viral genome.
2. The vector inserts the DNA into a new cell, which is grown to form a clone.
3. Large quantities of the gene product can be harvested from the clone.

Tools of Biotechnology (pp. 251–256)

Selection (p. 252)

1. Microbes with desirable traits are selected for culturing by artificial selection.

Mutation (p. 252)

1. Mutagens are used to cause mutations that might result in a microbe with desirable traits.

2. Site-directed mutagenesis is used to change a specific codon in a gene.

Restriction Enzymes (p. 252)

1. Prepackaged kits are available for many genetic engineering techniques.
2. A restriction enzyme recognizes and cuts only one particular nucleotide sequence in DNA.
3. Some restriction enzymes produce sticky ends, short stretches of single-stranded DNA at the ends of the DNA fragments.
4. Fragments of DNA produced by the same restriction enzyme will spontaneously join by base pairing. DNA ligase can covalently link the DNA backbones.

Vectors (pp. 253–254)

1. Shuttle vectors are plasmids that can exist in several different species.
2. A plasmid containing a new gene can be inserted into a cell by transformation.
3. A virus containing a new gene can insert the gene into a cell.

Polymerase Chain Reaction (pp. 254–256)

1. The polymerase chain reaction (PCR) is used to make multiple copies of a desired piece of DNA enzymatically.
2. PCR can be used to increase the amounts of DNA in samples to detectable levels. This may allow sequencing of genes, the diagnosis of genetic diseases, or the detection of viruses.

Techniques of Genetic Engineering (pp. 256–262)

Inserting Foreign DNA into Cells (pp. 256–257)

1. Cells can take up naked DNA by transformation. Chemical treatments are used to make cells that are not naturally competent to take up DNA.

2. Pores made in protoplasts and animal cells by electric current in the process of electroporation can provide entrance for new pieces of DNA.

3. Protoplast fusion is the joining of cells whose cell walls have been removed.

4. Foreign DNA can be introduced into plant cells by shooting DNA-coated particles into the cells.

5. Foreign DNA can be injected into animal cells by using a fine glass micropipette.

Obtaining DNA (pp. 257–259)

1. Gene libraries can be made by cutting up an entire genome with restriction enzymes and inserting the fragments into bacterial plasmids or phages.

2. cDNA made from mRNA by reverse transcription can be cloned in gene libraries.

3. Synthetic DNA can be made in vitro by a DNA synthesis machine.

Selecting a Clone (pp. 259–261)

1. Antibiotic-resistance markers on plasmid vectors are used to identify cells containing the engineered vector by direct selection.

2. In blue-white screening, the vector contains the genes for amp^R and β-galactosidase.

3. The desired gene is inserted into the β-galactosidase gene site, destroying the gene.

4. Clones containing the recombinant vector will be resistant to ampicillin and unable to hydrolyze X-gal (white colonies). Clones containing the vector without the new gene will be blue. Clones lacking the vector will not grow.

5. Clones containing foreign DNA can be tested for the desired gene product.

6. A short piece of labeled DNA called a DNA probe can be used to identify clones carrying the desired gene.

Making a Gene Product (pp. 261–262)

1. E. coli is used to produce proteins by genetic engineering because it is easily grown and its genomics are well understood.

2. Efforts must be made to ensure that E. coli's endotoxin does not contaminate a product intended for human use.

3. To recover the product, E. coli must be lysed or the gene must be linked to a gene that produces a naturally secreted protein.

4. Yeasts can be genetically engineered and are likely to continuously secrete a gene product.

5. Mammalian cells can be engineered to produce proteins such as hormones for medical use.

6. Plant cells can be engineered and used to produce plants with new properties.

Applications of Genetic Engineering (pp. 262–269)

1. Cloned DNA is used to produce products, study the cloned DNA, and alter the phenotype of an organism.

Therapeutic Applications (pp. 262–263)

1. Synthetic genes linked to the β-galactosidase gene (lacZ) in a plasmid vector were inserted into E. coli, allowing E. coli to produce and secrete the two polypeptides used to make human insulin.

2. Cells can be engineered to produce a pathogen's surface protein, which can be used as a subunit vaccine.

3. Animal viruses can be engineered to carry a gene for a pathogen's surface protein. When the virus is used as a vaccine, the host develops an immunity to the pathogen.

4. Gene therapy can be used to cure genetic diseases by replacing the defective or missing gene.

5. Genetic engineering techniques were used to map the human genome through the Human Genome Project.

6. This will provide tools for diagnosis and possibly the repair of genetic diseases.

Scientific Applications (pp. 264–267)

1. Recombinant DNA techniques can be used to increase understanding of DNA, for genetic fingerprinting, and for gene therapy.

2. DNA sequencing machines are used to determine the nucleotide base sequence of restriction fragments in random shotgun sequencing.

3. Bioinformatics is the use of computer applications to study genetic data; proteomics is the study of a cell's proteins.

4. Southern blotting can be used to locate a gene in a cell.

5. Southern blotting is used in DNA fingerprinting to identify bacterial or viral pathogens.

6. DNA probes can be used to quickly identify a pathogen in body tissue or food.

Agricultural Applications (pp. 267–269)

1. Cells from plants with desirable characteristics can be cloned to produce many identical cells. These cells can then be used to produce whole plants from which seeds can be harvested.

2. Plant cells can be engineered by using the Ti plasmid vector. The tumor-producing T genes are replaced with desired genes, and the recombinant DNA is inserted into Agrobacterium. The bacterium naturally transforms its plant hosts.

3. Genes for glyphosate resistance, Bt toxin, and pectinase suppression have been engineered into crop plants.

4. *Rhizobium* has been engineered for enhanced nitrogen fixation.

5. *Pseudomonas* has been engineered to produce *Bacillus thuringiensis* toxin against insects.

6. Bovine growth hormone is being produced by *E. coli*.

Safety Issues and the Ethics of Genetic Engineering (pp. 269–271)

1. Strict safety standards are used to avoid the accidental release of genetically engineered microorganisms.

2. Some microbes used in genetic engineering have been altered so that they cannot survive outside the laboratory.

3. Microorganisms intended for use in the environment may be engineered to contain suicide genes so that the organisms do not persist in the environment.

4. Genetic technology raises ethical questions such as: Should employers and insurance companies have access to a person's genetic records? Will some people be targeted for either breeding or sterilization? Will genetic counseling be available to everyone?

5. Genetically engineered crops must be safe for consumption and for release in the environment.

STUDY QUESTIONS

Review

1. Differentiate recombinant DNA from genetic engineering.

2. Compare and contrast the following terms:
   a. cDNA and gene
   b. restriction fragment and gene
   c. DNA probe and gene
   d. DNA polymerase and DNA ligase
   e. recombinant DNA and cDNA
   f. genome and proteome

3. How is each of the following used in genetic engineering?
   a. plasmid
   b. viral genome
   c. antibiotic-resistance genes
   d. restriction enzyme

4. Differentiate between a gene library and synthetic DNA.

5. Differentiate among the following terms. Which one is "hit and miss"—that is, does not add a specific gene to a cell?
   a. protoplast fusion
   b. gene gun
   c. microinjection
   d. electroporation

6. Some commonly used restriction enzymes are listed in the following table. The cutting site is indicated by ↓. Indicate which enzymes produce sticky ends. Of what value are sticky ends in making recombinant DNA?

Enzyme	Bacterial Source	Recognition Sequence
BamHI	*Bacillus amyloliquefaciens*	G↓G A T C C G C T A G↑G
EcoRI	*Escherichia coli*	G↓A A T T C C T T A A↑G
HaeIII	*Haemophilus aegyptius*	G G↓C C C C↑G G
HindIII	*Haemophilus influenzae*	A↓A G C T T T T C G A↑A

7. Suppose you want multiple copies of a gene you have synthesized. How would you obtain the necessary copies by cloning? By PCR?

8. Describe a genetic engineering experiment in two or three sentences. Use the following terms: intron, exon, DNA, mRNA, cDNA, RNA polymerase, reverse transcriptase.

9. List at least two examples of the use of genetic engineering in medicine and in agriculture.

10. You are attempting to insert a gene for saltwater tolerance into a plant by using the Ti plasmid. In addition to the desired gene, you add a gene for tetracycline resistance (tet^R) to the plasmid. What is the purpose of the tet^R gene?

Multiple Choice

1. Restriction enzymes were first discovered with the observation that
   a. DNA is restricted to the nucleus.
   b. phage DNA is destroyed in a host cell.
   c. foreign DNA is kept out of a cell.
   d. foreign DNA is restricted to the cytoplasm.
   e. all of the above

2. The DNA probe, 3'GGCTTA, will hybridize with DNA containing
   a. 5'CCGUUA.
   b. 5'CCGAAT.
   c. 5'GGCTTA.
   d. 3'CCGAAT.
   e. 3'GGCAAU.

3. Which of the following is the fourth basic step of genetic engineering?
   a. transformation
   b. ligation
   c. plasmid cleavage
   d. restriction-enzyme digestion of gene
   e. isolation of gene

4. The following steps are used to make cDNA. What is the second step?
   a. reverse transcription
   b. RNA processing to remove introns
   c. transcription
   d. translation

5. If you put a gene in a virus, the next step in genetic engineering would be
   a. insertion of a plasmid.   d. PCR.
   b. transformation.   e. Southern blotting.
   c. transduction.

6. You have a small gene that you want replicated by PCR. You add radioactively labeled nucleotides to the PCR thermalcycler. After three replication cycles, what percentage of the DNA single-strands are radioactively labeled?
   a. 0%
   b. 12.5%
   c. 50%
   d. 87.5%
   e. 100%

Match the following choices to the statements in questions 7 through 10.
   a. antisense
   b. clone
   c. library
   d. Southern blot
   e. vector

7. Pieces of human DNA stored in yeast cells.

8. A population of cells carrying a desired plasmid.

9. Self-replicating DNA for transmitting a gene from one organism to another.

10. A gene that hybridizes with mRNA.

Critical Thinking

1. Using the following map of plasmid pMICRO, give the number of restriction fragments that would result from digesting pMICRO with *Eco*RI, *Hind*III, and both enzymes together. Which enzyme makes the smallest fragment containing the tetracycline-resistance gene?

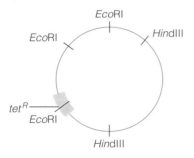

2. Design an experiment using vaccinia virus to make a vaccine against the AIDS virus (HIV).

3. Why did the use of DNA polymerase from the bacterium *Thermus aquaticus* allow researchers to add the necessary reagents to tubes in a preprogrammed heating block?

4. The following picture shows bacterial colonies growing on X-gal plus ampicillin in a blue-white screening test. Which colonies have the recombinant plasmid? The small satellite colonies do not have the plasmid. Why did they start growing on the medium 48 hours after the larger colonies?

Clinical Applications

1. PCR has been used to examine oysters for the presence of *Vibrio cholerae*. Oysters from different areas were homogenized, and DNA was extracted from the homogenates. The DNA was digested by the restriction enzyme *Hinc*II. A primer for the hemolysin gene of *V. cholerae* was used for the PCR reaction. After PCR, each sample was electrophoresed and stained with a probe for the hemolysin gene. Which of the oyster samples were (was) positive for *V. cholerae*? How can you tell? Why look for *V. cholerae* in oysters? What is the advantage of PCR over conventional biochemical tests to identify the bacteria?

2. Using the restriction enzyme *Eco*RI, the following gel electrophoresis patterns were obtained from digests of various DNA molecules from a transformation experiment. Can you conclude from these data that transformation occurred? Explain why or why not.

LEARNING WITH TECHNOLOGY

The Microbiology Place CD-ROM/website includes many resources to help you study and succeed in this course. In addition to self-quizzes, web links, and flashcards, you'll find the following activities for this chapter in the "Activities" section of The Microbiology Place:

  Cloning a Gene

  Cloning a Gene: Practice

  Cloning DNA Using PCR

  Identifying Organisms Using PCR

Chapter 10

Classification of Microorganisms

CHROMATIUM. *Purple sulfur bacteria use hydrogen sulfide as an electron donor and store globules of sulfur in their cells.*

The Study of Phylogenetic Relationships

Learning Objectives

- Define taxonomy, taxon, and phylogeny.
- Discuss the limitations of a two-kingdom classification system.
- Identify the contributions of Linnaeus, von Nägeli, Chatton, Whittaker, and Woese.

The science of classification, especially the classification of living forms, is called **taxonomy** (from the Greek for orderly arrangement). The objective of taxonomy is to classify living organisms—that is, to establish the relationships between one group of organisms and another and to differentiate between them. There may be as many as 100 million different living organisms, but fewer than 10% have been discovered, much less classified and identified.

Taxonomy also provides a common reference for identifying organisms already classified. For example, when a bacterium suspected of causing a specific disease is isolated from a patient, characteristics of that isolate are matched to lists of characteristics of previously classified bacteria to identify the isolate (see the box on page 286). Finally, taxonomy is a basic and necessary tool for scientists, providing a universal language of communication.

Modern taxonomy is an exciting and dynamic field. New techniques in molecular biology and genetics are providing new insights into classification and evolution. In this chapter, you will learn the various classification systems, the different criteria used for classification, and tests that are used to identify microorganisms that have already been classified.

In 2001, an international project called the All Species Inventory was launched. The project's purpose is to identify and record every species of life on Earth in the next 25 years. These researchers have undertaken a challenging goal: Whereas biologists have to date identified more than 1.5 million different organisms, it is estimated that the number of living species ranges from 7 to 100 million.

Among these many and diverse organisms, however, are many similarities. For example, all organisms are composed of cells surrounded by a plasma membrane, use ATP for energy, and store their genetic information in DNA. These similarities are the result of evolution, or descent from a common ancestor. In 1859, the English naturalist Charles Darwin proposed that natural selection was responsible for the similarities as well as the differences among organisms. The differences can be attributed to the survival of organisms with traits best suited to a particular environment.

To facilitate research, scholarship, and communication, we arrange organisms into taxonomic categories, or **taxa** (singular: *taxon*), to show degrees of similarities among organisms. These similarities are due to relatedness—all organisms are related through evolution. **Systematics,** or

phylogeny, is the study of the evolutionary history of organisms. The hierarchy of taxa reflects evolutionary or *phylogenetic* relationships.

From the time of Aristotle, living organisms were categorized in just two ways, as either plants or animals. In 1735, the Swedish botanist Carolus Linnaeus introduced a formal system of classification dividing living organisms into two kingdoms—Plantae and Animalia. He used latinized names to provide one common "language" for systematics. As the biological sciences developed, however, biologists began looking for a *natural* classification system—one that groups organisms based on ancestral relationships and allows us to see the order in life. In 1857, Carl von Nägeli, a contemporary of Pasteur, proposed that bacteria and fungi be placed in the plant kingdom. In 1866, Ernst Haeckel proposed the Kingdom Protista, to include bacteria, protozoa, algae, and fungi. Because of disagreements over the definition of protists, for the next 100 years biologists continued to follow von Nägeli's placement of bacteria and fungi in the plant kingdom. It is ironic that recent DNA sequencing places fungi closer to animals than plants. Fungi were placed in their own kingdom in 1959.

With the advent of electron microscopy, the physical differences between cells became apparent. The term *prokaryote* was introduced in 1937 by Edouard Chatton to distinguish cells having no nucleus from the nucleated cells of plants and animals. In 1961, Roger Stanier provided the current definition of prokaryotes: cells in which the nuclear material (nucleoplasm) is not surrounded by a nuclear membrane. In 1968, Robert G.E. Murray proposed the Kingdom Prokaryotae.

In 1969, Robert H. Whittaker founded the five-kingdom system in which prokaryotes were placed in the Kingdom Prokaryotae, or Monera, and eukaryotes comprised the other four kingdoms. The Kingdom Prokaryotae had been based on microscopic observations. Subsequently, new techniques in molecular biology revealed that there are actually two types of prokaryotic cells and one type of eukaryotic cell.

The Three Domains

Learning Objectives

- *Discuss the advantages of the three-domain system.*
- *List the characteristics of the Bacteria, Archaea, and Eukarya domains.*

The discovery of three cell types was based on the observations that ribosomes are not the same in all cells (see Chapter 4, page 92). Ribosomes provide a method of comparing cells because ribosomes are present in all cells.

Comparing the sequences of nucleotides in ribosomal RNA (rRNA) from different kinds of cells shows that there are three distinctly different cell groups: the eukaryotes and two different types of prokaryotes—the bacteria and the archaea.

In 1978, Carl R. Woese proposed elevating the three cell types to a level above kingdom, called domain. Woese believed that the archaea and the bacteria, although similar in appearance, should form their own separate domains on the evolutionary tree (Figure 10.1). In this widely accepted scheme, animals, plants, fungi, and protists are kingdoms in the Domain **Eukarya.** Organisms are classified by cell type in the three domain systems. In addition to differences in rRNA, the three domains differ in membrane lipid structure, transfer RNA molecules, and sensitivity to antibiotics (Table 10.1 on page 279).

The Domain **Bacteria** includes all of the pathogenic prokaryotes as well as many of the nonpathogenic prokaryotes found in soil and water. The photoautotrophic prokaryotes are also in this domain. The Domain **Archaea** includes prokaryotes that do not have peptidoglycan in their cell walls. They often live in extreme environments and carry out unusual metabolic processes. Archaea include three major groups:

1. The methanogens, strict anaerobes that produce methane (CH_4) from carbon dioxide and hydrogen.

2. Extreme halophiles, which require high concentrations of salt for survival.

3. Hyperthermophiles, which normally grow in hot environments.

The evolutionary relationship of the three domains is the subject of current research by biologists. Originally, archaea were thought to be the most primitive group, whereas bacteria were assumed to be more closely related to eukaryotes. However, studies of rRNA indicate that a universal ancestor split into three lineages. That split led to the Archaea, the Bacteria, and what eventually became the nucleoplasm of the eukaryotes. The oldest known fossils are the remains of prokaryotes that lived more than 3.5 billion years ago. Eukaryotic cells evolved more recently, about 1.4 billion years ago. According to the endosymbiotic theory, eukaryotic cells evolved from prokaryotic cells living inside one another, as endosymbionts (see Chapter 4, page 106). In fact, the similarities between prokaryotic cells and eukaryotic organelles provide striking evidence for this endosymbiotic relationship (Table 10.2 on page 279).

The original nucleoplasmic cell was prokaryotic. However, infoldings in its plasma membrane may have surrounded the nuclear region to produce a true nucleus

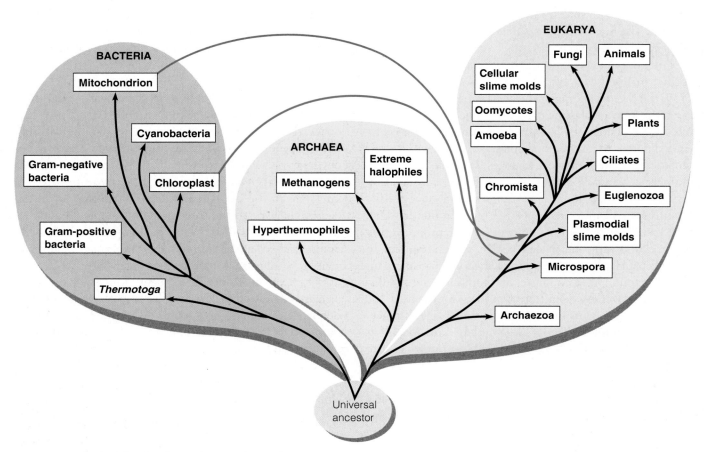

FIGURE 10.1 **The three-domain system.** The system recognizes the three types of cells. The Domain Eukarya includes the Kingdoms Fungi, Plantae, and Animalia. It also includes the protists discussed on page 284.

■ What characteristics are used to differentiate the three domains?

(Figure 10.2). That cell provided the original host in which endosymbiotic bacteria developed into organelles. An example of a modern prokaryote living in a eukaryotic cell is shown in Figure 10.3 on page 280. The cyanobacteriumlike cell and the eukaryotic host require each other for survival.

In sequencing the genome of a prokaryote called *Thermotoga maritima*, microbiologist Karen Nelson has discovered that this species has genes similar to members of both the Domain Bacteria and the Domain Archaea. Her findings suggest that *Thermotoga* is one of the earliest cells, arising before the Bacteria and Archaea split apart. For this reason, *Thermotoga* is referred to as one of the "deeply branching genera," that is, it is near the origin or "root" of the evolutionary tree.

Taxonomy provides tools for clarifying the evolution of organisms, as well as their interrelationships. New organisms are being discovered every day, and taxonomists continue to search for a natural classification system that reflects phylogenetic relationships.

A Phylogenetic Hierarchy

In a phylogenetic hierarchy, grouping organisms according to common properties implies that a group of organisms evolved from a common ancestor; each species retains some of the characteristics of the ancestor. Some of the information used to classify and determine phylogenetic relationships in higher organisms comes from fossils. Bones, shells, or stems that contain mineral matter or have left imprints in rock that was once mud are examples of fossils.

The structures of most microorganisms are not readily fossilized. Some exceptions are the following:

- A marine protist whose fossilized colonies form the White Cliffs of Dover, England.

- Stromatolites, the fossilized remains of filamentous bacteria and sediments that flourished between 0.5 and 2 billion years ago.

TABLE 10.1	**Some Characteristics of Archaea, Bacteria, and Eukarya**

	Archaea	**Bacteria**	**Eukarya**
	Methanosarcina SEM ⊢ 10 µm	*E. coli* SEM ⊢ 1 µm	*Amoeba* SEM ⊢ 10 µm
Cell Type	Prokaryotic	Prokaryotic	Eukaryotic
Cell Wall	Varies in composition; contains no peptidoglycan	Contains peptidoglycan	Varies in composition; contains carbohydrates
Membrane Lipids	Composed of branched carbon chains attached to glycerol by ether linkage	Composed of straight carbon chains attached to glycerol by ester linkage	Composed of straight carbon chains attached to glycerol by ester linkage
First Amino Acid in Protein Synthesis	Methionine	Formylmethionine	Methionine
Antibiotic Sensitivity	No	Yes	No
rRNA Loop*	Lacking	Present	Lacking
Common Arm of tRNA†	Lacking	Present	Present

*Binds to ribosomal protein; found in all bacteria.
†A sequence of bases in tRNA found in all eukaryotes and bacteria: guanine-thymine-pseudouridine-cytosine-guanine.

TABLE 10.2	**Prokaryotic Cells and Eukaryotic Organelles Compared**

	Prokaryotic Cell	**Eukaryotic Cell**	**Eukaryotic Organelles (Mitochondria and Chloroplasts)**
DNA	Circular	Linear	Circular
Histones	No	Yes	No
Ribosomes	70S	80S	70S
Growth	Binary fission	Mitosis	Binary fission

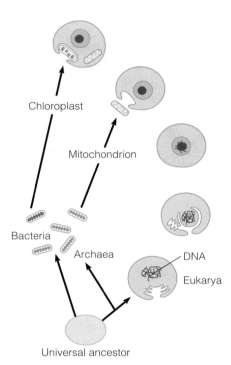

FIGURE 10.2 **A model of the origin of eukaryotes.** Invagination of the plasma membrane may have formed the nuclear envelope and endoplasmic reticulum. Similarities, including rRNA sequences, indicate that endosymbiotic prokaryotes gave rise to mitochondria and chloroplasts.

■ What features do chloroplasts, mitochondria, and bacteria have in common?

• Fossilized cyanobacteria found in rocks in Western Australia that are 3.0–3.5 billion years old. These are the oldest known fossils. Some fossils of prokaryotes are shown in Figure 10.4.

Because fossil evidence is not available for most prokaryotes, their phylogeny must be based on other evidence. But in one notable exception, scientists may have isolated living bacteria and yeast 25–40 million years old. In 1995, the American microbiologist Raul Cano and his colleagues reported growing *Bacillus sphaericus* and other as yet unidentified microorganisms that had survived embedded in amber (fossilized plant resin) for millions of years. If confirmed, this discovery should provide more information about the evolution of microorganisms.

Conclusions from rRNA sequencing and DNA hybridization studies (discussed on page 294) of selected orders and families of eukaryotes are in agreement with the fossil records. This has encouraged workers to use DNA hybridization and rRNA sequencing to gain an understanding of the evolutionary relationships among prokaryotic groups.

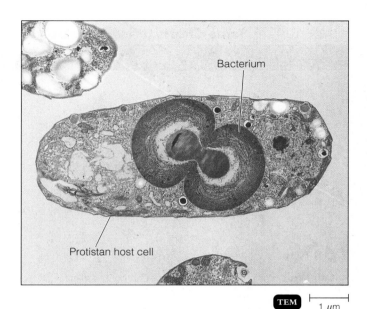

FIGURE 10.3 **Cyanophora paradoxa.** This organism, in which the eukaryotic host and the bacterium require each other for survival, provides a modern example of how eukaryotic cells might have evolved.

■ Eukaryotic organelles evolved from prokaryotic cells living inside host prokaryotic cells.

Classification of Organisms

Learning Objective

• Differentiate among eukaryotic, prokaryotic, and viral species.

Living organisms are grouped according to similar characteristics (classification), and each organism is assigned a unique scientific name. The rules for classifying and naming, which are used by biologists worldwide, are discussed next.

Scientific Nomenclature

Learning Objective

• Explain why scientific names are used.

In a world inhabited by millions of living organisms, biologists must be sure they know exactly which organism is being discussed. We cannot use common names, because the same name is often used for many different organisms in different locales. For example, there are two different organisms with the common name Spanish moss, and neither one is actually a moss. Plus, local languages are used for common names. Because common names can be misleading and are in different languages, a system of scientific names, referred to as *scientific nomenclature*, was developed in the eighteenth century.

(a) SEM ⊢——⊣ 2 μm (b) TEM ⊢——⊣ 10 μm

FIGURE 10.4 **Fossilized prokaryotes. (a)** Coccoid cyanobacteria from the Late Precambrian (850 million years ago) of central Australia. **(b)** Filamentous prokaryotes from the Early Precambrian (3.5 billion years ago) of western Australia.

■ Because fossil evidence is not available for most prokaryotes, their phylogeny must be based on other kinds of evidence.

Recall from Chapter 1 (page 2) that every organism is assigned two names, or a binomial. These names are the **genus** name and **specific epithet (species),** and both names are printed underlined or italicized. The genus name is always capitalized and is always a noun. The species name is lowercase and is usually an adjective. Because this system gives two names to each organism, the system is called **binomial nomenclature.**

Let's consider some examples. Our own genus and specific epithet are *Homo sapiens* (hō′mō sā′pē-ens). The noun, or genus, means man; the adjective, or specific epithet, means wise. A mold that contaminates bread is called *Rhizopus nigricans* (rī′zō-pŭs nī′gri-kans). *Rhizo-* (root) describes rootlike structures on the fungus; *nigr-* (black) identifies the color of its spore sacs. Table 10.3 contains more examples.

Binomials are used by scientists worldwide, regardless of their native language, which enables them to share knowledge efficiently and accurately. Several scientific entities are responsible for establishing rules governing the naming of organisms. Rules for assigning names for protozoa and parasitic worms are published in the *International Code of Zoological Nomenclature*. Rules for assigning names for fungi and algae are published in the *International Code of Botanical Nomenclature*. Rules for naming newly classified prokaryotes and for assigning prokaryotes to taxa are established by the International Committee on Systematic Bacteriology and are published in the *Bacteriological Code*. Descriptions of prokaryotes and evidence for their classifications are published in the *International Journal of Systematic Bacteriology* before being incorporated into a reference

called *Bergey's Manual*. According to the *Bacteriological Code*, scientific names are to be taken from Latin (a genus name can be taken from Greek) or latinized by the addition of the appropriate suffix. Suffixes for order and family are *-ales* and *-aceae*, respectively.

As new laboratory techniques make more detailed characterizations of microbes possible, two genera may be reclassified as a single genus, or a genus may be divided into two or more genera. For example, the genera "Diplococcus" and *Streptococcus* were combined in 1974; the only diplococcal species is now called *Streptococcus pneumoniae*. (The *Bacteriological Code* states that the older genus name should be retained in such cases.) In 1984, DNA hybridization studies indicated that "Streptococcus faecalis" and "Streptococcus faecium" were only distantly related to the other streptococcal species; consequently, a new genus called *Enterococcus* was created, and these species were renamed *E. faecalis* and *E. faecium* because the rules require that the original specific epithets be retained.

Making the transition to a new name can be confusing, so the old name is often written in parentheses. For example, a physician looking for information on the cause of a patient's fever and eye irritation (cat-scratch disease) would find the bacterial name *Bartonella (Rochalimaea) henselae* (bär tō-nel′lä rō-chä-lē′mä-ä hen′sel-ī).

Obtaining the name of the organism is important in order to know what treatment to use; antifungal drugs will not work against bacteria, and antibacterial drugs will not work against viruses.

TABLE 10.3	**Making Scientific Names Familiar**

Use the word roots guide inside the book's front and back covers to find out what the name means. The name will not seem so strange if you translate it. When you encounter a new name, practice saying it out loud. The exact pronunciation is not as important as the familiarity you will gain. Guidelines for pronunciation are given in Appendix E.

Following are some examples of microbial names you may encounter in the popular press as well as in the lab.

	Pronunciation	Source of Genus Name	Source of Specific Epithet
Klebsiella pneumoniae (bacterium)	kleb-sē-el′lä nü-mō′nē-ī	Honors bacteriologist Edwin Klebs	The disease it causes
Pfiesteria piscicida (alga)	fes-tėr′ē-ä pi′si-si-dä	Honors dinoflagellate biologist Lois Pfiester	Causes disease in fish (*piscis*)
Salmonella typhimurium (bacterium)	sal-mōn-el′lä tī-fi-mur′ē-um	Honors public health micro-biologist Daniel Salmon	Causes stupor (*typh-*) in mice (*muri-*)
Streptococcus pyogenes (bacterium)	strep-tō-kok′kus pī-äj′en-ēz	Appearance of cells in chains (*strepto-*)	Forms pus (*pyo-*)
Saccharomyces cerevisiae (yeast)	sak-ä-rō-mī′ses se-ri-vis′ē-ī	Fungus (*-myces*) that uses sugar (*saccharo-*)	Makes beer (*cerevisia*)
Penicillium notatum (fungus)	pen-i-sil′lē-um nō-tä′tum	Tuftlike or paintbrush (*penicill-*) appearance microscopically	Spores easily spread in air (*notus*, wind)
Trypanosoma cruzi (protozoan)	tri-pa-nō-sō′mä krūz′ē	Corkscrew (*trypano-*, borer; *soma-*, body)	Honors epidemiologist Oswaldo Cruz

The Taxonomic Hierarchy

Learning Objectives

- *List the major taxa.*
- *Differentiate between culture, clone, and strain.*

All organisms can be grouped into a series of subdivisions that make up the taxonomic hierarchy. Linnaeus developed this hierarchy for his classification of plants and animals. A **eukaryotic species** is a group of closely related organisms that breed among themselves. (Bacterial species will be discussed shortly.) A genus consists of species that differ from each other in certain ways but are related by descent. For example, *Quercus* (kwer′kus), the genus name for oak, consists of all types of oak trees (white oak, red oak, bur oak, velvet oak, and so on). Even though each species of oak differs from every other species, they are all related genetically. Just as a number of species make up a genus, related genera make up a **family.** A group of similar families constitutes an **order,** and a group of similar orders makes up a **class.** Related classes, in turn, make up a **phylum.** (In botany, the comparable term *division* has been used; it is equivalent to *phylum*.) Thus, a particular organism (or species) has a genus name and specific epithet and belongs to a family, order, class, and phylum or division.

All phyla or divisions that are related to each other make up a **kingdom,** and related kingdoms are grouped into a **domain.** (See Figure 10.5.)

Classification of Prokaryotes

The taxonomic classification scheme for prokaryotes is found in *Bergey's Manual of Systematic Bacteriology,* 2nd edition (see Appendix A). The first volume has been published, with the remaining four volumes to follow over the next few years. The contents of each volume are shown in Table 10.4. In *Bergey's Manual,* prokaryotes are divided into two domains: Bacteria and Archaea. Each domain is divided into phyla. Remember, the classification is based on similarities in nucleotide sequences in rRNA. Classes are divided into orders; orders, into families; families, into genera; and genera, into species.

A prokaryotic species is defined somewhat differently than a eukaryotic species, which is a group of closely related organisms that can interbreed. Unlike reproduction in eukaryotic organisms, cell division in bacteria is not directly tied to sexual conjugation, which is infrequent and does not always need to be species-specific. A **prokaryotic species,** therefore, is defined simply as a population of cells with similar characteristics. (The types of

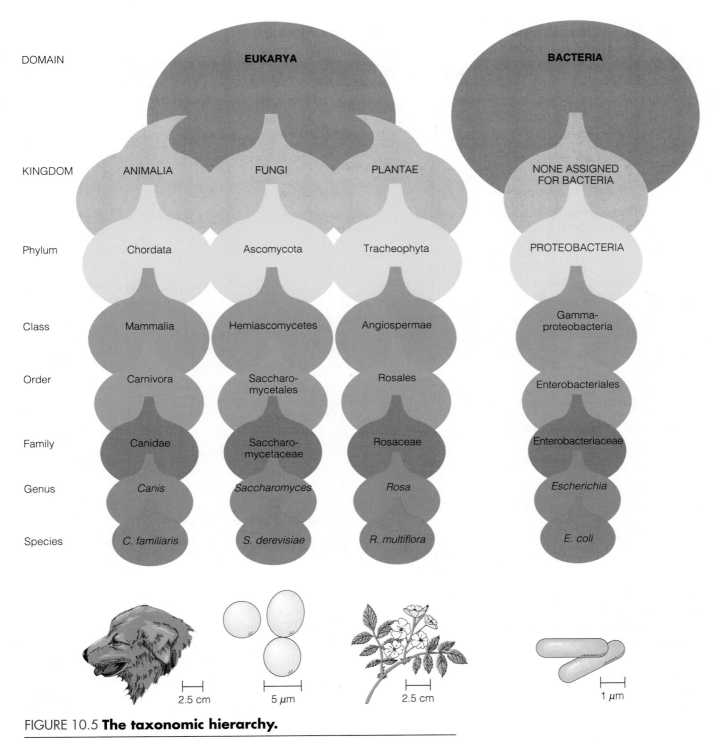

	EUKARYA			BACTERIA
DOMAIN	**EUKARYA**			**BACTERIA**
KINGDOM	ANIMALIA	FUNGI	PLANTAE	NONE ASSIGNED FOR BACTERIA
Phylum	Chordata	Ascomycota	Tracheophyta	PROTEOBACTERIA
Class	Mammalia	Hemiascomycetes	Angiospermae	Gamma-proteobacteria
Order	Carnivora	Saccharo-mycetales	Rosales	Enterobacteriales
Family	Canidae	Saccharo-mycetaceae	Rosaceae	Enterobacteriaceae
Genus	*Canis*	*Saccharomyces*	*Rosa*	*Escherichia*
Species	*C. familiaris*	*S. derevisiae*	*R. multiflora*	*E. coli*

2.5 cm   5 µm   2.5 cm   1 µm

FIGURE 10.5 **The taxonomic hierarchy.**

▪ What is the biological definition of a family?

characteristics will be discussed later in the chapter.) The members of a bacterial species are essentially indistinguishable from each other but are distinguishable from members of other species, usually on the basis of several features. As you know, bacteria grown at a given time in media are called a culture. A pure culture is often a **clone,** that is, a population of cells derived from a single parent cell. All cells in the clone should be identical. However, in some cases, pure cultures of the same species are not identical in all ways. Each such group is called a **strain.** Strains are identified by numbers, letters, or names that follow the specific epithet.

TABLE 10.4	**Classification of Prokaryotes (*Bergey's Manual of Systematic Bacteriology*, 2nd edition)**	
Volume	**Contents**	**Notes**
1	Archaea, cyanobacteria, phototrophs, and deeply branching genera	Includes the Domain Archaea and some gram-negative Bacteria
2	Proteobacteria	Phylum Proteobacteria; these are related gram-negative bacteria
3	Low G + C gram-positives	Phylum Firmicutes (gram-positive cell wall); Phylum Mycoplasmas (wall-less)
4	High G + C gram-positives	Includes actinomycetes
5	Chlamydiae, Spirochaetes, Bacteroidetes, and Fusobacteria	These are distinct phyla of bacteria, each with a unique rRNA sequence

Bergey's Manual provides a reference for identifying bacteria in the laboratory, as well as a classification scheme for bacteria. One scheme for the evolutionary relationships of bacteria is shown in Figure 10.6. Characteristics used to classify and identify bacteria are discussed in Chapter 11.

Classification of Eukaryotes

Learning Objectives

- List the major characteristics used to differentiate the three kingdoms of multicellular Eukarya.

- Define protist.

Some kingdoms in the Domain Eukarya are shown in Figure 10.1 (page 278).

Since 1969, simple eukaryotic organisms, mostly unicellular, have been grouped as the Kingdom **Protista,** a catchall kingdom for a variety of organisms. Historically, eukaryotic organisms that didn't fit into other kingdoms were placed in the Protista. Ribosomal RNA sequencing is making it possible to divide protists into groups based on their descent from common ancestors. Many biologists propose dividing Protista into five new kingdoms. For convenience, we will continue to use the term *protist* to refer to unicellular eukaryotes and their close relatives. These organisms will be discussed in Chapter 12.

Fungi, plants, and animals make up the three kingdoms of more complex eukaryotic organisms, most of which are multicellular.

The Kingdom **Fungi** includes the unicellular yeasts, multicellular molds, and macroscopic species such as mushrooms. To obtain raw materials for vital functions, a fungus absorbs dissolved organic matter through its plasma membrane. The cells of a multicellular fungus are commonly joined to form thin tubes called *hyphae*. The hyphae are usually divided into multinucleated units by cross-walls

that have holes, so that cytoplasm can flow between the cell-like units. Fungi develop from spores or from fragments of hyphae.

The Kingdom **Plantae** (plants) includes some algae and all mosses, ferns, conifers, and flowering plants. All members of this kingdom are multicellular. To obtain energy, a plant uses photosynthesis, the process that converts carbon dioxide and water into organic molecules used by the cell.

The kingdom of multicellular organisms called **Animalia** (animals) includes sponges, various worms, insects, and animals with backbones (vertebrates). Animals obtain nutrients and energy by ingesting organic matter through a mouth of some kind.

Classification of Viruses

Viruses are not classified as part of any of the three domains. Viruses are not composed of cells, and they use the anabolic machinery within living host cells to multiply. A viral genome can direct biosynthesis inside a host cell, and some viral genomes can become incorporated into the host genome. The ecological niche of a virus is its specific host cell, so viruses may be more closely related to their hosts than to other viruses. The International Committee on Taxonomy of Viruses defines a **viral species** as a population of viruses with similar characteristics (including morphology, genes, and enzymes) that occupies a particular ecological niche.

Viruses are obligatory intracellular parasites, so they must have evolved after a suitable host cell had evolved. There are two hypotheses on the origin of viruses: (1) They arose from independently replicating strands of nucleic acids (such as plasmids), and (2) they developed from degenerative cells that, through many generations, gradually lost the ability to survive independently but could survive when associated with another cell. Viruses will be discussed in Chapter 13.

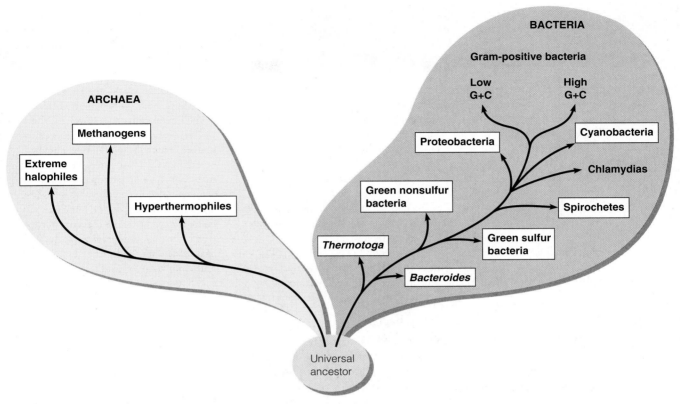

FIGURE 10.6 **Phylogenetic relationships of prokaryotes.** Arrows indicate major lines of descent of bacterial groups. Selected phyla are indicated by the white boxes.

■ Members of which phylum can be identified by Gram staining?

Methods of Classifying and Identifying Microorganisms

Learning Objectives

- *Compare and contrast classification and identification.*
- *Explain the purpose of* Bergey's Manual.

A classification scheme provides a list of characteristics and a means for comparison to aid in the identification of an organism. Once an organism is identified, it can be placed into a previously devised classification scheme. Microorganisms are *identified* for practical purposes—for example, to determine an appropriate treatment for an infection. They are not necessarily identified by the same techniques by which they are *classified*. Most identification procedures are easily performed in a laboratory and use as few procedures or tests as possible. Protozoa, parasitic worms, and fungi can usually be identified microscopically. Most prokaryotic organisms do not have distinguishing morphological features or even much variation in size and shape. Consequently, microbiologists have developed a va-

riety of methods to test metabolic reactions and other characteristics to identify prokaryotes.

Bergey's Manual of Determinative Bacteriology has been a widely used reference since the first edition was published in 1923. *Bergey's Manual of Determinative Bacteriology* (9th ed., 1994) does not classify bacteria according to evolutionary relatedness but provides identification (determinative) schemes based on such criteria as cell wall composition, morphology, differential staining, oxygen requirements, and biochemical testing.* The majority of Bacteria and Archaea have not been cultured, and scientists estimate that only 1% of these microbes have been discovered.

Medical microbiology (the branch of microbiology dealing with human pathogens) has dominated the interest in microbes, and this interest is reflected in many identification schemes. However, to put the pathogenic

---

*Both *Bergey's Manual of Systematic Bacteriology* (see page 282) and *Bergey's Manual of Determinative Biology* are referred to simply as *Bergey's Manual;* the complete titles are used when the information under discussion is found in one but not the other, for example, an identification table.

Mass Deaths of Marine Mammals Spur Veterinary Microbiology

A record number of manatees died in Florida waters during 1996. Over half the deaths were due to pneumonia. In 2000, Canadian biologists reported a high rate of opportunistic bacterial infections in beluga whales. Microbiologists are trying to determine the causes of these diseases.

Another potentially alarming instance of unusual disease in marine mammals occurred among the dolphin population. During an average year, 10 or 12 dead dolphins are found washed ashore on beaches between New Jersey and Virginia. However, in one recent year the death toll climbed to more than 200, and scientists believe that hundreds more may have died offshore. The current decline in the Southern sea otter population is the result of a 40% mortality rate due to a variety of infectious bacterial diseases. These mortality figures raise concerns that entire populations of marine mammals may ultimately be destroyed. Moreover, alarmed swimmers have asked whether they are also in danger from marine mammal diseases.

Large numbers of opportunistic pathogens, including 55 species of *Vibrio*, were found in the dolphins. These bacteria are a part of a dolphin's normal microbiota and the biota of coastal waters. They can cause disease only if the animals' immune system, their normal defense against infection, has been weakened. To find the ultimate cause of death, scientists must find out what has weakened the immune systems.

One possibility is chemical pollutants. Insecticides and polychlorinated biphenyls (PCBs) have been found in dolphins, manatees, and elephant seals. Daniel Martineau of Cornell University says that PCBs are strong immunosuppressants. Another possibility is a viral infection that affects the immune system. In 2000, the cause of thousands of Caspian seal deaths during a 5-month period was the canine distemper virus.

Information Is Scarce

Such questions are the concern of veterinary microbiology, which until recently has been a neglected branch of medical microbiology. Although the diseases of

such animals as cattle, chickens, and mink have been studied, partly because of their availability to researchers, the microbiology of wild animals, especially marine mammals, is a relatively newly emerging field. Gathering samples of animals that live in the open ocean and performing bacteriological analyses on them are very difficult. Currently, the animals being studied are those that live in captivity (see the photograph) and those that come onto the shore to breed, such as the northern fur sea lion.

The scientists are identifying bacteria in marine mammals by using conventional test batteries (see the figure) and genomic data on known species. The bacteria are compared with species described in *Bergey's Manual* to assign names or identify them. Perhaps new species of bacteria will be found in the marine mammals.

Veterinary microbiologists hope that increased study of the microbiology of wild animals, including marine mammals, will not only promote improved wildlife management but also provide models for the study of human diseases.

---

properties of bacteria in perspective, of the more than 2600 species listed in the *Approved Lists of Bacterial Names*, fewer than 10% are human pathogens.

We next discuss several criteria and methods for the classification of microorganisms and the routine identification of some of those organisms. In addition to properties of the organism itself, the source and habitat of a bacterial isolate is considered as part of the classification and identification processes.

Morphological Characteristics

Morphological (structural) characteristics have helped taxonomists classify organisms for 200 years. Higher organisms are frequently classified according to observed anatomical detail. But many microorganisms look too similar to be classified by their structures. Through a microscope, organisms that might differ in metabolic or physiological properties may look alike. Literally hundreds of bacterial species are small rods or small cocci.

Larger size and the presence of intracellular structures does not always mean easy classification, however. *Pneumocystis* (nü-mō-sis′tis) pneumonia is the most common opportunistic infection in immunosuppressed individuals and is a significant cause of death in AIDS patients. The causative agent of this infection, *P. jiroveci* (ye-rō′vet-zē) [formerly *P. carinii* (kär-i′nē-ē)] was not considered a human pathogen until the 1970s. *Pneumocystis* lacks structures that can be easily used for identification, and its taxonomic position has been uncertain since its discovery in 1909. Although it was provisionally classified as a protozoan, recent studies comparing its rRNA sequence with those of other protozoa, *Euglena*, cellular slime molds, plants, mammals, and fungi have shown that it is actually a member of the Kingdom Fungi. Researchers have not been able to culture *Pneumocystis*, but they have developed some useful treatments for *Pneumocystis* pneumonia. Perhaps as researchers take into account this organism's relatedness to fungi, appropriate culture methods and treatments will result.

Mass Deaths of Marine Mammals Spur Veterinary Microbiology
(continued)

Marine mammal researchers examine a Pacific bottlenosed dolphin.

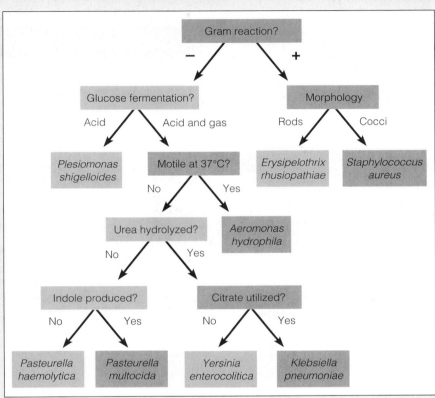

Biochemical tests used to identify selected species of human pathogens isolated from marine mammals.

- Assume you isolated a gram-negative rod that produces gas from glucose, is urease negative and indole positive. What is the bacterium?

Cell morphology tells us little about phylogenetic relationships. However, morphological characteristics are still useful in identifying bacteria. For example, differences in such structures as endospores or flagella can be helpful.

Differential Staining

Learning Objective

- *Describe how staining and biochemical tests are used to identify bacteria.*

Recall from Chapter 3 that one of the first steps in identifying bacteria is differential staining. Most bacteria are either gram-positive or gram-negative. Other differential stains, such as the acid-fast stain, can be useful for a more limited group of microorganisms. Recall that these stains are based on the chemical composition of cell walls and therefore are not useful in identifying either the wall-less bacteria or the archaea with unusual walls. Microscopic examination of a Gram stain or an acid-fast stain is used to obtain information quickly in the clinical environment. A physician can sometimes get sufficient information from a technician's lab report to begin appropriate treatment (Figure 10.7).

MICROBIOLOGY REQUISITION	Date:	Time:	Slip prepared by:
Lab: Date, time received:	Physician name:	Collected by:	Patient ID#:

↓ DO NOT WRITE BELOW THIS LINE ↓	**USE SEPARATE SLIP FOR EACH REQUEST**

GRAM STAIN REPORT

☐ GRAM POS. COCCI, GROUPS	☐ NO GROWTH
☐ GRAM POS. COCCI, PAIRS/CHAIN	☐ NO GROWTH IN ___DAYS
☐ GRAM POS. RODS	☐ MIXED MICROBIOTA
☒ GRAM NEG. COCCI	☐ SPECIMEN IMPROPERLY COLLECTED OR TRANSPORTED
☐ GRAM NEG. RODS	
☐ GRAM NEG. COCCOBACILLI	☐ ___DIFFERENT TYPES OF ORGANISMS
☐ YEAST	☐ NEGATIVE FOR *SALMONELLA, SHIGELLA,* AND *CAMPYLOBACTER*
☐ OTHER	☐ NO OVA, CYSTS, OR PARASITES SEEN
	☒ OXIDASE-POSITIVE GRAM-NEGATIVE DIPLOCOCCI
	☐ PRESUMPTIVE BETA STREP GROUP A BY BACITRACIN

SOURCE OF SPECIMEN

☐ BLOOD
☐ CEREBROSPINAL FLUID
☐ FLUID (Specify Source) _____
☐ THROAT
☐ SPUTUM, expectorated
☐ OTHER Respiratory (Describe) _____
☐ URINE, Clean Catch Midstream
☐ URINE, Indwelling Catheter
☐ URINE, Straight Catheter
☐ URINE, Entire First Morning
☐ URINE, Other (Describe) _____
☐ STOOL
☑ GU (Specify Source) ___*vag.*___
☐ ABSCESS (Specify Source) _____
☐ TISSUE (Specify Source) _____
☐ ULCER (Specify Source) _____
☐ WOUND (Specify Source) _____
☐ STERILIZER TEST

TEST(S) REQUESTED

Bacterial
☐ **Routine culture;** Gram stain, anaerobic culture, susceptibility testing. Throats done for Gp A Strep.
☐ *Legionella* culture
☐ *Bartonella*
☐ Blood Culture

Other Non-Routine Cultures
☐ *E. coli* 0157:H7
☐ *Vibrio*
☐ *Yersinia*
☑ *H. ducreyi*
☐ *B. pertussis*
☐ Other _____

Screening Cultures
☑ Gonococci
☐ Group B Strep
☐ Group A Strep
☐ Other _____

☐ **ACID-FAST BACILLI**

☐ **FUNGAL**

VIRAL
☐ Routine culture
☐ Herpes simplex
☐ Direct FA for _____

PARASITOLOGY
☐ Exam for intestinal ova and parasites
☐ *Giardia* immunoassay
☐ *Cryptosporidium*
☐ Pinworm prep
☐ Blood parasites
☐ Filaria concentration
☐ *Trichomonas*
☐ Other _____

TOXIN ASSAY
☐ *Clostridium difficile*

DIRECT (Antigen Detection)
☐ Cryptococcal antigen-CSF only
☐ Bacterial antigens (Specify) _____

SPECIAL
☐ Antimicrobial tests (MIC)

*filled out by one person   *filled out by different person

FIGURE 10.7 **A clinical microbiology lab report form.** In health care, morphology and differential staining are important in determining the proper treatment for microbial diseases. A clinician completes the form to identify the sample and specific tests. In this case, a genitourinary sample will be examined for sexually transmitted diseases. The lab technician reported the Gram stain and culture results. [Minimal inhibitory concentration (MIC) of antibiotics will be discussed in Chapter 20, page 579.]

■ A clinician requests information about patients with infectious diseases on a microbiology requisition.

Biochemical Tests

Enzymatic activities are widely used to differentiate bacteria. Even closely related bacteria can usually be separated into distinct species by subjecting them to biochemical tests, such as one to determine their ability to ferment an assortment of selected carbohydrates. For one example of the use of biochemical tests to identify bacteria (in this instance, in marine mammals), see the box on page 286. Moreover, biochemical tests can provide insight into a species' niche in the ecosystem. For example, a bacterium that can fix nitrogen gas or oxidize elemental sulfur will provide important nutrients for plants and animals.

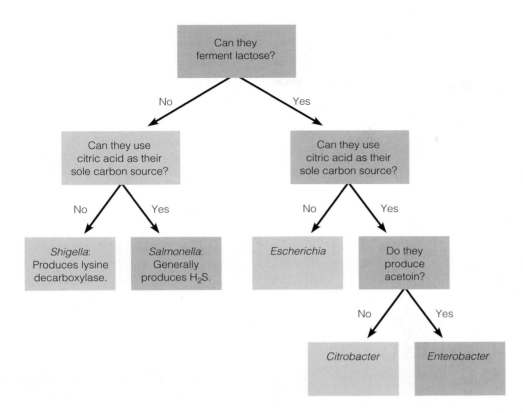

FIGURE 10.8 **The use of metabolic characteristics to identify selected genera of enteric bacteria.**

- Even closely related bacteria can usually be separated into distinct species by biochemical testing.

Enteric, gram-negative bacteria are a large heterogeneous group of microbes whose natural habitat is the intestinal tract of humans and other animals. This family contains several pathogens that cause diarrheal illness. A number of tests have been developed so that technicians can quickly identify the pathogen, a clinician can then provide appropriate treatment, and epidemiologists can locate the source of an illness. All members of the family Enterobacteriaceae are oxidase-negative. Among the enteric bacteria are members of the genera *Escherichia*, *Enterobacter* (en-te-rō-bak′tėr), *Shigella* (shi-gel′lä), *Citrobacter* (sit′rō-bak-tėr), and *Salmonella*. *Escherichia*, *Enterobacter*, and *Citrobacter*, which ferment lactose to produce acid and gas, can be distinguished from *Salmonella* and *Shigella*, which do not. Further biochemical testing, as represented in Figure 10.8, can differentiate among the genera.

The time needed to identify bacteria can be reduced considerably by the use of selective and differential media or by rapid identification methods. Recall from Chapter 6 (page 167) that selective media contain ingredients that suppress the growth of competing organisms and encourage the growth of desired ones, and that differential media allow the desired organism to form a colony that is somehow distinctive.

Bergey's Manual does not evaluate the relative importance of each biochemical test and does not always describe strains. In the clinical diagnosis of a disease, a particular species and even a particular strain must be identified in order to proceed with proper treatment. To this end, specific series of biochemical tests have been developed for fast identification in hospital laboratories. Rapid biochemical systems have been developed for yeasts and other fungi, as well as bacteria.

Rapid identification tools are manufactured for groups of medically important bacteria, such as the enterics. Such tools are designed to perform several biochemical tests simultaneously and can identify bacteria within 4–24 hours. This is sometimes called **numerical identification** because the results of each test are assigned a number. In the simplest form, a positive test would be assigned a value of 1, and a negative is assigned a value of 0. In most commercial testing kits, test results are assigned numbers ranging from 1 to 4 that are based on the relative reliability and importance of each test, and the resulting total is compared to a database of known organisms.

In the example shown in Figure 10.9,

❶ an unknown enteric bacterium is inoculated into a tube designed to perform 15 biochemical tests.

❷ After incubation, results in each compartment are observed.

❸ Positive results are marked on the scoring form. Notice that each test is assigned a value; the number derived from scoring all the tests is called the I.D.

FIGURE 10.9 **One type of rapid identification method for bacteria: Enterotube II from Becton Dickinson.**
❶ One tube containing media for 15 tests is inoculated. ❷ After incubation, the tube is observed for results. ❸ The value for each positive test is circled, and the numbers from each group of tests are added to give each portion of the I.D. value. ❹ Comparing the resultant I.D. value with a computerized listing shows that the organism in the tube is *Enterobacter cloacae*. This example shows results for a typical strain of *E. cloacae*; however, other strains may produce different test results, which are listed in the Atypical Test Results column. The V–P test is used to confirm an identification.

■ In rapid identification methods, test results are weighted according to their relative importance.

ID Value	Organism	Atypical Test Results	Confirmatory Test
32143	*Enterobacter cloacae*	Sorbitol⁻	–
	Enterobacter sakazakii	Urea⁺	+
32161	*Enterobacter cloacae*	None	V–P⁺
32162	*Enterobacter cloacae*	Citrate⁻	

value. Fermentation of glucose is important, and a positive reaction is valued at 2, compared with the production of acetoin (V–P test), which has no value.

❹ A computerized interpretation of the simultaneous test results is essential and is provided by the manufacturer. In this example, the test results indicate that the bacterium is *Enterobacter cloacae*, with the typical test results. A limitation of biochemical testing is that mutations and plasmid acquisition can result in strains with different characteristics. Unless a large number of tests are used, an organism could be incorrectly identified.

Serology

Learning Objectives

- *Differentiate Western blotting from Southern blotting.*
- *Explain how serological tests and phage typing can be used to identify an unknown bacterium.*

Serology is the science that studies blood serum and immune responses that are evident in serum (see Chapter 18). Microorganisms are antigenic; that is, microorganisms that enter an animal's body stimulate it to form antibodies. Antibodies are proteins that circulate in the blood and combine in a highly specific way with the bacteria that caused their production. For example, the immune system of a rabbit injected with killed typhoid bacteria (antigens) responds by producing antibodies against typhoid bacteria. Solutions of such antibodies used in the identification of many medically important microorganisms are commercially available; such a solution is called an **antiserum** (plural: *antisera*). If an unknown bacterium is isolated from a patient, it can be tested against known antisera and often identified quickly.

In a procedure called a **slide agglutination test,** samples of an unknown bacterium are placed in a drop of saline on each of several slides. Then a different known antiserum is added to each sample. The bacteria agglutinate (clump) when mixed with antibodies that were produced in response to that species or strain of bacterium; a positive test is indicated by the presence of agglutination. Positive and negative slide agglutination tests are shown in Figure 10.10.

(a) Positive test   **(b)** Negative test

FIGURE 10.10 **A slide agglutination test. (a)** In a positive test, the grainy appearance is due to the clumping (agglutination) of the bacteria. **(b)** In a negative test, the bacteria are still evenly distributed in the saline and antiserum.

■ Agglutination results when the bacteria are mixed with antibodies that were produced in response to the same strain.

Serological testing can differentiate not only among microbial species but also among strains within species. Strains with different antigens are called **serotypes, serovars,** or **biovars.** See the discussion of *Escherichia* and *Salmonella* serovars on page 314. As mentioned in Chapter 1, Rebecca Lancefield was able to classify serotypes of streptococci by studying serological reactions. She found that the different antigens in the cell walls of various serotypes of streptococci stimulate the formation of different antibodies. In contrast, because closely related bacteria also produce some of the same antigens, serological testing can be used to screen bacterial isolates for possible similarities. If an antiserum reacts with proteins from different bacterial species or strains, these bacteria can be tested further for relatedness.

Serological testing was used to determine whether the increase in number of cases of necrotizing fasciitis in the United States and England since 1987 was due to a common source of the infections. No common source was located, but there has been an increase in two serotypes of *Streptococcus pyogenes* that have been dubbed the flesh-eating bacteria.

A test called the **enzyme-linked immunosorbent assay (ELISA)** is widely used because it is fast and can be read by a computer scanner (Figure 10.11; see Figure 18.12, page 523). In a direct ELISA, known antibodies are placed in the wells of a microplate, and an unknown type of bacterium is added to each well. A reaction between the known antibodies and the bacteria provides identification of the bacteria. An ELISA is used in AIDS testing to detect the presence of antibodies against human immunodeficiency virus (HIV), the virus that causes AIDS (see Figure 19.12 on page 546).

Another serological test, **Western blotting,** is used to identify bacterial antigens in a patient's serum (Figure 10.12). HIV infection is confirmed by Western blotting,

(a)

(b)

FIGURE 10.11 **Using a computer scanner to read the results of an ELISA test. (a)** A technician uses a micropipette to add samples to a microplate for an ELISA. **(b)** ELISA results are then read by the computer scanner.

■ What are the similarities between the slide agglutination test and the ELISA test?

1 If Lyme disease is suspected in a patient, a sample of the patient's serum is taken. Electrophoresis is used to separate proteins in the serum sample. Each band consists of many molecules of a particular protein (antigen). The bands are not visible at this point.

2 The bands are transferred to a nitrocellulose filter by blotting.

3 The proteins (antigens) are positioned on the filter exactly as they were on the gel. The filter is then washed with antibodies against a particular antigen. The antibodies are tagged with an enzyme. The antibodies that combine with their specific antigen are visible (shown here in purple) when the enzyme's

4 The test is read. If the tagged antibodies stick to the filter, evidence of the presence of the microorganism in question—in this case, *Borrelia burgdorferi*—has been found in the patient's serum.

FIGURE 10.12 **The Western blot.** Proteins separated by electrophoresis can be detected by their reactions with antibodies.

■ Name two diseases that may be diagnosed by Western blotting.

and Lyme disease, caused by *Borrelia burgdorferi*, is often diagnosed by the Western blot.

1 Proteins (including bacterial proteins) in the patient's serum are separated by a process called electrophoresis (Chapter 9, page 267).

2 The proteins are then transferred to a filter by blotting.

3 Next, antibodies tagged with an enzyme are washed over the filter. If the specific antigen (in this case, *Borrelia* proteins) is present in the serum, the antibodies will combine with it and will be visible as a colored band on the filter after addition of the enzyme's substrate.

Phage Typing

Like serological testing, phage typing looks for similarities among bacteria. Both techniques are useful in tracing the origin and course of a disease outbreak. **Phage typing** is a test for determining which phages a bacterium is susceptible to. Recall from Chapter 8 (page 238) that bacteriophages (phages) are bacterial viruses and that they usually cause lysis of the bacterial cells they infect. They are highly specialized, in that they usually infect only members of a particular species, or even particular strains within a species. One bacterial strain might be susceptible to two different phages, whereas another strain of the same species might be susceptible to those two phages plus a third phage. Bacteriophages will be discussed further in Chapter 13.

The sources of food-associated infections can be traced by phage typing. One version of this procedure starts with a plate totally covered with bacteria growing on agar. A drop of each different phage type to be used in the test is then placed on the bacteria. Wherever the phages are able to infect and lyse the bacterial cells, clearings in the bacterial growth (called plaques) appear (Figure 10.13). Such a test might show, for instance, that bacteria isolated from a surgical wound have the same pattern of phage sensitivity as those isolated from the operating surgeon or surgical nurses. This establishes that the surgeon or a nurse is the source of infection.

Fatty Acid Profiles

Bacteria synthesize a wide variety of fatty acids, and in general, these fatty acids are constant for a particular species. A commercial system has been designed to separate cellular fatty acids to compare them to fatty acid profiles of known organisms. Fatty acid profiles often need to be backed up by biochemical testing, and they can be used only for identification, not to determine phylogenetic relatedness.

Flow Cytometry

Flow cytometry can be used to identify bacteria in a sample without culturing the bacteria. In a *flow cytometer*, a moving fluid containing bacteria is forced through a small opening (see Figure 18.11). The simplest method detects the presence of bacteria by detecting the difference in electrical conductivity between cells and the surrounding medium. If the fluid passing through the opening is illuminated by a laser, the scattering of light provides information about the cell size, shape, density, and surface, which is analyzed by a computer. Fluores-

FIGURE 10.13 **Phage typing of a strain of Salmonella enterica.** The tested strain was grown over the entire plate. Plaques, or areas of lysis, were produced by bacteriophages, indicating that the strain was sensitive to infection by these phages. Phage typing is used to distinguish *S. enterica* serotypes and *Staphylococcus aureus* types.

■ What is being identified in phage typing?

cence can be used to detect naturally fluorescent cells, such as *Pseudomonas*, or cells tagged with fluorescent dyes.

Milk can be a vehicle for disease transmission. A proposed test that uses flow cytometry to detect *Listeria* in milk could save time because the bacteria would not need to be cultured for identification. Antibodies against *Listeria* can be labeled with a fluorescent dye and added to the milk to be tested. The milk is passed through the flow cytometer, which records the light scattering of the antibody-labeled cells.

DNA Base Composition

Learning Objective

• *Describe how a newly discovered microbe can be classified by: DNA base composition, rRNA sequencing, DNA fingerprinting, PCR, and nucleic acid hybridization.*

A classification technique that has come into wide use among taxonomists is the determination of an organism's **DNA base composition.** This base composition is usually expressed as the percentage of guanine plus cytosine (G + C). The base composition of a single species is theoretically a fixed property; thus, a comparison of the G + C content in different species can reveal the degree of species relatedness. As we saw in Chapter 8, each guanine (G) in DNA has a complementary cytosine (C). Similarly, each adenine (A) in the DNA has a complementary thymine (T). Therefore, the percentage of DNA bases that are GC pairs also tells us the percentage that are AT pairs (GC + AT = 100%). Two organisms that are closely related and hence have many identical or similar genes will have similar amounts of the various bases in their DNA. However, if there is a difference of more than 10% in their percentage of GC pairs (for example, if one bacterium's DNA contains 40% GC and another bacterium has 60% GC), then these two organisms are probably not related. Of course, two organisms that have the same percentage of GC are not necessarily closely related; other supporting data are needed to draw conclusions about their phylogenetic relationship.

DNA Fingerprinting

Actually determining the entire sequence of bases in an organism's DNA is now possible with modern biochemical methods, but this is currently impractical for laboratory identification because of the great amount of time required. However, the use of restriction enzymes enables researchers to compare the base sequences of different organisms. Restriction enzymes cut a molecule of DNA everywhere a specific base sequence occurs, producing restriction fragments (as discussed in Chapter 9, page 252). For example, the enzyme *Eco*RI cuts DNA at the arrows in the sequence

$$...\text{G}^\downarrow\text{A A T T C}...$$
$$...\text{C T T A A}_\uparrow\text{G}...$$

In this technique, the DNA from two microorganisms is treated with the same restriction enzyme, and the restriction fragments produced are separated by electrophoresis on a thin layer of agar (see Chapter 9, page 267). A comparison of the number and sizes of restriction fragments that are produced from different organisms provides information about their genetic similarities and differences; the more similar the patterns, or *DNA fingerprints*, the more closely related the organisms are expected to be (Figure 10.14).

DNA fingerprinting is used to determine the source of hospital-acquired infections. In one hospital, patients undergoing coronary-bypass surgery developed infections caused by *Rhodococcus bronchialis* (rō-dō-kok′kus bronkē′al-is). The DNA fingerprints of the patients' bacteria and the bacteria of one nurse were identical. The hospi-

FIGURE 10.14 **DNA fingerprints.** Plasmids from seven different bacteria were digested with the same restriction enzyme. Each digest was put in a different well (origin) in the agarose gel. An electrical current was then applied to the gel to separate the fragments by size and electrical charge. The DNA was made visible by staining with ethidium bromide, which fluoresces under ultraviolet light. Comparison of the lanes shows that none of the DNA samples (and therefore none of the bacteria) is identical.

■ DNA fingerprinting is used to determine the source of hospital-acquired infections.

tal was thus able to break the chain of transmission of this infection by encouraging this nurse to use aseptic technique.

Ribosomal RNA Sequencing

Ribosomal RNA (rRNA) sequencing is currently being used to determine the diversity of organisms and the phylogenetic relationships among them. There are several advantages to using rRNA. First, all cells contain ribosomes. Two closely related organisms will have fewer different bases in their rRNA than two organisms that are distantly related. Another advantage is that RNA genes have undergone few changes over time. The rRNA used most often is a component of the smaller portion of ribosomes. A third advantage of rRNA sequencing is that cells do not have to be cultured in the laboratory.

DNA in a soil or water sample can be amplified by the polymerase chain reaction (PCR) using an rRNA primer (see below). The rRNA gene can be sequenced to determine evolutionary relationships between organisms, especially microorganisms.

The Polymerase Chain Reaction

When a microorganism cannot be cultured by conventional methods, the causative agent of an infectious disease might not be recognized. However, a technique called the **polymerase chain reaction (PCR)** can be used to increase the amount of microbial DNA to levels that can be tested by gel electrophoresis (see Chapter 9, page 254).

In 1992, researchers used PCR to determine the causative agent of Whipple's disease, which was previously an unknown bacterium now named *Tropheryma whippelii* (trō′fėr-ē-mä whip-pél′ē-ē). Whipple's disease was first described in 1907 by George Whipple as a gastrointestinal and nervous system disorder caused by an unknown bacillus. No one has been able to culture the bacterium to identify it, and thus PCR provides the only reliable methods of diagnosing and treating the disease.

In recent years, PCR has been used successfully in several discoveries that were not otherwise possible. For example, in 1992, Raul Cano used PCR to amplify DNA from *Bacillus* bacteria in amber that was 25–40 million years old. These primers were made from rRNA sequences in living *B. circulans* to amplify DNA coding for rRNA in the amber. These primers will cause amplification of DNA from other *Bacillus* species but do not cause amplification of DNA from other bacteria that might have been present, such as *Escherichia* or *Pseudomonas*. The DNA was sequenced after amplification. This information was used to determine the relationships between the ancient bacteria and modern bacteria.

In 1993, microbiologists identified a *Hantavirus* as the cause of an outbreak of hemorrhagic fever in the American Southwest using PCR. The identification was made in record time—less than 2 weeks. PCR was used in 1994 to identify the causative agent of a new tickborne disease (human granulocytic ehrlichiosis) as the bacterium *Ehrlichia chaffeensis* (ėr′lik-ē-ä chaf′ē-en-sis). PCR is used to identify the source of rabies viruses.

TaqMan is a commercial system that uses PCR to identify pathogenic *E. coli* in food and water. With this system, the newly amplified *E. coli* DNA fluoresces and can be detected using gel electrophoresis.

Nucleic Acid Hybridization

If a double-stranded molecule of DNA is subjected to heat, the complementary strands will separate as the hydrogen bonds between the bases break. If the single strands are then cooled slowly, they will reunite to form a double-stranded molecule identical to the original double strand. (This reunion occurs because the single strands have complementary sequences.) When this technique is applied to separated DNA strands from two different organisms, it is possible to determine the extent of similarity between the base sequences of the two organisms. This method is known as **nucleic acid hybridization.** The procedure assumes that if two species are similar or related, a major portion of their nucleic acid sequences will also be similar. The procedure measures the ability of DNA strands from one organism to hybridize (bind through complementary base pairing) with the DNA strands of another organism (Figure 10.15). The greater the degree of hybridization, the greater the degree of relatedness.

Recall from Chapter 8 (page 217) that RNA is single-stranded and is transcribed from one strand of DNA; a particular strand of RNA, therefore, is complementary to the strand of DNA from which it was transcribed and will hybridize with that separated strand of DNA. DNA-RNA hybridization can thus be used to determine relatedness between DNA from one organism and RNA from another organism in the same way that DNA-DNA hybridization is used.

Nucleic acid hybridization can be used to identify unknown microorganisms by **Southern blotting** (see Figure 9.12). In addition, rapid identification methods using **DNA probes** are being developed. One method involves breaking DNA extracted from *Salmonella* into fragments with a restriction enzyme, then selecting a specific fragment as the probe for *Salmonella* (Figure 10.16). This fragment must be able to hybridize with the DNA of all *Salmonella* strains, but not with the DNA of closely related enteric bacteria.

❶ The chosen DNA fragment is cloned in a plasmid in *E. coli*, producing hundreds of specific *Salmonella* DNA fragments.

❷ These fragments are tagged with radioactive isotopes or a fluorescent dye and are separated into single strands of DNA.

❸–❻ The resulting DNA probes can then be mixed with single-stranded DNA prepared from a food sample suspected of containing *Salmonella*.

❼ If *Salmonella* is present, the DNA probes will hybridize with the DNA of the *Salmonella*, and hybridization can be detected by the radioactivity or fluorescence of the probes.

An exciting new technology is the **DNA chip,** which will make it possible to quickly detect a pathogen in a host by identifying a gene that is unique to that pathogen (Figure 10.17). The DNA chip is composed of DNA probes. A sample containing DNA from an unknown organism is labeled with a fluorescent dye and added to the chip. Hybridization between the probe DNA and DNA in the sample is detected by fluoresence.

FIGURE 10.15 **DNA-DNA hybridization.** The greater the amount of pairing between DNA strands from different organisms (hybridization), the more closely the organisms are related.

■ What is the principle involved in DNA probes?

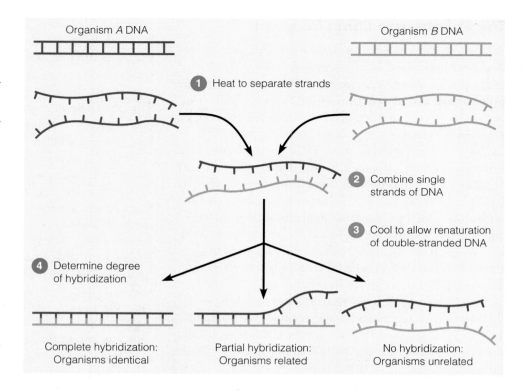

Organism *A* DNA

Organism *B* DNA

1 Heat to separate strands

2 Combine single strands of DNA

3 Cool to allow renaturation of double-stranded DNA

4 Determine degree of hybridization

Complete hybridization: Organisms identical

Partial hybridization: Organisms related

No hybridization: Organisms unrelated

FIGURE 10.16 **A DNA probe used to identify bacteria.**

■ Why do the DNA probe and cellular DNA hybridize?

Plasmid

Salmonella DNA fragment

1 A *Salmonella* DNA fragment is cloned in *E. coli.*

2 Cloned DNA fragments are marked with fluorescent dye and separated into single strands, forming DNA probes.

3 Unknown bacteria are collected on a filter.

4 The cells are lysed, and the DNA is released.

5 The DNA is separated into single strands.

6 DNA probes are added to the DNA from the unknown bacteria.

7 DNA probes hybridize with *Salmonella* DNA from sample. Then excess probe is washed off—fluorescence indicates presence of *Salmonella*.

Fluorescent probe

Salmonella DNA

DNA from other bacteria

(a) A DNA chip can be manufactured to contain hundreds of thousands of synthetic single-stranded DNA sequences. Assume that each DNA sequence was unique to a different bacterial species.

(b) Unknown DNA from a patient is separated into single strands, enzymatically cut, and labeled with a fluorescent dye.

(c) The unknown DNA is inserted into the chip and allowed to hybridize with the DNA on the chip.

(d) The tagged DNA will bind only to the complementary DNA on the chip. The bound DNA will be detected by its fluorescent dye and analyzed by a computer. The red light is a gene expressed in normal cells; green is a mutated gene expressed in tumor cells; and yellow, in both cells.

FIGURE 10.17 **DNA chip technology.** DNA chip technology will enable much more rapid, inexpensive analysis using many more probes. Chip technology is new; however, scientists soon hope to use DNA chips to detect microbes in a human or environmental sample, to identify cancer genes, to identify potential suspects in a crime or victims of a crime or catastrophe, to identify endangered species, and to match organ donors. If the DNA chip in (d) was constructed from different bacterial DNA, each spot would represent a bacterium, and the spot giving off fluorescent light would identify the bacterium in the sample.

Putting Classification Methods Together

Learning Objective

* Differentiate a dichotomous key from a cladogram.

Morphological characteristics, differential staining, and biochemical testing were the only identification tools available just a few years ago. Technological advancements are making it possible to use nucleic acid analysis techniques, once reserved for classification, for routine identification. A summary of taxonomic criteria and methods is provided in Table 10.5.

Information about microbes obtained by these methods is used to identify and classify the organisms. Two methods of using the information are described below.

Dichotomous Keys

Dichotomous keys are widely used for identification. In a dichotomous key, identification is based on successive questions, and each question has two possible answers (dichotomous means cut in two). After answering one question, the investigator is directed to another question until an organism is identified. Although these keys often have little to do with phylogenetic relationships, they are invaluable for identification. For example, a dichotomous key for bacteria could begin with an easily determined characteristic, such as cell shape, and move on to the ability to ferment a sugar. Dichotomous keys are shown in Figure 10.8 and in the box on page 286.

Cladograms

Cladograms are maps that show evolutionary relationships among organisms (*clado-* means branch). Cladograms are shown in Figures 10.1 and 10.6. Each branch point on the cladogram is defined by a feature shared by various species on that branch. Historically, cladograms for vertebrates were made using fossil evidence; however, rRNA sequences are now being used to confirm assumptions based

TABLE 10.5	**Taxonomic Criteria and Methods for Classifying and Identifying Bacteria**	
Criterion or Method	**Used for**	
	Classification	**Identification**
Morphological characteristics	No (yes for cyanobacteria)	Yes
Differential staining	Yes (for cell wall type)	Yes
Biochemical testing	No	Yes
Serology	No	Yes
Phage typing	No	Yes
Fatty acid profiles	No	Yes
Flow cytometry	No	Yes
DNA base composition	Yes	No
DNA fingerprinting	Yes	Yes
rRNA sequencing	Yes	No
PCR	Yes	Yes
Nucleic acid hybridization	Yes	Yes (DNA probes, DNA chips)

on fossils. As we said earlier, most microorganisms do not leave fossils; therefore, rRNA sequencing is primarily used to make cladograms for microorganisms. The small rRNA subunit used has 1500 bases, and computer programs do the calculations. The steps for constructing a cladogram are shown in Figure 10.18.

❶ Two rRNA sequences are aligned, and

❷ the percentage of similarity between the sequences is calculated.

❸ Then the horizontal branches are drawn in a length proportional to the calculated percent similarity. All species beyond a node (branch point) have similar rRNA sequences, suggesting that they arose from an ancestor at that node.

FIGURE 10.18 **Building a cladogram.**

▪ Why do *L. brevis* and *L. acidophilus* branch from the same node?

❶ Determine the sequence of bases in an rRNA molecule for each organism. Only a short sequence of bases is shown for this example.

Lactobacillus brevis	AGUCCAGAGC
L. sanfrancisco	GUAAAAGAGC
L. acidophilus	AGCGGAGAGC
L. plantarum	ACGUUAGAGC

❷ Calculate the percentage of similarity in the nucleotide bases between each species. For example, there is a 70% similarity between the sequences for *L. brevis* and *L. acidophilus*.

	Percent similarity
L. brevis ⟶ *L. sanfrancisco*	50%
L. brevis ⟶ *L. acidophilus*	70%
L. brevis ⟶ *L. plantarum*	60%
L. sanfrancisco ⟶ *L. acidophilus*	50%
L. sanfrancisco ⟶ *L. plantarum*	50%
L. plantarum ⟶ *L. acidophilus*	60%

❸ Construct a cladogram. The length of the horizontal lines corresponds to the percent similarity values. Each branch point, or node, in the cladogram represents an ancestor common to all species beyond that node. Each node is defined by a similarity in rRNA present in all species beyond that branch point.

STUDY OUTLINE

Introduction (p. 276)

1. Taxonomy is the science of the classification of organisms, with the goal of showing relationships among organisms.
2. Taxonomy also provides a means of identifying organisms.

The Study of Phylogenetic Relationships (pp. 276–280)

1. Phylogeny is the evolutionary history of a group of organisms.
2. The taxonomic hierarchy shows evolutionary, or phylogenetic, relationships among organisms.
3. Bacteria were separated into the Kingdom Prokaryotae in 1968.
4. Living organisms were divided into five kingdoms in 1969.

The Three Domains (pp. 277–278)

1. Living organisms are currently classified into three domains. A domain can be divided into kingdoms.
2. In this system, plants, animals, fungi, and protists belong to the Domain Eukarya.
3. Bacteria (with peptidoglycan) form a second domain.
4. Archaea (with unusual cell walls) are placed in the Domain Archaea.

A Phylogenetic Hierarchy (pp. 278–280)

1. Organisms are grouped into taxa according to phylogenetic relationships (from a common ancestor).
2. Some of the information for eukaryotic relationships is obtained from the fossil record.
3. Prokaryotic relationships are determined by rRNA sequencing.

Classification of Organisms (pp. 280–285)

Scientific Nomenclature (pp. 280–281)

1. According to scientific nomenclature, each organism is assigned two names, or binomial: a genus and a specific epithet, or species.
2. Rules for the assignment of names to bacteria are established by the International Committee on Systematic Bacteriology.
3. Rules for naming fungi and algae are published in the *International Code of Botanical Nomenclature*.
4. Rules for naming protozoa are found in the *International Code of Zoological Nomenclature*.

The Taxonomic Hierarchy (p. 282)

1. A eukaryotic species is a group of organisms that interbreeds but does not breed with individuals of another species.

2. Similar species are grouped into a genus; similar genera are grouped into a family; families, into an order; orders, into a class; classes, into a phylum; phyla, into a kingdom; and kingdoms, into a domain.

Classification of Prokaryotes (p. 282–284)

1. *Bergey's Manual of Systematic Bacteriology* is the standard reference on bacterial classification.
2. A group of bacteria derived from a single cell is called a strain.
3. Closely related strains constitute a bacterial species.

Classification of Eukaryotes (p. 284)

1. Eukaryotic organisms may be classified into the Kingdom Fungi, Plantae, or Animalia.
2. Protists are mostly unicellular organisms; these organisms are currently being assigned to kingdoms.
3. Fungi are absorptive chemoheterotrophs that develop from spores.
4. Multicellular photoautotrophs are placed in the Kingdom Plantae.
5. Multicellular ingestive heterotrophs are classified as Animalia.

Classification of Viruses (pp. 284–285)

1. Viruses are not placed in a kingdom. They are not composed of cells and cannot grow without a host cell.
2. A viral species is a population of viruses with similar characteristics that occupies a particular ecological niche.

Methods of Classifying and Identifying Microorganisms (pp. 285–298)

1. *Bergey's Manual of Determinative Bacteriology* is the standard reference for laboratory identification of bacteria.
2. Morphological characteristics are useful in identifying microorganisms, especially when aided by differential staining techniques.
3. The presence of various enzymes, as determined by biochemical tests, is used in identifying microorganisms.
4. Serological tests, involving the reactions of microorganisms with specific antibodies, are useful in determining the identity of strains and species, as well as relationships among organisms. ELISA and Western blotting are examples of serological tests.
5. Phage typing is the identification of bacterial species and strains by determining their susceptibility to various phages.
6. Fatty acid profiles can be used to identify some organisms.
7. Flow cytometry measures physical and chemical characteristics of cells.
8. The percentage of GC base pairs in the nucleic acid of cells can be used in the classification of organisms.

9. The number and sizes of DNA fragments, or DNA finger-prints, produced by restriction enzymes are used to determine genetic similarities.

10. The sequence of bases in ribosomal RNA can be used in the classification of organisms.

11. The polymerase chain reaction (PCR) can be used to detect small amounts of microbial DNA in a sample.

12. Single strands of DNA, or of DNA and RNA, from related organisms will hydrogen-bond to form a double-stranded molecule; this bonding is called nucleic acid hybridization.

13. Southern blotting, DNA probes, and DNA chips are examples of hybridization techniques.

14. Dichotomous keys are used for the identification of organisms. Cladograms show phylogenetic relationships among organisms.

STUDY QUESTIONS

Review

1. What is taxonomy?

2. Discuss the evidence that supports classifying organisms into three domains.

3. Explain why a two-, three-, or five-kingdom system is no longer acceptable for classification.

4. List and define the three kingdoms of multicellular eukaryotic organisms.

5. Compare and contrast the following terms:
   a. archaea and bacteria
   b. bacteria and eukarya
   c. archaea and eukarya

6. What is binomial nomenclature?

7. Why is binomial nomenclature preferable to the use of common names?

8. Using *Escherichia coli* and *Entamoeba coli* as examples, explain why the genus name must always be written out the first time you use it in a report.

9. Put the following terms in the correct sequence, from the most general to the most specific: order, class, genus, domain, species, phylum, family.

10. Find the gram-positive bacteria *Staphylococcus* in Appendix A. To what domain, phylum, class, order, and family does it belong? To which bacteria is it most closely related: *Gemella* or *Streptococcus*?

11. Define the following terms: eukaryotic species, bacterial species, and viral species.

12. List the twelve taxonomic criteria and methods discussed in this chapter for the classification of microorganisms. Separate your list into those tests used primarily for taxonomic classification and those used primarily for identifying microorganisms already classified.

13. Higher organisms are arranged into taxonomic groups on the basis of evolutionary relationships. Why is this type of classification only now being developed for bacteria?

14. Can you tell which of the following organisms are most closely related? Are any two the same species? On what did you base your answer?

Characteristic	A	B	C	D
Morphology	Rod	Coccus	Rod	Rod
Gram reaction	+	−	−	+
Glucose utilization	Fermentative	Oxidative	Fermentative	Fermentative
Cytochrome oxidase	Present	Present	Absent	Absent
GC moles %	48–52	23–40	50–54	49–53

Multiple Choice

1. *Bergey's Manual of Systematic Bacteriology* differs from *Bergey's Manual of Determinative Bacteriology* in that the former
   a. groups bacteria into species.
   b. groups bacteria according to phylogenetic relationships.
   c. groups bacteria according to pathogenic properties.
   d. groups bacteria into 19 species.
   e. all of the above

2. *Bacillus* and *Lactobacillus* are not in the same order. This indicates that which one of the following is *not* sufficient to assign an organism to a taxon?
   a. biochemical characteristics
   b. amino acid sequencing
   c. phage typing
   d. serology
   e. morphological characteristics

3. Which of the following is used to classify organisms into the Kingdom Fungi?
   a. ability to photosynthesize; possess a cell wall
   b. unicellular; possess cell wall; prokaryotic
   c. unicellular; lacking cell wall; eukaryotic
   d. absorptive; possess cell wall; eukaryotic
   e. ingestive; lacking cell wall; multicellular; prokaryotic

4. Which of the following is *not* true about scientific nomenclature?
   a. Each name is specific.
   b. Names vary with geographical location.
   c. The names are standardized.
   d. Each name consists of a genus and specific epithet.
   e. It was first designed by Linnaeus.

5. You could identify an unknown bacterium by all of the following *except*
   a. hybridizing a DNA probe from a known bacterium with the unknown's DNA.

b. making a fatty acid profile of the unknown.
c. specific antiserum agglutinating the unknown.
d. ribosomal RNA sequencing.
e. percentage of guanine + cytosine.

6. The wall-less mycoplasmas are considered to be related to gram-positive bacteria. Which of the following would provide the most compelling evidence for this?
    a. They share common rRNA sequences.
    b. Some gram-positive bacteria and some mycoplasmas produce catalase.
    c. Both groups are prokaryotic.
    d. Some gram-positive bacteria and some mycoplasmas have coccus-shaped cells.
    e. Both groups contain human pathogens.

Use the following choices to answer questions 7 and 8.
    a. Animalia
    b. Fungi
    c. Plantae
    d. Firmicutes (gram-positive bacteria)
    e. Proteobacteria (gram-negative bacteria)

7. Into which group would you place a multicellular organism that has a mouth and lives inside the human liver?

8. Into which group would you place a photosynthetic organism that lacks a nucleus and has a thin peptidoglycan wall surrounded by an outer membrane?

Use the following choices to answer questions 9 and 10.
    1. 9 + 2 flagella
    2. 70S ribosome
    3. fimbria
    4. nucleus
    5. peptidoglycan
    6. plasma membrane

9. Which is (are) found in all three domains?
    a. 2, 6
    b. 5
    c. 2, 4, 6
    d. 1, 3, 5
    e. all six

10. Which is (are) found *only* in prokaryotes?
    a. 1, 4, 6
    b. 3, 5
    c. 1, 2
    d. 4
    e. 2, 4, 5

Critical Thinking

1. Here is some additional information on the organisms in review question 13:

Organisms	% DNA Hybridization
A and B	5–15
A and C	5–15
A and D	70–90
B and C	10–20
B and D	2–5

Which of these organisms are most closely related? Compare this answer with your response to review question 13.

2. The GC content of *Micrococcus* is 66–75 moles %, and of *Staphylococcus*, 30–40 moles %. According to this information, would you conclude that these two genera are closely related?

3. Describe the use of a DNA probe and PCR for:
    a. rapid identification of an unknown bacterium.
    b. determining which of a group of bacteria are most closely related.

4. SF medium is a selective medium, developed in the 1940s, to test for fecal contamination of milk and water. Only certain gram-positive cocci can grow in this medium. Why is it named SF? Using this medium, which genus will you culture? (*Hint:* Refer to Chapter 11.)

Clinical Applications

1. A 55-year-old veterinarian was admitted to a hospital with a 2-day history of fever, chest pain, and cough. Gram-positive cocci were detected in his sputum, and he was treated for lobar pneumonia with penicillin. The next day, another Gram stain of his sputum revealed gram-negative rods, and he was switched to ampicillin and gentamicin. A sputum culture showed biochemically inactive gram-negative rods identified as *Enterobacter agglomerans*. After fluorescent-antibody staining and phage typing, *Yersinia pestis* was identified in the patient's sputum and blood, and chloramphenicol and tetracycline were administered. The patient died 3 days after admission to the hospital. Tetracycline was given to his 220 contacts (hospital personnel, family, and coworkers). What disease did the patient have? Discuss what went wrong in the diagnosis and how his death might have been prevented. Why were the 220 other people treated?

2. A 6-year-old girl was admitted to a hospital with endocarditis. Blood cultures showed a gram-positive, aerobic rod identified by the hospital laboratory as *Corynebacterium xerosis*. The girl died after 6 weeks of treatment with intravenous penicillin and chloramphenicol. The bacterium was tested by another laboratory and identified as *C. diphtheriae*. The following test results were obtained by each laboratory:

	Hospital Lab	Other lab
Catalase	+	+
Nitrate reduction	+	+
Urea	–	–
Esculin hydrolysis	–	–
Glucose fermentation	+	+
Sucrose fermentation	–	+
Serological test for toxin production	Not done	+

Provide a possible explanation for the incorrect identification. What are the potential public health consequences of misidentifying *C. diphtheriae*?

3. Use the information in the following table to construct a dichotomous key to these organisms. What is the purpose of a dichotomous key? Look up each genus in Chapter 11, and provide an example of why this organism is of interest to humans.

	Morphology	Gram Reaction	Acid from Glucose	Growth in Air (21% O_2)	Motile by Peritrichous Flagella	Presence of Cytochrome Oxidase	Produce Catalase
Staphylococcus aureus	Coccus	+	+	+	−	−	+
Streptococcus pyogenes	Coccus	+	+	+	−	−	−
Mycoplasma pneumoniae	Coccus	−	+	+ (Colonies < 1mm)	−	−	+
Clostridium botulinum	Rod	+	+	−	+	−	−
Escherichia coli	Rod	−	+	+	+	−	+
Pseudomonas aeruginosa	Rod	−	+	+	−	+	+
Campylobacter fetus	Vibrio	−	−	−	−	+	+
Listeria monocytogenes	Rod	+	+	+	+	−	+

4. Use the additional information shown at right to construct a cladogram for some of the organisms used in question 3. What is the purpose of a cladogram? How does your cladogram differ from a dichotomous key for these organisms?

	Similarity in rRNA Bases
P. aeruginosa—M. pneumoniae	52%
P. aeruginosa—C. botulinum	52%
P. aeruginosa—E. coli	79%
M. pneumoniae—C. botulinum	65%
M. pneumoniae—E. coli	52%
E. coli—C. botulinum	52%

5. Using the following information, create a dichotomous key for distinguishing these organisms.

	Unicellular?	Mitochondria?	Chlorophyll?	Nutritional Type?	Motile?	Cause Human Disease?
Euglena	+	+	+	Both	+	
Giardia				Hetero		
Nosema						
Pfiesteria				Auto		
Trichomonads						
Trypanosoma						

How would you define *protist*?

Using the additional information shown below, create a dichotomous key for these organisms.

	rRNA base #																			
	1	2	3	4	5	6	7	8	9	10	11	12	13	14	15	16	17	18	19	20
Euglena	C	C	A	G	G	U	U	G	U	U	C	C	A	G	U	U	U	U	A	A
Giardia	C	C	A	U	A	U	U	U	U	U	G	A	C	G	A	A	G	G	U	C
Nosema	C	C	A	U	A	U	U	U	U	U	A	A	C	G	A	A	G	G	C	C
Pfiesteria	C	C	A	A	C	U	U	A	U	U	C	C	A	G	U	U	U	C	A	G
Trichomonas	C	C	A	U	A	U	U	U	U	U	G	A	C	G	A	A	G	G	G	C
Trypanosoma	C	C	A	C	G	U	U	G	U	U	C	C	A	G	U	U	U	A	A	A

Do your two keys differ? Explain why. Which key is more useful for laboratory identification? For classification?

LEARNING WITH TECHNOLOGY

The Microbiology Place CD-ROM/website includes many resources to help you study and succeed in this course. In addition to self-quizzes, web links, and flashcards, you'll find the following activities for this chapter in the "Activities" section of The Microbiology Place:

DNA Fingerprints

The Three Domains

Building a Molecular Clock

Chapter 11

The Prokaryotes: Domains Bacteria and Archaea

STREPTOMYCES. Streptomyces *bacteria grow in filaments and make conidiospores. These bacteria produce most of the antibiotics used in medicine.*

When biologists first encountered microscopic bacteria, they were puzzled as to how to classify them. Bacteria were clearly not animals or rooted plants. At first, bacteria were included with the fungi, many of which were also microscopic. Attempts to build a taxonomic system for bacteria based on the phylogenetic system developed for plants and animals, as discussed in Chapter 10, failed. Bacteria were grouped by morphology (rod, coccus, helix), staining reactions, presence of endospores, and other obvious features. Although this system had its practical uses, microbiologists were aware that it had many imperfections. For example, lactic acid bacteria were placed in widely separated groups distinguished by either rod-shaped morphology or spherical morphology. Subsequent studies of their ribosomal RNA (rRNA) showed that the two groups are actually quite closely related.

In 2001, the first volume of a forthcoming five-volume second edition of Bergey's Manual was published. During the years since publication of the first edition, knowledge of bacteria at the molecular level has expanded to such a degree that it is possible to base this new edition on a phylogenetic system. One of the most important expressions of phylogenetic difference is rRNA. Above the species level, rRNA is especially useful. It is slow to change and performs the same functions in all organisms. It is now possible to begin constructing phylogenetic trees in which gene sequences can be compared. Organisms can then be grouped by a measure of differences in selected genes, sometimes expressed as evolutionary distance (see Figure 10.6, page 285). In general, there will be a tendency to greater variation as time allows differences in the genetic makeup to accumulate.

PROKARYOTIC GROUPS

In the second edition of *Bergey's Manual,* the prokaryotes are grouped into two domains, the Archaea and the Bacteria. Both domains consist of prokaryotic cells. Higher organisms are assigned to the Domain Eukarya, which contains all the unicellular and multicellular eukaryotic organisms. The grouping into domains is shown in Figure 10.1 and Table 10.1, and the differences between the two prokaryotic domains are summarized in Figure 10.6. Each domain is divided into phyla, each phylum into classes, and so on. The phyla discussed in this chapter are summarized in Table 11.1 on pages 307–309. (Also see Appendix A.)

DOMAIN BACTERIA

Most of us have been conditioned to think of bacteria as invisible, potentially harmful little creatures. Actually, relatively few species of bacteria cause disease in humans, animals, plants, or any other organisms. Indeed, once you have completed a course in microbiology, you will realize that without bacteria, much of life as we know it would not be possible. In fact, all organisms made up of eukaryotic cells probably evolved from bacterialike organisms, which were some of the earliest forms of life. In this chapter, you will learn how bacterial groups are differentiated from each other and how important bacteria are in the world of microbiology. Our discussion emphasizes bacteria that are considered to be of practical importance, those important in medicine, or those that illustrate biologically unusual or interesting principles.

The Proteobacteria

The proteobacteria, which includes most of the gram-negative, chemoheterotrophic bacteria, are presumed to have arisen from a common photosynthetic ancestor. They are now the largest taxonomic group of bacteria. However, few are now photosynthetic; other metabolic and nutritional capacities have arisen to replace this characteristic. The phylogenetic relationship in these groups is based upon rRNA studies. The name **Proteobacteria** was taken from the mythological Greek god Proteus, who could assume many shapes. Distinctive subgroups of the proteobacteria are designated by Greek letters.

The α (alpha) Proteobacteria

Learning Objective

The Learning Objectives in this chapter will help you to become familiar with these organisms and to look for similarities and differences between organisms. You will draw a dichotomous key to differentiate the bacteria described in each group. We'll draw the first one to get you started.

- *Make a dichotomous key to distinguish among the α-proteobacteria described in this chapter.*

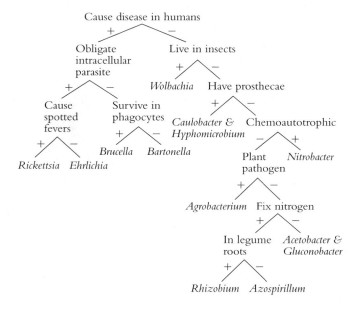

As a group, the α-proteobacteria includes most of the proteobacteria that are capable of growth at very low levels of nutrients. Some have unusual morphology, including protrusions such as stalks or buds known as **prosthecae.** The α-proteobacteria also include agriculturally important bacteria capable of inducing nitrogen fixation in symbiosis with plants, and several plant and human pathogens.

Azospirillum Agricultural microbiologists have been interested in members of the genus *Azospirillum* (ā-zō-spī′ril-lum), a soil bacterium that grows in close association with the roots of many plants, especially tropical grasses. It uses nutrients excreted by the plants and in return fixes nitrogen from the atmosphere. This form of nitrogen fixation is most significant in some tropical grasses and in sugar cane, although the organism can be isolated from the root system of many temperate-climate plants, such as corn. (The prefix *azo-* is frequently encountered in nitrogen-fixing genera of bacteria. It is derived from *a* (without) and *zo* (life), in reference to the early days of chemistry when oxygen was removed, by a burning candle, from an experimental atmosphere. Presumably, mostly nitrogen remained and mammalian life was found to

(a) Rickettsia   TEM |—————| 25 μm

(b) Rickettsias in chicken embryo cell   LM |—————| 50 μm

FIGURE 11.1 **Rickettsias.**

■ How are rickettsias transmitted from one host to another?

be not possible in this atmosphere. Hence, nitrogen came to be associated with absence of life.)

Acetobacter and Gluconobacter *Acetobacter* (ä′sē-tō-bak-tėr) and *Gluconobacter* (glü′kon-ō-bak-tėr) are industrially important aerobic organisms that convert ethanol into acetic acid (vinegar).

Rickettsia In the first edition of *Bergey's Manual*, the genera *Rickettsia*, *Coxiella*, and *Chlamydia* were grouped closely because they share the common characteristic of being obligate intracellular parasites—that is, they reproduce only within a mammalian cell. In the second edition they are now widely separated. A comparison of rickettsias, chlamydias, and viruses appears in Table 13.1, page 377.

The rickettsias are gram-negative rod-shaped bacteria, or coccobacilli (Figure 11.1a). One distinguishing feature of most rickettsias is that they are transmitted to humans by bites of insects and ticks, as are the *Coxiella* (discussed later with γ-proteobacteria). Rickettsia enter their host cell by inducing phagocytosis. They quickly enter the cytoplasm of the cell and begin reproducing by binary fission (Figure 11.1b). They can usually be cultivated artificially in cell culture or chick embryos (Chapter 13, pages 384–385).

The rickettsias are responsible for a number of diseases known as the spotted fever group. These include epidemic typhus, caused by *Rickettsia prowazekii* (ri-ket′sē-ä prou-wä-ze′kē-ē) and transmitted by lice; endemic murine typhus, caused by *R. typhi* (tī′fē) and transmitted by rat fleas, and Rocky Mountain spotted fever, caused by *R. rickettsii* (ri-ket′sē-ē) and transmitted by ticks. In humans, rickettsial infections damage the permeability of blood capillaries, which results in a characteristic spotted rash.

Ehrlichia Ehrlichiae are gram-negative, rickettsialike bacteria that live obligately within white blood cells. *Ehrlichia* (ėr′lik-ē-ä) species are transmitted by ticks to humans and cause ehrlichiosis, a sometimes fatal disease.

Caulobacter and Hyphomicrobium Members of the genus *Caulobacter* (kô-lō-bak′tėr) are found in low-nutrient aquatic environments, such as lakes. They feature stalks that anchor the organs to surfaces (Figure 11.2). This arrangement increases their nutrient uptake because they are exposed to a continuously changing flow of water and because the stalk increases the surface-to-volume ratio of the cell. Also, if the surface to which they anchor is a

SEM |—————| 2 μm

FIGURE 11.2 **Caulobacter.** Most of the *Caulobacter* in this photo are in the stalked stage.

■ Of what advantage are stalks?

TABLE 11.1	**Selected Prokaryotes from *Bergey's Manual of Systematic Bacteriology*, Second Edition***		
Phylum Class	**Order**	**Important Genera**	**Special Features**
DOMAIN BACTERIA **Proteobacteria** α (alpha) proteobacteria	Caulobacterales	*Caulobacter*	Stalked
	Rickettsiales	*Ehrlichia* *Rickettsia* *Wolbachia*	Obligately intracellular human pathogens Obligately intracellular human pathogens Symbionts of insects
	Rhizobiales	*Agrobacterium* *Bartonella* *Beijerinckia* *Bradyrhizobium* *Brucella* *Hyphomicrobium* *Nitrobacter* *Rhizobium*	Plant pathogens Human pathogens Free-living nitrogen fixers Symbiotic nitrogen fixers Human pathogens Budding Nitrifying Symbiotic nitrogen fixers
	Rhodospirillales	*Acetobacter* *Azospirillum* *Gluconobacter* *Rhodospirillum*	Acetic acid production Nitrogen fixers Acetic acid production Photosynthetic, anoxygenic
β (beta) proteobacteria	Burkholderiales	*Burkholderia* *Bordetella* *Sphaerotilus*	Opportunistic pathogens Human pathogens Sheathed
	Hydrogenophilales	*Thiobacillus*	Sulfur oxidizers
	Neisseriales	*Neisseria*	Human pathogens
	Nitrosomonadales	*Nitrosomonas* *Spirillum*	Nitrifying Found in stagnant freshwater
	Rhodocyclales	*Zoogloea*	Sewage treatment
γ (gamma) proteobacteria	Chromatiales	*Chromatium*	Photosynthetic, anoxygenic
	Thiotrichales	*Beggiatoa* *Thiomargarita* *Francisella*	Sulfur oxiders Giant bacterium Human pathogens
	Legionellales	*Legionella* *Coxiella*	Human pathogens Obligately intracellular human pathogens
	Pseudomonadales	*Azomonas* *Azotobacter* *Moraxella* *Pseudomonas*	Free-living nitrogen fixers Free-living nitrogen fixers Human pathogens Opportunistic pathogens
	Vibrionales	*Vibrio*	Human pathogens
	Enterobacteriales	*Citrobacter* *Enterobacter* *Erwinia* *Escherichia* *Klebsiella* *Proteus* *Salmonella*	Opportunistic pathogens Opportunistic pathogens Plant pathogens Normal intestinal bacteria, some pathogens Opportunistic pathogens Human intestinal bacteria, occasional pathogens Human pathogens

*See Appendix A for a complete taxonomic listing. This table includes prokaryotes mentioned in this text. Descriptions such as "pathogenic" mean that this trait is common in the genus but not that all members of the genus may have this trait.

TABLE 11.1	**Selected Prokaryotes from *Bergey's Manual of Systematic Bacteriology*, Second Edition** (continued)		
Phylum Class	**Order**	**Important Genera**	**Special Features**
		Serratia	Red pigment, opportunistic pathogens
		Shigella	Human pathogens
		Yersinia	Human pathogens
	Pasteurellales	*Pasteurella*	Human pathogens
		Haemophilus	Human pathogens
δ (delta) proteobacteria	Bdellovibrionales	*Bdellovibrio*	Parasites of bacteria
	Desulfovibrionales	*Desulfovibrio*	Sulfate reducers
	Myxococcales	*Myxococcus*	Gliding, fruiting
		Stigmatella	Gliding, fruiting
ε (epsilon) proteobacteria	Campylobacterales	*Campylobacter*	Human pathogens
		Helicobacter	Human pathogens, carcinogenic

Nonproteobacteria, Gram-Negative Bacteria
Cyanobacteria

		Anabaena	Photosynthetic, oxygenic
		Gloeocapsa	Photosynthetic, oxygenic

Chlorobi

		Chlorobium	Photosynthetic, anoxygenic

Chloroflexi

		Chloroflexus	Photosynthetic, anoxygenic

Firmicutes (The Low G + C Gram-Positive Bacteria)

	Clostridiales	*Clostridium*	Anaerobes, endospores, some human pathogens
		Epulopiscium	Giant bacterium
		Sarcina	Occur in cubical packets
	Mycoplasmatales[†]	*Mycoplasma*	No cell wall, human pathogens
		Spiroplasma	No cell wall, pleomorphic, plant pathogens
		Ureaplasma	No cell wall, ammonia from urea
	Bacillales	*Bacillus*	Endospores, some pathogens
		Listeria	Human pathogens
		Staphylococcus	Some human pathogens
	Lactobacillales	*Enterococcus*	Opportunistic pathogens
		Lactobacillus	Lactic acid producers
		Streptococcus	Many human pathogens

Actinobacteria (The High G + C Gram-Positive Bacteria)

	Actinomycetales	*Actinomyces*	Filamentous, branching, some human pathogens
		Corynebacterium	Human pathogens
		Frankia	Symbiotic nitrogen-fixers
		Gardnerella	Human pathogens
		Mycobacterium	Acid fast, human pathogens
		Nocardia	Filamentous, branching, opportunistic pathogens
		Propionibacterium	Produce propionic acid
		Streptomyces	Filamentous branching, many produce antibiotics

[†]The bacteria in the order Mycoplasmatales are genetically related to the low G + C gram-positive bacteria, but they lack a cell wall and stain gram-negative.

TABLE 11.1	(continued)			
Phylum Class		**Order**	**Important Genera**	**Special Features**

ASSORTED PHYLA DISCUSSED IN VOLUME FIVE OF *BERGEY'S MANUAL*

Chlamydiae

| | | Chlamydiales | *Chlamydia* | Intracellular parasites, human pathogens |

Spirochaetes

		Spirochaetales	*Borrelia*	Human pathogens
			Leptospira	Human pathogens
			Treponema	Human pathogens

Bacteroidetes

| | | Bacteroidales | *Bacteroides* | Human intestinal tract |
| | | | *Prevotella* | Human oral cavity |

Fusobacteria

| | | Fusobacteriales | *Fusobacterium* | Human intestinal tract |

DOMAIN ARCHAEA
Crenarchaeota (Gram-Negative)

| | | Desulfurococcales | *Pyrodictium* | Hyperthermophiles |
| | | Sulfolobales | *Sulfolobus* | Hyperthermophiles |

Euryarchaeota (Gram-Positive to Variable)

		Methanobacteriales	*Methanobacterium*	Methanogens
		Halobacteriales	*Halobacterium*	Require high salt concentration
			Halococcus	Require high salt concentration

living host, these bacteria can use the host's excretions as nutrients. When the nutrient concentration is exceptionally low, the size of the stalk increases, evidently to provide an even greater surface area for nutrient absorption.

Budding bacteria do not divide by binary fission into two nearly identical cells. The budding process resembles the asexual reproductive processes of many yeasts (Figure 12.3 on page 337). The parent cell retains its identity while the bud increases in size until it separates as a complete new cell. An example is the genus *Hyphomicrobium* (hī-fō-mī-krō′bē-um), as shown in Figure 11.3. These bacteria, like the caulobacteria, are found in low-nutrient aquatic environments and have even been found growing in laboratory water baths. Both *Caulobacter* and *Hyphomicrobium* produce prominent prosthecae.

Rhizobium and Agrobacterium The genus *Rhizobium* (rī-zō′bē-um) includes agriculturally important bacteria that specifically infect the roots of leguminous plants, such as beans, peas, or clover. The presence of the bacteria leads to formation of nodules in which the bacteria and plant form a symbiotic relationship, resulting in the fixation of nitrogen from the air for use by the plant.

Like *Rhizobium*, the genus *Agrobacterium* has the ability to invade plants. However, they do not induce root nodules

FIGURE 11.3 ***Hyphomicrobium*, a type of budding bacterium.**

■ Most bacteria do not reproduce by budding; what method do they use?

or fix nitrogen. Of particular interest is *Agrobacterium tumefaciens*. This is a plant pathogen that causes a disease called crown gall; the crown is the area of the plant where the roots and stem merge. The tumorlike gall is induced when *A. tumefaciens* inserts a plasmid containing bacterial genetic information into the plant's chromosomal DNA (see Figure 9.17, page 268). For this reason, microbial geneticists are very interested in this organism. Plasmids are the most common vector that genetic engineers use to carry new genes into a cell, and the thick wall of plants is especially difficult to penetrate (see Figure 9.18, page 268).

Bartonella The genus *Bartonella* contains several members that are human pathogens. The best known is *Bartonella henselae*, a gram-negative bacillus that causes cat-scratch disease.

Brucella *Brucella* (brü′sel-la) bacteria are small nonmotile coccobacilli. All species of *Brucella* are obligate parasites of mammals and cause the disease brucellosis. Of medical interest is the ability of *Brucella* to survive phagocytosis, an important element of the body's defense against bacteria (see Chapter 16, page 463).

Nitrobacter and Nitrosomonas *Nitrobacter* (ni-trō-bak′tėr) and *Nitrosomonas* (nī-trō-sō-mō′nas) are genera of nitrifying bacteria that are of great importance to the environment and to agriculture. They are chemoautotrophs capable of using inorganic chemicals as energy sources and carbon dioxide as the only source of carbon, from which they synthesize all of their complex chemical makeup. The energy sources of the genera *Nitrobacter* and *Nitrosomonas* (the latter a member of the β-proteobacteria) are reduced nitrogenous compounds. *Nitrobacter* species oxidize ammonium (NH_4^+) to nitrite (NO_2^-), which is in turn oxidized by *Nitrosomonas* species to nitrates (NO_3^-) in the process of *nitrification*. Nitrate is important to agriculture; it is a nitrogen form that is highly mobile in soil and therefore likely to be encountered and used by plants.

Wolbachia *Wolbachia* (wol-ba′kē-ä) are probably the most common infectious bacterial genus in the world. Little is known about *Wolbachia*; they live only inside the cells of their hosts, usually insects (a relationship known as endosymbiosis). Therefore, *Wolbachia* escape detection by the usual culture methods. This fascinating group of bacteria is described further in the box on page 323.

The β (beta) Proteobacteria

Learning Objective

- *Make a dichotomous key to distinguish among the β-proteobacteria described in this chapter.*

There is considerable overlap between the β-proteobacteria and the α-proteobacteria, for example among the nitrifying bacteria discussed earlier. The β-proteobacteria often use nutrient substances that diffuse away from areas of anaerobic decomposition of organic matter, such as hydrogen gas, ammonia, and methane. Several important pathogenic bacteria are found in this group.

Thiobacillus *Thiobacillus* (thī-ō-bä-sil′lus) species and other sulfur-oxidizing bacteria are important in the sulfur cycle. These chemoautotrophic bacteria are capable of obtaining energy by oxidizing the reduced forms of sulfur, such as hydrogen sulfide (H_2S), or elemental sulfur (S^0), into sulfates (SO_4^{2-}).

Spirillum The habitat of the genus *Spirillum* (spī-ril′lum) is mainly fresh water. An important morphological difference from the helical spirochetes (discussed later) is that *Spirillum* bacteria are motile by conventional polar flagella, rather than axial filaments. The spirilla are relatively large, gram-negative, aerobic bacteria. *Spirillum volutans* (vō-lū-tans) is often used as a demonstration slide when microbiology students are first introduced to the operation of the microscope (Figure 11.4).

Sphaerotilus Sheathed bacteria, which include *Sphaerotilus natans* (sfe-rä′ti-lus na′tans), are found in freshwater and in sewage. These gram-negative bacteria with polar flagella form a hollow, filamentous sheath in which to live (Figure 11.5). Sheaths are protective and also aid in nutrient accumulation. *Sphaerotilus* probably contributes to bulking, an important problem in sewage treatment.

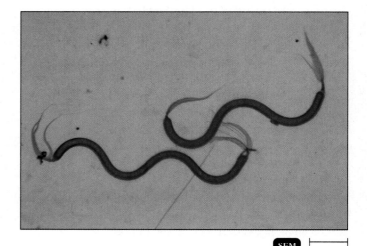

SEM |———| 10 μm

FIGURE 11.4 **Spirillum volutans.** These large helical bacteria are found in aquatic environments. Note the polar flagella.

■ Is this bacterium motile? How can you tell?

TEM ⊢——⊣ 2 μm

FIGURE 11.5 *Sphaerotilus natans.* These sheathed bacteria are found in dilute sewage and aquatic environments. It forms elongated sheaths in which the bacteria live. The bacteria have flagella (not visible here) and can eventually swim free of the sheath.

■ How does the sheath help the cell?

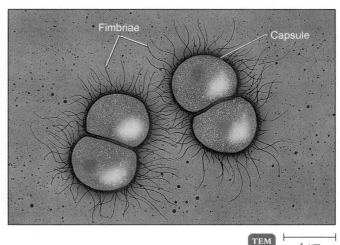

TEM ⊢——⊣ 1 μm

FIGURE 11.6 The gram-negative coccus *Neisseria gonorrhoeae.* Notice the paired arrangement (diplococci). The fimbriae enabled the organism to attach to mucous membranes and thus contribute to its pathogenicity. *N. gonorrhoeae* causes gonorrhea.

■ For what are fimbriae used?

Burkholderia The genus *Burkholderia* has only recently been reclassified; formerly it was grouped with the genus *Pseudomonas,* which is now classified under the γ-proteobacteria. Like the pseudomonads, almost all *Burkholderia* species are motile by a single polar flagellum or tuft of flagella. The best known species is the aerobic, gram-negative rod *Burkholderia cepacia* (berk′hōld-ér-ē-ä se-pā′se-ä). It has an extraordinary nutritional spectrum and is capable of degrading more than 100 different organic molecules. This capability is often a factor in the contamination of equipment and drugs in hospitals; these bacteria may actually grow in disinfectant solutions. This bacterium is also a problem for persons with the genetic lung disease cystic fibrosis, in whom it metabolizes accumulated respiratory secretions.

Bordetella Of special importance is the nonmotile, aerobic, gram-negative rod *Bordetella pertussis* (bȯr-de-tel′lä pėr-tus′sis). This serious pathogen is the cause of pertussis, or whooping cough.

Neisseria Bacteria of the genus *Neisseria* (nī-se′rē-ä) are aerobic, gram-negative cocci that usually inhabit the mucous membranes of mammals. Pathogenic species include the gonococcus bacterium *Neisseria gonorrhoeae* (go-nôr-rē′ī), the causative agent of gonorrhoea (Figure 11.6), and *N. meningitidis* (men-nin-ji′ti-dis), the agent of meningococcal meningitis.

Zoogloea The genus *Zoogloea* (zō′ō-glē-ä) is important in the context of aerobic sewage-treatment processes, such as the activated sludge system. As they grow, *Zoogloea* bacteria form fluffy, slimy masses that are essential to the proper operation of such systems.

The γ (gamma) Proteobacteria

Learning Objective

• *Make a dichotomous key to distinguish among the orders of γ-proteobacteria described in this chapter.*

The γ-proteobacteria constitute the largest subgroup of the proteobacteria and include a great variety of physiological types.

Beggiatoa *Beggiatoa alba* (bej′jē-ä-tō-ä al′ba), the only species of this unusual genus, grows in aquatic sediments at the interface between the aerobic and anaerobic layers. Morphologically, it resembles certain filamentous cyanobacteria (page 317), but it is not photosynthetic. Motility is by gliding. The exact mechanism of gliding motility is uncertain; there may actually be several mechanisms involved. The only common factor among the mechanisms is the production of slime, which attaches to the surface on which movement occurs and also provides lubrication allowing the organism to glide.

Nutritionally *Beggiatoa alba* uses hydrogen sulfide (H_2S) as an energy source and accumulates internal granules of sulfur.

TEM |———| 1 μm

FIGURE 11.7 *Pseudomonas.* This photo of a *Pseudomonas* bacterium shows polar flagella that are a characteristic of the genus. In some species only a single flagellum is present (see Figure 4.7a).

■ Pseudomonads are gram-negative rods.

Francisella *Francisella* (fran′sis-el′lä) is a genus of small, pleomorphic bacteria that grow only on complex media enriched with blood or tissue extracts. *Francisella tularensis* (tü′lär-en-sis) causes the disease tularemia.

Pseudomonadales

Members of the order Pseudomonadales are gram-negative aerobic rods or cocci. The most important genus in this group is *Pseudomonas*.

Pseudomonas A very important genus, *Pseudomonas* (sū-dō-mō′nas) consists of aerobic, gram-negative rods that are motile by polar flagella, either single or tufts (Figure 11.7). Pseudomonads are very common in soil and other natural environments.

Many species of pseudomonads excrete extracellular, water-soluble pigments that diffuse into their media. One species, *Pseudomonas aeruginosa* (ā-rü-ji-nō′sä), produces a soluble, blue-green pigmentation. Under certain conditions, particularly in weakened hosts, this organism can infect the urinary tract, burns, and wounds, and can cause blood infections (sepsis), abscesses, and meningitis. Other pseudomonads produce soluble fluorescent pigments that glow when illuminated by ultraviolet light. One species, *P. syringae* (sèr′in-gī), is an occasional plant pathogen. (Some species of *Pseudomonas* have been transferred, based upon rRNA studies, to the genus *Burkholderia*, which was discussed previously with the β-proteobacteria.)

Pseudomonads have almost as much genetic capacity as the eukaryotic yeasts and almost half as much as a fruit fly. While these bacteria are less efficient than some other heterotrophic bacteria in utilizing many of the common nutrients, they make use of their genetic capacity by compensating for this in other ways. For example, pseudomonads synthesize an unusually large number of enzymes and can metabolize a wide variety of substrates. Therefore, they probably contribute significantly to the decomposition of uncommon chemicals, such as pesticides, that are added to soil.

In hospitals and other places where pharmaceutical agents are prepared, the ability of pseudomonads to grow on minute traces of unusual carbon sources, such as soap residues or cap-liner adhesives found in a solution, has been unexpectedly troublesome. Pseudomonads are even capable of growth in some antiseptics, such as quaternary ammonium compounds. Their resistance to most antibiotics has also been a source of medical concern. This resistance is probably related to the characteristics of the cell wall porins, which control the entrance of molecules through the cell wall (see Chapter 4, page 86). The large genome of pseudomonads also codes for several very efficient efflux pump systems (page 582) that eject antibiotics from the cell before they can function. Pseudomonads are responsible for about one in ten nosocomial infections (hospital-acquired infections; see page 422), especially among infections in burn units. Persons with cystic fibrosis are also especially prone to infections by *Pseudomonas* and the closely related *Burkholderia*.

Although pseudomonads are classified as aerobic, some are capable of substituting nitrate for oxygen as a terminal electron acceptor. This process, anaerobic respiration, yields almost as much energy as aerobic respiration (see Chapter 5). In this way, pseudomonads cause important losses of valuable nitrogen in fertilizer and soil. Nitrate (NO_3^-) is the form of fertilizer nitrogen most easily used by plants. Under anaerobic conditions, as in water-logged soil, pseudomonads eventually convert this valuable nitrate into nitrogen gas (N_2), which is lost to the atmosphere.

Many pseudomonads can grow at refrigerator temperatures. This characteristic, combined with their ability to utilize proteins and lipids, makes them an important contributor to food spoilage.

Azotobacter and Azomonas Some nitrogen-fixing bacteria, such as *Azotobacter* (ā-zō-tō-bak′tèr) and *Azomonas* (ā-zō-mō′nas), are free-living in soil. These large, ovoid, heavily capsulated bacteria are frequently used in laboratory demonstrations of nitrogen fixation. However, to fix agriculturally significant amounts of nitrogen, they would require energy sources, such as carbohydrates, that are in limited supply in soil.

Moraxella Members of the genus *Moraxella* (mô-raks-el'lä) are strictly aerobic coccobacilli—that is, intermediate in shape between cocci and rods. *Moraxella lacunata* (la-kü-nä'tä) is implicated in conjunctivitis, an inflammation of the conjunctiva, the membrane that covers the eye and lines the eyelids.

Legionellales

The genera *Legionella* and *Coxiella* are closely associated in the second edition of *Bergey's Manual*, where both are placed in the same order, Legionellales. Because the *Coxiella* share an intracellular lifestyle with the rickettsial bacteria, they were previously considered rickettsial in nature and grouped with them. *Legionella* bacteria use binary fission and grow readily on suitable artificial media.

Legionella *Legionella* (lē-jä-nel'lä) bacteria were originally isolated during a search for the cause of an outbreak of pneumonia now known as legionellosis. The search was difficult because these bacteria did not grow on the usual laboratory isolation media then available. After intensive effort, special media were developed that enabled researchers to isolate and culture the first *Legionella*. Microbes of this genus are now known to be relatively common in streams, and they colonize such habitats as warm-water supply lines in hospitals and water in the cooling towers of air conditioning systems. An ability to survive and reproduce within aquatic amoebae often makes them difficult to eradicate in water systems.

Coxiella *Coxiella burnetii* (käks-ė-el'lä bėr-ne'tē-ē), which causes Q fever, was formerly grouped with the rickettsia. Like them, *Coxiella* bacteria require a mammalian host cell in order to reproduce. Unlike rickettsias, *Coxiella* are not transmitted among humans by insect or tick bites. Although cattle ticks harbor the organism, it is most commonly transmitted by aerosols or contaminated milk. A sporogenic cycle has been reported for *C. burnetii*. This might explain the bacterium's relatively high resistance to the stresses of airborne transmission and heat treatment.

Vibrionales

Members of this order are facultatively anaerobic gram-negative rods. Many are slightly curved. They are found mostly in aquatic habitats.

Vibrio Members of the genus *Vibrio* (vib'rē-ō) are rods that are often slightly curved (Figure 11.8). One important pathogen is *Vibrio cholerae* (kol'er-ī), the causative agent of cholera. The disease is characterized by a profuse and watery diarrhea. *V. parahaemolyticus* (pa-rä-hē-mō-li'ti-kus)

TEM |— 1 µm —|

FIGURE 11.8 **Vibrio cholerae.** Notice the slight curvature of these rods, which is a characteristic of the genus.

■ What disease does *Vibrio cholerae* cause?

causes a less serious form of gastroenteritis. Usually inhabiting coastal salt waters, it is transmitted to humans mostly by raw or undercooked shellfish.

Enterobacteriales

The members of the order Enterobacteriales are facultatively anaerobic, gram-negative rods that are, if motile, peritrichously flagellated. Morphologically, the rods are straight. They have simple nutritional requirements. This is an important bacterial group, often commonly called **enterics.** This reflects the fact that they inhabit the intestinal tracts of humans and other animals. Most enterics are active fermenters of glucose and other carbohydrates.

Because of the clinical importance of enterics, there are many techniques to isolate and identify them. An identification key to some enterics is shown in Figure 10.9, which incorporates a modern tool using 15 biochemical tests. Biochemical tests are especially important in clinical laboratory work and in food and water microbiology. Enterics can also be distinguished from each other according to antigens present on their surfaces, that is, by serology (see Chapter 18).

Enterics have fimbriae that help them adhere to surfaces or mucous membranes. Specialized sex pili aid in the exchange of genetic information between cells, which often includes antibiotic resistance (see Figures 8.26 and 8.27).

Enterics, like many bacteria, produce proteins called bacteriocins that cause the lysis of closely related species of bacteria. Bacteriocins may help maintain the ecological balance of various enterics in the intestines.

Escherichia The bacterial species *Escherichia coli* is one of the most common inhabitants of the human intestinal tract and is probably the most familiar organism in microbiology. Recall from previous chapters that a great deal is known about the biochemistry and genetics of *E. coli,* and it continues to be an important tool for basic biological research—many researchers consider it almost a laboratory pet. Its presence in water or food is an indication of fecal contamination. *E. coli* is not usually pathogenic. However, it can be a cause of urinary tract infections, and certain strains produce enterotoxins that cause traveler's diarrhea and occasionally cause very serious foodborne disease.

Salmonella Almost all members of the genus *Salmonella* (sal-mön-el'lä) are potentially pathogenic. Accordingly, there are extensive biochemical and serological tests to clinically isolate and identify salmonellae. Salmonellae are common inhabitants of the intestinal tracts of many animals, especially poultry and cattle. Under unsanitary conditions, they can contaminate food.

The nomenclature of the genus *Salmonella* is unusual. Instead of multiple species, members of the genus *Salmonella* that are infectious to warm-blooded animals can be considered for practical purposes to be a single species, *Salmonella enterica* (en-ter'i-kä). This species is divided into more than 2400 **serovars,** that is, *serological varieties.* The term **serotype** is often used to mean the same thing. By way of explanation of these terms, when salmonellae are injected into appropriate animals, their flagella, capsules, and cell walls serve as *antigens* that cause the animals to form *antibodies* in their blood that are specific for each of these structures. Thus, *serological* means are used to differentiate the microorganisms. Serology is discussed more fully in Chapter 18, but for now it will be sufficient to state that it can be used to differentiate and identify bacteria.

A serovar such as *Salmonella typhimurium* (tī-fi-mur' ē-um) is not a species and should be more properly written as "*Salmonella enterica* serovar Typhimurium." The convention now used by the Centers for Disease Control and Prevention (CDC) is to spell out the entire name at the first mention and then abbreviate as, for example, *Salmonella* Typhimurium. To minimize confusion, we will identify serovars of salmonellae in this text as we would species, that is, *S. typhimurium,* etc.

Specific antibodies, which are available commercially, can be used to differentiate *Salmonella* serovars by a system known as the Kauffmann-White scheme. This scheme designates an organism by numbers and letters that correspond to specific antigens on the organism's capsule, cell wall, and flagella, which are identified by the letters K, O,

and H, respectively. For example, the antigenic formula for the bacterium *S. typhimurium* is O1,4,[5],12:H,i,1,2.* Many salmonellae are named only by their antigenic formulas. Serovars can be further differentiated by special biochemical or physiological properties into **biovars,** or **biotypes.**

A recent taxonomic arrangement based upon the latest molecular technology adds another species, *Salmonella bongori* (bon'gôr-ē). This is a resident of "cold-blooded" animals—it was originally isolated from a lizard in the town of Bongor in the African desert nation of Chad—and is rarely found in humans.

Typhoid fever, caused by *Salmonella typhi* (tī'fē), is the most severe illness caused by any member of the genus *Salmonella.* A less severe gastrointestinal disease caused by other salmonellae is called salmonellosis. Salmonellosis is one of the most common forms of foodborne illness.

Shigella Species of *Shigella* (shi-gel'lä) are responsible for a disease called bacillary dysentery, or shigellosis. Unlike salmonellae, they are found only in humans. These organisms are second only to *E. coli* as a cause of traveler's diarrhea. Some strains of *Shigella* can cause life-threatening dysentery.

Klebsiella Members of the genus *Klebsiella* (kleb-sē-el'lä) are commonly found in soil or water. Many isolates are capable of fixing nitrogen from the atmosphere, which has been proposed as being a nutritional advantage in isolated populations with little protein nitrogen in their diet. The species *Klebsiella pneumoniae* (nü-mō'nē-ī) occasionally causes a serious form of pneumonia in humans.

Serratia *Serratia marcescens* (ser-rä'tē-ä mär-ses'sens) is a bacterial species distinguished by its production of red pigment. In hospital situations the organism can be found on catheters, in saline irrigation solutions, and in other supposedly sterile solutions. Such contamination is probably the cause of many urinary and respiratory tract infections in hospitals.

---

*The letters derive from the original German usage: K represents the German for capsule. (Salmonellae with capsules are identified serologically by a particular capsular antigen named Vi, for virulence.) Colonies that spread in a thin film over the agar surface were described by the German word for film, *hauch.* The motility needed to form a film implied the presence of flagella, and the letter H came to be assigned to the antigens of flagella. Nonmotile bacteria were described as *ohne hauch,* without film, and the O came to be assigned to the cell surface or body antigens. This terminology is also used in the naming of *E. coli* O157:H7, *Vibrio cholerae* O:1 and others.

(a) A gram-negative rod with peritrichous flagella

TEM |—————| 1 μm

(b) A swarming colony of *Proteus mirabilis;* note concentric rings of growth

|—————| 5 mm

FIGURE 11.9 ***Proteus mirabilis.***

■ The photo of the *Proteus* cell is probably a swarmer cell. How would you know?

Proteus Colonies of *Proteus* (prō′tē-us) bacteria growing on agar exhibit a swarming type of growth. Swarmer cells with many flagella (Figure 11.9a) move outward on the edges of the colony and then revert to normal cells with only a few flagella and reduced motility. Periodically, new generations of highly motile swarmer cells develop and the process is repeated. As a result a *Proteus* colony has the distinctive appearance of a series of concentric rings (Figure 11.9b). This genus of bacteria is implicated in many infections of the urinary tract and in wounds.

Yersinia *Yersinia pestis* (yĕr-sin′ē-ä pes′tis) causes plague, the Black Death of medieval Europe. Urban rats in some parts of the world and ground squirrels in the American Southwest carry these bacteria. Fleas usually transmit the organisms among animals and to humans, although contact with respiratory droplets from infected animals and people can be involved in transmission.

Erwinia *Erwinia* (ėr-wi′nē-ä) species are primarily plant pathogens; some cause plant soft-rot diseases. These species produce enzymes that hydrolyze the pectin between individual plant cells. This causes the plant cells to separate from each other, a disease that plant pathologists term plant rot.

Enterobacter Two *Enterobacter* (en-te-rō-bak′tėr) species, *E. cloacae* (klō-ā′kī), and *E. aerogenes* (ā-rä′jen-ēz) can cause urinary tract infections and hospital-acquired infections. They are widely distributed in humans and animals, as well as in water, sewage, and soil.

Pasteurellales

The bacteria in this order are nonmotile; they are best known as human and animal pathogens.

Pasteurella This genus is primarily known as a pathogen of domestic animals. It causes sepsis in cattle, fowl cholera in chickens and other fowl, and pneumonia in several types of animals. The best-known species is *Pasteurella* (pas-tyėr-el′lä) *multocida* (mul-tō′si-dä), which can be transmitted to humans by dog and cat bites.

Haemophilus *Haemophilus* (hē-mä′fil-us) is a very important genus of pathogenic bacteria. These organisms commonly inhabit the mucous membranes of the upper respiratory tract, mouth, vagina, and intestinal tract. The best-known species that affects humans is *Haemophilus influenzae* (in-flü-en′zī), named long ago because of the erroneous belief that it was responsible for influenza.

The name *Haemophilus* is derived from the bacteria's requirement for blood in their culture medium (*hemo* = blood). They are unable to synthesize important parts of the cytochrome system needed for respiration, and they obtain these substances from the heme fraction, known as the **X factor,** of blood hemoglobin. The culture medium must also supply the cofactor nicotinamide adenine dinucleotide (NAD^+ or $NADP^+$), which is known as **V factor.** Clinical laboratories use tests for the requirement of X and V factors to identify isolates as *Haemophilus* species.

Haemophilus influenzae is responsible for several important diseases. It has been a common cause of meningitis in young children and is a frequent cause of earaches. Other clinical conditions caused by *H. influenzae* include epiglotitis (a life-threatening condition in which the epiglottis becomes infected and inflamed), septic arthritis in children, bronchitis, and pneumonia. *Haemophilus ducreyi* (dü-krā′ē) is the cause of the sexually transmitted disease chancroid.

The δ (delta) Proteobacteria

Learning Objective

- *Make a dichotomous key to distinguish among the δ-proteobacteria described in this chapter.*

The δ-proteobacteria are distinctive in that they include some bacteria that are predators on other bacteria. Bacteria in this group are also important contributors to the sulfur cycle.

Bdellovibrio *Bdellovibrio* (del-lō-vib′rē-ō) is a particularly interesting genus. It attacks other gram-negative bacteria. It attaches tightly (*bdella* = leech), and after penetrating the outer layer of gram-negative bacteria, it reproduces within the periplasm. There, the cell elongates into a tight spiral, which then fragments almost simultaneously into several individual flagellated cells. The host cell then lyses, releasing them (see the box in Chapter 3, page 65).

Desulfovibrionales

Members of the order Desulfovibrionales are sulfur reducing bacteria. They are obligately anaerobic bacteria that use oxidized forms of sulfur, such as sulfates (SO_4^{2-}) or elemental sulfur (S^0), rather than oxygen as electron accep-

tors. The product of this reduction is hydrogen sulfide (H_2S). (Because the H_2S is not assimilated as a nutrient, this type of metabolism is termed *dissimilatory*.) The activity of these bacteria releases millions of tons of H_2S into the atmosphere every year and plays a key part in the sulfur cycle. Sulfur-oxidizing bacteria such as *Beggiatoa* are able to use H_2S either as part of photosynthesis or as an autotrophic energy source.

Desulfovibrio The best studied sulfur-reducing genus is *Desulfovibrio* (dē′sul-fō-vib′rē-ō), which is found in anaerobic sediments, and also in the intestinal tracts of humans and animals. Sulfur-reducing and sulfate-reducing bacteria use organic compounds such as lactate, ethanol, or fatty acids as electron donors. This reduces sulfur or sulfate to H_2S. When H_2S reacts with iron it forms insoluble FeS, which is responsible for the black color of many sediments.

Myxococcales

In the first edition of *Bergey's Manual* the Myxococcales were classified among the fruiting and gliding bacteria. They illustrate the most complex life cycle of all bacteria, part of which is predatory upon other bacteria.

Myxococcus Vegetative cells of the myxobacteria (*myxo* = nasal mucus) move by gliding and leave behind a slime trail (Figure 11.10a). *Myxococcus xanthus* (micks-ō-kok′kus zan′thus) and *M. fulvus* (ful′vus) are well-studied representatives of the genus. As they move, their source of nutrition is the bacteria they encounter, enzymatically lyse, and digest. Large numbers of these gram-negative microbes eventually

(a) Bacterial slime trails

Slime trails

Myxococcus bacteria

LM · 10 μm

(b) Myxobacterium with fruiting bodies that contain myxospores

LM · 100 μm

FIGURE 11.10 **Myxococcales.**

■ A few bacteria move by gliding over surfaces, usually leaving a trail.

aggregate. Where the moving cells aggregate, they differentiate and form a macroscopic stalked fruiting body that contains resting cells called *myxospores* (Figure 11.10b). Differentiation is usually triggered by low nutrients. Under proper conditions, usually a change in nutrients, the myxospores germinate and form new vegetative gliding cells. You might note the resemblance to the life cycle of the eukaryotic cellular slime molds in Figure 12.22 on page 360.

The ε (epsilon) Proteobacteria

Learning Objective

- *Make a dichotomous key to distinguish among the ε-proteobacteria described in this chapter.*

The ε-proteobacteria are slender gram-negative rods that are helical or vibrioid. Vibrioid is a term applied to helical bacteria that do not have a complete turn. We will discuss the two important genera, both of which are motile by means of flagella and are microaerophilic.

Campylobacter Members of the genus *Campylobacter* are microaerophilic vibrios; each cell has one polar flagellum. One species of *Campylobacter*, *C. fetus* (kam′pi-lō-bak-tėr fē′tus), causes spontaneous abortion in domestic animals. Another species, *C. jejuni* (je-ju′ni), is a leading cause of outbreaks of foodborne intestinal disease.

Helicobacter Members of the genus *Helicobacter* are microaerophilic curved rods with multiple flagella. The species *Helicobacter pylori* (hē-lik-ō-bak-tėr pī-lôr′ē) has been identified as the most common cause of peptic ulcers in humans and a cause of stomach cancer. See the box in Chapter 8 on page 235. (Figure 11.11).

The Nonproteobacteria Gram-Negative Bacteria

Learning Objective

- *Make a dichotomous key to distinguish among the gram-negative nonproteobacteria described in this chapter.*

There are a number of important gram-negative bacteria that are not closely related to the gram-negative proteobacteria. They include several physiologically and morphologically distinctive photosynthesizing bacteria, such as those included in the phyla Cyanobacteria (cyanobacteria), Chlorobi (green sulfur bacteria), and Chloroflexi (green nonsulfur bacteria). The cyanobacteria produce oxygen during photosynthesis (are oxygenic), and the green sulfur and green nonsulfur bacteria do not produce oxygen (are anoxygenic). For convenience we will divide this discussion of oxygenic photosynthetic bacteria and the anoxygenic photosynthetic bacteria. These groups are summarized in Table 11.2.

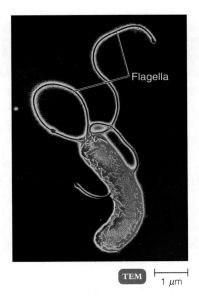

FIGURE 11.11 *Helicobacter pylori,* **an example of a helical bacterium that does not make a complete twist.**

■ How do helical bacteria differ from spirochetes?

Cyanobacteria (The Oxygenic Photosynthetic Bacteria)

The cyanobacteria, named for their characteristic blue-green (cyan) pigmentation, were once called blue-green algae. Although they resemble the eukaryotic algae and often occupy the same environmental niches, this is a misnomer because they are bacteria; algae are not. However, cyanobacteria do carry out oxygen-producing (oxygenic) photosynthesis, as do the eukaryotic plants and algae (see Chapter 12). Many of the cyanobacteria are capable of fixing nitrogen from the atmosphere. In most cases this activity is located in specialized cells called **heterocysts,** which contain enzymes that fix nitrogen gas (N_2) into ammonium (NH_4^+) that can be used by the growing cell (Figure 11.12a). Species that grow in water usually have gas vacuoles that provide buoyancy, helping the cell float at a favorable environment. Cyanobacteria that move about on solid surfaces use gliding motility.

Cyanobacteria are morphologically varied. They range from unicellular forms that divide by simple binary fission (Figure 11.12b), to colonial forms that divide by multiple fission, to filamentous forms that reproduce by fragmentation of the filaments. The filamentous forms usually exhibit some differentiation of cells that are often bound together within an envelope or sheath (Figure 11.12c).

Evidence indicates that oxygenic cyanobacteria played an important part in the development of life on Earth, which originally had very little free oxygen that would support life as we are familiar with it. Fossil evidence indicates that when

TABLE 11.2	**Selected Characteristics of Photosynthesizing Bacteria**					
Common Name	**Example**	**Phylum**	**Comments**	**Electron Donor for CO_2 Reduction**	**Oxygenic or Anoxygenic**
Cyanobacteria	*Anabaena*	Cyanobacteria	Plantlike photosynthesis; some use bacterial photosynthesis under anaerobic conditions	Usually H_2O	Usually oxygenic
Green nonsulfur bacteria	*Chloroflexus*	Chloroflexi	Grow chemoheterotrophically in aerobic environments	Organic compounds	Anoxygenic
Green sulfur bacteria	*Chlorobium*	Chlorobi	Deposit sulfur granules outside cells	Usually H_2S	Anoxygenic
Purple nonsulfur bacteria	*Rhodospirillum*	Proteobacteria	Can grow chemoheterotrophically as well	Organic compounds	Anoxygenic
Purple sulfur bacteria	*Chromatium*	Proteobacteria	Deposit sulfur granules outside cells	Usually H_2S	Anoxygenic

cyanobacteria first appeared, the atmosphere contained only about 0.1% free oxygen. When oxygen-producing eukaryotic plants appeared millions of years later, the concentration of oxygen was more than 10%. The increase presumably was a result of photosynthetic activity by cyanobacteria. The atmosphere we breathe today contains about 20% oxygen.

Cyanobacteria, especially those that fix nitrogen, are extremely important to the environment. They occupy environmental niches similar to those occupied by the eukaryotic algae (see Figure 12.11), but the ability of many of the cyanobacteria to fix nitrogen makes them even more adaptable in nutritionally poor environments.

Purple and Green Photosynthetic Bacteria (The Anoxygenic Photosynthetic Bacteria)

Learning Objective

• Compare and contrast purple and green photosynthetic bacteria with the cyanobacteria.

The photosynthetic bacteria are taxonomically confusing. The phyla Cyanobacteria, Chlorobi, and Chloroflexi are gram-negative, but they are not genetically included in the proteobacteria. The photosynthetic purple sulfur bacteria and purple nonsulfur bacteria are genetically included in, respectively, the α and γ-proteobacteria, although for simplicity we will discuss them at this point. These photosynthetic bacteria, which are not necessarily colored purple or green, are generally anaerobic. Their habitat is usually the deep sediments of lakes and ponds. Like plants, algae, and the cyanobacteria, purple and green bacteria carry out photosynthesis to make carbohydrates (CH_2O). Growing as they do in aquatic depths, these bacteria possess chlorophyll that makes use of parts of the visible spectrum not intercepted by photosynthetic organisms located at higher levels. Also, unlike plantlike photosynthesis, the photosynthesis of purple or green bacteria is *anoxygenic*—it does not produce oxygen.

Plants, algae, and the cyanobacteria, which we will discuss shortly, produce oxygen (O_2) from water (H_2O) as they carry out photosynthesis:

$$(1) \quad 2H_2O + CO_2 \xrightarrow{\text{light}} (CH_2O) + H_2O + O_2$$

The *purple sulfur* and *green sulfur bacteria* use reduced sulfur compounds, such as hydrogen sulfide (H_2S), instead of water, and they produce granules of sulfur (S^0) rather than oxygen, as follows:

$$(2) \quad 2H_2S + CO_2 \xrightarrow{\text{light}} (CH_2O) + H_2O + 2S^0$$

Chromatium (krō-mā′tē-um), shown in Figure 11.13, is a representative genus. At one time, an important question in biology concerned the source of the oxygen produced by

(a) Filamentous, with heterocysts   LM   10 μm

(b) Nonfilamentous   LM   10 μm

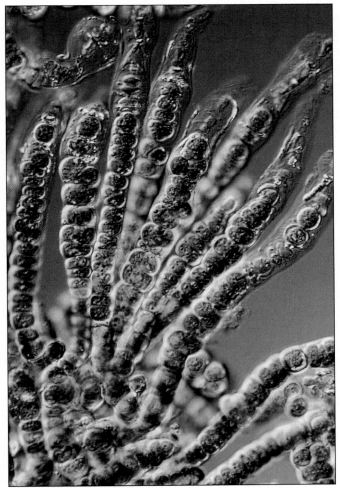

(c) Branching, filamentous   LM   10 μm

FIGURE 11.12 **Cyanobacteria. (a)** A filamentous cyanobacterium showing heterocysts, in which nitrogen-fixing activity is located. **(b)** A unicellular, nonfilamentous cyanobacterium, *Gloeocapsa*. Groups of these cells, which divide by binary fission, are held together by the surrounding glycocalyx. **(c)** A branching, filamentous cyanobacterium.

■ How does the photosynthesis of the cyanobacteria differ from that of the purple sulfur bacteria?

evidence that the oxygen source was from H_2O. It is important, also, to compare these two equations for an understanding of how reduced sulfur compounds, such as H_2S, can substitute for photosynthesis. (See the box in Chapter 6 on page 160.)

Other photoautotrophs, the *purple nonsulfur* and *green nonsulfur bacteria*, use organic compounds, such as acids and carbohydrates, for the photosynthetic reduction of carbon dioxide.

Morphologically, the photosynthetic bacteria are very diverse, with spirals, rods, cocci, and even budding forms.

The Gram-Positive Bacteria

The gram-positive bacteria can be divided into two groups: those that have a high G + C ratio, and those that have a low G + C ratio. To illustrate the variations in G + C ratio, the genus *Streptococcus* has a low G + C content of 33–44%; and the genus *Clostridium* has a low content of 21–54%. Included with the gram-positive, low G + C bacteria are the mycoplasmas, even though they lack a cell wall and therefore do not have a Gram reaction. Their G + C ratio is 23–40%.

plant photosynthesis; was it from CO_2 or from H_2O? Until the introduction of radioisotope tracers, which traced the oxygen in water and carbon dioxide and finally settled the question, comparison of equations 1 and 2 was the best

LM 10 μm

FIGURE 11.13 **Purple sulfur bacteria.** This photomicrograph of cells of the genus *Chromatium* shows the intracellular sulfur granules as multicolored refractile objects. The reason the sulfur accumulates can be surmised from inspection of equation 2 in the discussion.

■ Purple sulfur bacteria are anoxygenic photoautotrophs.

By contrast, filamentous actinomycetes of the genus *Streptomyces* have a high G + C content of 69–73%. Gram-positive bacteria of a more conventional morphology, such as the genera *Corynebacterium* and *Mycobacterium*, have, respectively, a G + C content of 51–63% and 62–70%.

These bacterial groups are placed into separate phyla, the Firmicutes (low G + C ratios) and Actinobacteria (high G + C ratios).

Firmicutes (Low G + C Gram-Positive Bacteria)

Learning Objective

- *Make a dichotomous key to distinguish among the low G + C gram-positive bacteria described in this chapter.*

Low G + C gram-positive bacteria are assigned to the phylum Firmicutes. This group includes important endospore-forming bacteria such as the genera *Clostridium* and *Bacillus*. Also of extreme importance in medical microbiology are the genera *Staphylococcus*, *Enterococcus*, and *Streptococcus*. In industrial microbiology the genus *Lactobacillus*, which produces lactic acid, is well known. The mycoplasma, which do not possess a cell wall, are also found in this phylum.

Clostridiales

Clostridium Members of the genus *Clostridium* (klôs-tri′dē-um) are obligate anaerobes. The rod-shaped cells

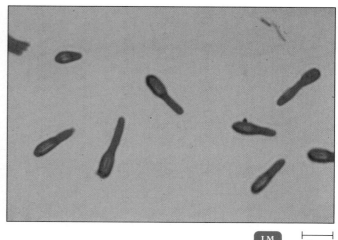

LM 5 μm

FIGURE 11.14 **Clostridium tetani.** The endospores of clostridia usually distend the cell wall as shown here.

■ Endospores are highly resistant bodies and an especially important factor in sterilization standards.

contain endospores that usually distend the cell (Figure 11.14). The formation of endospores by bacteria is important to both medicine and the food industry because of the endospore's resistance to heat and many chemicals. Diseases associated with clostridia include tetanus, caused by *C. tetani* (te′tan-e); botulism, caused by *C. botulinum* (bo-tū-lī′num); and gas gangrene, caused by *C. perfringens* (per-frin′jens) and other clostridia. *C. perfringens* is also the cause of a common form of foodborne diarrhea. *C. difficile* (dif′fi-sil-ē) is an inhabitant of the intestinal tract that may cause a serious diarrhea. This occurs only when antibiotic therapy alters the normal intestinal microbiota, allowing overgrowth by toxin-producing *C. difficile*.

Epulopiscium Biologists have long considered bacteria to be small by necessity because they lack the nutrient transport systems used by higher, eukaryotic organisms and because they depend on simple diffusion to obtain nutrients. These characteristics would seem to critically limit size. So, when a cigar-shaped organism living symbiotically in the gut of the Red Sea surgeonfish was first observed in 1985, it was considered to be a protozoan. Certainly, its size suggested this: the organism was as large as 80 μm × 600 μm—over half a millimeter in length—large enough to be seen with the unaided eye (Figure 11.15). Compared to the familiar bacterium *E. coli*, which is about 1 μm × 2 μm, this organism would be about a million times larger in volume.

Further investigation of the new organism showed that certain external structures thought to resemble the cilia of protozoa were actually similar to bacterial flagella, and it did not have a membrane-enclosed nucleus. Ribosomal

LM | 0.5 mm

FIGURE 11.15 **A giant prokaryote, *Epulopiscium fishelsoni.*** Several of these bacteria, each roughly a million times the volume of *Escherichia coli,* are shown on the head of a steel pin.

■ Why was this giant cell classified as a bacterium?

RNA analysis conclusively placed *Epulopiscium* (ep'ū-lō-pis-ē-um) with the prokaryotes. (The name means "guest at the banquet of a fish." It is literally bathed in semi-digested food.) It most closely resembles gram-positive bacteria of the genus *Clostridium.* Strangely, the species *Epulopiscium fishelsoni* (fish-el-sō'nē) does not reproduce by binary fission. Daughter cells formed within the cell are released through a slit opening in the parent cell. This may be related to the evolutionary development of sporulation.

This extraordinary bacterium has considerable significance for biologists. Because eukaryotes can be as small as 1–2 μm long, there is substantial overlap in size between prokaryotic and eukaryotic cells. Previously, scientists studying microfossils had considered small ones to be prokaryotic and large ones to be eukaryotic. The discovery of the giant prokaryote casts doubt on this view, which may ultimately affect how we view evolutionary development. It also raises questions about the capability of the nutrient transport systems of bacteria or the efficiency of simple diffusion in serving the needs of such a large cell. Recently it was discovered that this bacterium does not rely on diffusion to distribute nutrients. Instead, it makes use of its larger genetic capacity—it has 25 times as much DNA as a human cell and as many as 85,000 copies of at least one gene—to manufacture proteins at internal sites where they are needed. *E. fishelsoni* bacteria are large enough for researchers to insert microprobes and directly study such

aspects of bacterial physiology, something not possible with conventionally sized bacteria. (Near the conclusion of this chapter, we will describe another, more recently discovered, giant bacterium, *Thiomargarita* on page 330.)

Bacillales

This order includes several important genera of gram-positive rods and cocci.

Bacillus Bacteria of the genus *Bacillus* are typically rods that produce endospores. They are common in soil and only a few are pathogenic to humans. Several species produce antibiotics.

Bacillus anthracis (bä-sil'lus an-thrā'sis) causes anthrax, a disease of cattle, sheep, and horses that can be transmitted to humans. It is often mentioned as a possible agent of biological warfare. The anthrax bacillus is a nonmotile facultative anaerobe, often forming chains in culture. The centrally located endospore does not distend the walls. *Bacillus thuringiensis* (thur-in-jē-en'sis) is probably the best-known microbial insect pathogen (Figure 11.16a). It produces intracellular crystals when it sporulates. Commercial preparations containing endospores and crystalline toxin of this bacterium are sold in gardening supply shops to be sprayed on plants. *Bacillus cereus* (se'rē-us) (Figure 11.16b) is a common bacterium in the environment and occasionally is identified as a cause of food poisoning, especially in starchy foods such as rice.

The three species of the genus *Bacillus* that we have just described are dramatically different in important ways, especially their disease-causing properties. However, they are so closely related that taxonomists consider them to be variants of a single species, differing almost entirely in genes carried on plasmids, which are easily transferred from one bacterium to another.

Staphylococcus Staphylococci typically occur in grapelike clusters (Figure 11.17). The most important staphylococcal species is *Staphylococcus aureus* (staf-i-lō-kok'kus ô'rē-us), which is named for its yellow-pigmented colonies (*aureus* = golden). Members of this species are facultative anaerobes.

Some characteristics of the staphylococci account for their pathogenicity, which takes many forms. They grow comparatively well under conditions of high osmotic pressure and low moisture, which partially explains why they can grow and survive in nasal secretions (many of us carry the bacteria in our noses) and on the skin. This also explains how *S. aureus* can grow in some foods with high osmotic pressure (such as ham and other cured meats) or in low-moisture foods that tend to inhibit the growth of other organisms. The yellow pigment probably confers some protection from the antimicrobial effects of sunlight.

(a) *Bacillus thuringiensis*   TEM   1 μm

FIGURE 11.16 **Bacillus. (a)** *Bacillus thuringiensis;* the diamond-shaped crystal shown next to the endospore is toxic to insects that ingest it. This electron micrograph was made using the technique of shadow casting described on page 64. **(b)** A germinating cell of *Bacillus cereus.*

▪ What structure is made by both *Clostridium* and *Bacillus?*

(b) *Bacillus cereus* germinating   TEM   1 μm

S. *aureus* produces many toxins that contribute to the bacterium's pathogenicity by increasing its ability to invade the body or damage tissue. The infection of surgical wounds by S. *aureus* is a common problem in hospitals. And its ability to develop resistance quickly to such antibiotics as penicillin contributes to its danger to patients in hospital environments. S. *aureus* produces the toxin responsible for toxic shock syndrome, a severe infection characterized by high fever and vomiting, sometimes even death. S. *aureus* also produces an **enterotoxin** that causes vomiting and nausea when ingested; it is one of the most common causes of food poisoning.

Lactobacillales

Several important genera are found in the order Lactobacillales. The genus *Lactobacillus* is a representative of the industrially important lactic-acid producing bacteria. Most lack a cytochrome system and are unable to use oxygen as an electron acceptor. Unlike most obligate anaerobes, though, they are aerotolerant and capable of growth in the presence of oxygen. But compared to oxygen-utilizing microbes, they grow poorly. However, the production of lactic acid from simple carbohydrates inhibits the growth of competing organisms and allows them to grow competitively in spite of their inefficient metabolism. The genus *Streptococcus* shares the metabolic characteristics of the genus *Lactobacillus.* There are several industrially important species, but the streptococci are best known for their pathogenicity. The genera *Staphylococcus* and *Listeria* are

more conventional metabolically. Both are facultative anaerobes and several species are important pathogens.

Lactobacillus In humans, bacteria of the genus *Lactobacillus* (lak-tō-bä-sil'lus) are located in the vagina, intestinal tract, and oral cavity. Lactobacilli are used commercially in the production of sauerkraut, pickles, buttermilk, and yogurt. Typically, a succession of lactobacilli, each more acid tolerant than its predecessor, participates in these lactic acid fermentations.

SEM   1 μm

FIGURE 11.17 **Staphylococcus aureus.** Notice the grapelike clusters of these gram-positive cocci.

▪ Spherical bacteria tend to be resistant to environmental stresses.

Bacteria and Insect Sex

Wolbachia is quite possibly the most common infectious bacterial genus on Earth. Although they were first discovered in 1924, little had been known about them until the 1990s. They escape detection by the usual culture methods because they live as endosymbionts in the cells of insects and other invertebrates.

Wolbachia infect over a million species of insects as well as millipedes, mites, spiders, crustaceans, and nematodes. In all, as many as 75% of species of animals surveyed have been found to carry this bacterium. Wolbachia is essential to the nematodes. If the bacterium is killed with antibiotics, the host worm dies. And pea aphids infected with Wolbachia are not killed by a normally lethal parasitic wasp larva. The bacterium is harmless in the aphid but kills the wasp.

In some insects, Wolbachia destroys males of its host species. Wolbachia can turn males into females by interfering with the male hormone. If Wolbachia does this in an embryo, the developing gonads become ovaries. As shown in the figure, if a male and female insect are uninfected with Wolbachia, they produce offspring normally. If only the male is infected, the insects fail to reproduce. If

one or both insects of a mating pair are infected, only the infected females reproduce—and transmit Wolbachia in the cytoplasm of their eggs. Offspring produced without fertilization are female. The result is that the bacteria are transmitted to the next generation. This type of reproduction, called parthenogenesis, has been seen in a variety of insects and in some amphibians and reptiles. Thus a question arises: Is Wolbachia always responsible?

Eukaryotic species are defined as organisms that reproduce only with members of their own species. This reproductive isolation prevents the production of hybrids and thus maintains the uniqueness of each species. In the laboratory, researchers have found that after antibiotic treatment, wasps of one species will produce hybrid offspring with another species. This raises the question about the influence Wolbachia has had on the evolution of insects. Did insects that were not infected successfully reproduce outside of their species?

A virulent strain of Wolbachia called "popcorn" causes host cells to lyse, or "pop," which eventually kills the host insect. Researchers at the California Institute of Technology are investigating

whether the popcorn strain might be used to kill mosquitoes.

The unique biology of Wolbachia has attracted researchers interested in questions that range from the evolutionary implications of infection to the use of Wolbachia as a biological control of insect pests.

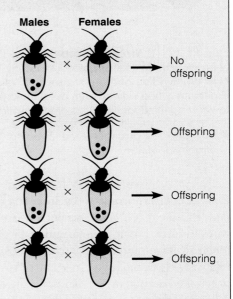

In an infected pair, only female hosts can reproduce.

Streptococcus Members of the genus *Streptococcus* (strep-tō-kok′kus) are spherical, gram-positive bacteria that typically appear in chains (Figure 11.18). They are a complex group, probably responsible for more illnesses and causing a greater variety of diseases than any other group of bacteria. The principal pathogen in the genus is *Streptococcus pyogenes* (pī-äj′en-ēz). Among the diseases caused by *S. pyogenes* are scarlet fever, pharyngitis (sore throat), erysipelas, impetigo, and rheumatic fever.

One basis for the classification of streptococci is their action on blood agar. *Alpha hemolytic* species produce a substance called alpha-hemolysin that reduces hemoglobin (red) to methemoglobin (green); this reduction causes a greenish zone to surround the colony. *Beta-hemolytic* species produce a hemolysin that forms a clear zone of hemolysis on blood agar (see Figure 6.8 on page 167). Some species have no apparent effect on red blood

cells; these are referred to as *nonhemolytic* (or less logically, gamma-hemolytic).

Streptococci produce several extracellular substances that contribute to their pathogenicity. Among them are products that destroy phagocytic cells, which are essential to the body's defenses. Enzymes produced by some streptococci spread infections by digesting connective tissue of the host, which may also result in extensive tissue destruction. Infections are also allowed to spread from sites of injury by enzymes that lyse the fibrin (a threadlike protein) of blood clots.

Other pathogenic streptococci include *S. mutans* (mū′tans), the cause of dental caries (cavities) and *S. pneumoniae* (nü-mō′nē-ī), probably the most common cause of pneumonia. This last species tends to grow as pairs, rather than in chains. A few nonpathogenic species are important in the production of dairy products.

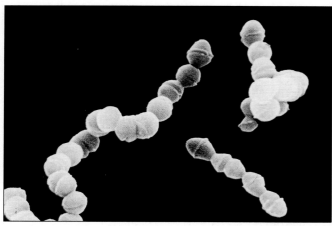

SEM ⊢ 1 μm

FIGURE 11.18 **Streptococcus.** Notice the chains of cells characteristic of most streptococci. Many of the spherical cells are dividing and are somewhat oval in appearance—especially when viewed with a light microscope, which has lower magnification than this electron micrograph.

■ How does the arrangement of *Streptococcus* differ from *Straphylococcus?*

Enterococcus The enterococci are adapted to areas of the body that are rich in nutrients but low in oxygen, such as the gastrointestinal tract, vagina, and oral cavity. They are also found in large numbers in human stool. Because they are relatively hardy microbes, they persist as contaminants in a hospital environment, on hands, bedding, and even as a fecal aerosol. In recent years they have become a leading cause of nosocomial infections, especially because of their high resistance to most antibiotics. Two species, *Enterococcus faecalis* and *Enterococcus faecium*, are responsible for much of the infections of surgical wounds and the urinary tract. In medical settings they frequently enter the bloodstream through invasive procedures, such as indwelling catheters.

Listeria The pathogenic species of the genus *Listeria*, *Listeria monocytogenes* (lis-te′rē-ä mo-nō-sī-to′je-nēz), can contaminate food, especially dairy products. Important characteristics of *L. monocytogenes* are that it survives within phagocytic cells and is capable of growth at refrigeration temperatures. If it infects a pregnant woman, the organism poses the threat of stillbirth or serious damage to the fetus.

Mycoplasmatales

The mycoplasmas are highly pleomorphic because they lack a cell wall (Figure 11.19) and can produce filaments that resemble fungi, hence their name (*mykes* = fungus, and *plasma* = formed). Cells of the genus *Mycoplasma* (mī-ko-plaz′ma) are very small, ranging in size from 0.1 to 0.25 μm,

with a cell volume that is only about 5% of that of a typical bacillus. Because their size and plasticity allowed them to pass through filters that retained bacteria, they were originally considered to be viruses. Mycoplasmas may represent the smallest self-replicating organisms that are capable of a cell-free existence. Studies of their DNA suggest that they are genetically related to the gram-positive bacterial group that includes the genera *Bacillus*, *Streptococcus*, and *Lactobacillus* but have gradually lost genetic material. The term *degenerative evolution* has been used to describe this process.

The most significant human pathogen among the mycoplasmas is M. *pneumoniae* (nu-mō-nē-ī), which is the cause of a common form of mild pneumonia. Other genera in the Order Mycoplasmatales are *Spiroplasma* (spī-rō-plaz′mä), cells with a tight corkscrew morphology that are serious plant pathogens and common parasites of plant-feeding insects, and *Ureaplasma* (ū-rē-ä-plaz′mä), so named because they can enzymatically split the urea in urine and are occasionally associated with urinary tract infections.

Mycoplasmas can be grown on artificial media that provide them with sterols (if necessary) and other special nutritional or physical requirements. Colonies are less than 1 mm in diameter and have a characteristic "fried egg" appearance when viewed under magnification. For many purposes, cell culture methods are often more satisfactory. In fact, mycoplasmas grow so well by this method that they are a frequent contamination problem in cell culture laboratories.

Actinobacteria (High G + C Gram-Positive Bacteria)

Learning Objective

- Make a dichotomous key to distinguish among the high G + C gram-positive bacteria described in this chapter.

High G + C gram-positive bacteria are in the phylum Actinobacteria. Many bacteria in this phylum are highly pleomorphic in their morphology; the genera *Corynebacterium* and *Gardnerella*, for example, and several genera such as *Streptomyces* grow only as extended, often branching filaments. Several important pathogenic genera are found in the Actinobacteria, such as the *Mycobacterium* species causing tuberculosis and leprosy. Although this genus is usually seen microscopically as a typical rod-shaped bacillus, it derives its name from the fact that it occasionally is observed growing as moldlike filaments. The genera *Streptomyces*, *Frankia*, *Actinomyces*, and *Nocardia* are often informally called actinomycetes from the Greek *actino*, because they have a radiate, or starlike, form of growth by reason of their often-branching filaments. Superficially, their morphology resembles that of filamentous fungi; however, the filaments of actinomycetes are prokaryotic cells with a diameter much

(a)　SEM ⊢ 1 μm　　**(b)**　SEM ⊢ 1 μm

FIGURE 11.19 ***Mycoplasma pneumoniae.*** Bacteria such as *M. pneumoniae* have no cell walls, and their morphology is irregular (pleomorphic). Mycoplasma can be grown only on highly specialized media, where they form tiny colonies with a "fried egg" appearance. **(a)** Individual cells of *M. pneumoniae.* Arrowheads indicate terminal structures that probably aid in attachment to eukaryotic cells, which then become infected. **(b)** This micrograph shows the filamentous growth of *M. pneumoniae.* Some individual cells can also be seen. The organism reproduces by fragmentation of the filaments at the bulges.

■ How can the cell structure of mycoplasmas account for their pleomorphism?

smaller than that of the eukaryotic molds. Some actinomycetes further resemble molds by their possession of externally carried asexual spores that are used for reproduction. Filamentous bacteria, like filamentous fungi, are very common inhabitants in soil, where a filamentous pattern of growth has advantages. The filamentous organism can bridge water-free gaps between soil particles to move to a new nutritional site. This morphology also gives the organism a much higher surface-to-volume ratio and improves its ability to absorb nutrients in the highly competitive soil environment.

Mycobacterium The mycobacteria are aerobic, non–endospore-forming rods. The name *myco*, meaning fungus-like, was derived from their occasional exhibition of filamentous growth. Many of the characteristics of mycobacteria, such as acid-fast staining, drug resistance, and pathogenicity, are related to their distinctive cell wall, which is structurally similar to gram-negative bacteria (see Figure 4.13c). However, the outermost lipopolysaccharide layer in mycobacteria is replaced by mycolic acids, which form a waxy, water-resistant layer. This makes the bacteria resistant to stresses such as drying, and few antimicrobial drugs are able to enter the cell. (See the box in Chapter 7 on page 200.) Also, nutrients enter the cell through this layer very slowly, which is a factor in the slow growth rate of mycobacteria; it sometimes takes weeks for visible colonies to appear. The mycobacteria include the impor-

tant pathogens *Mycobacterium tuberculosis* (mī-kō-bak-ti'rē-um tü-ber-kū-lō'sis), which causes tuberculosis, and *M. leprae* (lep'rī), which causes leprosy. A number of other mycobacteria species are found in soil and water and are occasional pathogens.

Corynebacterium The corynebacteria (*coryne* = club-shaped) tend to be pleomorphic, and their morphology often varies with the age of the cells. The best-known species is *Corynebacterium diphtheriae* (kôr'i-nē-bak-ti-rē-um dif-thi'rē-ī), the causative agent of diphtheria.

Propionibacterium The name of this genus is derived from the organism's ability to form propionic acid; some species are important in the fermentation of Swiss cheese. *Propionibacterium acnes* (prō-pē-on'ē-bak-ti-rē-um ak'nēz) are bacteria that are commonly found on human skin and are implicated as the primary bacterial cause of acne.

Gardnerella *Gardnerella vaginalis* (gard-ne-rel'la va-jin-al'is) is a bacterium that causes one of the most common forms of vaginitis. There has always been some difficulty in assigning a taxonomic position in this species, which is gram-variable, and which exhibits a highly pleomorphic morphology.

Frankia The genus *Frankia* (frank'ē-ä) causes nitrogen-fixing nodules to form in alder tree roots, much as *Rhizobium* causes nodules on the roots of legumes.

(a)

(b)   SEM ⊢ 1 μm

FIGURE 11.20 **Streptomyces. (a)** Drawing of a typical *Streptomyces* showing filamentous, branching growth with asexual reproductive conidiospores at the filament tips. **(b)** A chain of conidiospores is surrounded by filaments of the streptomycete.

■ Streptomycetes are filamentous bacteria that make up much of the bacterial population of soils.

Streptomyces The genus *Streptomyces* (strep-tō-mī'ses) is the best known of the actinomycetes, and is one of the bacteria most commonly isolated from soil (Figure 11.20). The reproductive asexual spores of *Streptomyces*, called *conidiospores*, are formed at the ends of aerial filaments. If each conidiospore lands on a suitable substrate, it is capable of germinating into a new colony. These organisms are strict aerobes. They often produce extracellular enzymes that enable them to utilize proteins; polysaccharides, such as starch; cellulose; and many other organic materials found in soil. *Streptomyces* characteristically produce a gaseous compound called *geosmin*, which gives fresh soil its typical musty odor. Species of *Streptomyces* are valuable because they produce most of our commercial antibiotics (see Table 20.1 on page 560). This has led to intensive

study of the genus—there are nearly 500 described species.

Actinomyces The genus *Actinomyces* (ak-tin-ō-mī'ses) consists of facultative anaerobes that are found in the mouth and throat of humans and animals. They occasionally form filaments that can fragment (Figure 11.21). One species, *Actinomyces israelii* (is-rā'lē-ē), causes actinomycosis, a tissue-destroying disease usually affecting the head, neck, or lungs.

Nocardia The genus *Nocardia* (nō-kär'dē-ä) morphologically resemble *Actinomyces*; however, they are aerobic. To reproduce, they form rudimentary filaments, which fragment into short rods. The structure of their cell wall resembles that of the mycobacteria; therefore, they are often acid-fast. *Nocardia* species are common in soil. Some species, such as *Nocardia asteroides* (as'ter-oi-dēz), occasionally cause a chronic, difficult-to-treat pulmonary infection. *N. asteroides* is also one of the causative agents of mycetoma, a localized destructive infection of the feet or hands.

* * *

Assorted Phyla Discussed in Volume Five of *Bergey's Manual*

The fifth and final volume of the second edition of *Bergey's Manual of Systematic Bacteriology* is scheduled to contain a varied assortment of phyla such as Chlamydiae, Spirochaetes, Bacteroidetes, and Fusobacteria. Several important pathogens such as the genera *Chlamydia*, *Borrelia*, and *Treponema* are included. Members of the genera *Bacteroides* and *Fusobacterium* are profuse and important inhabitants of the human intestinal tract.

Chlamydiae

Learning Objective

• *Make a dichotomous key to distinguish* Chlamydia, *spirochetes*, Cytophaga, Bacteroidetes, *and* Fusobacteria.

Members of the Phylum Chlamydiae are grouped with other genetically similar bacteria that do not contain peptidoglycan in their cell walls. We will discuss only the genus *Chlamydia*. Earlier editions of *Bergey's Manual of Systematic Bacteriology* grouped the chlamydia, which grow intracellularly within host cells, with the rickettsial bacteria, which also grow intracellularly. The rickettsia are now classified according to their genetic content with the α-proteobacteria.

Chlamydia Chlamydias have a unique developmental cycle that is perhaps their most distinguishing characteristic (Figure 11.22a). They are gram-negative coccoid bacteria (Figure 11.22b). The elementary body shown in Figure

FIGURE 11.21
Actinomyces. Notice the branched filamentous morphology.

■ Why are these bacteria not classified as fungi?

SEM 1 μm

11.22 is the infective agent. Unlike the rickettsias, chlamydias do not require insects or ticks for transmission. They are transmitted to humans by interpersonal contact or by airborne respiratory routes.

There are three species of chlamydias that are significant pathogens for humans. *Chlamydia trachomatis* (kla-mi′dē-ä trä-kō′mä-tis) is the causative agent of trachoma, a common cause of blindness in humans. *C. trachomatis* is considered to be the primary causative agent of both nongonococcal urethritis, which may be the most common sexually transmitted disease in the United States, and lymphogranuloma venereum, another sexually transmitted disease. *C. psittaci* (sit′tä-sē) is the causative agent of psittacosis (ornithosis). A third species, *C. pneumoniae* (nū-mō′nē-ī), is the cause of a mild form of pneumonia that is especially prevalent in young adults. Chlamydias can be cultivated in laboratory animals, cell cultures, or the yolk sac of embryonated chicken eggs.

Spirochaetes

The spirochetes have a coiled morphology, resembling a metal spring; some are more tightly coiled than others. The most distinctive characteristic of this order, however, is their method of motility, which makes use of two or more **axial filaments** (or *endoflagella*) enclosed in the space between an outer sheath and the body of the cell. One end of each axial filament is attached near a pole of the cell (see Figures 4.10 and 11.23). By rotating its axial filament, the cell rotates in the opposite direction, like a

corkscrew, which is very efficient in moving the organism through liquids. For bacteria, this is more difficult than it might seem. At the scale of a bacterium, water is as viscous as molasses is to an animal. However, a bacterium can typically move about 100 times its body length in a second (or about 50 μm/sec), whereas a large fish, such as a tuna, can move only about 10 times its body length in this time.

Many spirochetes are found in the human oral cavity and are probably among the first microorganisms described by van Leeuwenhoek in the 1600s that he found in saliva and tooth scrapings. An extraordinary location for spirochetes is on the surfaces of some of the cellulose-digesting protozoa found in termites, where they may function as substitutes for flagella.

Treponema The spirochetes include a number of important pathogenic bacteria. The best known is the genus *Treponema* (tre-pō-nē′mä), which includes *Treponema pallidum* (pal′li-dum), the cause of syphilis (Figure 11.23b).

Borrelia Members of the genus *Borrelia* (bor-rel′ē-ä) cause relapsing fever and Lyme disease, serious diseases that are usually transmitted by ticks or lice.

Leptospira Leptospirosis is a disease usually spread to humans by water contaminated by *Leptospira* (lep-tō-spī′ra) species. The bacteria are excreted in the urine of such animals as dogs, rats, and swine, so domestic dogs and cats are routinely immunized against leptospirosis.

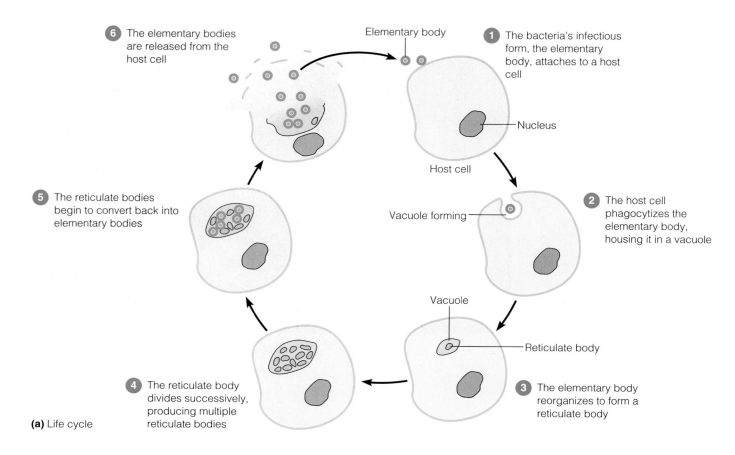

6 The elementary bodies are released from the host cell

Elementary body

1 The bacteria's infectious form, the elementary body, attaches to a host cell

Nucleus

Host cell

5 The reticulate bodies begin to convert back into elementary bodies

2 The host cell phagocytizes the elementary body, housing it in a vacuole

Vacuole forming

Vacuole

Reticulate body

4 The reticulate body divides successively, producing multiple reticulate bodies

3 The elementary body reorganizes to form a reticulate body

(a) Life cycle

Elementary body

Reticulate body

Intermediate body

(b) *Chlamydia psittaci*   TEM |—— 1 μm ——|

FIGURE 11.22 **Chlamydias. (a)** The generalized life cycle of chlamydia, which takes about 48 hours to complete. **(b)** A micrograph of *Chlamydia* in a slice of the cytoplasm of a host cell. The **elementary bodies,** which are the infectious stage, are dense, dark, and relatively small. **Reticulate bodies,** the form in which chlamydia reproduce within the host cell, are larger with a speckled appearance. **Intermediate bodies,** a stage between the change from reticulate to elementary bodies are dark centered.

■ Chlamydias reproduce only with a host cell; the elementary body is the infectious stage of the life cycle.

Axial filaments

Sheath

(a) Section of spirochete   LM   0.2 μm

Sheath

Axial filaments

(b) *Treponema pallidum*   TEM   0.2 μm

FIGURE 11.23 **Spirochetes.** Spirochetes are helical and have axial filaments under an outer sheath that enables them to move by a corkscrewlike rotation. **(a)** This section of a spirochete shows numerous axial filaments between the dark cell and the outer sheath. **(b)** This electron micrograph of a portion of *Treponema pallidum* shows the sheath, which has shrunk away from the cell during preparation for microscopy, and two axial filaments, which are attached near one end of the bacterial cell under the sheath.

■ How does a spirochete's motility differ from that of *Spirillum* (Figure 11.4)?

Bacteroidetes

The phylum Bacteroidetes includes several genera of anaerobic bacteria. Included are the genus *Bacteroides*, a common inhabitant of the human intestinal tract, and the genus *Prevotella*, found in the human mouth. Also included in the phylum Bacteroidetes are the important soil bacteria with gliding motility of the genus *Cytophaga*.

Bacteroides Bacteria of the genus *Bacteroides* (bak-tė-roi′dēz) live in the human intestinal tract in numbers approaching 1 billion per gram of feces. Some *Bacteroides* species also reside in anaerobic habitats such as the gingival crevice and are also frequently recovered from deep tissue infections. *Bacteroides* organisms are nonmotile and do not form endospores. Infections caused by *Bacteroides* often result from puncture wounds or surgery and are a frequent cause of peritonitis, an inflammation resulting from a perforated bowel.

Cytophaga Members of the genus *Cytophaga* (sī-täf′ag-a) are important in the degradation of cellulose and chitin, which are both abundant in soil. Gliding motility places the microbe in close contact with these substrates so that enzymatic action is very efficient.

Fusobacteria

The fusiform bacteria comprise another phylum of anaerobes. These bacteria are often pleomorphic but, as their name suggests, may be spindle-shaped (*fuso* = spindle).

SEM   1 μm

FIGURE 11.24 **Fusobacterium.** This is a common, anaerobic rod found in the human intestine. Notice the characteristic pointed ends.

■ In what other place in the human body do you often find *Fusobacterium*?

Fusobacterium Members of the genus *Fusobacterium* (fū-sō-bak-ti′rē-um) are long and slender, with pointed rather than blunt ends (Figure 11.24). In humans, they are found most often in the gingival crevice of the gums and may be responsible for some dental abscesses.

SEM | 1 μm

FIGURE 11.25 **Archaea.** *Pyrodictium abyssi,* an unusual member of the archaea found growing in deep ocean sediment at a temperature of 110°C. The cells are disk-shaped with a network of tubules (cannulae). Most archaea are more conventional in their morphology.

■ Do the terms included in the name, *Pyro* and *abyssi,* suggest a basis for the naming of this bacterium?

DOMAIN ARCHAEA

Learning Objective

• Name a habitat for each group of archaea.

In the late 1970s, a distinctive type of prokaryotic cell was discovered. So different from what were known then as the Eubacteria, it was believed to constitute almost a third form of life. Most strikingly, their cell walls lacked the peptidoglycan common to most bacteria. It soon became clear that they also shared many rRNA sequences, and the sequences were different from either those of the Domain Bacteria or the eukaryotic organisms.

This exceptionally interesting group of prokaryotes is highly diverse. Most are of conventional morphology, that is, rods, cocci, and helixes, but some are of very unusual morphology, as illustrated in Figure 11.25. Some are grampositive, others gram-negative; some may divide by binary fission, others by fragmentation or budding; a few lack cell walls. Organisms in this domain are physiologically diverse as well, ranging from aerobic, to facultatively anaerobic, to strictly anaerobic. Nutritionally, they include chemoautotrophs, photoautotrophs, and chemoheterotrophs. Of special interest to microbiologists is the fact that the archaea are frequent inhabitants of exceptionally extreme environments of heat, cold, acidity, and pressure.

Prominent among the archaea are the extreme halophiles, bacteria that survive in very high concentra-

tions of salt, such as the Great Salt Lake and solar evaporating ponds. Examples are *Halobacterium* (hā-lō-bak-ti′rē-um) (see the box in Chapter 5, page 143) and *Halococcus* (hā-lō-kok-kus), which live in high concentrations of sodium chloride (NaCl) and actually require such environmental conditions for growth.

Other archaea thrive in acidic, sulfur-rich hot springs. One such organism is *Sulfolobus* (sul′fō-lō-bus), which has a pH optimum of about 2 and a temperature optimum of more than 70°C. Archaea are also found in ocean depths near hydrothermal vents. Examples of hyperthermophiles, archaea that can thrive in extraordinarily high temperatures, are described in the box in Chapter 6, page 160.

The obligately anaerobic methane-producing members of the archaea, the genus *Methanobacterium* (meth-a-nō-bak-tér′ē-um), is of considerable economic importance. These bacteria are found in the human intestines and are used in sewage-treatment processes. These archaea derive energy from combining hydrogen (H_2) with carbon dioxide (CO_2) to form methane (CH_4). An essential part of treating sewage is encouraging the growth of these microbes in anaerobic digestion tanks to convert the sludge into CH_4.

MICROBIAL DIVERSITY

Learning Objective

• List two factors that contribute to the limits of our knowledge of microbial diversity.

Earlier in this chapter, we described the giant bacterium *Epulopiscium*. In 1999, another giant bacterium was discovered 100 meters deep in the sediments of the coastal waters off Namibia, on the southwestern coast of Africa. Named *Thiomargarita namibiensis* (thī′ō-mär-gär-ē-tà na′mi-bē-én-sis), meaning "sulfur pearl of Namibia," these spherical organisms, classified with the gamma-proteobacteria, are as large as 750 μm in diameter (Figure 11.26). This is a bit larger than the size of a period at the end of this sentence and even larger than *Epulopiscium*. As we have mentioned, a factor that limits the size of prokaryotic cells is that nutrients must enter the cytoplasm by simple diffusion. *T. namibiensis* minimizes this problem by resembling a fluid-filled balloon; the vacuole in the interior being surrounded by a relatively thin outer layer of cytoplasm. This cytoplasm is equal in volume to that of most other prokaryotes. Its nutrient supply is essentially hydrogen sulfide, which is plentiful in the sediments in which it is normally found, and nitrate, which it must extract intermittently from nitrate-rich seawaters when storms stir the loose sediment. The cell's interior vacuole, which makes up about 98% of the bacterium's volume, serves as a storage space to hold the nitrate between recharging of its supply. The cell's energy is derived from the oxidation of hydrogen sulfide; the nitrate, while a source of nutritional nitrogen, primarily serves as an electron acceptor in the absence of oxygen.

The discovery of uniquely large bacteria has raised the question of how large a prokaryotic cell can be and still absorb nutrients by diffusion. At the other extreme, there are reports of bacteria as small as $0.02-0.03$ μm (nanobacteria), found in deep rock formations. Many microbiologists think that nanobacteria are nonliving microscopic artifacts, but even so another question arises: Is there a lower size limit for microorganisms? (See the discussion of the genus *Mycoplasma* on page 324.) Some scientists have used theoretical considerations to calculate that a cell with a significant metabolism would have to have a diameter of at least 0.1 μm.

Until now, microbiologists have described only about 5000 bacterial species, of which about 3000 are listed in *Bergey's Manual*. The true number may be in the millions. Many bacteria in soil or water, or elsewhere in nature, cannot be cultivated with the media and conditions normally used for bacterial growth. Moreover, some bacteria are part of complex food chains and can only grow in the presence of other microbes that supply specific growth requirements. Recently, researchers have been using the polymerase chain reaction (PCR) to make millions of copies of genes found at random in a soil sample. By comparing the genes found in many rep-

LM   100 μm

FIGURE 11.26 ***Thiomargarita namibiensis.*** *Thiomargarita namibiensis* gets its energy from reduced sulfur compounds such as hydrogen sulfide.

■ The interior of this giant bacterium is mostly a fluid-filled vacuole; thus, it can obtain nutrients by diffusion, like much smaller bacteria.

etitions of this process, researchers can estimate the different bacterial species in such a sample. One report indicates that a single gram of soil may contain 10,000 or so bacterial types—about twice as many as have ever been described.

STUDY OUTLINE

Introduction (p. 305)

1. *Bergey's Manual* categorizes bacteria into taxa based on rRNA sequences.
2. *Bergey's Manual* lists identifying characteristics such as Gram stain reaction, cellular morphology, oxygen requirements, and nutritional properties.

Prokaryotic Groups (p. 305)

1. Prokaryotic organisms are classified into two domains: Archaea and Bacteria.

Domain Bacteria (pp. 305–329)

1. Bacteria are essential to life on Earth.

The Proteobacteria (pp. 305–317)

1. Members of the phylum Proteobacteria are gram-negative.

2. Alpha-proteobacteria includes nitrogen-fixing bacteria, chemoautotrophs, and chemoheterotrophs.
3. The β-proteobacteria include chemoautotrophs and chemoheterotrophs.
4. Pseudomonadales, Legionellales, Vibrionales, Enterobacteriales, and Pasteurellales are classified as γ-proteobacteria.
5. Purple and green photosynthetic bacteria are photoautotrophs that use light energy and CO_2 and do not produce O_2.
6. *Myxococcus* and *Bdellovibrio* in the δ-proteobacteria prey on other bacteria.
7. Epsilon-proteobacteria include *Campylobacter* and *Helicobacter*.

The Nonproteobacteria Gram-Negative Bacteria (pp. 317–319)

1. Several phyla of gram-negative bacteria are not related phylogenetically to the Proteobacteria.

2. Cyanobacteria are photoautotrophs that use light energy and CO_2 and do produce O_2.

3. Chemoheterotrophic examples are *Chlamydia*, spirochetes, *Bacteroides*, and *Fusobacterium*.

The Gram-Positive Bacteria (pp. 319–329)

1. In *Bergey's Manual*, gram-positive bacteria are divided into those that have low G + C ratio and those that have high G + C ratio.

2. Low G + C gram-positive bacteria include common soil bacteria, the lactic acid bacteria, and several human pathogens.

3. High G + C gram-positive bacteria include mycobacteria, corynebacteria, and actinomycetes.

Domain Archaea (pp. 329–330)

1. Extreme halophiles, extreme thermophiles, and methanogens are included in the archaea.

Microbial Diversity (pp. 330–331)

1. Few of the total number of different prokaryotes have been isolated and identified.

2. PCR can be used to uncover the presence of bacteria that can't be cultured in the laboratory.

STUDY QUESTIONS

Review

1. The following outline is a key that can be used to identify medically important bacteria. Fill in a representative genus in the space provided.

                                                    Name of
                                                    Representative
                                                    Genus

I. Gram-positive
  A. Endospore-forming rod
    1. Obligate anaerobe   _____
    2. Not obligate anaerobe   _____
  B. Nonendospore-forming
    1. Cells are rods
      a. Produce conidiospores   _____
      b. Acid-fast   _____
    2. Cells are cocci
      a. Lack cytochrome system   _____
      b. Use aerobic respiration   _____
II. Gram-negative
  A. Cells are helical or curved
    1. Axial filament   _____
    2. No axial filament   _____
  B. Cells are rods
    1. Aerobic, nonfermenting   _____
    2. Facultatively anaerobic   _____
III. Lack cell walls   _____
IV. Obligate intracellular parasites
  A. Transmitted by ticks   _____
  B. Reticulate bodies in host cells   _____

2. Compare and contrast each of the following:
  a. Cyanobacteria and algae
  b. Actinomycetes and fungi
  c. *Bacillus* and *Lactobacillus*
  d. *Pseudomonas* and *Escherichia*
  e. *Leptospira* and *Spirillum*
  f. *Escherichia* and *Bacteroides*
  g. *Rickettsia* and *Chlamydia*
  h. *Ureaplasma* and *Mycoplasma*

3. Matching:
  I. Gram-positive
    A. Nitrogen-fixing _____
  II. Gram-negative
    A. Phototrophic
      1. Anoxygenic _____
      2. Oxygenic _____
    B. Chemoautotrophic
      1. Oxidize NO_2^- _____
      2. Reduce CO_2 to CH_4 _____
    C. Chemoheterotrophic
      1. Cells inside a sheath _____
      2. Form myxospores _____
      3. Reduce sulfate to H_2S
        a. Anaerobic _____
        b. Thermophilic _____
      4. Long filaments, found in sewage _____
      5. Form projections from the cell _____

  a. Cyanobacteria
  b. *Cytophaga*
  c. *Desulfovibrio*
  d. *Frankia*
  e. *Hyphomicrobium*
  f. Methanogens
  g. Myxobacteria
  h. *Nitrobacter*
  i. Purple bacteria
  j. *Sphaerotilus*
  k. *Sulfolobus*

Multiple Choice

1. If you Gram-stained the bacteria that live in the human intestine, you would expect to find mostly
  a. gram-positive cocci.
  b. gram-negative rods.
  c. gram-positive, endospore-forming rods.
  d. gram-negative, nitrogen-fixing bacteria.
  e. all of the above

2. Which of the following does *not* belong with the others?
  a. Enterobacteriales
  b. Lactobacillales
  c. Legionellales
  d. Pasteurellales
  e. Vibrionales

3. Pathogenic bacteria can be
   a. motile.
   b. rods.
   c. cocci.
   d. anaerobic.
   e. all of the above

4. Which of the following is an intracellular parasite?
   a. *Rickettsia*
   b. *Mycobacterium*
   c. *Bacillus*
   d. *Staphylococcus*
   e. *Streptococcus*

5. Which of the following terms is the most specific?
   a. bacillus
   b. *Bacillus*
   c. gram-positive
   d. endospore-forming rods and cocci
   e. anaerobic

6. Which one of the following does *not* belong with the others?
   a. *Enterococcus*
   b. *Lactobacillus*
   c. *Staphylococcus*
   d. *Streptococcus*
   e. none of the above

7. Which of the following pairs is mismatched?
   a. anaerobic endospore-forming gram-positive rods— *Clostridium*
   b. facultatively anaerobic gram-negative rods—*Escherichia*
   c. facultatively anaerobic gram-negative rods—*Shigella*
   d. pleomorphic gram-positive rods—*Corynebacterium*
   e. spirochete—*Helicobacter*

8. *Spirillum* is not classified as a spirochete because spirochetes
   a. do not cause disease.
   b. possess axial filaments.
   c. possess flagella.
   d. are prokaryotes.
   e. none of the above

9. When *Legionella* was newly discovered, it was classified with the pseudomonads because
   a. it is a pathogen.
   b. it is an aerobic gram-negative rod.
   c. it is difficult to culture.
   d. it is found in water.
   e. none of the above

10. Cyanobacteria differ from purple and green phototrophic bacteria because cyanobacteria
    a. produce oxygen during photosynthesis.
    b. do not require light.
    c. use H_2S as an electron donor.
    d. have a membrane-enclosed nucleus.
    e. all of the above

Critical Thinking

1. Place each phylum listed in Table 11.1 in the appropriate category:

a. typical gram-positive cell wall
b. typical gram-negative cell wall
c. no peptidoglycan in cell wall
d. no cell wall

2. To which of the following is the photosynthetic bacterium *Chromatium* most closely related? Briefly explain why.
   a. cyanobacteria
   b. *Chloroflexus*
   c. *Escherichia*

3. Identify the genus that best fits each of the following descriptions:
   a. This organism can produce a fuel used for home heating and for generating electricity.
   b. This gram-positive genus presents the greatest source of bacterial damage to the beekeeping industry.
   c. This gram-positive rod is used in dairy fermentations.
   d. This γ-proteobacterial genus is well suited to degrade hydrocarbons in an oil spill.

Clinical Applications

1. After contact with a patient's spinal fluid, a lab technician developed fever, nausea, and purple lesions on her neck and extremities. A throat culture grew gram-negative diplococci. What is the genus of the bacteria?

2. Between April 1 and May 15 of one year, 22 children in three states developed diarrhea, fever, and vomiting. The children had each received pet ducklings. Gram-negative, facultatively anaerobic bacteria were isolated from both the patients' and the ducks' feces; the bacteria were identified as serovar C2. What is the genus of these bacteria?

3. A woman complaining of lower abdominal pain with a temperature of 39°C gave birth soon after to a stillborn baby. Blood cultures from the infant revealed gram-positive rods. The woman had a history of eating unheated hot dogs during her pregnancy. Which organism is most likely involved?

LEARNING WITH TECHNOLOGY

The Microbiology Place CD-ROM/website includes many resources to help you study and succeed in this course. In addition to self-quizzes, web links, and flashcards, you'll find the following activities for this chapter in the "Activities" section of The Microbiology Place:

Investigating the Archaea

Taxonomy of Prokaryotes

Chapter 12

The Eukaryotes: Fungi, Algae, Protozoa, and Helminths

PENICILLIUM. *Fungi such as* Penicillium *are important decomposers that recycle organic matter in the soil.*

Over half of the world's population is infected with eukaryotic pathogens. The World Health Organization (WHO) ranks six parasitic diseases among the top 20 microbial causes of death in the world. Every year, there are over 5 million cases each of malaria, schistosomiasis, amoebiasis, hookworm, African trypanosomiasis, and intestinal parasites reported in developing countries. Emerging eukaryotic pathogens in developed countries include Pneumocystis, the leading cause of death in AIDS patients; the protozoan Cryptosporidium,

which caused disease in 400,000 people in Milwaukee in 1993; a new protozoan parasite Cyclospora, discovered in the United States in 1993; and new and increased poisonings due to algae. Increases in mold-related illness have led to debate about the passage of laws regarding safe levels of exposure to molds.

In this chapter, we examine the eukaryotic microorganisms that affect humans: fungi, algae, protozoa, and parasitic helminths, or worms. (For a comparison of their characteristics, see Table 12.1.)

Fungi

Over the last 10 years, the incidence of serious fungal infections has been increasing. These infections are occurring as nosocomial infections and in people with compromised immune systems. In addition, thousands of fungal diseases afflict economically important plants, costing more than one billion dollars annually.

Fungi are also beneficial. They are important in the food chain because they decompose dead plant matter, thereby recycling vital elements. Through the use of extracellular enzymes such as cellulases, fungi are the primary decomposers of the hard parts of plants, which cannot be digested by animals. Nearly all plants depend on symbiotic fungi, known as **mycorrhizae,** which help their roots absorb minerals and water from the soil. Fungi are also valuable to animals. Fungi-farming ants cultivate fungi in order to break down cellulose and lignin from plants so the ants can then digest them. Fungi are used by humans for food (mushrooms) and to produce foods (bread and citric acid) and drugs (alcohol and penicillin). Of the more than 100,000 species of fungi, only about 200 are pathogenic to humans and animals.

The study of fungi is called **mycology.** We will first look at the structures that are the basis of fungal identification in a clinical laboratory, then we will explore their life cycles. Recall from Chapter 10 that identification of a pathogen is often required in order to properly treat a disease and to prevent its spread.

We will also examine nutritional needs. All fungi are chemoheterotrophs, requiring organic compounds for energy and carbon. Fungi are aerobic or facultatively anaerobic; only a few anaerobic fungi are known.

Table 12.2 lists the basic differences between fungi and bacteria.

TABLE 12.1	**Major Differences Among Eukaryotic Microorganisms: Fungi, Algae, Protozoa, and Helminths**			
	Fungi	**Algae**	**Protozoa**	**Helminths**
Kingdom	Fungi	Protist	Protist	Animalia
Nutritional type	Chemoheterotroph	Photoautotroph	Chemoheterotroph	Chemoheterotroph
Multicellularity	All, except yeasts	Some	None	All
Cellular arrangement	Unicellular, filamentous, fleshy (such as mushrooms)	Unicellular, colonial, filamentous; tissues	Unicellular	Tissues and organs
Food acquisition method	Absorptive	Absorptive	Absorptive; ingestive (cytostome)	Ingestive (mouth); absorptive
Characteristic features	Sexual and asexual spores	Pigments	Motility; some form cysts	Many have elaborate life cycles, including egg, larva, and adult
Embryo formation	None	None	None	All

TABLE 12.2	**Selected Features of Fungi and Bacteria Compared**	
	Fungi	**Bacteria**
Cell type	Eukaryotic	Prokaryotic
Cell membrane	Sterols present	Sterols absent, except in *Mycoplasma*
Cell wall	Glucans; mannans; chitin (no peptidoglycan)	Peptidoglycan
Spores	Produce a wide variety of sexual and asexual reproductive spores	Endospores (not for reproduction); some asexual reproductive spores
Metabolism	Limited to heterotrophic; aerobic, facultatively anaerobic	Heterotrophic, chemoautotrophic, photoautotrophic; aerobic, facultatively anaerobic, anaerobic
Sensitivity to antibiotics	Often sensitive to polyenes, imidazoles, and griseofulvin	Often sensitive to penicillins, tetracyclines, and aminoglycosides

SOURCE: After B. D. Davis et al., *Microbiology,* 4th ed. Philadelphia: J. B. Lippincott, 1990, p. 746.

Characteristics of Fungi

Learning Objectives
- List the defining characteristics of fungi.
- Differentiate asexual from sexual reproduction, and describe each of these processes in fungi.

Yeast identification, like bacterial identification, involves biochemical tests. However, multicellular fungi are identified on the basis of physical appearance, including colony characteristics and reproductive spores.

Vegetative Structures

Fungal colonies are described as **vegetative** structures because they are composed of the cells involved in catabolism and growth.

Molds and Fleshy Fungi The **thallus** (body) of a mold or fleshy fungus consists of long filaments of cells joined together; these filaments are called **hyphae** (singular: **hypha**). Hyphae can grow to immense proportions. The hyphae of a single fungus in Michigan extend across 40 acres and are estimated to weigh over 10 tons.

FIGURE 12.1 **Characteristics of fungal hyphae.** (a) Septate hyphae have cross-walls, or septa, dividing the hyphae into cell-like units. (b) Coenocytic hyphae lack septa. (c) Hyphae grow by elongating at the tips.

■ What is a hypha? A mycelium?

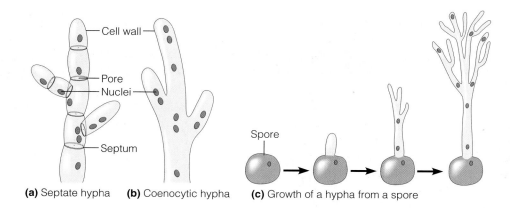

(a) Septate hypha   (b) Coenocytic hypha   (c) Growth of a hypha from a spore

In most molds, the hyphae contain cross-walls called **septa** (singular: *septum*), which divide them into distinct, uninucleate (one-nucleus) cell-like units. These hyphae are called **septate hyphae** (Figure 12.1a). In a few classes of fungi, the hyphae contain no septa and appear as long, continuous cells with many nuclei. These are called **coenocytic hyphae** (Figure 12.1b). Even in fungi with septate hyphae, there are usually openings in the septa that make the cytoplasm of adjacent "cells" continuous; these fungi are actually coenocytic organisms, too.

Hyphae grow by elongating at the tips (Figure 12.1c). Each part of a hypha is capable of growth, and when a fragment breaks off, it can elongate to form a new hypha. In the laboratory, fungi are usually grown from fragments obtained from a fungal thallus.

The portion of a hypha that obtains nutrients is called the *vegetative hypha*; the portion concerned with reproduction is the *reproductive* or *aerial hypha,* so named because it projects above the surface of the medium on which the fungus is growing. Aerial hyphae often bear reproductive spores (Figure 12.2a), discussed later. When environmental conditions are suitable, the hyphae grow to form a filamentous mass called a **mycelium,** which is visible to the unaided eye (Figure 12.2b).

Yeasts Yeasts are nonfilamentous, unicellular fungi that are typically spherical or oval. Like molds, yeasts are widely distributed in nature; they are frequently found as a white powdery coating on fruits and leaves. **Budding yeast,** such as *Saccharomyces* (sak-ä-rō-mī′sēs), divide unevenly.

In budding (Figure 12.3), the parent cell forms a protuberance (bud) on its outer surface. As the bud elongates, the parent cell's nucleus divides, and one nucleus migrates into the bud. Cell wall material is then laid down between the bud and parent cell, and the bud eventually breaks away.

One yeast cell can in time produce up to 24 daughter cells by budding. Some yeasts produce buds that fail to detach themselves; these buds form a short chain of cells

called a **pseudohypha.** *Candida albicans* (kan′did-ä al′bi-kanz) attaches to human epithelial cells as a yeast but usually requires pseudohyphae to invade deeper tissues.

Fission yeasts, such as *Schizosaccharomyces* (skiz-ō-sak-ä-rō-mī′sēs), divide evenly to produce two new cells. During fission, the parent cell elongates, its nucleus divides, and two daughter cells are produced. Increases in the number of yeast cells on a solid medium produce a colony similar to a bacterial colony.

Yeasts are capable of facultative anaerobic growth. Yeasts can use oxygen or an organic compound as the final electron acceptor; this is a valuable attribute because it allows these fungi to survive in various environments. If given access to oxygen, yeasts perform aerobic respiration to metabolize carbohydrates to carbon dioxide and water; denied oxygen, they ferment carbohydrates and produce ethanol and carbon dioxide. This fermentation is used in the brewing, wine-making, and baking industries. Species of *Saccharomyces* produce ethanol in brewed beverages and carbon dioxide for leavening bread dough.

Dimorphic Fungi Some fungi, most notably the pathogenic species, exhibit **dimorphism**—two forms of growth. Such fungi can grow either as a mold or as a yeast. The moldlike forms produce vegetative and aerial hyphae; the yeastlike forms reproduce by budding. Dimorphism in pathogenic fungi is temperature-dependent: At 37°C, the fungus is yeastlike, and at 25°C, it is moldlike. However, the appearance of the dimorphic (in this instance, non-pathogenic) fungus shown in Figure 12.4 changes with CO_2 concentration.

Life Cycle

Filamentous fungi can reproduce asexually by fragmentation of their hyphae. In addition, both sexual and asexual reproduction in fungi occurs by the formation of **spores.** In fact, fungi are usually identified by spore type.

(a) *Aspergillus niger*

LM 20 μm

(b) *A. niger* on agar

FIGURE 12.2 **Aerial and vegetative hyphae. (a)** A photomicrograph of aerial hyphae, showing reproductive spores. **(b)** A colony of *Aspergillus niger* grown on a glucose agar plate, showing both vegetative and aerial hyphae.

■ How do fungal colonies differ from bacterial colonies?

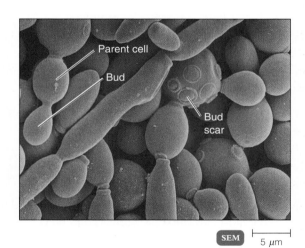

SEM 5 μm

FIGURE 12.3 **A budding yeast.** A micrograph of *Saccharomyces cerevisiae* in various stages of budding.

■ Following mitosis, some yeast cells produce a small bud that grows into a full-size cell.

LM 50 μm

FIGURE 12.4 **Fungal dimorphism.** Dimorphism in the fungus *Mucor rouxii* depends on CO_2 concentration. On the agar surface, *Mucor* exhibits yeastlike growth, but in the agar it is moldlike.

■ What is fungal dimorphism?

Fungal spores, however, are quite different from bacterial endospores. Bacterial endospores allow a bacterial cell to survive adverse environmental conditions (see Chapter 4). A single vegetative bacterial cell forms one endospore, which eventually germinates to produce a single vegetative bacterial cell. This process is not reproduction because it does not increase the total number of bacterial cells. But after a mold forms a spore, the spore detaches from the parent and germinates into a new mold (see Figure 12.1c). Unlike the bacterial endospore, this is a true reproductive spore; a second organism grows from the spore. Although fungal spores can survive for extended periods in dry or hot environments, most do not exhibit the extreme tolerance and longevity of bacterial endospores.

Spores are formed from aerial hyphae in a number of different ways, depending on the species. Fungal spores can be either asexual or sexual. **Asexual spores** are formed by the hyphae of one organism. When these spores germinate, they become organisms that are genetically identical to the parent. **Sexual spores** result from the fusion of nuclei from two opposite mating strains of the same species of fungus. Fungi produce sexual spores less frequently than asexual spores. Organisms that grow from sexual spores will have genetic characteristics of both parental strains. Because spores are of considerable importance in the identification of fungi, we will next look at some of the various types of asexual and sexual spores.

Asexual Spores Asexual spores are produced by an individual fungus through mitosis and subsequent cell division; there is no fusion of the nuclei of cells. Several types of asexual spores are produced by fungi. One type is a **conidiospore** or **conidium,** a unicellular or multicellular spore that is not enclosed in a sac (Figure 12.5a). Conidia are produced in a chain at the end of a **conidiophore.** Such spores are produced by *Aspergillus.* One type of conidium, an **arthrospore,** is formed by the fragmentation of a septate hypha into single, slightly thickened cells (see Figure 12.5b). One species that produces such spores is *Coccidioides immitis* (kok-sid-ē-oi′dēz im′mi-tis). Another type of conidium, **blastoconidia,** consists of buds coming off the parent cell (Figure 12.5c). Such spores are found in some yeasts, such as *Candida albicans* and *Cryptococcus.* A second type of spore is a **chlamydospore,** a thick-walled spore formed by rounding and enlargement within a hyphal segment (Figure 12.5d). A fungus that produces chlamydospores is the yeast *C. albicans.*

A third type of asexual spore is a **sporangiospore,** formed within a **sporangium,** or sac, at the end of an aerial hypha called a **sporangiophore.** The sporangium can contain hundreds of sporangiospores (Figure 12.5e). Such spores are produced by *Rhizopus.*

Sexual Spores A fungal sexual spore results from sexual reproduction, consisting of three phases:

1. **Plasmogamy.** A haploid nucleus of a donor cell (+) penetrates the cytoplasm of a recipient cell (−).

2. **Karyogamy.** The (+) and (−) nuclei fuse to form a diploid zygote nucleus.

3. **Meiosis.** The diploid nucleus gives rise to haploid nuclei (sexual spores), some of which may be genetic recombinants.

The sexual spores produced by fungi characterize the phyla. In laboratory settings, most fungi exhibit only asexual spores. Consequently, clinical identification is based on microscopic examination of asexual spores.

Nutritional Adaptations

Fungi are generally adapted to environments that would be hostile to bacteria. Fungi are chemoheterotrophs, and, like bacteria, they absorb nutrients rather than ingesting them as animals do. However, fungi differ from bacteria in certain environmental requirements and in the following nutritional characteristics:

- Fungi usually grow better in an environment with a pH of about 5, which is too acidic for the growth of most common bacteria.

- Almost all molds are aerobic. Most yeasts are facultative anaerobes.

- Most fungi are more resistant to osmotic pressure than bacteria; most can therefore grow in relatively high sugar or salt concentrations.

- Fungi can grow on substances with a very low moisture content, generally too low to support the growth of bacteria.

- Fungi require somewhat less nitrogen than bacteria for an equivalent amount of growth.

- Fungi are often capable of metabolizing complex carbohydrates, such as lignin (a component of wood), that most bacteria cannot use for nutrients.

These characteristics enable fungi to grow on such unlikely substrates as bathroom walls, shoe leather, and discarded newspapers.

(a) Conidia

SEM |— 5 μm

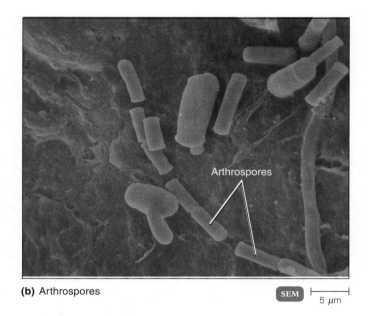

(b) Arthrospores

SEM |— 5 μm

(c) Blastoconidia

SEM |— 10 μm

(d) Chlamydospores

SEM |— 10 μm

FIGURE 12.5 **Representative asexual spores.**
(a) Conidia are arranged in chains at the end of a conidiophore on this *Aspergillus flavus*. **(b)** Fragmentation of
hyphae results in the formation of arthroconidia in this
Coccidioides immitis. **(c)** Blastoconidia are formed from the
buds of a parent cell of *Candida albicans*. **(d)** Chlamydospores are thick-walled cells within hyphae of this *C. albicans*.
(e) Sporangiospores are formed within a sporangium
(spore sac) of this *Rhizopus*.

■ What are the green powdery structures on moldy food?

(e) Sporangiospores

SEM |— 10 μm

Medically Important Phyla of Fungi

Learning Objective

- List the defining characteristics of the three phyla of fungi described in this chapter.

This section provides an overview of medically important phyla of fungi. Note that not all fungi cause disease.

The genera named in the following phyla include many that are readily found as contaminants in foods and in laboratory bacterial cultures. Although these genera are not all of primary medical importance, they are typical examples of their respective groups.

Zygomycota

The Zygomycota, or conjugation fungi, are saprophytic molds that have coenocytic hyphae. An example is *Rhizopus nigricans*, the common black bread mold. The asexual spores of *Rhizopus* are sporangiospores (Figure 12.6, upper right). The dark sporangiospores inside the sporangium give *Rhizopus* its descriptive common name. When the sporangium breaks open, the sporangiospores are dispersed. If they fall on a suitable medium, they will germinate into a new mold thallus.

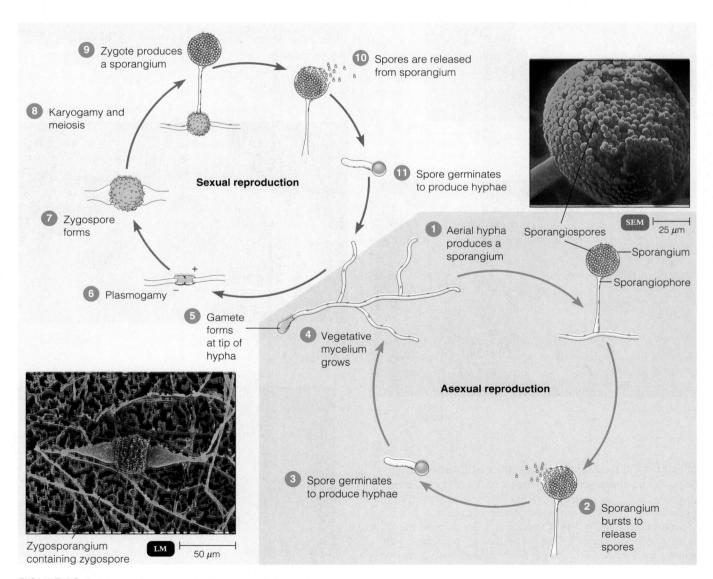

FIGURE 12.6 The life cycle of *Rhizopus*, a zygomycete. This fungus will reproduce asexually most of the time. Two opposite mating strains (designated + and −) are necessary for sexual reproduction.

- What is an opportunistic mycosis?

FIGURE 12.7 **The life cycle of _Talaromyces,_ an ascomycete.** Occasionally, when two opposite mating cells from two different strains (+ and −) fuse, sexual reproduction occurs.

■ Name one ascomycete that can infect humans.

The sexual spores are zygospores. A **zygospore** is a large spore enclosed in a thick wall (Figure 12.6, lower left). This type of spore results from the fusion of the nuclei of two cells that are morphologically similar to each other.

Ascomycota

The Ascomycota, or sac fungi, include molds with septate hyphae and some yeasts. Their asexual spores are usually conidia produced in long chains from the conidiophore. The term _conidia_ means dust, and these spores freely detach from the chain at the slightest disturbance and float in the air like dust.

An **ascospore** results from the fusion of the nuclei of two cells that can be either morphologically similar or dissimilar. These spores are produced in a saclike structure called an **ascus** (Figure 12.7, upper left). The members of this phylum are called sac fungi because of the ascus.

Basidiomycota

The Basidiomycota, or club fungi, also possess septate hyphae. This phylum includes fungi that produce mushrooms. **Basidiospores** are formed externally on a base pedestal called a **basidium** (Figure 12.8). (The common name of the fungus is derived from the club shape of the basidium.)

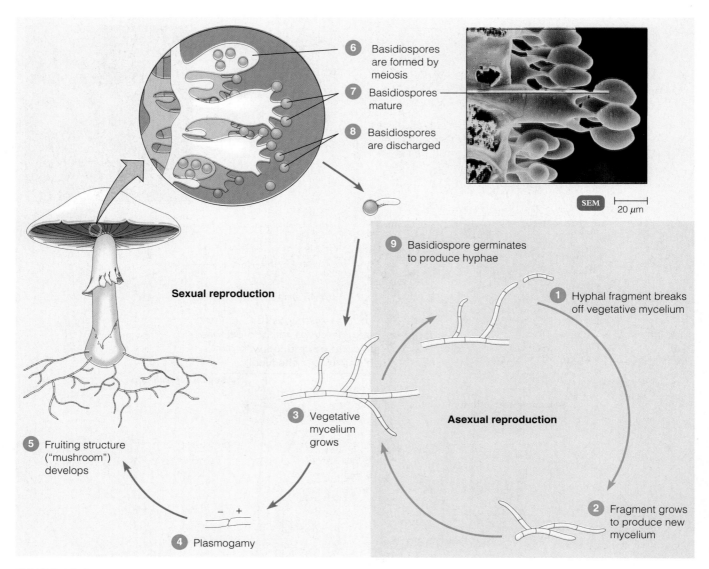

FIGURE 12.8 **A generalized life cycle of a basidiomycete.**
Mushrooms appear after cells from two mating strains (+ and −) have fused.

■ On what basis are fungi classified into phyla?

There are usually four basidiospores per basidium. Some of the basidiomycota produce asexual conidiospores. Representative basidiomycetes are shown in Figure 12.9.

The fungi we have looked at thus far are **teleo-morphs;** that is, they produce both sexual and asexual spores. Some ascomycetes have lost the ability to reproduce sexually. These asexual fungi are called **ana-morphs.** *Penicillium* is an example of an anamorph that arose from a mutation in a teleomorph. Historically, fungi whose sexual cycle had not been observed were put in a "holding category" called **Deuteromycota.** Now, mycologists are using rRNA sequencing to classify these organisms. Most of these previously unclassified deu-

teromycetes are anamorph phases of Ascomycota, and a few are basidiomycetes.

Table 12.3 on pages 344–345 lists some fungi that cause human diseases. Two generic names are given for some of the fungi because medically important fungi that are well known by their anamorph, or asexual, name are often referred to by that name.

Fungal Diseases

Any fungal infection is called a **mycosis.** Mycoses are generally chronic (long-lasting) infections because fungi grow slowly. Mycoses are classified into five groups according to the degree of tissue involvement and mode of entry into

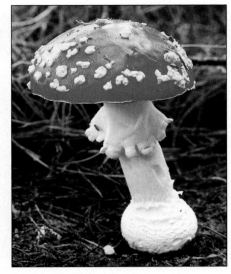

(a) *Crucibulum vulgare*   **(b)** *Amanita muscaria*

FIGURE 12.9 **Representative basidiomycetes. (a)** A bird's nest fungus (*Crucibulum vulgare*) growing on a twig. The basidiospores visible in one of the cups will pop out when the cup is hit by a raindrop. **(b)** *Amanita muscaria* grows in close association with plant roots (mycorrhiza) and produces a neurotoxin and a possible antitumor chemical.

■ What is the primary role of fungi in the ecosystem?

the host: systemic, subcutaneous, cutaneous, superficial, or opportunistic. In Chapter 10, we saw that fungi are related to animals. Consequently, drugs that affect fungal cells may also affect animal cells. This fact makes fungal infections of humans and other animals often difficult to treat.

Systemic mycoses are fungal infections deep within the body. They are not restricted to any particular region of the body but can affect a number of tissues and organs. Systemic mycoses are usually caused by fungi that live in the soil. Inhalation of spores is the route of transmission; these infections typically begin in the lungs and then spread to other body tissues. They are not contagious from animal to human or from human to human.

Subcutaneous mycoses are fungal infections beneath the skin caused by saprophytic fungi that live in soil and on vegetation. Sporotrichosis is a subcutaneous infection acquired by gardeners and farmers. Infection occurs by direct implantation of spores or mycelial fragments into a puncture wound in the skin.

Fungi that infect only the epidermis, hair, and nails are called **dermatophytes,** and their infections are called dermatomycoses or **cutaneous mycoses.** Dermatophytes secrete keratinase, an enzyme that degrades **keratin,** a protein found in hair, skin, and nails. Infection is transmitted from human to human or from animal to human by direct contact or by contact with infected hairs and epidermal cells (as from barber shop clippers or shower room floors).

The fungi that cause **superficial mycoses** are localized along hair shafts and in superficial (surface) epidermal cells. These infections are prevalent in tropical climates.

An **opportunistic pathogen** is generally harmless in its normal habitat but can become pathogenic in a host who is seriously debilitated or traumatized, who is under treatment with broad-spectrum antibiotics, whose immune system is suppressed by drugs or by an immune disorder, or who has a lung disease.

Pneumocystis is an opportunistic pathogen seen in individuals with compromised immune systems and is the most common life-threatening infection in AIDS patients. It was first classified as a protozoan, but recent studies of its RNA indicate it is a unicellular anamorphic fungus. Another example of an opportunistic pathogen is the fungus *Stachybotrys* (sta'ke-bo-tris), which normally grows on cellulose found in dead plants but in recent years has been found growing on water-damaged walls of homes. One preliminary study reported that its toxic spores can cause fatal pulmonary hemorrhage in infants. Additional research is being designed to determine whether *Stachybotrys* is a health hazard.

Mucormycosis is an opportunistic mycosis caused by *Rhizopus* and *Mucor* (mū-kôr); the infection occurs mostly in patients with diabetes mellitus, with leukemia, or

TABLE 12.3	**Characteristics of Some Pathogenic Fungi**		
Phylum	**Growth Characteristics**	**Asexual Spore Types**	**Human Pathogens**
Zygomycota	Nonseptate hyphae	Sporangiospores	*Rhizopus* *Mucor*
Ascomycota	Dimorphic	Conidia	*Aspergillus* *Blastomyces* (Ajellomyces†) dermatitidis* *Histoplasma* (Ajellomyces†) capsulatum*
	Septate hyphae, strong affinity for keratin	Conidia Arthroconidia	*Microsporum* *Trichophyton* (Arthroderma†)*
Anamorphs		Conidia	*Epidermophyton*
	Dimorphic	Conidia Arthroconidia	*Sporothrix schenckii, Stachybotrys* *Coccidioides immitis*
	Yeastlike, pseudohyphae	Chlamydospores	*Candida albicans*
	Unknown	Unknown	*Pneumocystis*
Basidiomycota	Septate hyphae; includes rusts and smuts, and plant pathogens; yeastlike encapsulated cells	Conidia	*Cryptococcus neoformans* (Filobasidiella)†*

*Anamorph name.
†Teleomorph name.

undergoing treatment with immunosuppressive drugs. Aspergillosis is also an opportunistic mycosis; it is caused by *Aspergillus* (see Figure 12.2). This disease occurs in people who have debilitating lung diseases or cancer and have inhaled *Aspergillus* spores. Opportunistic infections by *Cryptococcus* and *Penicillium* can cause fatal diseases in AIDS patients. These opportunistic fungi may be transmitted from one person to an uninfected person but do not usually infect immunocompetent people. **Yeast infection,** or candidiasis, is most frequently caused by *Candida albicans* and may occur as vulvovaginal candidiasis or thrush, a mucocutaneous candidiasis. Candidiasis frequently occurs in newborns, in people with AIDS, and in people being treated with broad-spectrum antibiotics.

Some fungi cause disease by producing toxins. These toxins are discussed in Chapter 15.

Economic Effects of Fungi

Learning Objective

- Identify two beneficial and two harmful effects of fungi.

Fungi have been used in biotechnology for many years. *Aspergillus niger*, for example, has been used to produce citric acid for foods and beverages since 1914. The yeast *Saccharomyces cerevisiae* is used to make bread and wine. It is also genetically engineered to produce a variety of proteins, including hepatitis B vaccine. *Saccharomyces* and another yeast, *Torulopsis*, are used as protein supplements for humans and cattle. *Trichoderma* is used commercially to produce the enzyme cellulase, which is used to remove plant cell walls to produce a clear fruit juice. When the anticancer drug taxol, which is produced by yew trees, was discovered, there was concern that the yew forests of the U.S. Northwest coast would be decimated to harvest the drug. However, in 1993, Andrea and Donald Stierle saved the yews by discovering that the fungus *Taxomyces* also produces taxol.

Fungi are used as biological controls of pests. In 1990, the fungus *Entomorphaga* unexpectedly proliferated and killed gypsy moths that were destroying trees in the eastern United States. Scientists are investigating whether this fungus can be used in place of chemical insecticides. Annually, 25–50% of harvested fruits and vegetables are ruined by fungi. Chemical fungicides cannot be used to prevent this decay because of safety and environmental concerns. However, another fungus, *Candida oleophila*, can be and is used to prevent fungal growth on harvested fruits. This process of biocontrol works because the *C. oleophila* grows on the fruit surface before spoilage fungi grow.

TABLE 12.3	(continued)	
Habitat	**Type of Mycosis**	**Clinical Notes**
Ubiquitous	Systemic	Opportunistic pathogen
Ubiquitous	Systemic	Opportunistic pathogen
Ubiquitous	Systemic	Opportunistic pathogen
Unknown	Systemic	Inhalation
Soil	Systemic	Inhalation
Soil, animals	Cutaneous	Tinea capitis (ringworm)
Soil, animals	Cutaneous	Tinea pedis (athlete's foot)
Soil, humans	Cutaneous	Tinea cruris (jock itch), tinea unguium (of fingernails or toenails)
Soil	Subcutaneous	Puncture wound
Soil	Systemic	Inhalation
Human normal microbiota	Cutaneous, systemic, mucocutaneous	Opportunistic pathogen
Ubiquitous	Systemic	Opportunistic pathogen
Soil, bird feces	Systemic	Inhalation

In contrast to these beneficial effects, fungi can have undesirable effects for industry and agriculture because of their nutritional adaptations. As most of us have observed, mold spoilage of fruits, grains, and vegetables is relatively common, but bacterial spoilage of such foods is not. There is little moisture on the unbroken surfaces of such foods, and the interiors of fruits are too acidic for many bacteria to grow there. Jams and jellies also tend to be acidic, and they have a high osmotic pressure from the sugars they contain. These factors all discourage bacterial growth but readily support the growth of molds. A paraffin layer on top of a jar of homemade jelly helps deter mold growth because molds are aerobic and the paraffin layer keeps out the oxygen. However, fresh meats and certain other foods are such good substrates for bacterial growth that bacteria not only will outgrow molds but also will actively suppress mold growth in these foods.

The spreading chestnut tree, of which Longfellow wrote, no longer grows in the United States except in a few widely isolated locations; a fungal blight killed virtually all of them. This blight was caused by the ascomycete *Cryphonectria parasitica* (kri-fō-nek'trē-ä par-ä-si'ti-kä), which was introduced from China around 1904. The fungus allows the tree roots to live and put forth shoots regularly, but then it kills the shoots just as regularly.

Cryphonectria-resistant chestnuts are being developed. Another imported fungal plant disease is Dutch elm disease, caused by *Ceratocystis ulmi* (sē-rä-tō-sis'tis ul'mē). Carried from tree to tree by a bark beetle, the fungus blocks the afflicted tree's circulation. The disease has devastated the American elm population.

Lichens

Learning Objectives

- *List the distinguishing characteristics of lichens, and describe their nutritional needs.*
- *Describe the roles of the fungus and the alga in a lichen.*

A **lichen** is a combination of a green alga (or a cyanobacterium) and a fungus. Lichens are placed in the Kingdom Fungi and are classified according to the fungal partner, most often an ascomycete. The two organisms exist in a *mutualistic* relationship, in which each partner benefits. The lichen is very different from either the alga or fungus growing alone, and if the partners are separated, the lichen no longer exists. Approximately 13,500 species of lichens occupy quite diverse habitats. Because they can inhabit areas in which neither fungi nor algae could survive alone, lichens are often the first life forms to colonize newly exposed soil or rock. Lichens secrete organic acids that

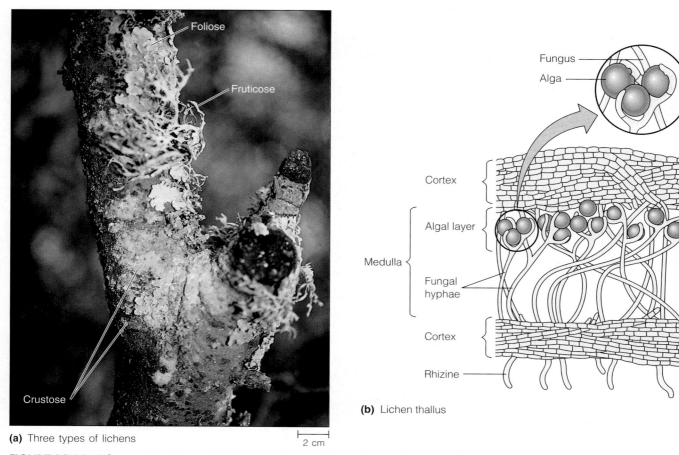

(a) Three types of lichens

2 cm

(b) Lichen thallus

FIGURE 12.10 **Lichens.** The lichen medulla is composed of fungal hyphae surrounding the algal layer. The protective cortex is a layer of fungal hyphae that covers the surface and sometimes the bottom of the lichen.

■ In what ways are lichens unique?

chemically weather rock, and they accumulate nutrients needed for plant growth. Also found on trees, concrete structures, and rooftops, lichens are some of the slowest-growing organisms on Earth.

Lichens can be grouped into three morphologic categories (Figure 12.10a). *Crustose lichens* grow flush or encrusting onto the substratum, *foliose lichens* are more leaflike, and *fruticose lichens* have fingerlike projections. The lichen's thallus, or body, forms when fungal hyphae grow around algal cells to become the **medulla** (Figure 12.10b). Fungal hyphae project below the lichen body to form **rhizines,** or holdfasts. Fungal hyphae also form a **cortex,** or protective covering, over the algal layer and sometimes under it as well. After incorporation into a lichen thallus, the alga continues to grow, and the growing hyphae can incorporate new algal cells.

When the algal partner is cultured separately in vitro, about 1% of the carbohydrates produced during photosynthesis are released into the culture medium; however, when the alga is associated with a fungus, the algal plasma membrane is more permeable, and up to 60% of the products of photosynthesis are released to the fungus or are found as end-products of fungus metabolism. The fungus clearly benefits from this association. The alga, while giving up valuable nutrients, is in turn compensated; it receives from the fungus both protection from desiccation (cortex) and attachment (holdfast).

Lichens had considerable economic importance in ancient Greece and other parts of Europe as dyes for clothing. Usnic acid from *Usnea* is used as an antimicrobial agent in China. Erythrolitmin, the dye used in litmus paper to indicate changes in pH, is extracted from a variety of lichens. Some lichens or their acids can cause allergic contact dermatitis in humans.

Populations of lichens readily incorporate cations (positively charged ions) into their thalli. Therefore, the concentrations and types of cations in the atmosphere can be determined by chemical analyses of lichen thalli. In addition, the presence or absence of species that are quite sensitive to pollutants can be used to ascertain air quality. A 1985 study in the Cuyahoga Valley in Ohio revealed that 81% of the 172 lichen species that were present in 1917 were gone. Because this area is severely affected by air pollution, the inference is that air pollutants, primarily sulfur dioxide (the major contributor to acid rain), caused the death of sensitive species.

Lichens are the major food for tundra herbivores such as caribou and reindeer. After the 1986 Chernobyl nuclear disaster, 70,000 reindeer in Lapland that had been raised for food had to be destroyed because of high levels of radiation. The lichens on which the reindeer fed had absorbed radioactive cesium-137, which had spread in the air.

Algae

Algae are familiar as the large brown kelp in coastal waters, the green scum in a puddle, and the green stains on soil or on rocks. A few algae are responsible for food poisonings. Some algae are unicellular; others form chains of cells (are filamentous); and a few have thalli.

Algae are mostly aquatic, although some are found in soil or on trees when sufficient moisture is available there. Unusual algal habitats include the hair of both the sedentary South American sloth and the polar bear. Water is necessary for physical support, reproduction, and the diffusion of nutrients. Generally, algae are found in cool temperate waters, although the large floating mats of the brown alga *Sargassum* (sär-gas′sum) are found in the subtropical Sargasso Sea. Some species of brown algae grow in antarctic waters.

Characteristics of Algae

Learning Objective

- List the defining characteristics of algae.

Algae are relatively simple eukaryotic photoautotrophs that lack the tissues (roots, stem, and leaves) of plants. The identification of unicellular and filamentous algae requires microscopic examination. Most algae are found in the ocean. Their locations depend on the availability of appropriate nutrients, wavelengths of light, and surfaces on which to grow. Probable locations for representative algae are shown in Figure 12.11a.

Vegetative Structures

The body of a multicellular alga is called a thallus. Thalli of the larger multicellular algae, those commonly called seaweeds, consist of branched **holdfasts** (which anchor the alga to a rock), stemlike and often hollow **stipes,** and

leaflike **blades** (Figure 12.11b). The cells covering the thallus can carry out photosynthesis. The thallus lacks the conductive tissue (xylem and phloem) characteristic of vascular plants; algae absorb nutrients from the water over their entire surface. The stipe is not lignified or woody, so it does not offer the support of a plant's stem; instead, the surrounding water supports the algal thallus; some algae are also buoyed by a floating, gas-filled bladder called a *pneumatocyst*.

Life Cycle

All algae can reproduce asexually. Multicellular algae with thalli and filamentous forms can fragment; each piece is capable of forming a new thallus or filament. When a unicellular alga divides, its nucleus divides (mitosis), and the two nuclei move to opposite parts of the cell. The cell then divides into two complete cells (cytokinesis).

Sexual reproduction occurs in algae (Figure 12.12 on page 349). In some species, asexual reproduction may occur for several generations and then, under different conditions, the same species reproduce sexually. Other species alternate generations so that the offspring resulting from sexual reproduction reproduce asexually, and the next generation then reproduces sexually.

Nutrition

"Algae" is a common name that includes several phyla (Table 12.4, page 349). Most algae are photosynthetic; however, the oomycotes, or fungal-like algae, are chemoheterotrophs. Photosynthetic algae are found throughout the photic (light) zone of bodies of water. Chlorophyll *a* (a light-trapping pigment) and accessory pigments involved in photosynthesis are responsible for the distinctive colors of many algae.

Algae are classified according to their rRNA sequences, structures, pigments, and other qualities (Table 12.4). Following are descriptions of some phyla of algae.

Selected Phyla of Algae

Learning Objective

- List the outstanding characteristics of the five phyla of algae discussed in this chapter.

The *brown algae*, or kelp, are macroscopic; some reach lengths of 50 m (see Figure 12.11b). Most brown algae are found in coastal waters. Brown algae have a phenomenal growth rate. Some grow at rates exceeding 20 cm per day and therefore can be harvested regularly. **Algin,** a thickener used in many foods (such as ice cream and cake decorations), is extracted from their cell walls. Algin is also used in the production of a wide variety of

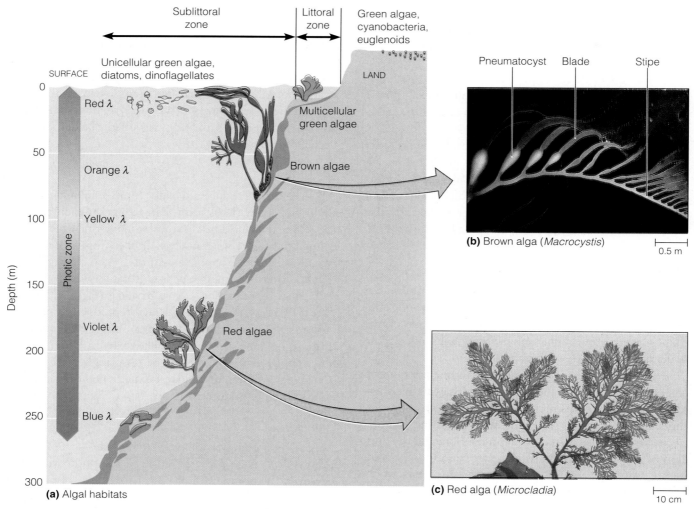

FIGURE 12.11 **Algae and their habitats. (a)** Although unicellular and filamentous algae can be found on land, they frequently exist in marine and freshwater environments as plankton. Multicellular green, brown, and red algae require a suitable attachment site, adequate water for support, and light of the appropriate wavelengths. **(b)** *Macrocystis porifera*, a brown alga. The hollow stipe and gas-filled pneumatocysts hold the thallus upright to ensure that sufficient sunlight is received for growth. **(c)** *Microcladia*, a red alga. The delicately branched red algae get their color from phycobiliprotein accessory pigments.

■ What red alga is toxic for humans?

nonfood goods, including rubber tires and hand lotion. The brown alga *Laminaria japonica* is used to induce vaginal dilation before surgical entry into the uterus through the vagina.

Most *red algae* have delicately branched thalli and can live at greater ocean depths than other algae (see Figure 12.11c). The thalli of a few red algae form crustlike coatings on rocks and shells. The red pigments enable red algae to absorb the blue light that penetrates deepest into the ocean. The agar used in microbiological

media is extracted from many red algae. Another gelatinous material, carrageenan, comes from a species of red algae commonly called Irish moss. Carrageenan and agar can be a thickening ingredient in evaporated milk, ice cream, and pharmaceutical agents. *Gracilaria* species, which grow in the Pacific Ocean, are used by humans for food. However, members of this genus can produce a lethal toxin.

Green algae have cellulose cell walls, contain chlorophyll *a* and *b*, and store starch, as plants do (see Figure

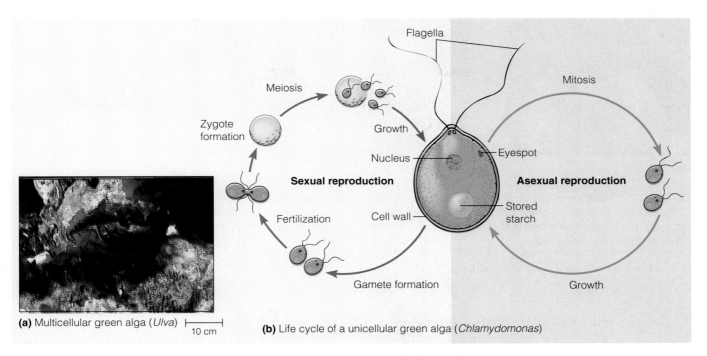

(a) Multicellular green alga (*Ulva*)   10 cm

(b) Life cycle of a unicellular green alga (*Chlamydomonas*)

FIGURE 12.12 **Green algae. (a)** The multicellular green alga *Ulva*. **(b)** The life cycle of the unicellular green alga *Chlamydomonas*. Two whiplike flagella propel this cell.

■ What is the primary role of algae in the ecosystem?

TABLE 12.4	**Characteristics of Selected Phyla of Algae**					
	Brown Algae	**Red Algae**	**Green Algae**	**Diatoms**	**Dinoflagellates**	**Water Molds**
Phylum	Phaeophyta	Rhodophyta	Chlorophyta	Bacillariophyta	Dinoflagellata	Oomycota
Color	Brownish	Reddish	Green	Brownish	Brownish	Colorless, white
Cell Wall	Cellulose and alginic acid	Cellulose	Cellulose	Pectin and silica	Cellulose in membrane	Cellulose
Cell Arrangement	Multicellular	Most are multicellular	Unicellular and multicellular	Unicellular	Unicellular	Multicellular
Photosynthetic Pigments	Chlorophyll *a* and *c*, xanthophylls	Chlorophyll *a* and *d*, phycobili-proteins	Chlorophyll *a* and *b*	Chlorophyll *a* and *c*, carotene, xanthophylls	Chlorophyll *a* and *c*, carotene, xanthins	None
Sexual Reproduction	Yes	Yes	Yes	Yes	In a few (?)	Yes (similar to the Zygomycota)
Storage Material	Carbohydrate	Glucose polymer	Glucose polymer	Oil	Starch	None

(a)

SEM | 50 μm

(b) Asexual reproduction of a diatom

FIGURE 12.13 **Diatoms. (a)** In this micrograph of *Isthmia nervosa,* notice how the two parts of the cell wall fit together at the arrows. **(b)** Asexual reproduction in a diatom. During mitosis, each daughter cell retains one-half of the cell wall from the parent (green) and must synthesize the remaining half (blue).

■ What human disease is caused by diatoms?

12.12a). Green algae are believed to have given rise to terrestrial plants. Most green algae are microscopic, although they may be either unicellular or multicellular. Some filamentous kinds form grass-green scum in ponds.

Diatoms, dinoflagellates, and water molds are grouped into the Kingdom Chromista or Stramenopila. *Diatoms* (Figure 12.13) are unicellular or filamentous algae with complex cell walls that consist of pectin and a layer of silica. The two parts of the wall fit together like the halves of a Petri dish. The distinctive patterns of the walls are a useful tool in diatom identification. Diatoms store energy captured through photosynthesis in the form of oil.

The first reported outbreak of a neurological disease caused by diatoms was reported in 1987 in Canada. Affected people ate mussels that had been feeding on diatoms. The diatoms produced *domoic acid,* a toxin that was then concentrated in the mussels. Symptoms included diarrhea and memory loss. The fatality rate was less than 4%. Since 1991, hundreds of marine birds and sea lions have died from the same **domoic acid intoxication** in California.

Dinoflagellates are unicellular algae collectively called **plankton,** or free-floating organisms (Figure 12.14). Their rigid structure is due to cellulose embedded in the plasma membrane. Some dinoflagellates produce neurotoxins. In the last 20 years, a worldwide increase in toxic marine algae has killed millions of fish, hundreds of marine mammals, and even some humans. When fish swim through large numbers of the dinoflagellate *Gymnodinium breve* (jim'nō-din-ē-um brev'ē), the algae trapped in the gills of the fish release a neurotoxin that stops the fish from breathing. Dinoflagellates in the genus *Alexandrium* (al-eg'zan-drē-um) produce neurotoxins (called **saxitoxins**) that cause **paralytic shellfish poisoning (PSP).** The toxin is concentrated when large numbers of dinoflagel-

lates are eaten by mollusks, such as mussels or clams. Humans who eat these mollusks develop PSP. Large concentrations of *Alexandrium* give the ocean a deep red color, from which the name **red tide** originates. Mollusks should not be harvested for consumption during a red tide. A disease called **ciguatera** occurs when the dinoflagellate *Gambierdiscus toxicus* (gam'bē-ėr-dis-kus toks'i-kus) passes up the food chain and is concentrated in large fish. Ciguatera is endemic (constantly present) in the south Pacific Ocean and the Caribbean Sea. An emerging disease associated with dinoflagellates is discussed in the box on page 351.

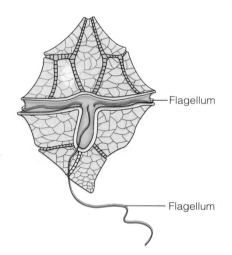

Flagellum

Flagellum

FIGURE 12.14 *Peridinium,* **a dinoflagellate.** Like some other dinoflagellates, *Peridinium* has two flagella in perpendicular, opposing grooves. When the two flagella beat simultaneously, they cause the cell to spin.

■ What human diseases are caused by dinoflagellates?

MMWR

Fish Killer Emerges as Human Pathogen

The periodic proliferation of dinoflagellates has long been linked to human death. When the dinoflagellates' toxins become concentrated in animals that are eaten by humans and fish, deaths due to these dinoflagellates are not uncommon. However, the number of these algal blooms has increased during the last 50 years. Researchers suggest the increase is due to several factors:

- Increases in nitrogen and phosphate from fertilizers and animal feces in runoff from fields.

- Ships' introducing algae into new waters with their ballast water

- Dams that affect the flow and nutrient levels of rivers, which in turn alters the chemistry of the river's estuary

- Global warming and increased ultraviolet radiation that may affect algal growth

In 1991, a billion fish died in North Carolina's Neuse Estuary, and in 1997, thousands of fish died in Chesapeake Bay estuaries. The 1997 fish kill was unusual because a woman became sick only after handling a fish from the poisoned waters. The CDC calls this possible estuary-associated syndrome (PEAS), and it can include memory loss, confusion, acute skin burning, headaches, skin rash, eye irritation, upper respiratory irritation, muscle cramps, and gastrointestinal complaints (e.g., nausea, vomiting, diarrhea, and abdominal cramps). Elevated liver enzyme levels in the blood of one patient also suggest liver damage.

SEM   15 μm

Pfiesteria piscicida

The culprit is a dinoflagellate, named *Pfiesteria* (see the figure), that was first discovered in 1988 in fish cultures at the North Carolina School of Veterinary Medicine by JoAnn Burkholder. Since then it has been implicated as the cause of half of the major fish kills from Delaware Bay to the Gulf Coast. The dinoflagellate was named in honor of the dinoflagellate biologist Lois Pfiester.

Most of the information on human disease was obtained from researchers working with toxic cultures of the dinoflagellates. They inadvertently discovered that *Pfiesteria* is unusual in that its toxic effects can be contracted by inhaling cultures containing 2000 or more cells per milliliter. It is unclear whether people exposed to *Pfiesteria* while swimming or boating are at risk for developing the illness. PEAS is not infectious and has not been associated with eating fish or mollusks caught in waters where *Pfiesteria* has been found. CDC recommends avoiding areas with large numbers of diseased, dying, or dead fish.

The CDC is conducting multistate surveillance and epidemiologic studies for PEAS and has established that the following conditions may represent adverse consequences to *Pfiesteria* exposure.

1. 20% of a sample of 50 fish with *Pfiesteria* lesions or toxicity

2. A fish kill involving fish with *Pfiesteria* lesions

3. A fish kill involving fish without *Pfiesteria* lesions if there is no alternative reason for the fish kill

Life Cycle

Pfiesteria has a complex life cycle that may include as many as 24 stages. Like other dinoflagellates, *Pfiesteria* can photosynthesize—but unlike other dinoflagellates, it must steal chloroplasts from algae. Without stolen chloroplasts, amoeboid and flagellated forms of *Pfiesteria* feed on bacteria, algae, and small animals in the water column. When large numbers of fish are in an area, their excreta causes *Pfiesteria* to become toxic. It may be that the *Pfiesteria* population increases by feeding on the excreta, thereby increasing the likelihood that fish will encounter *Pfiesteria* cells. It is unclear whether *Pfiesteria* produce a neurotoxin like some other algae or *Pfiesteria* kill fish by attaching to the outside of fish and eating their tissues.

SOURCE: Adapted from MMWR 49(17): 372–373 (5/5/00).

Most *water molds*, or *oomycota*, are decomposers. They form the cottony masses on dead algae and animals, usually in fresh water (Figure 12.15). Asexually, the oomycotes resemble the zygomycete fungi in that they produce spores in a sporangium (spore sac). However, oomycote spores, called **zoospores,** have two flagella; fungi do not have flagella. Because of their superficial similarity to fungi, oomycotes were previously classified with the fungi. Their cellulose cell walls always raised the question about their relationship to algae, and recent DNA analyses have confirmed that oomycotes are more closely related to diatoms and dinoflagellates than to fungi. Many of the terrestrial oomycotes are plant parasites. The U.S.D.A. inspects imported plants for white rust and other parasites. Often travelers, or even commercial plant importers, do not realize that one little blossom or seedling could carry a pest that is

FIGURE 12.15 **An oomycote.** Note the fuzzy mass on this fish. The funguslike algae are common decomposers in aquatic environments.

■ Is this oomycote more closely related to *Penicillium* or to diatoms?

capable of causing millions of dollars' worth of damage to U.S. agriculture.

In Ireland during the mid-1800s, 1 million people died when the country's potato crop failed. The alga that caused the great potato blight, *Phytophthora infestans* (fī-tof'thô-rä in-fes'tans), was one of the first microorganisms to be associated with a disease. Today, *Phytophthora* infects soybeans, potatoes, and cocoa.

In Australia, *P. cinnamomi* has infected about 20% of one species of *Eucalyptus*. *Phytophthora* was introduced into the United States in the 1990s and caused widespread damage to fruit and vegetable crops. When California oak trees suddenly starting dying in 1995, University of California scientists identified the cause of this "sudden oak death" to be a new species, *P. ramorum*. *P. ramorum* also infects redwood trees.

Roles of Algae in Nature

Algae are an important part of any aquatic food chain because they fix carbon dioxide into organic molecules that can be consumed by chemoheterotrophs. Using the energy produced in photophosphorylation, algae convert carbon dioxide in the atmosphere into carbohydrates. Molecular oxygen (O_2) is a by-product of their photosynthesis. The top few meters of any body of water contain planktonic algae. As 75% of the Earth is covered with water, it is estimated that 80% of the Earth's O_2 is produced by planktonic algae.

Seasonal changes in nutrients, light, and temperature cause fluctuations in algal populations; periodic increases in numbers of planktonic algae are called **algal blooms.** Blooms of dinoflagellates are responsible for seasonal red tides. Blooms of a certain few species indicate that the water in which they grow is polluted because these algae thrive in high concentrations of organic materials that exist in sewage or industrial wastes. When algae die, the decomposition of the large numbers of cells associated with an algal bloom depletes the level of dissolved oxygen in the water.

Much of the world's petroleum was formed from diatoms and other planktonic organisms that lived several million years ago. When such organisms died and were buried by sediments, the organic molecules they contained did not decompose to be returned to the carbon cycle as CO_2. Heat and pressure resulting from the Earth's geologic movements altered the oil stored in the cells, as well as the cell membranes. Oxygen and other elements were eliminated, leaving a residue of hydrocarbons in the form of petroleum and natural gas deposits.

Many unicellular algae are symbionts in animals. The giant clam *Tridacna* (trī-dak'nä) has evolved special organs that host dinoflagellates. As the clam sits in shallow water, the algae proliferate in these organs when they are exposed to the sun. The algae release glycerol into the clam's bloodstream, thus supplying the clam's carbohydrate requirement. In addition, evidence suggests that the clam gets essential proteins by phagocytizing old algae.

Protozoa

Protozoa are unicellular, eukaryotic chemoheterotrophic organisms. Among the protozoa are many variations on this cell structure, as we shall see. Protozoa inhabit water and soil. The feeding and growing stage, or **trophozoite,** feeds upon bacteria and small particulate nutrients. Some protozoa are part of the normal microbiota of animals. Of the nearly 20,000 species of protozoa, relatively few cause disease.

Characteristics of Protozoa

Learning Objective

• List the defining characteristics of protozoa.

The term *protozoan* means "first animal," which generally describes its animal-like nutrition. In addition to getting food, a protozoan must reproduce, and parasitic species must be able to get from one host to another.

Life Cycle

Protozoa reproduce asexually by fission, budding, or schizogony. **Schizogony** is multiple fission; the nucleus undergoes multiple divisions before the cell divides. After many nuclei are formed, a small portion of cytoplasm concentrates around each nucleus, and then the single cell separates into daughter cells.

Sexual reproduction has been observed in some protozoa. The ciliates, such as *Paramecium*, reproduce sexually by **conjugation** (Figure 12.16), which is very different from the bacterial process of the same name (see Figure 8.26 on page 238). During protozoan conjugation, two cells fuse, and a haploid nucleus (the micronucleus) from each cell migrates to the other cell. This haploid micronucleus fuses with the haploid macronucleus within the cell. The parent cells separate, each now a fertilized cell. When the cells later divide, they produce daughter cells with recombined DNA. Some protozoa produce **gametes (gametocytes)**, haploid sex cells. During reproduction, two gametes fuse to form a diploid zygote.

Encystment Under certain adverse conditions, some protozoa produce a protective capsule called a **cyst.** A cyst permits the organism to survive when food, moisture, or oxygen are lacking, when temperatures are not suitable, or when toxic chemicals are present. A cyst also enables a parasitic species to survive outside a host. This is important because parasitic protozoa may have to be excreted from one host in order to get to a new host. The cyst form in members of the phylum Apicomplexa is called an **oocyst.** It is a reproductive structure in which new cells are produced asexually.

Nutrition

Protozoa are mostly aerobic heterotrophs, although many intestinal protozoa are capable of anaerobic growth. Two chlorophyll-containing groups, dinoflagellates and euglenoids, are often studied with algae.

All protozoa live in areas with a large supply of water. Some protozoa transport food across the plasma membrane. However, some have a protective covering, or *pellicle*, and thus require specialized structures to take in food. Ciliates take in food by waving their cilia toward a mouth-like opening called a **cytostome.** Amoebas engulf food by surrounding it with pseudopods and phagocytizing it. In all protozoa, digestion takes place in membrane-enclosed **vacuoles,** and waste may be eliminated through the plasma membrane or through a specialized **anal pore.**

Medically Important Phyla of Protozoa

Learning Objectives

- *Describe the outstanding characteristics of the seven phyla of protozoa discussed in this chapter, and give an example of each.*
- *Differentiate an intermediate host from a definitive host.*

LM |—— 25 μm

FIGURE 12.16 Conjugation in the ciliate protozoan *Paramecium.* Sexual reproduction in ciliates is by conjugation. Each cell has two nuclei: a micronucleus and a macronucleus. The micronucleus is haploid and is specialized for conjugation. One micronucleus from each cell will migrate to the other cell during conjugation. Both cells will then go on to produce two daughter cells with recombined DNA. Condensed chromosomes are visible in the micronuclei.

■ Some protozoa can reproduce sexually as well as asexually.

The biology of protozoa is discussed in this chapter. Diseases caused by protozoa are described in Part Four.

Protozoa are a large and diverse group. Current schemes of classifying protozoan species into phyla are based on rRNA sequencing. Researchers have begun sorting out groups within the protists based on their evolutionary history; that is, members in a group derived from a single ancestor. At present, the following groups are phyla of protists. As more information is obtained, some of these groups may be classified as kingdoms in the Domain Eukarya.

Archaezoa

The **Archaezoa** are eukaryotes that lack mitochondria. The absence of mitochondria suggests that they may have evolved before the endosymbiotic event that gave rise to mitochondria, or that they have lost the ability to make mitochondria. Many archaezoans live as symbionts in the digestive tracts of animals. Archaezoans are typically spindle-shaped, with flagella projecting from the front end (Figure 12.17a). Most have two or more flagella, which move in a whiplike manner that pulls the cells through their environment.

An example of an archaezoan that is a human parasite is *Trichomonas vaginalis* (trik-ō-mōn'as va-jin-al'is), shown

(a) *Chilomastix*   5 μm

1st–3rd flagella
4th flagellum
Nucleus
Oral groove
Food vacuole

Flagella
Undulating membrane
Axostyle

(b) *Trichomonas vaginalis*   SEM   10 μm

(c) *Giardia lamblia*   LM   15 μm

(d) *Giardia lamblia* cyst   LM   15 μm

FIGURE 12.17 **Archaezoa. (a)** *Chilomastix*. This flagellate, found in the human intestine, may be mildly pathogenic. The cysts survive for months outside a human host. The fourth flagellum is used to move food into the oral groove, where food vacuoles are formed. **(b)** *Trichomonas vaginalis*. This flagellate causes urinary and genital tract infections. Notice the small undulating membrane. This flagellate does not have a cyst stage. **(c)** *Giardia lamblia*. The trophozoite of this intestinal parasite has eight flagella and two prominent nuclei, giving it a distinctive appearance. **(d)** The *G. lamblia* cyst provides protection from the environment before it is ingested by a new host.

■ How do archaezoans obtain energy without mitochondria?

in Figure 12.17b. Like some other flagellates, *T. vaginalis* has an **undulating membrane,** which consists of a membrane bordered by a flagellum. *T. vaginalis* does not have a cyst stage and must be transferred from host to host quickly before desiccation occurs. *T. vaginalis* is found in the vagina and in the male urinary tract. It is usually transmitted by sexual intercourse but can also be transmitted by toilet facilities or towels.

Another parasitic archaezoan is *Giardia lamblia* (jē-är′dē-ä lam′lē-ä) (see Figure 12.17c for the vegetative

trophozoite and Figure 12.17d for the cyst stage). The parasite is found in the small intestine of humans and other mammals. It is excreted in the feces as a cyst and survives in the environment before being ingested by the next host. Diagnosis of giardiasis, the disease caused by *G. lamblia*, is often based on the identification of cysts in feces.

Microspora

Microspora, like the Archaezoa, are unusual eukaryotes because they lack mitochondria. Microspora do not have microtubules (see Chapter 4, page 98), and they are obligate intracellular parasites. Microsporidial protozoa have been reported since 1984 to be the cause of a number of human diseases, including chronic diarrhea and keratoconjunctivitis (inflammation of the conjunctiva near the cornea), most notably in AIDS patients.

Rhizopoda

The **rhizopoda** or amoebas move by extending blunt, lobelike projections of the cytoplasm called **pseudopods** (Figure 12.18a). Any number of pseudopods can flow from one side of the amoeba, and the rest of the cell will flow toward the pseudopods.

Entamoeba histolytica (en-ta-me′ba his-to-li′ti-ka) is the only pathogenic amoeba found in the human intestine. As many as 10% of the human population may be colonized by this amoeba. New techniques, including DNA analyses and lectin binding, have revealed that the amoeba thought to be *E. histolytica* are actually two distinct species. The nonpathogenic species, *E. dispar* (dis′par) is most common. The invasive *E. histolytica* (Figure 12.18b) causes amoebic dysentery. In the human intestine, *E. histolytica* uses proteins called lectins to attach to the galactose of the plasma membrane and causes cell lysis. *E. dispar* does not have galactose-binding lectins. *Entamoeba* is transmitted between humans through ingestion of the cysts that are excreted in the feces of the infected person. *Acanthamoeba* growing in water, including tap water, can infect the cornea and cause blindness.

Since 1990, *Balamuthia* (bal′am-üth-ē-ä) has been reported as the cause of brain abscesses called granulomatous amoebic encephalitis in the United States and other countries. The amoeba most often infects immunocompromised people. Like *Acanthamoeba*, *Balamuthia* is a free-living amoeba found in water and is not transmitted from human to human.

Apicomplexa

The **Apicomplexa** are not motile in their mature forms and are obligate intracellular parasites. Apicomplexans are characterized by the presence of a complex of special organelles at the apexes (tips) of their cells (hence the phylum name). The organelles in these apical complexes contain enzymes that penetrate the host's tissues.

Apicomplexans have a complex life cycle that involves transmission between several hosts. An example of an apicomplexan is *Plasmodium* (plaz-mō′dē-um), the causative agent of malaria. The complex life cycle makes it difficult to develop a vaccine against malaria (see the box in Chapter 18, page 513).

(a) *Amoeba proteus*

(b) *Entamoeba histolytica*

FIGURE 12.18 **Rhizopoda. (a)** To move and to engulf food, amoebas (such as this *Amoeba proteus*) extend cytoplasmic structures called pseudopods. Food vacuoles are created when pseudopods surround food and bring it into the cell. **(b)** *Entamoeba histolytica*. The presence of ingested red blood cells is diagnostic for *Entamoeba*.

■ How do amoebic dysentery and bacillary dysentery differ?

Plasmodium grows by sexual reproduction in the *Anopheles* (an-of′el-ēz) mosquito (Figure 12.19).

❶ When an *Anopheles* carrying the infective stage of *Plasmodium*, called a **sporozoite,** bites a human, sporozoites can be injected into the human. The sporozoites are carried by the blood to the liver.

❷ They undergo schizogony in liver cells and produce thousands of progeny called **merozoites.**

❸ Merozoites enter the bloodstream and infect red blood cells.

❹ The young trophozoite looks like a ring in which the nucleus and cytoplasm are visible. This is called a **ring stage.**

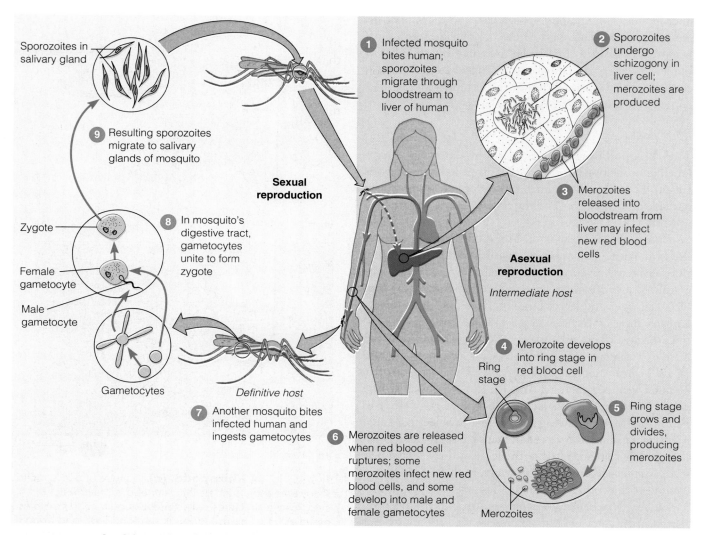

FIGURE 12.19 **The life cycle of *Plasmodium vivax,* the apicomplexan that causes malaria.** Asexual reproduction (schizogony) of the parasite takes place in the liver and in the red blood cells of a human host. Sexual reproduction occurs in the intestine of an *Anopheles* mosquito after the mosquito has ingested gametocytes.

■ What is the definitive host for *Plasmodium?*

⑤ The ring stage enlarges and divides repeatedly, and

⑥ the red blood cells eventually rupture and release more merozoites. Upon release of the merozoites, their waste products, which cause fever and chills, are also released. Most of the merozoites infect new red blood cells and perpetuate their cycle of asexual reproduction. However, some develop into male and female sexual forms (gametocytes). Even though the gametocytes themselves cause no further damage,

⑦ they can be picked up by the bite of another *Anopheles* mosquito; they then enter the mosquito's intestine and begin their sexual cycle.

⑧ Here the male and female gametocytes unite to form a zygote. The zygote forms an oocyst, in which cell division occurs, and asexual sporozoites are formed.

⑨ When the oocyst ruptures, the sporozoites migrate to the salivary glands of the mosquito. They can then be injected into a new human host by the biting mosquito.

The mosquito is the **definitive host** because it harbors the sexually reproducing stage of *Plasmodium.* The host in which the parasite undergoes asexual reproduction (in this case, the human) is the **intermediate host.**

Malaria is diagnosed in the laboratory by microscopic observation of thick blood smears for the presence of *Plas-*

modium. A peculiar characteristic of malaria is that the interval between periods of fever caused by the release of merozoites is always the same for a given species of *Plasmodium* and is always a multiple of 24 hours. The reason and mechanism for such precision have intrigued scientists. After all, why should a parasite need a biological clock? *Plasmodium's* development is regulated by the host's body temperature, which normally fluctuates over a 24-hour period. The parasite's careful timing ensures that gametocytes are mature at night, when *Anopheles* mosquitoes are feeding, to facilitate transmission of the parasite to a new host.

Another apicomplexan parasite of red blood cells is *Babesia microti* (ba-bē-sē-ä mī-krō'tē). *Babesia* causes fever and anemia in immunosuppressed individuals. In the United States, it is transmitted by the tick *Ixodes scapularis*.

Toxoplasma gondii (toks-ō-plaz'mä gon'dē-ē) is another apicomplexan intracellular parasite of humans. The life cycle of this parasite involves domestic cats. The trophozoites, called **tachyzoites,** reproduce sexually and asexually in an infected cat, and **oocysts,** each containing eight sporozoites, are excreted with feces. If the oocysts are ingested by humans or other animals, the sporozoites emerge as trophozoites, which can reproduce in the tissues of the new host. *T. gondii* is dangerous to pregnant women, as it can cause congenital infections in utero. Tissue examination and observation of *T. gondii* are used for diagnosis. Antibodies may be detected by ELISA and by indirect fluorescent-antibody tests (see Chapter 18).

Cryptosporidium (krip-tō-spô-ri'dē-um) is a newly recognized parasite of humans. In AIDS patients and other immunosuppressed people, *Cryptosporidium* can cause respiratory and gallbladder infections and may be a major cause of death. The organism, which lives inside the cells lining the small intestine, can be transmitted to humans through the feces cows, rodents, dogs, and cats. Waterborne and nosocomial infections have also been reported. Inside the host cell, each *Cryptosporidium* organism forms four oocysts, each containing four sporozoites. When the oocyst ruptures, sporozoites may infect new cells in the host or be released with the feces. The disease is diagnosed by acid-fast staining or fluorescent-antibody tests.

During the 1980s, epidemics of waterborne diarrhea were identified on every continent except Antarctica. The causative agent was misidentified as a cyanobacterium because the outbreaks occurred during warm months, and the disease agent looked like a prokaryotic cell. In 1993, the organism was identified as an apicomplexan similar to *Cryptosporidium.* In 1996, the new parasite, named *Cyclospora cayetanensis* (sī-klō-spô-rä kī'ē-tan-en-sis), was responsible for 850 cases of diarrhea associated with raspberries in the United States and Canada.

Ciliophora

Members of the phylum **Ciliophora** or ciliates have cilia that are similar to but shorter than flagella. The cilia are arranged in precise rows on the cell (Figure 12.20). They are moved in unison to propel the cell through its environment and to bring food particles to the mouth.

The only ciliate that is a human parasite is *Balantidium coli* (bal-an-tid'ē-um kō'lī), the causative agent of a severe, though rare, type of dysentery. When cysts are ingested by the host, they enter the large intestine, into which the

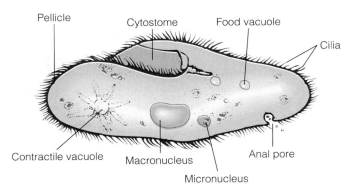

Pellicle · Cytostome · Food vacuole · Cilia
Contractile vacuole · Macronucleus · Micronucleus · Anal pore

(a) *Paramecium*

Stalk · Cytostome · Cilia

LM | 50 µm

(b) *Vorticella*

FIGURE 12.20 **Ciliates. (a)** *Paramecium* is covered with rows of cilia. It has specialized structures for ingestion (mouth), elimination of wastes (anal pore), and the regulation of osmotic pressure (contractile vacuoles). The macronucleus is involved with protein synthesis and other ongoing cellular activities. The micronucleus functions in sexual reproduction. **(b)** *Vorticella* attaches to objects in water by the base of its stalk. The springlike stalk can expand so *Vorticella* can feed in different areas. Cilia surround its cytostome.

■ What ciliate can cause disease in humans?

trophozoites are released. The trophozoites produce proteases and other substances that destroy host cells. The trophozoite feeds on host cells and tissue fragments. Its cysts are excreted with feces.

Ciliates, apicomplexans, and dinoflagellates (page 350) may be placed in their own phylum or kingdom, called **Alveolata** because they all have membrane-bound cavities (alveoli) under the cell surface and rRNA sequences in common.

Euglenozoa

Two groups of flagellated cells are included in the **Euglenozoa** based on common rRNA sequences, disk-shaped mitochondria, and absence of sexual reproduction.

Euglenoids are photoautotrophs (Figure 12.21). Euglenoids have a semirigid plasma membrane called a pellicle, and they move by means of a flagellum at the anterior end. Most euglenoids also have a red *eyespot* at the anterior end. This carotenoid-containing organelle senses light and directs the cell in the appropriate direction by using a *preemergent flagellum*. Some euglenoids are facultative chemoheterotrophs. In the dark, they ingest organic matter through a cytostome. Euglenoids are frequently studied with algae because they can photosynthesize.

The **hemoflagellates** (blood parasites) are transmitted by the bites of blood-feeding insects and are found in the circulatory system of the bitten host. To survive in this viscous fluid, hemoflagellates have long, slender bodies and an undulating membrane. The genus *Trypanosoma* (tri-pa′nō-sō-mä) includes the species that causes African sleeping sickness, *T. brucei gambiense* (brüs′ē gam-bē-ens′), which is transmitted by the tsetse fly. *T. cruzi* (kruz′ē), the causative agent of Chagas' disease, is transmitted by the "kissing bug," so named because it bites on the face (see Figure 12.33d). After entering the insect, the trypanosome rapidly multiplies by fission. If the insect then defecates while biting a human, it can release trypanosomes that can contaminate the bite wound.

Table 12.5 lists some typical parasitic protozoa and the diseases they cause.

Slime Molds

Learning Objective
* Compare and contrast cellular slime molds and plasmodial slime molds.

Slime molds have both fungal and amoebal characteristics; they are probably more closely related to amoeba. There are two phyla of slime molds: cellular and plasmodial. **Cellular slime molds** are typical eukaryotic cells that resemble amoebas. In the life cycle of cellular slime molds (Figure 12.22),

1 the amoeboid cells live and grow by ingesting fungi and bacteria by phagocytosis. Cellular slime molds are of interest to biologists who study cellular migration and aggregation, because when conditions are unfavorable,

2–3 large numbers of amoeboid cells aggregate to form a single structure. This aggregation occurs because some individual amoebas produce the chemical cyclic AMP (cAMP), toward which the other amoebas migrate.

LM ⊢ 4 μm

FIGURE 12.21 **Euglena.** Euglenoids are photoautotrophs. Semirigid rings supporting the pellicle allow *Euglena* to change shape.

■ Why is *Euglena* classified with the hemoflagellates?

TABLE 12.5	**Some Representative Parasitic Protozoa**					
Phylum	**Human Pathogens**	**Distinguishing Features**	**Disease**	**Source of Human Infections**	**Figure/ Table Reference**	**Page Reference**
Archaezoa	Giardia lamblia	Two nuclei, eight flagella	Giardial enteritis	Fecal contamination of drinking water	—	—
	Trichomonas vaginalis	No encysting stage	Urethritis, vaginitis	Contact with vaginal-urethral discharge	—	—
Microspora	Nosema	Unknown	Diarrhea, kerato-conjunctivitis, conjunctivitis	Other animals	—	—
Rhizopoda	Acanthamoeba	Pseudopods	Keratitis	Water	—	—
	Entamoeba histolytica, E. dispar		Amoebic dysentery	Fecal contamination of drinking water	—	
	Balamuthia		Encephalitis	Water		
Alveolata (Proposed) Apicomplexa	Babesia microti	Complex	Babesiosis	Domestic animals, ticks	—	—
	Cryptosporidium	Life cycles may require more than one host	Diarrhea	Humans, other animals, water	—	—
	Cyclospora	—	Diarrhea	Water	—	—
	Isospora	—	Coccidiosis	Domestic animals	Table 19.5	550
	Plasmodium	—	Malaria	Bite of Anopheles mosquito	12.19	356
	Toxoplasma gondii	—	Toxoplasmosis	Cats, beef; congenital	—	—
Dinoflagellates	Alexandrium, Pfiesteria	Photosynthetic, see Table 12.4	Paralytic shellfish poisoning; ciguatera	Ingestion of dinoflagellates in mollusks, fish	12.14	350
Ciliophora	Balantidium coli	Only parasitic ciliate of humans	Balantidial dysentery	Fecal contamination of drinking water	—	—
Euglenozoa	Leishmania	Flagellated form in sand fly; ovoid form in vertebrate host	Leishmaniasis	Bite of sand fly (Phlebotomus)	—	—
	Naegleria fowleri	Flagellated and amoeboid forms	Meningo-encephalitis	Water in which people swim	—	—
	Trypanosoma cruzi	Undulating membrane	Chagas' disease	Bite of Triatoma (kissing bug)	—	—
	T. brucei gambiense, T.b. rhodesiense		African trypanosomiasis	Bite of tsetse fly	—	—

❹ The aggregated amoebas are enclosed in a slimy sheath called a *slug*. The slug migrates as a unit toward light.

❺ After a period of hours, the slug ceases to migrate and begins to form differentiated structures.

❻ Some of the amoeboid cells form a stalk; others swarm up the stalk to form a spore cap, and

❼ most of these differentiate into spores.

❽ When spores are released under favorable conditions,

❾ they germinate to form single amoebas.

In 1973, a Dallas resident discovered a pulsating red blob in his backyard. The news media claimed that a "new life form" had been found. For some people, the "creature"

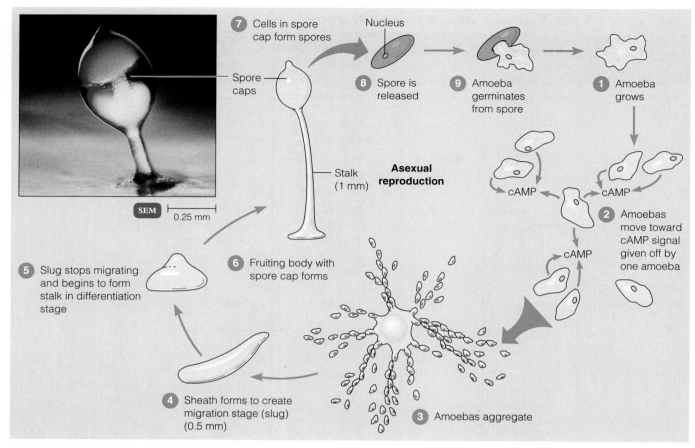

FIGURE 12.22 The generalized life cycle of a cellular slime mold.
The micrograph shows a spore cap of *Dictyostelium*.

■ Slime molds have both fungal and protozoal characteristics.

evoked spine-chilling recollections of an old science fiction movie. Before imaginations got carried away too far, biologists calmed everyone's worst fears (or highest hopes). The amorphous mass was merely a plasmodial slime mold, they explained. But its unusually large size—46 cm in diameter—startled even scientists.

Plasmodial slime molds were first scientifically reported in 1729. They belong to a separate phylum. A plasmodial slime mold exists as a mass of protoplasm with many nuclei (it is multinucleated). This mass of protoplasm is called a **plasmodium** (Figure 12.23).

❶ The entire plasmodium moves as a giant amoeba; it engulfs organic debris and bacteria. Biologists have found that musclelike proteins forming microfilaments account for the movement of the plasmodium.

❷ When plasmodial slime molds are grown in laboratories, a phenomenon called **cytoplasmic streaming** is observed, during which the protoplasm within the

plasmodium moves and changes both its speed and direction so that the oxygen and nutrients are evenly distributed.

The plasmodium continues to grow as long as there is enough food and moisture for it to thrive.

❸ When either is in short supply, the plasmodium separates into many groups of protoplasm;

❹ each of these groups forms a stalked sporangium,

❺ in which spores (a resistant, resting form of the slime mold) develop.

❻ Nuclei within these spores undergo meiosis and form uninucleate haploid cells.

❼ The spores are then released.

❽ When conditions improve, these spores germinate,

❾ fuse to form diploid cells, and

❿ develop into a multinucleated plasmodium.

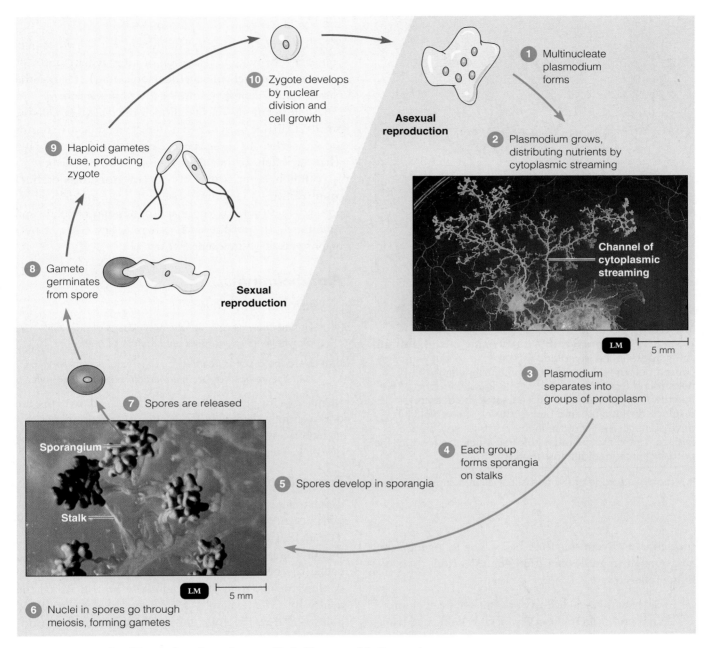

FIGURE 12.23 **The life cycle of a plasmodial slime mold.** Pictured in the photomicrographs is *Physarum*.

■ How do cellular and acellular slime molds differ?

Helminths

A number of parasitic animals spend part or all of their lives in humans. Most of these animals belong to two phyla: Platyhelminthes (flatworms) and Nematoda (roundworms). These worms are commonly called **helminths.** There are also free-living species in these phyla, but we will limit our discussion to the parasitic species. Diseases caused by parasitic worms are discussed in Part Four.

Characteristics of Helminths

Learning Objectives

- List the distinguishing characteristics of parasitic helminths.
- Provide a rationale for the elaborate life cycle of parasitic worms.

Helminths are multicellular eukaryotic animals that generally possess digestive, circulatory, nervous, excretory, and reproductive systems. Parasitic helminths must be highly

FIGURE 12.24 Infection by a parasitic platyhelminth.
An increase in the trematode *Ribeiroia* in recent years has caused deformed frogs. Frogs with multiple limbs have been found from Minnesota to California. Cercaria of the trematode infect tadpoles. The encysted metacercariae displace developing limb buds, causing abnormal limb development. The increase in the parasite may be due to fertilizer runoff that increases algae, which are food for the parasite's intermediate host snail.

■ What tailed stage of the parasite lives in a snail?

specialized to live inside their hosts. The following generalizations distinguish parasitic helminths from their free-living relatives:

1. They may *lack* a digestive system. They can absorb nutrients from the host's food, body fluids, and tissues.

2. Their nervous system is *reduced*. They do not need an extensive nervous system because they do not have to search for food or respond much to their environment. The environment within a host is fairly constant.

3. Their means of locomotion is occasionally *reduced* or *completely lacking*. Because they are transferred from host to host, they do not need to search actively for a suitable habitat.

4. Their reproductive system is often complex; an individual produces large numbers of eggs, by which a suitable host is infected.

Life Cycle

The life cycle of parasitic helminths can be extremely complex, involving a succession of intermediate hosts for completion of each **larval** (developmental) stage of the parasite and a definitive host for the adult parasite.

Adult helminths may be **dioecious;** male reproductive organs are in one individual, and female reproductive organs are in another. In those species, reproduction occurs only when two adults of the opposite sex are in the same host.

Adult helminths may also be **monoecious** or **hermaphroditic**—one animal has both male and female reproductive organs. Two hermaphrodites may copulate and simultaneously fertilize each other. A few types of hermaphrodites fertilize themselves.

Platyhelminths

Learning Objectives

- List the characteristics of the two classes of parasitic platyhelminths, and give an example of each.

- Describe a parasitic infection in which humans serve as a definitive host, as an intermediate host, and as both.

Members of the phylum Platyhelminthes, the **flatworms,** are dorsoventrally flattened. The classes of parasitic flatworms include the trematodes and cestodes. These parasites cause disease in a wide variety of animals (Figure 12.24).

Trematodes

Trematodes, or **flukes,** often have flat, leaf-shaped bodies with a ventral sucker and an oral sucker (Figure 12.25). The suckers hold the organism in place. Flukes obtain food by absorbing it through their nonliving outer covering, called the **cuticle.** Flukes are given common names according to the tissue of the definitive host in which the adults live (for example, lung fluke, liver fluke, blood fluke). The Asian liver fluke *Clonorchis sinensis* (klo-nôr′kis si-nen′sis) is occasionally seen in immigrants in the United States, but it cannot be transmitted because its intermediate hosts are not in the United States.

To exemplify a fluke's life cycle, let's look at the lung fluke, *Paragonimus westermani* (pār-ä-gōn′e-mus we-ster-ma′nē), shown in Figure 12.26 on page 364. The intermediate hosts for this fluke, and therefore the fluke itself, occur throughout the world, including the United States and Canada. The adult lung fluke lives in the bronchioles of humans and other mammals, and is approximately 6 mm wide and 12 mm long.

❶ The hermaphroditic adults liberate eggs into the bronchi. Because sputum that contains eggs is

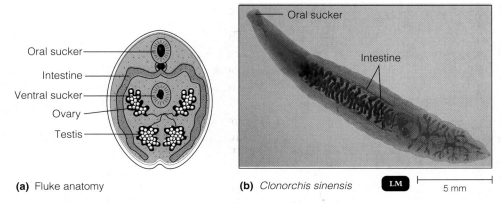

(a) Fluke anatomy

Oral sucker
Intestine
Ventral sucker
Ovary
Testis

Oral sucker
Intestine

(b) *Clonorchis sinensis*   LM   5 mm

FIGURE 12.25 **Flukes. (a)** General anatomy of an adult fluke, shown in cross section. The oral and ventral suckers attach the fluke to the host. The mouth is located in the center of the oral sucker. Flukes are hermaphroditic; each animal contains both testes and ovaries. **(b)** The Asian liver fluke *Clonorchis sinensis*. Notice the incomplete digestive system. Heavy infestations may block bile ducts from the liver.

■ The flatworm digestive system has one opening that functions as the mouth and anus.

frequently swallowed, the eggs are usually excreted in feces of the definitive host. If the life cycle is to continue,

❷ the eggs must reach a body of water.

❸ Inside the egg, a miracidial larva, or **miracidium,** then develops. When the egg hatches,

❹ the larva enters a suitable snail. Only certain species of aquatic snails can be this first intermediate host.

❺ Inside the snail, the miracidium develops into a **redia,** which then undergoes asexual reproduction and produces more rediae. Each redia produces **cercariae** that

❻ bore out of the snail and penetrate the cuticle of a crayfish.

❼ The parasite encysts as a **metacercaria** in the muscles and other tissues of the crayfish.

❽ When the crayfish is eaten by a human, the metacercaria is freed in the small intestine. It bores out and wanders around until it penetrates the lungs, enters the bronchioles, and develops into an adult lung fluke.

In a laboratory diagnosis, sputum and feces are examined microscopically for fluke eggs. Infection results from eating undercooked crayfish, and the disease can be prevented by thoroughly cooking crayfish.

The cercariae of the blood fluke *Schistosoma* (shis-tō-sō'ma) are not ingested. Instead, they burrow through the skin of the human host and enter the circulatory system. The adults are found in certain abdominal and pelvic veins. The disease schistosomiasis is a major world health problem.

Cestodes

Cestodes, or **tapeworms,** are intestinal parasites. Their structure is shown in Figure 12.27 on page 365. The head, or **scolex** (plural: *scoleces*), has suckers for attaching to the intestinal mucosa of the definitive host; some species also have small hooks for attachment. Tapeworms do not ingest the tissues of their hosts; in fact, they completely lack a digestive system. To obtain nutrients from the small intestine, they absorb food through their cuticle. The body consists of segments called **proglottids.** Proglottids are continually produced by the neck region of the scolex, as long as the scolex is attached and alive. Each mature proglottid contains both male and female reproductive organs. The proglottids farthest away from the scolex are the mature ones containing eggs. Mature proglottids are essentially bags of eggs, each of which is infective to the proper intermediate host.

Humans as Definitive Hosts The adults of *Taenia saginata* (te'nē-ä sa-ji-nä'tä), the beef tapeworm, live in humans and can reach a length of 6 m. The scolex is about 2 mm long and is followed by a thousand or more proglottids. The feces of an infected human contain mature proglottids, each of which contains thousands of eggs. As the proglottids wriggle away from the fecal material, they increase their chances of being ingested by an animal that is grazing. Upon ingestion by cattle, the larvae

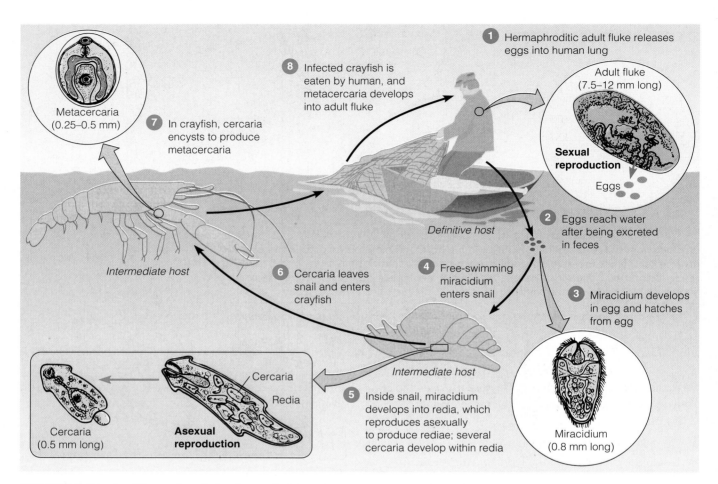

FIGURE 12.26 **The life cycle of the lung fluke *Paragonimus westermani*.** The trematode reproduces sexually in a human and asexually in a snail, its first intermediate host. Larvae encysted in the second intermediate host, a crayfish, infect humans when ingested.

■ Parasites often have complex life cycles that ensure that their progeny get into a suitable host.

hatch from the eggs and bore through the intestinal wall. The larvae migrate to muscle (meat), in which they encyst as **cysticerci.** When the cysticerci are ingested by humans, all but the scolex is digested. The scolex anchors itself in the small intestine and begins producing proglottids.

Diagnosis of tapeworm infection in humans is based on the presence of mature proglottids and eggs in feces. Cysticerci can be seen macroscopically in meat; their presence is referred to as "measly beef." Inspecting beef that is intended for human consumption for "measly" appearance is one way to prevent infections by beef tapeworm. Another method of prevention is to avoid the use of untreated human sewage as fertilizer in grazing pastures.

Humans are the only known definitive host of the pork tapeworm, *Taenia solium*. Adult worms living in the human intestine produce eggs, which are passed out in feces. When eggs are eaten by pigs, the larval helminth encysts in the pig's muscles; humans become infected when they eat undercooked pork. The human-pig-human cycle of *T. solium* is common in Latin America, Asia, and Africa. In the United States, however, *T. solium* is virtually nonexistent in pigs; the parasite is transmitted from human to human. Eggs shed by one person and ingested by another person hatch, and the larvae encyst in the brain and other parts of the body, causing cysticercosis. The human hosting *T. solium*'s larvae is serving as an intermediate host. Approximately 7% of the few hundred cases reported in recent years were acquired by people who had never been outside the United States. They may have become infected through household contact with people who were born in or had traveled in other countries.

Humans as Intermediate Hosts Humans are the intermediate hosts for *Echinococcus granulosus* (ē-kīn-ō-kok′kus gra-nū-lō′sus), shown in Figure 12.28. Dogs and coyotes are the definitive hosts for this minute (2–8 mm) tapeworm.

❶ Eggs are excreted with feces and

❷ are ingested by deer, sheep, or humans. Humans can also become infected by contaminating their hands with dog feces or saliva from a dog that has licked itself.

❸ The eggs hatch in the human's small intestine, and the larvae migrate to the liver or lungs.

❹ The larva develops into a **hydatid cyst.** The cyst contains "brood capsules," from which thousands of scoleces might be produced.

❺ Humans are a dead-end for the parasite, but in the wild, the cysts might be in a deer that is eaten by a wolf.

❻ The scoleces would be able to attach themselves in the wolf's intestine and produce proglottids.

Diagnosis of hydatid cysts is frequently made only on autopsy, although X rays can detect the cysts.

Nematodes

Learning Objectives

- *List the characteristics of parasitic nematodes, and give an example of infective eggs and infective larvae.*
- *Compare and contrast platyhelminths and nematodes.*

Members of the Phylum Nematoda, the **roundworms,** are cylindrical and tapered at each end. Roundworms have a *complete* digestive system, consisting of a mouth, an intestine, and an anus. Most species are dioecious. Males are smaller than females and have one or two hardened **spicules** on their posterior ends. Spicules are used to guide sperm to the female's genital pore.

Some species of nematodes are free-living in soil and water, and others are parasites on plants and animals. Some nematodes pass their entire life cycle, from egg to mature adult, in a single host.

Nematode infections of humans can be divided into two categories: those in which the egg is infective, and those in which the larva is infective.

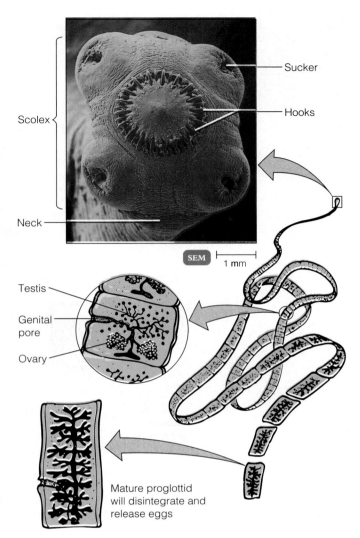

FIGURE 12.27 **General anatomy of an adult tapeworm.** The scolex, shown in the micrograph, consists of suckers and hooks that attach to the host's tissues. The body lengthens as new proglottids form at the neck. Each mature proglottid contains both testes and ovaries.

■ What are the similarities between tapeworms and flukes?

Eggs Infective for Humans

The pinworm *Enterobius vermicularis* (en-te-rō′bē-us ver-mi-kū-lar′is) spends its entire life in a human host (Figure 12.29 on page 367). Adult pinworms are found in the large intestine. From there, the female pinworm migrates to the anus to deposit her eggs on the perianal skin. The eggs can be ingested by the host or by another person exposed through contaminated clothing or bedding. Pinworm infections are diagnosed by the Graham sticky-tape method. A piece of transparent tape is placed on the perianal skin

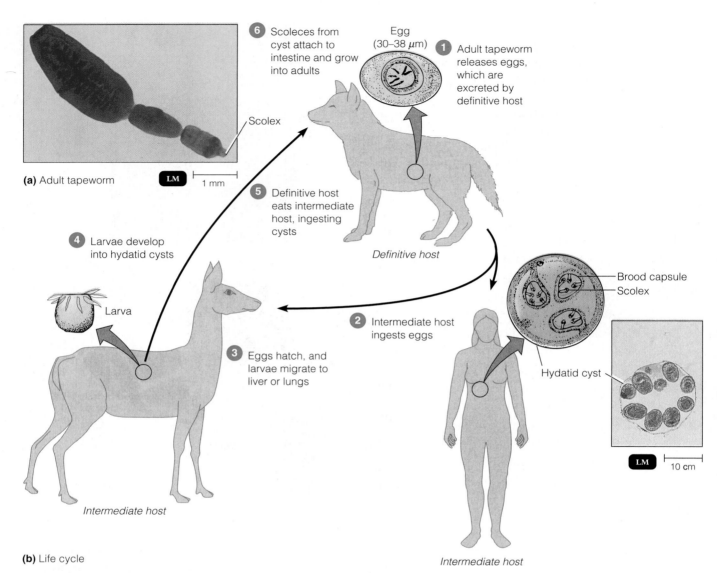

(a) Adult tapeworm   LM   ⊢—⊢ 1 mm

6 Scoleces from cyst attach to intestine and grow into adults

Scolex

Egg (30–38 μm)

1 Adult tapeworm releases eggs, which are excreted by definitive host

5 Definitive host eats intermediate host, ingesting cysts

Definitive host

4 Larvae develop into hydatid cysts

Larva

3 Eggs hatch, and larvae migrate to liver or lungs

Intermediate host

2 Intermediate host ingests eggs

Brood capsule
Scolex

Hydatid cyst

LM   ⊢—⊢ 10 cm

(b) Life cycle

Intermediate host

FIGURE 12.28 **The tapeworm *Echinococcus granulosus.*** This tiny tapeworm is found in the intestines of dogs, wolves, and foxes. **(a)** The adult of the closely related *Echinococcus multilocularis.* **(b)** Life cycle. The photomicrograph shows a hydatid cyst. The parasite can complete its life cycle only if the cysts are ingested by a definitive host that eats the intermediate host.

■ A human serving as the intermediate host for a tapeworm is a dead end for the parasite, unless the human is eaten by an animal.

in such a way that the sticky side picks up eggs that were deposited earlier. The tape is then microscopically examined for the presence of eggs.

Ascaris lumbricoides (as′kar-is lum-bri-koi′dēz) is a large nematode (30 cm in length) that infects over one billion people worldwide. It is dioecious with **sexual dimorphism;** that is, the male and female worms look distinctly

different, the male being smaller with a curled tail. The adult *Ascaris* lives in the small intestines of humans exclusively; it feeds primarily on semidigested food. Eggs, excreted with feces, can survive in the soil for long periods until accidentally ingested by another host. The eggs hatch in the small intestine of the host. The larvae then burrow out of the intestine and enter the blood. They are

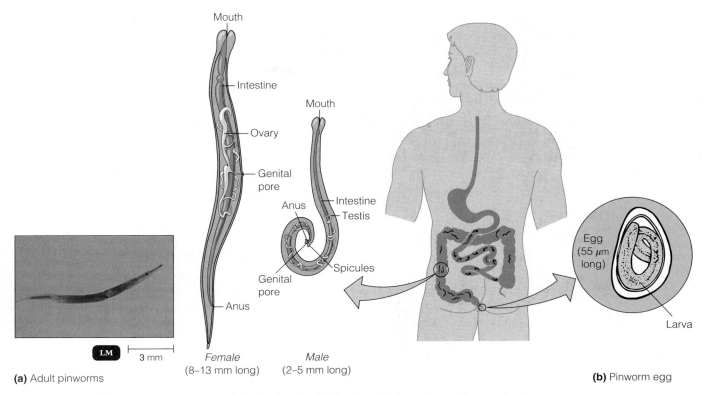

(a) Adult pinworms

(b) Pinworm egg

FIGURE 12.29 **The pinworm *Enterobius vermicularis*. (a)** Adult pinworms live in the large intestine of humans. Most roundworms are dioecious, and the female (left and photomicrograph) is often distinctly larger than the male (right). **(b)** Pinworm eggs are deposited by the female on the perianal skin at night.

■ Roundworms have a complete digestive system with a mouth, intestine, and anus.

carried to the lungs, where they grow. The larvae will then be coughed up, swallowed, and returned to the small intestine, where they mature into adults.

Diagnosis is frequently made when the adult worms are excreted with feces. *Ascaris* eggs remain in soil for 10 years. Children become infected by putting their hands and toys in their mouths. Preventing infection in humans is managed by proper sanitary habits.

Larvae Infective for Humans

Adult hookworms, *Necator americanus* (ne-kā-tôr ä-me-ri-ka′nus) and *Ancylostoma duodenale* (an-sil-os′toma dü′o-den-al-ē), live in the small intestine of humans (Figure 12.30); the eggs are excreted in feces. The larvae hatch in the soil, where they feed on bacteria. A larva enters its host by penetrating the host's skin. It then enters a blood or lymph vessel, which carries it to the lungs. It is coughed up in sputum, swallowed, and finally carried to the small intestine. Diagnosis is based on the presence of eggs in feces. People can avoid hookworm infections by wearing shoes.

Worldwide, *Trichinella spiralis* (trik-in-el′lä spī-ra′lis) infections, called trichinosis, are usually acquired by eating encysted larvae in poorly cooked pork. In the United States, trichinosis is more often acquired from eating game animals, such as bears. In the human digestive tract, the larvae are freed from the cysts. They mature into adults in the small intestine and sexually reproduce there. Eggs develop in the female, and she gives birth to live larvae. The larvae enter lymph and blood vessels in the intestines and migrate from there throughout the body. They encyst in muscles and other tissues and remain there until ingested by another host.

Diagnosis of trichinosis is made by microscopic examination for larvae in a muscle biopsy. Trichinosis can be prevented by thoroughly cooking meat prior to consumption.

Four genera of roundworms called anisakines, or wriggly worms, can be transmitted to humans from infected fish

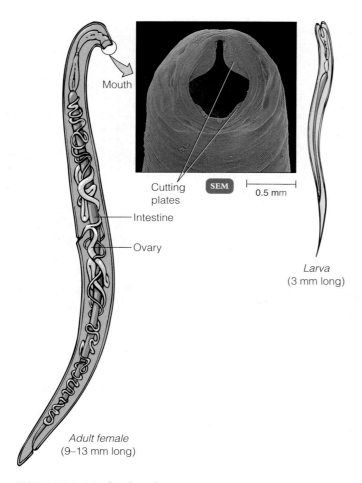

FIGURE 12.30 **The hookworm *Necator americanus.*** Notice the hook shape of the adult. The cutting plates around the mouth are used for attaching to and removing food from the host tissue. The free-living larvae inhabit the soil and infect their definitive host (humans) by penetrating the skin.

■ How do roundworms and flatworms differ?

and squid. Anisakine larvae are in the fish's intestinal mesenteries and migrate to the muscle when the fish dies. Freezing or thorough cooking will kill the larvae.

Table 12.6 lists representative parasitic helminths of each phylum and class and the diseases they cause.

Arthropods as Vectors

Learning Objectives
- *Define arthropod vector.*
- *Differentiate between a tick and a mosquito, and name a disease transmitted by each.*

Arthropods are animals characterized by segmented bodies, hard external skeletons, and jointed legs. With nearly

1 million species, this is the largest phylum in the animal kingdom. Although not microbes themselves, we will briefly describe arthropods here because a few suck the blood of humans and other animals and can transmit microbial diseases while doing so. Arthropods that carry pathogenic microorganisms are called **vectors.** Scabies and pediculosis are diseases that are caused by arthropods.

Representative classes of arthropods include the following:

- Arachnida (eight legs): spiders, mites, ticks
- Crustacea (four antennae): crabs, crayfish
- Insecta (six legs): bees, flies, lice

Table 12.7 lists those arthropods that are important vectors, and Figures 12.31, 12.32, and 12.33 illustrate some of them. These insects and ticks reside on an animal only when they are feeding. An exception to this is the louse, which spends its entire life on its host and cannot survive for long away from a host.

Some vectors are just a mechanical means of transport for a pathogen. For example, houseflies lay their eggs on decaying organic matter such as feces. While doing so, a housefly can pick up a pathogen on its feet or body and transport the pathogen to our food.

Some parasites multiply in their vectors. When this happens, the parasites can accumulate in the vector's feces or saliva. Large numbers of parasites can then be deposited on or in the host while the vector is feeding there. The spirochete that causes Lyme disease is transmitted by ticks in this manner, and the West Nile virus is transmitted in the same way by mosquitoes.

As discussed earlier, *Plasmodium* is an example of a parasite that requires that its vector also be the definitive host. *Plasmodium* can sexually reproduce only in the gut of an *Anopheles* mosquito. *Plasmodium* is introduced into a human host with the mosquito's saliva, which acts as an anticoagulant to keep blood flowing.

To eliminate vectorborne diseases (such as African sleeping sickness), health workers focus on eradicating the vectors.

| TABLE 12.6 | | **Representative Parasitic Helminths** | | | | | | |

Phylum	Class	Human Parasites	Intermediate Host	Definitive Host Site	Stage Passed to Humans; Method	Disease	Location in Humans	Figure Reference
Platyhelminthes	Trematodes	*Paragonimus westermani*	Freshwater snails and crayfish	Humans; lungs	Metacercaria in crayfish; ingested	Paragonimiasis (lung fluke)	Lungs	12.26
		Schistosoma	Freshwater snails	Humans	Cercariae; through skin	Schistosomiasis	Veins	— —
	Cestodes	*Taenia saginata*	Cattle	Humans; small intestine	Cysticerci in beef; ingested	Tapeworm	Small intestine	—
		Taenia solium	Humans; pigs	Humans	Eggs; ingested	Neurocysticercosis	Brain; any tissue	—
		Echinococcus granulosus	Humans	Dogs and other animals; intestines	Eggs from other animals; ingested	Hydatidosis	Lungs, liver, brain	12.28 —
Nematoda		*Ascaris lumbricoides*	—	Humans; small intestine	Eggs; ingested	Ascariasis	Small intestine	—
		Enterobius vermicularis	—	Humans; large intestine	Eggs; ingested	Pinworm	Large intestine	12.29
		Necator americanus	—	Humans; small intestine	Larvae; through skin	Hookworm	Small intestine	12.30
		Ancylostoma duodenale	—	Humans; small intestine	Larvae; through skin	Hookworm	Small Intestine	—
		Trichinella spiralis	—	Humans, swine, and other mammals; small intestine	Larvae; ingested	Trichinosis	Muscles	—
		Anisakines	Marine fish and squid	Marine mammals	Larvae in fish; ingested	Anisakiasis (sashimi worms)	Gastrointestinal tract	—

| TABLE 12.7 | | **Important Arthropod Vectors of Human Diseases** | | |

Class	Order	Vector	Disease	Figure Reference
Arachnida	Mites and ticks	*Dermacentor* (tick)	Rocky Mountain spotted fever	—
		Ixodes (tick)	Lyme disease, babesiosis, ehrlichiosis	12.32
		Ornithodorus (tick)	Relapsing fever	—
Insecta	Sucking lice	*Pediculus* (human louse)	Epidemic typhus, relapsing fever	12.33a
	Fleas	*Xenopsylla* (rat flea)	Endemic murine typhus, plague	12.33b
	True flies	*Chrysops* (deer fly)	Tularemia	12.33c
		Aedes (mosquito)	Dengue fever, yellow fever	—
		Anopheles (mosquito)	Malaria	12.31
		Culex (mosquito)	Arboviral encephalitis	—
		Glossina (tsetse fly)	African trypanosomiasis	—
	True bugs	*Triatoma* (kissing bug)	Chagas' disease	12.33d

(a) Female mosquito

(b) *Anopheles*

FIGURE 12.31 **Mosquitoes. (a)** A female mosquito sucking blood from human skin. Mosquitoes transmit several diseases from person to person, including the yellow fever and West Nile viruses. **(b)** The *Anopheles* mosquito transmits malaria.

■ Mosquitoes are one family of flies.

FIGURE 12.32 **Ticks.** *Ixodes pacificus* is the Lyme disease vector on the West Coast.

■ Why aren't ticks classified as insects?

LM | 1 mm

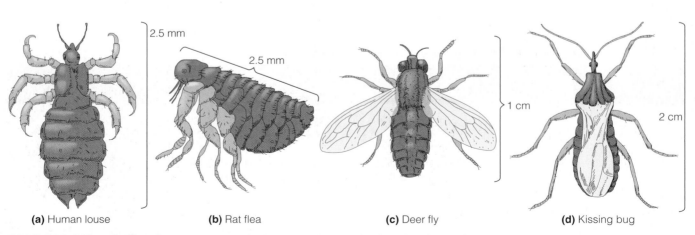

(a) Human louse   **(b)** Rat flea   **(c)** Deer fly   **(d)** Kissing bug

FIGURE 12.33 **Arthropod vectors. (a)** The human louse, *Pediculus.* **(b)** The rat flea, *Xenopsylla.* **(c)** The deer fly, *Chrysops.* **(d)** The kissing bug, *Triatoma.*

■ Name one disease carried by each of these vectors.

STUDY OUTLINE

Fungi (pp. 334–345)

1. Mycology is the study of fungi.

2. The number of serious fungal infections is increasing.

3. Fungi are aerobic or facultatively anaerobic chemoheterotrophs.

4. Most fungi are decomposers, and a few are parasites of plants and animals.

Characteristics of Fungi (pp. 335–339)

1. A fungal thallus consists of filaments of cells called hyphae; a mass of hyphae is called a mycelium.

2. Yeasts are unicellular fungi. To reproduce, fission yeasts divide symmetrically, whereas budding yeasts divide asymmetrically.

3. Buds that do not separate from the mother cell form pseudohyphae.

4. Pathogenic dimorphic fungi are yeastlike at 37°C and mold-like at 25°C.

5. These spores can be produced asexually: sporangiospores, conidiospores, and chlamydospores.

6. Fungi are classified according to the type of sexual spore that they form.

7. Sexual spores are usually produced in response to special circumstances, often changes in the environment.

8. Fungi can grow in acidic, low-moisture, aerobic environments.

9. They are able to metabolize complex carbohydrates.

Medically Important Phyla of Fungi (pp. 340–342)

1. The Zygomycota have coenocytic hyphae and produce sporangiospores and zygospores.

2. The Ascomycota have septate hyphae and produce ascospores and frequently conidiospores.

3. Basidiomycota have septate hyphae and produce basidiospores; some produce conidiospores.

4. Teleomorphic fungi produce sexual and asexual spores; anamorphic fungi produce asexual spores only.

Fungal Diseases (pp. 342–345)

1. Systemic mycoses are fungal infections deep within the body that affect many tissues and organs.

2. Subcutaneous mycoses are fungal infections beneath the skin.

3. Cutaneous mycoses affect keratin-containing tissues such as hair, nails, and skin.

4. Superficial mycoses are localized on hair shafts and superficial skin cells.

5. Opportunistic mycoses are caused by normal microbiota or fungi that are not usually pathogenic.

6. Opportunistic mycoses include *Pneumocystis* pneumonia; aspergillosis, caused by *Aspergillus*; and candidiasis, caused by *Candida*.

7. Opportunistic mycoses can infect any tissues. However, they are usually systemic.

Economic Effects of Fungi (p. 345)

1. *Saccharomyces* and *Trichoderma* are used in the production of foods.

2. Fungi are used for the biological control of pests.

3. Mold spoilage of fruits, grains, and vegetables is more common than bacterial spoilage of these products.

4. Many fungi cause diseases in plants.

Lichens (pp. 345–347)

1. A lichen is a mutualistic combination of an alga (or a cyanobacterium) and a fungus.

2. The alga photosynthesizes, providing carbohydrates for the lichen; the fungus provides a holdfast.

3. Lichens colonize habitats that are unsuitable for either the alga or the fungus alone.

4. Lichens may be classified on the basis of morphology as crustose, foliose, or fruticose.

5. Lichens are used for their pigments and as air quality indicators.

Algae (pp. 347–352)

1. Algae are unicellular, filamentous, or multicellular (thallic).

2. Most algae live in aquatic environments.

Characteristics of Algae (p. 347)

1. Algae are eukaryotic; most are photoautotrophs.

2. The thallus (body) of multicellular algae usually consists of a stipe, a holdfast, and blades.

3. Algae reproduce asexually by cell division and fragmentation.

4. Many algae reproduce sexually.

5. Photoautotrophic algae produce oxygen.

6. Algae are classified according to their structures and pigments.

Selected Phyla of Algae (pp. 347–352)

1. Brown algae (kelp) may be harvested for algin.

2. Red algae grow deeper in the ocean than other algae because their red pigments can absorb the blue light that penetrates to deeper levels.

3. Green algae have cellulose and chlorophyll *a* and *b* and store starch.

4. Diatoms are unicellular and have pectin and silica cell walls; some produce a neurotoxin.

5. Dinoflagellates produce neurotoxins that cause paralytic shellfish poisoning and ciguatera.

6. The oomycotes are heterotrophic; they include decomposers and plant parasites.

Roles of Algae in Nature (p. 352)

1. Algae are the primary producers in aquatic food chains.

2. Planktonic algae produce most of the molecular oxygen in the Earth's atmosphere.

3. Petroleum is the fossil remains of planktonic algae.

4. Unicellular algae are symbionts in such animals as *Tridacna*.

Protozoa (pp. 352–358)

1. Protozoa are unicellular, eukaryotic chemoheterotrophs.

2. Protozoa are found in soil and water and as normal microbiota in animals.

Characteristics of Protozoa (pp. 352–353)

1. The vegetative form is called a trophozoite.

2. Asexual reproduction is by fission, budding, or schizogony.

3. Sexual reproduction is by conjugation.

4. During ciliate conjugation, two haploid nuclei fuse to produce a zygote.

5. Some protozoa can produce a cyst that provides protection during adverse environmental conditions.

6. Protozoa have complex cells with a pellicle, a cytostome, and an anal pore.

Medically Important Phyla of Protozoa (pp. 353–358)

1. Archaezoa lack mitochondria and have flagella; they include *Trichomonas* and *Giardia*.

2. Microsporidia lack mitochondria and microtubales; microsporans cause diarrhea in AIDS patients.

3. Rhizopoda are amoeba; they include *Entamoeba* and *Acanthamoeba*.

4. Apicomplexa have apical organelles for penetrating host tissue; they include *Plasmodium* and *Cryptosporidium*.

5. Ciliophora move by means of cilia; *Balantidium coli* is the human parasitic ciliate.

6. Euglenozoa move by means of flagella and lack sexual reproduction; they include *Trypanosoma*.

Slime Molds (pp. 358–360)

1. Cellular slime molds resemble amoebas and ingest bacteria by phagocytosis.

2. Plasmodial slime molds consist of a multinucleated mass of protoplasm that engulfs organic debris and bacteria as it moves.

Helminths (pp. 361–368)

1. Parasitic flatworms belong to the Phylum Platyhelminthes.

2. Parasitic roundworms belong to the Phylum Nematoda.

Characteristics of Helminths (pp. 361–362)

1. Helminths are multicellular animals; a few are parasites of humans.

2. The anatomy and life cycle of parasitic helminths are modified for parasitism.

3. The adult stage of a parasitic helminth is found in the definitive host.

4. Each larval stage of a parasitic helminth requires an intermediate host.

5. Helminths can be monoecious or dioecious.

Platyhelminths (pp. 362–365)

1. Flatworms are dorsoventrally flattened animals; parasitic flatworms may lack a digestive system.

2. Adult trematodes, or flukes, have an oral and ventral sucker with which they attach to host tissue.

3. Eggs of trematodes hatch into free-swimming miracidia that enter the first intermediate host; two generations of rediae develop in the first intermediate host; the rediae become cercariae that bore out of the first intermediate host and penetrate the second intermediate host; cercariae encyst as metacercariae in the second intermediate host; after they are ingested by the definitive host, the metacercariae develop into adults.

4. A cestode, or tapeworm, consists of a scolex (head) and proglottids.

5. Humans serve as the definitive host for the beef tapeworm, and cattle are the intermediate host.

6. Humans serve as the definitive host and can be an intermediate host for the pork tapeworm.

7. Humans serve as the intermediate host for *Echinococcus granulosus*; the definitive hosts are dogs, wolves, and foxes.

Nematodes (pp. 365–368)

1. Roundworms have a complete digestive system.

2. The nematodes that infect humans with their eggs are *Enterobius vermicularis* (pinworm) and *Ascaris lumbricoides*.

3. The nematodes that infect humans with their larvae are *Necator americanus*, *Trichinella spiralis*, and anisakine worms.

Arthropods as Vectors (pp. 368–370)

1. Jointed-legged animals, including ticks and insects, belong to the Phylum Arthropoda.

2. Arthropods that carry diseases are called vectors.

3. Elimination of vectorborne diseases is best done by the control or eradication of the vectors.

STUDY QUESTIONS

Review

1. Contrast the mechanisms of conidiospore and ascospore formation by a fungus.

2. Fill in the following table.

Phylum	Spore Type(s)	
	Sexual	**Asexual**
Zygomycota		
Ascomycota		
Basidiomycota		

3. Following is a list of fungi, their methods of entry into the body, and sites of infections they cause. Categorize each type of mycosis as cutaneous, opportunistic, subcutaneous, superficial, or systemic.

Genus	Method of Entry	Site of Infection	Mycosis
Blastomyces	Inhalation	Lungs	
Sporothrix	Puncture	Ulcerative lesions	
Microsporum	Contact	Fingernails	
Trichosporon	Contact	Hair shafts	
Aspergillus	Inhalation	Lungs	

4. A mixed culture of *Escherichia coli* and *Penicillium chrysogenum* is inoculated onto the following culture media. On which medium would you expect each to grow? Why?
   a. 0.5% peptone in tap water
   b. 10% glucose in tap water

5. What is the role of the alga in a lichen? What is the role of the fungus?

6. Briefly discuss the importance of lichens in nature. Briefly discuss the importance of algae.

7. Complete the following table.

Division	Cell Wall Composition	Special Features/ Importance
Oomycotes		
Dinoflagellates		
Diatoms		
Red algae		
Brown algae		
Green algae		

Indicate which divisions consist primarily of unicellular forms. Which division could you include in the plant kingdom? Why?

8. Differentiate between cellular and plasmodial slime molds. How does each survive adverse environmental conditions?

9. Complete the following table.

Phylum	Method of Motility	One Human Parasite
Archaezoa		
Microspora		
Rhizopoda		
Apicomplexa		
Ciliophora		
Euglenozoa		

10. Why is it significant that *Trichomonas* does not have a cyst stage? Name a protozoan parasite that does have a cyst stage.

11. Recall the life cycle of *Plasmodium*. Where does asexual reproduction occur? Where does sexual reproduction occur? Identify the definitive host. Identify the vector.

12. By what means are helminthic parasites transmitted to humans?

13. To what phylum and class does this animal belong?

List two characteristics that put it in this phylum. Name the body parts. What is the name of the encysted larva of this animal?

14. Most nematodes are dioecious. What does this term mean? To what phylum do nematodes belong?

15. Vectors can be divided into three major types, according to the roles they play for the parasite. List the three types of vectors and a disease transmitted by each.

Multiple Choice

1. How many phyla are represented in the following list of organisms: *Echinococcus, Cyclospora, Aspergillus, Taenia, Toxoplasma, Trichinella?*
   a. 1
   b. 2
   c. 3
   d. 4
   e. 5

Use the following choices to answer questions 2 and 3:
- (1) metacercaria
- (2) redia
- (3) adult
- (4) miracidium
- (5) cercaria

2. Put the above stages in order of development, beginning with the egg.
   - a. 5, 4, 1, 2, 3
   - b. 4, 2, 5, 1, 3
   - c. 2, 5, 4, 3, 1
   - d. 3, 4, 5, 1, 2
   - e. 2, 4, 5, 1, 3

3. If a snail is the first intermediate host of a parasite with these stages, which stage would be found in the snail?
   - a. 1
   - b. 2
   - c. 3
   - d. 4
   - e. 5

4. Which of the following statements about yeasts are true?
   - (1) Yeasts are fungi.
   - (2) Yeasts can form pseudohyphae.
   - (3) Yeasts reproduce asexually by budding.
   - (4) Yeasts are facultatively anaerobic.
   - (5) All yeasts are pathogenic.
   - (6) All yeasts are dimorphic.
   - a. 1, 2, 3, 4
   - b. 3, 4, 5, 6
   - c. 2, 3, 4, 5
   - d. 1, 3, 5, 6
   - e. 2, 3, 4

5. Which of the following events follows cell fusion in an ascomycete?
   - a. conidiophore formation
   - b. conidiospore germination
   - c. ascus opening
   - d. ascospore formation
   - e. conidiospore release

6. The definitive host for *Plasmodium vivax* is
   - a. human.
   - b. *Anopheles*.
   - c. a sporocyte.
   - d. a gametocyte.

7. Fleas are the intermediate host for *Dipylidium caninum* tapeworm and dogs are the definitive host. Which stage of the parasite could be found in the flea?
   - a. cysticerus larva
   - b. proglottids
   - c. scolex
   - d. adult

Use the following choices to answer questions 8–10:
   - a. Apicomplexa
   - b. Ciliophora
   - c. Dinoflagellates
   - d. Microspora

8. These are obligate intracellular parasites that lack mitochondria.

9. These are nonmotile parasites with special organelles for penetrating host tissue.

10. These photosynthetic organisms can cause paralytic shellfish poisoning.

Critical Thinking

1. A generalized life cycle of the liver fluke *Clonorchis sinensis* is shown below. Identify the intermediate host(s). Identify the definitive host(s). To what phylum and class does this animal belong?

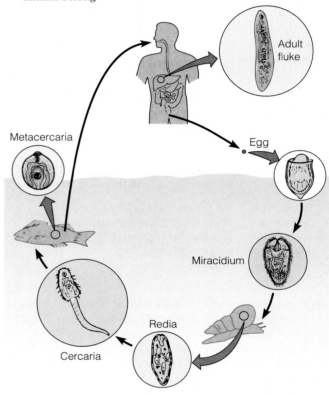

2. The size of a cell is limited by its surface-to-volume ratio; that is, if the volume becomes too great, internal heat cannot be dissipated, and nutrients and wastes cannot be efficiently transported. How do plasmodial slime molds manage to circumvent the surface-to-volume rule?

3. The life cycle of the fish tapeworm *Diphyllobothrium* is similar to that of *Taenia saginata*, except that the intermediate

host is fish. Describe the life cycle and method of transmission to humans. Why are freshwater fish more likely to be a source of tapeworm infection than marine fish?

4. *Trypanosoma brucei gambiense* (part a in the figure below) is the causative agent of African sleeping sickness. To what phylum does it belong? Part b shows a simplified life cycle for *T. b. gambiense*. Identify the host and vector of this parasite.

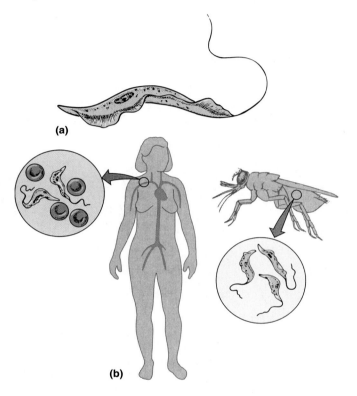

(a)

(b)

Clinical Applications

1. A girl developed generalized seizures. A CAT scan revealed a single brain lesion consistent with a tumor. Biopsy of the lesion showed a cysticercus. The patient lived in South Carolina and had never traveled outside the state. What parasite caused her disease? How is this disease transmitted? How might it be prevented?

2. A California farmer developed a low-grade fever, myalgia, and cough. A chest X ray revealed an infiltrate in the lung. Microscopic examination of the sputum revealed round, budding cells. A sputum culture grew mycelia and arthro-

conidia. What organism is most likely the cause of the symptoms? What is causing the man's disease? How is this disease transmitted? How might it be prevented?

3. A teenaged male in California complained of remittent fever, chills, and headaches. A blood smear revealed ring-shaped cells in his white blood cells. He was successfully treated with primaquine and chloroquine. The patient lives near the San Luis Rey River and has no history of foreign travel, blood transfusion, or IV drug use. What is the disease? How was it acquired?

4. Seventeen patients in ten hospitals had cutaneous infections caused by *Rhizopus*. In all 17 patients, Elastoplast bandages were placed over sterile gauze pads to cover wounds. Fourteen of the patients had surgical wounds, two had venous line insertion sites, and one had a bite wound. Lesions present when the bandages were removed ranged from vesiculopustular eruptions to ulcerations and skin necrosis requiring debridement.
   a. How did the wounds most likely get contaminated?
   b. Why is a fungus a more likely contaminant than a bacterium in this instance?

5. In mid-December, a female who had been on prednisone and is an insulin-dependent diabetic fell and received an abrasion on the dorsal side of her right hand. She was placed on penicillin. By the end of January, the ulcer had not healed, and she was referred to a plastic surgeon. On January 30, a swab of the wound was cultured at 35°C on blood agar. On the same day, a smear was made for Gram staining. The Gram stain showed fungal elements. What would you do next? Brownish, waxy colonies grew on the blood agar. Slide cultures set up on February 1 and incubated at 25°C showed septate hyphae and single conidia. What caused the infection? What treatment do you recommend? (See Chapter 20.)

LEARNING WITH TECHNOLOGY

The Microbiology Place CD-ROM/website includes many resources to help you study and succeed in this course. In addition to self-quizzes, web links, and flashcards, you'll find the following activities for this chapter in the "Activities" section of The Microbiology Place:

Taxonomy of Eukaryotes

Case Study: The Unfortunate Alaskan Fishing Trip

Chapter 13

Viruses, Viroids, and Prions

MASTADENOVIRUS. *This virus causes upper respiratory tract infections in humans.*

Viruses are too small to be seen with a light microscope and cannot be cultured outside their hosts. Therefore, although viral diseases are not new, the viruses themselves could not be studied until the twentieth century. In 1886, the Dutch chemist Adolf Mayer showed that tobacco mosaic disease (TMD) was transmissible from a diseased plant to a healthy plant. He could not satisfy Koch's postulates because he was not able to culture the infectious agent. In 1892, in an attempt to isolate the cause of TMD, the Russian bacteriologist Dimitri Iwanowski filtered the sap of diseased plants through a porcelain filter that

was designed to retain bacteria. He expected to find the microbe trapped in the filter; instead, he found that the infectious agent had passed through the minute pores of the filter. When he infected healthy plants with the filtered fluid, they contracted TMD. This led to a host of experiments by other scientists to isolate filterable agents of disease. The first human disease associated with a filterable agent was yellow fever.

These early workers could not imagine submicroscopic particles, and thus they described the infectious agent as contagium vivum fluidum—a contagious fluid. By the 1930s, scientists had begun

using the word virus, the Latin word for poison, to describe these filterable agents. The nature of viruses, however, remained elusive until 1935, when Wendell Stanley, an American chemist, isolated tobacco mosaic virus, making it possible for the first time to carry out chemical and structural studies on a purified virus. At about the same time, the invention of the electron microscope made it possible to see viruses.

We now know that viruses are found as parasites in all types of organisms. Although viruses must live inside a host cell, not all of them cause disease. In 1997, Japanese researchers discov-

ered a new virus they named TT virus (TTV), for the initials of the patient from whom it was isolated. Since then TTV has been found in 2% of healthy humans. It appears that TTV is a harmless symbiont. Perhaps there are other viruses that have adapted to live peacefully without causing disease.

Advances in molecular biology techniques in the 1980s and 1990s have led to the recognition of several other new human viruses. Human immunodeficiency virus (HIV), hepatitis C virus, Hantavirus *Sin Nombre*, and West Nile virus are a few examples. In this chapter, we will study the biology of viruses.

General Characteristics of Viruses

Learning Objective

- Differentiate a virus from a bacterium.

The question of whether viruses are living organisms has an ambiguous answer. Life can be defined as a complex set of processes resulting from the actions of proteins specified by nucleic acids. The nucleic acids of living cells are in action all the time. Because viruses are inert outside living host cells, in this sense they are not considered to be living organisms. However, once viruses enter a host cell, the viral nucleic acids become active, and viral multiplication results. In this sense, viruses are alive when they multiply in the host cells they infect. From a clinical point of view, viruses can be considered alive because they cause infection and disease, just as pathogenic bacteria, fungi, and protozoa do. Depending on one's viewpoint, a virus may be regarded as an exceptionally complex aggregation of nonliving chemicals, or as an exceptionally simple living microorganism.

How, then, do we define a virus? Viruses were originally distinguished from other infectious agents because they are especially small (filterable) and because they are **obligatory intracellular parasites**—that is, they absolutely require living host cells in order to multiply. However, both of these properties are shared by certain small bacteria, such as some rickettsias. Viruses and bacteria are compared in Table 13.1.

The truly distinctive features of viruses are now known to relate to their simple structural organization and their mechanism of multiplication. Accordingly, **viruses** are entities that

- Contain a single type of nucleic acid, either DNA or RNA.
- Contain a protein coat (sometimes itself enclosed by an envelope of lipids, proteins, and carbohydrates) that surrounds the nucleic acid.
- Multiply inside living cells by using the synthesizing machinery of the cell.
- Cause the synthesis of specialized structures that can transfer the viral nucleic acid to other cells.

Viruses have few or no enzymes of their own for metabolism; for example, they lack enzymes for protein synthesis and ATP generation. To multiply, viruses must take over the metabolic machinery of the host cell. This fact has considerable medical significance for the development of antiviral drugs, because most drugs that would interfere with viral multiplication would also interfere with the functioning of the host cell and therefore are too toxic for clinical use. (Antiviral drugs are discussed in Chapter 20.)

TABLE 13.1	Viruses and Bacteria Compared		
	Bacteria		Viruses
	Typical Bacteria	Rickettsias/ Chlamydias	
Intracellular parasite	No	Yes	Yes
Plasma membrane	Yes	Yes	No
Binary fission	Yes	Yes	No
Pass through bacteriological filters	No	No/Yes	Yes
Possess both DNA and RNA	Yes	Yes	No
ATP-generating metabolism	Yes	Yes/No	No
Ribosomes	Yes	Yes	No
Sensitive to antibiotics	Yes	Yes	No
Sensitive to interferon	No	No	Yes

Host Range

The **host range** of a virus is the spectrum of host cells the virus can infect. There are viruses that infect invertebrates, vertebrates, plants, protists, fungi, and bacteria. However, most viruses are able to infect specific types of cells of only one host species. In this chapter we are concerned mainly with viruses that infect either humans or bacteria. Viruses that infect bacteria are called **bacteriophages,** or **phages.**

The particular host range of a virus is determined by the virus's requirements for its specific attachment to the host cell and the availability within the potential host of cellular factors required for viral multiplication. For the virus to infect the host cell, the outer surface of the virus must chemically interact with specific receptor sites on the surface of the cell. The two complementary components are held together by weak bonds, such as hydrogen bonds. For some bacteriophages, the receptor site is part of the cell wall of the host; in other cases, it is part of the fimbriae or flagella. For animal viruses, the receptor sites are on the plasma membranes of the host cells.

The potential to use viruses to treat diseases is intriguing because of their narrow host range and their ability to kill their host cells. The idea of *phage therapy*—using bacteriophage to treat bacterial infections, has been around for 75

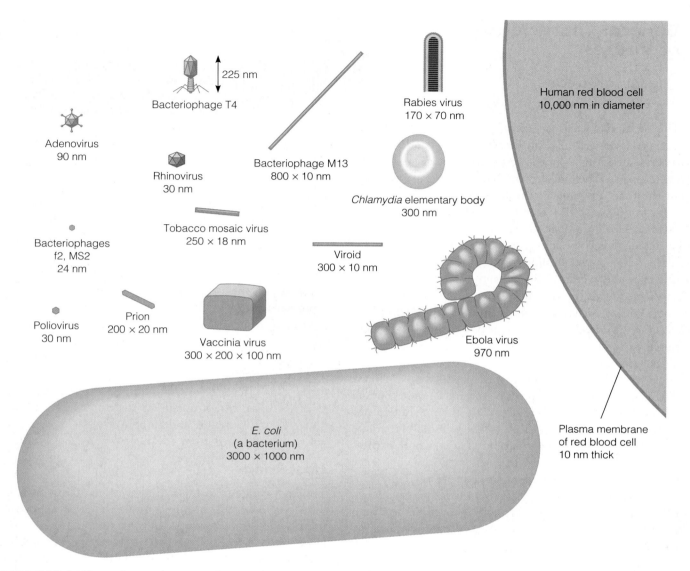

FIGURE 13.1 **Virus sizes.** The sizes of several viruses (teal blue) and bacteria (pink) are compared with a human red blood cell, shown to the right of the microbes. Dimensions are given in nanometers (nm) and are either diameters or length by width.

■ How do viruses differ from bacteria?

years. Unfortunately, the immune system of a treated person or animal quickly eliminates the administered phage.

Experimentally induced viral infections in cancer patients during the 1920s suggested that viruses might have antitumor activity. These tumor destroying or *oncolytic viruses* may selectively infect and kill tumor cells or cause an immune response against tumor cells. Some viruses naturally infect tumor cells and other viruses can be genetically engineered to infect tumor cells. At present several studies are underway to determine the killing mechanism of oncolytic viruses and the safety of using viral therapy.

Viral Size

Viral sizes are determined with the aid of electron microscopy. Different viruses vary considerably in size. Although most are quite a bit smaller than bacteria, some of the larger viruses (such as the vaccinia virus) are about the same size as some very small bacteria (such as the mycoplasmas, rickettsias, and chlamydias). Viruses range from 20 to 1000 nm in length. The comparative sizes of several viruses and bacteria are shown in Figure 13.1.

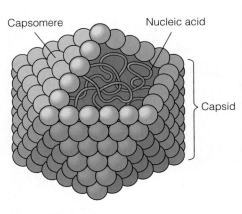

Capsomere — Nucleic acid

Capsid

(a) A polyhedral virus

(b) *Mastadenovirus*

TEM ⊢——⊣ 40 nm

FIGURE 13.2 **Morphology of a nonenveloped polyhedral virus. (a)** A diagram of a polyhedral (icosahedral) virus. **(b)** A micrograph of the adenovirus *Mastadenovirus*. Individual capsomeres in the protein coat are visible.

■ A viral capsid is composed of capsomeres that often form an icosahedron.

Viral Structure

Learning Objective

• Describe the chemical and physical structure of both an enveloped and a nonenveloped virus.

A **virion** is a complete, fully developed, infectious viral particle composed of nucleic acid and surrounded by a protein coat that protects it from the environment and is a vehicle of transmission from one host cell to another. Viruses are classified by differences in the structures of these coats.

Nucleic Acid

In contrast to prokaryotic and eukaryotic cells, in which DNA is always the primary genetic material (and RNA plays an auxiliary role), a virus can have either DNA or RNA, but never both. The nucleic acid of a virus can be single-stranded or double-stranded. Thus, there are viruses with the familiar double-stranded DNA, with single-stranded DNA, with double-stranded RNA, and with single-stranded RNA. Depending on the virus, the nucleic acid can be linear or circular. In some viruses (such as the influenza virus), the nucleic acid is in several separate segments.

The percentage of nucleic acid in relation to protein is about 1% for the *Influenzavirus* and about 50% for certain bacteriophages. The total amount of nucleic acid varies from a few thousand nucleotides (or pairs) to as many as 250,000 nucleotides. (*E. coli*'s chromosome consists of approximately 4 million nucleotide pairs.)

Capsid and Envelope

The nucleic acid of a virus is surrounded by a protein coat called the **capsid** (Figure 13.2a). The structure of the capsid is ultimately determined by the viral nucleic acid and accounts for most of the mass of a virus, especially of small

ones. Each capsid is composed of protein subunits called **capsomeres.** In some viruses, the proteins composing the capsomeres are of a single type; in other viruses, several types of protein may be present. Individual capsomeres are often visible in electron micrographs (see Figure 13.2b for an example). The arrangement of capsomeres is characteristic of a particular type of virus.

In some viruses, the capsid is covered by an **envelope** (Figure 13.3a), which usually consists of some combination of lipids, proteins, and carbohydrates. Some animal viruses are released from the host cell by an extrusion process that coats the virus with a layer of the host cell's plasma membrane; that layer becomes the viral envelope. In many cases, the envelope contains proteins determined by the viral nucleic acid and materials derived from normal host cell components.

Depending on the virus, envelopes may or may not be covered by **spikes,** which are carbohydrate-protein complexes that project from the surface of the envelope. Some viruses attach to host cells by means of spikes. Spikes are such a reliable characteristic of some viruses that they can be used as a means of identification. The ability of certain viruses, such as the *Influenzavirus*, to clump red blood cells is associated with spikes. Such viruses bind to red blood cells and form bridges between them. The resulting clumping is called *hemagglutination* and is the basis for several useful laboratory tests.

Viruses whose capsids are not covered by an envelope are known as **nonenveloped viruses** (see Figure 13.2). The capsid of a nonenveloped virus protects the nucleic acid from nuclease enzymes in biological fluids and promotes the virus's attachment to susceptible host cells.

When the host has been infected by a virus, the host immune system is stimulated to produce antibodies (proteins that react with the surface proteins of the virus). This interaction between host antibodies and virus proteins should inactivate the virus and stop the infection. However, some

FIGURE 13.3 **Morphology of an enveloped helical virus. (a)** A diagram of an enveloped helical virus. **(b)** A micrograph of *Influenzavirus* A2. Notice the halo of spikes projecting from the outer surface of each envelope.

■ What is the nucleic acid in a virus?

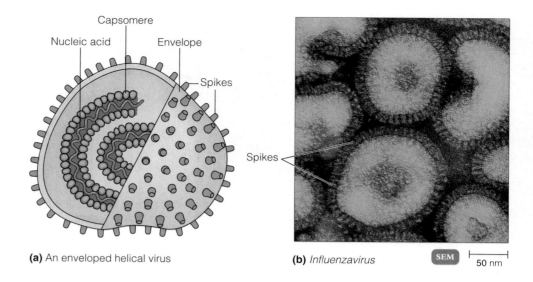

(a) An enveloped helical virus

(b) *Influenzavirus*

FIGURE 13.4 **Morphology of a helical virus. (a)** A diagram of a portion of a helical virus. Several rows of capsomeres have been removed to reveal the nucleic acid. **(b)** A micrograph of Ebola virus, a filovirus showing helical rods.

■ Helical viruses look like long or coiled threads.

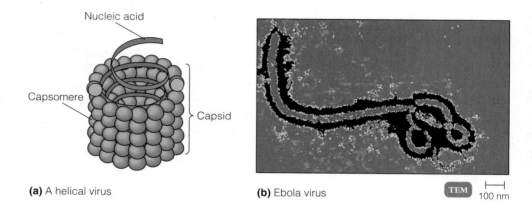

(a) A helical virus

(b) Ebola virus

viruses can escape antibodies because regions of the genes that code for these viruses' surface proteins are susceptible to mutations. The progeny of mutant viruses have altered surface proteins, such that the antibodies are not able to react with them. *Influenzavirus* frequently undergoes such changes in its spikes. This is why you can get influenza more than once. Although you may have produced antibodies to one *Influenzavirus*, the virus can mutate and infect you again.

General Morphology

Viruses may be classified into several different morphological types on the basis of their capsid architecture. The structure of these capsids has been revealed by electron microscopy and a technique called X-ray crystallography.

Helical Viruses

Helical viruses resemble long rods that may be rigid or flexible. The viral nucleic acid is found within a hollow, cylindrical capsid that has a helical structure (Figure 13.4). The

viruses that cause rabies and Ebola hemorrhagic fever are helical viruses.

Polyhedral Viruses

Many animal, plant, and bacterial viruses are polyhedral, or many-sided, viruses. The capsid of most polyhedral viruses is in the shape of an *icosahedron*, a regular polyhedron with 20 triangular faces and 12 corners (see Figure 13.2a). The capsomeres of each face form an equilateral triangle. An example of a polyhedral virus in the shape of an icosahedron is the adenovirus (shown in Figure 13.2b). Another icosahedral virus is the poliovirus.

Enveloped Viruses

As noted earlier, the capsid of some viruses is covered by an envelope. Enveloped viruses are roughly spherical. When helical or polyhedral viruses are enclosed by envelopes, they are called *enveloped helical* or *enveloped polyhedral viruses*. An example of an enveloped helical virus is the *Influenzavirus* (see Figure 13.3b). An example of an enveloped polyhedral

(b) *Orthopoxvirus*   TEM 200 nm

FIGURE 13.5 **Morphology of complex viruses. (a)** A diagram and micrograph of a T-even bacteriophage. **(b)** A micrograph of variola virus, a species in the genus *Orthopoxvirus*, which causes smallpox.

■ What is the chemical composition of a capsid?

(icosahedral) virus is the herpes simplex virus (genus *Simplexvirus*) (see Figure 13.14 on page 391).

Complex Viruses

Some viruses, particularly bacterial viruses, have complicated structures and are called **complex viruses.** One example of a complex virus is a bacteriophage. Some bacteriophages have capsids to which additional structures are attached (Figure 13.5a). In this figure, notice that the capsid (head) is polyhedral and the tail sheath is helical. The head contains the nucleic acid. Later in the chapter we will discuss the functions of the other structures, such as the tail sheath, tail fibers, plate, and pin. Another ex-

ample of complex viruses are poxviruses, which do not contain clearly identifiable capsids but have several coats around the nucleic acid (Figure 13.5b).

Taxonomy of Viruses

Learning Objectives
- *Define viral species.*
- *Give an example of a family, genus, and common name for a virus.*

Just as we need taxonomic categories of plants, animals, and bacteria, we need viral taxonomy to help us organize and understand newly discovered organisms. The oldest classification of viruses is based on symptomatology, such as for diseases that affect the respiratory system. This system was convenient but not scientifically acceptable because the same virus may cause more than one disease, depending on the tissue affected. In addition, this system artificially grouped viruses that do not infect humans.

Virologists began addressing the problem of viral taxonomy in 1966 with the formation of the International Committee on the Taxonomy of Viruses (ICTV). Since then, the ICTV has been grouping viruses into families based on (1) nucleic acid type, (2) strategy for replication, and (3) morphology. The suffix *-virus* is used for genus names; family names end in *-viridae;* and order names end in *-ales.* In formal usage, the family and genus names are used in the following manner: Family Herpesviridae, genus *Simplexvirus*, human herpes virus 2.

A **viral species** is a group of viruses sharing the same genetic information and ecological niche (host range). Specific epithets for viruses are not used. Thus, viral species are designated by descriptive common names, such as human immunodeficiency virus (HIV), with subspecies (if any) designated by a number (HIV-1). Table 13.2 presents a summary of the classification of viruses that infect humans.

Isolation, Cultivation, and Identification of Viruses

The fact that viruses cannot multiply outside a living host cell complicates their detection, enumeration, and identification. Viruses must be provided with living cells instead of a fairly simple chemical medium. Living plants and animals are difficult and expensive to maintain, and pathogenic viruses that grow only in higher primates and human hosts cause additional complications. However, viruses that use bacterial cells as a host (bacteriophages) are rather easily grown on bacterial cultures. This is one reason so much of our understanding of viral multiplication has come from bacteriophages.

TABLE 13.2	**Families of Viruses That Affect Humans**		
Characteristics/ Dimensions	**Viral Family**	**Important Genera**	**Clinical or Special Features**
Single-stranded DNA nonenveloped 18–25 nm	Parvoviridae	Human parvovirus B19	Fifth disease; anemia in immunocompromised patients.
Double-stranded DNA nonenveloped 70–90 nm	Adenoviridae	*Mastadenovirus*	Medium-sized viruses that cause various respiratory infections in humans; some cause tumors in animals.
40–57 nm	Papovaviridae	*Papillomavirus* (human wart virus) *Polyomavirus*	Small viruses that induce tumors; the human wart virus (papilloma) and certain viruses that produce cancer in animals (polyoma and simian) belong to this family.
Double-stranded DNA enveloped 200–350 nm	Poxviridae	*Orthopoxvirus* (vaccinia and smallpox viruses) *Molluscipoxvirus*	Very large, complex, brick-shaped viruses that cause diseases such as smallpox (variola), molluscum contagiosum (wartlike skin lesion), and cowpox.
150–200 nm	Herpesviridae	*Simplexvirus* (HHV-1 and 2) *Varicellovirus* (HHV-3) *Lymphocryptovirus* (HHV-4) *Cytomegalovirus* (HHV-5) *Roseolovirus* (HHV-6) HHV-7 Kaposi's sarcoma (HHV-8)	Medium-sized viruses that cause various human diseases, such as fever blisters, chickenpox, shingles, and infectious mononucleosis; causes a type of human cancer called Burkitt's lymphoma.
42 nm	Hepadnaviridae	*Hepadnavirus* (hepatitis B virus)	After protein synthesis, hepatitis B virus uses reverse transcriptase to produce its DNA from mRNA; causes hepatitis B and liver tumors.
Single-stranded RNA, + strand nonenveloped 28–30 nm	Picornaviridae	*Enterovirus* *Rhinovirus* (common cold virus) Hepatitis A virus	At least 70 human enteroviruses are known, including the polio-, coxsackie-, and echoviruses; more than 100 rhinoviruses exist and are the most common cause of colds.
35–40 nm	Caliciviridae	Hepatitis E virus *Norovirus*	Includes causes of gastroenteritis and one cause of human hepatitis.
Single-stranded RNA, + strand enveloped 60–70 nm	Togaviridae	*Alphavirus* *Rubivirus* (rubella virus)	Included are many viruses transmitted by arthropods (*Alphavirus*); diseases include eastern equine encephalitis (EEE) and western equine encephalitis (WEE). Rubella virus is transmitted by the respiratory route.

TABLE 13.2	(continued)		
Characteristics/ Dimensions	**Viral Family**	**Important Genera**	**Clinical or Special Features**
40–50 nm	Flaviviridae	*Flavivirus* *Pestivirus* Hepatitis C virus	Can replicate in arthropods that transmit them; diseases include yellow fever, dengue, St. Louis encephalitis, and West Nile virus.
Nidovirales 80–160 nm	Coronaviridae	*Coronavirus*	Associated with upper respiratory tract infections and the common cold; SARS virus.
Mononegavirales – strand, one strand of RNA 70–180 nm	Rhabdoviridae	*Vesiculovirus* (vesicular stomatatis virus) *Lyssavirus* (rabies virus)	Bullet-shaped viruses with a spiked enve-lope; cause rabies and numerous animal diseases.
80–14,000 nm	Filoviridae	*Filovirus*	Enveloped, helical viruses; Ebola and Marburg viruses are filoviruses.
150–300 nm	Paramyxoviridae	*Paramyxovirus* *Morbillivirus* (measleslike virus)	Paramyxoviruses cause parainfluenza, mumps, and Newcastle disease in chick-ens.
– strand, one strand of RNA 32 nm	Deltaviridae	Hepatitis D	Depend on coinfection with hepadnavirus.
– strand, multiple strands of RNA 80–200 nm	Orthomyxoviridae	*Influenzavirus* (influenza viruses A and B) Influenza C virus	Envelope spikes can agglutinate red blood cells.
90–120 nm	Bunyaviridae	*Bunyavirus* (California encephalitis virus) *Hantavirus*	Hantaviruses cause hemorrhagic fevers such as Korean hemorrhagic fever and *Hantavirus* pulmonary syndrome; associ-ated with rodents.
110–130 nm	Arenaviridae	*Arenavirus*	Helical capsids contain RNA-containing granules; cause lymphocytic choriomenin-gitis, Venezuelan hemorrhagic fever, and Lassa fever.
Produce DNA 100–120 nm	Retroviridae	Oncoviruses *Lentivirus* (HIV)	Includes all RNA tumor viruses. Oncoviruses cause leukemia and tumors in animals; the *Lentivirus* HIV causes AIDS.
Double-stranded RNA nonenveloped 60–80 nm	Reoviridae	*Reovirus* *Rotavirus*	Involved in mild respiratory infections and gastroenteritis; an unclassified species causes Colorado tick fever.

Growing Bacteriophages in the Laboratory

Learning Objective

• *Describe how bacteriophages are cultured.*

Bacteriophages can be grown either in suspensions of bacteria in liquid media or in bacterial cultures on solid media. The use of solid media makes possible the *plaque method* for detecting and counting viruses. A sample of bacteriophage is mixed with host bacteria and melted agar. The agar containing the bacteriophages and host bacteria is then poured into a Petri plate containing a hardened layer of agar growth medium. The virus-bacteria mixture solidifies into a thin top layer that contains a layer of bacteria approximately one cell thick. Each virus infects a bacterium, multiplies, and releases several hundred new viruses. These newly produced viruses infect other bacteria in the immediate vicinity, and more new viruses are produced. Following several viral multiplication cycles, all the bacteria in the area surrounding the original virus are destroyed. This produces a number of clearings, or **plaques,** visible against a lawn of bacterial growth on the surface of the agar (Figure 13.6). While the plaques form, uninfected bacteria elsewhere in the Petri plate multiply rapidly and produce a turbid background.

Each plaque theoretically corresponds to a single virus in the initial suspension. Therefore, the concentrations of viral suspensions measured by the number of plaques are usually given in terms of **plaque-forming units (pfu).**

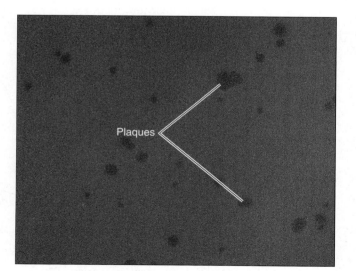

Plaques

FIGURE 13.6 **Viral plaques formed by bacteriophages.** Clear viral plaques of various sizes have been formed by bacteriophage λ (lambda) on a lawn of *E. coli.*

■ What is a plaque-forming unit?

Growing Animal Viruses in the Laboratory

Learning Objective

• *Describe how animal viruses are cultured.*

In the laboratory, three methods are commonly used for culturing animal viruses. These methods involve using living animals, embryonated eggs, or cell cultures.

In Living Animals

Some animal viruses can be cultured only in living animals, such as mice, rabbits, and guinea pigs. Most experiments to study the immune system's response to viral infections must also be performed in virally infected live animals. Animal inoculation may be used as a diagnostic procedure for identifying and isolating a virus from a clinical specimen. After the animal is inoculated with the specimen, the animal is observed for signs of disease, or is killed so that infected tissues can be examined for the virus.

Some human viruses cannot be grown in animals, or can be grown but do not cause disease. The lack of natural animal models for AIDS has slowed our understanding of its disease process and prevented experimentation with drugs that inhibit growth of the virus in vivo. Chimpanzees can be infected with one subspecies of human immunodeficiency virus (HIV-1, genus *Lentivirus*), but because they do not show symptoms of the disease, they cannot be used to study the effects of viral growth and disease treatments. AIDS vaccines are presently being tested in humans, but the disease progresses so slowly in humans that it can take years to determine the effectiveness of these vaccines. In 1986, simian AIDS, an immune deficiency disease of green monkeys, was reported, followed in 1987 by feline AIDS, an immune deficiency disease of domestic cats. These diseases are caused by lentiviruses, which are closely related to HIV, and the diseases develop within a few months, thus providing a model for studying viral growth in different tissues. In 1990, a way to infect mice with human AIDS was found when immunodeficient mice were grafted to produce human T cells and human gamma globulin. The mice provide a reliable model for studying viral replication, although they do not provide models for vaccine development.

In Embryonated Eggs

If the virus will grow in an *embryonated egg,* this can be a fairly convenient and inexpensive form of host for many animal viruses. A hole is drilled in the shell of the embryonated egg, and a viral suspension or suspected virus-containing tissue is injected into the fluid of the egg. There are several membranes in an egg, and the virus is injected near the one most appropriate for its growth (Figure 13.7).

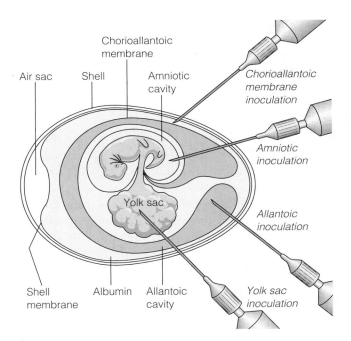

FIGURE 13.7 **Inoculation of an embryonated egg.** The injection site determines the membrane on which the viruses will grow.

■ Viruses must be grown in living host cells.

Viral growth is signaled by the death of the embryo, by embryo cell damage, or by the formation of typical pocks or lesions on the egg membranes. This method was once the most widely used method of viral isolation and growth, and it is still used to grow viruses for some vaccines. For this reason, you may be asked if you are allergic to eggs before receiving a vaccination, because egg proteins may be present in the viral vaccine preparations. (Allergic reactions will be discussed in Chapter 19.)

In Cell Cultures

Cell cultures have replaced embryonated eggs as the preferred type of growth medium for many viruses. Cell cultures consist of cells grown in culture media in the laboratory. Because these cultures are generally rather homogeneous collections of cells and can be propagated and handled much like bacterial cultures, they are more convenient to work with than whole animals or embryonated eggs.

Cell culture lines are started by treating a slice of animal tissue with enzymes that separate the individual cells (Figure 13.8). These cells are suspended in a solution that provides the osmotic pressure, nutrients, and growth factors needed for the cells to grow. Normal cells tend to adhere to the glass or plastic container and reproduce to form a monolayer. Viruses infecting such a monolayer sometimes cause the cells of the monolayer to deteriorate as they multiply. This cell deterioration is called **cytopathic effect (CPE),** illustrated in Figure 13.9. CPE can be detected and counted in much the same way as plaques caused by bacteriophages on a lawn of bacteria.

Viruses may be grown in primary or continuous cell lines. **Primary cell lines,** derived from tissue slices, tend to die out after only a few generations. Certain cell lines, called **diploid cell lines,** developed from human embryos can be maintained for about 100 generations and are widely used for culturing viruses that require a human host. Cell lines developed from embryonic human cells are used to culture rabies virus (genus *Lyssavirus*) for a rabies vaccine called human diploid culture vaccine.

When viruses are routinely grown in a laboratory, **continuous cell lines** are used. These are transformed (cancerous) cells that can be maintained through an indefinite number of generations, and they are sometimes called immortal cell lines (see the discussion of transformation on page 400). One of these, the HeLa cell line, was isolated from the cancer of a woman who died in 1951. After years of laboratory cultivation, many such cell lines have lost almost all the original characteristics of the cell, but these changes have not interfered with the use of the cells for viral propagation. In spite of the success of cell culture in viral isolation and growth, there are still some viruses that have never been successfully cultivated in cell culture.

① A tissue is treated with enzymes to separate the cells.

② Cells are suspended in culture medium.

③ Normal cells or primary cells grow in a monolayer across the glass or plastic container. Transformed cells or continuous cell cultures do not grow in a monolayer.

FIGURE 13.8 **Cell cultures.** Transformed cells can be grown indefinitely in laboratory culture.

(a) |———| 20 μm (b) SEM |———| 0.5 μm

FIGURE 13.9 **The cytopathic effect of viruses.**
(a) Uninfected mouse cells align next to each other forming a monolayer. **(b)** The same cells 24 hours after infection with vesicular stomatitis virus (VSV) (see Figure 13.18b). Notice the cells pile up and "round up."

■ How did VSV infection affect the cells?

The idea of cell culture dates back to the end of the nineteenth century, but it was not a practical laboratory technique until the development of antibiotics in the years following World War II. A major problem with cell culture is that the cell lines must be kept free of microbial contamination. The maintenance of cell culture lines requires trained technicians with considerable experience working on a full-time basis. Because of these difficulties, most hospital laboratories and many state health laboratories do not isolate and identify viruses in clinical work. Instead, the tissue or serum samples are sent to central laboratories that specialize in such work.

Viral Identification

Learning Objective

• List three techniques used to identify viruses.

Identifying viral isolates is not an easy task. For one thing, viruses cannot be seen at all without the use of an electron microscope. Serological methods, such as Western blotting, are the most commonly used means of identification (see Figure 10.12, page 292). In these tests, the virus is detected and identified by its reaction with antibodies. We will discuss antibodies in detail in Chapter 17 and a number of immunological tests for identifying viruses in Chapter 18. Observation of cytopathic effects, described in Chapter 15 on page 450, is also useful for the identification of a virus.

Virologists can identify and characterize viruses by using such modern molecular methods as restriction fragment length polymorphisms (RFLPs) and the polymerase chain reaction (PCR) (Chapter 9, pages 254 and 267). PCR was used to amplify RNA from birds and humans to identify the virus in the 1999 West Nile encephalitis outbreak in the United States.

Viral Multiplication

The nucleic acid in a virion contains only a few of the genes needed for the synthesis of new viruses. These include genes for the virion's structural components, such as the capsid proteins, and genes for a few of the enzymes used in the viral life cycle. These enzymes are synthesized and functional only when the virus is within the host cell. Viral enzymes are almost entirely concerned with replicating or processing viral nucleic acid. Enzymes needed for protein synthesis, ribosomes, tRNA, and energy production are supplied by the host cell and are used for synthesizing viral proteins, including viral enzymes. Although the smallest nonenveloped virions do not contain any preformed enzymes, the larger virions may contain one or a few enzymes, which usually function in helping the virus penetrate the host cell or replicate its own nucleic acid.

Thus, for a virus to multiply, it must invade a host cell and take over the host's metabolic machinery. A single virion can give rise to several or even thousands of similar viruses in a single host cell. This process can drastically change the host cell and can even cause its death.

Multiplication of Bacteriophages

Learning Objectives

• Describe the lytic cycle of T-even bacteriophages.

• Describe the lysogenic cycle of bacteriophage lambda.

Although the means by which a virus enters and exits a host cell may vary, the basic mechanism of viral multiplication is similar for all viruses. The best-understood viral life cycles are those of the bacteriophages. Phages can multiply by two alternative mechanisms: the lytic cycle or the lysogenic cycle. The **lytic cycle** ends with the lysis and death of the host cell, whereas the host cell remains alive in the **lysogenic cycle.** Because the *T-even bacteriophages* (T2, T4, and T6) have been studied most extensively, we will describe the multiplication of T-even bacteriophages in their host, *E. coli*, as an example of the lytic cycle.

T-Even Bacteriophages: The Lytic Cycle

The virions of T-even bacteriophages are large, complex, and nonenveloped, with a characteristic head-and-tail structure shown in Figures 13.5a and 13.10. The length of DNA contained in these bacteriophages is only about 6% of that contained in *E. coli*, yet the phage has enough DNA for over 100 genes. The multiplication cycle of these phages, like that of all viruses, occurs in five distinct stages: attachment, penetration, biosynthesis, maturation, and release (Figure 13.10).

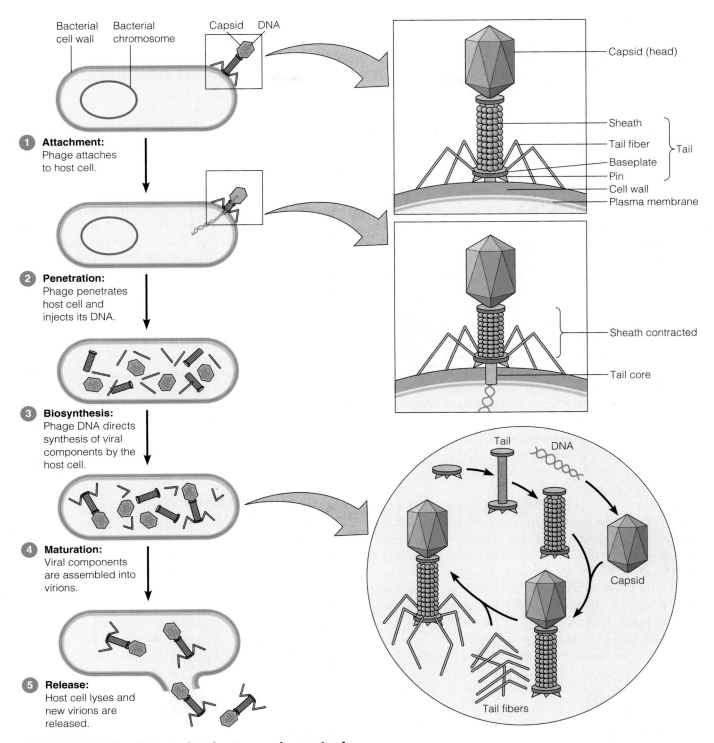

1 Attachment:
Phage attaches
to host cell.

2 Penetration:
Phage penetrates
host cell and
injects its DNA.

3 Biosynthesis:
Phage DNA directs
synthesis of viral
components by the
host cell.

4 Maturation:
Viral components
are assembled into
virions.

5 Release:
Host cell lyses and
new virions are
released.

Bacterial
cell wall

Bacterial
chromosome

Capsid   DNA

Capsid (head)

Sheath
Tail fiber   Tail
Baseplate
Pin
Cell wall
Plasma membrane

Sheath contracted

Tail core

Tail   DNA

Capsid

Tail fibers

FIGURE 13.10 **The lytic cycle of a T-even bacteriophage.**

■ The lytic cycle results in lysis of the host cell and the release of new phage
particles.

Attachment ❶ After a chance collision between phage particles and bacteria, *attachment,* or *adsorption,* occurs. During this process, an attachment site on the virus attaches to a complementary receptor site on the bacterial cell. This attachment is a chemical interaction in which weak bonds are formed between the attachment and receptor sites. T-even bacteriophages use fibers at the end of the tail as attachment sites. The complementary receptor sites are on the bacterial cell wall.

Penetration ❷ After attachment, the T-even bacteriophage injects its DNA (nucleic acid) into the bacterium. To do this, the bacteriophage's tail releases an enzyme, phage **lysozyme,** which breaks down a portion of the bacterial cell wall. During the process of *penetration,* the tail sheath of the phage contracts, and the tail core is driven through the cell wall. When the tip of the core reaches the plasma membrane, the DNA from the bacteriophage's head passes through the tail core, through the plasma membrane, and enters the bacterial cell. The capsid remains outside the bacterial cell. Therefore, the phage particle functions like a hypodermic syringe to inject its DNA into the bacterial cell.

Biosynthesis ❸ Once the bacteriophage DNA has reached the cytoplasm of the host cell, the biosynthesis of viral nucleic acid and protein occurs. Host protein synthesis is stopped by virus-induced degradation of the host DNA, viral proteins that interfere with transcription, or the repression of translation.

Initially, the phage uses the host cell's nucleotides and several of its enzymes to synthesize many copies of phage DNA. Soon after, the biosynthesis of viral proteins begins. Any RNA transcribed in the cell is mRNA transcribed from phage DNA for the biosynthesis of phage enzymes and capsid proteins. The host cell's ribosomes, enzymes, and amino acids are used for translation. Genetic controls regulate when different regions of phage DNA are transcribed into mRNA during the multiplication cycle. For example, early messages are translated into early phage

proteins, the enzymes used in the synthesis of phage DNA. Also, late messages are translated into late phage proteins for the synthesis of capsid proteins.

For several minutes following infection, complete phages cannot be found in the host cell. Only separate components—DNA and protein—can be detected. The period during viral multiplication when complete, infective virions are not yet present is called the **eclipse period.**

Maturation ❹ In the next sequence of events, *maturation* occurs. In this process, bacteriophage DNA and capsids are assembled into complete virions. The viral components essentially assemble into a viral particle spontaneously, eliminating the need for many nonstructural genes and gene products. The phage heads and tails are separately assembled from protein subunits, and the head is filled with phage DNA and attached to the tail.

Release ❺ The final stage of viral multiplication is the *release* of virions from the host cell. The term **lysis** is generally used for this stage in the multiplication of T-even phages because in this case, the plasma membrane actually breaks open (lyses). Lysozyme, which is coded for by a phage gene, is synthesized within the cell. This enzyme causes the bacterial cell wall to break down, and the newly produced bacteriophages are released from the host cell. The released bacteriophages infect other susceptible cells in the vicinity, and the viral multiplication cycle is repeated within those cells.

A One-Step Growth Experiment

The time that elapses from phage attachment to release is known as **burst time** and averages 20–40 minutes. The number of newly synthesized phage particles released from a single cell is referred to as **burst size** and usually ranges from about 50 to 200.

The various stages involved in the multiplication of phages can be demonstrated experimentally in what is known as a *one-step growth experiment* (Figure 13.11). In this procedure, a phage suspension is diluted until a

FIGURE 13.11 **A bacteriophage one-step growth curve.** No new infective phage particles are found in a culture until after biosynthesis and maturation have taken place.

■ What can be found in the cell during biosynthesis and maturation?

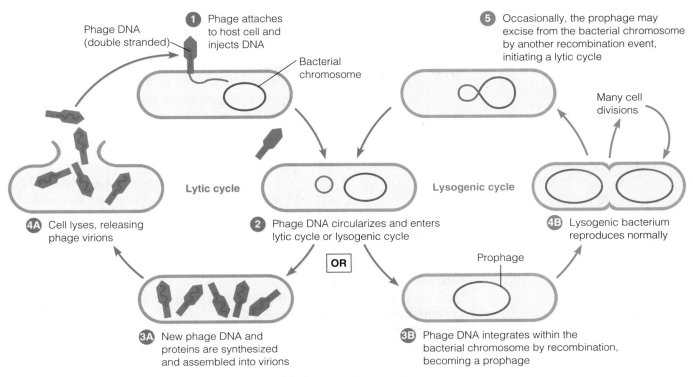

① Phage attaches to host cell and injects DNA

Phage DNA (double stranded)

Bacterial chromosome

Lytic cycle

Lysogenic cycle

⑤ Occasionally, the prophage may excise from the bacterial chromosome by another recombination event, initiating a lytic cycle

Many cell divisions

④A Cell lyses, releasing phage virions

② Phage DNA circularizes and enters lytic cycle or lysogenic cycle

OR

④B Lysogenic bacterium reproduces normally

Prophage

③A New phage DNA and proteins are synthesized and assembled into virions

③B Phage DNA integrates within the bacterial chromosome by recombination, becoming a prophage

FIGURE 13.12 **The lysogenic cycle of bacteriophage λ in E. coli.**

■ How does lysogeny differ from the lytic cycle?

sample containing only a few phage particles is obtained. These particles are then introduced into a culture of host cells. Periodically, samples of phage particles are removed from the culture and inoculated onto a plate culture of susceptible host cells; the plaque method is used to determine the number of infective phage particles on this culture (see Figure 13.6). A few minutes after attachment, no infective particles are present. However, phage nucleic acid is found inside the infected cells, and capsid proteins will be synthesized. The time required for maturation is the interval between the appearance of phage nucleic acid and the synthesis of mature phages. After a few minutes, the number of infective phage particles found in subcultures begins to rise. The burst size is determined once the number of infective phage particles remains constant, indicating that no further phage multiplication will occur.

Bacteriophage Lambda: The Lysogenic Cycle

In contrast to T-even bacteriophages, some viruses do not cause lysis and death of the host cell when they multiply. These *lysogenic phages* (also called *temperate phages*) may indeed proceed through a lytic cycle, but they are also capable of incorporating their DNA into the host cell's DNA to begin a lysogenic cycle. In **lysogeny,** the phage remains latent (inactive). The participating bacterial host cells are known as *lysogenic cells*.

We will use the bacteriophage **λ** (lambda), a well-studied lysogenic phage, as an example of the lysogenic cycle (Figure 13.12).

❶ Upon penetration into an *E. coli* cell,

❷ the originally linear phage DNA forms a circle.

❸A This circle can multiply and be transcribed,

❹A leading to the production of new phage and to cell lysis (the lytic cycle).

❸B Alternatively, the circle can recombine with and become part of the circular bacterial DNA (the lysogenic cycle). The inserted phage DNA is now called a **prophage.** Most of the prophage genes are repressed by two repressor proteins that are the products of phage genes. These repressors stop transcription of all the other phage genes by binding to operators. Thus, the phage genes that would otherwise direct the synthesis and release of new virions are turned off, in much the same way that the genes of the *E. coli lac* operon are turned off by the *lac* repressor (Figure 8.14, page 225).

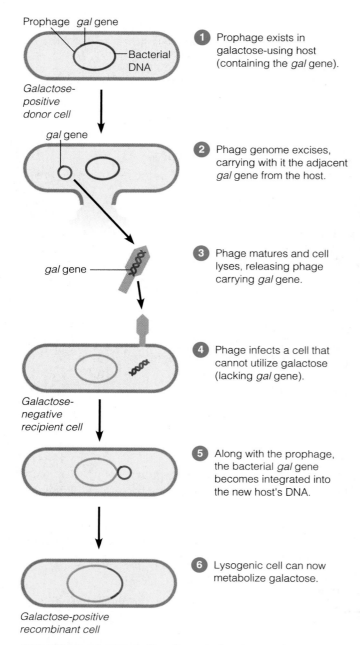

Prophage *gal* gene

Bacterial DNA

Galactose-positive donor cell

gal gene

gal gene

Galactose-negative recipient cell

Galactose-positive recombinant cell

1 Prophage exists in galactose-using host (containing the *gal* gene).

2 Phage genome excises, carrying with it the adjacent *gal* gene from the host.

3 Phage matures and cell lyses, releasing phage carrying *gal* gene.

4 Phage infects a cell that cannot utilize galactose (lacking *gal* gene).

5 Along with the prophage, the bacterial *gal* gene becomes integrated into the new host's DNA.

6 Lysogenic cell can now metabolize galactose.

FIGURE 13.13 Specialized transduction. When a prophage is excised from its host chromosome, it can take with it a bit of the adjacent DNA from the bacterial chromosome.

■ How does specialized transduction differ from the lytic cycle?

Every time the host cell's machinery replicates the bacterial chromosome, **4B** it also replicates the prophage DNA. The prophage remains latent within the progeny cells.

5 However, a rare spontaneous event, or the action of UV light or certain chemicals, can lead to the excision (popping-out) of the phage DNA, and to initiation of the lytic cycle.

There are three important results of lysogeny. First, the lysogenic cells are immune to reinfection by the same phage. (However, the host cell is not immune to infection by other phage types.) The second result of lysogeny is **phage conversion;** that is, the host cell may exhibit new properties. For example, the bacterium *Corynebacterium diphtheriae*, which causes diphtheria, is a pathogen whose disease-producing properties are related to the synthesis of a toxin. The organism can produce toxin only when it carries a temperate phage, because the prophage carries the gene coding for the toxin. As another example, only streptococci carrying a temperate phage are capable of producing the toxin associated with scarlet fever. The toxin produced by *Clostridium botulinum*, which causes botulism, is encoded by a prophage gene, as is the cholera toxin produced by pathogenic strains of *Vibrio cholerae*.

The third result of lysogeny is that it makes **specialized transduction** possible. Recall from Chapter 8 that bacterial genes can be picked up in a phage coat and transferred to another bacterium in a process called generalized transduction (see Figure 8.28 on page 240). Any bacterial genes can be transferred by generalized transduction because the host chromosome is broken down into fragments, any of which can be packaged into a phage coat. In specialized transduction, however, only certain bacterial genes can be transferred.

Specialized transduction is mediated by a lysogenic phage, which packages bacterial DNA *along with* its own DNA in the same capsid. When a prophage is excised from the host chromosome, adjacent genes from either side may remain attached to the phage DNA. In Figure 13.13, bacteriophage λ has picked up the *gal* gene for galactose fermentation from its galactose-positive host. The phage carries this gene to a galactose-negative cell, which then becomes galactose-positive.

Certain animal viruses can undergo processes very similar to lysogeny. Animal viruses that can remain latent in cells for long periods without multiplying or causing disease may become inserted in a host chromosome or remain separate from host DNA in a repressed state (as some lysogenic phages). Cancer-causing viruses may also be latent, as will be discussed later in the chapter.

Multiplication of Animal Viruses

Learning Objective

- Compare and contrast the multiplication cycle of DNA- and RNA-containing animal viruses.

The multiplication of animal viruses follows the basic pattern of bacteriophage multiplication but has several

TABLE 13.3	**Bacteriophage and Viral Multiplication Compared**	
Stage	**Bacteriophages**	**Animal Viruses**
Attachment	Tail fibers attach to cell wall proteins	Attachment sites are plasma membrane proteins and glycoproteins
Penetration	Viral DNA injected into host cell	Capsid enters by endocytosis or fusion
Uncoating	Not required	Enzymatic removal of capsid proteins
Biosynthesis	In cytoplasm	In nucleus (DNA viruses) or cytoplasm (RNA viruses)
Chronic infection	Lysogeny	Latency; slow viral infections; cancer
Release	Host cell lysed	Enveloped viruses bud out; nonenveloped viruses rupture plasma membrane

differences, summarized in Table 13.3. Animal viruses differ from phages in their mechanism of entering the host cell. Also, once the virus is inside, the synthesis and assembly of the new viral components are somewhat different, partly because of the differences between prokaryotic cells and eukaryotic cells. Animal viruses may have certain types of enzymes not found in phages. Finally, the mechanisms of maturation and release, and the effects on the host cell, differ in animal viruses and phages.

In the following discussion of the multiplication of animal viruses, we will consider the processes that are shared by both DNA- and RNA-containing animal viruses. These processes are attachment, penetration, uncoating, and release. We will also examine how DNA- and RNA-containing viruses differ with respect to their processes of biosynthesis.

Attachment

Like bacteriophages, animal viruses have attachment sites that attach to complementary receptor sites on the host cell's surface. However, the receptor sites of animal cells are proteins and glycoproteins of the plasma membrane (Figure 13.14a). Moreover, animal viruses do not possess appendages like the tail fibers of some bacteriophages. The

(a) Attachment **(b)** Endocytosis **(c)** Penetration **(d)** Uncoating

TEM ⊢————⊣ 100 nm

FIGURE 13.14 **The entry of herpes simplex virus (*Simplexvirus*) into an animal cell. (a)** Attachment of the viral envelope to the plasma membrane. **(b)** The cell's plasma membrane folds inward, forming a vesicle around the virus; this results in loss of the envelope. **(c)** The nonenveloped capsid penetrates the cytoplasm of the cell from the vesicle. **(d)** The nucleic acid core is uncoated by digestion of the capsid.

■ Viruses enter animal cells by endocytosis.

attachment sites of animal viruses are distributed over the surface of the virus. The sites themselves vary from one group of viruses to another. In adenoviruses, which are icosahedral viruses, the attachment sites are small fibers at the corners of the icosahedron (see Figure 13.2b). In many of the enveloped viruses, such as *Influenzavirus*, the attachment sites are spikes located on the surface of the envelope (see Figure 13.3b). As soon as one spike attaches to a host receptor, additional receptor sites on the same cell migrate to the virus. Attachment is completed when many sites are bound.

Receptor sites are inherited characteristics of the host. Consequently, the receptor for a particular virus can vary from person to person. This could account for the individual differences in susceptibility to a particular virus. Understanding the nature of attachment can lead to the development of drugs that prevent viral infections. Monoclonal antibodies (discussed in Chapter 17) that combine with a virus's attachment site or the cell's receptor site may soon be used to treat some viral infections. One experimental treatment for AIDS involved injecting the patient with numerous receptor molecules. If the virus attaches to these, it cannot attach to natural receptor sites on cells.

Penetration

Following attachment, penetration occurs. Penetration of viruses into eukaryotic cells occurs by **endocytosis,** an active cellular process by which nutrients and other molecules are brought into a cell (see Chapter 4, page 99). A cell's plasma membrane continuously folds inward to form vesicles. These vesicles contain elements that originate outside the cell and are brought into the interior of the cell to be digested. If a virion attaches to a small outfolding (called a microvillus) on the plasma membrane of a potential host cell, the host cell will enfold the virion into a fold of plasma membrane, forming a vesicle (Figure 13.14b and c). Once the virion is enclosed within the vesicle, its viral envelope is destroyed. The capsid is digested when the cell attempts to digest the vesicle's contents, or the nonenveloped capsid may be released into the cytoplasm of the host cell (Figure 13.14d).

Enveloped viruses can penetrate by an alternative method called **fusion,** in which the viral envelope fuses with the plasma membrane and releases the capsid into the cell's cytoplasm. For example, HIV penetrates cells by this method.

Uncoating

Uncoating is the separation of the viral nucleic acid from its protein coat. This poorly understood process apparently varies with the type of virus. Some animal viruses accomplish uncoating by the action of lysosomal enzymes contained inside phagocytic vacuoles and coated vesicles of the host cell. These enzymes degrade the proteins of the viral capsid. The uncoating of poxviruses is completed by a specific enzyme encoded by the viral DNA and synthesized soon after infection. For other viruses, uncoating appears to be exclusively caused by enzymes in the host cell cytoplasm (Figure 13.14d). For at least one virus, the poliovirus, uncoating seems to begin while the virus is still attached to the host cell's plasma membrane.

The Biosynthesis of DNA Viruses

Generally, DNA-containing viruses replicate their DNA in the nucleus of the host cell by using viral enzymes, and they synthesize their capsid and other proteins in the cytoplasm by using host cell enzymes. Then the proteins migrate into the nucleus and are joined with the newly synthesized DNA to form virions. These virions are transported along the endoplasmic reticulum to the host cell's membrane for release. Herpesviruses, papovaviruses, adenoviruses, and hepadnaviruses all follow this pattern of biosynthesis (Table 13.4). Poxviruses are an exception because all of their components are synthesized in the cytoplasm.

As an example of the multiplication of a DNA virus, the sequence of events in papovavirus is shown in Figure 13.15. After

❶ attachment,

❷ penetration, and uncoating, the viral DNA is released into the nucleus of the host cell.

❸ Transcription of a portion of the viral DNA—the "early" genes occurs next. Translation follows. The products of these genes are enzymes that are required for the multiplication of viral DNA. In most DNA viruses, early transcription is carried out with the host's transcriptase (RNA polymerase); poxviruses, however, contain their own transcriptase.

❹ Sometime after the initiation of DNA replication, transcription and translation of the remaining "late" viral genes occur. Late proteins include capsid and other structural proteins.

❺ This leads to the synthesis of capsid proteins, which occurs in the cytoplasm of the host cell.

❻ After the capsid proteins migrate into the nucleus of the host cell, maturation occurs; the viral DNA and capsid proteins assemble to form complete viruses, which

❼ are then released from the host cell.

Some DNA viruses are described on the page 394.

TABLE 13.4	The Biosynthesis of DNA and RNA Viruses Compared	
Viral Nucleic Acid	**Virus Family**	**Special Features of Biosynthesis**
DNA, single-stranded	Parvoviridae	Cellular enzyme transcribes viral DNA in nucleus
DNA, double-stranded	Herpesviridae Papovaviridae	Cellular enzyme transcribes viral DNA in nucleus
	Poxviridae	Viral enzyme transcribes viral DNA in virion, in cytoplasm
DNA, reverse transcriptase	Hepadnaviridae	Cellular enzyme transcribes viral DNA in nucleus; reverse transcriptase copies mRNA to make viral DNA
RNA, + strand	Picornaviridae Togaviridae	Viral RNA functions as a template for synthesis of RNA polymerase which copies − strand RNA to make mRNA in cytoplasm
RNA, − strand	Rhabdoviridae	Viral enzyme copies viral RNA to make mRNA in cytoplasm
RNA, double-stranded	Reoviridae	Viral enzyme copies − strand RNA to make mRNA in cytoplasm
RNA, reverse transcriptase	Retroviridae	Viral enzyme copies viral RNA to make DNA in cytoplasm; DNA moves to nucleus

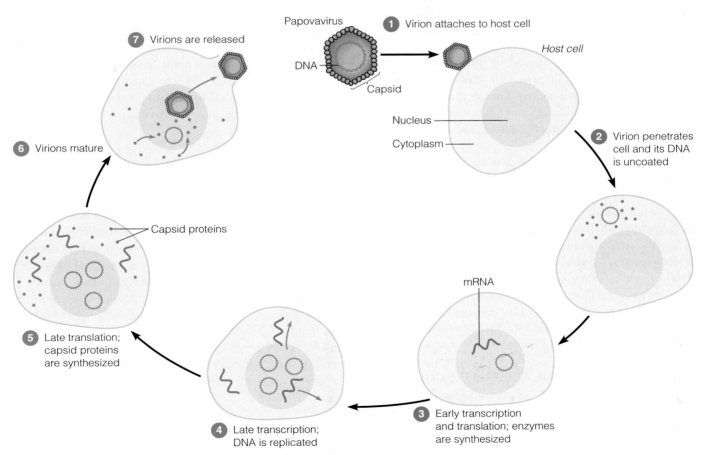

FIGURE 13.15 Multiplication of *Papovavirus*, a DNA-containing virus.

■ Why is mRNA made?

(a) *Mastadenovirus*   SEM ⊢——⊣ 100 nm

(b) *Herpesvirus*   TEM ⊢——⊣ 50 nm

FIGURE 13.16 DNA-containing animal viruses. (a) Negatively stained adenoviruses that have been concentrated in a centrifuge gradient. The individual capsomeres are clearly visible. **(b)** The envelope around this herpes simplex virus capsid has broken, giving a "fried egg" appearance.

■ What is the morphology of these viruses?

Adenoviridae Named after adenoids, from which they were first isolated, adenoviruses cause acute respiratory diseases—the common cold (Figure 13.16a).

Poxviridae All diseases caused by poxviruses, including smallpox and cowpox, include skin lesions. *Pox* refers to pus-filled lesions. Viral multiplication is started by viral transcriptase; the viral components are synthesized and assembled in the cytoplasm of the host cell.

Herpesviridae Nearly 100 herpesviruses are known (Figure 13.16b). They are named after the spreading (*herpetic*) appearance of cold sores. Species of human herpesviruses (HHV) include HHV-1 and HHV-2, both in the genus *Simplexvirus*, which causes cold sores; HHV-3, genus *Varicellavirus*, which causes chickenpox; HHV-4, genus *Lymphocryptovirus*, which causes infectious mononucleosis; HHV-5, genus *Cytomegalovirus*, which causes CMV inclusion disease; HHV-6, genus *Roseolovirus*, which causes roseola; HHV-7, which infects most infants, causing measleslike rashes; and HHV-8, which causes Kaposi's sarcoma, primarily in AIDS patients.

Papovaviridae Papovaviruses are named for *pa*pillomas (warts), *po*lyomas (tumors), and *va*cuolation (cytoplasmic vacuoles produced by some of these viruses). Warts are caused by members of the genus *Papillomavirus*. Some *Papillomavirus* species are capable of transforming cells and causing cancer. Viral DNA is replicated in the host cell's nucleus along with host cell chromosomes. Host cells may proliferate, resulting in a tumor.

Hepadnaviridae Hepadnaviridae are so named because they cause *hep*atitis and contain *DNA*. The only genus in this family causes hepatitis B. (Hepatitis A, C, D, E, F, and G viruses, although not related to each other, are RNA viruses.) Hepadnaviruses differ from other DNA viruses because they synthesize DNA by copying RNA, using viral reverse transcriptase. This enzyme is discussed later with the retroviruses, the only other family with reverse transcriptase.

The Biosynthesis of RNA Viruses

The multiplication of RNA viruses is essentially the same as that of DNA viruses, except that several different mechanisms of mRNA formation occur among different groups of RNA viruses (see Table 13.4). Although the details of these mechanisms are beyond the scope of this text, for comparative purposes we will trace the multiplication cycles of the four nucleic acid types of RNA viruses (three of which are shown in Figure 13.17). RNA viruses multiply in the host cell's cytoplasm. The major differences among the multiplication processes of these viruses lie in how mRNA and viral RNA are produced. Once viral RNA and viral proteins are synthesized, maturation occurs by similar means among all animal viruses, as will be discussed shortly.

Picornaviridae Picornaviruses, such as poliovirus, are single-stranded RNA viruses. They are the smallest viruses; and the prefix *pico-* (small) plus *RNA* gives these viruses their name. The RNA within the virion is called a **sense strand** (or **+ strand**), because it can act as mRNA.

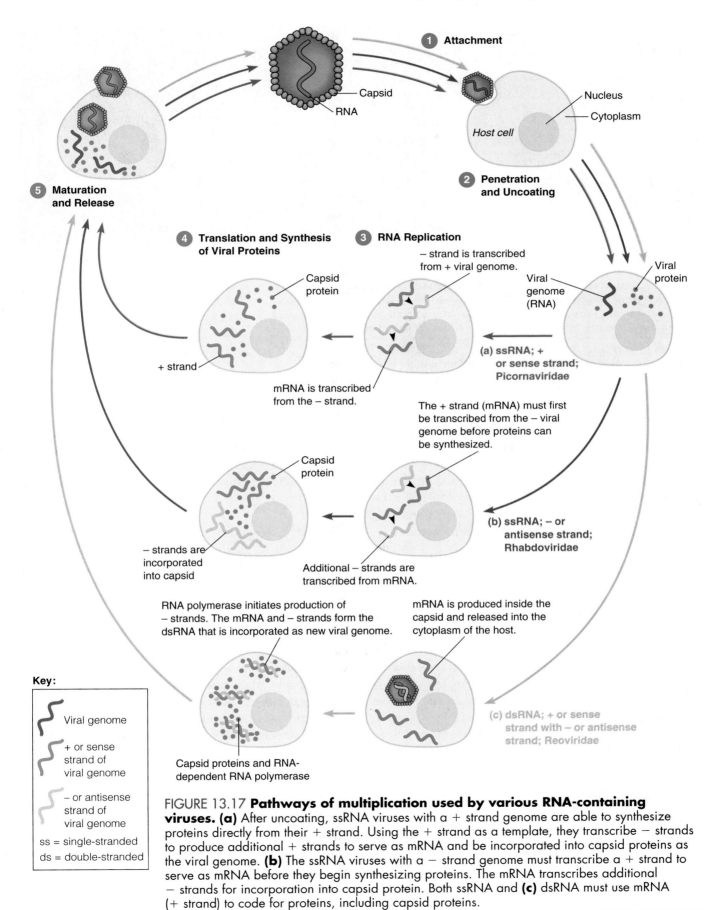

FIGURE 13.17 **Pathways of multiplication used by various RNA-containing viruses. (a)** After uncoating, ssRNA viruses with a + strand genome are able to synthesize proteins directly from their + strand. Using the + strand as a template, they transcribe − strands to produce additional + strands to serve as mRNA and be incorporated into capsid proteins as the viral genome. **(b)** The ssRNA viruses with a − strand genome must transcribe a + strand to serve as mRNA before they begin synthesizing proteins. The mRNA transcribes additional − strands for incorporation into capsid protein. Both ssRNA and **(c)** dsRNA must use mRNA (+ strand) to code for proteins, including capsid proteins.

■ Why is − strand RNA made by picornaviruses and reoviruses? By rhabdoviruses?

(a) TEM |—| 10 nm **(b)** TEM |—| 100 nm **(c)** TEM |—| 100 nm

FIGURE 13.18 **RNA-containing animal viruses. (a)** A single rubella virus (*Rubivirus*), which belongs to the togavirus family. **(b)** Particles of vesicular stomatitis virus (*Vesiculovirus*), a member of the Family Rhabdoviridae. **(c)** Mouse mammary tumor virus, a Retroviridae, causes tumors in mice.

■ The RNA of + strand RNA viruses serves as mRNA; − strand RNA viruses must make + strand RNA to serve as mRNA.

After attachment, penetration, and uncoating are completed, the single-stranded viral RNA (Figure 13.17a) is translated into two principal proteins, which inhibit the host cell's synthesis of RNA and protein and which form an enzyme called *RNA-dependent RNA polymerase.* This enzyme catalyzes the synthesis of another strand of RNA, which is complementary in base sequence to the original infecting strand. This new strand, called an **antisense strand** (or **− strand**), serves as a template to produce additional + strands. The + strands may serve as mRNA for the translation of capsid proteins, may become incorporated into capsid proteins to form a new virus, or may serve as a template for continued RNA multiplication. Once viral RNA and viral protein are synthesized, maturation occurs.

Togaviridae Togaviruses, which include *a*rthropod-*bo*rne *arbo*viruses or alphaviruses, also contain a single + strand of RNA. Togaviruses are enveloped viruses; their name is from the Latin word for covering, *toga* (Figure 13.18a). Keep in mind that these are not the only enveloped viruses. After a − strand is made from the + strand, two types of mRNA are transcribed from the − strand. One type of mRNA is a short strand that codes for envelope proteins; the other, longer strand serves as mRNA for capsid proteins and can become incorporated into a capsid.

Rhabdoviridae Rhabdoviruses, such as rabiesvirus (genus *Lyssavirus*), are usually bullet-shaped (Figure 13.18b). *Rhabdo* is from the Greek word for rod, which is not really an accu-

rate description of their morphology. They contain a single − strand of RNA (Figure 13.17b). They also contain an RNA-dependent RNA polymerase that uses the − strand as a template from which to produce a + strand. The + strand serves as mRNA and as a template for synthesis of new viral RNA.

Reoviridae Reoviruses were named for their habitats: the respiratory and enteric (digestive) systems of humans. They were not associated with any diseases when first discovered, so they were considered orphan viruses. Their name comes from the first letters of *r*espiratory, *e*nteric, and *o*rphan. Three serotypes are now known to cause respiratory tract and intestinal tract infections.

The capsid containing the double-stranded RNA is digested upon entering a host cell. Viral mRNA is produced in the cytoplasm, where it is used to synthesize more viral proteins (Figure 13.17c). One of the newly synthesized viral proteins acts as RNA-dependent RNA polymerase to produce more − strands of RNA. The mRNA (+) and − strands form the double-stranded RNA that is then surrounded by capsid proteins.

Retroviridae Many retroviruses infect vertebrates (Figure 13.18c). One genus of retrovirus, *Lentivirus*, includes the subspecies HIV-1 and HIV-2, which cause AIDS (see the box on page 398 and in Chapter 19, page 546). The retroviruses that cause cancer will be discussed later in this chapter.

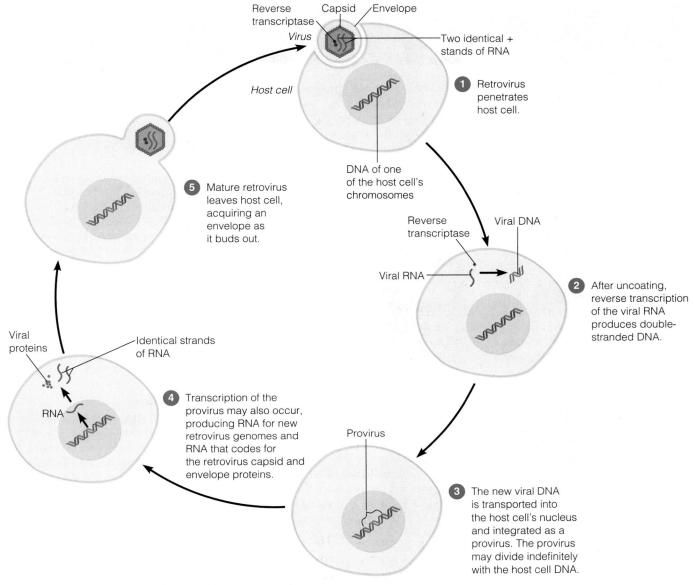

FIGURE 13.19 **Multiplication and inheritance processes of the Retroviridae.** A retrovirus may become a provirus that replicates in a latent state, and it may produce new retroviruses.

- How does the biosynthesis of a retrovirus differ from that of other RNA viruses?

The formation of mRNA and RNA for new retrovirus virions is shown in Figure 13.19.

1 These viruses carry their own polymerase, which is an RNA-dependent DNA polymerase called **reverse transcriptase,** so named because it carries out a reaction (RNA → DNA) that is exactly the reverse of the familiar transcription of DNA → RNA. The name *retrovirus* is derived from the first letters of *reverse transcriptase.*

2 Reverse transcriptase uses the RNA of the virus to synthesize a complementary strand of DNA, which in turn is replicated to form double-stranded DNA. This enzyme also degrades the original viral RNA. The formation of complete viruses requires that DNA be transcribed back into the RNA that will serve as mRNA for viral protein synthesis and be incorporated into new virions.

M M W R

AIDS: The Risk to Health Care Workers

With the advent of the AIDS epidemic, health care workers are understandably concerned about the risk of contracting AIDS after exposure to the body fluids of infected patients. However, when precautions are observed, the risk to workers is very small, even for those treating AIDS patients.

Understanding the Risk

The first protection for health care workers is a clear understanding of how HIV can (and cannot) be transmitted in the course of their work. To date, direct inoculation of infected material is the only proven method of transmission in the health care environment. Infected materials that can transmit HIV are blood, semen, vaginal secretions, and breast milk. The most common route of transmission is through accidental needle sticks. However, inoculation is also possible if infected material contacts mucous membranes or a break in the health care worker's skin. There is no evidence of HIV transmission by aerosols, the fecal-oral route, mouth-to-mouth or casual contact, or contact with environmental surfaces such as floors, walls, chairs, and toilets. Although HIV has been detected in fluids such as saliva, tears, cerebrospinal fluid, amniotic fluid, and urine, HIV is unlikely to be transmitted through exposure to these fluids.

The number of health care workers who have become infected, and the sources of their infections, are being carefully monitored by the Centers for Disease Control and Prevention (CDC). In the United States, by mid-2001, 57

health care workers who denied other risk factors and had seroconverted to HIV positive had contracted HIV infections, 50 of these following needlestick injuries. Twenty-six of them have AIDS.

The CDC estimates that the probability of infection following a needlestick injury with HIV-infected blood is 3 out of 1000 exposures, or 0.3%. The probability of infection from mucous membrane exposure to blood is 0.1%. The probability of acquiring hepatitis B from a needlestick injury with blood containing HBV (hepatitis B virus) is 6–30%.

Precautions

Health care workers should be vaccinated against HBV. Avoidance of exposure is the health care worker's first line of defense against HIV. The CDC has developed the strategy of "universal precautions," which should be followed in *all* health care settings. These are described below.

Gloves Disposable gloves should be used for direct exposure to infected blood, other body fluids, and tissues. Double-gloving is recommended during invasive surgical procedures. Personnel should not work when they have open skin lesions, dermatitis exuding fluids, or cutaneous wounds.

Gowns, Masks, and Goggles Masks and protective eyewear are recommended when splashes are expected, such as during airway manipulation, endoscopy, and dental procedures, and in the laboratory.

Needles To minimize the risk of needlesticks, needles should not be resheathed

and should be put in a puncture-proof container for sterilization and disposal. More expensive, safer needle devices that minimize the risk of needlesticks are available.

Disinfection Routine housekeeping in health care settings should include washing floors, walls, and other areas not normally associated with disease transmissions with a 1:100 dilution of household bleach. A 1:10 dilution is recommended for disinfecting a spill.

Preventive Treatment after Exposure

Accidental exposure cannot always be prevented. Studies indicate that exposed individuals can reduce their risk by the prophylactic use of two or three antiviral drugs for 4 weeks.

The Risk to Patients

Transmission of HBV from health care workers to patients during invasive dental procedures (tooth extractions) has been documented. The only documented cases of HIV transmission from a health care worker to patients involved a dentist in Florida and a surgeon in France. The risk of HIV transmission from an infected health care worker to patients is small and can be reduced with universal precautions.

The risk of HIV transmission from health care workers to patients is greatest during invasive procedures; restriction of patient care by infected workers who perform such procedures is made on an individual basis.

SOURCE: Adapted from MMWR 38(S-2) (5/12/89); MMWR 47(RR-7)(5/15/98); CDC Fact Sheet (2/02).

❸ However, before transcription can take place, the viral DNA must be integrated into the DNA of a host cell chromosome. In this integrated state, the viral DNA is called a **provirus.** Unlike a prophage, the provirus never comes out of the chromosome. As a provirus, HIV is protected from the host's immune system and antiviral drugs. Once the provirus is integrated into the host cell's DNA, several things can happen.

Sometimes the provirus simply remains in a latent state and replicates when the DNA of the host cell replicates.

❹ In other cases, the provirus is expressed and

❺ produces new viruses, which may infect adjacent cells. Mutagens such as gamma radiation can induce expression of a provirus. The provirus can also convert the host cell into a tumor cell; possible mechanisms for this phenomenon will be discussed later.

Maturation and Release

The first step in viral maturation is the assembly of the protein capsid; this assembly is usually a spontaneous process. The capsids of many animal viruses are enclosed by an envelope consisting of protein, lipid, and carbohydrate, as noted earlier. Examples of such viruses include orthomyxoviruses and paramyxoviruses. The envelope protein is encoded by the viral genes and is incorporated into the plasma membrane of the host cell. The envelope lipid and carbohydrate are encoded by host cell genes and are present in the plasma membrane. The envelope actually develops around the capsid by a process called **budding** (Figure 13.20).

After the sequence of attachment, penetration, uncoating, and biosynthesis of viral nucleic acid and protein, the assembled capsid containing nucleic acid pushes through the plasma membrane. As a result, a portion of the plasma membrane, now the envelope, adheres to the virus. This extrusion of a virus from a host cell is one method of release. Budding does not immediately kill the host cell, and in some cases the host cell survives.

Nonenveloped viruses are released through ruptures in the host cell plasma membrane. In contrast to budding, this type of release usually results in the death of the host cell.

Viruses and Cancer

Learning Objectives
- *Define oncogene and transformed cell.*
- *Discuss the relationship between DNA- and RNA-containing viruses and cancer.*

Several types of cancer are now known to be caused by viruses. Molecular biological research shows that the mechanisms of the diseases are similar, even when a virus does not cause the cancer.

The relationship between cancers and viruses was first demonstrated in 1908, when virologists Wilhelm Ellerman and Olaf Bang, working in Denmark, were trying to isolate the causative agent of chicken leukemia. They found that leukemia could be transferred to healthy chickens by cell-free filtrates that contained viruses. Three years later, F. Peyton Rous, working at the Rockefeller Institute in New York, found that a chicken **sarcoma** (cancer of connective tissue) can be similarly transmitted. Virus-induced **adeno-carcinomas** (cancers of glandular epithelial tissue) in

(a) Release by budding

(b) *Alphavirus*

FIGURE 13.20 **Budding of an enveloped virus.** **(a)** A diagram of the budding process. **(b)** The small "bumps" (at arrow) seen on this freeze-fractured plasma membrane are Sindbis virus (*Alphavirus*) particles caught in the act of budding out from an infected cell.

■ Most enveloped viruses take part of the host's plasma membrane for their envelope.

mice were discovered in 1936. At that time, it was clearly shown that mouse mammary gland tumors are transmitted from mother to offspring through the mother's milk. A human cancer-causing virus was discovered and isolated in 1972 by American bacteriologist Sarah Stewart.

The viral cause of cancer can often go unrecognized for several reasons. First, most of the particles of some viruses infect cells but do not induce cancer. Second, cancer might not develop until long after viral infection. Third, cancers do not seem to be contagious, as viral diseases usually are.

The Transformation of Normal Cells into Tumor Cells

Almost anything that can alter the genetic material of a eukaryotic cell has the potential to make a normal cell cancerous. These cancer-causing alterations to cellular DNA affect parts of the genome called **oncogenes.** Oncogenes were first identified in cancer-causing viruses and were thought to be a part of the normal viral genome. However, American microbiologists J. Michael Bishop and Harold E. Varmus received the 1989 Nobel Prize in Medicine for proving that the cancer-inducing genes carried by viruses are actually derived from animal cells. Bishop and Varmus showed that the cancer-causing *src* gene in avian sarcoma viruses is derived from a normal part of chicken genes.

Oncogenes can be activated to abnormal functioning by a variety of agents, including mutagenic chemicals, high-energy radiation, and viruses. Viruses capable of inducing tumors in animals are called **oncogenic viruses,** or *oncoviruses.* Approximately 10% of cancers are known to be virus-induced. An outstanding feature of all oncogenic viruses is that their genetic material integrates into the host cell's DNA and replicates along with the host cell's chromosome. This mechanism is similar to the phenomenon of lysogeny in bacteria, and it can alter the host cell's characteristics in the same way.

Tumor cells undergo **transformation;** that is, they acquire properties that are distinct from the properties of uninfected cells or from infected cells that do not form tumors. After being transformed by viruses, many tumor cells contain a virus-specific antigen on their cell surface, called **tumor-specific transplantation antigen (TSTA),** or an antigen in their nucleus, called the **T antigen.** Transformed cells tend to be less round than normal cells, and they tend to exhibit certain chromosomal abnormalities, such as unusual numbers of chromosomes and fragmented chromosomes.

DNA Oncogenic Viruses

Oncogenic viruses are found within several families of DNA-containing viruses. These groups include the Adenoviridae, Herpesviridae, Poxviridae, Papovaviridae, and Hepadnaviridae. Among the papovaviruses, papillomaviruses cause uterine (cervical) cancer.

A genus of the Herpesviridae, *Lymphocryptovirus,* includes the Epstein-Barr virus (EB virus), which causes infectious mononucleosis and two human cancers, Burkitt's lymphoma and nasopharyngeal carcinoma. Burkitt's lymphoma is a rare cancer of the lymphatic system; it affects mostly children in certain parts of Africa. Nasopharyngeal carcinoma, a cancer of the nose and throat, is found worldwide. Some researchers have also suggested that EB virus is involved in Hodgkin's disease, a cancer of the lymphatic system.

About 90% of the U.S. population probably carry the latent stage of the EB virus in their lymphocytes but have no signs of disease. This latent stage is indicated by the presence of antibodies to the virus in blood serum. Although infection with this virus leads to mild symptoms in healthy children, it can cause infectious mononucleosis, mostly in teenagers.

EB virus was isolated from Burkitt's lymphoma cells in 1964 by Michael Epstein and Yvonne Barr. The proof that EB virus can cause cancer was accidentally demonstrated in 1985 when a 12-year-old boy known only as David received a bone marrow transplant. David was born without a functioning immune system and was kept from all microbes inside a sterile "bubble" his entire life; the news media called him "the bubble boy." (The lymphocytes that provide immunity are produced in bone marrow.) Several months after the transplant, he developed signs of infectious mononucleosis, and a few months after that, he died of cancer. An autopsy revealed that the virus had been unwittingly introduced into the boy with the bone marrow transplant. The case supported suspicions that EB virus can induce cancer in immunosuppressed individuals.

Another DNA virus that causes cancer is hepatitis B virus (HBV, genus *Hepadnavirus*). Many animal studies have been performed that have clearly indicated the causal role of HBV in liver cancer. In one human study, virtually all people with liver cancer had previous HBV infections. Moreover, results of this study indicated that the risk of getting liver cancer is over 98 times higher in people with prior HBV infection than in those with no prior infection.

RNA Oncogenic Viruses

Among the RNA viruses, only the oncoviruses in the family Retroviridae cause cancer. The human T-cell leukemia viruses (HTLV-1 and HTLV-2) are retroviruses that cause adult T-cell leukemia and lymphoma in humans. (T cells are a type of white blood cell involved in the immune response.)

Sarcoma viruses of cats, chickens, and rodents, and the mammary tumor viruses of mice, are also retroviruses. Another retrovirus, feline leukemia virus (FeLV), causes leukemia in cats and is transmissible among cats. There is a test to detect the virus in cat serum.

TABLE 13.5	Examples of Persistent Viral Infections in Humans	
Disease	**Primary Effect**	**Causative Virus**
Subacute sclerosing panencephalitis (SSPE)	Mental deterioration	Measles virus (*Morbillivirus*)
Progressive encephalitis	Rapid mental deterioration	Rubella virus (*Rubivirus*)
Progressive multifocal leukoencephalopathy	Brain degeneration	Papovavirus (family Papovaviridae)
AIDS dementia complex	Brain degeneration	HIV (*Lentivirus*)
Persistent enterovirus infection	Mental deterioration associated with AIDS	Echoviruses

The ability of retroviruses to induce tumors is related to their production of a reverse transcriptase by the mechanism described earlier (see Figure 13.19). The provirus, which is the double-stranded DNA molecule synthesized from the viral RNA, becomes integrated into the host cell's DNA; new genetic material is thereby introduced into the host's genome, and this is the key reason retroviruses can contribute to cancer. Some retroviruses contain oncogenes; others contain promoters that turn on oncogenes or other cancer-causing factors.

Latent Viral Infections

Learning Objective

• *Provide an example of a latent viral infection.*

A virus can remain in equilibrium with the host and not actually produce disease for a long period, often many years. All of the human herpesviruses can remain in host cells throughout the life of an individual. When herpesviruses are reactivated by immunosuppression (for example, AIDS), the resulting infection may be fatal. The classic example of such a **latent infection** in viruses is the infection of the skin by herpes simplex virus, which produces cold sores. This virus inhabits the host's nerve cells but causes no damage until it is activated by a stimulus such as fever or sunburn—hence the term *fever blister*.

In some individuals, viruses are produced, but symptoms never appear. Even though a large percentage of the human population carries the herpes simplex virus, only 10–15% of people carrying the virus exhibit the disease. The virus of some latent infections can exist in a lysogenic state within host cells.

The chickenpox virus (genus *Varicellovirus*) can also exist in a latent state. Chickenpox (varicella) is a skin disease that is usually acquired in childhood. The virus gains access to the skin via the blood. From the blood, some viruses may enter nerves, where they remain latent. Later,

changes in the immune (T-cell) response can activate these latent viruses, causing shingles (zoster). The shingles rash appears on the skin along the nerve in which the virus was latent. Shingles occurs in 10–20% of people who have had chickenpox.

Persistent Viral Infections

Learning Objective

• *Differentiate persistent viral infections from latent viral infections.*

The term *slow viral infection* was coined in the 1950s to refer to a disease process that occurs gradually over a long period and was thought to be caused by a virus. A better term to describe these prolonged diseases is **persistent viral infection.** Typically, persistent viral infections are fatal.

A number of persistent viral infections have in fact been shown to be caused by conventional viruses. For example, several years after causing measles, the measles virus can be responsible for a rare form of encephalitis called subacute sclerosing panencephalitis (SSPE). A persistent viral infection is apparently different from a latent viral infection in that, in most persistent viral infections, detectable infectious virus gradually builds up over a long period, rather than appearing suddenly.

Several examples of persistent viral infections caused by conventional viruses are listed in Table 13.5.

Prions

Learning Objective

• *Discuss how a protein can be infectious.*

Other infectious diseases that have not been found to have a viral cause might be caused by prions. In 1982, American neurobiologist Stanley Prusiner proposed that infectious proteins caused a neurological disease in sheep called scrapie. The infectivity of scrapie-infected brain tissue is reduced by treatment with proteases but not by treatment with radiation, suggesting that the infectious agent is pure protein. Prusiner coined the name **prion** for *pr*oteinaceous *in*fectious particle.

Nine animal diseases now fall into this category, including the "mad cow disease" that emerged in cattle in Great Britain in 1987. All nine are neurological diseases called spongiform encephalopathies because large vacuoles develop in the brain. The human diseases are kuru, Creutzfeldt-Jakob disease (CJD), Gerstmann-Sträussler-Scheinker syndrome, and fatal familial insomnia. These diseases run in families, which indicates a possible genetic cause. However, they cannot be purely inherited, because mad cow disease arose from feeding scrapie-infected sheep meat to cattle, and the new (bovine) variant was transmitted to humans who ate undercooked beef from infected cattle (see Chapter 1, page 19). Additionally, CJD has been transmitted with transplanted nerve tissue and contaminated surgical instruments.

These diseases are caused by the conversion of a normal host glycoprotein called PrP^C, for cellular prion protein, into an infectious form called PrP^{Sc}, for scrapie protein. The cell-surface location of PrP^C suggests the protein might be associated with cell adhesion and recognition or the uptake of an element. One hypothesis for how an infectious agent that lacks any nucleic acid can reproduce is shown in Figure 13.21.

❶–❷ The gene for PrP^C is located on chromosome 20 in humans. PrP^C produced by cells is secreted to the cell surface.

❸–❺ PrP^{Sc} reacts with PrP^C on the cell surface converting the PrP^C to PrP^{Sc}.

❻–❽ PrP^{Sc} is taken in by endocytosis and accumulates in lysosomes.

The actual cause of cell damage is not known. Fragments of PrP^{Sc} molecules accumulate in the brain, forming plaques; these plaques are used for postmortem diagnosis, but they do not appear to be the cause of cell damage.

Plant Viruses and Viroids

Learning Objectives
- *Differentiate virus, viroid, and prion.*
- *Name a virus that causes a plant disease.*

Plant viruses resemble animal viruses in many respects: Plant viruses are morphologically similar to animal viruses, and they have similar types of nucleic acid (Table 13.6). In fact, some plant viruses can multiply inside insect cells.

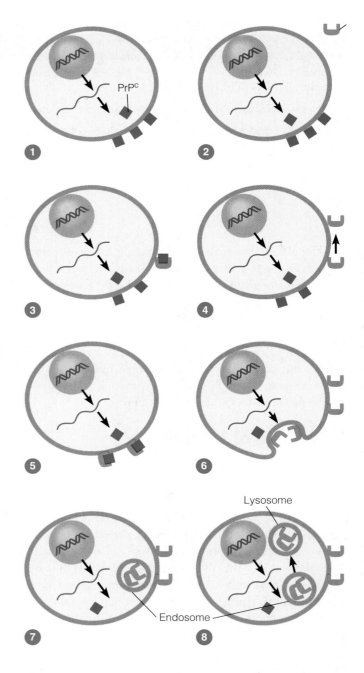

FIGURE 13.21 **How a protein can be infectious.** If an abnormal prion protein (PrP^{Sc}) enters a cell, it changes a normal prion protein to PrP^{Sc} which now can change another normal PrP, resulting in an accumulation of the abnormal PrP^{Sc}.

■ How do prions differ from viruses?

Plant viruses cause many diseases of economically important crops, including beans (bean mosaic virus), corn and sugarcane (wound tumor virus), and potatoes (potato yel-

TABLE 13.6	**Classification of Some Major Plant Viruses**			
Characteristic	**Viral Family**	**Viral Genus or Unclassified Members**	**Morphology**	**Method of Transmission**
Double-stranded DNA, nonenveloped	Papovaviridae	Cauliflower mosaic virus		Aphids
Single-stranded RNA, + strand, nonenveloped	Picornaviridae	Bean mosaic virus		Pollen
	Tetraviridae	*Tobamovirus*		Wounds
Single-stranded RNA, − strand, enveloped	Rhabdoviridae	Potato yellow dwarf virus		Leafhoppers and aphids
Double-stranded RNA, nonenveloped	Reovirus	Wound tumor virus		Leafhoppers

low dwarf virus). Viruses can cause color change, deformed growth, wilting, and stunted growth in their plant hosts. Some hosts, however, remain symptomless and only serve as reservoirs of infection.

Plant cells are generally protected from disease by an impermeable cell wall. Viruses must enter through wounds or be assisted by other plant parasites, including nematodes, fungi, and, most often, insects that suck the plant's sap. Once one plant is infected, it can spread infection to other plants in its pollen and seeds.

In laboratories, plant viruses are cultured in protoplasts (plant cells with the cell walls removed) and in insect cell cultures.

Some plant diseases are caused by **viroids,** short pieces of naked RNA, only 300–400 nucleotides long, with no protein coat. The nucleotides are often internally paired, so the molecule has a closed, folded, three-dimensional structure that presumably helps protect it from attack by cellular enzymes. The RNA does not code for any proteins. Thus far, viroids have been conclusively identified as pathogens only of plants. Annually, infections by viroids, such as potato spindle tuber viroid, result in losses of millions of dollars from crop damage (Figure 13.22).

Current research on viroids has revealed similarities between the base sequences of viroids and introns. Recall from Chapter 8 (page 221) that introns are sequences of

TEM | 200 nm

FIGURE 13.22 **Linear and circular potato spindle tuber viroid (PSTV).**

■ A viroid is infectious RNA; a prion is infectious protein.

genetic material that do not code for polypeptides. This observation has led to the hypothesis that viroids evolved from introns, leading to speculation that future researchers may discover animal viroids.

STUDY OUTLINE

General Characteristics of Viruses (pp. 377–378)

1. Depending on one's viewpoint, viruses may be regarded as exceptionally complex aggregations of nonliving chemicals or as exceptionally simple living microbes.

2. Viruses contain a single type of nucleic acid (DNA or RNA) and a protein coat, sometimes enclosed by an envelope composed of lipids, proteins, and carbohydrates.

3. Viruses are obligatory intracellular parasites. They multiply by using the host cell's synthesizing machinery to cause the synthesis of specialized elements that can transfer the viral nucleic acid to other cells.

Host Range (pp. 377–378)

1. Host range refers to the spectrum of host cells in which a virus can multiply.

2. Most viruses infect only specific types of cells in one host species.

3. Host range is determined by the specific attachment site on the host cell's surface and the availability of host cellular factors.

Viral Size (p. 378)

1. Viral size is ascertained by electron microscopy.

2. Viruses range from 20 to 1000 nm in length.

Viral Structure (pp. 379–381)

1. A virion is a complete, fully developed viral particle composed of nucleic acid surrounded by a coat.

Nucleic Acid (p. 379)

1. Viruses contain either DNA or RNA, never both, and the nucleic acid may be single- or double-stranded, linear or circular, or divided into several separate molecules.

2. The proportion of nucleic acid in relation to protein in viruses ranges from about 1% to about 50%.

Capsid and Envelope (pp. 379–380)

1. The protein coat surrounding the nucleic acid of a virus is called the capsid.

2. The capsid is composed of subunits, capsomeres, which can be a single type of protein or several types.

3. The capsid of some viruses is enclosed by an envelope consisting of lipids, proteins, and carbohydrates.

4. Some envelopes are covered with carbohydrate-protein complexes called spikes.

General Morphology (pp. 380–381)

1. Helical viruses (for example, Ebola virus) resemble long rods, and their capsids are hollow cylinders surrounding the nucleic acid.

2. Polyhedral viruses (for example, adenovirus) are many-sided. Usually the capsid is an icosahedron.

3. Enveloped viruses are covered by an envelope and are roughly spherical but highly pleomorphic. There are also enveloped helical viruses (for example, *Influenzavirus*) and enveloped polyhedral viruses (for example, *Simplexvirus*).

4. Complex viruses have complex structures. For example, many bacteriophages have a polyhedral capsid with a helical tail attached.

Taxonomy of Viruses (p. 381)

1. Classification of viruses is based on type of nucleic acid, strategy for replication, and morphology.

2. Virus family names end in *-viridae*; genus names end in *-virus*.

3. A viral species is a group of viruses sharing the same genetic information and ecological niche.

Isolation, Cultivation, and Identification of Viruses (pp. 381–386)

1. Viruses must be grown in living cells.

2. The easiest viruses to grow are bacteriophages.

Growing Bacteriophages in the Laboratory (p. 384)

1. The plaque method mixes bacteriophages with host bacteria and nutrient agar.

2. After several viral multiplication cycles, the bacteria in the area surrounding the original virus are destroyed; the area of lysis is called a plaque.

3. Each plaque originates with a single viral particle; the concentration of viruses is given as plaque-forming units.

Growing Animal Viruses in the Laboratory (pp. 384–386)

1. Cultivation of some animal viruses requires whole animals.

2. Simian AIDS and feline AIDS provide models for studying human AIDS.

3. Some animal viruses can be cultivated in embryonated eggs.

4. Cell cultures are cells growing in culture media in the laboratory.

5. Primary cell lines and embryonic diploid cell lines grow for a short time in vitro.

6. Continuous cell lines can be maintained in vitro indefinitely.

7. Viral growth can cause cytopathic effects in the cell culture.

Viral Identification (p. 386)

1. Serological tests are used most often to identify viruses.

2. Viruses may be identified by RFLPs and PCR.

Viral Multiplication (pp. 386–399)

1. Viruses do not contain enzymes for energy production or protein synthesis.

2. For a virus to multiply, it must invade a host cell and direct the host's metabolic machinery to produce viral enzymes and components.

Multiplication of Bacteriophages (pp. 386–390)

1. During a lytic cycle, a phage causes the lysis and death of a host cell.

2. Some viruses can either cause lysis or have their DNA incorporated as a prophage into the DNA of the host cell. The latter situation is called lysogeny.

3. The T-even bacteriophages that infect *E. coli* have been studied extensively.

4. During the attachment phase of the lytic cycle, sites on the phage's tail fibers attach to complementary receptor sites on the bacterial cell.

5. In penetration, phage lysozyme opens a portion of the bacterial cell wall, the tail sheath contracts to force the tail core through the cell wall, and phage DNA enters the bacterial cell. The capsid remains outside.

6. In biosynthesis, transcription of phage DNA produces mRNA coding for proteins necessary for phage multiplication. Phage DNA is replicated, and capsid proteins are produced. During the eclipse period, separate phage DNA and protein can be found.

7. During maturation, phage DNA and capsids are assembled into complete viruses.

8. During release, phage lysozyme breaks down the bacterial cell wall, and the multiplied phages are released.

9. The time from phage adsorption to release is called burst time (20–40 minutes). Burst size, the number of newly synthesized phages produced from a single infected cell, ranges from 50 to 200.

10. During the lysogenic cycle, prophage genes are regulated by a repressor coded for by the prophage. The prophage is replicated each time the cell divides.

11. Exposure to certain mutagens can lead to excision of the prophage and initiation of the lytic cycle.

12. Because of lysogeny, lysogenic cells become immune to reinfection with the same phage and may undergo phage conversion.

13. A lysogenic phage can transfer bacterial genes from one cell to another through transduction. Any genes can be transferred in generalized transduction, and specific genes can be transferred in specialized transduction.

Multiplication of Animal Viruses (pp. 390–399)

1. Animal viruses attach to the plasma membrane of the host cell.

2. Penetration occurs by endocytosis or fusion.

3. Animal viruses are uncoated by viral or host cell enzymes.

4. The DNA of most DNA viruses is released into the nucleus of the host cell. Transcription of viral DNA and translation produce viral DNA and, later, capsid proteins. Capsid proteins are synthesized in the cytoplasm of the host cell.

5. DNA viruses include members of the families Adenoviridae, Poxviridae, Herpesviridae, Papovaviridae, and Hepadnaviridae.

6. Multiplication of RNA viruses occurs in the cytoplasm of the host cell. RNA-dependent RNA polymerase synthesizes a double-stranded RNA.

7. Picornaviridae + strand RNA acts as mRNA and directs the synthesis of RNA-dependent RNA polymerase.

8. Togaviridae + strand RNA acts as a template for RNA-dependent RNA polymerase, and mRNA is transcribed from a new − RNA strand.

9. Rhabdoviridae − strand RNA is a template for viral RNA-dependent RNA polymerase, which transcribes mRNA.

10. Reoviridae are digested in host cell cytoplasm to release mRNA for viral biosynthesis.

11. Retroviridae reverse transcriptase (RNA-dependent DNA polymerase) transcribes DNA from RNA.

12. After maturation, viruses are released. One method of release (and envelope formation) is budding. Naked viruses are released through ruptures in the host cell membrane.

Viruses and Cancer (pp. 399–401)

1. The earliest relationship between cancer and viruses was demonstrated in the early 1900s, when chicken leukemia and chicken sarcoma were transferred to healthy animals by cell-free filtrates.

The Transformation of Normal Cells into Tumor Cells (p. 400)

1. When activated, oncogenes transform normal cells into cancerous cells.

2. Viruses capable of producing tumors are called oncogenic viruses.

3. Several DNA viruses and retroviruses are oncogenic.

4. The genetic material of oncogenic viruses becomes integrated into the host cell's DNA.

5. Transformed cells lose contact inhibition, contain virus-specific antigens (TSTA and T antigen), exhibit chromosome abnormalities, and can produce tumors when injected into susceptible animals.

DNA Oncogenic Viruses (p. 400)

1. Oncogenic viruses are found among the Adenoviridae, Herpesviridae, Poxviridae, and Papovaviridae.
2. The EB virus, a herpesvirus, causes Burkitt's lymphoma and nasopharyngeal carcinoma. *Hepadnavirus* causes liver cancer.

RNA Oncogenic Viruses (pp. 400–401)

1. Among the RNA viruses, only retroviruses seem to be oncogenic.
2. HTLV-1 and HTLV-2 have been associated with human leukemia and lymphoma.
3. The virus's ability to produce tumors is related to the production of reverse transcriptase. The DNA synthesized from the viral RNA becomes incorporated as a provirus into the host cell's DNA.
4. A provirus can remain latent, can produce viruses, or can transform the host cell.

Latent Viral Infections (p. 401)

1. A latent viral infection is one in which the virus remains in the host cell for long periods without producing an infection.
2. Examples are cold sores and shingles.

Persistent Viral Infections (p. 401)

1. Persistent viral infections are disease processes that occur over a long period and are generally fatal.
2. Persistent viral infections are caused by conventional viruses; viruses accumulate over a long period.

Prions (pp. 401–402)

1. Prions are infectious proteins first discovered in the 1980s.
2. Prion diseases, such as CJD and mad cow disease, all involve the degeneration of brain tissue.
3. Prion diseases are the result of an altered protein; the cause can be a mutation in the normal gene for PrP or contact with an altered protein (PrP^{Sc}).

Plant Viruses and Viroids (pp. 402–403)

1. Plant viruses must enter plant hosts through wounds or with invasive parasites, such as insects.
2. Some plant viruses also multiply in insect (vector) cells.
3. Viroids are infectious pieces of RNA that cause some plant diseases, such as potato spindle tuber viroid disease.

STUDY QUESTIONS

Review

1. Viruses were first detected because they are filterable. What do we mean by the term *filterable*, and how could this property have helped researchers detect viruses before the invention of the electron microscope?
2. Why do we classify viruses as obligatory intracellular parasites?
3. List the four properties that define a virus. What is a virion?
4. Describe the four morphological classes of viruses, then diagram and give an example of each.
5. Describe how bacteriophages are detected and enumerated by the plaque method.
6. Why are continuous cell lines of more practical use than primary cell lines for culturing viruses? What is unique about continuous cell lines?
7. *Vibrio cholerae* produces toxin and is capable of causing cholera only when it is lysogenic. What does this mean?
8. Describe the principal events of attachment, penetration, uncoating, biosynthesis, maturation, and release of an enveloped DNA-containing virus.
9. Recall from Chapter 1 that Koch's postulates are used to determine the etiology of a disease. Why is it difficult to determine the etiology of
   a. a viral infection, such as influenza?
   b. cancer?

10. Persistent viral infections such as _____ might be caused by _____ that are _____ .
11. The DNA of DNA-containing oncogenic viruses can become integrated into the host DNA. When integrated, the DNA is called a _____ . How does this process result in transformation of the cell? Describe the changes of transformation. How can an RNA-containing virus be oncogenic?
12. Contrast viroids and prions. Name a disease caused by each.
13. Plant viruses cannot penetrate intact plant cells because _____ ; therefore, they enter cells by _____ . Plant viruses can be cultured in _____ .

Multiple Choice

1. Place the following in the most likely order for biosynthesis of a bacteriophage: (1) phage lysozyme; (2) mRNA; (3) DNA; (4) viral proteins; (5) DNA polymerase.
   a. 5, 4, 3, 2, 1
   b. 1, 2, 3, 4, 5
   c. 5, 3, 4, 2, 1
   d. 3, 5, 2, 4, 1
   e. 2, 5, 3, 4, 1
2. The molecule serving as mRNA can be incorporated in the newly synthesized virus capsids of all of the following *except*
   a. + strand RNA picornaviruses.
   b. + strand RNA togaviruses.

c. − strand RNA rhabdoviruses.
d. double-stranded RNA reoviruses.
e. double-stranded DNA herpesviruses.

3. A virus with RNA-dependent RNA polymerase
   a. synthesizes DNA from an RNA template.
   b. synthesizes double-stranded RNA from an RNA template.
   c. synthesizes double-stranded RNA from a DNA template.
   d. transcribes mRNA from DNA.
   e. none of the above

4. Which of the following would be the first step in the biosynthesis of a virus with reverse transcriptase?
   a. A complementary strand of RNA must be synthesized.
   b. Double-stranded RNA must be synthesized.
   c. A complementary strand of DNA must be synthesized from an RNA template.
   d. A complementary strand of DNA must be synthesized from a DNA template.
   e. none of the above

5. An example of lysogeny in animals could be
   a. slow viral infections.
   b. latent viral infections.
   c. T-even bacteriophages.
   d. infections resulting in cell death.
   e. none of the above

6. The ability of a virus to infect an organism is regulated by
   a. the host species.
   b. the type of cells.
   c. the availability of an attachment site.
   d. cell factors necessary for viral replication.
   e. all of the above

7. Which of the following statements is *not* true?
   a. Viruses contain DNA or RNA.
   b. The nucleic acid of a virus is surrounded by a protein coat.
   c. Viruses multiply inside living cells using viral mRNA, tRNA, and ribosomes.
   d. Viruses cause the synthesis of specialized infectious elements.
   e. Viruses multiply inside living cells.

8. Place the following in the order in which they are found in a host cell: (1) capsid proteins; (2) infective phage particles; (3) phage nucleic acid.
   a. 1, 2, 3
   b. 3, 2, 1
   c. 2, 1, 3
   d. 3, 1, 2
   e. 1, 3, 2

9. Which of the following does *not* initiate DNA synthesis?
   a. A double-stranded DNA virus (Poxviridae)
   b. A DNA virus with reverse transcriptase (Hepadnaviridae)
   c. An RNA virus with reverse transcriptase (Retroviridae)
   d. A single-stranded RNA virus (Togaviridae)
   e. none of the above

10. A viral species is not defined on the basis of the disease symptoms it causes. The best example of this is
    a. polio.
    b. rabies.

c. hepatitis.
d. chickenpox and shingles.
e. measles.

Critical Thinking

1. Discuss the arguments for and against the classification of viruses as living organisms.

2. In some viruses, capsomeres function as enzymes as well as structural supports. Of what advantage is this to the virus?

3. Why was the discovery of simian AIDS and feline AIDS important?

4. Prophages and proviruses have been described as being similar to bacterial plasmids. What similar properties do they exhibit? How are they different?

Clinical Applications

1. A 40-year-old male who was seropositive for HIV experienced abdominal pain, fatigue, and low-grade fever (38°C) for 2 weeks. A chest X-ray examination revealed lung infiltrates. Gram and acid-fast stains were negative. A viral culture revealed the cause of his symptoms: a large, enveloped polyhedral virus with double-stranded DNA. What is the disease? Which virus causes it? Why was a viral culture done after the Gram and acid-fast stain results were obtained?

2. A newborn female developed extensive vesicular and ulcerative lesions over her face and chest. What is the most likely cause of her symptoms? How would you determine the viral cause of this disease without doing a viral culture?

3. Thirty-two people in the same town reported to their physicians with fever (40°C), jaundice, and tender abdomen. All 32 had eaten an ice-slush beverage purchased from a local convenience store. Liver function tests were abnormal. Over the next several months, the symptoms subsided and liver function returned to normal. What is the disease? This disease could be caused by a member of the Picornaviridae, Hepadnaviridae, or Flaviviridae. Differentiate among these families by method of transmission, morphology, nucleic acid, and type of replication.

LEARNING WITH TECHNOLOGY

The Microbiology Place CD-ROM/website includes many resources to help you study and succeed in this course. In addition to self-quizzes, web links, and flashcards, you'll find the following activities for this chapter in the "Activities" section of The Microbiology Place:

Lytic Cycle

Lysogenic Cycle

DNA Viruses

RNA Viruses

Viruses

Case Study: Encephalitis, Texas

Chapter 14

Principles of Disease and Epidemiology

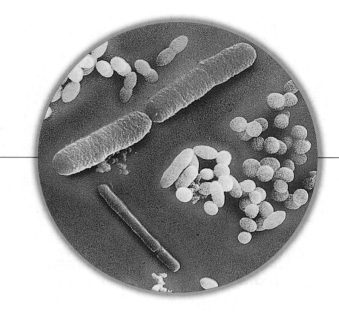

INTESTINAL BACTERIA. *Millions of bacteria live in the human large intestine and are essential for good health.*

Now that you have a basic understanding of the structures and functions of microorganisms and some idea of the variety of microorganisms that exist, we can consider how the human body and various microorganisms interact in terms of health and disease.

We all have our defense mechanisms to keep us healthy. For instance, unbroken skin and mucous membranes are effective barriers against microbial invasion. Within the body, certain cells and certain specialized proteins called antibodies can work together to destroy microbes. In spite of our defenses, however, we are still susceptible to **pathogens** (disease-causing microorganisms). Some bacteria can invade our tissues and resist our defenses by producing capsules or enzymes; other bacteria release toxins that can seriously affect our health. A rather delicate balance exists between our defenses and the pathogenic mechanisms of microorganisms. When our defenses resist these pathogenic capabilities, we maintain our health. But when the pathogen's capability overcomes our defenses, disease results. After the disease has become established, an infected person may recover completely, suffer temporary or permanent damage, or die. The outcome depends on many factors.

Infectious disease is the primary cause of death worldwide. In the United States infectious disease mortality decreased markedly during most of the twentieth century. However, between 1980 and 1992, the death rate from infectious disease increased 58% (including only those people for whom the primary cause of death was an infectious disease). The sharp increase in infectious disease deaths in 1918 and 1919 was caused by the influenza pandemic, which killed more than 20 million people worldwide and over 500,000 people in the United States. This episode illustrates the volatility of infectious disease death rates.

In Part Three we examine some of the principles of infection and disease, the mechanisms by which pathogens cause disease, the body's defenses against disease, and the ways that microbial diseases can be prevented by immunization and controlled by drugs. This first chapter discusses the general principles of disease, starting with a discussion of the meaning and scope of pathology. In the last section of this chapter, "Epidemiology," you will learn how these principles are useful in studying and controlling disease.

Pathology, Infection, and Disease

Learning Objective

• *Define pathology, etiology, infection, and disease.*

Pathology is the scientific study of disease (*pathos* = suffering; *logos* = science). Pathology is first concerned with the cause, or **etiology,** of disease. Second, it deals with **pathogenesis,** the manner in which a disease develops. Third, pathology is concerned with the structural and functional changes brought about by disease and with their final effects on the body.

Although the terms *infection* and *disease* are sometimes used interchangeably, they differ somewhat in meaning. **Infection** is the invasion or colonization of the body by pathogenic microorganisms; **disease** occurs when an infection results in any change from a state of health. Disease is an abnormal state in which part or all of the body is not properly adjusted or incapable of performing its normal functions. An infection may exist in the absence of detectable disease. For example, the body may be infected with the virus that causes AIDS, but there may be no symptoms of the disease.

The presence of a particular type of microorganism in a part of the body where it is not normally found is also called an infection—and may lead to disease. For example, although large numbers of *E. coli* are normally present in the healthy intestine, their infection of the urinary tract usually results in disease.

Few microorganisms are pathogenic. In fact, the presence of some microorganisms can even benefit the host. Therefore, before we discuss the role of microorganisms in causing disease, let's examine the relationship of the microorganisms to the healthy human body.

Normal Microbiota

Learning Objective

• *Define normal and transient microbiota.*

Animals, including humans, are generally free of microbes in utero. At birth, however, normal and characteristic microbial populations begin to establish themselves. Just before a woman gives birth, lactobacilli in her vagina multiply rapidly. The newborn's first contact with microorganisms is usually with these lactobacilli, and they become the predominant organisms in the newborn's intestine. More microorganisms are introduced to the newborn's body from the environment when breathing begins and feeding starts. After birth, *E. coli* and other bacteria acquired from foods begin to inhabit the large intestine.

These microorganisms remain there throughout life and, in response to altered environmental conditions, may increase or decrease in number and contribute to disease.

Many other usually harmless microorganisms establish themselves inside other parts of the normal adult body and on its surface. A typical human body contains 1×10^{13} body cells, yet harbors an estimated 1×10^{14} bacterial cells. This gives you an idea of the abundance of microorganisms that normally reside in the human body. The microorganisms that establish more or less permanent residence (colonize) but that do not produce disease under normal conditions are members of the body's **normal microbiota** or **normal flora** (Figure 14.1). Others, called **transient microbiota,** may be present for several days, weeks, or months and then disappear. Microorganisms are not found throughout the entire human body but are localized in certain regions, as shown in Figure 14.2.

The principal normal microbiota in different regions of the body and some distinctive features of each region are listed in Table 14.1 on page 411. Normal microbiota are also discussed more specifically in Part Four.

Relationships Between the Normal Microbiota and the Host

Learning Objective

• *Compare commensalism, mutualism, and parasitism, and give an example of each.*

Once established, the normal microbiota can benefit the host by preventing the overgrowth of harmful microorganisms. This phenomenon is called **microbial antagonism.** Microbial antagonism involves competition among microbes. One consequence of this competition is that the normal microbiota protect the host against colonization by potentially pathogenic microbes by competing for nutrients, producing substances harmful to the invading microbes, and affecting conditions such as pH and available oxygen. When this balance between normal microbiota and pathogenic microbes is upset, disease can result. For example, the normal bacterial microbiota of the adult human vagina maintains a local pH of 3.5–4.5. The presence of normal microbiota inhibits the overgrowth of the yeast *Candida albicans*, which cannot grow under these conditions. If the bacterial population is eliminated by antibiotics, excessive douching, or deodorants, the pH of the vagina reverts to nearly neutral, and *C. albicans* can flourish and become the dominant microorganism there. This condition can lead to a form of vaginitis (vaginal infection).

Another example of microbial antagonism is seen in the mouth, where streptococci produce compounds that

(a)   SEM ⊢—⊣ 2 μm

(b)   SEM ⊢—⊣ 2 μm

(c)   SEM ⊢—⊣ 2 μm

FIGURE 14.1 Representative normal microbiota for different regions of the body. (a) Bacteria on the surface of the skin. **(b)** Plaque on enamel near the gums. The bacteria that cause dental plaque, although part of the normal microbiota, will cause gum disease and caries (cavities) unless removed frequently. **(c)** Bacteria of the large intestine.

■ Microbes that establish more or less permanent residence but do not produce disease under normal conditions are referred to as normal microbiota.

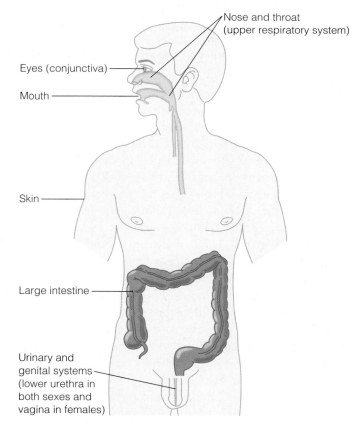

Nose and throat
(upper respiratory system)

Eyes (conjunctiva)

Mouth

Skin

Large intestine

Urinary and
genital systems
(lower urethra in
both sexes and
vagina in females)

FIGURE 14.2 Locations of normal microbiota on and in the human body.

■ Why is the normal microbiota essential for good health?

prevent the growth of most gram-positive and gram-negative cocci. And in the large intestine, *E. coli* cells produce *bacteriocins*, proteins that inhibit the growth of other bacteria of the same or closely related species, such as pathogenic *Salmonella* and *Shigella*. A bacterium that makes a particular bacteriocin is not killed by that bacteriocin, but may be killed by other ones. Bacteriocins are used in medical microbiology to help identify different strains of bacteria. Such identification helps determine whether several outbreaks of an infectious disease are caused by one or more strains of a bacterium.

A final example involves another bacterium, *Clostridium difficile*, also in the large intestine. The normal microbiota of the large intestine effectively inhibit *C. difficile*, possibly by making host receptors unavailable, competing for available nutrients, or producing bacteriocins. However, if the normal microbiota are eliminated, for example, by antibiotics, *C. difficile* can become a problem. This microbe is responsible for nearly all gastrointestinal infections that follow antibiotic therapy, from mild diarrhea to severe or even fatal colitis (inflammation of the colon). The microbe invades the intestinal lining and releases toxins that induce diarrhea, fever, and abdominal pain. See the discussion of probiotics in the box on page 412.

The relationship between the normal microbiota and the host is called **symbiosis** (living together). In the symbiotic relationship called **commensalism,** one of the organisms is benefited and the other is unaffected. Many of

TABLE 14.1	**Representative Members of the Normal Microbiota by Body Region***	
Region	**Principal Components**	**Comments**
Skin	*Propionibacterium acnes, Staphylococcus epidermidis, Staphylococcus aureus, Corynebacterium xerosis, Pityrosporum* spp. (fungus), *Candida* spp. (fungus)	Most of the microbes in direct contact with skin do not become residents because secretions from sweat and oil glands have antimicrobial properties.
Eyes (conjunctiva)	*S. epidermidis, S. aureus,* diphtheroids	The conjunctiva, a continuation of the skin or mucous membrane, contains basically the same microbiota found on the skin.
Nose and throat (upper respiratory system)	*S. aureus, S. epidermidis,* and aerobic diphtheroids in the nose; *S. epidermidis, S. aureus,* diphtheroids, *Streptococcus pneumoniae, Haemophilus,* and *Neisseria* in the throat	Although some normal microbiota are potential pathogens, their ability to cause disease is reduced by microbial antagonism.
Mouth	Various species of *Streptococcus, Lactobacillus, Actinomyces, Bacteroides, Fusobacterium, Treponema, Corynebacterium,* and *Candida* (fungus)	Abundant moisture, warmth, and the constant presence of food make the mouth an ideal environment that supports very large and diverse microbial populations on the tongue, cheeks, teeth, and gums.
Large intestine	*Bacteroides, Fusobacterium, Lactobacillus, Enterococcus, Bifidobacterium, Escherichia coli, Enterobacter, Citrobacter, Proteus, Klebsiella, Candida* (fungus)	The large intestine contains the largest numbers of resident microbiota in the body because of its available moisture and nutrients.
Urogenital system	*Staphylococcus epidermidis,* aerobic micrococci, *Enterococcus, Lactobacillus,* aerobic diphtheroids, *Pseudomonas, Klebsiella,* and *Proteus* in urethra; lactobacilli, aerobic diphtheroids, *Streptococcus, Staphylococcus, Bacteroides, Clostridium, Candida albicans* (fungus), and *Trichomonas vaginalis* (protozoan) in vagina	The lower urethra in both sexes has a resident population; the vagina has its acid-tolerant population of microbes because of the nature of its secretions.

*This table lists some microbiota that are not discussed in this chapter. Unless indicated, the organisms are bacteria.

the microorganisms that make up our normal microbiota are commensals; these include the corynebacteria that inhabit the surface of the eye and certain saprophytic mycobacteria that inhabit the ear and external genitals. These bacteria live on secretions and sloughed-off cells, and they bring no apparent benefit or harm to the host.

Mutualism is a type of symbiosis that benefits both organisms. For example, the large intestine contains bacteria, such as *E. coli*, that synthesize vitamin K and some B vitamins. These vitamins are absorbed into the bloodstream and distributed for use by body cells. In exchange, the large intestine provides nutrients used by the bacteria, resulting in their survival. In still another kind of symbiosis, one organism is benefited at the expense of the other; this relationship is called **parasitism.** Many disease-causing bacteria are parasites.

Opportunistic Microorganisms

Learning Objective

- Contrast normal and transient microbiota with opportunistic microorganisms.

Although categorizing symbiotic relationships by type is convenient, keep in mind that under certain conditions the relationship can change. For example, given the proper circumstances, a mutualistic organism, such as *E. coli*, can become harmful. *E. coli* is generally harmless as long as it remains in the large intestine; but if it gains access to other body sites, such as the urinary bladder, lungs, spinal cord, or wounds, it may cause urinary tract infections, pulmonary infections, meningitis, or abscesses, respectively. Microbes such as *E. coli* are called **opportunistic pathogens**. They ordinarily do not cause disease

APPLICATIONS OF MICROBIOLOGY

Bacteria Help the Body

There is a very large and diverse microbial population inhabiting the human body. This normal microbiota includes about 100 trillion bacteria. In fact, all animals beyond the embryo stage are hosts for a myriad of microbes. Impressed by this fact, Pasteur speculated that microbes are essential for animal life.

We now know that animals with no microbiota whatsoever can be reared in the laboratory. Such germfree animals are very useful for studying the functions of microorganisms in the body, as well as the interactions among microorganisms in the body. (Germfree animals that are colonized with known microorganisms are called gnotobiotic.)

Germfree mammals can be obtained by surgically removing the fetus from the mother's uterus under aseptic conditions and maintaining it in a sterile environment. It is also possible to breed germfree animals in a sterile environment, and today most germfree mammals used in research are obtained in that way.

On the one hand, research with germfree animals has shown that microbes are not absolutely essential to animal life. On the other hand, this research has also shown that germfree animals are at a number of disadvantages compared with normal animals. Germfree animals have undeveloped immune systems; they have little lymphoid tissue, and the levels of immune proteins in their body fluids are very low. When they do come in contact with microorganisms, these animals are unusually susceptible to infection and serious disease. In addition, germfree animals have higher requirements for certain nutrients than other animals, in particular for vitamins normally produced by intestinal bacteria. Clearly, normal

microbiota benefit the host both directly and indirectly.

One hundred years ago, the microbiologist who discovered phagocytes, Elie Metchnikoff, believed that aging and even death were due to poisoning from intestinal bacteria. He believed that eating a diet rich in sour milk could add 100 years to one's life. He isolated the lactic acid bacteria (LAB) responsible for fermenting milk, producing yogurt. Several people got rich selling his "Bulgarian bacillus" to health-conscious consumers. Recently, a renewed interest in normal microbiota has led to the study of probiotics. Probiotics (*pro* = for, *bios* = life) are live microbial cultures applied to or ingested into the body that are intended to exert a beneficial effect on health. Probiotics may be administered with prebiotics, which are chemicals that selectively promote the growth of beneficial bacteria.

Normal bacteria can prevent the growth of pathogens by competitive exclusion. The large numbers of normal bacteria compete for attachment sites and nutrients and may stimulate the immune response. Antibiotics taken for a nonintestinal infection can kill normal intestinal microbiota, leading to diarrhea. Moreover, *Salmonella enterica* can colonize more easily in the absence of normal bacteria. Several studies have shown that ingestion of certain LAB such as *Lactobacillus acidophilus* and *Bifidobacterium bifidum* can alleviate diarrhea and prevent colonization by *Salmonella enterica* during antibiotic therapy. If these LAB colonize the large intestine, the lactic acid and bacteriocins they produce can inhibit the growth of certain pathogenic bacteria. Researchers are also investigating whether recombinant LAB can be used to deliver vaccines, antimi-

crobials, and digestive enzymes. Many therapeutic proteins cannot be taken orally because they are destroyed by stomach acid. However, the LAB can survive passage through the stomach to deliver molecules for adsorption in the intestine.

Staphylococcus aureus is a common cause of surgical wound infections. In one study, researchers tested whether probiotic LAB could prevent infection by *S. aureus*. They inoculated surgical wounds on nine (control) mice with *S. aureus*; on nine other mice, they inoculated surgical wounds with both *S. aureus* and LAB. None of the animals inoculated with LAB developed infections, whereas five of the control animals developed staphylococcal infections.

One study at Baylor University tested the use of a probiotic *E. coli* inoculated into the urinary bladders of 44 patients susceptible to recurrent bladder infections. These patients had an average of two bladder infections per year. The *E. coli* colonized in 30 patients for more than a year, during which only two bladder infections occurred.

Vaginal infections occur when the normal LAB population in the vagina is disrupted. In one preliminary study, the use of vaginal suppositories containing LAB from healthy women restored the normal vaginal bacteria.

A recent study done at the Institute of Food Research in the United Kingdom raised a question about the potential for transfer of virulence genes from a few pathogens in one's intestine to the 10 billion probiotic cells that a person might consume. Probiotics need to be investigated further—for their potential therapeutic value as well as for risks associated with their use.

in their normal habitat in a healthy person but may do so in a different environment. For example, microbes that gain access through broken skin or mucous membranes can cause opportunistic infections. Or, if the host is already weakened or compromised by infection, microbes

that are usually harmless can cause disease. AIDS is often accompanied by a common opportunistic infection, *Pneumocystis* pneumonia, caused by the opportunistic organism *Pneumocystis jiroveci* (formerly *Pneumocystis carinii*). This secondary infection can develop in AIDS patients because their immune systems are suppressed. Before the AIDS epidemic, this type of pneumonia was rare. Opportunistic pathogens possess other features that contribute to their ability to cause disease. For example, they are present in or on the body or in the external environment in relatively large numbers. In addition, some opportunistic pathogens may be found in locations in or on the body that are somewhat protected from the body's defenses. Also, some opportunistic pathogens are resistant to antibiotics.

In addition to the usual symbionts, many people carry other microorganisms that are generally regarded as pathogenic but that may not cause disease in those people. Among the pathogens that are frequently carried in healthy individuals are echoviruses (*echo* comes from *enteric cytopathogenic human orphan*), which can cause intestinal diseases; and adenoviruses, which can cause respiratory diseases. *Neisseria meningitidis*, which often resides benignly in the respiratory tract, can cause meningitis, a disease that inflames the coverings of the brain and spinal cord. *Streptococcus pneumoniae*, a normal resident of the nose and throat, can cause a type of pneumonia.

Cooperation Among Microorganisms

In addition to competition among microbes in causing disease, there are situations in which cooperation among microbes is a factor in causing disease. One example of cooperation among microbes in the development of disease involves oral streptococci that colonize the teeth. Pathogens that cause periodontal disease and gingivitis have been found to have receptors for the streptococci, rather than for the teeth.

The Etiology of Infectious Diseases

Learning Objective
• List Koch's postulates.

Some diseases—such as polio, Lyme disease, and tuberculosis—have a well-known etiology. Some have an etiology that is not completely understood—for example, the relationship between ulcers and *Helicobacter pylori*. For still others, such as Alzheimer's disease, the etiology is unknown. Of course, not all diseases are caused by microorganisms. For example, the disease hemophilia is an *inherited (genetic) disease*; osteoarthritis and cirrhosis are considered *degenerative diseases*. There are several other categories of disease, but here we will discuss only *infectious diseases*, those caused by microorganisms. To see how microbiologists determine the etiology of an infectious disease, we will discuss in greater detail the work of Robert Koch, which was introduced in Chapter 1 on page 9.

Koch's Postulates

In the historical overview of microbiology presented in Chapter 1, we briefly discussed Koch's famous postulates. Recall that Koch was a German physician who played a major role in establishing that microorganisms cause specific diseases. In 1877, he published some early papers on anthrax, a disease of cattle that can also occur in humans. Koch demonstrated that certain bacteria, today known as *Bacillus anthracis*, were always present in the blood of animals that had the disease and were not present in healthy animals. He knew that the mere presence of the bacteria did not prove that they had caused the disease; the bacteria could have been there as a result of the disease. Thus, he experimented further.

He took a sample of blood from a diseased animal and injected it into a healthy one. The second animal developed the same disease and died. He repeated this procedure many times, always with the same results. (A key criterion in the validity of any scientific proof is that experimental results be repeatable.) Koch also cultivated the microorganism in fluids outside the animal's body, and he demonstrated that the bacterium would cause anthrax even after many culture transfers.

Koch showed that a specific infectious disease (anthrax) is caused by a specific microorganism (*B. anthracis*) that can be isolated and cultured on artificial media. He later used the same methods to show that the bacterium *Mycobacterium tuberculosis* is the causative agent of tuberculosis.

Koch's research provides a framework for the study of the etiology of any infectious disease. Today, we refer to Koch's experimental requirements as **Koch's postulates** (Figure 14.3). They are summarized as follows:

1. The same pathogen must be present in every case of the disease.

2. The pathogen must be isolated from the diseased host and grown in pure culture.

3. The pathogen from the pure culture must cause the disease when it is inoculated into a healthy, susceptible laboratory animal.

4. The pathogen must be isolated from the inoculated animal and must be shown to be the original organism.

Exceptions to Koch's Postulates

Although Koch's postulates are useful in determining the causative agent of most bacterial diseases, there are some

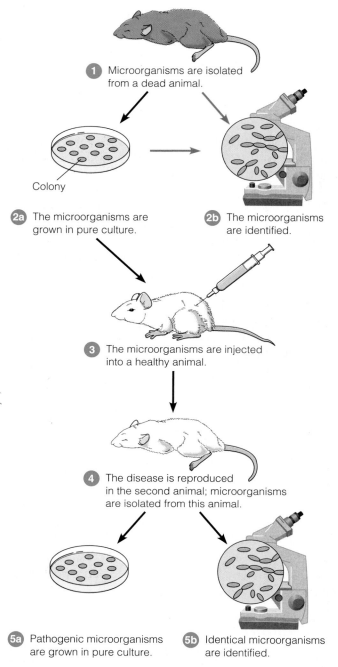

1 Microorganisms are isolated from a dead animal.

Colony

2a The microorganisms are grown in pure culture.

2b The microorganisms are identified.

3 The microorganisms are injected into a healthy animal.

4 The disease is reproduced in the second animal; microorganisms are isolated from this animal.

5a Pathogenic microorganisms are grown in pure culture.

5b Identical microorganisms are identified.

FIGURE 14.3 **Application of Koch's postulates.**

■ Koch's postulates provide a framework for the study of the etiology of any infectious disease.

exceptions. For example, some microbes have unique culture requirements. The bacterium *Treponema pallidum* is known to cause syphilis, but virulent strains have never been cultured on artificial media. The causative agent of leprosy, *Mycobacterium leprae*, has also never been grown on artificial media. Moreover, many rickettsial and viral pathogens cannot be cultured on artificial media because they multiply only within cells.

The discovery of microorganisms that cannot grow on artificial media has necessitated some modifications of Koch's postulates and the use of alternative methods of culturing and detecting certain microbes. For example, when researchers looking for the microbial cause of legionellosis (Legionnaires' disease) were unable to isolate the microbe directly from a victim, they took the alternative step of inoculating a victim's lung tissue into guinea pigs. These guinea pigs developed the disease's pneumonialike symptoms, whereas guinea pigs inoculated with tissue from an unafflicted person did not. Then tissue samples from the diseased guinea pigs were cultured in yolk sacs of chick embryos, a method (see Figure 13.7, page 385) that reveals the growth of extremely small microbes. After the embryos were incubated, electron microscopy revealed rod-shaped bacteria in the chick embryos. Finally, modern immunological techniques (which will be discussed in Chapter 18) were used to show that the bacteria in the chick embryos were the same bacteria as those in the guinea pigs and in afflicted humans.

In a number of situations, a human host exhibits certain signs and symptoms that are associated only with a certain pathogen and its disease. For example, the pathogens responsible for diphtheria and tetanus cause distinguishing signs and symptoms (described shortly) that can be produced by no other microbe. They are unequivocally the only organisms that produce their respective diseases. But some infectious diseases are not as clear-cut and provide another exception to Koch's postulates. For example, nephritis (inflammation of the kidneys) can involve any of several different pathogens, all of which cause the same signs and symptoms. Thus, it is often difficult to know which particular microorganism is causing a disease. Other infectious diseases that sometimes have poorly defined etiologies are pneumonia, meningitis, and peritonitis (inflammation of the peritoneum, the membrane that lines the abdomen and pelvis and covers the organs within them).

Still another exception to Koch's postulates results because some pathogens can cause several disease conditions. *Mycobacterium tuberculosis*, for example, is implicated in diseases of the lungs, skin, bones, and internal organs. *Streptococcus pyogenes* can cause sore throat, scarlet fever, skin infections (such as erysipelas), and osteomyelitis (inflammation of bone), among other diseases. When clinical signs and symptoms are used together with laboratory methods, these infections can usually be distinguished from infections of the same organs by other pathogens.

Ethical considerations may also impose an exception to Koch's postulates. For example, some agents that cause disease in humans have no other known host. An example is human immunodeficiency virus (HIV), the cause of AIDS. This poses the ethical question of whether humans can be intentionally inoculated with infectious agents. In 1721, King George I told condemned prisoners they could be inoculated with smallpox to test a smallpox vaccine (see Chapter 18). He promised their freedom if they lived. Human experiments with untreatable diseases are not acceptable today. Sometimes accidental inoculation does occur. A contaminated bone marrow transplant provided the third Koch's postulate to prove that a herpesvirus caused cancer.

Classifying Infectious Diseases

Learning Objective

- *Differentiate a communicable from a noncommunicable disease.*

Every disease that affects the body alters body structures and functions in particular ways, and these alterations are usually indicated by several kinds of evidence. For example, the patient may experience certain **symptoms,** or changes in body function, such as pain and *malaise* (a vague feeling of body discomfort). These *subjective* changes are not apparent to an observer. The patient can also exhibit **signs,** which are *objective* changes the physician can observe and measure. Frequently evaluated signs include lesions (changes produced in tissues by disease), swelling, fever, and paralysis. A specific group of symptoms or signs may always accompany a particular disease; such a group is called a **syndrome.** The diagnosis of a disease is made by evaluation of the signs and symptoms, together with the results of laboratory tests.

Diseases are often classified in terms of how they behave within a host and within a given population. Any disease that spreads from one host to another, either directly or indirectly, is said to be a **communicable disease.** Chickenpox, measles, genital herpes, typhoid fever, and tuberculosis are examples. Chickenpox and measles are also examples of **contagious diseases,** that is, diseases that are *easily* spread from one person to another. A **noncommunicable disease** is not spread from one host to another. These diseases are caused by microorganisms that normally inhabit the body and only occasionally produce disease, or by microorganisms that reside outside the body and produce disease only when introduced into the body. An example is tetanus: *Clostridium tetani* produces disease only when it is introduced into the body via abrasions or wounds.

Occurrence of a Disease

Learning Objective

- *Categorize diseases according to frequency of occurrence.*

To understand the full scope of a disease, we should know something about its occurrence. The **incidence** of a disease is the number of people in a population who develop a disease during a particular time period. It is an indicator of the spread of the disease. The **prevalence** of a disease is the number of people in a population who develop a disease at a specified time, regardless of when it first appeared. Prevalence takes into account both old and new cases. It's an indicator of how seriously and how long a disease affects a population. For example, the incidence of AIDS in the United States in 1999 was 45,000, while the prevalence in that same year was estimated to be about 700,000. Knowing the incidence and the prevalence of a disease in different populations (for example, in populations representing different geographic regions or different racial groups) enables scientists to estimate the range of the disease's occurrence and its tendency to affect some groups of people more than others.

Frequency of occurrence is another criterion that is used in the classification of diseases. If a particular disease occurs only occasionally, it is called a **sporadic disease;** typhoid fever in the United States is such a disease. A disease constantly present in a population is called an **endemic disease;** an example of such a disease is the common cold. If many people in a given area acquire a certain disease in a relatively short period, it is called an **epidemic disease;** influenza is an example of a disease that often achieves epidemic status. Figure 14.4 shows the epidemic incidence of AIDS in the United States. Some authorities consider gonorrhea and certain other sexually transmitted diseases to be epidemic at this time as well. An epidemic disease that occurs worldwide is called a **pandemic disease.** We experience pandemics of influenza from time to time. Some authorities also consider AIDS to be pandemic.

Severity or Duration of a Disease

Learning Objectives

- *Categorize diseases according to severity.*
- *Define herd immunity.*

Another useful way of defining the scope of a disease is in terms of its severity or duration. An **acute disease** is one that develops rapidly but lasts only a short time; a good example is influenza. A **chronic disease** develops more slowly, and the body's reactions may be less severe, but the

FIGURE 14.4 **Reported AIDS cases in the United States.** Notice that the first 250,000 cases occurred over a 12-year period, whereas the second and third 250,000 cases in this epidemic occurred in just 3 and 5 years, respectively. Much of the increase shown for 1993 is due to an expanded definition of AIDS cases adopted in that year.
SOURCE: CDC.

■ Distinguish among sporadic, endemic, epidemic, and pandemic diseases.

disease is likely to be continual or recurrent for long periods. Infectious mononucleosis, tuberculosis, and hepatitis B fall into this category. A disease that is intermediate between acute and chronic is described as a **subacute disease;** an example is subacute sclerosing panencephalitis, a rare brain disease characterized by diminished intellectual function and loss of nervous function. A **latent disease** is one in which the causative agent remains inactive for a time but then becomes active to produce symptoms of the disease; an example is shingles, one of the diseases caused by varicella-zoster virus.

The rate at which a disease, or an epidemic, spreads, and the number of individuals involved, are determined in part by the immunity of the population. Vaccination can provide long-lasting and sometimes lifelong protection of an individual against certain diseases. People who are immune to an infectious disease will not be carriers, thereby reducing the occurrence of the disease. Immune individuals act as a barrier to the spread of infectious agents. Even though a highly communicable disease may cause an epidemic, many nonimmune people will be protected because of the unlikelihood of their coming into contact with an infected person. A great advantage of vaccination is that enough individuals in a population will be protected from a disease to prevent its rapid spread to those in the population who are not vaccinated. When many immune people are present in a community, **herd immunity** exists.

Extent of Host Involvement

Infections can also be classified according to the extent to which the host's body is affected. A **local infection** is one in which the invading microorganisms are limited to a relatively small area of the body. Some examples of local infections are boils and abscesses. In a **systemic (generalized) infection,** microorganisms or their products are spread throughout the body by the blood or lymph. Measles is an example of a systemic infection. Very often, agents of a local infection enter a blood or lymphatic vessel and spread to other specific parts of the body, where they are confined to specific areas of the body. This condition is called a **focal infection.** Focal infections can arise from infections in areas such as the teeth, tonsils, or sinuses.

The presence of bacteria in the blood is known as **bacteremia,** and if the bacteria actually multiply in the blood, the condition is called **sepsis. Toxemia** refers to the presence of toxins in blood (as occurs in tetanus), and **viremia** refers to the presence of viruses in blood.

The state of host resistance also determines the extent of infections. A **primary infection** is an acute infection that causes the initial illness. A **secondary infection** is one caused by an opportunistic pathogen after the primary infection has weakened the body's defenses. Secondary infections of the skin and respiratory tract are common and are sometimes more dangerous than the primary infections. *Pneumocystis* pneumonia as a consequence of AIDS is an example of a secondary infection; streptococcal bronchopneumonia following influenza is an example of a secondary infection that is more serious than the primary infection. A **subclinical (inapparent) infection** is one that does not cause any noticeable illness. Poliovirus and hepatitis A virus, for example, can be carried by people who never develop the illness.

Patterns of Disease

A definite sequence of events usually occurs during infection and disease. As you will learn shortly, there must be a reservoir of infection as a source of pathogens for an infectious disease to occur. Next, the pathogen must be transmitted to a susceptible host by direct contact, by indirect contact, or by vectors. Transmission is followed by invasion, in which the microorganism enters the host and multiplies. Following invasion, the microorganism injures the host through a process called pathogenesis (discussed further in the next chapter). The extent of injury depends on the degree to which host cells are damaged, either directly or by toxins. Despite the effects of all these factors, the occurrence of disease ultimately depends on the resistance of the host to the activities of the pathogen.

Predisposing Factors

Learning Objective

- *Identify four predisposing factors for disease.*

Certain predisposing factors also affect the occurrence of disease. A **predisposing factor** is one that makes the body more susceptible to a disease and may alter the course of the disease. Gender is sometimes a predisposing factor; for example, females have a higher incidence of urinary tract infections than males, whereas males have higher rates of pneumonia and meningitis. Other aspects of genetic background may play a role as well. For example, sickle-cell disease is a severe, life-threatening form of anemia that occurs when the genes for the disease are inherited from both parents. Individuals who carry only one sickle-cell gene have a condition called sickle-cell trait, and are normal unless specially tested. However, they are relatively resistant to the most serious form of malaria. The potential that individuals in a population might inherit a life-threatening disease is more than counterbalanced by protection from malaria among carriers of the gene for sickle-cell trait. Of course, in countries where malaria is not present, sickle-cell trait is an entirely negative condition.

Climate and weather seem to have some effect on the incidence of infectious diseases. In temperate regions, the incidence of respiratory diseases increases during the winter. This increase may be related to the fact that when

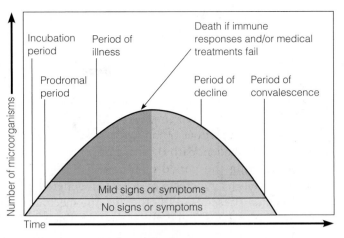

FIGURE 14.5 **The stages of a disease.**

■ During which period can a disease be transmitted?

people stay indoors, the closer contact with one other facilitates the spread of respiratory pathogens.

Other predisposing factors include inadequate nutrition, fatigue, age, environment, habits, lifestyle, occupation, preexisting illness, chemotherapy, and emotional disturbances. It is often difficult to know the exact relative importance of the various predisposing factors.

Development of Disease

Learning Objective

- *Put the following in proper sequence, according to the pattern of disease: period of decline, period of convalescence, period of illness, prodromal period, incubation period.*

Once a microorganism overcomes the defenses of the host, development of the disease follows a certain sequence that tends to be similar whether the disease is acute or chronic (Figure 14.5).

Incubation Period

The **incubation period** is the time interval between the initial infection and the first appearance of any signs or symptoms. In some diseases, the incubation period is always the same; in others, it is quite variable. The time of incubation depends on the specific microorganism involved, its virulence (degree of pathogenicity), the number of infecting microorganisms, and the resistance of the host. (See Table 15.1 on page 439 for the incubation periods of a number of microbial diseases.)

Prodromal Period

The **prodromal period** is a relatively short period that follows the period of incubation in some diseases. The prodromal period is characterized by early, mild symptoms of disease, such as general aches and malaise.

Period of Illness

During the period of illness, the disease is most acute. The person exhibits overt signs and symptoms of disease, such as fever, chills, muscle pain (myalgia), sensitivity to light (photophobia), sore throat (pharyngitis), lymph node enlargement (lymphadenopathy), and gastrointestinal disturbances. During the period of illness, the number of white blood cells may increase or decrease. Generally, the patient's immune response and other defense mechanisms overcome the pathogen, and the period of illness ends. When the disease is not successfully overcome (or successfully treated), the patient dies during this period.

Period of Decline

During the period of decline, the signs and symptoms subside. The fever decreases, and the feeling of malaise diminishes. During this phase, which may take from less than 24 hours to several days, the patient is vulnerable to secondary infections.

Period of Convalescence

During the **period of convalescence,** the person regains strength and the body returns to its prediseased state. Recovery has occurred.

We all know that during the period of illness, people can serve as reservoirs of disease and can easily spread infections to other people. However, you should also know that people can spread infection during incubation and convalescence as well. This is especially true of diseases such as typhoid fever and cholera, in which the convalescing person carries the pathogenic microorganism for months or even years.

The Spread of Infection

Now that you have an understanding of normal microbiota, the etiology of infectious diseases, and the types of infectious diseases, we will examine the sources of pathogens and how diseases are transmitted.

Reservoirs of Infection

Learning Objectives

- *Define reservoir of infection.*
- *Contrast human, animal, and nonliving reservoirs, and give one example of each.*

For a disease to perpetuate itself, there must be a continual source of the disease organisms. This source can be either a living organism or an inanimate object that provides a pathogen with adequate conditions for survival and multiplication and an opportunity for transmission. Such a source is called a **reservoir of infection.** These reservoirs may be human, animal, or nonliving.

Human Reservoirs

The principal living reservoir of human disease is the human body itself. Many people harbor pathogens and transmit them directly or indirectly to others. People with signs and symptoms of a disease may transmit the disease; in addition, some people can harbor pathogens and transmit them to others without exhibiting any signs of illness. These people, called **carriers,** are important living reservoirs of infection. Some carriers have inapparent infections for which no signs or symptoms are ever exhibited. Other people, such as those with latent diseases, carry a disease during its symptom-free stages—during the incubation period (before symptoms appear) or during the convalescent period (recovery). Human carriers play an important role in the spread of such diseases as AIDS, diphtheria, typhoid fever, hepatitis, gonorrhea, amoebic dysentery, and streptococcal infections.

Animal Reservoirs

Both wild and domestic animals are living reservoirs of microorganisms that can cause human diseases. Diseases that occur primarily in wild and domestic animals and can be transmitted to humans are called **zoonoses** (singular: **zoonosis**). Rabies (found in bats, skunks, foxes, dogs, and cats), and Lyme disease (found in field mice) are examples of zoonoses. Other representative zoonoses are presented in Table 14.2.

About 150 zoonoses are known. The transmission of zoonoses to humans can occur via one of many routes: by direct contact with infected animals; by direct contact with domestic pet waste (such as cleaning a litter box or bird cage); by contamination of food and water; by air from contaminated hides, fur, or feathers; by consuming infected animal products; or by insect vectors (insects that transmit pathogens).

Nonliving Reservoirs

The two major nonliving reservoirs of infectious disease are soil and water. Soil harbors such pathogens as fungi, which cause mycoses including ringworm and systemic infections; *Clostridium botulinum*, the bacterium that causes botulism; and *C. tetani*, the bacterium that causes tetanus. Because both species of clostridia are part of the normal intestinal microbiota of horses and cattle, the bacteria are found especially in soil where animal feces are used as fertilizer.

TABLE 14.2	**Selected Zoonoses**

Disease	Causative Agent	Reservoir	Method of Transmission
Viral			
Influenza (some types)	*Influenzavirus*	Swine, waterfowl	Direct contact
Rabies	*Lyssavirus*	Bats, skunks, foxes, dogs, cats	Direct contact (bite)
Western equine encephalitis	*Alphavirus*	Horses, birds	*Culex* mosquito bite
Hantavirus pulmonary syndrome (HPS)	*Hantavirus*	Rodents (primarily deer mice)	Direct contact with rodent saliva, feces, or urine
Bacterial			
Anthrax	*Bacillus anthracis*	Domestic livestock	Direct contact with contaminated hides or animals; air; food
Brucellosis	*Brucella* spp.	Domestic livestock	Direct contact with contaminated milk, meat, or animals
Bubonic plague	*Yersinia pestis*	Rodents	Flea bites
Cat-scratch disease	*Bartonella henselae*	Domestic cats	Direct contact
Ehrlichiosis	*Ehrlichia* spp.	Deer, rodents	Tick bite
Leptospirosis	*Leptospira*	Wild mammals, domestic dogs and cats	Direct contact with urine, soil, water
Lyme disease	*Borrelia burgdorferi*	Field mice	Tick bites
Psittacosis (ornithosis)	*Chlamydia psittaci*	Birds, especially parrots	Direct contact
Rocky Mountain spotted fever	*Rickettsia rickettsii*	Rodents	Tick bites
Salmonellosis	*Salmonella* spp.	Poultry, rats, turtles, reptiles	Ingestion of contaminated food and water and putting hands in mouth
Endemic typhus	*Rickettsia typhi*	Rodents	Flea bites
Fungal			
Ringworm	*Trichophyton Microsporum Epidermophyton*	Domestic mammals	Direct contact; fomites (nonliving objects)
Protozoan			
Malaria	*Plasmodium* spp.	Monkeys	*Anopheles* mosquito bite
Toxoplasmosis	*Toxoplasma gondii*	Cats and other mammals	Ingestion of contaminated meat or by direct contact with infected tissues or fecal matter
Helminthic			
Tapeworm (pork)	*Taenia solium*	Pigs	Ingestion of undercooked contaminated pork
Trichinosis	*Trichinella spiralis*	Pigs, bears	Ingestion of undercooked contaminated pork

(a)

(b)

FIGURE 14.6 **Means of disease transmission.**
(a) Contact, **(b)** vehicles, and **(c)** vectors.

■ What is a reservoir of infection? What is a disease vector?

(c)

Water that has been contaminated by the feces of humans and other animals is a reservoir for several pathogens, notably those responsible for gastrointestinal diseases. These include *Vibrio cholerae*, which causes cholera, and *Salmonella typhi*, which causes typhoid fever. Other nonliving reservoirs include foods that are improperly prepared or stored. They may be sources of diseases such as trichinosis and salmonellosis.

Transmission of Disease

Learning Objective

* *Explain three methods of disease transmission.*

The causative agents of disease can be transmitted from the reservoir of infection to a susceptible host by three principal routes: contact, vehicles, and vectors (Figure 14.6).

Contact Transmission

Contact transmission is the spread of an agent of disease by direct contact, indirect contact, or droplet transmission. **Direct contact transmission,** also known as *person-to-person transmission*, is the direct transmission of an agent by physical contact between its source and a susceptible host; no intermediate object is involved (Figure 14.6a). The most common forms of direct contact transmission are touching, kissing, and sexual intercourse. Among the diseases that can be transmitted by direct contact are viral respiratory tract diseases (the common cold and influenza), staphylococcal infections, hepatitis A, measles, scarlet fever, and sexually transmitted diseases (syphilis, gonorrhea, and genital herpes). Direct contact is also one way of spreading AIDS and infectious mononucleosis. To guard against person-to-person transmission, health care workers use gloves and other protective measures (Figure 14.7). Potential pathogens can also be transmitted by direct contact from animals (or animal products) to humans. Examples are the pathogens causing rabies and anthrax.

Indirect contact transmission occurs when the agent of disease is transmitted from its reservoir to a susceptible host by means of a nonliving object. The general term for any nonliving object involved in the spread of an infection is a **fomite.** Examples of fomites are tissues, handkerchiefs, towels, bedding, diapers, drinking cups, eating utensils, toys, money, and thermometers. Contaminated syringes serve as fomites in the transmission of AIDS and hepatitis B. Other fomites may transmit diseases such as tetanus.

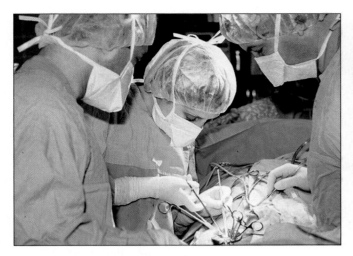

FIGURE 14.7 **Preventing direct contact transmission.** Health care workers use masks, gloves, and other protective measures to safeguard against the transmission of pathogens.

■ How do direct contact, indirect contact, and droplet transmission differ?

FIGURE 14.8 **Droplet transmission.** This high-speed photograph shows the spray of small droplets that come from the mouth during a sneeze.

■ In droplet transmission, microbes are spread in mucus droplets that are discharged into the air by coughing, sneezing, laughing, or talking and travel less than 1 meter from reservoir to host.

Droplet transmission is a third type of contact transmission in which microbes are spread in *droplet nuclei* (mucus droplets) that travel only short distances (Figure 14.8). These droplets are discharged into the air by coughing, sneezing, laughing, or talking and travel less than 1 meter from the reservoir to the host. In one sneeze, 20,000 droplets may be produced. Disease agents that travel such short distances are not regarded as airborne (airborne transmission is discussed shortly). Examples of diseases spread by droplet transmission are influenza, pneumonia, and pertussis (whooping cough).

Vehicle Transmission

Vehicle transmission is the transmission of disease agents by a medium, such as water, food, or air (Figure 14.6b). Other media include blood and other body fluids, drugs, and intravenous fluids. Here we will discuss water, food, and air as vehicles of transmission.

In *waterborne transmission*, pathogens are usually spread by water contaminated with untreated or poorly treated sewage. Diseases transmitted via this route include cholera, waterborne shigellosis, and leptospirosis. In *foodborne transmission*, pathogens are generally transmitted in foods that are incompletely cooked, poorly refrigerated, or prepared under unsanitary conditions. Foodborne pathogens cause diseases such as food poisoning and tapeworm infestation.

Airborne transmission refers to the spread of agents of infection by droplet nuclei in dust that travel more than 1 meter from the reservoir to the host. For example, microbes are spread by droplets, which may be discharged in a fine spray from the mouth and nose during coughing and sneezing (see Figure 14.8). These droplets are small enough to remain airborne for prolonged periods. The virus that causes measles and the bacterium that causes tuberculosis can be transmitted via airborne droplets. Dust particles can harbor various pathogens. Staphylococci and streptococci can survive on dust and be transmitted by the airborne route. Spores produced by certain fungi are also transmitted by the airborne route and can cause such diseases as histoplasmosis, coccidioidomycosis, and blastomycosis.

Vectors

Arthropods are the most important group of disease **vectors**—animals that carry pathogens from one host to another. (Insects and other arthropod vectors are discussed in Chapter 12 on page 368.) Arthropod vectors transmit disease by two general methods. **Mechanical transmission** is the passive transport of the pathogens on the insect's feet or other body parts (Figure 14.6c). If the insect makes contact with a host's food, pathogens can be transferred to the food and later swallowed by the host. Houseflies, for instance, can transfer the pathogens of typhoid fever and bacillary dysentery (shigellosis) from the feces of infected people to food.

TABLE 14.3	Representative Arthropod Vectors and the Diseases They Transmit	
Disease	**Causative Agent**	**Arthropod Vector**
Malaria	*Plasmodium* spp.	*Anopheles* (mosquito)
African trypanosomiasis	*Trypanosoma brucei gambiense* and *T. b. rhodesiense*	*Glossina* sp. (tsetse fly)
American trypanosomiasis	*T. cruzi*	*Triatoma* sp. (kissing bug)
Yellow fever	*Alphavirus* (yellow fever virus)	*Aedes* (mosquito)
Dengue	*Alphavirus* (dengue fever virus)	*A. aegypti* (mosquito)
Arthropod-borne encephalitis	*Alphavirus* (encephalitis virus)	*Culex* (mosquito)
Ehrlichiosis	*Ehrlichia* spp.	*Ixodes* spp. (tick)
Epidemic typhus	*Rickettsia prowazekii*	*Pediculus humanus* (louse)
Endemic murine typhus	*R. typhi*	*Xenopsylla cheopis* (rat flea)
Rocky Mountain spotted fever	*R. rickettsii*	*Dermacentor andersoni* and other species (tick)
Plague	*Yersinia pestis*	*Xenopsylla cheopis* (rat flea)
Relapsing fever	*Borrelia* spp.	*Ornithodorus* spp. (soft tick)
Lyme disease	*B. burgdorferi*	*Ixodes* spp. (tick)

Biological transmission is an active process and is more complex. The arthropod bites an infected person or animal and ingests some of the infected blood. The pathogens then reproduce in the vector, and the increase in the number of pathogens increases the possibility that they will be transmitted to another host. Some parasites reproduce in the gut of the arthropod; these can be passed with feces. If the arthropod defecates or vomits while biting a potential host, the parasite can enter the wound. Other parasites reproduce in the vector's gut and migrate to the salivary gland; these are directly injected into a bite. Some protozoan and helminthic parasites use the vector as a host for a developmental stage in their life cycle.

Table 14.3 lists a few important arthropod vectors and the diseases they transmit.

Nosocomial (Hospital-Acquired) Infections

Learning Objective

• *Define nosocomial infections, and explain their importance.*

A **nosocomial infection** is one that does not show any evidence of being present or incubating at the time of admission to a hospital; it is acquired as a result of a hospital stay. (The word *nosocomial* is derived from the Greek word for hospital; the term also includes infections acquired in nursing homes and other health care facilities.) The Centers for Disease Control and Prevention (CDC) estimates that 5–15% of all hospital patients acquire some type of nosocomial infection. The work of pioneers in aseptic techniques such as Lister and Semmelweis (Chapter 1, page 9) decreased the rate of nosocomial infections considerably. However, despite modern advances in sterilization techniques and disposable materials, the rate of nosocomial infections has increased 36% during the last 20 years. In the United States, about 2 million people per year contract nosocomial infections, and over 100,000 die as a result. Nosocomial infections represent the fourth leading cause of death in the United States behind heart disease, cancer, and strokes.

Nosocomial infections result from the interaction of several factors: (1) microorganisms in the hospital environment, (2) the compromised (or weakened) status of the host, and (3) the chain of transmission in the hospital. Figure 14.9 illustrates that the presence of any one of these factors alone is generally not enough to cause infection; it is the interaction of all three factors that poses a significant risk of nosocomial infection.

Microorganisms in the Hospital

Although every effort is made to kill or check the growth of microorganisms in the hospital, the hospital environment

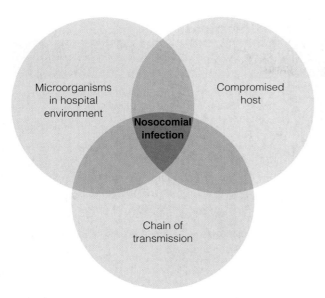

FIGURE 14.9 **Nosocomial infections.**

■ The interaction of microorganisms in the hospital environment, a compromised (weakened) host, and the chain of transmission in the hospital contributes to nosocomial infections.

is a major reservoir for a variety of pathogens. One reason is that certain normal microbiota of the human body are opportunistic and present a particularly strong danger to hospital patients. In fact, most of the microbes that cause nosocomial infections do not cause disease in healthy people but are pathogenic only for individuals whose defenses have been weakened by illness or therapy (see the box in Chapter 7, page 200).

In the 1940s and 1950s, most nosocomial infections were caused by gram-positive microbes. At one time, the gram-positive *Staphylococcus aureus* was the primary cause of nosocomial infections. In the 1970s, gram-negative rods, such as *E. coli* and *Pseudomonas aeruginosa*, were the most common causes of nosocomial infections. During the 1980s, antibiotic-resistant gram-positive bacteria, *Staphylococcus aureus*, coagulase-negative staphylococci, and *Enterococcus* spp., emerged as nosocomial pathogens. By the 1990s, these gram-positive bacteria accounted for 34% of nosocomial infections and four gram-negative pathogens accounted for 32%. In the 2000s, antibiotic resistance in nosocomial infections is a major concern. The principal microorganisms involved in nosocomial infections are summarized in Table 14.4.

In addition to being opportunistic, some microorganisms in the hospital become resistant to antimicrobial drugs, which are commonly used there. For example, *P. aeruginosa* and other such gram-negative bacteria tend to be difficult to control with antibiotics because of their

R factors, which carry genes that determine resistance to antibiotics (see Chapter 8, page 241). As the R factors recombine, new and multiple resistance factors are produced. These strains become part of the microbiota of patients and hospital personnel and become progressively more resistant to antibiotic therapy. In this way, people become part of the reservoir (and chain of transmission) for antibiotic-resistant strains of bacteria. Usually, if the host's resistance is high, the new strains are not much of a problem. But if disease, surgery, or trauma has weakened the host's defenses, secondary infections may be difficult to treat.

Compromised Host

Learning Objective

• *Define compromised host.*

A **compromised host** is one whose resistance to infection is impaired by disease, therapy, or burns. Two principal conditions can compromise the host: broken skin or mucous membranes, and a suppressed immune system.

As long as the skin and mucous membranes remain intact, they provide formidable physical barriers against most pathogens. Burns, surgical wounds, trauma (such as accidental wounds), injections, invasive diagnostic procedures, ventilators, intravenous therapy, and urinary catheters (used to drain urine) can all break the first line of defense and make a person more susceptible to disease in hospitals. Burn patients are especially susceptible to nosocomial infections because their skin is no longer an effective barrier to microorganisms.

The risk of infection is also related to other invasive procedures, such as administering anesthesia, which may alter breathing and contribute to pneumonia, and tracheotomy, in which an incision is made into the trachea to assist breathing. Patients who require invasive procedures usually have a serious underlying disease, which further increases susceptibility to infections. Invasive devices provide a pathway for microorganisms in the environment to enter the body; they also help transfer microbes from one part of the body to another. Pathogens can also proliferate on the devices themselves.

In healthy individuals, white blood cells called T lymphocytes (T cells) provide resistance to disease by killing pathogens directly, mobilizing phagocytes and other lymphocytes, and secreting chemicals that kill pathogens. White blood cells called B lymphocytes (B cells), which develop into antibody-producing cells, also protect against infection. Antibodies provide immunity by such actions as neutralizing toxins, inhibiting the attachment of a pathogen to host cells, and helping to lyse pathogens.

TABLE 14.4	**Microorganisms Involved in Most Nosocomial Infections**			
Microorganism	**Percentage of Total Infections**	**Percentage Resistant to Antibiotics**		**Infections Caused**
Staphylococcus aureus, coagulase-negative staphylococci, enterococci	34%	25–87%		Surgical wound infections, pneumonia, sepsis, urinary tract infections
Escherichia coli, Pseudomonas aeruginosa, Enterobacter spp., and *Klebsiella pneumoniae*	32%	3–34%		Pneumonia and surgical wound infections
Clostridium difficile	17%			Causes nearly half of all nosocomial diarrhea
Fungi (mostly *Candida albicans*)	10%			Urinary tract infections and sepsis
Other gram-negative bacteria (*Acinetobacter, Citrobacter, Haemophilus*)	7%			Urinary tract infections and surgical wound infections

DATA SOURCE: CDC, National Nosocomial Infections Surveillance.

Drugs, radiation therapy, steroid therapy, burns, diabetes, leukemia, kidney disease, stress, and malnutrition can all adversely affect the actions of T and B lymphocytes and compromise the host. In addition, the AIDS virus destroys certain T cells.

A summary of the principal sites of nosocomial infections is presented in Figure 14.10 and Table 14.5.

Chain of Transmission

Given the variety of pathogens (and potential pathogens) in the hospital and the compromised state of the host, routes of transmission are a constant concern. The principal routes of transmission of nosocomial infections are direct contact transmission from hospital staff to patient and from patient to patient, and indirect contact transmission through fomites and the hospital's ventilation system (airborne transmission).

Because hospital personnel are in direct contact with patients, they can often transmit disease. For example, a physician or nurse may transmit microbiota to a patient when changing a dressing, or a kitchen worker who carries *Salmonella* can contaminate a food supply.

Certain areas of a hospital are reserved for specialized care; these include the burn, hemodialysis, recovery, intensive care, and oncology units. Unfortunately, these units also group patients together and provide environments for the epidemic spread of nosocomial infections from patient to patient.

Many diagnostic and therapeutic hospital procedures provide a fomite route of transmission. The urinary catheter used to drain urine from the urinary bladder is a fomite in many nosocomial infections. Intravenous catheters, which pass through the skin and into a vein to provide fluids, nutrients, or medication, can also transmit nosocomial infections. Respiratory aids can introduce contaminated fluids into the lungs. Needles may introduce pathogens into muscle or blood, and surgical dressings can become contaminated and promote disease.

Control of Nosocomial Infections

Learning Objectives

- *List several methods of disease transmission in hospitals.*
- *Explain how nosocomial infections can be prevented.*

Control measures aimed at preventing nosocomial infections vary from one institution to another, but certain procedures are generally implemented. It is important to reduce the number of pathogens to which patients are exposed by using aseptic techniques, handling contaminated

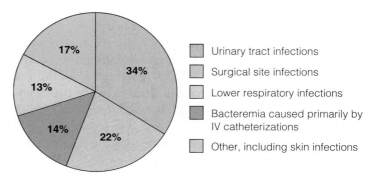

FIGURE 14.10 Relative frequency of nosocomial infections.

- Urinary tract infections
- Surgical site infections
- Lower respiratory infections
- Bacteremia caused primarily by IV catheterizations
- Other, including skin infections

■ Urinary tract infections account for most nosocomial infections.

TABLE 14.5	**Principal Sites of Nosocomial Infections**
Type of Infection	**Comment**
Urinary tract infections	Most common, usually accounts for about 34% of all nosocomial infections. Typically related to urinary catheterization.
Surgical site infections	Ranks second in infection incidence (about 17%). An estimated 5–12% of all surgical patients develop postoperative infections; the percentage can reach 30% for certain surgeries, such as colon surgery and amputations.
Bacteremia, caused primarily by intravenous catheterizations	Bacteremias rank third in incidence and account for about 14% of nosocomial infections. Intravenous catheterization is implicated in nosocomial infections of the bloodstream, particularly infections caused by bacteria and fungi.
Lower respiratory infections	Nosocomial pneumonias account for about 13% and have high mortality rates (13–55%). Most of these pneumonias are related to respiratory devices that aid breathing or administer medications.
Cutaneous infections	Among the least common of all nosocomial infections. However, newborns have a high rate of susceptibility to skin and eye infections.

DATA SOURCE: CDC, National Nosocomial Infection Surveillance.

materials carefully, insisting on frequent and thorough hand washing, educating staff members about basic infection control measures, and using isolation rooms and wards.

According to the CDC, hand washing is the single most important means of preventing the spread of infection. Nevertheless, in 1997, a CDC report revealed that in one long-term care facility, health care workers washed their hands before interacting with patients only 27% of the time. And in 1996, researchers identified hand washing rates as low as 31% in one emergency department.

In addition to hand washing, tubs used to bathe patients should be disinfected between uses so that bacteria from the previous patient will not contaminate the next one. Respirators and humidifiers provide both a suitable growth environment for some bacteria and a method of airborne transmission. These sources of nosocomial infections must be kept scrupulously clean and disinfected, and materials used for bandages and intubation (insertion of tubes into organs, such as the trachea) should be single-use disposable or sterilized before use. Packaging used to maintain sterility should be removed aseptically. Physicians can help improve patients' resistance to infection by prescribing antibiotics only when necessary, avoiding invasive procedures if possible, and minimizing the use of immunosuppressive drugs.

Accredited hospitals should have an infection control committee. Most hospitals have at least an infection control nurse or epidemiologist (an individual who studies disease in populations). The role of these staff members is to identify problem sources, such as antibiotic-resistant strains of bacteria and improper sterilization techniques. The infection control officer should make periodic examinations of hospital equipment to determine the extent of microbial contamination. Samples should be taken from tubing, catheters, respirator reservoirs, and other equipment.

Emerging Infectious Diseases

Learning Objective

- List several probable reasons for emerging infectious diseases, and name one example for each reason.

As noted in Chapter 1, **emerging infectious diseases (EIDs)** are ones that are new or changing, showing an increase in incidence in the recent past, or a potential to increase in the near future. An emerging disease can be caused by a virus, a bacterium, a fungus, a protozoan, or a helminth. Several criteria are used for identifying an EID. For example, some diseases present symptoms that are clearly distinctive from all other diseases. Some are recognized because improved diagnostic techniques allow the identification of a new pathogen. Others are identified when a local disease becomes widespread, a rare disease becomes common, a mild disease becomes more severe, or an increase in life span permits a slow disease to develop. Examples of emerging infectious diseases are listed in Table 14.6.

A variety of factors contribute to the emergence of new infectious diseases:

- A new serovar, such as *Vibrio cholerae* O139, may result from changes in or the evolution of existing microorganisms.

- The widespread, and sometimes unwarranted, use of antibiotics and pesticides encourages the growth of more resistant populations of microbes and the insects (mosquitoes and lice) and ticks that carry them.

TABLE 14.6	**Emerging Infectious Diseases**	
Microorganism	**Year of Emergence**	**Disease Caused**
Bacteria		
Bacillus anthracis	2001	Anthrax
Streptococcus pneumoniae	1995	Antibiotic-resistant pneumonia
Streptococcus pyogenes	1995	Streptococcal toxic shock syndrome
Corynebacterium diphtheriae	1994	Diphtheria epidemic, eastern Europe
Vibrio cholerae O139	1992	New serovar of cholera, Asia
Ehrlichia chaffeenis	1986	Human monocytic ehrlichiosis (tickborne)
Bartonella henselae	1983	Cat-scratch disease
Escherichia coli O157:H7	1982	Hemorrhagic diarrhea
Staphylococcus aureus	1978	Toxic shock syndrome
Legionella pneumophila	1976	Legionnaires' disease
Borrelia burgdorferi	1975	Lyme disease
Fungi		
Coccidioides immitis	1993	Coccidioidomycosis
Pneumocystis jiroveci	1981	Pneumonia in immunocompromised patients
Protozoa		
Plasmodium spp.	1986	Malaria in United States; 40% of the world's population is at risk; insecticide-resistant mosquitoes and drug-resistant protozoa
Cyclospora cayetanensis	1993	Severe diarrhea and wasting syndrome
Cryptosporidium parvum	1976	Cryptosporidiosis
Helminths		
Baylisascaris procyonis	2001	Raccoon roundworm encephalitis, in humans
Viruses		
Ebola virus	2002, 1995, 1975	Ebola hemorrhagic fever
West Nile virus	1999	West Nile encephalitis
Nipah virus	1998	Encephalitis, Malaysia
Hendra virus	1994	Encephalitis-like symptoms, Australia
Hantavirus	1993	Hantavirus pulmonary syndrome
Venezuelan hemorrhagic fever	1991	Hemorrhagic fever, South America
Hepatitis E virus	1990	Hepatitis
Hepatitis C virus	1989	Hepatitis
Dengue fever virus	1984	Dengue fever and dengue hemorrhagic fever, South and Central America and the Caribbean
HIV	1983	AIDS
Prions		
Bovine spongiform encephalitis agent	1996	Mad cow disease, Great Britain

- Global warming and changes in weather patterns may increase the distribution and survival of reservoirs and vectors, resulting in the introduction and dissemination of diseases such as malaria, and *Hantavirus* pulmonary syndrome disease.

- Known diseases, such as cholera and West Nile virus, may spread to new geographic areas by modern transportation. This was less likely 100 years ago when travel took so long that infected travelers either died or recovered during passage.

- Previously unrecognized infections may appear in individuals living or working in regions undergoing ecological changes brought about by natural disaster, construction, wars, and expanding human settlement. In California, the incidence of coccidioidomycosis increased tenfold following the Northridge earthquake of 1994. Workers clearing South American forests are now contracting Venezuelan hemorrhagic fever.

- Even animal control measures may affect the incidence of a disease. The increase in Lyme disease in recent years could be due to rising deer populations resulting from the killing of deer predators.

- Failures in public health measures may be a contributing factor to the emergence of previously controlled infections. For example, the failure of adults to get a diphtheria booster vaccination led to a diphtheria epidemic in the newly independent republics of the former Soviet Union in the 1990s.

The CDC, the National Institutes of Health (NIH), and the World Health Organization (WHO) have developed plans to address issues relating to emerging infectious diseases. Their priorities include the following:

1. To detect, promptly investigate, and monitor emerging infectious pathogens, the diseases they cause, and factors that influence their emergence

2. To expand basic and applied research on ecological and environmental factors, microbial changes and adaptations, and host interactions that influence EIDs

3. To enhance the communication of public health information and the prompt implementation of prevention strategies regarding EIDs

4. To establish plans to monitor and control EIDs worldwide

The importance of emerging infectious diseases to the scientific community resulted in a new publication, *Emerging Infectious Diseases,* devoted exclusively to the topic. It was first published in January 1995. In addition, as a result of a global focus of attention, January 1996 was proclaimed Emerging Infection Month, in which 36 scientific journals worldwide documented the occurrence, causes, and consequences of emerging and reemerging infectious diseases.

Epidemiology

Learning Objective

- *Define epidemiology, and describe three types of epidemiologic investigations.*

In today's crowded, overpopulated world, in which frequent travel and the mass production and distribution of food and other goods are a way of life, diseases can spread rapidly. A contaminated food or water supply, for example, can affect many thousands of people very quickly. Identifying the causative agent is desirable so that a disease can be effectively controlled and treated. It is also desirable to understand the mode of transmission and geographical distribution of the disease. The science that studies when and where diseases occur and how they are transmitted in populations is called **epidemiology.**

Modern epidemiology began in the mid-1800s with three now-famous investigations. John Snow, a British physician, conducted a series of investigations related to outbreaks of cholera in London. As the cholera epidemic of 1848–1849 raged, Snow analyzed the death records attributed to cholera, gathered information about the victims, and interviewed survivors who lived in the neighborhood. Using the information he compiled, Snow made a map showing that most individuals who died of cholera drank or brought water from the Broad Street pump; those who used other pumps (or drank beer, like the workers at a nearby brewery) did not get cholera. He concluded that contaminated water from the Broad Street pump was the source of the epidemic. When the pump's handle was removed and people could no longer get water from this location, the number of cholera cases dropped significantly.

Between 1846 and 1848, Ignaz Semmelweis meticulously recorded the number of births and maternal deaths at Vienna General Hospital. The First Maternity Clinic had become a source of gossip throughout Vienna because the death rate due to puerperal sepsis ranged between 13% and 18%, four times that of the Second Maternity Clinic. Puerperal sepsis (childbirth fever) is a nosocomial infection that begins in the uterus as a result of childbirth or abortion. It is frequently caused by *Streptococcus pyogenes.* The infection progresses to the abdominal cavity (peritonitis) and in many cases to sepsis (proliferation of microbes in the blood). Wealthy women did not go to the clinic, and poor women had learned they had a better chance of surviving childbirth if they gave birth elsewhere before going to the hospital. Looking at his data, Semmelweis realized that wealthy women and the poor women who had given birth prior to entering the clinic were not examined by the medical students, who had spent their mornings dissecting cadavers. In May 1847, he ordered all medical students to wash their hands before entering the delivery room, and the mortality rate dropped to under 2%.

Florence Nightingale recorded statistics on epidemic typhus in the English civilian and military populations. In 1858, she published a thousand-page report using statistical comparisons to demonstrate that diseases, poor food, and unsanitary conditions were killing the soldiers. Her work resulted in reforms in the British Army and to her admission to the Statistical Society, their first female member.

These three careful analyses of where and when a disease occurs and how it is transmitted within a population constituted a new approach to medical research and demonstrated the importance of epidemiology. The works of Snow, Semmelweis, and Nightingale resulted in changes that lowered the incidence of diseases even though there was limited knowledge of the causes of infectious disease. Most physicians believed that the symptoms they saw were the causes of the disease, not the result of disease. Koch's work on the germ theory of disease was still 30 years in the future.

An epidemiologist not only determines the etiology of a disease but also identifies other possibly important factors and patterns concerning the people affected. An important part of the epidemiologist's work is assembling and analyzing such data as age, sex, occupation, personal habits, socioeconomic status, history of immunization, presence of any other diseases, and the common history of affected individuals (such as eating the same food or visiting the same doctor's office). Also important for the prevention of future outbreaks is knowledge of the site at which a susceptible host came into contact with the agent of infection. In addition, the epidemiologist considers the period during which the disease occurs, either on a seasonal basis (to indicate whether the disease is prevalent during the summer or winter) or on a yearly basis (to indicate the effects of immunization or an emerging or reemerging disease).

An epidemiologist is also concerned with various methods for controlling a disease. The strategies controlling diseases include the use of drugs (chemotherapy) and vaccines (immunization). Other methods include the control of human, animal, and nonliving reservoirs of infection, water treatment, proper sewage disposal (enteric diseases), cold storage, pasteurization, food inspection, and adequate cooking (foodborne diseases), improved nutrition to bolster host defenses, changes in personal habits, and screening of transfused blood and transplanted organs.

Figure 14.11 contains graphs indicating the incidence of selected diseases. Such graphs provide information about whether disease outbreaks are sporadic or epidemic,

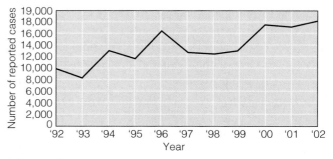

(a) Lyme disease cases, 1992–2002

(b) Lyme disease by month, 2002

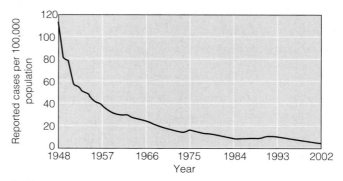

(c) Reported tuberculosis cases, 1948–2002

FIGURE 14.11 **Epidemiological graphs. (a)** A graph of Lyme disease cases, showing the annual occurrence of the disease during the covered period. **(b)** This bar graph represents Lyme disease from a different perspective, enabling epidemiologists to begin drawing some conclusions about the disease. Over a period of years, researchers saw that the disease was transmitted by a vector that was active during summer and fall. This graph records the number of cases per 100,000 people, rather than the total number of cases. **(c)** A graph of the incidence of tuberculosis shows a rapid decrease in the rate of decline from 1948 to 1957. SOURCE: CDC.

■ Graphs representing the incidence of diseases indicate whether disease outbreaks are sporadic or epidemic, how a disease might spread, and how effectively a disease is controlled.

and, if epidemic, how the disease might have spread. By establishing the frequency of a disease in a population and identifying the factors responsible for its transmission, an epidemiologist can provide physicians with information that is important in determining the prognosis and treatment of a disease. Epidemiologists also evaluate how effectively a disease is being controlled in a community—by a vaccination program, for example. Finally, epidemiologists can provide data to help in evaluating and planning overall health care for a community.

Epidemiologists use three basic types of investigations when analyzing the occurrence of a disease: descriptive, analytical, and experimental.

Descriptive Epidemiology

Descriptive epidemiology entails collecting all data that describe the occurrence of the disease under study. Relevant information usually includes information about the affected individuals and the place and period in which the disease occurred. Snow's search for the cause of the cholera outbreak in London is an example of descriptive epidemiology.

Such a study is generally *retrospective* (looking backward after the episode has ended). In other words, the epidemiologist backtracks to the cause and source of the disease. The search for the cause of toxic shock syndrome is an example of a fairly recent retrospective study. In the initial phase of an epidemiological study, retrospective studies are more common than *prospective* (looking forward) studies, in which an epidemiologist chooses a group of people who are free of a particular disease to study. The group's subsequent disease experiences are then recorded for a given period. Prospective studies were used to test the Salk polio vaccine in 1954 and 1955.

Analytical Epidemiology

Analytical epidemiology analyzes a particular disease to determine its probable cause. This study can be done in two ways. With the *case control method,* the epidemiologist looks for factors that might have preceded the disease. A group of people who have the disease is compared with another group of people who are free of the disease. For example, one group with meningitis and one without the disease might be matched by age, sex, socioeconomic status, and location. These statistics are compared to determine which of all the possible factors—genetic, environmental, nutritional, and so forth—might be responsible for the meningitis. Nightingale's work was an example of analytical epidemiology, in which she compared disease in soldiers and civilians. With the *cohort method,* the epidemiologist studies two populations: one that has had contact with the agent causing a disease and another

that has not (both groups are called *cohort groups*). For example, a comparison of one group composed of people who have received blood transfusions and one composed of people who have not could reveal an association between blood transfusions and the incidence of hepatitis B virus.

Experimental Epidemiology

Experimental epidemiology begins with a hypothesis about a particular disease; experiments to test the hypothesis are then conducted with a group of people. One such hypothesis could be the assumed effectiveness of a drug. A group of infected individuals is selected and divided randomly so that some receive the drug and others receive a placebo, a substance that has no effect. If all other factors are kept constant between the two groups, and if those people who received the drug recover more rapidly than those who received the placebo, it can be concluded that the drug was the experimental factor (variable) that made the difference.

Case Reporting

We noted earlier in this chapter that establishing the chain of transmission for a disease is extremely important. Once known, the chain can be interrupted in order to slow down or stop the spread of the disease.

An effective way to establish the chain of transmission is *case reporting,* a procedure that requires health care workers to report specified diseases to local, state, and national health officials. Examples of such diseases are AIDS, measles, gonorrhea, tetanus, and typhoid fever. Case reporting provides epidemiologists with an approximation of the incidence and prevalence of a disease. This information helps officials decide whether or not to investigate a given disease.

Case reporting provided epidemiologists with valuable leads regarding the origin and spread of AIDS. In fact, one of the first clues about AIDS came from reports of young males with Kaposi's sarcoma, formerly a disease of older males. Using these reports, epidemiologists began various studies of the patients. If an epidemiological study shows that a large enough segment of the population is affected by a disease, an attempt is then made to isolate and identify its causative agent. Identification is accomplished by a number of different microbiological methods. Identification of the causative agent often provides valuable information regarding the reservoir for the disease.

Once the chain of transmission is discovered, it is possible to apply control measures to stop the disease from spreading. These might include elimination of the source of infection, isolation and segregation of infected people, the development of vaccines, and, as in the case of AIDS, education.

TABLE 14.7	**Nationally Notifiable Diseases, 2002**
Acquired immunodeficiency syndrome (AIDS)	Meningococcal disease
Anthrax	Mumps
Botulism	Pertussis
Brucellosis	Plague
Chancroid	Poliomyelitis, paralytic
Chlamydia trachomatis, genital infections	Psittacosis
Cholera	Q fever
Coccidioidomycosis	Rabies, animal
Cryptosporidiosis	Rabies, human
Cyclosporiasis	Rocky Mountain spotted fever
Diphtheria	Rubella
Ehrlichiosis	Rubella, congenital syndrome
Encephalitis, arboviral	Salmonellosis
Enterohemorrhagic *Escherichia coli*	Shigellosis
Giardiasis	Streptococcal disease, invasive, group A
Gonorrhea	Streptococcal toxic shock syndrome
Haemophilus influenzae, invasive	*Streptococcus pneumoniae,* drug-resistant
Hansen's disease (Leprosy)	*S. pneumoniae,* invasive, children
Hantavirus pulmonary syndrome	Syphilis
Hemolytic uremic syndrome, postdiarrheal	Syphilis, congenital
Hepatitis A	Tetanus
Hepatitis B	Toxic shock syndrome
Hepatitis C (non-A, non-B hepatitis)	Trichinosis
HIV infection	Tuberculosis
Legionellosis	Tularemia
Listeriosis	Typhoid fever
Lyme disease	Varicella (deaths only)
Malaria	Yellow fever
Measles	

The Centers for Disease Control and Prevention (CDC)

Learning Objectives

- Identify the function of the CDC.
- Define the following terms: morbidity, mortality, and notifiable disease.

Epidemiology is a major concern of state and federal public health departments. The **Centers for Disease Control and Prevention (CDC),** a branch of the U.S. Public Health Service located in Atlanta, Georgia, is a central source of epidemiological information in the United States.

The CDC issues a publication called the *Morbidity and Mortality Weekly Report* (www.cdc.gov). The **MMWR,** as it is called, is read by microbiologists, physicians, and other hospital and public health professionals. The MMWR contains data on **morbidity,** the incidence of specific notifiable diseases, and **mortality,** the number of deaths from these diseases. These data are usually organized by state. **Notifiable diseases,** shown in Table 14.7, are those for which physicians are required by law to report cases to the U.S. Public Health Service. As of 2002, a total of 57 infectious diseases were reported at the national level. **Morbidity rate** is the number of people affected by a disease in a given period of time in relation to the total population. **Mortality rate** is the number of deaths resulting from a disease in a population in a given period of time in relation to the total population.

MMWR articles include reports of disease outbreaks, case histories of special interest, and summaries of the status of particular diseases during a recent period. These articles often include recommendations for procedures for

diagnosis, immunization, and treatment. Several graphs and other data in this textbook are from the MMWR, and the case history boxes are adapted from reports from this publication.

* * *

In the next chapter, we consider the mechanisms of pathogenicity. We will discuss in more detail the methods by which microorganisms enter the body and cause disease, the effects of disease on the body, and the means by which pathogens leave the body.

STUDY OUTLINE

Introduction (p. 408)

1. Disease-causing microorganisms are called pathogens.
2. Pathogenic microorganisms have special properties that allow them to invade the human body or produce toxins.
3. When a microorganism overcomes the body's defenses, a state of disease results.

Pathology, Infection, and Disease (p. 409)

1. Pathology is the scientific study of disease.
2. Pathology is concerned with the etiology (cause), pathogenesis (development), and effects of disease.
3. Infection is the invasion and growth of pathogens in the body.
4. A host is an organism that shelters and supports the growth of pathogens.
5. Disease is an abnormal state in which part or all of the body is not properly adjusted or is incapable of performing normal functions.

Normal Microbiota (pp. 409–413)

1. Animals, including humans, are usually germ-free in utero.
2. Microorganisms begin colonization in and on the surface of the body soon after birth.
3. Microorganisms that establish permanent colonies inside or on the body without producing disease make up the normal microbiota.
4. Transient microbiota are microbes that are present for various periods and then disappear.

Relationships Between the Normal Microbiota and the Host (pp. 409–411)

1. The normal microbiota can prevent pathogens from causing an infection; this phenomenon is known as microbial antagonism.

2. Normal microbiota and the host exist in symbiosis (living together).
3. The three types of symbiosis are commensalism (one organism benefits and the other is unaffected), mutualism (both organisms benefit), and parasitism (one organism benefits and one is harmed).

Opportunistic Microorganisms (pp. 411–413)

1. Opportunistic pathogens do not cause disease under normal conditions but cause disease under special conditions.

Cooperation Among Microorganisms (p. 413)

1. In some situations, one microorganism makes it possible for another to cause a disease or produce more severe symptoms.

The Etiology of Infectious Diseases (pp. 413–415)

Koch's Postulates (p. 413)

1. Koch's postulates are criteria for establishing that specific microbes cause specific diseases.
2. Koch's postulates have the following requirements: (a) the same pathogen must be present in every case of the disease; (b) the pathogen must be isolated in pure culture; (c) the pathogen isolated from pure culture must cause the same disease in a healthy, susceptible laboratory animal; and (d) the pathogen must be reisolated from the inoculated laboratory animal.

Exceptions to Koch's Postulates (pp. 413–415)

1. Koch's postulates are modified to establish etiologies of diseases caused by viruses and some bacteria, which cannot be grown on artificial media.
2. Some diseases, such as tetanus, have unequivocal signs and symptoms.

3. Some diseases, such as pneumonia and nephritis, may be caused by a variety of microbes.

4. Some pathogens, such as *S. pyogenes,* cause several different diseases.

5. Certain pathogens, such as HIV, cause disease in humans only.

Classifying Infectious Diseases (pp. 415–417)

1. A patient may exhibit symptoms (subjective changes in body functions) and signs (measurable changes), which a physician uses to make a diagnosis (identification of the disease).

2. A specific group of symptoms or signs that always accompanies a specific disease is called a syndrome.

3. Communicable diseases are transmitted directly or indirectly from one host to another.

4. A contagious disease is one that is easily spread from one person to another.

5. Noncommunicable diseases are caused by microorganisms that normally grow outside the human body and are not transmitted from one host to another.

Occurrence of a Disease (p. 415)

1. Disease occurrence is reported by incidence (number of people contracting the disease) and prevalence (number of cases at a particular time).

2. Diseases are classified by frequency of occurrence: sporadic, endemic, epidemic, and pandemic.

Severity or Duration of a Disease (pp. 415–416)

1. The scope of a disease can be defined as acute, chronic, subacute, or latent.

2. Herd immunity is the presence of immunity to a disease in most of the population.

Extent of Host Involvement (pp. 416–417)

1. A local infection affects a small area of the body; a systemic infection is spread throughout the body via the circulatory system.

2. A secondary infection can occur after the host is weakened from a primary infection.

3. An inapparent, or subclinical, infection does not cause any signs of disease in the host.

Patterns of Disease (pp. 417–418)

Predisposing Factors (p. 417)

1. A predisposing factor is one that makes the body more susceptible to disease or alters the course of a disease.

2. Examples include gender, climate, age, fatigue, and inadequate nutrition.

Development of Disease (pp. 417–418)

1. The incubation period is the time interval between the initial infection and the first appearance of signs and symptoms.

2. The prodromal period is characterized by the appearance of the first mild signs and symptoms.

3. During the period of illness, the disease is at its height, and all disease signs and symptoms are apparent.

4. During the period of decline, the signs and symptoms subside.

5. During the period of convalescence, the body returns to its prediseased state, and health is restored.

The Spread of Infection (pp. 418–422)

Reservoirs of Infection (pp. 418–420)

1. A continual source of infection is called a reservoir of infection.

2. People who have a disease or are carriers of pathogenic microorganisms are human reservoirs of infection.

3. Zoonoses are diseases that affect wild and domestic animals and can be transmitted to humans.

4. Some pathogenic microorganisms grow in nonliving reservoirs, such as soil and water.

Transmission of Disease (pp. 420–422)

1. Transmission by direct contact involves close physical contact between the source of the disease and a susceptible host.

2. Transmission by fomites (inanimate objects) constitutes indirect contact.

3. Transmission via saliva or mucus in coughing or sneezing is called droplet transmission.

4. Transmission by a medium such as water, food, or air is called vehicle transmission.

5. Airborne transmission refers to pathogens carried on water droplets or dust for a distance greater than 1 meter.

6. Arthropod vectors carry pathogens from one host to another by both mechanical and biological transmission.

Nosocomial (Hospital-Acquired) Infections (pp. 422–425)

1. A nosocomial infection is any infection that is acquired during the course of stay in a hospital, nursing home, or other health care facility.

2. About 5–15% of all hospitalized patients acquire nosocomial infections.

Microorganisms in the Hospital (pp. 422–423)

1. Certain normal microbiota are often responsible for nosocomial infections when they are introduced into the body through such medical procedures as surgery and catheterization.

2. Opportunistic, drug-resistant gram-negative bacteria are the most frequent causes of nosocomial infections.

Compromised Host (pp. 423–424)

1. Patients with burns, surgical wounds, and suppressed immune systems are the most susceptible to nosocomial infections.

Chain of Transmission (p. 424)

1. Nosocomial infections are transmitted by direct contact between staff members and patients and between patients.

2. Fomites such as catheters, syringes, and respiratory devices can transmit nosocomial infections.

Control of Nosocomial Infections (pp. 424–425)

1. Aseptic techniques can prevent nosocomial infections.

2. Hospital infection control staff members are responsible for overseeing the proper cleaning, storage, and handling of equipment and supplies.

Emerging Infectious Diseases (pp. 425–427)

1. New diseases and diseases with increasing incidences are called emerging infectious diseases (EIDs).

2. EIDs can result from the use of antibiotics and pesticides, climatic changes, travel, the lack of vaccinations, and improved case reporting.

3. The CDC, NIH, and WHO are responsible for surveillance and responses to emerging infectious diseases.

Epidemiology (pp. 427–431)

1. The science of epidemiology is the study of the transmission, incidence, and frequency of disease.

2. Modern epidemiology began in in the mid-1800s with the works of Snow, Semmelweis, and Nightingale.

3. Data about infected people are collected and analyzed in descriptive epidemiology.

4. In analytical epidemiology, a group of infected people is compared with an uninfected group.

5. Controlled experiments designed to test hypotheses are performed in experimental epidemiology.

6. Case reporting provides data on incidence and prevalence to local, state, and national health officials.

7. The Centers for Disease Control and Prevention (CDC) is the main source of epidemiologic information in the United States.

8. The CDC publishes the *Morbidity and Mortality Weekly Report* to provide information on morbidity (incidence) and mortality (deaths).

STUDY QUESTIONS

Review

1. Differentiate the terms in each of the following pairs:
a. etiology and pathogenesis
b. infection and disease
c. communicable disease and noncommunicable disease

2. What is meant by normal microbiota? How do they differ from transient microbiota?

3. Define symbiosis. Differentiate among commensalism, mutualism, and parasitism, and give an example of each.

4. What is a reservoir of infection? Match the following diseases with their reservoirs:
_____ influenza   **a.** nonliving
_____ rabies   **b.** human
_____ botulism   **c.** animal

5. Describe how Koch's postulates establish the etiology of many infectious diseases. Why don't Koch's postulates apply to all infectious diseases?

6. Describe the various ways diseases can be transmitted in each of the following categories. Name one disease transmitted by each method.
a. transmission by direct contact
b. transmission by indirect contact
c. transmission by arthropod vectors
d. droplet transmission
e. vehicle transmission
f. airborne transmission

7. List four predisposing factors to disease.

8. Indicate whether each of the following conditions is typical of subacute, chronic, or acute infections.
a. The patient experiences a rapid onset of malaise; symptoms last 5 days.
b. The patient experiences cough and breathing difficulty for months.
c. The patient has no apparent symptoms and is a known carrier.

9. Of all the hospital patients with infections, one-third do not enter the hospital with an infection. How do they acquire these infections? What is the method of transmission of these infections? What is the reservoir of infection?

10. Differentiate between endemic and epidemic states of infectious disease.

11. What is epidemiology? What is the role of the Centers for Disease Control and Prevention (CDC)?

12. Distinguish between symptoms and signs as signals of disease.

13. How can a local infection become a systemic infection?

14. Why are some organisms that constitute the normal microbiota described as commensals, whereas others are described as mutualistic?

15. Put the following in the correct order to describe the pattern of disease: period of convalescence, prodromal period, period of decline, incubation period, period of illness.

Multiple Choice

1. The emergence of new infectious diseases is probably due to all of the following *except*
   a. the need of bacteria to cause disease.
   b. the ability of humans to travel by air.
   c. changing environments (e.g., flood, drought, pollution).
   d. a pathogen crossing the species barrier.
   e. the increasing human population.

2. All members of a group of ornithologists studying barn owls in the wild have had salmonellosis (*Salmonella* gastroenteritis). One birder is experiencing her third infection. What is the most likely source of their infections?
   a. The ornithologists are eating the same food.
   b. They are contaminating their hands while handling the owls and nests.
   c. One of the workers is a *Salmonella* carrier.
   d. Their drinking water is contaminated.

3. Which of the following statements is *not* true?
   a. *E. coli* never causes disease.
   b. *E. coli* provides vitamin K for its host.
   c. *E. coli* often exists in a mutual relationship with humans.
   d. *E. coli* gets nutrients from intestinal contents.

4. Which of the following is *not* one of Koch's postulates?
   a. The same pathogen must be present in every case of the disease.
   b. The pathogen must be isolated and grown in pure culture from the diseased host.
   c. The pathogen from pure culture must cause the disease when inoculated into a healthy, susceptible laboratory animal.
   d. The disease must be transmitted from a diseased animal to a healthy, susceptible animal by some form of contact.
   e. The pathogen must be isolated in pure culture from an experimentally infected lab animal.

Use the following information to answer questions 5–7.

On September 6, a 6-year-old boy experienced fever, chills, and vomiting. On September 7, he was hospitalized with diarrhea and swollen lymph nodes under both arms. On September 3, the boy had been scratched and bitten by a cat. The cat was found dead on September 5, and *Yersinia pestis* was isolated from the cat. Chloramphenicol was administered to the boy from September 7, when *Y. pestis* was isolated from him. On September 17, the boy's temperature returned to normal, and on September 22, he was released from the hospital.

5. Identify the incubation period for this case of bubonic plague.
   a. September 3–5.
   b. September 3–6.

   c. September 6–7.
   d. September 6–17.

6. Identify the prodromal period for this disease.
   a. September 3–5.
   b. September 3–6.
   c. September 6–7.
   d. September 6–17.

7. Identify the crisis during this disease.
   a. September 6.
   b. September 7.
   c. September 6–17.
   d. September 17.

Use the following information to answer questions 8–10.

A Maryland woman was hospitalized with dehydration; *Vibrio cholerae* and *Plesiomonas shigelloides* were isolated from the patient. She had neither traveled outside the United States nor eaten raw shellfish during the preceding month. She had attended a party 2 days before her hospitalization. Two other people at the party had acute diarrheal illness and elevated levels of serum antibodies against *Vibrio*. Everyone at the party ate crabs and rice pudding with coconut milk. Crabs left over from this party were served at a second party. One of the 20 people at the second party had onset of mild diarrhea; specimens from 14 of these people were negative for vibriocidal antibodies.

8. This is an example of
   a. vehicle transmission.
   b. airborne transmission.
   c. transmission by fomites.
   d. direct contact transmission.
   e. nosocomial transmission.

9. The etiologic agent of the disease is
   a. *Plesiomonas shigelloides*.
   b. crabs.
   c. *Vibrio cholerae*.
   d. coconut milk.
   e. rice pudding.

10. The source of the disease was
    a. *Plesiomonas shigelloides*.
    b. crabs.
    c. *Vibrio cholerae*.
    d. coconut milk.
    e. rice pudding.

Critical Thinking

1. Ten years before Robert Koch published his work on anthrax, Anton De Bary showed that potato blight was caused by the alga *Phytophthora infestans*. Why do you suppose we use Koch's postulates instead of something called "De Bary's postulates"?

2. Florence Nightingale gathered the following data in 1855.

Population Sampled	Deaths from Contagious Diseases
Englishmen	0.2%
English soldiers (in England)	18.7%
English soldiers (in Crimean War)	42.7%
English soldiers (in Crimean War) after Nightingale's sanitary reforms	2.2%

Discuss how Nightingale used the three basic types of epidemiologic investigation. The contagious diseases were primarily cholera and typhus; how are these diseases transmitted and prevented?

3. Name the method of transmission of each of the following diseases:

   a. malaria
   b. tuberculosis
   c. nosocomial infections
   d. salmonellosis
   e. streptococcal pharyngitis
   f. mononucleosis
   g. measles
   h. hepatitis A
   i. tetanus
   j. hepatitis B
   k. chlamydial urethritis

4. The following graph shows the incidence of typhoid fever in the United States from 1954 through 2002. Mark the graph to show when this disease occurred sporadically and epidemically. What appears to be the endemic level? What would have to be shown to indicate a pandemic of this disease? How is typhoid fever transmitted?

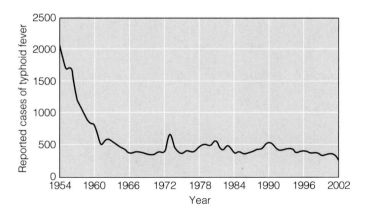

5. Cholera is reemerging in the Western Hemisphere. In 1991, a cholera epidemic caused by *Vibrio cholerae* O1 Inaba was discovered in Peru. The map that follows shows the location of cholera cases in the continental United States caused by *V. cholerae* O1 Inaba (gold), as well as where this strain has been found in coastal waters (blue).

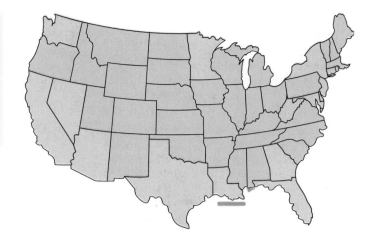

How is cholera transmitted? How did *V. cholerae* O1 Inaba get into the United States? Explain how eight people who did not travel outside the U.S. contracted *V. cholerae* O1 Inaba.

Clinical Applications

1. Three days before a nurse developed meningococcemia, she assisted with intubation of a patient with a *Neisseria meningitidis* infection. Of the 24 medical personnel involved, only this nurse became ill. The nurse recalled that she was exposed to nasopharyngeal secretions and did not receive antibiotic prophylaxis. What two mistakes did the nurse make? How is meningitis transmitted?

2. Three patients in a large hospital acquired infections of *Pseudomonas cepacia* during their stay. All three patients received cryoprecipitate, which is prepared from blood that has been frozen in a standard plastic blood transfer pack. The transfer pack is then placed in a water bath to thaw. What is the probable origin of the nosocomial infections? What characteristics of *Pseudomonas* would allow it to be involved in this type of infection?

3. Following is a case history of a 49-year-old man. Identify each period in the pattern of disease that he experienced. On February 7, he handled a parakeet with a respiratory illness. On March 9, he experienced intense pain in his legs, followed by severe chills and headaches. On March 16, he had chest pains, cough, and diarrhea, and his temperature was 40°C. Appropriate antibiotics were administered on March 17, and his fever subsided within 12 hours. He continued taking antibiotics for 14 days. (*Note:* The disease is psittacosis. Can you find the etiology?)

4. Twenty-one percent of the patients in a large hospital acquired *Clostridium difficile* diarrhea and colitis during their hospital stay. These patients required longer hospital stays than uninfected patients. Epidemiological studies provided

the following information. What is the most likely mode of transmission of this bacterium in hospitals? How can transmission be prevented?

Rate of infection for patients:

Single room	7%
Double room	17%
Triple room	26%

Rate of environmental isolations of C. *difficile*:

Bed rail	10%
Commode	1%
Floor	18%
Call button	6%
Toilet	3%

Hands of hospital personnel after contact with patients that were culture-positive for C. *difficile*:

Used gloves	0%
Did not use gloves	59%
Had C. *difficile* before patient contact	3%
Washed with nondisinfectant soap	40%
Washed with disinfectant soap	3%
Did not wash hands	20%

5. *Mycobacterium avium-intracellulare* is prevalent in AIDS patients. In an effort to determine the source of this infection, hospital water systems were sampled. The water contained chlorine.

Percentage of samples with M. *avium*:

Hot water		Cold water	
February	88%	February	22%
June	50%	June	11%

What is the usual method of transmission for *Mycobacterium*? What is a probable source of infection in hospitals? How can such nosocomial infections be prevented?

LEARNING WITH TECHNOLOGY

The Microbiology Place CD-ROM/website includes many resources to help you study and succeed in this course. In addition to self-quizzes, web links, and flashcards, you'll find the following activities for this chapter in the "Activities" section of The Microbiology Place:

Investigating AIDS and HIV-II

Case Study: An Outbreak of Food Poisoning, San Francisco

Chapter 15

Microbial Mechanisms of Pathogenicity

SALMONELLA. Salmonella *bacteria enter host cells by causing the host's cytoskeleton to form a basket around the bacterial cells.*

Now that you have a basic understanding of how microorganisms cause disease, we will take a look at some of the specific properties of microorganisms that contribute to **pathogenicity,** the ability to cause disease by overcoming the defenses of a host, and **virulence,** the degree or extent of pathogenicity. (As discussed throughout the chapter, the term host usually refers to humans.) Microbes don't try to cause disease; the microbial cells are getting food and defending themselves. The box on page 440 describes the balance between survival of the host and microbe. Keep in mind that many of the properties contributing to microbial pathogenicity and virulence are unclear or unknown. We do know, however, that if the microbe overpowers the host defenses, disease results.

How Microorganisms Enter a Host

To cause disease, most pathogens must gain access to the host, adhere to host tissues, penetrate or evade host defenses, and damage the host tissues. However, some microbes do not cause disease by directly damaging host tissue. Instead, disease is due to the accumulation of microbial waste products. Some microbes, such as those that cause dental caries and acne, can cause disease without penetrating the body. Pathogens can gain entrance to the human body and other hosts through several avenues, which are called **portals of entry.**

Portals of Entry

Learning Objective
- *Identify the principal portals of entry.*

The portals of entry for pathogens are mucous membranes, skin, and direct deposition beneath the skin or membranes (the parenteral route).

Mucous Membranes

Many bacteria and viruses gain access to the body by penetrating mucous membranes lining the respiratory tract, gastrointestinal tract, genitourinary tract, and conjunctiva, a delicate membrane that covers the eyeballs and lines the eyelids. Most pathogens enter through the mucous membranes of the gastrointestinal and respiratory tracts.

The respiratory tract is the easiest and most frequently traveled portal of entry for infectious microorganisms. Microbes are inhaled into the nose or mouth in drops of moisture and dust particles. Diseases that are commonly contracted via the respiratory tract include the common cold, pneumonia, tuberculosis, influenza, measles, and smallpox.

Microorganisms can gain access to the gastrointestinal tract in food and water and via contaminated fingers. (See the box on page 440, which discusses how these and other human behaviors can, over time, influence the virulence of microorganisms.) Most microbes that enter the body in these ways are destroyed by hydrochloric acid (HCl) and enzymes in the stomach, or by bile and enzymes in the small intestine. Those that survive can cause disease. Microbes in the gastrointestinal tract can cause poliomyelitis, hepatitis A, typhoid fever, amoebic dysentery, giardiasis, shigellosis (bacillary dysentery), and cholera. These pathogens are then eliminated with feces and can be transmitted to other hosts via contaminated water, food, or fingers.

The genitourinary tract is a portal of entry for pathogens that are contracted sexually. Some microbes that cause sexually transmitted diseases (STDs) may penetrate an unbroken mucous membrane. Others require a cut or abrasion of some type. Examples of STDs are HIV infection, genital warts, chlamydia, herpes, syphilis, and gonorrhea.

Skin

The skin is one of the largest organs of the body in terms of surface area and is an important defense against disease. Unbroken skin is impenetrable by most microorganisms. Some microbes gain access to the body through openings in the skin, such as hair follicles and sweat gland ducts. In addition, larvae of the hookworm actually bore through intact skin, and some fungi grow on the keratin in skin or infect the skin itself.

The Parenteral Route

Other microorganisms gain access to the body when they are deposited directly into the tissues beneath the skin or into mucous membranes when these barriers are penetrated or injured. This route is called the **parenteral route.** Punctures, injections, bites, cuts, wounds, surgery, and splitting due to swelling or drying can all establish parenteral routes.

The Preferred Portal of Entry

Even after microorganisms have entered the body, they do not necessarily cause disease. The occurrence of disease depends on several factors, only one of which is the portal of entry. Many pathogens have a preferred portal of entry that is a prerequisite to their being able to cause disease. If they gain access to the body by another portal, disease might not occur. For example, the bacteria of typhoid

fever, *Salmonella typhi*, produce all the signs and symptoms of the disease when swallowed (preferred route), but if the same bacteria are rubbed on the skin, no reaction (or only a slight inflammation) occurs. Streptococci that are inhaled (preferred route) can cause pneumonia; those that are swallowed generally do not produce signs or symptoms. Some pathogens, such as *Yersinia pestis*, the microorganism that causes plague, can initiate disease from more than one portal of entry. The preferred portals of entry for some common pathogens are listed in Table 15.1.

Numbers of Invading Microbes

Learning Objective

- Define LD_{50} and ID_{50}.

If only a few microbes enter the body, they will probably be overcome by the host's defenses. However, if large numbers of microbes gain entry, the stage is probably set for disease. Thus, the likelihood of disease increases as the number of pathogens increases.

The virulence of a microbe is often expressed as the **ID_{50}** (infectious dose for 50% of a sample population). The 50 is not an absolute value; rather, it is used to compare relative virulence under experimental conditions. *Bacillus anthracis* can cause infection via three different portals of entry. The ID_{50} through the skin (cutaneous anthrax) is 10–50 endospores; the ID_{50} for inhalation anthrax is inhalation of 10,000–20,000 endospores; and the ID_{50} for gastrointestinal anthrax is ingestion of 250,000–1,000,000 endospores. These data show that cutaneous anthrax is significantly easier to acquire than either the inhalation or the gastrointestinal forms. A study of *Vibrio cholerae* showed that the ID_{50} is 10^8 cells; but if stomach acid is neutralized with bicarbonate, the number of cells required to cause an infection decreases significantly.

The potency of a toxin is often expressed as the **LD_{50}** (lethal dose for 50% of a sample population). For example, the LD_{50} for botulinum toxin in mice is 0.03 ng/kg and for Shiga toxin, 250 ng/kg, and staphylococcal enterotoxin, 1350 ng/kg. In other words, compared to the other two toxins, a much smaller dose of botulinum toxin is needed to cause symptoms.

Adherence

Learning Objective

- Using examples, explain how microbes adhere to host cells.

Once pathogens gain entry to a host, almost all of them have some means of attaching themselves to host tissues. For most pathogens, this attachment, called **adherence,** is

TABLE 15.1	**Portals of Entry for the Pathogens of Some Common Diseases**		
Portal of Entry	**Pathogen***	**Disease**	**Incubation Period**
Mucous Membranes			
Respiratory tract	*Streptococcus pneumoniae*	Pneumococcal pneumonia	Variable
	Mycobacterium tuberculosis[†]	Tuberculosis	Variable
	Bordetella pertussis	Whooping cough (pertussis)	12–20 days
	Influenza virus *(Influenzavirus)*	Influenza	18–36 hours
	Measles virus *(Morbillivirus)*	Measles (rubeola)	11–14 days
	Rubella virus *(Rubivirus)*	German measles (rubella)	2–3 weeks
	Epstein-Barr virus *(Lymphocryptovirus)*	Infectious mononucleosis	2–6 weeks
	Varicella-zoster virus *(Varicellovirus)*	Chickenpox (varicella) (primary infection)	14–16 days
	Histoplasma capsulatum (fungus)	Histoplasmosis	5–18 days
Gastrointestinal tract	*Shigella* spp.	Bacillary dysentery (shigellosis)	1–2 days
	Brucella spp.	Brucellosis (undulant fever)	6–14 days
	Vibrio cholerae	Cholera	1–3 days
	Salmonella enterica	Salmonellosis	7–22 hours
	Salmonella typhi	Typhoid fever	14 days
	Hepatitis A virus *(Hepatovirus)*	Hepatitis A	15–50 days
	Mumps virus *(Rubulavirus)*	Mumps	2–3 weeks
	Trichinella spiralis (helminth)	Trichinosis	2–28 days
Genitourinary tract	*Neisseria gonorrhoeae*	Gonorrhea	3–8 days
	Treponema pallidum	Syphilis	9–90 days
	Chlamydia trachomatis	Nongonococcal urethritis	1–3 weeks
	Herpes simplex virus type 2	Herpes virus infections	4–10 days
	Human immunodeficiency virus (HIV)[‡]	AIDS	10 years
	Candida albicans (fungus)[‡]	Candidiasis	2–5 days
Skin or Parenteral Route	*Clostridium perfringens*	Gas gangrene	1–5 days
	Clostridium tetani	Tetanus	3–21 days
	Rickettsia rickettsii	Rocky Mountain spotted fever	3–12 days
	Hepatitis B virus *(Hepadnavirus)*[†]	Hepatitis B	6 weeks–6 months
	Rabiesvirus *(Lyssavirus)*	Rabies	10 days–1 year
	Plasmodium spp. (protozoan)	Malaria	2 weeks

*All pathogens are bacteria, unless indicated otherwise. For viruses, the viral species and/or genus name is given.
[†]These pathogens can also cause disease after entering the body via the gastrointestinal tract.
[‡]These pathogens can also cause disease after entering the body via the parenteral route.

a necessary step in pathogenicity. (Of course, non-pathogens also have structures for attachment.) The attachment between pathogen and host is accomplished by means of surface molecules on the pathogen called **adhesins** or **ligands** that bind specifically to complementary surface **receptors** on the cells of certain host tissues (Figure 15.1 on page 441). Adhesins may be located on a microbe's glycocalyx or on other microbial surface structures, such as pili, fimbriae, and flagella (see Chapter 4).

The majority of adhesins on the microorganisms studied so far are glycoproteins or lipoproteins. The receptors on host cells are typically sugars, such as mannose. Adhesins on different strains of the same species of pathogen can vary in structure. Different cells of the same host can also have different receptors that vary in structure. If adhesins, receptors, or both can be altered to interfere with adherence, infection can often be prevented (or at least controlled).

The following examples illustrate the diversity of adhesins. *Streptococcus mutans*, a bacterium that plays a key role in tooth decay, attaches to the surface of teeth by its glycocalyx. An enzyme produced by *S. mutans*, called glucosyltransferase, converts glucose (derived from sucrose or table sugar) into a sticky polysaccharide called dextran, which forms the glycocalyx. *Actinomyces* bacterial cells have fimbriae that adhere to the glycocalyx of *S. mutans*. The combination of *S. mutans*, *Actinomyces*, and dextran make up dental plaque and contribute to dental caries (tooth decay).

How Human Behavior Influences the Evolution of Virulence in Microorganisms

According to human logic, a parasite that kills its host is harming itself. Thus, it seems reasonable that some parasites have evolved to a sort of commensalism with their human hosts. In this respect, tapeworms might be considered the "perfect parasite." Tapeworms cause no obvious symptoms in most patients, thus ensuring that their hosts will be able to move around and shed tapeworm eggs that will be ingested by intermediate hosts.

However, other diseases such as cholera, antibiotic-resistant tuberculosis, viral cancers, and syphilis sometimes do kill their hosts. Does it make sense for a parasite to kill its host? Remember that nature does not have a plan for evolution; the genetic variations that give rise to evolution are due to random mutations, not logic. However, according to natural selection, organisms best adapted to their environments will survive and reproduce. Coevolution between a parasite and its host seems to occur: The behavior of one influences that of the other.

Some parasites, such as the cholera bacterium, get transmitted before their host dies. *Vibrio cholerae* quickly induces diarrhea, threatening the host's life resulting from a loss of fluids and salts, but providing a way to transmit the parasite to another person through the ingestion of contaminated water. This cycle continues in countries where war and poverty have prevented the implementa-

tion of water purification systems. However, the virulent form of cholera has been selected against in countries where water-treatment systems prevent the bacteria from getting to a new host. After water purification was instituted in India, the milder agent, *V. cholerae* eltor, replaced the deadly *V. cholerae*.

When we interfere with a pathogen's survival by using antibiotics, we inadvertently select for antibiotic-resistant mutants. The rapid rise of multidrug-resistant *Mycobacterium tuberculosis*, documented in the popular and scientific media, demonstrates this. In an antibiotic-laden environment, the "fittest" bacterium is one that is resistant to antibiotics. Antibiotic resistance is selected for when chemotherapy is not completed and when antibiotics are not used correctly. For example, tuberculosis (TB) chemotherapy protocols are very long (up to 2 years). In the United States, only 75% of TB patients complete the prescribed chemotherapy. When a person doesn't take the complete regimen of the prescribed antibiotic, the level of the drug in the body decreases, and some resistant bacteria survive. These bacteria reproduce, and some of the new population are able to survive in even higher levels of antibiotics.

Comparisons of human T-cell leukemia virus type 1 (HTLV-1) infections in Japan and Jamaica also suggest that human sexual behavior influences the virulence of sexually transmitted dis-

eases. HTLV-1 is a sexually transmitted retrovirus that causes a type of adult leukemia. In Japan, the average age of onset of cancers caused by HTLV-1 is 60, whereas in Jamaica, onset occurs at about age 45. In Japan, barrier contraceptives are more widely used than birth control pills, so the virus needs a healthy host for a longer period of time. In Jamaica, barrier contraceptives are not widely used and HTLV-1 is more virulent, perhaps because it has a greater potential for sexual transmission.

Human behavior may also have influenced the evolution of syphilis. *Treponema pallidum* has a long incubation period and goes into noninfectious latent periods between the primary, secondary, and tertiary stages of the disease. Although it cannot be transmitted during the incubation and latent periods, these periods allow the bacteria to stay alive until the infected host changes sex partners. Partner changes might take several years in a monogamous society.

Paul Ewald of Amherst College reminds us that commensalism is not the inevitable end point of evolution. In order to control diseases, we must modify our behavior to make commensalism the most favorable outcome for the pathogen. Some examples of these modifications are ensuring completion of the prescribed antibiotic protocol for tuberculosis, and the use of condoms to prevent the transmission of sexually transmitted diseases.

Microbes have the ability to come together in masses, cling to surfaces, and take in and share available nutrients. These communities, which constitute masses of microbes and their extracellular products that can attach to living and nonliving surfaces, are called **biofilms.** Examples of biofilms include the dental plaque on teeth, the algae on the walls of swimming pools, and the scum that builds on shower doors. A biofilm forms when microbes adhere to a particular surface that is typically moist and contains organic matter. The first microbes to attach are usually bacte-

ria. Once they adhere to the surface, they multiply and secrete a glycocalyx that further attaches the bacteria to each other and to the surface. In some cases biofilms can be several layers thick and may contain several types of microbes. Biofilms represent another method of adherence and are important because they resist disinfectants and antibiotics. This characteristic is significant, especially when biofilms colonize structures such as teeth, medical catheters, stents, heart valves, hip replacements, and contact lenses. Dental plaque is actually a biofilm that mineralizes over time. It is

TEM |———| 1 μm

TEM |———| 1 μm

(a) Surface molecules on a pathogen, called adhesins or ligands, bind specifically to complementary surface receptors on cells of certain host tissues.

(b) *E. coli* bacteria (green) on human bladder cells.

(c) *Streptococcus* bacteria adhering to human skin.

FIGURE 15.1 **Adherence.**

■ Most adhesins are glycoproteins or lipoproteins.

estimated that biofilms are involved in 65% of all human bacterial infections.

Enteropathogenic strains of *E. coli* (those responsible for gastrointestinal disease) have adhesins on fimbriae that adhere only to specific kinds of cells in certain regions of the small intestine. After adhering, *Shigella* and *E. coli* induce endocytosis as a vehicle to enter host cells and then multiply within them. *Treponema pallidum*, the causative agent of syphilis, uses its tapered end as a hook to attach to host cells. *Listeria monocytogenes*, which causes meningitis, spontaneous abortions, and stillbirths, produces an adhesin for a specific receptor on host cells. *Neisseria gonorrhoeae* (nī-se′rē-ä go-nôr-rē′i), the causative agent of gonorrhea, also has fimbriae containing adhesins, which in this case permit attachment to cells with appropriate receptors in the genitourinary tract, eyes, and pharynx. *Staphylococcus aureus*, which can cause skin infections, binds to skin by a mechanism of adherence that resembles viral attachment (see Chapter 13).

How Bacterial Pathogens Penetrate Host Defenses

Learning Objective

- *Explain how capsules and cell wall components contribute to pathogenicity.*

Although some pathogens can cause damage on the surface of tissues, most must penetrate tissues to cause disease.

Here we will consider several factors that contribute to the ability of bacteria to invade a host.

Capsules

Recall from Chapter 4 that some bacteria make glycocalyx material that forms capsules around their cell walls; this property increases the virulence of the species. The capsule resists the host's defenses by impairing phagocytosis, the process by which certain cells of the body engulf and destroy microbes (see Chapter 16, page 463). The chemical nature of the capsule appears to prevent the phagocytic cell from adhering to the bacterium. However, the human body can produce antibodies against the capsule, and when these antibodies are present on the capsule surface, the encapsulated bacteria are easily destroyed by phagocytosis.

One bacterium that owes its virulence to the presence of a polysaccharide capsule is *Streptococcus pneumoniae*, the causative agent of pneumococcal pneumonia. Some strains of this organism have capsules, and others do not. Strains with capsules are virulent, but strains without capsules are avirulent because they are susceptible to phagocytosis. Other bacteria that produce capsules related to virulence are *Klebsiella pneumoniae*, a causative agent of bacterial pneumonia; *Haemophilus influenzae*, a cause of pneumonia and meningitis in children; *Bacillus anthracis*, the cause of anthrax; and *Yersinia pestis*, the causative agent of bubonic plague. Keep in mind that capsules are not the only cause of virulence. Many nonpathogenic bacteria produce capsules, and the virulence of some pathogens is not related to the presence of a capsule.

Components of the Cell Wall

The cell walls of certain bacteria contain chemical substances that contribute to virulence. For example, *Streptococcus pyogenes* produces a heat-resistant and acid-resistant protein called **M protein.** This protein is found on both the cell surface and fimbriae. The M protein mediates attachment of the bacterium to epithelial cells of the host and helps the bacterium resist phagocytosis by white blood cells. The protein thereby increases the virulence of the microorganism. Immunity to *S. pyogenes* depends on the body's production of an antibody specific to M protein. *Neisseria gonorrhoeae* grows inside human epithelial cells and leukocytes. These bacteria use **fimbriae** and an outer membrane protein called **Opa** to attach to host cells. Following attachment by both Opa and fimbriae, the host cells take in the bacteria. (Bacteria that produce Opa form *opa*que colonies in culture media.) The **waxes** that make up the cell wall of *Mycobacterium tuberculosis* also increase virulence by resisting digestion by phagocytes. In fact, *M. tuberculosis* can even multiply inside phagocytes.

Enzymes

Learning Objective

- Compare the effects of coagulases, kinases, hyaluronidase, and collagenase.

The virulence of some bacteria is thought to be aided by the production of extracellular enzymes (exoenzymes) and related substances. These chemicals can dissolve materials between cells and form or dissolve blood clots, among other functions.

Coagulases are bacterial enzymes that coagulate (clot) the fibrinogen in blood. Fibrinogen, a plasma protein produced by the liver, is converted by coagulases into fibrin, the threads that form a blood clot. The fibrin clot may protect the bacterium from phagocytosis and isolate it from other defenses of the host. Coagulases are produced by some members of the genus *Staphylococcus*; they may be involved in the walling-off process in boils produced by staphylococci. However, some staphylococci that do not produce coagulases are still virulent. (Capsules may be more important to their virulence.)

Bacterial **kinases** are bacterial enzymes that break down fibrin and thus dissolve clots formed by the body to isolate the infection. One of the better-known kinases is *fibrinolysin (streptokinase)*, which is produced by such streptococci as *Streptococcus pyogenes*. Another kinase, *staphylokinase*, is produced by such staphylococci as *Staphylococcus aureus*. Injected directly into the blood, streptokinase has been used successfully to dissolve some

types of blood clots in cases of heart attacks due to obstructed coronary arteries.

Hyaluronidase is another enzyme secreted by certain bacteria, such as streptococci. It hydrolyzes hyaluronic acid, a type of polysaccharide that holds together certain cells of the body, particularly cells in connective tissue. This digesting action is thought to be involved in the tissue blackening of infected wounds and to help the microorganism spread from its initial site of infection. Hyaluronidase is also produced by some clostridia that cause gas gangrene. For therapeutic use, hyaluronidase may be mixed with a drug to promote the spread of the drug through a body tissue.

Another enzyme, **collagenase,** produced by several species of *Clostridium*, facilitates the spread of gas gangrene. Collagenase breaks down the protein collagen, which forms the connective tissue of muscles and other body organs and tissues.

As a defense against adherence of pathogens to mucosal surfaces, the body produces a class of antibodies called IgA antibodies. There are some pathogens with the ability to produce enzymes, called **IgA proteases,** that can destroy these antibodies. *N. gonorrhoeae* has this ability, as does *N. meningitidis* (me-nin-ji′ti-dis), the causative agent of meningococcal meningitis, and other microbes that infect the central nervous system.

Antigenic Variation

Learning Objective

- Define and give an example of antigenic variation.

In Chapter 17 you will learn that specific resistance (immunity) refers to a specific defensive response of the body to an infection or to antigens. In the presence of antigens the body produces proteins called antibodies, which bind to the antigens and inactivate or destroy them. However, some pathogens can alter their surface antigens, by a process called **antigenic variation.** Thus, by the time the body mounts an immune response against a pathogen, the pathogen has already altered its antigens and is unaffected by the antibodies. Some microbes can activate alternative genes resulting in antigenic changes. For example, *N. gonorrhoeae* has several copies of the Opa-encoding gene, resulting in cells with different antigens and in cells that express different antigens over time.

A wide range of bacteria and protozoa are capable of antigenic variation. Examples include *Influenzavirus*, the causative agent of influenza (flu); *Neisseria gonorrhoeae*, the causative agent of gonorrhea; and *Trypanosoma brucei gambiense* (tri-pa′nō-sō-mä brüs′ē gam-bē-ens′), the causative agent of African trypanosomiasis (sleeping sickness).

SEM ⊢———⊣ 3 μm

FIGURE 15.2 *Salmonella* entering epithelial cells.

■ Some microbes produce surface proteins called invasins that rearrange actin filaments of the cytoskeleton.

Penetration into the Host Cell Cytoskeleton

Learning Objective

• *Describe how bacteria use the host cell's cytoskeleton to enter the cell.*

As previously noted, microbes attach to host cells by adhesins. The interaction triggers signals in the host cell that activate factors that can result in the entrance of some bacteria. The actual mechanism is provided by the host cell cytoskeleton. Recall from Chapter 4 that eukaryotic cytoplasm has a complex internal structure consisting of protein filaments called microfilaments, intermediate filaments, and microtubules that comprise the cytoskeleton.

A major component of the cytoskeleton is a protein called actin, which is used by some microbes to penetrate host cells and by others to move through and between host cells.

Salmonella strains and *E. coli* make contact with the host cell plasma membrane. This leads to dramatic changes in the membrane at the point of contact. The microbes produce surface proteins called **invasins** that rearrange nearby actin filaments of the cytoskeleton. This causes a cytoplasmic structure to project from the host cell like a pedestal under the *Salmonella*. The structure supports the bacterial cells, then forms an actin basket that appears to cuddle the *Salmonella* and move it into the cell (Figure 15.2).

Once inside the host cell, certain bacteria such as *Shigella* species and *Listeria* species can actually use actin to propel themselves through the host cell cytoplasm and from one host cell to another. The condensation of actin

$$\begin{array}{c} O \\ \diagdown \\ Fe^{3+} \\ \diagup \\ O \end{array}$$

C=O
|
NH
|
$(CO-CH-CH_2-O)_3$

FIGURE 15.3 **Structure of enterobactin, one type of bacterial siderophore.** Note where the iron (Fe^{3+}) is attached to the siderophore.

on one end of the bacteria propels them through the cytoplasm. The bacteria also make contact with membrane junctions that form part of a transport network between host cells. The bacteria use a glycoprotein called cadherin, which bridges the junctions, to move from cell to cell.

The study of the numerous interactions between microbes and host cell cytoskeleton is a very intense area of investigation on virulence mechanisms.

How Bacterial Pathogens Damage Host Cells

When a microorganism invades a body tissue, it initially encounters phagocytes of the host. If the phagocytes are successful in destroying the invader, no further damage is done to the host. But if the pathogen overcomes the host defense, then the microorganism can damage host cells in four basic ways: (1) by using the host's nutrients; (2) by causing direct damage in the immediate vicinity of the invasion; (3) by producing toxins, transported by blood and lymph, that damage sites far removed from the original site of invasion; and (4) by inducing hypersensitivity reactions. This fourth mechanism is considered in detail in Chapter 19. For now, we will discuss only the first three mechanisms.

Using the Host's Nutrients

Learning Objective

• *Describe the function of siderophores.*

Iron is required for the growth of most pathogenic bacteria. However, the concentration of free iron in the human body is fairly low because most of the iron is tightly bound to iron-transport proteins, such as lactoferrin, transferrin, and ferritin, as well as hemoglobin. In order to obtain free iron, some pathogens secrete proteins called **siderophores** (Figure 15.3). When iron is needed by a pathogen, siderophores are released into the medium where they take the iron away from iron-transport proteins by binding the iron even more tightly. Once the

FIGURE 15.4 **Exotoxins and endotoxins.**

■ What are the three principal types of exotoxins based on their mode of action?

Cell wall

Exotoxin

Endotoxin

(a) **Exotoxins** are produced inside mostly gram-positive bacteria as part of their growth and metabolism. They are then secreted or released following lysis into the surrounding medium.

(b) **Endotoxins** are part of the outer portion of the cell wall (lipid A; see Figure 4.13c) of gram-negative bacteria. They are liberated when the bacteria die and the cell wall breaks apart.

iron-siderophore complex is formed, it is taken up by siderophore receptors on the bacterial surface. Then the iron is brought into the bacterium. In some cases the iron is released from the complex to enter the bacterium; in other cases the iron enters as part of the complex.

As an alternative to iron acquisition by siderophores, some pathogens have receptors that bind directly to iron-transport proteins and hemoglobin. Then these are taken into the bacterium directly along with the iron. Also, it is possible that some bacteria produce toxins (described shortly) when iron levels are low. The toxins kill host cells, releasing their iron, which is then available to the bacteria.

Direct Damage

Learning Objective

• Provide an example of direct damage, and compare this to toxin production.

Once pathogens attach to host cells, they can cause direct damage as the pathogens use the host cell for nutrients and produce waste products. As pathogens metabolize and multiply in cells, the cells usually rupture. Many viruses and some intracellular bacteria and protozoa that grow in host cells are released when the host cell ruptures. Following their release, pathogens that rupture cells can spread to other tissues in even greater numbers. Some bacteria, such as *E. coli, Shigella, Salmonella,* and *Neisseria gonorrhoeae,* can induce host epithelial cells to engulf them by a process that resembles phagocytosis. These pathogens can disrupt host cells as they pass through and can then be extruded from the host cells by a reverse phagocytosis process, enabling them to enter other host cells. Some bacteria can also penetrate host cells by excreting enzymes and by their own motility;

such penetration can itself damage the host cell. Most damage by bacteria, however, is done by toxins.

The Production of Toxins

Learning Objectives

• Contrast the nature and effects of exotoxins and endotoxins.

• Outline the mechanisms of action of A-B toxins, membrane-disrupting toxins, and superantigens. Classify diphtheria toxin, erythrogenic toxin, botulinum toxin, tetanus toxin, Vibrio enterotoxin, and staphylococcal enterotoxin.

• Identify the importance of the LAL assay.

Toxins are poisonous substances that are produced by certain microorganisms. They are often the primary factor contributing to the pathogenic properties of those microbes. The capacity of microorganisms to produce toxins is called **toxigenicity.** Toxins transported by the blood or lymph can cause serious, and sometimes fatal, effects. Some toxins produce fever, cardiovascular disturbances, diarrhea, and shock. Toxins can also inhibit protein synthesis, destroy blood cells and blood vessels, and disrupt the nervous system by causing spasms. Of the 220 or so known bacterial toxins, nearly 40% cause disease by damaging eukaryotic cell membranes. The term **toxemia** refers to the presence of toxins in the blood. Toxins are of two general types, based on their position relative to the microbial cell: exotoxins and endotoxins.

Exotoxins

Exotoxins are produced inside some bacteria as part of their growth and metabolism and are secreted by the bacterium into the surrounding medium or released following lysis (Figure 15.4a). Exotoxins are proteins, and many are enzymes that catalyze only certain biochemical reactions. Because of the enzymatic nature of most exotoxins, even small amounts are quite harmful because they can be used over and over again. Bacteria that produce exotoxins may be gram-positive or gram-negative. The genes for most (perhaps all) exotoxins are carried on bacterial plasmids or phages. Because exotoxins are soluble in body fluids, they can easily diffuse into the blood and are rapidly transported throughout the body.

Exotoxins work by destroying particular parts of the host's cells or by inhibiting certain metabolic functions. They are highly specific in their effects on body tissues. Exotoxins are among the most lethal substances known. Only 1 mg of the botulinum exotoxin is enough to kill 1 million guinea pigs. Fortunately, only a few bacterial species produce such potent exotoxins.

Diseases caused by bacteria that produce exotoxins are often caused by minute amounts of exotoxins, not by the bacteria themselves. It is the exotoxins that produce the specific signs and symptoms of the disease. Thus, exotoxins are disease-specific. For example, botulism is usually due to ingesting the exotoxin, not a bacterial infection. Likewise, staphylococcal food poisoning is an *intoxication*, not an infection.

The body produces antibodies called **antitoxins** that provide immunity to exotoxins. When exotoxins are inactivated by heat or by formaldehyde, iodine, or other chemicals, they no longer cause the disease but can still stimulate the body to produce antitoxins. Such altered exotoxins are called **toxoids.** When toxoids are injected into the body as a vaccine, they stimulate antitoxin production so that immunity is produced. Diphtheria and tetanus can be prevented by toxoid vaccination.

Exotoxins are divided into three principal types on the basis of their structure and function: (1) A-B toxins, (2) membrane-disrupting toxins, and (3) superantigens.

A-B Toxins

A-B toxins, also called *type III toxins*, were the first toxins to be studied intensively and are so named because they consist of two parts designated A and B, both of which are polypeptides. Most exotoxins are A-B toxins. The A part is the active (enzyme) component, and the B part is the binding component. An example of an A-B toxin is the diphtheria toxin, which is illustrated in Figure 15.5.

❶ In the first step, the A-B toxin is released from the bacterium.

❷ The B part binds to a surface receptor on the host cell, usually a carbohydrate. Following binding, the A-B toxin is transported across the plasma membrane into the host cell cytoplasm.

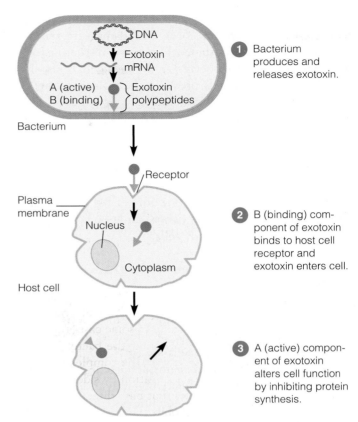

FIGURE 15.5 **The action of an exotoxin.** A proposed model for the mechanism of action of diphtheria toxin.

■ Diphtheria toxin consists of two different polypeptides: A (active) and B (binding); B is required to make A active.

❸ Within the cell, the A-B components separate. The A part inhibits protein synthesis and kills the host cell while the B part is released from the cell.

Membrane-Disrupting Toxins

Membrane-disrupting toxins, also called *type II toxins*, cause lysis of host cells by disrupting their plasma membranes. Some do this by forming protein channels in the plasma membrane; others disrupt the phospholipid portion of the membrane. The cell-lysing exotoxin of *Staphylococcus aureus* is an example of an exotoxin that forms protein channels, whereas that of *Clostridium perfringens* is an example of an exotoxin that disrupts the phospholipids. Membrane-disrupting toxins contribute to virulence by killing host cells, especially phagocytes, and by aiding the escape of bacteria from sacs within phagocytes (phagosomes) into the host cell's cytoplasm.

Membrane-disrupting toxins that kill phagocytic leukocytes (white blood cells) are called **leukocidins.** They act by forming protein channels. Leukocidins are also active against macrophages, phagocytes present in tissues. Most leukocidins are produced by staphylococci, streptococci, and pneumococci. The damage to phagocytes decreases host resistance. Membrane-disrupting toxins that destroy erythrocytes (red blood cells), also by forming protein channels, are called **hemolysins.** Important producers of hemolysins include staphylococci and streptococci. Hemolysins produced by streptococci are called *streptolysins*. One kind, called *streptolysin O (SLO)*, is so named because it is inactivated by atmospheric oxygen. Another kind of streptolysin is called *streptolysin S (SLS)* because it is stable in an oxygen environment. Both streptolysins can cause lysis not only of red blood cells, but also of white blood cells (whose function is to kill the streptococci) and other body cells.

Superantigens **Superantigens,** or *type I toxins,* are antigens that provoke a very intense immune response. They are bacterial proteins. Through a series of interactions with various cells of the immune system, superantigens nonspecifically stimulate the proliferation of immune cells called T cells. These cells are types of white blood cells (lymphocytes) that act against foreign organisms and tissues (transplants) and regulate the activation and proliferation of other cells of the immune system. In response to superantigens, T cells are stimulated to release enormous amounts of chemicals called cytokines (see Chapter 17, page 495). Cytokines are small protein hormones that stimulate or inhibit many normal cell functions. The excessively high levels of cytokines released by T cells enter the bloodstream and give rise to a number of symptoms, including fever, nausea, vomiting, diarrhea, and sometimes shock and even death. Bacterial superantigens include the staphylococcal toxins that cause food poisoning and toxic shock syndrome.

Exotoxins are named on the basis of several characteristics. One is the type of host cell that is attached. For example, *neurotoxins* attack nerve cells, *cardiotoxins* attack heart cells, *hepatotoxins* attack liver cells, *leukotoxins* attack leukocytes, *enterotoxins* attack the lining of the gastrointestinal tract, and *cytotoxins* attack a wide variety of cells. Some exotoxins are named for the diseases with which they are associated. Examples include diphtheria toxin (cause of diphtheria) and tetanus toxin (cause of tetanus). Other exotoxins are named for the specific bacterium that produces them, for example, botulinum toxin (*Clostridium botulinum*) and *Vibrio* enterotoxin (*Vibrio cholerae*).

Next we briefly describe a few of the more notable exotoxins (antitoxins will be discussed further in Chapter 18).

Diphtheria Toxin Corynebacterium diphtheriae produces the *diphtheria toxin* only when it is infected by a lysogenic phage carrying the *tox* gene. This cytotoxin inhibits protein synthesis in eukaryotic cells. The mechanism by which it does this is an A-B toxin, which is illustrated in Figure 15.5.

Erythrogenic Toxins Streptococcus pyogenes has the genetic material to synthesize three types of cytotoxins, designated A, B, and C. These *erythrogenic* (*erythro* = red; *gen* = producing) *toxins* damage the plasma membranes of blood capillaries under the skin and produce a red skin rash. Scarlet fever, caused by *S. pyogenes* exotoxins, is named for this characteristic rash.

Botulinum Toxin Botulinum toxin is produced by *Clostridium botulinum*. Although toxin production is associated with the germination of endospores and the growth of vegetative cells, little of the toxin appears in the medium until it is released by lysis late in growth. Botulinum toxin is an A-B neurotoxin; it acts at the neuromuscular junction (the junction between nerve cell and muscle cell) and prevents the transmission of impulses from the nerve cell to the muscle. The toxin accomplishes this by binding to the nerve cell and inhibiting the release of a neurotransmitter called acetylcholine. As a result, botulinum toxin causes paralysis in which muscle tone is lacking (flaccid paralysis). *C. botulinum* produces several different types of botulinum toxin, and each possesses a different potency.

Tetanus Toxin Clostridium tetani produces tetanus neurotoxin, also known as *tetanospasmin*. This A-B toxin reaches the central nervous system and binds to nerve cells that control the contraction of various skeletal muscles. These nerve cells normally send inhibiting impulses that prevent random contractions and terminate completed contractions. The binding of tetanospasmin blocks this relaxation pathway. The result is uncontrollable muscle contractions, producing the convulsive symptoms (spasmodic contractions) of tetanus, or "lockjaw."

Vibrio Enterotoxin Vibrio cholerae produces an enterotoxin called *cholera toxin*. Subunit B binds to epithelial cells, and subunit A causes cells to secrete large amounts of fluids and electrolytes (ions). Normal muscular contractions are disturbed, leading to severe diarrhea that may be accompanied by vomiting. *Heat-labile enterotoxin* (so named because it is more sensitive to heat than are most toxins), produced by some strains of *E. coli*, has an action identical to that of cholera toxin.

Staphylococcal Enterotoxin Staphylococcus aureus produces an enterotoxin that affects the intestines in the same way as cholera toxin. A strain of *S. aureus* also produces

TABLE 15.2	**Diseases Caused by Exotoxins**	
Disease	**Bacterium**	**Mechanism**
Botulism	*Clostridium botulinum*	Neurotoxin prevents the transmission of nerve impulses; flaccid paralysis results.
Tetanus	*Clostridium tetani*	Neurotoxin blocks nerve impulses to muscle relaxation pathway that results in uncontrollable muscle contractions.
Gas gangrene and food poisoning	*Clostridium perfringens* and other species of *Clostridium*	One exotoxin (cytotoxin) causes massive red blood cell destruction (hemolysis); another exotoxin (enterotoxin) is related to food poisoning and causes diarrhea.
Diphtheria	*Corynebacterium diphtheriae*	Cytotoxin inhibits protein synthesis, especially in nerve, heart, and kidney cells.
Scalded skin syndrome, food poisoning, and toxic shock syndrome	*Staphylococcus aureus*	One exotoxin causes skin layers to separate and slough off (scalded skin); another exotoxin (enterotoxin) produces diarrhea and vomiting; still another exotoxin produces symptoms associated with TSS.
Cholera	*Vibrio cholerae*	Enterotoxin causes secretion of large amounts of fluids and electrolytes that result in diarrhea.
Scarlet fever	*Streptococcus pyogenes*	Cytotoxins cause vasodilation that results in the characteristic rash.
Traveler's diarrhea	Enterotoxigenic *Escherichia coli* and *Shigella* spp.	Enterotoxin causes secretion of large amounts of fluids and electrolytes that result in diarrhea.

enterotoxins that result in the symptoms associated with toxic shock syndrome. A summary of diseases produced by exotoxins is shown in Table 15.2.

Endotoxins

Endotoxins differ from exotoxins in several ways. Endotoxins are part of the outer portion of the cell wall of gram-negative bacteria (see Figure 15.4b). Recall from Chapter 4 that gram-negative bacteria have an outer membrane surrounding the peptidoglycan layer of the cell wall. This outer membrane consists of lipoproteins, phospholipids, and lipopolysaccharides (LPS) (see Figure 4.13c on page 84). The lipid portion of LPS, called **lipid A,** is the endotoxin. Thus, endotoxins are lipopolysaccharides, whereas exotoxins are proteins.

Endotoxins are released when gram-negative bacteria die and their cell walls undergo lysis, thus liberating the endotoxin. (Endotoxins are also released during bacterial multiplication.) Antibiotics used to treat diseases caused by gram-negative bacteria can lyse the bacterial cells; this reaction releases endotoxin and may lead to an immediate worsening of the symptoms, but the condition usually improves as the endotoxin breaks down. Endotoxins exert their effects by stimulating macrophages to release cytokines in very high concentrations. At these levels, cytokines are toxic. All endotoxins produce the same signs and symptoms, regardless of the species of microorganism, although not to the same degree. These include chills, fever, weakness, generalized aches, and, in some cases, shock and even death. Endotoxins can also induce miscarriage.

Another consequence of endotoxins is the activation of blood-clotting proteins, causing the formation of small blood clots. These blood clots obstruct capillaries, and the resulting decreased blood supply induces the death of tissues. This condition is referred to as *disseminated intravascular clotting.*

The fever (pyrogenic response) caused by endotoxins is believed to occur as depicted in Figure 15.6.

❶ When gram-negative bacteria are ingested by phagocytes and

❷ degraded in vacuoles, the LPS of the bacterial cell wall are released. These endotoxins cause macrophages to produce a cytokine called **interleukin-1 (IL-1),** formerly called *endogenous pyrogen,*

❸ which is carried via the blood to the hypothalamus, a temperature control center in the brain.

❹ IL-1 induces the hypothalamus to release lipids called prostaglandins, which reset the thermostat in the hypothalamus at a higher temperature. The result is a fever.

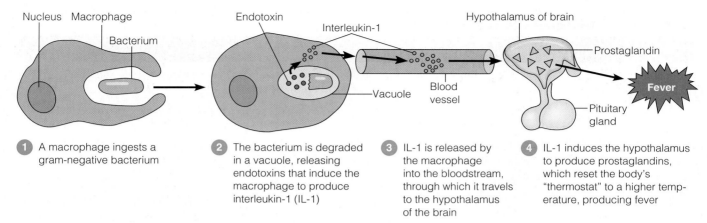

① A macrophage ingests a gram-negative bacterium

② The bacterium is degraded in a vacuole, releasing endotoxins that induce the macrophage to produce interleukin-1 (IL-1)

③ IL-1 is released by the macrophage into the bloodstream, through which it travels to the hypothalamus of the brain

④ IL-1 induces the hypothalamus to produce prostaglandins, which reset the body's "thermostat" to a higher temperature, producing fever

FIGURE 15.6 **Endotoxins and the pyrogenic response.** The proposed mechanism by which endotoxins cause fever.

■ All endotoxins produce the same signs and symptoms, regardless of the species of microbe.

Bacterial cell death caused by lysis or antibiotics can also produce fever by this mechanism. Both aspirin and acetaminophen reduce fever by inhibiting the synthesis of prostaglandins. (The function of fever in the body is discussed in Chapter 16 on page 471.)

Shock refers to any life-threatening loss of blood pressure. Shock caused by bacteria is called **septic shock.** Gram-negative bacteria cause endotoxic shock. Like fever, the shock produced by endotoxins is related to the secretion of a cytokine by macrophages. Phagocytosis of gram-negative bacteria causes the phagocytes to secrete a polypeptide called **tumor necrosis factor (TNF),** or *cachectin.* TNF binds to many tissues in the body and alters their metabolism in a number of ways. One effect of TNF is damage to blood capillaries; their permeability is increased, and they lose large amounts of fluid. The result is a drop in blood pressure that results in shock. Low blood pressure has serious effects on the kidneys, lungs, and gastrointestinal tract. Another mechanism that causes septic shock is discussed in the box in Chapter 16 (page 472). In addition, the presence of gram-negative bacteria such as *Haemophilus influenzae* type b in cerebrospinal fluid causes the release of IL-1 and TNF. These, in turn, cause a weakening of the blood–brain barrier that normally protects the central nervous system from infection. The barrier is weakened to let phagocytes in, but this also lets more bacteria enter from the bloodstream. In the United States, 500,000 cases of septic shock occur each year. One-third of the patients die within a month, and nearly half die within 6 months.

Endotoxins do not promote the formation of effective antitoxins against the carbohydrate component of an endotoxin. Antibodies are produced, but they tend not to counter the effect of the toxin; sometimes, in fact, they actually enhance its effect.

Representative microorganisms that produce endotoxins are *Salmonella typhi* (the causative agent of typhoid fever), *Proteus* spp. (frequently the causative agents of urinary tract infections), and *Neisseria meningitidis* (the causative agent of meningococcal meningitis).

It is important to have a sensitive test to identify the presence of endotoxins in drugs, medical devices, and body fluids. Materials that have been sterilized may contain endotoxins, even though no bacteria can be cultured from them. One such laboratory test is called the **Limulus amoebocyte lysate (LAL) assay,** which can detect even minute amounts of endotoxin. The hemolymph (blood) of the Atlantic coast horseshoe crab, *Limulus polyphemus,* contains white blood cells called amoebocytes, which have large amounts of a protein (lysate) that causes clotting. In the presence of endotoxin, amoebocytes in the crab hemolymph lyse and liberate their clotting protein. The resulting gel-clot (precipitate) is a positive test for the presence of endotoxin. The degree of the reaction is measured by a spectrophotometer, an instrument that measures degree of turbidity.

A comparison of exotoxins and endotoxins appears in Table 15.3.

Plasmids, Lysogeny, and Pathogenicity

Learning Objective

• Using examples, describe the roles of plasmids and lysogeny in pathogenicity.

Recall from Chapters 4 (page 92) and 8 (page 240) that plasmids are small, circular DNA molecules that are not

TABLE 15.3	**Exotoxins and Endotoxins**	
Property	**Exotoxin**	**Endotoxin**
Bacterial source	Mostly from gram-positive bacteria	Gram-negative bacteria
Relation to microorganism	Metabolic product of growing cell	Present in LPS of outer membrane of cell wall and released with destruction of cell or during cell division
Chemistry	Proteins, usually with two parts (A-B)	Lipid portion (lipid A) of LPS of outer membrane (lipopolysaccharide).
Pharmacology (effect on body)	Specific for a particular cell structure or function in the host (mainly affects cell functions, nerves, and gastrointestinal tract)	General, such as fever, weaknesses, aches, and shock; all produce the same effects
Heat stability	Unstable; can usually be destroyed at 60–80°C (except staphylococcal enterotoxin)	Stable; can withstand autoclaving (121°C for 1 hour)
Toxicity (ability to cause disease)	High	Low
Fever-producing	No	Yes
Immunology (relation to antibodies)	Can be converted to toxoids to immunize against toxin; neutralized by antitoxin	Not easily neutralized by antitoxin; therefore, effective toxoids cannot be made to immunize against toxin
Lethal dose	Small	Considerably larger
Representative diseases	Gas gangrene, tetanus, botulism, diphtheria, scarlet fever	Typhoid fever, urinary tract infections, and meningococcal meningitis

connected to the main bacterial chromosome and are capable of independent replication. One group of plasmids, called R (resistance) factors, is responsible for the resistance of some microorganisms to antibiotics. In addition, a plasmid may carry the information that determines a microbe's pathogenicity. Examples of virulence factors that are encoded by plasmid genes are tetanospasmin, heat-labile enterotoxin, and staphylococcal enterotoxin. Other examples are dextransucrase, an enzyme produced by *Streptococcus mutans* that is involved in tooth decay; adhesins and coagulase produced by *Staphylococcus aureus*; and a type of fimbria specific to enteropathogenic strains of *E. coli*.

In Chapter 13, we noted that some bacteriophages (viruses that infect bacteria) can incorporate their DNA into the bacterial chromosome, becoming a prophage, and thus remain latent and not cause lysis of the bacterium. Such a state is called *lysogeny*, and cells containing a prophage are said to be lysogenic. One outcome of

lysogeny is that the host bacterial cell and its progeny may exhibit new properties coded for by the bacteriophage DNA. Such a change in the characteristics of a microbe due to a prophage is called **lysogenic conversion.** As a result of lysogenic conversion, the bacterial cell is immune to infection by the same type of phage. In addition, lysogenic cells are of medical importance because some bacterial pathogenesis is caused by the prophages they contain.

Among the bacteriophage genes that contribute to pathogenicity are the genes for diphtheria toxin, erythrogenic toxins, staphylococcal enterotoxin and pyrogenic toxin, botulinum neurotoxin, and the capsule produced by *Streptococcus pneumoniae*. Pathogenic strains of *Vibrio cholerae* carry lysogenic phages. These phages can transmit the cholera toxin gene to nonpathogenic *V. cholerae* strains, increasing the number of pathogenic bacteria.

Pathogenic Properties of Viruses

Learning Objective
- *List nine cytopathic effects of viral infections.*

The pathogenic properties of viruses depend on their gaining access to a host, evading the host's defenses, and then causing damage to or death of the host cell while reproducing themselves.

Viral Mechanisms for Evading Host Defenses

Viruses have a variety of mechanisms that enable them to evade destruction by the host's immune response (see Chapter 17). For example, viruses can penetrate and grow inside of host cells, where components of the immune system cannot reach them. Viruses gain access to cells because they have attachment sites for receptors on their target cells. When such an attachment site is brought together with an appropriate receptor, the virus can bind to and penetrate the cell. Some viruses gain access to host cells because their attachment sites mimic substances useful to those cells. For example, the attachment sites of rabies virus can mimic the neurotransmitter acetylcholine. As a result, the virus can enter the host cell along with the neurotransmitter.

The AIDS virus (HIV) goes further by hiding its attachment sites from the immune response and by attacking components of the immune system directly. Like some viruses, HIV is cell-specific; that is, it attacks only particular body cells. HIV attacks only those cells that have a surface marker called the CD4 protein, most of which are cells of the immune system called T lymphocytes (T cells). Binding sites on HIV are complementary to the CD4 protein. The surface of the virus is folded to form ridges and valleys, and the HIV binding sites are located on the floors of the valleys. CD4 proteins are long enough and slender enough to reach these binding sites, whereas antibody molecules made against HIV are too large to make contact with the sites. As a result, it is difficult for these antibodies to destroy HIV.

Cytopathic Effects of Viruses

Infection of a host cell by an animal virus usually kills the host cell (see Chapter 12). Death can be caused by the accumulation of large numbers of multiplying viruses, by the effects of viral proteins on the permeability of the host cell's plasma membrane, or by inhibition of host DNA, RNA, or protein synthesis. The visible effects of viral infection are known as **cytopathic effects (CPE).** Those cytopathic effects that result in cell death are called *cytocidal effects;* those that result in cell damage but not cell death are called *noncytocidal effects.* CPEs are used to diagnose many viral infections.

Cytopathic effects vary with the virus. One difference is the point in the viral infection cycle at which the effects occur. Some viral infections result in early changes in the host cell; in other infections, changes are not seen until a much later stage. A virus can produce one or more of the following cytopathic effects:

1. At some stage in their multiplication, cytocidal viruses cause the macromolecular synthesis within the host cell to stop. Some viruses, such as herpes simplex virus, irreversibly stop mitosis.

2. When a cytocidal virus infects a cell, it causes the cell's lysosomes to release their enzymes, resulting in destruction of intracellular contents and host cell death.

3. **Inclusion bodies** are granules found in the cytoplasm or nucleus of some infected cells (Figure 15.7a). These granules are sometimes viral parts—nucleic acids or proteins in the process of being assembled into virions. The granules vary in size, shape, and staining properties, according to the virus. Inclusion bodies are characterized by their ability to stain with an acidic stain (acidophilic) or with a basic stain (basophilic). Other inclusion bodies arise at sites of earlier viral synthesis but do not contain assembled viruses or their components. Inclusion bodies are important because their presence can help identify the virus causing an infection. For example, rabies virus produces inclusion bodies (Negri bodies) in the cytoplasm of nerve cells, and their presence in the brain tissue of animals suspected of being rabid has been used as a diagnostic tool for rabies. Diagnostic inclusion bodies are also associated with measles virus, vaccinia virus, smallpox virus, herpesvirus, and adenoviruses.

4. At times, several adjacent infected cells fuse to form a very large multinucleate cell called a **syncytium** (Figure 15.7b). Such giant cells are produced from infections by viruses that cause diseases, such as measles, mumps, and the common cold (paramyxoviruses); some herpesviruses; and other viruses.

5. Some viral infections result in changes in the host cell's functions with no visible changes in the infected cells. For example, when measles virus attaches to its receptor called CD46, the CD46 prompts the cell to reduce production of an im-

(a)

TEM   .25 μm

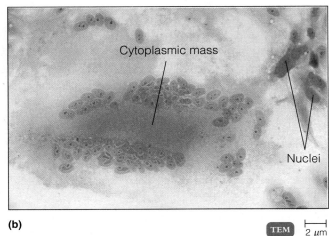

Cytoplasmic mass

Nuclei

(b)

TEM   2 μm

FIGURE 15.7 **Some cytopathic effects of viruses. (a)** Cytoplasmic inclusion body in a brain cell from a person who died of rabies. **(b)** Portion of a syncytium (giant cell) formed in a cell infected with measles. The cytoplasmic mass is probably the Golgi complexes of fused cells.

■ The visible effects that can lead to death or damage of a host cell are called cytopathic effects.

mune substance called IL-12, reducing the host's ability to fight the infection.

6. Some virus-infected cells produce substances called **interferons.** Viral infection induces cells to produce interferons, but the host cell's DNA actually codes for the interferon. This protects neighboring uninfected cells from viral infection. (Interferons will be discussed further in Chapter 16 on page 476.)

7. Many viral infections induce antigenic changes on the surface of the infected cells. These antigenic changes elicit a host antibody response against the infected cell, and thus they target the cell for destruction by the host's immune system.

8. Some viruses induce chromosomal changes in the host cell. For example, some viral infections result in chromosomal damage to the host cell, most often chromosomal breakage. Frequently oncogenes (cancer-causing genes) may be contributed or activated by a virus.

9. Most normal cells cease growing in vitro when they come close to another cell, a phenomenon known as **contact inhibition.** Viruses capable of causing cancer *transform* host cells, as discussed in Chapter 13. Transformation results in an abnormal, spindle-shaped cell that does not recognize contact inhibition (Figure 15.8). The loss of contact inhibition results in unregulated cell growth.

LM   100 μm

FIGURE 15.8 **Transformed cells in culture.** In the center of this photomicrograph is a cluster of chick embryo cells transformed by Rous sarcoma virus. Such a cluster results from the multiplication of a single cell infected with a transforming virus. Notice how the transformed cells appear dark, in contrast to the monolayer of light, flat, normal cells around them. This appearance is caused by their spindle shapes and their uninhibited growth due to the absence of contact inhibition.

■ What is contact inhibition?

TABLE 15.4	Cytopathic Effects of Selected Viruses	
Virus (Genus)	**Cytopathic Effect**
Poliovirus (Enterovirus)	Cytocidal (cell death)
Papovavirus (family Papovaviridae)	Acidophilic inclusion bodies in nucleus
Adenovirus (Mastadenovirus)	Basophilic inclusion bodies in nucleus
Rhabdovirus (family Rhabdoviridae)	Acidophilic inclusion bodies in cytoplasm
Cytomegalovirus	Acidophilic inclusion bodies in nucleus and cytoplasm
Measles virus (Morbillivirus)	Cell fusion
Polyomavirus	Transformation
HIV (Lentivirus)	Destruction of T cells

Some representative viruses that cause cytopathic effects are presented in Table 15.4. In subsequent chapters, we will discuss the pathological properties of viruses in more detail.

Pathogenic Properties of Fungi, Protozoa, Helminths, and Algae

Learning Objective

- Discuss the causes of symptoms in fungal, protozoan, helminthic, and algal diseases.

This section describes some general pathological effects of fungi, protozoa, helminths, and algae that cause human disease.

Fungi

Although fungi cause disease, they do not have a well-defined set of virulence factors. Some fungi have metabolic products that are toxic to human hosts. In such cases, however, the toxin is only an indirect cause of disease, as the fungus is already growing in or on the host. Chronic fungal infections, such as from molds growing in homes, can also provoke an allergic response in the host.

Trichothecenes are fungal toxins that inhibit protein synthesis in eukaryotic cells. Ingestion of these toxins causes headaches, chills, severe nausea, vomiting, and vi-sual disturbances. These toxins are produced by *Fusarium* (fu′sãr-ē-um) growing on grains, such as wheat and rice.

There is evidence that some fungi do have virulence factors. Two fungi that can cause skin infections, *Candida albicans* and *Trichophyton* (trik-ō-fī′ton), secrete proteases. These enzymes may modify host cell membranes to allow attachment of the fungi. *Cryptococcus neoformans* (krip-tō-kok′kus nē-ō-fôr′manz) is a fungus that causes a type of meningitis; it produces a capsule that helps it resist phagocytosis. Some fungi have become resistant to antifungal drugs by decreasing their synthesis of receptors for these drugs.

The disease called ergotism, which was common in Europe during the Middle Ages, is caused by a toxin produced by an ascomycete plant pathogen, *Claviceps purpurea* (kla′vi-seps p̆ur-p̆u-rē′ä), that grows on grains. The toxin is contained in **sclerotia,** highly resistant portions of the mycelia of the fungus that can detach. The toxin itself, **ergot,** is an alkaloid that can cause hallucinations resembling those produced by LSD; in fact, ergot is a natural source of LSD. Ergot also constricts capillaries and can cause gangrene of the limbs by preventing proper blood circulation in the body. Although *C. purpurea* still occasionally occurs on grains, modern milling usually removes the sclerotia.

Several other toxins are produced by fungi that grow on grains or other plants. For example, peanut butter is occasionally recalled because of excessive amounts of **aflatoxin,** a toxin that has carcinogenic properties. Aflatoxin is produced by the growth of the mold *Aspergillus flavus*. When ingested, the toxin might be altered in a human body to a mutagenic compound.

A few mushrooms produce toxins called **mycotoxins** (toxins produced by fungi). Examples are **phalloidin** and **amanitin,** produced by *Amanita phalloides* (am-an-ī′ta fal-loi′dēz), commonly known as the death angel. These neurotoxins are so potent that ingestion of the *Amanita* mushroom may result in death.

Protozoa

The presence of protozoa and their waste products often produces disease symptoms in the host (see Table 12.5, page 359). Some protozoa, such as *Plasmodium*, the causative agent of malaria, invade host cells and reproduce within them, causing their rupture. *Toxoplasma* attaches to macrophages and gains entry by phagocytosis. The parasite prevents normal acidification and digestion; thus, it can grow in the phagocytic vacuole. Other protozoa, such as *Giardia lamblia*, the causative agent of giardiasis, attach to host cells and digest the cells and tissue fluids.

Some protozoa can evade host defenses and cause disease for very long periods of time. For example, *Giardia*, which causes diarrhea, and *Trypanosoma*, which causes African trypanosomiasis (sleeping sickness), both have a mechanism that enables them to stay one step ahead of the host's immune system. The immune system is alerted to recognize foreign substances called antigens; the presence of antigens causes the immune system to produce antibodies designed to destroy them (see Chapter 17). When *Trypanosoma* is introduced into the bloodstream by a tsetse fly, it produces and displays a specific antigen. In response, the body produces antibodies against that antigen. However, within 2 weeks, the microbe stops displaying the original antigen and instead produces and displays a different one. Thus, the original antibodies are no longer effective. Because the microbe can make up to 1000 different antigens, such an infection can last for decades.

Helminths

The presence of helminths also often produces disease symptoms in a host (see Table 12.6, page 369). Some of these organisms actually use host tissues for their own growth or produce large parasitic masses; the resulting cellular damage evokes the symptoms. An example is the roundworm *Wuchereria bancrofti* (vū-kėr-ār′ē-ä ban-krof′tē), the causative agent of elephantiasis. This parasite blocks lymphatic circulation, leading to an accumulation of lymph and eventually causing grotesque swelling of the legs and other body parts. Waste products of the metabolism of these parasites can also contribute to the symptoms of a disease.

Algae

A few species of algae produce neurotoxins. For example, some genera of dinoflagellates, such as *Alexandrium*, are important medically because they produce a neurotoxin called **saxitoxin.** Although mollusks that feed on the dinoflagellates that produce saxitoxin show no symptoms of disease, people who eat the mollusks develop symptoms similar to botulism called paralytic shellfish poisoning. Public health agencies frequently prohibit human consumption of mollusks during red tides.

Portals of Exit

Learning Objective
• Compare and contrast portal of entry and portal of exit.

In the beginning of the chapter, you learned how microbes enter the body through a preferred route, or portal of entry.

Microbes also leave the body via specific routes called **portals of exit** in secretions, excretions, discharges, or tissue that has been shed. In general, portals of exit are related to the part of the body that has been infected. Thus, in general, a microbe uses the same portal for entry and exit. By using various portals of exit, pathogens can spread through a population by moving from one susceptible host to another. As you learned in Chapter 14, this type of information about the dissemination of a disease is very important to epidemiologists.

The most common portals of exit are the respiratory and gastrointestinal tracts. For example, many pathogens living in the respiratory tract exit in discharges from the mouth and nose; these discharges are expelled during coughing or sneezing. These microorganisms are found in droplets formed from mucus. Pathogens that cause tuberculosis, whooping cough, pneumonia, scarlet fever, meningococcal meningitis, chickenpox, measles, mumps, smallpox, and influenza are discharged through the respiratory route. Other pathogens exit via the gastrointestinal tract in feces or saliva. Feces may be contaminated with pathogens associated with salmonellosis, cholera, typhoid fever, shigellosis, amoebic dysentery, and poliomyelitis. Saliva can also contain pathogens, such as those that cause rabies, mumps, and infectious mononucleosis.

Another important route of exit is the genitourinary tract. Microbes responsible for sexually transmitted diseases are found in secretions from the penis and vagina. Urine can also contain the pathogens responsible for typhoid fever and brucellosis, which can exit via the urinary tract. Skin or wound infections are other portals of exit. Infections transmitted from the skin include yaws, impetigo, skin ringworm, herpes simplex, and warts. Drainage from wounds can spread infections to another person directly or by contact with a contaminated fomite. Infected blood can be removed and reinjected by biting insects and contaminated needles and syringes to spread infection within a population. Examples of diseases transmitted by biting insects are yellow fever, plague, tularemia, and malaria. AIDS and hepatitis B may be transmitted by contaminated needles and syringes.

* * *

In the next chapter, we will examine a group of nonspecific defenses of the host against disease. But before proceeding, examine Figure 15.9 carefully. It summarizes some key concepts of the microbial mechanisms of pathogenicity we have discussed in this chapter.

FIGURE 15.9 **Microbial mechanisms of pathogenicity.** A summary of how microorganisms cause disease.

STUDY OUTLINE

Introduction (p. 437)

1. Pathogenicity is the ability of a pathogen to produce a disease by overcoming the defenses of the host.

2. Virulence is the degree of pathogenicity.

How Microorganisms Enter a Host (pp. 437–441)

1. The specific route by which a particular pathogen gains access to the body is called its portal of entry.

Portals of Entry (pp. 437–438)

1. Many microorganisms can penetrate mucous membranes of the conjunctiva and the respiratory, gastrointestinal, and genitourinary tracts.

2. Microorganisms that are inhaled with droplets of moisture and dust particles gain access to the respiratory tract.

3. The respiratory tract is the most common portal of entry.

4. Microorganisms that gain access via the genitourinary tract can enter the body through mucous membranes.

5. Microorganisms enter the gastrointestinal tract via food, water, and contaminated fingers.

6. Most microorganisms cannot penetrate intact skin; they enter hair follicles and sweat ducts.

7. Some microorganisms can gain access to tissues by inoculation through the skin and mucous membranes in bites, injections, and other wounds. This route of penetration is called the parenteral route.

The Preferred Portal of Entry (p. 438)

1. Many microorganisms can cause infections only when they gain access through their specific portal of entry.

Numbers of Invading Microbes (p. 438)

1. Virulence can be expressed as LD_{50} (lethal dose for 50% of the inoculated hosts) or ID_{50} (infectious dose for 50% of the inoculated hosts).

Adherence (pp. 438–441)

1. Surface projections on a pathogen called adhesins (ligands) adhere to complementary receptors on the host cells.

2. Adhesins can be glycoproteins or lipoproteins and are frequently associated with fimbriae.

3. Mannose is the most common receptor.

4. Biofilms provide attachment and resistance to antimicrobial agents.

How Bacterial Pathogens Penetrate Host Defenses (pp. 441–443)

Capsules (p. 441)

1. Some pathogens have capsules that prevent them from being phagocytized.

Components of the Cell Wall (p. 442)

1. Proteins in the cell wall can facilitate adherence or prevent a pathogen from being phagocytized.

2. Some microbes can reproduce inside phagocytes.

Enzymes (p. 442)

1. Leukocidins destroy neutrophils and macrophages.

2. Local infections can be protected in a fibrin clot caused by the bacterial enzyme coagulase.

3. Bacteria can spread from a focal infection by means of kinases (which destroy blood clots), hyaluronidase (which destroys a mucopolysaccharide that holds cells together), and collagenase (which hydrolyzes connective tissue collagen).

4. IgA proteases destroy IgA antibodies.

Antigenic Variation (p. 442)

1. Some microbes vary expression of antigens, thus avoiding the host's antibodies.

Penetration into the Host Cell Cytoskeleton (p. 443)

1. *Salmonella* bacteria produce invasins, proteins that cause the actin of the host cell's cytoskeleton to form a basket to carry the bacteria into the cell.

How Bacterial Pathogens Damage Host Cells (pp. 443–449)

Using the Host's Nutrients (pp. 443–444)

1. Bacteria get iron from the host using siderophores.

Direct Damage (p. 444)

1. Host cells can be destroyed when pathogens metabolize and multiply inside the host cells.

The Production of Toxins (pp. 444–448)

1. Poisonous substances produced by microorganisms are called toxins; toxemia refers to the presence of toxins in the blood. The ability to produce toxins is called toxigenicity.

2. Exotoxins are produced by bacteria and released into the surrounding medium. Exotoxins, not the bacteria, produce the disease symptoms.

3. Antibodies produced against exotoxins are called antitoxins.

4. Exotoxins occur as A-B toxins, membrane-disrupting toxins, and superantigens.

5. Cytotoxins include diphtheria toxin (which inhibits protein synthesis) and erythrogenic toxins (which damage capillaries).

6. Neurotoxins include botulinum toxin (which prevents nerve transmission) and tetanus toxin (which prevents inhibitory nerve transmission).

7. *Vibrio cholerae* toxin and staphylococcal enterotoxin are enterotoxins, which induce fluid and electrolyte loss from host cells.

8. Endotoxins are lipopolysaccharides (LPS), the lipid A component of the cell wall of gram-negative bacteria.

9. Bacterial cell death, antibiotics, and antibodies may cause the release of endotoxins.

10. Endotoxins cause fever (by inducing the release of interleukin-1) and shock (because of a TNF-induced decrease in blood pressure).

11. Endotoxins allow bacteria to cross the blood–brain barrier.

12. The *Limulus* amoebocyte lysate (LAL) assay is used to detect endotoxins in drugs and on medical devices.

Plasmids, Lysogeny, and Pathogenicity (pp. 448–449)

1. Plasmids may carry genes for antibiotic resistance, toxins, capsules, and fimbriae.

2. Lysogenic conversion can result in bacteria with virulence factors, such as toxins or capsules.

Pathogenic Properties of Viruses (pp. 450–451)

1. Viruses avoid the host's immune response by growing inside cells.

2. Viruses gain access to host cells because they have attachment sites for receptors on the host cell.

3. Visible signs of viral infections are called cytopathic effects (CPE).

4. Some viruses cause cytocidal effects (cell death), and others cause noncytocidal effects (damage but not death).

5. Cytopathic effects include the stopping of mitosis, lysis, the formation of inclusion bodies, cell fusion, antigenic changes, chromosomal changes, and transformation.

Pathogenic Properties of Fungi, Protozoa, Helminths, and Algae (pp. 452–453)

1. Symptoms of fungal infections can be caused by capsules, toxins, and allergic responses.

2. Symptoms of protozoan and helminthic diseases can be caused by damage to host tissue or by the metabolic waste products of the parasite.

3. Some protozoa change their surface antigens while growing in a host so that the host's antibodies don't kill the protozoa.

4. Some algae produce neurotoxins that cause paralysis when ingested by humans.

Portals of Exit (pp. 453–454)

1. Just as pathogens have preferred portals of entry, they also have definite portals of exit.

2. Three common portals of exit are the respiratory tract via coughing or sneezing, the gastrointestinal tract via saliva or feces, and the urogenital tract via secretions from the vagina or penis.

3. Arthropods and syringes provide a portal of exit for microbes in blood.

STUDY QUESTIONS

Review

1. List three portals of entry, and describe how microorganisms gain access through each.

2. Compare pathogenicity with virulence.

3. Explain how drugs that bind each of the following would affect pathogenicity:
   a. mannose on human cell membranes
   b. *Neisseria gonorrhoeae* fimbriae
   c. *Streptococcus pyogenes* M protein

4. Define cytopathic effects, and give five examples.

5. Compare and contrast the following aspects of endotoxins and exotoxins: bacterial source, chemistry, toxicity, and pharmacology. Give an example of each toxin.

6. How are capsules and cell wall components related to pathogenicity? Give specific examples.

7. Describe how hemolysins, leukocidins, coagulase, kinases, hyaluronidase, siderophores, and IgA proteases might contribute to pathogenicity.

8. Describe the factors contributing to the pathogenicity of fungi, protozoa, and helminths.

9. Which of the following genera is the most infectious?

Genus	ID_{50}	Genus	ID_{50}
Legionella	1 cell	Shigella	200 cells
Salmonella	10^5 cells	Treponema	52 cells

10. The LD_{50} of botulinum toxin is 0.000025 μg. The LD_{50} of *Salmonella* toxin is 200 μg. Which of these is the more potent toxin? How can you tell from the LD_{50} values?

11. Food poisoning can be divided into two categories: food infection and food intoxication. On the basis of toxin production by bacteria, explain the difference between these two categories.

12. How can viruses and protozoa avoid being killed by the host's immune response?

Multiple Choice

1. The removal of plasmids reduces virulence in which of the following organisms?
   a. *Clostridium tetani*
   b. *Escherichia coli*
   c. *Staphylococcus aureus*
   d. *Streptococcus mutans*
   e. *Clostridium botulinum*

2. What is the LD_{50} for the bacterial toxin tested in the example below?

Dilution	No. of Animals Died	No. of Animals Survived
a. 6 μg/kg	0	6
b. 12.5 μg/kg	0	6
c. 25 μg/kg	3	3
d. 50 μg/kg	4	2
e. 100 μg/kg	6	0

3. Which of the following is *not* a portal of entry for pathogens?
   a. mucous membranes of the respiratory tract
   b. mucous membranes of the gastrointestinal tract
   c. skin
   d. blood
   e. parenteral route

4. All of the following can occur during bacterial infection. Which would prevent all of the others?
   a. vaccination against fimbriae
   b. phagocytosis
   c. inhibition of phagocytic digestion
   d. destruction of adhesins
   e. alteration of cytoskeleton

5. The ID_{50} for *Campylobacter* sp. is 500 cells; the ID_{50} for *Cryptosporidium* sp. is 100 cells. Which of the following statements is *not* true?
   a. Both microbes are pathogens.
   b. Both microbes produce infections in 50% of the inoculated hosts.
   c. *Cryptosporidium* is more virulent than *Campylobacter*.
   d. *Campylobacter* and *Cryptosporidium* are equally virulent; they cause infections in the same number of test animals.
   e. The severity of infections caused by *Campylobacter* and *Cryptosporidium* cannot be determined by the information provided.

6. An encapsulated bacterium can be virulent because the capsule
   a. resists phagocytosis.
   b. is an endotoxin.
   c. destroys host tissues.
   d. interferes with physiological processes.
   e. has no effect; since many pathogens do not have capsules, capsules do not contribute to virulence.

7. A drug that binds to mannose on human cells would prevent
   a. the entrance of *Vibrio* enterotoxin.
   b. the attachment of pathogenic *E. coli*.
   c. the action of botulinum toxin.
   d. streptococcal pneumonia.
   e. the action of diphtheria toxin.

8. The earliest smallpox vaccines were infected tissue rubbed into the skin of a healthy person. The recipient of such a vaccine usually developed a mild case of smallpox, recovered, and was immune thereafter. The most likely reason this vaccine did not kill more people is:
   a. Skin is the wrong portal of entry for smallpox.
   b. The vaccine consisted of a mild form of the virus.
   c. Smallpox is normally transmitted by skin-to-skin contact.

d. Smallpox is a virus.

e. The virus mutated.

9. Which of the following does *not* represent the same mechanism for avoiding host defenses as the others?

a. Rabiesvirus attaches to the receptor for the neurotransmitter acetylcholine.

b. *Salmonella* attaches to the receptor for epidermal growth factor.

c. EB virus binds to the host receptor for C3.

d. Surface protein genes in *Neisseria gonorrhoeae* mutate frequently.

e. none of the above

10. Which of the following statements is true?

a. The primary goal of a pathogen is to kill its host.

b. Evolution selects for the most virulent pathogens.

c. A successful pathogen doesn't kill its host before it is transmitted.

d. A successful pathogen never kills its host.

Critical Thinking

1. How can plasmids and lysogeny turn the normally harmless *E. coli* into a pathogen? Using the graph below, showing confirmed cases of enteropathogenic *E. coli*, what is the usual portal of entry for this bacterium?

2. The cyanobacterium *Microcystis aeruginosa* produces a peptide that is toxic to humans. According to the graph below, when is this bacterium most toxic?

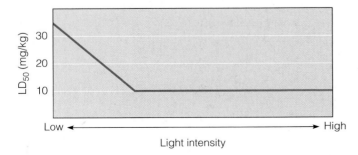

3. When injected into rats, the ID_{50} for *Salmonella typhimurium* is 10^6 cells. If sulfonamides are injected with the salmonellae, the ID_{50} is 35 cells. Explain the change in ID_{50} value.

4. How do each of the following strategies contribute to the virulence of the pathogen? What disease does each organism cause?

Strategy	Pathogen
Changes its cell wall after entry into host	*Yersinia pestis*
Uses urea to produce ammonia	*Helicobacter pylori*
Causes host to make more receptors	*Rhinovirus*
Binds to site for epidermal growth factor on human cells	*Salmonella*

Clinical Applications

1. On July 8, a woman was given an antibiotic for presumptive sinusitis. However, her condition worsened, and she was unable to eat for 4 days because of severe pain and tightness of the jaw. On July 12, she was admitted to a hospital with severe facial spasms. She reported that on July 5 she had incurred a puncture wound at the base of her big toe; she cleaned the wound but did not seek medical attention. What caused her symptoms? Was her condition due to an infection or an intoxication? Can she transmit this condition to another person?

2. Explain whether each of the following examples is a food infection or intoxication. What is the probable etiological agent in each case?

a. Eighty-two people who ate shrimp at a dinner in Port Allen, Louisiana, developed diarrhea, cramps, weakness, nausea, chills, headache, and fever from 4 hours to 2 days after eating.

b. Two people in Vermont who ate barracuda caught in Florida developed malaise, nausea, blurred vision, breathing difficulty, and numbness 3–6 hours after eating.

3. Washwater containing *Pseudomonas* was sterilized and used to wash cardiac catheters. Three patients undergoing cardiac catheterization developed fever, chills, and hypotension. The water and catheters were sterile. Why did the patients show these reactions?

4. Cancer patients undergoing chemotherapy are normally *more* susceptible to infections. However, a patient receiving an antitumor drug that inhibited cell division was resistant to *Salmonella*. Provide a possible mechanism for the resistance.

LEARNING WITH TECHNOLOGY

The Microbiology Place CD-ROM/website includes many resources to help you study and succeed in this course. In addition to self-quizzes, web links, and flashcards, you'll find the following activities for this chapter in the "Activities" section of The Microbiology Place:

Host and Pathogen

Case Study: Food Poisoning, New Mexico

Chapter 16

Nonspecific Defenses of the Host

A PHAGOCYTE. *Phagocytes provide protection against pathogens. This macrophage is engulfing bacteria which will prevent the bacteria from colonizing and causing disease.*

*F*rom our discussion to this point, you can see that pathogenic microorganisms are endowed with special properties that enable them to cause disease if given the right opportunity. If microorganisms never encountered resistance from the host, we would constantly be ill and would eventually die of various diseases. But in most cases, our body's defenses prevent this from happening.

Some of these defenses are designed to keep out microorganisms altogether, other defenses remove the microorganisms if they do get in, and still others combat them if they remain inside. Our ability to ward off disease through our defenses is called **resistance.** Vulnerability or lack of resistance is known as **susceptibility.**

In discussing resistance, we will divide our body defenses into two general kinds: nonspecific and specific (Figure 16.1). **Nonspecific resistance** refers to defenses that protect us against any pathogen, regardless of species. These include a first line of defense (skin and mucous membranes) and a second line of defense (phagocytes, inflammation, fever, and antimicrobial substances). **Specific resistance,** or **immunity,** is a third line of defense the body provides against particular pathogens. Specific defenses are based on specialized cells of the immune system called lymphocytes (a type of white blood cell) and the production of specific proteins called antibodies. Chapter 17 deals with specific resistance. In this chapter, we consider nonspecific resistance.

Nonspecific Resistance		Specific Resistance (Responses of the Immune System, Chapter 17)
First line of defense	**Second line of defense**	**Third line of defense**
• Intact skin • Mucous membranes and their secretions • Normal microbiota	• Phagocytic white blood cells • Inflammation • Fever • Antimicrobial substances	• Specialized lymphocytes: B cells and T cells • Antibodies

FIGURE 16.1 **An overview of the body's defenses.** Nonspecific resistance refers to defenses against any pathogen, regardless of species; specific resistance refers to defenses against a specific pathogen.

Skin and Mucous Membranes

Learning Objectives

- *Describe the role of the skin and mucous membranes in nonspecific resistance.*

- *Differentiate mechanical from chemical factors, and list five examples of each.*

- *Describe the role of normal microbiota in nonspecific resistance.*

The skin and mucous membranes are the body's first line of defense against pathogens. This function results from both mechanical and chemical factors. Whereas mechanical factors include physical barriers to entry or processes that remove microbes from the body's surface, chemical factors include substances made by the body that inhibit microbial growth or destroy them.

Mechanical Factors

The intact **skin** is one of the human body's largest organs in terms of surface area. It consists of two distinct portions: the dermis and the epidermis (Figure 16.2). The **dermis,** the skin's inner, thicker portion, is composed of connective tissue. The **epidermis,** the outer, thinner portion, is in direct contact with the external environment. The epidermis consists of many layers of continuous sheets of tightly packed epithelial cells with little or no material between the cells. The top layer of epidermal cells is dead and contains a protective protein called **keratin.** In addition, the dryness of the skin is a major factor in inhibiting microbial growth on the skin. Although normal microbiota and other microbes are present on the entire skin, they are most numerous on moist areas of the skin. When the skin is moist, as in hot, humid climates, skin infections are quite common, especially fungal infections such as athlete's foot. These fungi hydrolyze keratin when water is available. In addition, periodic shedding of the epidermis helps remove microbes on the skin's surface.

If we consider the closely packed cells, continuous layering, the presence of keratin and the dryness and shedding of the skin, we can see why the intact skin provides such a formidable barrier to the entrance of microorganisms. The intact surface of healthy epidermis is rarely, if ever, penetrated by microorganisms. But when the epithelial surface

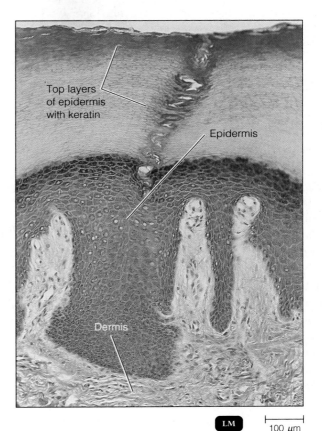

FIGURE 16.2 **A section through human skin.** The thin layers at the top of this photomicrograph contain keratin. These layers and the darker purple cells beneath them make up the epidermis. The lighter purple material near the bottom is the dermis.

■ The intact skin and mucous membranes are components of the first line of defense against pathogens.

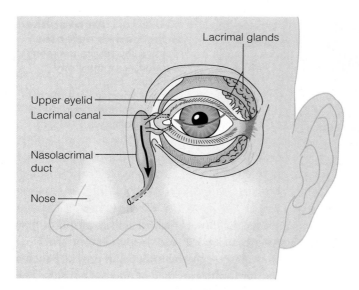

Lacrimal glands

Upper eyelid

Lacrimal canal

Nasolacrimal duct

Nose

FIGURE 16.3 **The lacrimal apparatus.** The washing action of the tears is shown by the arrow passing over the surface of the eyeball. Tears produced by the lacrimal glands pass across the surface of the eyeball into two small holes that convey the tears into the lacrimal canals and the nasolacrimal duct.

■ The lacrimal apparatus protects the eye by continually producing tears that wash away microorganisms.

is broken, a subcutaneous (below-the-skin) infection often develops. The bacteria most likely to cause infection are the staphylococci that normally inhabit the epidermis, hair follicles, and sweat and oil glands of the skin. Infections of the skin and underlying tissues frequently occur as a result of burns, cuts, stab wounds, or other conditions that break the skin.

Epithelial cells that line blood and lymphatic vessels are not closely packed like those of the epidermis. Although this arrangement permits defensive cells to move from blood into tissues during inflammation, it also permits microbes to move into and out of blood and lymph.

Mucous membranes also consist of an epithelial layer and an underlying connective tissue layer. Although mucous membranes inhibit the entrance of many microorganisms, they offer less protection than the skin. Mucous membranes line the entire gastrointestinal, respiratory, and genitourinary tracts. The epithelial layer of a mucous membrane secretes a fluid called **mucus,** a slightly viscous (thick) glycoprotein produced by goblet cells of a mucous membrane. Among other functions, mucus prevents the tracts from drying out. Some pathogens that can thrive on the moist secretions of a mucous membrane are able to penetrate the membrane if the microorganism is present in sufficient numbers. *Treponema pallidum* is such a pathogen.

This penetration may be facilitated by toxic substances produced by the microorganism, prior injury by viral infection, or mucosal irritation.

Besides the physical barrier presented by the skin and mucous membranes, several other mechanical factors help protect certain epithelial surfaces. One such mechanism that protects the eyes is the **lacrimal apparatus,** a group of structures that manufactures and drains away tears (Figure 16.3). The lacrimal glands, located toward the upper, outermost portion of each eye socket, produce the tears and pass them under the upper eyelid. From here, tears pass toward the corner of the eye near the nose and into two small holes that lead through tubes (lacrimal canals) to the nose. The tears are spread over the surface of the eyeball by blinking. Normally, the tears evaporate or pass into the nose as fast as they are produced. This continual washing action helps keep microorganisms from settling on the surface of the eye. If an irritating substance or large numbers of microorganisms come in contact with the eye, the lacrimal glands start to secrete heavily, and the tears accumulate more rapidly than they can be carried away. This excessive production is a protective mechanism because the excess tears dilute and wash away the irritating substance or microorganisms.

In a cleansing action very similar to that of tears, **saliva** is produced by the salivary glands. Saliva helps dilute the numbers of microorganisms and wash them from both the surface of the teeth and the mucous membrane of the mouth. This helps prevent colonization by microbes.

The respiratory and gastrointestinal tracts have many mechanical forms of defense. Mucus traps many of the microorganisms that enter the respiratory and gastrointestinal tracts. The mucous membrane of the nose also has mucus-coated hairs that filter inhaled air and trap microorganisms, dust, and pollutants. The cells of the mucous membrane of the lower respiratory tract are covered with cilia. By moving synchronously, these cilia propel inhaled dust and microorganisms that have become trapped in mucus upward toward the throat. This so-called **ciliary escalator** (Figure 16.4) keeps the mucus blanket moving toward the throat at a rate of 1–3 cm per hour; coughing and sneezing speed up the escalator. (Some substances in cigarette smoke are toxic to cilia and can seriously impair the functioning of the ciliary escalator by inhibiting or destroying the cilia.) Microorganisms are also prevented from entering the lower respiratory tract by a small lid of cartilage called the **epiglottis,** which covers the larynx (voicebox) during swallowing.

The cleansing of the urethra by the flow of **urine** is another mechanical factor that prevents microbial colonization in the genitourinary tract. **Vaginal secretions** likewise move microorganisms out of the female body.

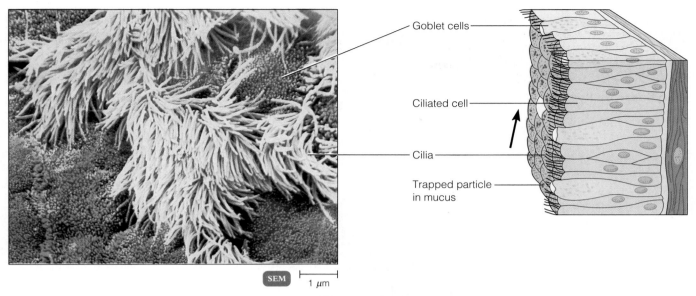

FIGURE 16.4 **The ciliary escalator.** Microorganisms in the lower respiratory tract are trapped in mucus produced by goblet cells, then propelled upward by the synchronized beating of cilia.

Chemical Factors

Mechanical factors alone do not account for the high degree of resistance of skin and mucous membranes to microbial invasion. Certain chemical factors also play important roles.

Sebaceous (oil) glands of the skin produce an oily substance called **sebum** that prevents hair from drying and becoming brittle. Sebum also forms a protective film over the surface of the skin. One of the components of sebum is unsaturated fatty acids, which inhibit the growth of certain pathogenic bacteria and fungi. The low pH of the skin, between pH 3 and 5, is caused in part by the secretion of fatty acids and lactic acid. The skin's acidity probably discourages the growth of many other microorganisms.

Bacteria that live commensally on the skin decompose sloughed-off skin cells, and the resultant organic molecules and the end-products of their metabolism produce body odor. Certain bacteria commonly found on the skin metabolize sebum, and this metabolism forms free fatty acids that cause the inflammatory response associated with acne. Isotretinoin (Accutane), a derivative of vitamin A that prevents sebum formation, is a treatment for a very severe type of acne called cystic acne.

The sweat glands of the skin produce **perspiration,** which helps maintain body temperature, eliminate certain wastes, and flush microorganisms from the surface of the skin. Perspiration also contains **lysozyme,** an enzyme capable of breaking down cell walls of gram-positive bacteria and, to a lesser extent, gram-negative bacteria (see Figure 4.13 on page 84). Specifically, lysozyme breaks chemical bonds on peptidoglycan, which destroys the cell walls. Lysozyme is also found in tears, saliva, nasal secretions, and tissue fluids, where it exhibits its antimicrobial activity. Alexander Fleming was studying lysozyme in 1929 when he accidentally discovered the antimicrobial effects of penicillin (see Figure 20.1 on page 560).

Gastric juice is produced by the glands of the stomach. It is a mixture of hydrochloric acid, enzymes, and mucus. The very high acidity of gastric juice (pH 1.2–3.0) is sufficient to destroy bacteria and most bacterial toxins, except those of *Clostridium botulinum* and *Staphylococcus aureus*. However, many enteric pathogens are protected by food particles and can enter the intestines via the gastrointestinal tract. The bacterium *Helicobacter pylori* neutralizes stomach acid which allows the bacterium to grow in the stomach. Its growth results in ulcers and gastritis.

Vaginal secretions are also slightly acidic, which discourages bacterial growth. Blood also contains antimicrobial chemicals. For example, iron-binding proteins, called **transferrins,** inhibit bacterial growth by reducing the amount of available iron. Iron is not only required for microbial growth (the synthesis of cytochromes and certain enzymes), it also suppresses chemotaxis and phagocytosis. Iron overload increases the risk of infection.

Normal Microbiota and Nonspecific Resistance

In Chapter 14, several relationships between normal microbiota and host cells were described. Some of these

TABLE 16.1	**Formed Elements in Blood**		
Type of Cell		**Numbers per Microliter (μl) or Cubic mm (mm^3)**	**Function**
Erythrocytes (Red Blood Cells)		4.8–5.4 million	Transport of O_2 and CO_2
Leukocytes (White Blood Cells) A. Granulocytes (stained) 1. Neutrophils (PMNs) (60–70% of leukocytes)		5000–10,000	Phagocytosis
2. Basophils (0.5–1%)			Production of histamine
3. Eosinophils (2–4%)			Production of toxic proteins against certain parasites; some phagocytosis

relationships assist in preventing the overgrowth of pathogens and thus may be considered components of nonspecific defense. For example, in microbial antagonism, the normal microbiota prevent pathogens from colonizing the host by competing with them for nutrients (competitive exclusion), by producing substances that are harmful to the pathogens, and by altering conditions that affect the survival of the pathogens, such as pH and oxygen availability. The presence of normal microbiota in the vagina, for example, alters pH to prevent overpopulation by *Candida albicans*, a pathogenic yeast that causes vaginitis. In the large intestine, *E. coli* produce bacteriocins that inhibit the growth of *Salmonella* and *Shigella*.

In commensalism, one organism uses the body of a larger organism as its physical environment and may make use of the body to obtain nutrients. Thus in commensalism, one organism benefits while the other is unaffected. Most microbes that are part of the commensal microbiota are found on the skin and in the gastrointestinal tract. The majority of such microbes are bacteria that have highly specialized attachment mechanisms and precise environmental requirements for survival. Normally, such microbes are harmless, but may cause disease if their environmental conditions change. These opportunistic pathogens include *E. coli*, *Staphylococcus aureus*, *S. epidermidis*, *Enterococcus faecalis*, *Pseudomonas aeruginosa*, and oral streptococci.

Type of Cell		Numbers per Microliter (μl) or Cubic mm (mm^3)	Function
B. Agranulocytes (stained) 1. Monocytes (3–8%)			Phagocytosis (when they mature into macrophages)
2. Lymphocytes (20–25%)			Antibody production (descendants of B lymphocytes); cell-mediated immunity (T lymphocytes)*
Platelets		150,000–400,000	Blood clotting

TABLE 16.1 | *(continued)*

*Discussed in Chapter 17.

Phagocytosis

Learning Objective

• *Define phagocytosis and phagocyte.*

Phagocytosis (from Greek words meaning eat and cell) is the ingestion of a microorganism or other particles such as debris by a cell. We have previously mentioned phagocytosis as the method of nutrition of certain protozoa. In this chapter, phagocytosis is discussed as a means by which cells in the human body counter infection—the second line of defense. The cells that perform this function are collectively called **phagocytes,** all of which are types of white blood cells or derivatives of white blood cells. Before we look at phagocytes specifically, let's consider the various components of the blood more generally.

Formed Elements in Blood

Learning Objectives

• *Classify phagocytic cells, and describe the roles of granulocytes and monocytes.*

• *Define differential white blood cell count.*

Blood consists of a fluid called **plasma,** which contains **formed elements**—that is, cells and cell fragments (Table 16.1). Of the cells listed in Table 16.1, those that concern us at present are the **leukocytes,** or white blood cells (Figure 16.5).

During many kinds of infections, especially bacterial infections, the total number of white blood cells increases as a protective response to combat the microbes; this increase is called *leukocytosis.* During the active stage of infection, the leukocyte count might double, triple, or quadruple, depending on the severity of the infection. Diseases that might cause such an elevation in the leukocyte count include meningitis, infectious mononucleosis, appendicitis, pneumococcal pneumonia, and gonorrhea. Other diseases, such

Leukocytes

Phagocytes

Granulocytes

| Basophil | Neutrophil | Eosinophil | Monocyte → Macrophage | Lymphocyte (T cell and B cell) |

(a)   **(b)**   **(c)**

FIGURE 16.5 **Principal kinds of leukocytes. (a)** Granulocytes can be distinguished by staining. **(b)** Monocytes mature into actively phagocytic macrophages. Neutrophils, eosinophils, and macrophages are the phagocytic cells of nonspecific resistance. **(c)** Lymphocytes are important in specific resistance.

■ All phagocytes are leukocytes or cells derived from them.

as salmonellosis and brucellosis, and some viral and rickettsial infections, may cause a *decrease* in the leukocyte count, called *leukopenia*. Leukopenia may be related to either impaired white blood cell production or the effect of increased sensitivity of white blood cell membranes to damage by complement, antimicrobial plasma proteins discussed later in the chapter. Leukocyte increase or decrease can be detected by a **differential white blood cell count,** which is a calculation of the percentage of each kind of white cell in a sample of 100 white blood cells. The percentages in a normal differential white blood cell count are shown in parentheses in the first column of Table 16.1.

Leukocytes are divided into two main categories based on their appearance under a light microscope: granulocytes and agranulocytes. **Granulocytes** owe their name to the presence of large granules in their cytoplasm that can be seen under a light microscope after staining. They are differentiated into three types on the basis of how the granules stain. The granules of **neutrophils** stain pale lilac with a mixture of acidic and basic dyes, those of **basophils** stain blue-purple with the basic dye methylene blue, and those of **eosinophils** stain red or orange with the acidic dye eosin.

Neutrophils are also commonly called *polymorphonuclear leukocytes (PMNs)*, or *polymorphs*. (The term *polymorphonuclear* refers to the fact that the nuclei of neutrophils contain two to five lobes.) Neutrophils, which are highly phagocytic and motile, are active in the initial stages of an infection. They have the ability to leave the blood, enter an infected tissue, and destroy microbes and foreign particles.

The role of basophils is not clear. However, they release substances, such as histamine, that are important in inflammation and allergic responses.

Eosinophils are somewhat phagocytic and also have the ability to leave the blood. Their major function is to produce toxic proteins against certain parasites, such as helminths. Although eosinophils are physically too small to ingest and destroy helminths, they can attach to the outer surface of the parasites and discharge peroxide ions that destroy them. (See Figure 17.18 on page 502.) Their number increases significantly during certain parasitic worm infections and hypersensitivity (allergy) reactions.

Agranulocytes also have granules in their cytoplasm but they are not visible under the light microscope after staining. **Monocytes** are not actively phagocytic until they leave circulating blood, enter body tissues, and mature into **macrophages.** In fact, the maturation and proliferation of macrophages (along with lymphocytes) is one factor responsible for the swelling of lymph nodes during an infection. As blood and lymph that contain microorganisms pass through organs with macrophages, the microorganisms are removed by phagocytosis. Macrophages also dispose of worn out blood cells.

Lymphocytes (T cells and B cells) are not phagocytic but play a key role in specific immunity (see Chapter 17). They occur in lymphoid tissues of the lymphatic system: in the tonsils, spleen, thymus, thoracic duct, red bone marrow, appendix, Peyer's patches of the small intestine, and lymph nodes in the respiratory, gastrointestinal, and reproductive tracts (Figure 16.6). They also circulate in the blood.

Phagocytes may be activated by components of bacteria such as lipid A or lipopolysaccharides (LPS). Among the more important activators are small protein hormones secreted by phagocytes and other cells involved in immunity called cytokines (see Chapter 17).

Tonsil

Right lymphatic duct

Right subclavian vein

Thymus

Lymphatic vessel

Left subclavian vein

Thoracic duct

Spleen

Small intestine

Peyer's patch

Lymph node

Red bone marrow

Lymphatic vessel

(a) Components of lymphatic system

Venule

Tissue cell

Blood

Interstitial fluid

Lymph

Blood capillary

Arteriole

Blood

Lymphatic capillary

(b) Relationship of lymphatic capillaries to tissue cells and blood capillaries

FIGURE 16.6 **The lymphatic system. (a)** Components of the lymphatic system. **(b)** Fluid circulating between tissue cells (interstitial fluid) is picked up by lymphatic capillaries. The fluid is then called lymph. Next, the lymph is passed into lymphatic vessels. Macrophages in lymph nodes at various intervals along lymphatic vessels remove microorganisms from the lymph. Lymph is ultimately collected in the right lymphatic duct and thoracic duct. Eventually, the lymph is returned to the blood from these ducts at the subclavian veins, just before the blood enters the heart.

■ Why do lymph nodes swell during an infection?

Bacteria

SEM   1 μm

FIGURE 16.7 **A macrophage engulfing rod-shaped bacteria.** Macrophages in the mononuclear phagocytic system remove microorganisms after the initial phase of infection.

■ Monocytes enlarge and develop into macrophages.

Actions of Phagocytic Cells

When an infection occurs, both granulocytes (especially neutrophils) and monocytes migrate to the infected area. During this migration, monocytes enlarge and develop into actively phagocytic macrophages (Figure 16.7). These cells leave the blood and migrate into tissues where they enlarge and develop into macrophages. Some macrophages, called **fixed macrophages** or *histiocytes,* are located in certain tissues and organs of the body. Fixed macrophages are found in the liver (Kupffer's cells), lungs (alveolar macrophages), nervous system (microglial cells), bronchial tubes, spleen, lymph nodes, red bone marrow, and the peritoneal cavity surrounding abdominal organs. Other macrophages are called **wandering macrophages,** which roam the tissues and gather at sites of infection or inflammation. The various macrophages of the body constitute the **mononuclear phagocytic (reticuloendothelial) system.**

During the course of an infection, a shift occurs in the type of white blood cell that predominates in the bloodstream. Granulocytes, especially neutrophils, dominate during the initial phase of bacterial infection, at which time they are actively phagocytic; this dominance is indicated by their increased number in a differential white blood cell count. However, as the infection progresses, the macrophages dominate; they scavenge and phagocytize remaining living bacteria and dead or dying bacteria. The increased number of monocytes (which develop into

TABLE 16.2	**Classification and Functions of Phagocytes**	
Type of Phagocyte	**Functional Cells**	**Functions**
Granulocytes	Neutrophils and eosinophils	Phagocytic against microbes during the initial phase of infection
Agranulocytes	Mononuclear phagocytic system: wandering macrophages and fixed macrophages developed from monocytes	Phagocytic against microbes as infection progresses and against worn-out blood cells as infection subsides; also involved in cell-mediated immunity (see Chapter 17)

macrophages) is also reflected in a differential white blood cell count. In viral and fungal infections, macrophages predominate in all phases of defense.

Table 16.2 summarizes the different types of phagocytic cells and their functions.

The Mechanism of Phagocytosis

Learning Objective

• Describe the process of phagocytosis, and include the stages of adherence and ingestion.

How does phagocytosis occur? For the convenience of study, we will divide phagocytosis into four main phases: chemotaxis, adherence, ingestion, and digestion (Figure 16.8).

Chemotaxis

❶ **Chemotaxis** is the chemical attraction of phagocytes to microorganisms. (The mechanism of chemotaxis is discussed in Chapter 4 on page 80.) Among the chemotactic chemicals that attract phagocytes are microbial products, components of white blood cells and damaged tissue cells, and peptides derived from complement, a system of host defense discussed later in the chapter.

Adherence

As it pertains to phagocytosis, **adherence** is the attachment of the phagocyte's plasma membrane to the surface of the microorganism or other foreign material. In some instances, adherence occurs easily, and the microorganism is readily phagocytized. Microorganisms can be more readily phagocytized if they are first coated with certain serum pro-

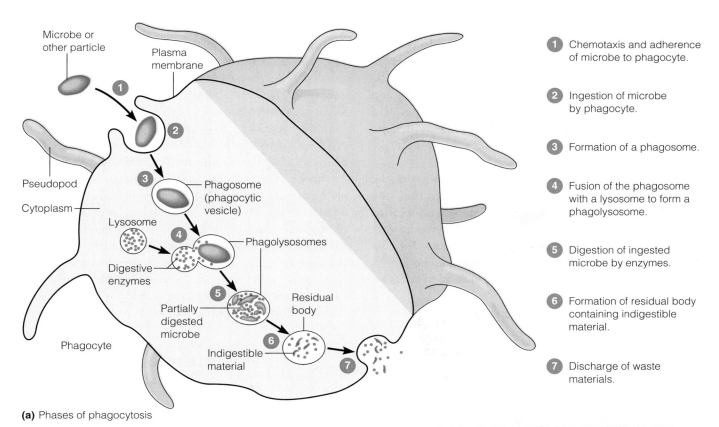

Microbe or
other particle

Plasma
membrane

Pseudopod

Cytoplasm

Lysosome

Digestive
enzymes

Phagocyte

Phagosome
(phagocytic
vesicle)

Phagolysosomes

Partially
digested
microbe

Residual
body

Indigestible
material

1 Chemotaxis and adherence
of microbe to phagocyte.

2 Ingestion of microbe
by phagocyte.

3 Formation of a phagosome.

4 Fusion of the phagosome
with a lysosome to form a
phagolysosome.

5 Digestion of ingested
microbe by enzymes.

6 Formation of residual body
containing indigestible
material.

7 Discharge of waste
materials.

(a) Phases of phagocytosis

FIGURE 16.8 **The mechanism of phagocytosis in a phagocyte.** The phases of phagocytosis are chemotaxis, adherence, ingestion, and digestion.

■ What does *phagocyte* mean?

Bacteria   Phagocyte

(b) A phagocyte engulfing bacteria SEM ├──────┤
(*Neisseria gonorrhoeae*)   4 μm

teins that promote attachment of the microorganisms to the phagocyte. This coating process is called **opsonization.** The proteins that act as *opsonins* include some components of the complement system and antibody molecules (described later in this chapter and in Chapter 17).

Ingestion

❷ Following adherence, **ingestion** occurs. During this process, the plasma membrane of the phagocyte extends projections called **pseudopods** that engulf the microorganism. ❸ Once the microorganism is surrounded, the pseudopods meet and fuse, surrounding the microorganism with a sac called a **phagosome** or *phagocytic vesicle*. The membrane of a phagosome has enzymes that pump protons (H^+) into the phagosome, reducing the pH to about 4. At this pH, hydrolytic enzymes are activated.

Digestion

In this phase of phagocytosis, the phagosome pinches off from the plasma membrane and enters the cytoplasm. Within the cytoplasm, it contacts lysosomes that contain

digestive enzymes and bactericidal substances (see Chapter 4 on page 102). ❹ Upon contact, the phagosome and lysosome membranes fuse to form a single, larger structure called a **phagolysosome.** ❺ The contents of the phagolysosome take only 10–30 minutes to kill most types of bacteria.

Lysosomal enzymes that attack microbial cells directly include lysozyme, which hydrolyzes peptidoglycan in bacterial cell walls. A variety of other enzymes, such as lipases, proteases, ribonuclease, and deoxyribonuclease, hydrolyze other macromolecular components of microorganisms. Lysosomes also contain enzymes that can produce toxic oxygen products such as superoxide radical (O_2^-), hydrogen peroxide (H_2O_2), singlet oxygen (O_2), and hydroxyl radical (OH·) (see Chapter 6 on page 162). Toxic oxygen products are produced by a process called an *oxidative burst*. Other enzymes can make use of these toxic oxygen products in killing ingested microorganisms. For example, the enzyme myeloperoxidase converts chloride (Cl⁻) ions and hydrogen peroxide into highly toxic hypochlorous acid (HOCl). The acid contains hypochlorous ions, which are found in household bleach and account for its antimicrobial activity (see Chapter 7 on page 196).

❻ After enzymes have digested the contents of the phagolysosome brought into the cell by ingestion, the phagolysosome contains indigestible material and is called a *residual body*. ❼ This residual body then moves toward the cell boundary and discharges its wastes outside the cell. The box on page 472 describes another mechanism by which phagocytes can kill microorganisms and tumor cells.

Microbial Evasion of Phagocytosis

Learning Objective

• *Identify six methods of avoiding destruction by phagocytes.*

Some bacteria have structures that inhibit adherence, such as the M protein and capsules. As mentioned in Chapter 15 on page 442, the M protein of *Streptococcus pyogenes* inhibits the attachment of phagocytes to their surfaces and makes adherence more difficult. Organisms with large capsules include *Streptococcus pneumoniae* and *Haemophilus influenzae* type b. Heavily encapsulated microorganisms like these can be phagocytized only if the phagocyte traps the microorganism against a rough surface, such as a blood vessel, blood clot, or connective tissue fiber, from which the microbe cannot slide away.

Other microbes may be ingested but not killed. For example, *Staphylococcus* produces leukocidins that may kill phagocytes by causing the release of the phagocyte's own lysosomal enzymes into its cytoplasm. Streptolysin released by streptococci has a similar mechanism.

A number of intracellular pathogens secrete membrane attack complexes (described shortly) that lyse phagocyte cell membranes once inside the phagocyte. For example, *Trypanosoma cruzi* (the causative agent of American trypanosomiasis), and *Listeria monocytogenes* (the causative agent of listeriosis), produce membrane attack complexes that lyse phagolysosome membranes and release the microbes into the cytoplasm of the phagocyte, where they propagate. Later the microbes secrete more membrane attack complexes that lyse the plasma membrane and release the microbes from the phagocyte, resulting in lysis of the phagocyte and infection of neighboring cells by the microbe.

Still other microbes have the ability to survive inside phagocytes. *Coxiella burnetti*, the causative agent of Q fever,

actually requires the low pH inside a phagolysosome to replicate. *Listeria monocytogenes* (the causative agent of listeriosis), *Shigella* (the causative agent of shigellosis), and *Rickettsia* (the causative agent of Rocky Mountain spotted fever and typhus) have the ability to escape from a phagosome before it fuses with a lysosome. *Mycobacterium tuberculosis* (the causative agent of tuberculosis), HIV (the causative agent of AIDS), *Chlamydia* (the causative agent of trachoma, nongonococcal urethritis, and lymphogranuloma venereum), *Leishmania* (the causative agent of leishmaniasis), and *Plasmodium* (malarial parasites) can prevent both the fusion of a phagosome with a lysosome and the proper acidification of digestive enzymes. The microbes then multiply within the phagocyte, almost completely filling it. In most cases, the phagocyte dies and the microbes are released by autolysis to infect other cells. Still other microbes, such as the causative agents of tularemia and brucellosis, can remain dormant within phagocytes for months or years at a time.

*   *   *

In addition to providing nonspecific resistance for the host, phagocytosis plays a role in immunity. Macrophages help T and B lymphocytes perform vital immune functions. In Chapter 17, we will discuss in more detail how phagocytosis supports immunity.

In the next section, we will see how phagocytosis often occurs as part of another nonspecific mechanism of resistance: inflammation.

Inflammation

Learning Objective

• *List the stages of inflammation.*

Damage to the body's tissues triggers a defensive response called **inflammation.** The damage can be caused by microbial infection, physical agents (such as heat, radiant energy, electricity, or sharp objects), or chemical agents (acids, bases, and gases). Inflammation is usually characterized by four signs and symptoms: *redness, pain, heat,* and *swelling.* Sometimes a fifth, *loss of function,* is present; its occurrence depends on the site and extent of damage.

If the cause of an inflammation is removed in a relatively short period of time, the inflammatory response is intense and is referred to as an *acute inflammation.* An example is the response to a boil caused by *S. aureus.* If, instead, the cause of an inflammation is difficult or impossible to remove, the inflammatory response is longer lasting but less intense (although overall more destructive). This type of inflammation is referred to as a *chronic inflammation.* An example is the response to a chronic infection such as tuberculosis, caused by *M. tuberculosis.*

Inflammation has the following functions: (1) to destroy the injurious agent, if possible, and to remove it and its by-products from the body; (2) if destruction is not possible, to limit the effects on the body by confining or walling off the injurious agent and its by-products; and (3) to repair or replace tissue damaged by the injurious agent or its by-products.

During inflammation, there is an activation and increased concentration of a group of proteins in the blood called **acute-phase proteins.** Some acute-phase proteins are produced by the liver; others are present in the blood in an inactive form and are converted to an active form during inflammation. Acute-phase proteins induce both localized and systemic responses and include complement (page 472), the cytokines (Chapter 17, page 495), and several specialized proteins such as fibrinogen for clotting, and kinins for vasodilation.

For purposes of our discussion, we will divide the process of inflammation into three stages: vasodilation and increased permeability of blood vessels, phagocyte migration and phagocytosis, and tissue repair.

Vasodilation and Increased Permeability of Blood Vessels

Learning Objective

- *Describe the roles of vasodilation, kinins, prostaglandins, and leukotrienes in inflammation.*

Immediately following tissue damage, blood vessels dilate (increase in diameter) in the area of damage, and their permeability increases (Figure 16.9a and b). Dilation of blood vessels, called **vasodilation,** increases blood flow to the damaged area and is responsible for the redness (erythema) and heat associated with inflammation.

Increased permeability permits defensive substances normally retained in the blood to pass through the walls of the blood vessels and enter the injured area. The increase in permeability, which permits fluid to move from the blood into tissue spaces, is responsible for the **edema** (swelling) of inflammation. The pain of inflammation can be caused by nerve damage, irritation by toxins, or the pressure of edema. ❶ Vasodilation and the increase in permeability of blood vessels are caused by chemicals released by damaged cells in response to injury. One such substance is **histamine,** a chemical present in many cells of the body, especially in mast cells in connective tissue, circulating basophils, and blood platelets. Histamine is released in direct response to the injury of cells that contain it; it is also released in response to stimulation by certain components of the complement system (discussed later). Phagocytic granulocytes attracted to the site of injury can also produce chemicals that cause the release of histamine.

Kinins are another group of substances that cause vasodilation and increased permeability of blood vessels. Kinins are present in blood plasma, and once activated, they play a role in chemotaxis by attracting phagocytic granulocytes, chiefly neutrophils, to the injured area.

Prostaglandins, substances released by damaged cells, intensify the effects of histamine and kinins and help phagocytes move through capillary walls. **Leukotrienes** are substances produced by mast cells (cells especially numerous in the connective tissue of the skin and respiratory system, and in blood vessels) and basophils (one of the types of white blood cells). Leukotrienes cause increased permeability of blood vessels and help attach phagocytes to pathogens. Various components of the complement system stimulate the release of histamine, attract phagocytes, and promote phagocytosis.

Vasodilation and the increased permeability of blood vessels also help deliver clotting elements of blood into the injured area. ❷ The blood clots that form around the site of activity prevent the microbe (or its toxins) from spreading to other parts of the body. ❸ As a result, there may be a localized collection of **pus,** a mixture of dead cells and body fluids, in a cavity formed by the breakdown of body tissues. This focus of infection is called an **abscess.** Common abscesses include pustules and boils.

The next stage in inflammation involves the migration of phagocytes to the injured area.

Phagocyte Migration and Phagocytosis

Learning Objective

- *Describe phagocyte migration.*

Generally, within an hour after the process of inflammation is initiated, phagocytes appear on the scene (Figure 16.9c). ❹ As the flow of blood gradually decreases, phagocytes (both neutrophils and monocytes) begin to stick to the inner surface of the endothelium (lining) of blood vessels. This sticking process is called **margination.** ❺ Then the collected phagocytes begin to squeeze between the endothelial cells of the blood vessel to reach the damaged area. This migration, which resembles amoeboid movement, is called **emigration** (*diapedesis*); the migratory process can take as little as 2 minutes. ❻ The phagocytes then begin to destroy invading microorganisms by phagocytosis.

As mentioned earlier, certain chemicals attract neutrophils to the site of injury (chemotaxis). These include chemicals produced by microorganisms and even other neutrophils; other chemicals are kinins, leukotrienes, chemokines, and components of the complement system. Chemokines are cytokines that are chemotactic for phagocytes and T cells and thus stimulate both the inflammatory response and an immune response. The availability of a

(a) Tissue damage

① Chemicals such as histamine, kinins, prostaglandins, and leukotrienes (represented as blue dots) are released by damaged cells

② Blood clot forms

③ Abscess starts to form (yellow area)

(b) Vasodilation and increased permeability of blood vessels

④ Margination—phagocytes stick to endothelium

⑤ Emigration—phagocytes squeeze between endothelial cells

⑥ Phagocytosis of invading bacteria

(c) Phagocyte migration and phagocytosis

(d) Tissue repair

FIGURE 16.9 **The process of inflammation. (a)** Damage to otherwise healthy tissue—in this case, skin. **(b)** Vasodilation and increased permeability of blood vessels. **(c)** Phagocyte migration and phagocytosis of bacteria and cellular debris by macrophages and neutrophils. Macrophages develop from monocytes. **(d)** The repair of damaged tissue.

■ The signs and symptoms of inflammation are redness, pain, heat, and swelling, and sometimes loss of function.

steady stream of neutrophils is ensured by the production and release of additional granulocytes from bone marrow.

As the inflammatory response continues, monocytes follow the granulocytes into the infected area. Once the monocytes are contained in the tissue, they undergo changes in biological properties and become wandering macrophages. The granulocytes predominate in the early stages of infection but tend to die off rapidly. Macrophages enter the picture during a later stage of the infection, once granulocytes have accomplished their function. They are several times more phagocytic than granulocytes and are large enough to phagocytize tissue that has been destroyed, granulocytes that have been destroyed, and invading microorganisms.

After granulocytes or macrophages engulf large numbers of microorganisms and damaged tissue, they themselves eventually die. As a result, pus forms, and its formation usually continues until the infection subsides. At times, the pus pushes to the surface of the body or into an internal cavity for dispersal. On other occasions the pus remains even after the infection is terminated. In this case, the pus is gradually destroyed over a period of days and is absorbed by the body.

As effective as phagocytosis is in contributing to nonspecific resistance, there are times when the mechanism becomes less functional in response to certain conditions. For example, some individuals are born with an inability to produce phagocytes. And with age, there is a progressive decline in the efficiency of phagocytosis. Recipients of heart or kidney transplants have impaired nonspecific defenses as a result of receiving drugs that prevent the rejection of the transplant. Radiation treatments can also depress nonspecific immune responses by damaging bone marrow. Even certain diseases such as AIDS and cancer can cause defective functioning of nonspecific defenses.

Tissue Repair

The final stage of inflammation is tissue repair, the process by which tissues replace dead or damaged cells (Figure 16.9d). Repair begins during the active phase of inflammation, but it cannot be completed until all harmful substances have been removed or neutralized at the site of injury. The ability of a tissue to regenerate, or repair itself, depends on the type of tissue. For example, skin has a high capacity for regeneration, whereas cardiac muscle tissue does not regenerate at all.

A tissue is repaired when its stroma or parenchyma produces new cells. The *stroma* is the supporting connective tissue, and the *parenchyma* is the functioning part of the tissue. For example, the capsule around the liver that encloses and protects it is part of the stroma because it is not involved in the functions of the liver; liver cells (hepatocytes) that perform the functions of the liver are part of the parenchyma. If only parenchymal cells are active in repair, a perfect or near-perfect reconstruction of the tissue occurs. A familiar example of perfect reconstruction is a minor skin cut, in which parenchymal cells are more active in repair. However, if repair cells of the stroma of the skin are more active, scar tissue is formed.

Fever

Learning Objective

• *Describe the cause and effects of fever.*

Inflammation is a local response of the body to injury. There are also systemic, or overall, responses; one of the most important is **fever,** an abnormally high body temperature. The most frequent cause of fever is infection from bacteria (and their toxins) or viruses.

Body temperature is controlled by a part of the brain called the hypothalamus. The hypothalamus is sometimes called the body's thermostat, and it is normally set at 37°C (98.6°F). It is believed that certain substances affect the hypothalamus by setting it at a higher temperature. Recall from Chapter 15 that when phagocytes ingest gram-negative bacteria, the lipopolysaccharides (LPS) of the cell wall (endotoxins) are released, causing the phagocytes to release the cytokine interleukin-1 (formerly called endogenous pyrogen). IL-1 causes the hypothalamus to release prostaglandins that reset the hypothalamic thermostat at a higher temperature, thereby causing fever (see Figure 15.6 on page 448). Another cytokine called *alpha tumor necrosis factor,* which is produced by macrophages and mast cells, also induces fever.

Assume that the body is invaded by pathogens and that the thermostat setting is increased to 39°C (102.2°F). To adjust to the new thermostat setting, the body responds with blood vessel constriction, increased rate of metabolism, and **shivering,** all of which raise body temperature. Even though body temperature is climbing higher than normal, the skin remains cold, and shivering occurs. This condition, called a *chill,* is a definite sign that body temperature is rising. When body temperature reaches the setting of the thermostat, the chill disappears. But the body will continue to maintain its temperature at 39°C until the IL-1 is eliminated. The thermostat is then reset at 37°C. As the infection subsides, heat-losing mechanisms such as vasodilation and sweating go into operation. The skin becomes warm and the person begins to sweat. This phase of the fever, called the **crisis,** indicates that body temperature is falling.

Up to a certain point, fever is considered a defense against disease. Interleukin-1 helps step up the production

Macrophages Say NO

In 1998, three scientists received the Nobel Prize for Physiology or Medicine for their discovery that nitric oxide (NO) is a powerful signaling molecule in the human body. NO is unusual for a chemical messenger in humans. Chemical messengers such as interferon or epinephrine are large organic molecules, but NO is composed of only two atoms, nitrogen and oxygen ($\cdot N = O$). Nitric oxide is not the same compound as the anesthetic nitrous oxide (N_2O), called laughing gas.

Robert Furchgott of the State University of New York Health Sciences Center, Ferid Murad of the University of Texas, Houston Health Science Center, and Louis Ignarro of the University of California, Los Angeles discovered that NO, released by endothelial cells lining blood vessels, is the molecule that relaxes the smooth muscle of the blood vessel to allow increased blood flow. Murad discovered that nitroglycerin dilates blood vessels because the nitroglycerin releases NO. Since its discovery, NO has been found to be in-volved in a wide variety of processes from memory to immunity to impotence to blood pressure regulation.

High nitrate concentrations have been detected in the urine of people with certain types of infections. The reason is that macrophages also produce NO. The macrophage enzyme, NO synthase, produces NO from the amino acid arginine, and the NO is oxidized to nitrates and nitrites, which are excreted in urine.

Macrophages that are activated by γ interferon from T cells or bacterial endotoxin produce NO synthase, which makes NO. NO appears to kill microorganisms as well as tumor cells by inhibiting ATP production. Evidence for the role of NO is provided by experiments demonstrating that macrophages can't kill tumor cells or parasites if NO production is inhibited. Although NO is quickly destroyed after it is made, large amounts of NO will dilate blood vessels, which causes blood pressure to drop and the patient to go into shock. This contributes to the symptoms of gram-negative septic shock. A future treatment for septic shock may involve inhibiting NO synthase.

NO Suppresses Autoimmunity
Autoimmune diseases, such as lupus erythematosus and rheumatoid arthritis, in which a person makes antibodies against their own molecules, appear to be uncommon in less developed countries. Researchers in Australia suggest that the reason lies in the higher incidence of childhood infections in these countries. We now know that NO production is stimulated by infection and that NO can activate the suppressor cells that are responsible for preventing an immune response to self. Therefore, Bill Cowden at the Australian National University reasons that boosting NO production may prevent autoimmune diseases. He and his colleagues are developing a vaccine made from dead malarial parasites in the hope that the vaccine will prevent autoimmune reactions.

of T lymphocytes. High body temperature intensifies the effect of interferons (antiviral proteins we will discuss shortly). They are believed to inhibit the growth of some microorganisms by decreasing the amount of iron available to them. Also, because the high temperature speeds up the body's reactions, it may help body tissues repair themselves more quickly.

Among the complications of fever are tachycardia (rapid heart rate), which may compromise elderly persons with cardiopulmonary disease; increased metabolic rate, which may produce acidosis; dehydration; electrolyte imbalances; seizures in young children; and delirium and coma. As a rule, death results if body temperature rises above 44–46°C (112–114°F).

Antimicrobial Substances

The body produces certain antimicrobial substances in addition to the chemical factors mentioned earlier. The recently discovered antimicrobial action of nitric oxide is described in the box above. Among the most important of these are the proteins of the complement system and interferons.

The Complement System

Learning Objectives
- *List the components of the complement system.*
- *Describe three pathways of activating complement.*
- *Describe three consequences of complement activation.*

The **complement system** is a defensive system consisting of over 30 proteins produced by the liver and found circulating in blood serum* and within tissues throughout the body. Together, proteins of the complement system destroy

---
****Blood serum** is the straw-colored liquid remaining after blood is allowed to clot. Plasma is the liquid remaining after formed elements are removed from unclotted blood, for example, by centrifugation.

Opsonization:
Enhancement of phagocytosis
by coating with C3b

Histamine

Mast cell

Inflammation:
Increase of blood vessel
permeability and chemotactic
attraction of phagocytes
(see also Figure 16.12)

Microbial plasma
membrane

C5b C6 C7 C8   C9

Cytolysis:
Loss of cellular contents through
transmembrane channel formed
by membrane attack complex
C5–C9 (see also Figure 16.11)

FIGURE 16.10 **Outcomes of complement activation.**

■ The complement system consists of over 30 proteins
produced by the liver and found circulating in serum.

microbes by cytolysis, by inflammation, and by phagocyto-sis and also prevent excessive damage to host tissues. Com-plement proteins are usually designated by an uppercase letter C and are inactive until they are split into products. The proteins are numbered C1 through C9, named for the order in which they were discovered. The products are activated proteins and are indicated by the lowercase letters *a* and *b*. For example, inactive complement protein C3 is split into two activated fragments, C3a and C3b. The activated fragments have new physiological or enzy-matic functions.

Complement proteins act in a *cascade*, that is, one re-action triggers another which, in turn, triggers another, and so on. Also, as part of the cascade, more product is formed with each succeeding reaction so that the effect is amplified as the reactions continue.

The Result of Complement Activation

C3 can be activated by three mechanisms that will be de-scribed shortly. Activation of C3 (Figure 16.10) is very im-portant because it starts a cascade that results in cytolysis, inflammation, and phagocytosis.

SEM   1 μm

FIGURE 16.11 **Cytolysis caused by complement.** Micrographs of a rod-shaped bacterium before cytolysis (left) and after cytolysis (right). Reprinted from Schreiber, R. D. et al. "Bactericidal Activity of the Alternative Complement Pathway Generated from 11 Isolated Plasma Proteins." *Journal of Experimental Medicine,* 149:870–882, 1979.

■ What is the relationship of complement to fighting infections?

❶ When C3 is activated, it splits into C3a and C3b.

❷ C3b enhances phagocytosis. C3b binds to the surface of an invading microbe, and receptors on phagocytes attach to the C3b. Recall that this enhancement of phagocytosis by coating a microbe is called *opsonization* or *immune adherence.* It promotes attachment of a phagocyte to a microbe.

❸ C3b also initiates a series of reactions that result in cytolysis. First, C3b splits C5. Fragment C5b then binds to C6 and C7, which attach to the invading cell's plasma membrane. Next, C8 and several C9 molecules join the other complement proteins and together form a cylinder-shaped **membrane attack complex (MAC),** which inserts itself through the membrane.

❹ The MAC creates transmembrane channels (holes) in the membrane. This causes water to enter the cell and ions to leave the cell, resulting in cytolysis (Figure 16.11).

❺ C3a and C5a bind to mast cells and cause them to release histamine and other chemicals that increase blood vessel permeability during inflammation (Figure 16.12). C5a also functions as a very powerful chemotactic factor that attracts phagocytes to the site of an infection.

Host cell plasma membranes contain proteins that protect against lysis by preventing attachment of the MAC proteins to their surfaces. Also, the MAC forms the basis for the complement fixation test used to diagnose some diseases. This is explained in Chapter 18 (see Figure 18.9 on page 520). Gram-negative bacteria are more susceptible to cytolysis because they have only one or very few layers of peptidoglycan to protect the plasma membrane from the effects of complement. The many layers of peptidoglycan of gram-positive bacteria limit access of complement to the plasma membrane and thus interfere with cytolysis.

The cascade of complement proteins that occurs during an infection is called **complement activation** and may occur in three pathways.

The Classical Pathway

The **classical pathway** (Figure 16.13 on page 476), so named because it was the first to be discovered, is initiated by an antigen-antibody reaction as follows:

❶ A pair of antibody molecules attaches to an antigen, for example, proteins or large polysaccharides on the surface of a bacterium or other cell, forming an antigen-antibody complex. The antigen-antibody complex binds and activates C1.

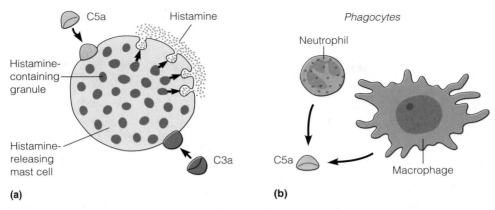

FIGURE 16.12 **Inflammation stimulated by complement. (a)** C3a and C5a bind to mast cells, basophils, and platelets to trigger the release of histamine, which increases blood vessel permeability. **(b)** C5a functions as a chemotactic factor that attracts phagocytes to the site of complement activation.

■ How is complement inactivated once it has completed its functions?

❷ Next, activated C1 activates C4 and C2 by splitting them. C4 is split into fragments called C4a and C4b, and C2 into fragments called C2a and C2b.

❸ C4b and C2b combine and together they activate C3 by splitting it into C3a and C3b. The C3 fragments then initiate cytolysis, inflammation, and opsonization (see Figure 16.10).

The Alternative Pathway

The **alternative pathway** is so named because it was discovered after the classical pathway. Unlike the classical pathway, the alternative pathway does not involve antibodies. The alternative pathway is activated by contact between certain complement proteins and a pathogen. C3 is constantly present in the blood. It combines with complement proteins called factor B, factor D, and factor P (properdin) on the surface of a pathogen (Figure 16.14). The complement proteins are attracted to microbial cell surface material (mostly lipid-carbohydrate complexes of certain bacteria and fungi). Once the complement proteins combine and interact, C3 is split into fragments C3a and C3b. As in the classical pathway, C3a participates in inflammation, and C3b functions in cytolysis and opsonization (see Figure 16.10).

The Lectin Pathway

The **lectin pathway** is the most recently discovered pathway for complement activation. When macrophages ingest bacteria, viruses, and other foreign matter by phagocytosis, they release chemicals that stimulate the liver to produce **lectins,** proteins that bind to carbohydrates (Figure 16.15 on page 477).

❶ One such lectin, **mannose-binding lectin (MBL),** binds to the carbohydrate mannose. MBL binds to many pathogens because the MBL molecules recognize a distinctive pattern of carbohydrates that includes mannose, which is found in bacterial cell walls and on some viruses. As a result of binding, MBL functions as an opsonin to enhance phagocytosis and

❷ activates C2 and C4;

❸ C2a and C4b activate C3 (see Figure 16.10).

Inactivation of Complement

Once complement is activated, its destructive capabilities usually cease very quickly in order to minimize the destruction of host cells. This is accomplished by various regulatory proteins in the host's blood and on certain cells, such as blood cells. The proteins bring about the breakdown of activated complement and function as inhibitors and destructive enzymes.

Complement and Disease

In addition to its importance in defense, the complement system assumes a role in causing disease as a result of inherited deficiencies. For example, deficiencies of C1, C2, or C4 cause collagen vascular disorders that result in hypersensitivity (anaphylaxis); deficiency of C3, though rare, results in increased susceptibility to recurrent infections with pyogenic microbes; and C5–C9 defects result in increased susceptibility to *Neisseria meningitidis* and *N. gonorrhoeae* infections.

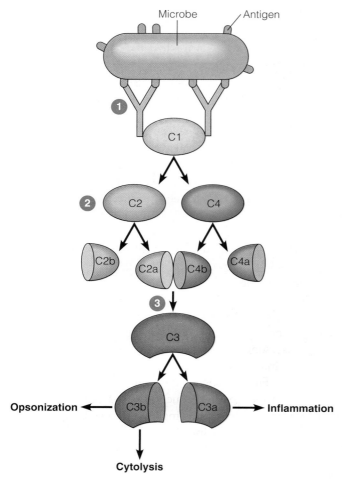

FIGURE 16.13 **Classical pathway of complement activation.** This pathway is initiated by an antigen-antibody reaction. The splitting of C3 into C3a and C3b starts a cascade that results in cytolysis, inflammation, and phagocytosis. (See Figure 16.10)

■ What happens after C3 is cleaved into C3a and C3b?

FIGURE 16.14 **Alternative pathway of complement activation.** This pathway is initiated by contact between certain complement proteins (C3 and factors B, D, and P) and a pathogen. There are no antibodies involved. Once C3a and C3b are formed, they participate in cytolysis, inflammation, and phagocytosis per the classical pathway (see Figure 16.10).

■ How is the alternative pathway similar to the classical pathway?

Interferons

Learning Objectives

- Define interferons.
- Compare and contrast the actions of α-IFN and β-IFN with γ-IFN.

Because viruses depend on their host cells to provide many functions of viral multiplication, it is difficult to inhibit viral multiplication without affecting the host cell itself. One way the infected host counters viral infections is with interferons. **Interferons (IFNs)** are a class of similar antiviral proteins produced by certain animal cells after viral stimulation. One of the principal functions of interferons is to interfere with viral multiplication.

An interesting feature of interferons is that they are host-cell-specific but not virus-specific. Interferons produced by human cells protect human cells but produce little antiviral activity for cells of other species, such as mice or chickens. However, the interferons of a species are active against a number of different viruses.

Just as different animal species produce different interferons, different types of cells in the same animal also produce different interferons. Human interferons are of three

Evading the Complement System

Some bacteria evade the complement system by means of their capsules, which prevent complement activation. For example, some capsules contain large amounts of a substance called sialic acid, which discourages opsonization and MAC formation. Some gram-negative bacteria can lengthen their surface lipid-carbohydrate complexes to prevent MAC formation. Bacteria that are not killed by the MAC are said to be *serum resistant*, a property of many gram-negative bacteria that cause systemic infections. Gram-positive cocci release an enzyme that breaks down C5a, the fragment that serves as a chemotactic factor that attracts phagocytes.

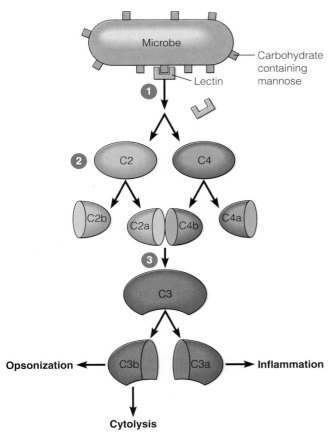

FIGURE 16.15 The lectin pathway of complement activation. When mannose-binding lectin binds to mannose on the surface of microbes, MBL functions as an opsin that enhances phagocytosis and activates complement via the classical and alternative pathways.

■ How do the lectin and alternative pathways differ from the classical pathway?

principal types: *alpha interferon* (α-IFN), *beta interferon* (β-IFN), and *gamma interferon* (γ-IFN). There are also various subtypes of interferons within each of the principal groups. In the human body, interferons are produced by fibroblasts in connective tissue and by lymphocytes and other leukocytes. Each of the three types of interferons produced by these cells can have a slightly different effect on the body.

All interferons are small proteins, with molecular weights between 15,000 and 30,000. They are quite stable at low pH and are fairly resistant to heat.

Both α-IFN and β-IFN are produced by virus-infected host cells only in very small quantities and diffuse to uninfected neighboring cells (Figure 16.16). They react with plasma or nuclear membrane receptors, inducing the uninfected cells to manufacture mRNA for the synthesis of **antiviral proteins (AVPs).** These proteins are enzymes that disrupt various stages of viral multiplication. For example, one AVP, called *oligoadenylate synthetase*, degrades viral mRNA.

Another, called *protein kinase*, inhibits protein synthesis.

Gamma-IFN is produced by lymphocytes; it induces neutrophils and macrophages to kill bacteria by phagocytosis. Neutrophils and macrophages in individuals with an inherited condition called chronic granulomatous disease (CGD) do not kill bacteria. When these people take a recombinant γ-IFN, their neutrophils and macrophages kill bacteria. Gamma-IFN is not a cure and must be taken for the life of the CGD individual.

Because of their beneficial properties, they would seem to be ideal antiviral substances. But certain problems do exist. For one thing, interferons are effective for only short periods; they do not remain stable for long periods of time in the body. And when injected, interferons have side effects, such as nausea, fatigue, headache, vomiting, weight loss, and fever. High concentrations of interferons are toxic to the heart, liver, kidneys, and red bone marrow. They typically play a major role in infections that are acute and short-term, such as colds and influenza. Another problem is that they have no effect on viral multiplication in cells already infected. Also, some viruses, such as adenoviruses (which cause respiratory infections), have resistance mechanisms that inhibit AVPs. Further, some viruses, such as the hepatitis B virus, do not induce the production of sufficient amounts of interferon in host cells following viral stimulation.

The importance of interferons in protecting the body against viruses, as well as their potential as anticancer agents, has made their production in large quantities a top health priority. Several groups of scientists have successfully applied recombinant DNA technology in inducing certain species of bacteria to produce interferons. (This technique is described in Chapter 9.) The interferons produced with recombinant DNA techniques, called *recombinant interferons (rIFNs)*, are important for two reasons: They are pure, and they are plentiful.

In clinical trials, IFNs have exhibited no effects against some types of tumors and only limited effects against others. Alpha-IFN (Intron A) is approved in the United States for treating several virus-associated disorders. One is Kaposi's sarcoma, a cancer that often occurs in patients infected with HIV. Other approved uses for α-IFN include treating genital herpes caused by herpesvirus; hepatitis B and C, caused by the hepatitis B and C viruses; and hairy cell leukemia. Alpha-IFN also is being tested to see if it can slow the development of AIDS in HIV-infected people. A form of β-IFN (Betaseron) slows the progression of multiple sclerosis (MS) and lessens the frequency and severity of MS attacks.

* * *

In Chapter 17, we will discuss specific immune responses. We will also look at the principal factors that contribute to immunity.

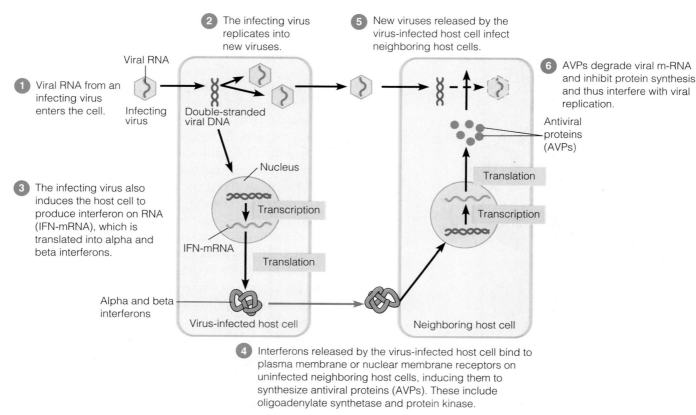

① Viral RNA from an infecting virus enters the cell.

② The infecting virus replicates into new viruses.

③ The infecting virus also induces the host cell to produce interferon on RNA (IFN-mRNA), which is translated into alpha and beta interferons.

⑤ New viruses released by the virus-infected host cell infect neighboring host cells.

⑥ AVPs degrade viral m-RNA and inhibit protein synthesis and thus interfere with viral replication.

④ Interferons released by the virus-infected host cell bind to plasma membrane or nuclear membrane receptors on uninfected neighboring host cells, inducing them to synthesize antiviral proteins (AVPs). These include oligoadenylate synthetase and protein kinase.

FIGURE 16.16 Antiviral action of alpha and beta interferons (IFNs). Interferons are host-cell-specific but not virus-specific.

■ How does interferon stop viruses?

STUDY OUTLINE

Introduction (p. 458)

1. The ability to ward off disease through body defenses is called resistance.

2. Lack of resistance is called susceptibility.

3. Nonspecific resistance refers to all body defenses that protect the body against any kind of pathogen.

4. Specific resistance (immunity) refers to defenses (antibodies) against specific microorganisms.

Skin and Mucous Membranes (pp. 459–462)

Mechanical Factors (pp. 459–460)

1. The structure of intact skin and the waterproof protein keratin provide resistance to microbial invasion.

2. Some pathogens, if present in large numbers, can penetrate mucous membranes.

3. The lacrimal apparatus protects the eyes from irritating substances and microorganisms.

4. Saliva washes microorganisms from teeth and gums.

5. Mucus traps many microorganisms that enter the respiratory and gastrointestinal tracts; in the lower respiratory tract, the ciliary escalator moves mucus up and out.

6. The flow of urine moves microorganisms out of the urinary tract, and vaginal secretions move microorganisms out of the vagina.

Chemical Factors (p. 461)

1. Sebum contains unsaturated fatty acids, which inhibit the growth of pathogenic bacteria. Some bacteria commonly found on the skin can metabolize sebum and cause the inflammatory response associated with acne.

2. Perspiration washes microorganisms off the skin.

3. Lysozyme is found in tears, saliva, nasal secretions, and perspiration.

4. The high acidity (pH 1.2–3.0) of gastric juice prevents microbial growth in the stomach.

5. Transferrins bind iron.

6. Normal microbiota prevent the growth of many pathogens.

Normal Microbiota and Nonspecific Resistance (pp. 461–462)

1. Normal microbiota change the environment, which can prevent the growth of pathogens.

Phagocytosis (pp. 463–468)

1. Phagocytosis is the ingestion of microorganisms or particulate matter by a cell.
2. Phagocytosis is performed by phagocytes, certain types of white blood cells or their derivatives.

Formed Elements in Blood (pp. 463–465)

1. Blood consists of plasma (fluid) and formed elements (cells and cell fragments).
2. Leukocytes (white blood cells) are divided into three categories: granulocytes (neutrophils, basophils, and eosinophils), lymphocytes, and monocytes.
3. During many infections, the number of leukocytes increases (leukocytosis); some infections are characterized by leukopenia (decrease in leukocytes).

Actions of Phagocytic Cells (p. 466)

1. Among the granulocytes, neutrophils are the most important phagocytes.
2. Enlarged monocytes become wandering macrophages and fixed macrophages.
3. Fixed macrophages are located in selected tissues and are part of the mononuclear phagocytic system.
4. Granulocytes predominate during the early stages of infection, whereas monocytes predominate as the infection subsides.

The Mechanism of Phagocytosis (pp. 466–468)

1. Chemotaxis is the process by which phagocytes are attracted to microorganisms.
2. The phagocyte then adheres to the microbial cells; adherence may be facilitated by opsonization—coating the microbe with serum proteins.
3. Pseudopods of phagocytes engulf the microorganism and enclose it in a phagocytic vesicle to complete ingestion.
4. Many phagocytized microorganisms are killed by lysosomal enzymes and oxidizing agents.

Microbial Evasion of Phagocytosis (p. 468)

1. Some microbes are not killed by phagocytes and can even reproduce in phagocytes.
2. Evasion mechanisms include M protein, capsules, leukocidins, membrane attack complexes, and prevention of phagolysosome formation.

Inflammation (pp. 468–471)

1. Inflammation is a bodily response to cell damage; it is characterized by redness, pain, heat, swelling, and sometimes the loss of function.
2. Acute inflammation is a short, intense response to infection; chronic inflammation is a prolonged response.

Vasodilation and Increased Permeability of Blood Vessels (p. 469)

1. The release of histamine, kinins, and prostaglandins causes vasodilation and increased permeability of blood vessels.
2. Blood clots can form around an abscess to prevent dissemination of the infection.

Phagocyte Migration and Phagocytosis (pp. 469–471)

1. Phagocytes have the ability to stick to the lining of the blood vessels (margination).
2. They also have the ability to squeeze through blood vessels (emigration).
3. Pus is the accumulation of damaged tissue and dead microbes, granulocytes, and macrophages.

Tissue Repair (p. 471)

1. A tissue is repaired when the stroma (supporting tissue) or parenchyma (functioning tissue) produces new cells.
2. Stromal repair by fibroblasts produces scar tissue.

Fever (pp. 471–472)

1. Fever is an abnormally high body temperature produced in response to a bacterial or viral infection.
2. Bacterial endotoxins, interleukin-1, and α tumor necrosis factor can induce fever.
3. A chill indicates a rising body temperature; crisis (sweating) indicates that the body's temperature is falling.

Antimicrobial Substances (pp. 472–478)

The Complement System (pp. 472–476)

1. The complement system consists of a group of serum proteins that activate one another to destroy invading microorganisms. Serum is the liquid remaining after blood plasma is clotted.
2. Complement proteins are activated in a cascade.
3. C3 activation can result in cell lysis, inflammation, and opsonization.
4. Complement is activated via the classical pathway, the alternative pathway, and the lectin pathway.
5. Complement is deactivated by host-regulatory proteins.
6. Complement deficiencies can result in an increased susceptibility to disease.

7. Some bacteria evade destruction by complement by means of capsules, surface lipid-carbohydrate complexes, and enzymatic destruction of C5a.

Interferons (pp. 476–478)

1. Interferons (IFNs) are antiviral proteins produced in response to viral infection.

2. There are three types of human interferon: α-IFN, β-IFN, and γ-IFN. Recombinant interferons have been produced.

3. The mode of action of α-IFN and β-IFN is to induce uninfected cells to produce antiviral proteins (AVPs) that prevent viral replication.

4. Interferons are host-cell-specific but not virus-specific.

5. Gamma-IFN activates neutrophils and macrophages to kill bacteria.

STUDY QUESTIONS

Review

1. Define the following terms:
   a. resistance
   b. susceptibility
   c. nonspecific resistance

2. Identify at least one mechanical and one chemical factor that prevent microbes from entering the body through each of the following:
   a. skin
   b. eyes
   c. digestive tract
   d. respiratory tract
   e. urinary tract
   f. reproductive tract

3. Describe the five different white blood cells, and name a function for each cell type.

4. Define phagocytosis.

5. Compare the structures and functions of granulocytes and monocytes in phagocytosis.

6. How do fixed and wandering macrophages differ?

7. Diagram the following processes that result in phagocytosis: margination, emigration, adherence, and ingestion.

8. Define inflammation, and list its characteristics.

9. Why is inflammation beneficial to the body?

10. How is fever related to nonspecific resistance?

11. What causes the periods of chill and crisis during fever?

12. What is complement? List the steps of complement activation via (a) the classical pathway, (b) the alternative pathway, and (c) the lectin pathway.

13. Summarize the major outcomes of complement activation.

14. How is the complement system activated by bacterial endotoxin in the bloodstream? How does endotoxic shock result in massive host cell destruction?

15. What are interferons? Discuss their roles in nonspecific resistance. Why do α-IFN and β-IFN share the same receptor on target cells yet γ-IFN has a different receptor?

Multiple Choice

1. *Legionella* uses C3b receptors to enter monocytes. This
   a. prevents phagocytosis.
   b. degrades complement.
   c. inactivates complement.
   d. prevents inflammation.
   e. prevents cytolysis.

2. *Chlamydia* can prevent the formation of phagolysosomes, and therefore *Chlamydia* can
   a. avoid being phagocytized.
   b. avoid destruction by complement.
   c. prevent adherence.
   d. avoid being digested.
   e. none of the above

3. If the following are placed in the order of occurrence, which would be the *third* step?
   a. Emigration
   b. Digestion
   c. Formation of a phagosome
   d. Formation of a phagolysosome
   e. Margination

4. If the following are placed in the order of occurrence, which would be the *third* step?
   a. Activation of C5–C9
   b. Cell lysis
   c. Antigen–antibody reaction
   d. Activation of C3
   e. Activation of C2–C4

5. A human host can prevent a pathogen from getting enough iron by
   a. reducing dietary intake of iron.
   b. binding iron with transferrin.
   c. binding iron with hemoglobin.
   d. excreting excess iron.
   e. binding iron with siderophores.

6. A decrease in the production of C3 would result in
   a. increased susceptibility to infection.
   b. increased numbers of white blood cells.
   c. increased phagocytosis.
   d. activation of C5–C9.
   e. none of the above

7. In 1884, Elie Metchnikoff observed blood cells collected around a splinter inserted in a sea star embryo. This was the discovery of
   a. blood cells.
   b. sea stars.
   c. phagocytosis.
   d. immunity.
   e. none of the above

8. *Helicobacter pylori* uses the enzyme urease to counteract a chemical defense in the human organ in which it lives. This chemical defense is
   a. lysozyme.
   b. hydrochloric acid.
   c. superoxide radicals.
   d. sebum.
   e. complement.

9. Which of the following statements about α-IFN is *not* true?
   a. It interferes with viral replication.
   b. It is host specific.
   c. It is released by fibroblasts.
   d. It is virus-specific.
   e. It is released by lymphocytes.

10. Which of the following does not stimulate phagocytes?
    a. Cytokines
    b. γ-IFN
    c. C3b
    d. Lipid A
    e. Histamine

Critical Thinking

1. Why do serum levels of iron increase during an infection? What can a bacterium do to respond to high levels of transferrin?

2. A variety of drugs with the ability to reduce inflammation are available. Comment on the danger of misuse of these anti-inflammatory drugs.

3. To be a successful parasite, a microbe must avoid destruction by complement. The following list provides examples of complement-evading techniques. For each microbe, identify the disease it causes, and describe how its strategy enables it to avoid destruction by complement.

Pathogen	Strategy
Group A streptococci	C3 does not bind to M protein
Haemophilus influenzae type b	Has a capsule
Pseudomonas aeruginosa	Sheds cell wall polysaccharides
Trypanosoma cruzi	Degrades C1

4. The list below identifies a virulence factor for a selected microorganism. Describe the effect of each factor listed. Name a disease caused by each organism.

Microorganism	Virulence Factor
Influenzavirus	Causes release of lysosomal enzymes
Mycobacterium tuberculosis	Inhibits lysosome fusion
Toxoplasma gondii	Prevents phagosome acidification
Trichophyton	Secretes keratinase
Trypanosoma cruzi	Lyses phagosomal membrane

Clinical Applications

1. People with *Rhinovirus* infections of the nose and throat have an 80-fold increase in kinins and no increase in histamine. What do you expect for rhinoviral symptoms? What disease is caused by rhinoviruses?

2. A hematologist often performs a differential white blood cell count on a blood sample. Such a count determines the relative numbers of white blood cells. Why are these numbers important? What do you think a hematologist would find in a differential white blood cell count of a patient with mononucleosis? With neutropenia? With eosinophilia?

3. Leukocyte adherence deficiency (LAD) is an inherited disease resulting in the inability of neutrophils to recognize C3b-bound microorganisms. What are the most likely consequences of LAD?

4. The neutrophils of individuals with Chediak-Higashi syndrome (CHS) have fewer than normal chemotactic receptors and lysosomes that spontaneously rupture. What are the consequences of CHS?

5. Consider the following.
   a. In laboratory experiments, plant lectins can bind to mannose in human plasma membranes. Why doesn't human mannose-binding lectin bind to human cells?
   b. About 4% of the human population have a mannose-binding lectin deficiency. How might this deficiency affect a person?

LEARNING WITH TECHNOLOGY

The Microbiology Place CD-ROM/website includes many resources to help you study and succeed in this course. In addition to self-quizzes, web links, and flashcards, you'll find the following activities for this chapter in the "Activities" section of The Microbiology Place:

Host and Pathogen

Case Study: Complement Evasion

Chapter 17

Specific Defenses of the Host: The Immune Response

MACROPHAGES. *Macrophages are phagocytic cells usually found in a resting state. When exposed to a specific antigen they become activated, such as the three cells with ruffled surfaces shown here.*

In Chapter 16, we discussed nonspecific defenses of the body against harmful invaders in the environment. These defenses include the skin and mucous membranes, phagocytosis, and inflammation. Humans, and indeed most animals, even the simplest, are born with such defenses, which are always present to provide instant protection, or immunity. **Im-munity** is a term derived from the Latin word immunis, meaning to exempt. Taken altogether, these defenses are termed **innate (nonspecific) immunity.** (For a review of the body's defense systems, see Figure 16.1, page 459.)

Innate immunity seems to have an inherited component. For example, measles usually has a relatively mild effect on individuals of European ancestry, but the disease devastated the populations of Pacific Islanders when they were first exposed to it by European explorers. The reason probably lies in natural selection. For the Europeans, many generations of exposure to the measles virus presumably led to the selection of genes that conferred some resistance to the virus. Body temperature and other factors also make humans resistant to many infectious diseases of other animals, such as canine distemper and hog cholera.

Higher animals also are protected by a more specialized type of immunity called adaptive or **acquired (specific) immunity.** This aspect of the body's defenses is the focus of this chapter.

Immunity

Learning Objectives

- *Differentiate between innate and acquired immunity.*
- *Differentiate between immunity and nonspecific resistance.*
- *Contrast the four types of acquired immunity.*

The human immune system recognizes foreign substances as not belonging to the body and develops a *specific* immune response against them. This is clearly different from the *nonspecific* defenses that were discussed in Chapter 16. Substances that provoke a specific response are called **antigens.** This immune response involves the production of proteins called **antibodies** and certain specialized lymphocytes. (Antigens and antibodies are discussed in detail later in the chapter.) Both antibodies and specialized lymphocytes are specifically targeted for the antigens that cause their formation, and they can destroy or inactivate those antigens if they encounter them again.

Invading organisms might be pathogenic bacteria, viruses, fungi, protozoa, or helminths; other substances include pollen, insect venom, and transplanted tissue. Body cells that become cancerous are also recognized as foreign and may be eliminated. (However, if they survive to become established as solid tumors, the immune system is no longer effective against them.) We will see in this chapter and in Chapters 18 and 19 that the immune system is essential to survival, but that it can also cause harm when its efforts are misdirected.

Types of Acquired Immunity

Acquired immunity refers to the protection an animal develops against certain types of microbes or foreign substances. Acquired immunity is developed during an individual's lifetime. The various types of acquired immunity are summarized in Figure 17.1.

Immunity can be acquired either actively or passively. Immunity is acquired *actively* when a person is exposed to microorganisms or foreign substances and the immune system responds. Immunity is acquired *passively* when antibodies are transferred from one person to another. Passive immunity in the recipient lasts only as long as the antibodies are present—in most cases, a few weeks or months. Both actively acquired immunity and passively acquired immunity can be obtained by natural or artificial means.

Naturally Acquired Immunity

Naturally acquired active immunity is obtained when a person is exposed to antigens in the course of daily life. Once acquired, immunity is lifelong for some diseases, such

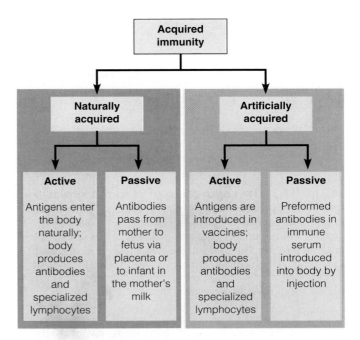

FIGURE 17.1 Types of acquired immunity.

- Which type of immunity, active or passive, lasts longer?

as measles and chickenpox. For other diseases, especially intestinal diseases, the immunity may last for only a few years. *Subclinical infections* (those that produce no noticeable symptoms or signs of illness) can also confer immunity.

Naturally acquired passive immunity involves the natural transfer of antibodies from a mother to her infant. A pregnant woman's body passes some of her antibodies to her fetus across the placenta, a mechanism known as *transplacental transfer*. If the mother is immune to diphtheria, rubella, or polio, the newborn will be temporarily immune to these diseases as well. Certain antibodies are also passed from the mother to her nursing infant in breast milk, especially in the first secretions called *colostrum*. In the infant, immunity generally lasts only as long as the transmitted antibodies are present—usually a few weeks or months. These maternal antibodies are essential for providing immunity to the infant until its own immune system matures.

Colostrum is even more important to some other mammals. Calves, for example, do not have antibodies that cross the placenta and rely on colostrum ingested during the first day of life for this type of immunity. Researchers often specify fetal calf serum for certain experimental or clinical uses because it does not contain maternal antibodies.

Artificially Acquired Immunity

Artificially acquired active immunity results from vaccination. **Vaccination,** also called **immunization,** introduces specially prepared antigens called **vaccines** into the

FIGURE 17.2 **The separation of serum proteins by gel electrophoresis.** In this procedure, serum is placed in a trough cut into a gel. In response to an electrical current, the negatively charged proteins of the serum migrate through the gel from the negatively charged end (cathode) to the positively charged end (anode).

■ What serum fraction contains the most antibodies?

body. Vaccines may be inactivated bacterial toxins (toxoids), killed microorganisms, living but attenuated (weakened) microorganisms, or parts of microorganisms such as capsules. These substances can no longer cause disease, but they can still stimulate an immune response, much as naturally acquired pathogens do. Vaccines are discussed in more detail in Chapter 18.

Artificially acquired passive immunity involves the introduction of antibodies (rather than antigens) into the body. These antibodies come from an animal or person who is already immune to the disease.

The antibodies are found in the serum (page 472) of the immune animal or individual. Because most of the antibodies remain in the serum, **antiserum** has become a generic term for blood-derived fluids containing antibodies. Hence, the study of reactions between antibodies and antigens is called **serology.** When serum is subjected to an electrical current in the laboratory during gel electrophoresis (discussed in Chapter 9), the proteins within it move at different rates, as shown in Figure 17.2. The proteins separate into fractions, that are termed alpha (α), beta (β), and gamma (γ) for their relative locations. These proteins are members of a class of soluble proteins called **globulins.** The fraction that contains most of the antibodies (proteins that are able to specifically target the foreign antigens that caused their formation), are in the γ fraction of the sample. Therefore, the blood fraction that is rich in antibodies has come to be called **gamma globulin.** Antibodies are also known collectively as **immunoglobulins (Ig).**

When immune serum globulin from an individual who is immune to a disease is injected into the body, it confers immediate protection against the disease. However, although artificially acquired passive immunity is immediate, it is short-lived because antibodies are degraded by the recipient. The half-life of an injected antibody (the time required for half of the antibodies to disappear) is typically about 3 weeks.

The Duality of the Immune System

Learning Objective

• Differentiate humoral (antibody-mediated) from cell-mediated immunity.

The first Nobel prize for physiology or Medicine was awarded in 1901 to the German bacteriologist Emil von Behring for his discovery that immunity to diphtheria could be transferred from an animal immunized against diphtheria to another animal. The factors involved in this immunity were known as *humoral immunity,* because they were found in body fluids (formerly called *humors* in medical parlance). These factors were identified much later, in the 1930s, as antibodies.

Even before von Behring's work, Russian biologist Elie Metchnikoff had shown the importance of phagocytic cells in the body's nonspecific defenses. He also observed that these cells were much more effective in immunized animals. (Metchnikoff received a Nobel prize for his work in immunology in 1908.) Specific immunity of this type was called *cell-mediated immunity.* During the 1940s and 1950s, lymphocytes were identified as the responsible cells. In experiments similar to those of von Behring, it was shown that lymphocytes could transfer immunity to certain diseases, such as tuberculosis, between animals. As you will soon see, cell-mediated and humoral immunity are closely interrelated.

Humoral (Antibody-Mediated) Immunity

Humoral immunity, or **antibody-mediated immunity,** involves the production of antibodies that act against foreign organisms and substances. These antibodies are found in extracellular fluids, such as blood plasma, lymph, and mucus secretions. Cells called **B cells** (or B lymphocytes) are responsible for the production of antibodies. The humoral immune response defends primarily against bacteria, bacterial toxins, and viruses that are circulating freely in the body's fluids. It is also a factor in some reactions against transplanted tissue.

Cell-Mediated Immunity

Cell-mediated immunity involves specialized lymphocytes called **T cells** (or T lymphocytes) that act against foreign organisms or tissues. T cells also regulate the activation and proliferation of other immune system cells, such as macrophages. The cell-mediated immune response is most effective against bacteria and viruses located within phagocytic or infected host cells, and against fungi, protozoa, and helminths. This system is the primary responder to transplanted tissue, such as a foreign skin graft. Cell-mediated immunity mounts a response to reject the foreign tissue. It is also an important factor in our defense against cancer.

Antigens and Antibodies

Learning Objective

• *Define antigen and hapten.*

Antigens and antibodies play key roles in the response of the immune system. Antigens (sometimes more descriptively called *immunogens*) provoke a highly specific immune response in an organism.

The Nature of Antigens

Normally, the immune system recognizes components of the body it protects as "self" and foreign matter as "nonself." This recognition is the reason people's defense systems do not usually produce antibodies against their own body tissues (although as we will see in Chapter 19, this sometimes does occur).

Most antigens are either proteins or large polysaccharides. Lipids and nucleic acids usually are antigenic only when combined with proteins and polysaccharides. Anti-

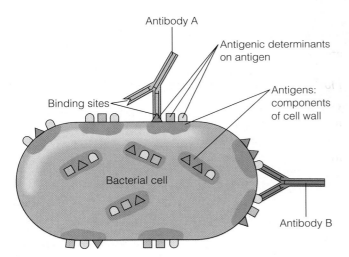

FIGURE 17.3 **Antigenic determinants (epitopes).** In this diagram, the antigenic molecules are components of the bacterial cell wall. Each antigen carries more than one antigenic determinant. Each antibody molecule has at least two binding sites that can attach to a specific kind of determinant site on an antigen. An antibody can also bind to identical determinants on two different cells at the same time (see Figure 18.4 on page 516), which can cause neighboring cells to aggregate.

■ Molecules that elicit an antibody response are called antigens.

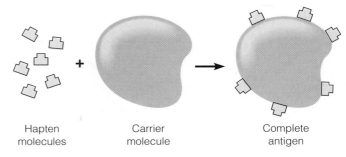

FIGURE 17.4 **Haptens.** A hapten is a molecule too small to stimulate antibody formation by itself. However, when the hapten is combined with a larger carrier molecule, usually a serum protein, the hapten and its carrier together function as an antigen and can stimulate an immune response.

■ A hapten is a partial antigen.

genic compounds are often components of invading microbes, such as the capsules, cell walls, flagella, fimbriae, and toxins of bacteria; the coats of viruses; or the surfaces of other types of microbes. Nonmicrobial antigens include pollen, egg white, blood cell surface molecules, serum proteins from other individuals or species, and surface molecules of transplanted tissues and organs.

Generally, antibodies recognize and interact with specific regions on antigens called **antigenic determinants,** or **epitopes** (Figure 17.3). The nature of this interaction depends on the size, shape, and chemical nature of the antigenic determinant, as well as the chemical structure of the binding site on the antibody molecule.

Most antigens have a molecular weight of 10,000 or higher. A foreign substance that has a low molecular weight is often not antigenic unless it is attached to a carrier molecule. These small compounds are called **haptens** (from the Greek *hapto,* to grasp; Figure 17.4). Once an antibody against the hapten has been formed, the hapten will react with the antibody independent of the carrier molecule. Penicillin is a good example of a hapten. This drug is not antigenic by itself, but some people develop an allergic reaction to it. (Allergic reactions are a type of immune response.) In these people, when penicillin combines with serum proteins, the resulting combined molecule initiates an immune response.

The Nature of Antibodies

Learning Objectives

• *Explain the function of antibodies, and describe their structural and chemical characteristics.*

• *Name one function for each of the five classes of antibodies.*

Antibodies are proteins (immunoglobulins) that are made in response to an antigen and can recognize and bind to

(a) Antibody molecule

(b) Enlarged antigen-binding site bound to an antigenic determinant

(c) Antibody symbol

(d) Computer graphic model of an antibody molecule

FIGURE 17.5 The structure of a typical antibody molecule. The Y-shaped molecule is composed of two light chains and two heavy chains linked by disulfide bridges (S—S). Most of the molecule is made up of constant regions (C), which are the same for all antibodies of the same class. The amino acid sequences of the variable regions (V), which form the two antigen-binding sites, differ from molecule to molecule.

■ What is responsible for the specificity of each different antibody?

that antigen. Antibodies can therefore help neutralize or destroy that antigen. Antibodies are highly specific in recognizing the antigen that stimulated their original formation. An antigen such as a bacterium or virus will probably have several antigenic determinant sites that cause the production of different antibodies.

Each antibody has at least two identical sites that bind to antigenic determinants. These sites are known as **antigen-binding sites.** The number of antigen-binding sites on an antibody is called the **valence** of that antibody. For example, most human antibodies have two binding sites; therefore, they are bivalent.

Antibody Structure

Because a bivalent antibody has the simplest molecular structure, it is called a **monomer.** A typical antibody

monomer has four protein chains: two identical *light (L) chains* and two identical *heavy (H) chains.* ("Light" and "heavy" refer to the relative molecular weights.) The chains are joined by disulfide links (Figure 2.15c on page 46) and other bonds to form a Y-shaped molecule (Figure 17.5). The Y-shaped molecule is flexible and can assume a T shape (notice the hinge region in Figure 17.5a).

The two sections located at the ends of the Y's arms are called *variable (V) regions.* The amino acid sequences, and therefore the three-dimensional structure, of these two variable regions are identical on any one antibody. Their structure reflects the nature of the antigen for which they are specific. These are the two antigen-binding sites found on each antibody monomer. The stem of the antibody monomer and the lower parts of the Y's arms are called *constant (C) regions.* There are five major types of C regions,

TABLE 17.1	**A Summary of Immunoglobulin Classes**				
Characteristics	**IgG**	**IgM**	**IgA**	**IgD**	**IgE**
Structure	Monomer	Pentamer	Dimer (with secretory component)	Monomer	Monomer
Percentage of total serum antibody	80%	5–10%	10–15%*	0.2%	0.002%
Location	Blood, lymph, intestine	Blood, lymph, B cell surface (as monomer)	Secretions (tears, saliva, mucus, intestine, milk), blood lymph	B cell surface, blood, lymph	Bound to mast and basophil cells throughout body, blood
Molecular weight	150,000	970,000	405,000	175,000	190,000
Half-life in serum	23 days	5 days	6 days	3 days	2 days
Complement fixation	Yes	Yes	No†	No	No
Placental transfer	Yes	No	No	No	No
Known functions	Enhances phagocytosis; neutralizes toxins and viruses; protects fetus and newborn	Especially effective against microorganisms and agglutinating antigens; first antibodies produced in response to initial infection	Localized protection on mucosal surfaces	Serum function not known; presence on B cells functions in initiation of immune response	Allergic reactions; possibly lysis of parasitic worms

*Percentage in serum only; if mucous membranes and body secretions are included, percentage is much higher.
†May be yes via alternate pathway.

which accounts for the five major classes of immunoglobulins, which will be described shortly.

The stem of the Y-shaped antibody monomer is called the *Fc region*, so named because in the early days when antibody structure was being identified, it was a fragment (F) that crystallized (c) in cold storage.

These Fc regions are often important in immunological reactions. If left exposed after both antigen-binding sites attach to an antigen such as a bacterium, the Fc regions of adjacent antibodies can bind complement. This leads to the destruction of the bacterium, as shown in Figure 16.11, page 474. Conversely, the Fc region may bind to a cell, leaving the antigen-binding sites of adjacent antibodies free to react with antigens, as shown in Figure 19.1, page 531.

Immunoglobulin Classes

The five classes of immunoglobulins (Igs) are designated IgG, IgM, IgA, IgD, and IgE. Each class plays a different role in the immune response. The structures of IgG, IgD, and IgE molecules resemble the structure shown in Figure 17.5a; molecules of IgA and IgM are usually two or five monomers, respectively, joined together by disulfide bonds. The structures and characteristics of the immunoglobulin classes are summarized in Table 17.1.

IgG **IgG** antibodies account for about 80% of all antibodies in serum. These monomer antibodies readily cross the walls of blood vessels and enter tissue fluids. Maternal IgG antibodies, for example, can cross the placenta and confer passive immunity to a fetus.

IgG antibodies protect against circulating bacteria and viruses, neutralize bacterial toxins, trigger the complement system, and, when bound to antigens, enhance the effectiveness of phagocytic cells.

IgM Antibodies of the **IgM** (from *macro,* reflecting their large size) class make up 5–10% of the antibodies in serum. IgM has a pentamer structure consisting of five monomers held together by a polypeptide called a J (*joining) chain* (see Table 17.1). The large size of the molecule prevents IgM from moving about as freely as IgG does. IgM antibodies generally remain in blood vessels and do not enter the surrounding tissues.

IgM is the predominant type of antibody involved in the response to the ABO blood group antigens on the surface of red blood cells (see Table 19.2, page 534). It is also effective in aggregating antigens and in reactions involving complement. Finally, it can enhance the ingestion of target cells by phagocytic cells, as does IgG.

IgM antibodies are the first antibodies to appear in response to initial exposure to an antigen. A second exposure to antigen results mostly in increased production of IgG.

The fact that IgM appears first in response to a primary infection and is relatively short-lived makes it uniquely valuable in the diagnosis of disease. If high concentrations of IgM against a pathogen are detected in a patient, it is likely that the disease observed is caused by that pathogen. The detection of IgG, which is relatively long-lived, may indicate only that immunity against a particular pathogen was acquired in the more distant past.

IgA **IgA** accounts for only about 10–15% of the antibodies in serum, but it is by far the most common form in mucous membranes and in body secretions such as mucus, saliva, tears, and breast milk. Taking this into consideration, IgA is the most abundant immunoglobulin in the body. (IgG is the most abundant in serum.)

So-called *serum IgA* circulates in the serum mostly as a monomer. The most effective form of IgA is a dimer, called *secretory IgA,* consisting of two monomers connected by a J chain. It is produced in this form by plasma cells in the mucous membranes. Each dimer then enters and passes through a mucosal cell, where it acquires a polypeptide called a *secretory component* that protects it from enzymatic degradation. The main function of secretory IgA is probably to prevent the attachment of pathogens, particularly viruses and certain bacteria, to mucosal surfaces. This is especially important in resistance to respiratory pathogens. And because IgA immunity is relatively shorter-lived, the length of immunity to many respiratory infections is corre-

spondingly short. IgA's presence in colostrum probably helps protect infants from gastrointestinal infections.

IgD **IgD** antibodies make up only about 0.2% of the total serum antibodies. Their structure resembles that of IgG molecules. IgD antibodies are found in blood and lymph and on the surfaces of B cells. IgD molecules have no known function in serum; they act as antigen receptors on the surfaces of B cells.

IgE Antibodies of the **IgE** class are slightly larger than IgG molecules, but they constitute only 0.002% of the total serum antibodies. IgE molecules bind tightly by their Fc (stem) regions to receptors on mast cells and basophils, specialized cells that participate in allergic reactions (see Chapter 19). When an antigen such as pollen reacts with the IgE antibodies attached to a mast cell or basophil, that cell releases histamine and other chemical mediators. These chemicals provoke a response—for example, an allergic reaction such as hay fever. However, the response can be protective as well, for it attracts IgG, complement, and phagocytic cells. This is especially useful when the antibodies bind to parasitic worms.

B Cells and Humoral Immunity

Learning Objectives
- *Name the function of B cells.*
- *Define apoptosis, and give a potential medical application.*

As mentioned earlier, the humoral (antibody-mediated) response is carried out by antibodies, which are produced by a special group of lymphocytes, activated B cells. The process that leads to the production of antibodies starts when B cells are exposed to free, or extracellular, antigens. In response, the B cell becomes activated, and divides and differentiates into a clone of many effector cells called **plasma cells.** These plasma cells then produce antibodies that are directed against the specific antigen that activated the original B cell.

B cells develop from **stem cells** (see also page 541) located in the red bone marrow in adults and in the liver of fetuses (Figure 17.6). After maturation, the mature B cells migrate to lymphoid organs, such as the lymph nodes or the spleen (see Figure 16.6 on page 465). Once in these organs, B cells recognize antigens by means of the **antigen receptors** found on their cell surfaces.

The process by which B cells are stimulated to make antibodies is intimately associated with the assistance of T cells of the cell-mediated immune system. A detailed discussion is presented later in this chapter (see Figure 17.16).

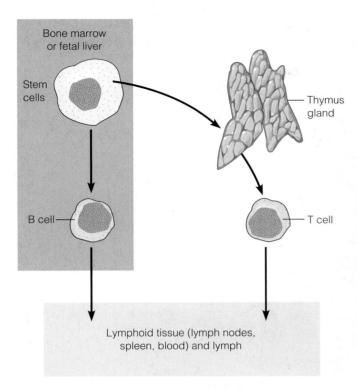

FIGURE 17.6 **Differentiation of T cells and B cells.** Both B cells and T cells originate from stem cells in adult red bone marrow or in the fetal liver. (Red blood cells, macrophages, neutrophils, and other white blood cells also originate from these same stem cells.) Some cells pass through the thymus and emerge as mature T cells. Other cells probably remain in the bone marrow and become B cells. Both types of cells then migrate to lymphoid tissues, such as the lymph nodes or spleen.

■ B cells are responsible for humoral immunity, and T cells are responsible for cellular immunity.

Apoptosis

The human body makes about 100 million lymphocytes (T cells and B cells) each day, so an equivalent number must die. If not, the result would be the proliferation of lymphocytes that characterizes the disease leukemia. B cells, for example, that do not encounter a stimulating antigen soon undergo **apoptosis** (from the Greek for falling away like leaves). Several diseases are linked to defects in apoptosis, ranging from AIDS, in which there is an excess; to cancer, in which there is inadequate apoptosis. Nature accomplishes apoptosis, also called *programmed cell death*, primarily by means of a family of enzymes called *caspases*. Caspases are especially interesting because they represent a potential target for the regulation of apoptosis.

Cells that die from apoptosis first cut their genome into fragments, and the external membranes bulge outward in a manner called *blebbing* (Figure 17.7). Signals are also dis-

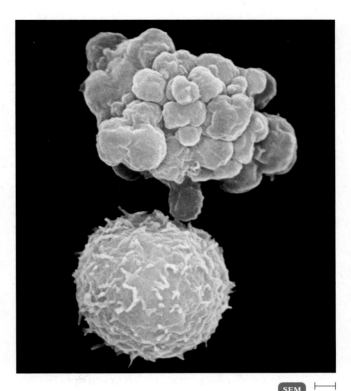

SEM ⊢ 4 μm

FIGURE 17.7 **Apoptosis.** A normal B cell is shown at the bottom of the photo. A B cell undergoing apoptosis is at the top. Notice the bubble-like blebs.

played on the cell's surface that attract circulating phagocytes to digest the remains before any significant leakage of contents occurs. By contrast, cells that meet a violent end, called *necrotic death* or *necrosis*, from tissue injury or infection swell, burst, and release components that trigger an inflammatory response.

Apoptosis rids the body of other unneeded cells. For example, the elimination of activated macrophages helps terminate immune responses. Skin cells migrate from deeper layers and undergo apoptosis, leaving dead cells on the skin's outer layer. It is even thought that humans might be born with webbed digits if apoptosis did not eliminate the webs during embryonic life. The death of a *Bacillus* bacterium in order to release an endospore is a form of apoptosis.

As our understanding of apoptosis improves, we can expect many practical applications. For example, the viruses in a virus-infected cell cannot reproduce effectively during rapid apoptosis. The host cells are continually destroyed before the virus can reproduce. Preventing the inhibition of apoptosis would, therefore, be a potent antiviral weapon. Cancer causes cells to proliferate uncontrolled at the same time that it blocks apoptosis. This

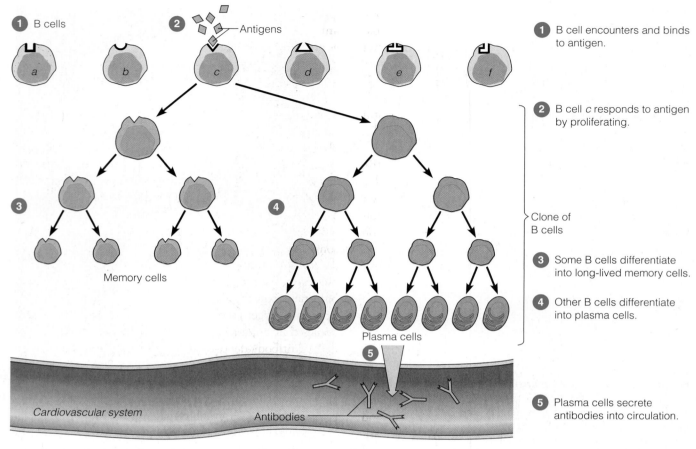

① B cells

② Antigens

① B cell encounters and binds to antigen.

② B cell *c* responds to antigen by proliferating.

Clone of B cells

③ Some B cells differentiate into long-lived memory cells.

④ Other B cells differentiate into plasma cells.

⑤ Plasma cells secrete antibodies into circulation.

Memory cells

Plasma cells

Cardiovascular system

Antibodies

FIGURE 17.8 Clonal selection and differentiation of B cells. B cells can recognize an almost infinite number of antigens, but each particular cell recognizes only one type of antigen. An encounter with a particular antigen triggers the proliferation of a cell that is specific for that antigen (here, B cell *c*) into a clone of cells with the same specificity, hence the term clonal selection.

■ What is selected for in clonal selection?

suggests that restoring apoptosis would open a new avenue for anticancer therapy.

Activation of Antibody-Producing Cells by Clonal Selection

Learning Objective

• Describe the clonal selection theory.

Although the total population of B cells in a person can produce antibodies against an enormous assortment of antigens, each B cell can produce antibodies against only one specific antigen. How, then, do B cells recognize their complementary antigens? In fact, IgM and IgD antibodies bound to the surface of B cells act as very specific antigen receptors. By the time a B cell is mature, it may carry as many as 100,000 such antibody molecules embedded in its surface membrane.

The basic principle that allows a B cell to recognize an appropriate antigen is that the antibodies embedded in its surface (the antigen receptors) will bind only one specific antigen. We will focus here on the simplest case of B cell activation, in which antigens bind directly to antigen receptors on B cells. When the appropriate antigen binds to the antigen receptors on a B cell, the B cell proliferates into a large clone of cells, a phenomenon called **clonal selection** (Figure 17.8). Thus, an antigen *selects* the lymphocyte that will multiply to form a *clone* of cells with the same immunological specificity. (The same process occurs with T cells.)

According to some estimates, an individual may be able to respond to as many as 100 million different antigens. The mechanism by which so many antigens can be made is analogous to the generation of huge numbers of words from a limited alphabet. This "alphabet" is found in the genetic makeup of the variable regions of the antibody molecule. For his work in showing how such remarkable diversity is possible, the Japanese immunologist Susumu Tonegawa received a Nobel prize in 1987.

As mentioned previously, some of these activated B cells differentiate into plasma cells, which secrete antibodies against the antigen. Each plasma cell lives only for a few days but can produce about 2000 antibody molecules per second.

Stimulation of a B cell by an antigen also results in the production of a population of cells called *memory cells,* which function in long-term immunity (discussed shortly).

An important question raised by the clonal selection theory is why the immune system does not react to the body's own cells and macromolecules, a phenomenon called **self-tolerance,** or the differentiation between *self* and *nonself.* The exact answer is still a mystery, but apparently the B and T cells that interact with self antigens are somehow destroyed during fetal development, most probably during passage through the thymus. This mechanism is called **clonal deletion.** A more recent proposal is that the body selectively learns to discriminate between the dangerous and the nondangerous rather than between self and nonself. Normal cell death, or apoptosis, leads to toleration. Anything that violently kills cells signals danger and triggers an inflammatory response. For example, it is possible that the immune response that causes the rejection of an organ transplant is not against nonself, but rather is initiated by the cell damage associated with the transplant surgery.

Antigen–Antibody Binding and Its Results

Learning Objective

- *Explain how an antibody reacts with an antigen, and identify the consequences of the reaction.*

When an antibody encounters an antigen for which it is specific, an **antigen–antibody complex** rapidly forms. An antibody binds to an antigen at the *antigen-binding site* (see Figure 17.5a, b). Sometimes, especially when large amounts of antigen are present, some antigens attach to antigen-binding sites for which they are less than a perfect fit. Because of this, some antibodies are poorer matches for the antigen; these antibodies are said to have less *affinity.*

The binding of an antibody to an antigen basically protects the host by tagging foreign cells and molecules for destruction by phagocytes and complement. The antibody molecule itself is not damaging to the antigen. Foreign organisms and toxins are rendered harmless by only a few mechanisms, as summarized in Figure 17.9.

These are agglutination, opsonization, neutralization, antibody-dependent cell-mediated cytotoxicity, and the action of complement, resulting in cell lysis and inflammation (see Figure 16.10 on page 473).

In **agglutination,** antibodies cause antigens to clump together. For example, the two antigen-binding sites of an IgG antibody can combine with antigenic determinants on two different foreign cells, aggregating the cells into clumps that are more easily ingested by phagocytes. Because of its numerous binding sites, IgM is more effective at cross-linking and aggregating particulate antigens (see Figure 18.4 on page 516). IgG requires 100 to 1000 times as many molecules for the same results. (In Chapter 18, we will see how agglutination is important in the diagnosis of some diseases.)

In neutralization, IgG antibodies inactivate viruses by blocking their attachment to host cells, and they neutralize bacterial toxins by blocking their active sites. In opsonization, the antigen, such as a bacterium, is coated with antibodies that enhance its ingestion and lysis by phagocytic cells. Antibody-dependent cell-mediated cytotoxicity (see page 501 and Figure 17.18) resembles opsonization in that the target organism is coated with antibodies. However, destruction of the target cell is by nonspecific immune system cells that remain external to the target cell.

Finally, both IgG and IgM antibodies trigger the complement system, so named because it complements the action of antibodies. **Inflammation** is caused by infection or tissue injury (see Figure 16.9 on page 470). One aspect of inflammation is that it will often cause microbes in the inflamed area to become coated with certain reactive proteins. This leads to the attachment and activation of complement on the microbe's surface, which is then lysed. **Lysis** of the microbe attracts phagocytes and other defensive immune system cells to the area.

As we will see in Chapter 19, the action of antibodies can also damage the host. For example, immune complexes of antibody, antigen, and complement can damage host tissue. Antigens combining with IgE on mast cells can initiate allergic reactions, and antibodies can react with host cells and cause autoimmune disorders.

Immunological Memory

Learning Objective

- *Distinguish a primary from a secondary immune response.*

The intensity of the humoral response is reflected by the **antibody titer,** the amount of antibody in the serum. After the initial contact with an antigen, the exposed person's serum contains no detectable antibodies for several days.

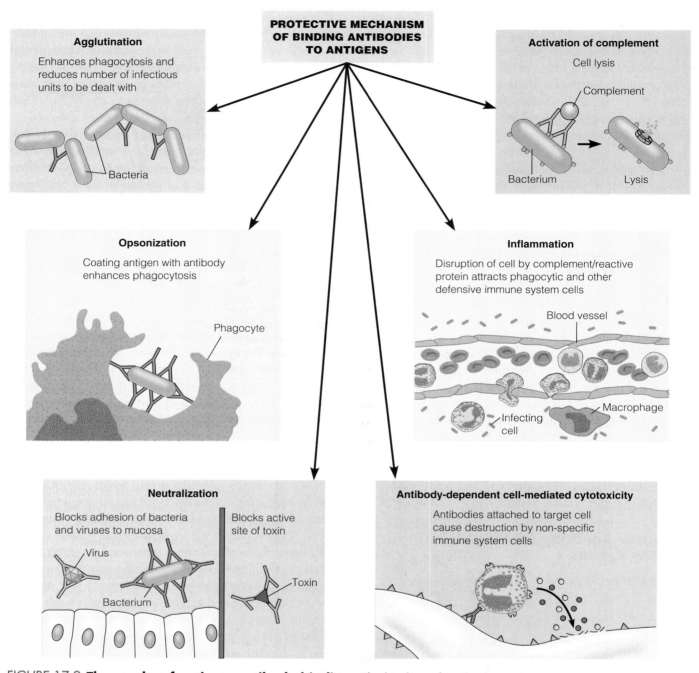

FIGURE 17.9 **The results of antigen–antibody binding.** The binding of antibodies to antigens to form antigen–antibody complexes tags foreign cells and molecules for destruction by phagocytes and complement.

■ What are some of the possible outcomes of an antigen–antibody reaction?

Then there is a slow rise in antibody titer; first IgM antibodies are produced, followed by IgG (Figure 17.10). Finally, a gradual decline in antibody titer occurs. This pattern is characteristic of a **primary response** to an antigen.

The immune responses of the host intensify after a second exposure to an antigen. This **secondary response** is also called the **memory,** or **anamnestic, response.** As we saw in Figure 17.8, some activated B lymphocytes do not become antibody-producing plasma cells but persist as long-lived **memory cells.** Years (perhaps decades) later, if these cells are stimulated by the same antigen, they rapidly differentiate into antibody-producing plasma cells.

This is the basis of the secondary immune response shown in Figure 17.10.

A similar response occurs with T cells, which, as we will see in Chapter 19, is necessary for establishing the lifelong memory for distinguishing between self and nonself.

Monoclonal Antibodies and Their Uses

Learning Objective

- *Define monoclonal antibodies, and identify their advantage over conventional antibody production.*

We have been discussing antibodies primarily in the context of their protective effects against disease. Antibodies are also useful in clinical medicine, especially in diagnosing diseases, as we will emphasize in Chapter 18.

A known antibody can be used, for example, to identify an unknown pathogen. Before monoclonal antibodies became available, antibodies were generally produced in animals by a procedure similar to natural infection. An antigen such as a toxin or microbial pathogen was injected into an animal. If large amounts were needed, a large animal was used. However, the amounts were still quite limited, and the serum contained many other antibodies produced by the animal. Ideally, if an antibody-producing B cell could be grown by standard cell-culture methods, it would produce the desired antibody in nearly unlimited amounts without contamination by other antibodies. Unfortunately, a B cell reproduces only a few times under such conditions.

However, in 1984, three immunologists shared the Nobel prize for their discovery of a method to prolong the culture life of antibody-producing B cells. Niels Jerne, Georges Köhler, and César Milstein made this discovery in 1975. Scientists have long observed that antibody-producing B cells may become cancerous. In this case, their proliferation is unchecked and they are called *myelomas.* These cancerous B cells can be isolated and propagated indefinitely in cell culture. Cancer cells, in this sense, are "immortal." The breakthrough came in combining an "immortal" cancerous B cell with an antibody-producing normal B cell. When fused, this combination is termed a **hybridoma.**

When a hybridoma is grown in culture, its genetically identical cells continue to produce the type of antibody characteristic of the ancestral B cell. The importance of the technique is that clones of the antibody-secreting cells now can be maintained indefinitely in cell culture and can produce immense amounts of identical antibody molecules. Because all of these antibody molecules are produced by a single hybridoma clone, they are called **monoclonal antibodies,** or **Mabs** (Figure 17.11).

Monoclonal antibodies are useful for three reasons: They are uniform; they are highly specific; and they can be produced readily in large quantities. Because of these qual-

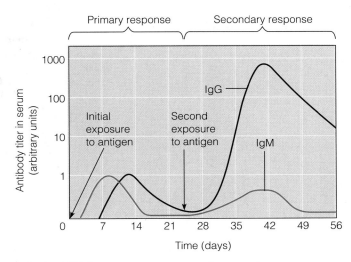

FIGURE 17.10 The primary and secondary immune responses to an antigen. IgM appears first in response to the initial exposure. IgG follows and provides longer-term immunity. The second exposure to the same antigen stimulates the memory cells formed at the time of initial exposure to rapidly produce a large amount of antibody. The antibodies produced in response to this second exposure are mostly IgG.

- Why do many diseases, such as measles, occur only once in a person, yet others, such as colds, occur more than once?

ities, Mabs have assumed enormous importance as diagnostic tools. For instance, commercial kits use Mabs to recognize chlamydial and streptococcal bacteria, and nonprescription pregnancy tests use Mabs to indicate the presence of a hormone excreted only in the urine of a pregnant woman (see Figure 18.13 on page 524).

Monoclonal antibodies are also being used therapeutically to overcome unwanted effects of the immune system. For example, *muromonab (Orthoclone OKT3)* has been used since 1986 to minimize rejection of kidney transplants. For these purposes, monoclonals are prepared that react with the T cells that are responsible for rejection of the transplanted tissue. The Mabs suppress the T cell activity.

The use of Mabs is revolutionizing the treatment of many illnesses. The FDA has approved several for the treatment of specific diseases, such as *alemtuzumab (Campath)* for a form of leukemia, *infliximab (Remicade)* for Crohn's disease and rheumatoid arthritis, and *rituximab (Rituxan)* for non-Hodgkin lymphoma (note that the generic names of those approved to date end in *-mab*). More than 90 other Mabs are in clinical trials.

Cancer is a particularly attractive target for Mab therapy. Theoretically at least, it is possible to combine Mabs that are specifically targeted against cancer cells with a radioactive isotope or a toxin intended to kill the cancer cell

① A mouse is injected with a specific antigen that will induce antibodies against that antigen.

Antigen

Spleen

② The spleen of the mouse is removed, and a suspension is made. The suspension includes B cells that produce antibodies against the injected antigen.

Suspension of spleen cells

Cultured myeloma cells (cancerous B cells)

Suspension of myeloma cells

Unfused spleen cells

Unfused myeloma cells

Fused hybrid cells

③ The spleen cells are then mixed with *myeloma cells* that are capable of continuous growth in culture but have lost the ability to produce antibodies. Some of the antibody-producing spleen cells and myeloma cells fuse to form hybrid cells. These hybrid cells are now capable of growing continuously in culture while producing antibodies.

Hybrid cell

④ The mixture of cells is placed in a selective medium that allows only hybrid cells to grow.

Hybridoma

⑤ Hybrid cells proliferate into clones called *hybridomas*. The hybridomas are screened for production of the desired antibody.

Desired monoclonal antibodies

⑥ The selected hybridomas are then cultured to produce large amounts of *monoclonal antibodies*.

FIGURE 17.11 The production of monoclonal antibodies.

■ Monoclonal antibody production provides pure preparations of antibodies in large quantities.

and leave the healthy, normal cells untouched. These Mabs are termed **immunotoxins** or **conjugated monoclonal antibodies.** One such product, *trastuzumab (Herceptin)* has proven useful for a particular type of advanced breast cancer that bears a receptor on its cells called HER-2.

The therapeutic use of Mabs had been limited because these antibodies once were produced only by mouse (murine) cells. The immune systems of patients reacted against the foreign mouse proteins, leading to rashes, swelling, and even occasional kidney failure, plus the destruction of the Mabs. For example, the success of *muromonab* in minimizing tissue rejection was severely limited by side effects related to the administration of the foreign (murine) fraction of the monoclonal antibody.

In recognition of this problem, researchers are developing new generations of Mabs that are less likely to cause side effects due to their "foreignness." Essentially, the more human the antibody, the more successful it is likely to be. Researchers are exploring several approaches.

Chimeric monoclonal antibodies use genetically manipulated mice to make a human-murine hybrid. The variable part of the antibody molecule, including the antigen-binding sites (see Figure 17.5a), is murine. The remainder of the antibody molecule, the constant region, has been derived from a human source. These Mabs are about 66% human. An example is rituximab.

Humanized antibodies are constructed so that the murine portion is limited to the antigen-binding sites. The balance of the variable region and all of the constant region are derived from human sources. Such Mabs are about 90% human. An example is alemtuzumab.

The eventual goal is to develop **fully human antibodies.** One approach is to genetically engineer mice to contain human antibody genes. The mice would produce antibodies that are fully human; in some cases, it might even be possible to produce an antibody that is an exact match to the patient.

It is also possible that Mab therapies may succeed so well that it would be difficult to produce them in sufficient volumes. Several potential solutions to this problem are under investigation. For example, the use of mice could be avoided entirely by using bacteriophages to insert desired genes into bacteria, which would be able to produce the desired mabs on an industrial scale. Another approach to the problem is to genetically engineer animals that can secrete the mabs in their milk. Genetic alteration of plants to produce mabs is another possible avenue to large-scale production.

T Cells and Cell-Mediated Immunity

Learning Objective

- *Identify at least one function of each of the following in cell-mediated immunity: cytokines, interleukins, interferons.*

As briefly mentioned earlier, antigens that stimulate cell-mediated immunity are mostly intracellular in nature. This type of immunity also generally requires the continued presence of the antigen to maintain its effectiveness. Unlike humoral immunity, cell-mediated immunity is not transferred to the fetus via the placenta.

Cell-mediated immunity is based on the activity of certain specialized lymphocytes, primarily T cells.

Chemical Messengers of Immune Cells: Cytokines

As immunologists began working with cultures of immune system cells, they found that the fluids supporting these cultures contained a confusing array of soluble "factors"— chemical messengers—that regulated many other cells of the immune system. In effect, the cells communicated with each other, triggering their activity or causing them to differentiate. These factors were assigned names such as B cell activating factor or B cell differentiation factor. Only after decades of work did the picture begin to be clarified.

Many of these variously named factors, now known as **cytokines,** were ultimately determined to be the same. Cytokines function as chemical messengers within the immune system. Also, certain cytokines may carry different meanings in different contexts; for example, a cytokine might *inhibit* a response under one set of circumstances and *stimulate* a response under other circumstances.

Cytokines that serve as communicators between leukocytic (white cell) populations are now known as **interleukins** (between leukocytes). When enough information, including the amino acid sequence, is known, the cytokine is assigned an interleukin number by an international committee. At present, we know of at least 26 interleukins. For example, *interleukin-1 (IL-1)* is produced by every nucleated cell, especially macrophages, and affects an array of activities of practically all cell types. It has had at least eight descriptive names. *Interleukin-2 (IL-2)* is produced mostly by CD4 T_H cells (which we will be discussing shortly), and is important because of its role in T-cell proliferation.

As more cytokines were identified, there developed an inclination to group them by function, as was done with interleukins. One family of cytokines induces the migration of leukocytes into infected areas. These are called **chemokines,** from *chemotaxis.* The best known of at least 30 identified chemokines is IL-8 (Table 17.2).

Interferons that help protect against the viral infection of cells are cytokines. Other named cytokines are tumor necrosis factor (TNF), which is important in inflammatory reactions. TNF was so named because tumor cells were observed to be one of its targets. Cytokines known as colony stimulating factors (CSF) stimulate the formation of various blood cells. Cytokines of this type (GM-CSF) are used to increase the numbers of protective macrophages and granulocytes in patients undergoing bone marrow transplants.

The role of cytokines in the stimulation of the immune system has suggested their use as therapeutic agents (see the box on page 497).

TABLE 17.2	A Summary of Some Important Cytokines	
Cytokine	**Representative Activity**	
Interleukin-1 (IL-1)	Stimulates T_H cells in presence of antigens; chemically attracts phagocytes in inflammatory response	
Interleukin-2 (IL-2)	Involved in proliferation of antigen-stimulated T_H cells, proliferation and differentiation of B cells, and activation of T_C cells and NK cells	
Interleukin-8 (IL-8)	Chemoattractant for immune system cells and phagocytes to site of inflammation	
Interleukin-12 (IL-12)	Mainly involved in differentiation of CD4-type T cells	
Gamma-interferon (γ-IFN)	Inhibits intracellular viral replication; increases activity of macrophages against microbes and tumor cells	
Tumor necrosis factor-beta (TNF-β)	Cytotoxic to tumor cells; enhances activity of phagocytic cells	
Granulocyte-macrophage colony-stimulating factor (GM-CSF)	Stimulates formation of red and white blood cells from stem cells	

Cellular Components of Immunity

Learning Objective

- Describe at least one function for each of the following: T_H1 cell, T_H2 cell, T_C cell, T_D cell, T_S cell, APC, MHC, activated macrophage, NK cell.

T cells are the key cellular component of immunity. Like B cells and all other cells involved in the immune response, T cells develop from stem cells in the bone marrow (see Figure 17.6). The precursors of the T cells migrate from the bone marrow and reach maturity in the thymus, a lymphoid organ in the upper chest (the *T* in *T cell* is for *thymus*). Next, T cells migrate to various lymphoid organs, where they are apt to encounter antigens.

Most pathogens first enter the gastrointestinal tract or lungs, where they encounter a barrier of epithelial cells. Normally they can pass this barrier in the gastrointestinal

tract only by way of a scattered number of gateway cells called **microfold cells,** or **M cells.** (Instead of the myriad of fingerlike microvilli found on the surface of absorptive epithelial cells of the intestinal tract, M cells have microfolds). M cells are located over **Peyer's patches,** which are secondary lymphoid organs (see Figure 16.6 on page 465) located in the intestinal wall. M cells are well adapted to take up antigens from the intestinal tract and allow their transfer to the lymphocytes and dendritic cells (page 497) of the immune system in the Peyer's patches. It is there that antibodies, mostly IgA essential for mucosal immunity, are formed and migrate to the mucosal lining.

Like the B cells of humoral immunity, T cells use antigen receptors to recognize and react to an antigen. Also, like B cells, each T cell can react specifically with only a single antigen. As in humoral immunity, clonal selection is the mechanism by which T cells with particular antigen receptors are stimulated to be differentiated into the effector T cells that carry out cell-mediated immunity and proliferate. As with B cells, some of the *effector (antigen-stimulated) T cells* become memory cells, which provide the basis for a secondary immune response if they encounter the same antigen later. The body's ability to make new T cells decreases with age, beginning in late adolescence. Eventually, the T-cell-producing thymus gland becomes less active, and bone marrow produces fewer B cells. As a

FIGURE 17.12 **A dendritic cell.** These antigen-presenting cells are named for their long arms, or dendrites. They are especially plentiful in the skin and mucous membranes.

MICROBIOLOGY IN THE NEWS

Is IL-12 the Next "Magic Bullet"?

Worldwide, HIV/AIDS-related illnesses kill 3 million people, and measles kills 1 million people annually. If laboratory tests are any indication of the future, the cytokine IL-12 (interleukin-12) could be the "magic bullet" against these diseases.

A dozen research teams in the United States and Italy are investigating the potential for IL-12 to treat disease. Since its discovery in the 1980s, IL-12 proved to be different from the other cytokines. For one thing, it is made from the products of two different genes instead of the usual one gene. IL-12 is released by B cells and macrophages in response to infections. It inhibits the humoral response and activates the cell-mediated response by activating T_H cells and recruiting NK cells. Gamma-interferon, γ-IFN, causes T_H cells and NK cells to produce more γ-IFN, which activates more T_H cells and NK cells.

Scientists at the National Institute of Allergy and Infectious Diseases (NIAID)

have found that treating mice with IL-12 can help activate phagocytes to kill *Histoplasma* fungi, *Leishmania* protozoa, *Cryptosporidium* protozoa, *Toxoplasma gondii* protozoa, and *Mycobacterium avium*. The last three are common opportunistic infections in people with late-stage AIDS. The NIAID recently began recruiting people with both AIDS and *Mycobacterium avium* complex into a clinical trial to test IL-12.

Researchers have shown that HIV and measles virus decrease the production of IL-12, which may make the patient more susceptible to secondary infections. When treated with IL-12, however, T_H cells taken from HIV-positive people responded to viruses, including HIV.

IL-12 is known to inhibit about 20 kinds of tumors in mice by inhibiting blood vessel growth to tumors. The Genetics Institute of Massachusetts is currently conducting clinical trials to test the effectiveness of IL-12 in patients

with advanced kidney cancer. (A safe dosage for IL-12 was determined in the Phase I trial.)

IL-12 could also provide the boost needed for DNA vaccines and gene therapy to be successful. Viruses are used as vectors in gene therapy to insert desirable genes into animal cells. Unfortunately, in this case, the introduction of a virus causes the recipient to produce antibodies, thereby preventing the administration of more than one dose. This is especially significant because many patients, such as those with cystic fibrosis, need more than one dose of the new genes. James Wilson of the University of Pennsylvania Medical Centers has observed that antibodies are not made against the gene-carrying virus if it is administered with IL-12.

Will IL-12 be a panacea? Further studies are needed to determine whether the exaggerated responses caused by IL-12 could have adverse effects, such as autoimmune disease.

---

result, the immune system is relatively weak in the elderly. However, sufficient T and B memory cells survive to make immunization effective for such diseases as influenza and pneumococcal pneumonia.

Types of T Cells

There seem to be four main functional types of T cells: helper T (T_H) cells, cytotoxic T (T_C) cells, delayed hypersensitivity T (T_D) cells, and perhaps, suppressor T (T_S) cells. Mature T cells of these four types can be identified by their characteristic cell-surface molecules.

Another classification of T cells is based on the type of cell-surface receptor called CD, for *clusters of differentiation*. Two types of these receptors are CD4 and CD8. **CD4 cells** are primarily helper T cells and include the long-lived memory cells. **CD8 cells** include both cytotoxic and suppressor T cells. The cause of the pathogenicity of the AIDS virus was identified by using monoclonal antibodies to show that a population of CD4 cells was decimated by this infection. These CD4 cells are vital to a vigorous immune response.

Individual T cells interact specifically with only a single antigen, which must be displayed on a cell surface.

These cells are known as **antigen-presenting cells (APCs).** The primary APCs are macrophages (see Figure 16.7 on page 466) or **dendritic cells** (Figure 17.12). Dendritic cells have no other function. The APC ingests and processes the antigen; it then displays fragments of the antigen, each of about 10 amino acids, on the cell surface. A T cell will recognize an antigenic fragment on an APC only if it is in close association with certain cell-surface self molecules. These self molecules are components of the **major histocompatibility complex (MHC).** The particular MHC proteins a person carries are unique to that individual. Thus, they serve as signals by which the immune system distinguishes self from nonself. The MHC in humans will be discussed in more detail in Chapter 19 in relation to autoimmune disease conditions.

Helper T Cells Immune system cells known as **helper T (T_H) cells** play a central roll in the immune response. Once activated by an antigen, T_H cells use their cytokines to influence the activity of other immune system cells. T_H cells differentiate into two major subpopulations: T_H1 and

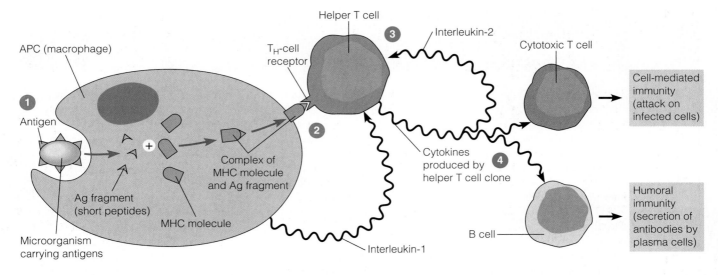

① An antigen-presenting cell (APC) encounters and ingests a microorganism. Antigen fragments (short peptides) from the microorganism combine internally with MHC (self molecules) and the complex of MHC molecules and antigen fragments is presented on the surface of the APC.

② A helper T (T_H) cell receptor binds to the complex, stimulating the APC to secrete interleukin-1.

③ This interleukin-1 stimulates the helper T cell to produce interleukin-2, which then stimulates that helper T cell to form a clone of helper T cells.

④ The cells of this clone in turn produce cytokines, stimulating cells of both immune systems.

FIGURE 17.13 The central role of helper T cells.

■ What is the function of T_H cells?

T_H2, which produce unique groups of cytokines. The cytokines produced by **T_H1 cells** mostly activate cells related to cell-mediated immunity. Among these are macrophages, CD8 T cells, and natural killer cells (see page 499). **T_H2 cells** produce cytokines that activate eosinophils (page 464), which are protective against certain parasites. They also support allergic reactions and stimulate B cells to produce relatively large amounts of IgM and IgE.

In helper T cell activation (Figure 17.13),

❶ an antigen-presenting cell first encounters and processes an antigen.

❷ Binding of a T_H cell to an antigen-MHC complex on the APC stimulates the APC to secrete the cytokine IL-1.

❸ The IL-1 in turn activates the T_H cell, which begins to synthesize interleukin-2 (IL-2). The IL-2 secreted by the T_H cell returns to receptors on the surface of the same cell, which then begins to proliferate and differentiate into mature T_H cells. Only T_H cells that have been stimulated by an antigen have receptors for IL-2, so although the cytokine IL-2 is nonspecific, the resulting T_H cells are specific to that antigen.

❹ These activated T_H cells then begin to produce IL-2 and other cytokines. This IL-2 in turn stimulates other T_H cells specific to that antigen to proliferate and mature.

Cytotoxic T Cells **Cytotoxic T (T_C) cells** destroy target cells on contact. Because viruses, and some bacteria, reproduce within host cells, they cannot be attacked there by antibodies. The T_C cell attacks instead—for example, the virus-infected host cell shown in Figure 17.14.

❶ The T_C cell binds with the MHC-antigen complex on the cell's surface and then

❷ releases a protein called *perforin*. The perforin forms a pore in the membrane of the target cell, causing

❸ lysis. T_C cells continue their activity as long as the stimulating antigen persists; when it disappears, they undergo apoptosis. (Chapter 19 discusses how cytotoxic T cells recognize and lyse cancerous cells.)

Delayed Hypersensitivity T Cells **Delayed hypersensitivity T (T_D) cells** are the cells for which cell-mediated immunity was originally named, because the transfer of these cells between animals was found to transfer immu-

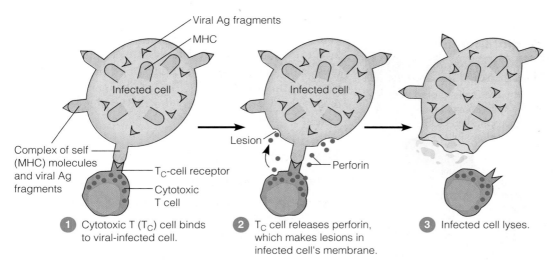

1 Cytotoxic T (T$_C$) cell binds to viral-infected cell.

2 T$_C$ cell releases perforin, which makes lesions in infected cell's membrane.

3 Infected cell lyses.

FIGURE 17.14 **Cell-mediated cytotoxicity.** A cytotoxic T (T$_C$) cell binds to the MHC-antigen complex on the surface of a viral-infected cell. The T$_C$ cell then discharges a protein called perforin, which lyses the infected cell.

■ What is the function of T$_C$ cells?

nity to tuberculosis. However, T$_D$ cells are probably not a separate population, but mostly T$_H$ cells and a few T$_C$ cells that are involved in these types of immune functions. They are associated with certain allergic reactions, such as to poison ivy, and with the rejection of transplanted tissues (discussed fully in Chapter 19).

Suppressor T Cells **Suppressor T (T$_S$) cells** are not well understood. They are generally thought to be T cells that regulate the immune response by turning it off when an antigen is no longer present. Most immunologists consider it likely that T$_S$ cells are not a distinct population but represent temporary suppressive activity by the T$_H$ and T$_C$ cell populations.

Nonspecific Cellular Components

T cells, which are generally directed at specific antigens, are the primary warriors in cell-mediated defense. Other important elements are activated macrophages and natural killer cells, which are less specific in their activity.

Activated Macrophages Macrophages are phagocytic cells usually found in a resting state. Their phagocytic capabilities are greatly increased when they are stimulated to become **activated macrophages.** This stimulation is primarily by ingestion of antigenic material. However, cytokines from antigenically activated helper T cells may also activate macrophages. Activated macrophages are more effective, and their appearance becomes recognizably different as well; they are larger and become ruffled (Figure 17.15).

Activated macrophages are especially valued for their enhanced ability to eliminate certain virus-infected cells and pathogenic intracellular bacteria, such as the tubercle bacillus. Of great importance is their ability to attack and destroy many cancerous cells. Furthermore, they function well as antigen-presenting cells.

Natural Killer Cells Certain lymphocytes called **natural killer (NK) cells** are capable of destroying other cells, especially virus-infected cells and tumor cells. They can also attack large parasites, as illustrated in Figure 17.18.

FIGURE 17.15 **Activated macrophages.** When activated, macrophages became larger and ruffled.

■ How do macrophages become activated?

TABLE 17.3	**Principal Cells That Function in Cell-Mediated Immunity**	
Cell	**Function**	
Helper T (T$_H$1) cell	Activates cells related to cell-mediated immunity: macrophages, CD8 T cells, and natural killer cells	
Helper T (T$_H$2) cell	Stimulates production of eosinophils, IgM, and IgE	
Cytotoxic T (T$_C$) cell	Destroys target cells on contact	
Delayed hypersensitivity T (T$_D$) cell	Provides protection against infectious agents; causes inflammation associated with allergic reactions and tissue transplant rejection	
Suppressor T (T$_S$) cell	Regulates immune response and helps maintain tolerance	
Activated macrophage	Enhanced phagocytic activity, attacks cancer cells	
Natural killer (NK) cell	Attacks and destroys target cells; participates in antibody-dependent cell-mediated cytotoxicity	

In contrast to cytotoxic T cells, NK cells are not immunologically specific; that is, they do not need to be stimulated by an antigen. They are not phagocytic but must contact the target cell to lyse it.

The functions of natural killer cells and the other principal cells involved in cell-mediated immunity are briefly summarized in Table 17.3.

The Interrelationship of Cell-Mediated and Humoral Immunity

Learning Objective

• Compare and contrast cell-mediated and humoral immunity.

Although cell-mediated and humoral immunity are considered separate branches of the specific immune system, they are closely interrelated. In our earlier discussion of the mechanisms by which humoral antibodies are protective (see Figure 17.9), we saw instances in which an-

tibodies of the humoral immune system and cells of the cell-mediated immune system worked in concert. In fact, it is not even possible to describe the production of most antibodies—the core concept of humoral immunity—without introducing the involvement of helper T cells, which are central to the cell-mediated immune system.

The Production of Antibodies

Learning Objectives

• Compare and contrast T-dependent antigens and T-independent antigens.

• Describe the role of antibodies and NK cells in antibody-dependent cell-mediated cytotoxicity.

The production of antibodies against certain antigens requires the assistance of helper T cells; this type of antigen is therefore known as a **T-dependent antigen.** T-dependent antigens are mainly proteins, such as those found on viruses, bacteria, foreign red blood cells, and combinations of haptens and carrier molecules.

Figure 17.16 illustrates the process by which a T-dependent antigen stimulates the formation of an antibody.

❶ The antigen is ingested and processed by an APC. Fragments of the antigen are presented on the surface of the APC in close proximity to MHC molecules.

❷ A T$_H$ cell specific for the antigen reacts with the MHC-antigen complex. (The need to recognize the complex rather than the antigen alone minimizes the possibility of producing antibodies against host tissues.)

❸ The T$_H$ cell reacting with the MHC-antigen complex then produces the cytokine IL-2 that influences a B cell to

❹ differentiate into a plasma cell that begins to produce humoral antibodies. (Notice that, as shown in Figure 17.13, a similar mechanism activates T cells to become T$_C$ cells.)

Antigens that can stimulate B cells directly, without the help of T cells, are called **T-independent antigens.** This type of antigen is usually composed of polysaccharides or lipopolysaccharides characterized by repeating subunits. Bacterial capsules are often good examples of T-independent antigens. The repeating subunits, as shown in Figure 17.17, can bind to multiple B-cell receptors, which is probably why they do not require T-cell assistance. T-independent antigens generally provoke a weaker immune response than T-dependent antigens, and the immune system of infants may not be stimulated

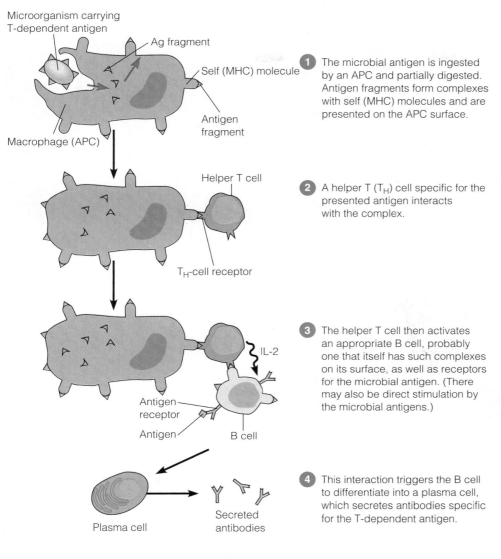

Microorganism carrying T-dependent antigen

Ag fragment

Self (MHC) molecule

Antigen fragment

Macrophage (APC)

Helper T cell

T$_H$-cell receptor

IL-2

Antigen receptor

Antigen

B cell

Plasma cell

Secreted antibodies

1 The microbial antigen is ingested by an APC and partially digested. Antigen fragments form complexes with self (MHC) molecules and are presented on the APC surface.

2 A helper T (T$_H$) cell specific for the presented antigen interacts with the complex.

3 The helper T cell then activates an appropriate B cell, probably one that itself has such complexes on its surface, as well as receptors for the microbial antigen. (There may also be direct stimulation by the microbial antigens.)

4 This interaction triggers the B cell to differentiate into a plasma cell, which secretes antibodies specific for the T-dependent antigen.

FIGURE 17.16 How helper T cells may activate B cells to make antibodies against T-dependent antigens. B cells require the help of antigen-presenting cells (APCs) and specialized helper T cells to produce antibodies against T-dependent antigens.

■ To make antibodies against certain antigens, B cells require T$_H$ cells.

by them until about age 2. These factors will be discussed further in the context of vaccines in Chapter 18.

Antibody-Dependent Cell-Mediated Cytotoxicity

With the help of antibodies produced by the humoral immune system, the cell-mediated immune system can stimu-late NK cells and cells of the nonspecific defense system to kill targeted cells. In this way, an organism, such as a protozoan or helminth, that is too large to be phagocytized can be attacked by immune system cells. This process is referred to as **antibody-dependent cell-mediated cytotoxicity (ADCC).** In this process,

T-independent antigen

FIGURE 17.17 T-independent antigens. T-independent antigens have repeating units that can cross-link several antigen receptors on the same B cell. These antigens stimulate the B cell to make antibodies without the aid of helper T cells. The polysaccharides of bacterial capsules are examples of this type of antigen.

■ Differentiate T-dependent from T-independent antigens.

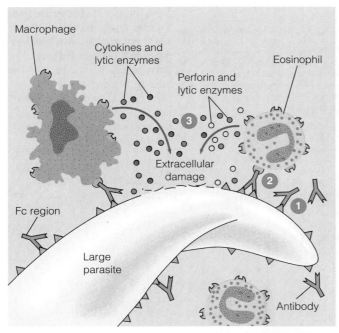

(a) Organisms, such as many parasites, are too large for ingestion by phagocytic cells. They must be attacked externally.

(b) Eosinophils adhering to the larval stage of a parasitic fluke. LM ⊢——⊣ 20 μm

FIGURE 17.18 **Antibody-dependent cell-mediated cytotoxicity (ADCC).** If an organism, such as a parasitic worm, is too large for ingestion and destruction by phagocytosis, it can be attacked by immune system cells that remain external to it. ❶ The parasite first becomes coated with antibodies, leaving their Fc regions extended. ❷ Immune system cells such as eosinophils, macrophages, and NK cells (not shown) tightly attach to the parasite. ❸ They then secrete lytic enzymes and other factors that destroy the parasite.

■ Why is ADCC important protection against parasitic protozoa and helminths?

❶ the target cell must first be coated with antibodies, leaving their Fc (stem) regions pointing outward.

❷ In addition to NK cells, a variety of cells, including macrophages, neutrophils, and eosinophils, have receptors that will bind to these protruding Fc regions, and thus to the target cell.

❸ The target cell is then lysed by substances secreted by the attacking cells. By this process the immune systems can successfully attack many relatively large organisms, such as the various life-cycle stages of parasitic helminths (Figure 17.18).

* * *

Figure 17.19 on page 503 summarizes the duality of the immune system, with its interconnected humoral (antibody-mediated) and cell-mediated responses. In the next chapter, we will discuss immunization and how it stimulates the immune system to prevent disease, as well as some common tests used to diagnose diseases.

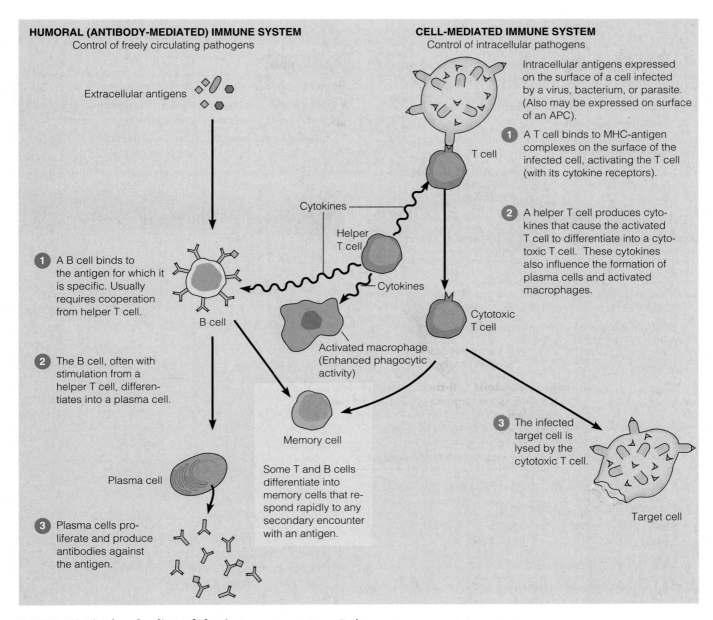

HUMORAL (ANTIBODY-MEDIATED) IMMUNE SYSTEM
Control of freely circulating pathogens

Extracellular antigens

1 A B cell binds to the antigen for which it is specific. Usually requires cooperation from helper T cell.

B cell

2 The B cell, often with stimulation from a helper T cell, differentiates into a plasma cell.

Plasma cell

3 Plasma cells proliferate and produce antibodies against the antigen.

Memory cell

Some T and B cells differentiate into memory cells that respond rapidly to any secondary encounter with an antigen.

CELL-MEDIATED IMMUNE SYSTEM
Control of intracellular pathogens

Intracellular antigens expressed on the surface of a cell infected by a virus, bacterium, or parasite. (Also may be expressed on surface of an APC).

1 A T cell binds to MHC-antigen complexes on the surface of the infected cell, activating the T cell (with its cytokine receptors).

T cell

Cytokines

Helper T cell

Cytokines

Activated macrophage (Enhanced phagocytic activity)

2 A helper T cell produces cytokines that cause the activated T cell to differentiate into a cytotoxic T cell. These cytokines also influence the formation of plasma cells and activated macrophages.

Cytotoxic T cell

3 The infected target cell is lysed by the cytotoxic T cell.

Target cell

FIGURE 17.19 The duality of the immune system. Pathogens can come into contact with antibodies only when they circulate in the blood or spaces outside cells. However, viruses and some bacterial pathogens and parasites reproduce only inside living cells, where circulating antibodies cannot reach them. Elimination of these intracellular pathogens requires cell-mediated immune responses that depend on T cells, especially cytotoxic T cells.

■ When does a B cell require stimulation by a T_H cell?

STUDY OUTLINE

Introduction (p. 482)

1. An individual's genetically predetermined resistance to certain diseases is called innate (nonspecific) immunity or resistance.

2. Individual resistance is affected by gender, age, nutritional status, and general health.

Immunity (pp. 483–485)

1. Immunity is the ability of the body to specifically counteract foreign organisms or substances called antigens.

2. Immunity results from the production of specialized lymphocytes and antibodies.

Types of Acquired Immunity (pp. 483–484)

1. Acquired immunity is specific resistance to infection developed during the life of the individual.
2. A person may develop or acquire immunity after birth.

Naturally Acquired Immunity (p. 483)

1. Immunity resulting from infection is called naturally acquired active immunity; this type of immunity may be long-lasting.
2. Antibodies transferred from a mother to a fetus (transplacental transfer) or to a newborn in colostrum results in naturally acquired passive immunity in the newborn; this type of immunity can last up to a few months.

Artificially Acquired Immunity (pp. 483–484)

1. Immunity resulting from vaccination is called artificially acquired active immunity and can be long-lasting.
2. Vaccines can be prepared from attenuated, inactivated, or killed microorganisms and toxoids.
3. Artificially acquired passive immunity refers to humoral antibodies acquired by injection; this type of immunity can last for a few weeks.
4. Antibodies made by a human or other mammal may be injected into a susceptible individual.
5. Serum containing antibodies is often called antiserum.
6. When serum is separated by gel electrophoresis, antibodies are found in the gamma fraction of the serum and are termed immune serum globulin, or gamma globulin.

The Duality of the Immune System (pp. 484–485)

1. Humoral immunity is in body fluids.
2. Cell-mediated immunity is due to certain types of lymphocytes.

Humoral (Antibody-Mediated) Immunity (p. 484)

1. The humoral immune system involves antibodies produced by B cells in response to a specific antigen.
2. Antibodies primarily defend against bacteria, viruses, and toxins in blood plasma and lymph.

Cell-Mediated Immunity (pp. 484–485)

1. The cell-mediated immune system depends on T cells and does not involve antibody production.
2. Cellular immunity is primarily a response to intracellular bacteria and viruses, multicellular parasites, transplanted tissue, and cancer cells.

Antigens and Antibodies (pp. 485–488)

The Nature of Antigens (p. 485)

1. An antigen (or immunogen) is a chemical substance that causes the body to produce specific antibodies or sensitized T cells.

2. As a rule, antigens are foreign substances; they are not part of the body's chemistry.
3. Most antigens are components of invading microbes: proteins, nucleoproteins, lipoproteins, glycoproteins, or large polysaccharides with a molecular weight greater than 10,000.
4. Antibodies are formed against specific regions on the surface of an antigen called antigenic determinants.
5. Most antigens have many different determinants.
6. A hapten is a low-molecular-weight substance that cannot cause the formation of antibodies unless combined with a carrier molecule.

The Nature of Antibodies (pp. 485–488)

1. An antibody, or immunoglobulin, is a protein produced by B cells in response to the presence of an antigen and capable of combining specifically with that antigen.
2. An antibody has at least two identical antigen-binding (valence) sites.

Antibody Structure (pp. 486–487)

1. A single bivalent antibody unit is a monomer.
2. Typical monomers consist of four polypeptide chains. Two are heavy chains, and two are light chains.
3. Within each chain is a variable (V) region, where antigen binding occurs, and a constant (C) region, which serves as a basis for distinguishing the classes of antibodies.
4. An antibody monomer is Y- or T-shaped; the variable regions form the tips, and the constant regions form the base and Fc (stem) region.
5. The Fc region can attach to a host cell or complement.

Immunoglobulin Classes (pp. 487–488)

1. IgG antibodies are the most prevalent in serum; they provide naturally acquired passive immunity, neutralize bacterial toxins, participate in complement fixation, and enhance phagocytosis.
2. IgM antibodies consist of five monomers held by a joining chain; they are involved in agglutination and complement fixation.
3. Serum IgA antibodies are monomers; secretory IgA antibodies are dimers that protect mucosal surfaces from invasion by pathogens.
4. IgD antibodies are antigen receptors on B cells.
5. IgE antibodies bind to mast cells and basophils and are involved in allergic reactions.

B Cells and Humoral Immunity (pp. 488–495)

1. Humoral immunity involves antibodies that are produced by B cells.
2. Bone marrow stem cells give rise to B cells.
3. Mature B cells migrate to lymphoid organs.
4. A mature B cell recognizes an antigen with antigen receptors.

Apoptosis (pp. 489–490)

1. Lymphocytes that are not needed undergo apoptosis, or programmed cell death, and are destroyed by phagocytes.

Activation of Antibody-Producing Cells by Clonal Selection (pp. 490–491)

1. According to the clonal selection theory, a B cell becomes activated when an antigen reacts with antigen receptors on its surface.

2. The activated B cell produces a clone of plasma cells and memory cells.

3. Plasma cells secrete antibodies. Memory cells recognize pathogens from previous encounters.

4. T cells and B cells that react with self antigens are destroyed during fetal development; this is called clonal deletion.

Antigen–Antibody Binding and Its Results (p. 491)

1. An antigen binds to the antigen-binding site (variable region) of an antibody to form an antigen–antibody complex.

2. IgG antibodies inactivate viruses and also neutralize bacterial toxins.

3. Agglutination of cellular antigens occurs when an IgG or IgM antibody combines with two cells.

4. Antigen–antibody complexes involving IgG and IgM antibodies can fix complement, resulting in the lysis of a bacterial (antigenic) cell.

Immunological Memory (pp. 491–493)

1. The amount of antibody in serum is called the antibody titer.

2. The response of the body to the first contact with an antigen is called the primary response. It is characterized by the appearance of IgM followed by IgG.

3. Subsequent contact with the same antigen results in a very high antibody titer and is called the secondary, anamnestic, or memory response. The antibodies are primarily IgG.

Monoclonal Antibodies and Their Uses (pp. 493–495)

1. Hybridomas are produced in the laboratory by fusing a cancerous cell with an antibody-secreting plasma cell.

2. A hybridoma cell culture produces large quantities of the plasma cell's antibody, called monoclonal antibodies.

3. Monoclonal antibodies are used in serologic identification tests, to prevent tissue rejections, and to make immunotoxins to treat cancer.

4. Immunotoxins can be made by combining a monoclonal antibody and a toxin; the toxin will then kill a specific antigen.

T Cells and Cell-Mediated Immunity (pp. 495–500)

1. Cell-mediated immunity involves specialized lymphocytes, primarily T cells, that respond to intracellular antigens.

Chemical Messengers of Immune Cells: Cytokines (p. 495)

1. Cells of the immune system communicate with each other by means of chemicals called cytokines.

2. Interleukins (IL) are cytokines that serve as communicators between leukocytes.

3. Interferons protect cells against viruses.

4. Chemokines cause leukocytes to move to the site of infection.

Cellular Components of Immunity (pp. 496–500)

1. T cells are responsible for cell-mediated immunity.

2. After differentiation in the thymus gland, T cells migrate to lymphoid tissue.

3. T cells differentiate into effector T cells when they are stimulated by an antigen.

4. Some effector T cells become memory cells.

Types of T Cells (pp. 497–499)

1. T cells are classified according to their functions and cell-surface receptors called CDs.

2. The antigen must be processed by an antigen-presenting cell (APC) and positioned on the surface of the APC.

3. The major histocompatibility complex (MHC) consists of cell-surface proteins that are unique to each individual and provide self molecules.

4. A T cell recognizes antigens in association with MHC on an APC, causing the APC to release IL-1.

5. After binding to an APC, helper T (T_H) or CD4 cells secrete cytokines that activate other T cells and B cells.

6. Cytotoxic T (T_C) or CD8 cells release perforin to lyse cells carrying the target antigen and MHC.

7. Delayed hypersensitivity T (T_D) cells are associated with certain types of allergic reactions and transplant rejection.

8. Suppressor T (T_S) cells appear to regulate the immune response.

Nonspecific Cellular Components (pp. 499–500)

1. Macrophages that are stimulated by ingesting an antigen or by cytokines become activated to have enhanced phagocytic ability.

2. Natural killer (NK) cells lyse virus-infected and tumor cells. They are not T cells and are not antigenically specific.

The Interrelationship of Cell-Mediated and Humoral Immunity (pp. 500–503)

1. T_H cells activate B cells to produce antibodies against T-dependent antigens.

2. Antigens that directly activate B cells are called T-independent antigens.

3. In antibody-dependent cell-mediated cytotoxicity (ADCC), NK cells, macrophages, and other leukocytes lyse antibody-coated cells.

4. ADCC is useful against helminthic parasites.

STUDY QUESTIONS

Review

1. Define *immunity*.

2. Contrast the terms in the following pairs:
   a. innate and acquired immunity
   b. humoral and cell-mediated immunity
   c. active and passive immunity
   d. T_H1 and T_H2 cells
   e. natural and artificial immunity
   f. T-dependent and T-independent antigens
   g. CD4 and CD8

3. Classify the following examples of immunity as naturally acquired active immunity, naturally acquired passive immunity, artificially acquired active immunity, or artificially acquired passive immunity:
   a. Immunity following the injection of diphtheria toxoid
   b. Immunity following an infection
   c. A newborn's immunity to yellow fever
   d. Immunity following an injection of antirabies serum

4. Explain what an antigen is. Distinguish an antigen from a hapten.

5. Explain what an antibody is by describing the characteristics of antibodies. Diagram the structure of a typical antibody; label the heavy chain, light chain, and constant, variable, and Fc regions.

6. Discuss the clonal selection mechanism.

7. By means of a diagram, explain the role of T cells and B cells in immunity.

8. Explain a function for the following types of cells: T_C, T_D, T_H, and T_S. What is a cytokine?

9. a. In the graph below, at time A, the host was injected with tetanus toxoid. At time B, the host was given a booster dose. Explain the meaning of the areas of the curve marked a and b.
   b. Identify in the graph the antibody response of this same individual to exposure to a new antigen indicated at time B.

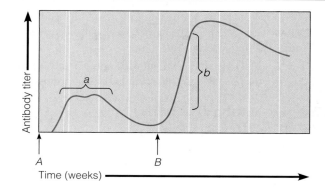

10. What effect does an antibody have on an antigen?

11. How does a T cell recognize an antigen?

12. What are natural killer cells?

13. How would each of the following prevent infection?
    a. antibodies against *Neisseria gonorrhoeae* fimbriae
    b. antibodies against host cell mannose

14. How are monoclonal antibodies produced?

15. Explain why a person who recovers from a disease can attend others with the disease without fear of contracting the disease.

16. Pooled human immune serum globulin is sometimes administered to a patient after exposure to hepatitis A. What is human immune serum globulin? What type of immunity might this confer on the patient?

Multiple Choice

Match the following choices to questions 1–3:
   a. innate resistance
   b. naturally acquired active immunity
   c. naturally acquired passive immunity
   d. artificially acquired active immunity
   e. artificially acquired passive immunity

1. The type of protection provided by the injection of a toxoid.

2. The type of protection provided by the injection of an antitoxin.

3. The type of protection resulting from recovery from an infection.

Match the following choices to the statements in questions 4–6:

- **a.** IgA
- **b.** IdD
- **c.** IgE
- **d.** IgG
- **e.** IgM

4. Antibodies that protect the fetus and newborn.

5. The first antibodies synthesized; especially effective against microorganisms.

6. Antibodies that are bound to mast cells and involved in allergic reactions.

7. Put the following in the correct sequence to elicit an antibody response: (1) T_H cell recognizes B cell; (2) APC phagocytizes antigen; (3) antigen fragment goes to surface of APC; (4) T_H recognizes antigen digest and MHC; (5) B cell recognizes antigen.

- **a.** 1, 2, 3, 4, 5
- **b.** 5, 4, 3, 2, 1
- **c.** 3, 4, 5, 1, 2
- **d.** 2, 3, 4, 1, 5
- **e.** 4, 5, 3, 1, 2

Use the following choices to answer questions 8 and 9:

- **a.** antigen
- **b.** hapten
- **c.** IL-1
- **d.** IL-2
- **e.** perforin

8. Responsible for the differentiation of T and NK cells.

9. Activates a T cell to bind IL-2.

10. Patients with Chediak-Higashi syndrome suffer from various types of cancer. These patients are most likely lacking which of the following:

- **a.** T_D cells
- **b.** T_H cells
- **c.** B cells
- **d.** NK cells
- **e.** T_S cells

Critical Thinking

1. Injections of T_C cells completely removed all hepatitis B viruses from infected mice, but they killed only 5% of the infected liver cells. Explain how T_C cells cured the mice.

2. Provide an explanation for the following:

- **a.** IL-2 has been used to treat pancreatic cancer.
- **b.** IL-2 would exacerbate autoimmune diseases.

3. A positive tuberculin skin test shows cell-mediated immunity to *Mycobacterium tuberculosis*. How could a person acquire this immunity?

Clinical Applications

1. A woman had life-threatening salmonellosis that was successfully treated with monoclonal antibodies. Why did this treatment work, when antibiotics and her own immune system failed?

2. A patient with AIDS has a low T_H/T_S cell ratio. What clinical manifestations would this cause?

3. A patient with chronic diarrhea was found to lack IgA in his secretions, although he had a normal level of serum IgA. What was this patient found to be unable to produce?

4. Newborns (under 1 year) who contract dengue fever have a higher chance of dying from it if their mothers had dengue fever prior to pregnancy. Explain why.

5. A woman died from a *Capnocytophaga* bacterial infection introduced by a dog bite. *Capnocytophaga* kills only people who lack a spleen. What is the relationship between infection and the spleen?

LEARNING WITH TECHNOLOGY

The Microbiology Place CD-ROM/website includes many resources to help you study and succeed in this course. In addition to self-quizzes, web links, and flashcards, you'll find the following activities for this chapter in the "Activities" section of The Microbiology Place:

B cells

T cells

Case Study: Jules Bordet's Discovery

Chapter 18

Practical Applications of Immunology

IMMUNOFLUORESCENCE. *These streptococci fluoresce under UV light because dye-tagged antibodies are attached to them.*

Vaccines

Learning Objective

- *Define vaccine.*

Long before the invention of vaccines, it was known that people who recovered from certain diseases, such as smallpox, were immune to the disease thereafter. As discussed in Chapter 1 on page 11, Chinese physicians may have been the first to try to exploit this phenomenon to prevent disease when they had children inhale dried smallpox scabs.

In 1717, Mary Montagu reported from her travels there that in Turkey, an "old woman comes with a nutshell full of the matter of the best sort of smallpox and asks what veins you please to have opened, and puts into the vein as much venom as can lie upon the head of her needle." This practice usually led to a week of mild illness, and the person was subsequently protected from smallpox. Called **variolation,** this procedure became commonplace in England. Unfortunately, however, it sometimes resulted in a serious case of smallpox. In eighteenth-century England, the mortality rate associated with variolation was about 1%, still a significant improvement over the 50% mortality rate that could be expected from smallpox.

One person who received this treatment, at the age of 8, was Edward Jenner. Later in life, as a physician, Jenner was intrigued by a dairymaid's assertion that she had no fear of smallpox because she had already had cowpox. Cowpox is a mild disease that causes lesions on cows' udders; dairymaids' hands often became infected during milking. Motivated by his childhood memory of variolation, Jenner began a series of experiments in 1798 in which he

In Chapter 17, we learned the basics of the immune system, in which the body recognizes foreign microbes, toxins, or tissues. In response, the body forms antibodies and activates other immune system cells that are programmed to recognize and neutralize or destroy this foreign material if the body encounters it again. This specific immunity is an important element of our defenses against foreign pathogens.

In this chapter, we will discuss some useful tools that have been developed from knowledge of the basics of the immune system. Vaccines were briefly mentioned in the previous chapter; in this chapter we will expand our discussion of this important field of immunology. The diagnosis of disease frequently depends on tests that make use of the specificity of the immune system. Antibodies, especially monoclonal antibodies, are of special value in many of these diagnostic tests.

TABLE 18.1	Principal Vaccines Used in the United States to Prevent Bacterial Diseases in Humans		
Disease	**Vaccine**	**Recommendation**	**Booster**
Diphtheria	Purified diphtheria toxoid	See Table 18.3	Every 10 years for adults
Meningococcal meningitis	Purified polysaccharide from *Neisseria meningitidis*	For people with substantial risk of infection	Need not established
Pertussis (whooping cough)	Killed whole or acellular fragments of *Bordetella pertussis*	Children prior to school age; see Table 18.3	For high-risk adults
Pneumococcal pneumonia	Purified polysaccharide from *Streptococcus pneumoniae*	For adults with certain chronic diseases; people over 65; children 2–23 months	
Tetanus	Purified tetanus toxoid	See Table 18.3	Every 10 years for adults
Haemophilus influenzae type b meningitis	Polysaccharide from *Haemophilus influenzae* type b conjugated with protein to enhance effectiveness	Children prior to school age; see Table 18.3	None recommended

deliberately inoculated people with cowpox in an attempt to prevent smallpox. To honor Jenner's work, the term *vaccination* (from *vacca* = cow) was coined by Louis Pasteur. A **vaccine** is a suspension of organisms or fractions of organisms that is used to induce immunity. Two centuries later, the disease of smallpox has been eliminated worldwide by vaccination, and two other viral diseases, measles and polio, are also targeted for elimination.

Principles and Effects of Vaccination

Learning Objective

- *Explain why vaccination works.*

We now know that Jenner's inoculations worked because the cowpox virus, which is not a serious pathogen, is closely related to the smallpox virus. The injection, by skin scratches, provoked a primary immune response in the recipients, leading to the formation of antibodies and long-term memory cells. Later, when the recipient encountered the smallpox virus, the memory cells were stimulated, producing a rapid, intense secondary immune response (see Figure 17.10 on page 493). This response mimics the immunity gained by recovering from the disease. The cowpox vaccine was soon replaced by a vaccinia virus vaccine. The vaccinia virus also confers immunity to smallpox. Although, strangely, little is known with certainty about the origin of this important virus, it is probably a hybrid that arose long ago from an accidental mixing of cowpox and smallpox viruses. The development of vaccines based on the model of the small-

pox vaccine is the single most important application of immunology.

Many communicable diseases can be controlled by behavioral and environmental methods. For example, proper sanitation can prevent the spread of cholera, and the use of condoms can slow the spread of sexually transmitted diseases. If prevention fails, bacterial diseases can often be treated with antibiotics. Viral diseases, however, often cannot be effectively treated once contracted. Therefore, vaccination is frequently the only feasible method of controlling viral disease. Controlling a disease does not necessarily require that everyone be immune to it. If most of the population is immune, a phenomenon called *herd immunity*, outbreaks are limited to sporadic cases because there are not enough susceptible individuals to support the spread of epidemics.

The principal vaccines used to prevent bacterial and viral diseases in the United States are listed in Tables 18.1 and 18.2. Recommendations for childhood immunizations against some of these diseases are given in Table 18.3. American travelers who might be exposed to cholera, yellow fever, or other diseases not endemic in the United States can obtain current inoculation recommendations from the U.S. Public Health Service and local public health agencies.

Experience has shown that vaccines against enteric bacterial pathogens, such as those causing cholera and typhoid, are not nearly as effective or long-lived as those against viral diseases, such as measles and smallpox. The chief reason is that the polysaccharide components of the outer surfaces of these bacterial pathogens do not

TABLE 18.2	**Principal Vaccines Used in the United States to Prevent Viral Diseases in Humans**		
Disease	**Vaccine**	**Recommendation**	**Booster**
Influenza	Injected vaccine, inactivated virus (nasally administered vaccine with attenuated virus available soon)	For chronically ill people, especially with respiratory diseases, or for healthy people over 65	Annual
Measles	Attenuated virus	For infants age 15 months	See Table 18.3
Mumps	Attenuated virus	For infants age 15 months	(Duration of immunity not known)
Rubella	Attenuated virus	For infants age 15 months; for females of childbearing age who are not pregnant	(Duration of immunity not known)
Chickenpox	Attenuated virus	For infants age 12 months	(Duration of immunity not known)
Poliomyelitis	Killed virus	For children, see Table 18.3; for adults, as risk to exposure warrants	(Duration of immunity not known)
Rabies	Killed virus	For field biologists in contact with wildlife in endemic areas; for veterinarians; for people exposed to rabies virus by bites	Every 2 years
Hepatitis B	Antigenic fragments of virus	For children, see Table 18.3; for adults, especially health care workers, homosexual males, injecting drug users, heterosexual people with multiple partners, and household contacts of hepatitis B carriers	Duration of protection at least 7 years; need for boosters uncertain
Hepatitis A	Inactivated virus	Mostly for travel to endemic areas and protecting contacts during outbreaks	Duration of protection estimated at about 10 years
Smallpox	Live vaccinia virus	Certain military and healthcare personnel	Duration of protection estimated at about 3 to 5 years

stimulate production of effective antibodies as well as protein-coated viruses.

Types of Vaccines and Their Characteristics

Learning Objectives

- Differentiate among the following, and provide an example of each: attenuated, inactivated, toxoid, subunit, and conjugated vaccines.

- Contrast subunit vaccines and nucleic acid vaccines.

There are now several basic types of vaccine. Some of the newer vaccines take full advantage of knowledge and technology developed in recent years.

Attenuated whole-agent vaccines use living, but attenuated (weakened), microbes. Live vaccines more

closely mimic an actual infection. Lifelong immunity, especially with viruses, is often achieved without booster immunizations, and an effectiveness rate of 95% is not unusual. This long-term effectiveness probably occurs because the attenuated viruses replicate in the body, increasing the original dose and acting as a series of secondary (booster) immunizations.

Examples of attenuated vaccines are the Sabin polio vaccine and those used against measles, mumps, and rubella (MMR). The widely used vaccine against the tuberculosis bacillus and certain of the newly introduced, orally administered typhoid vaccines contain attenuated bacteria. Attenuated microbes are usually derived from mutations accumulated during long-term culture. A danger of such vaccines is that the live microbes can backmutate to a virulent form (discussed later in the chapter).

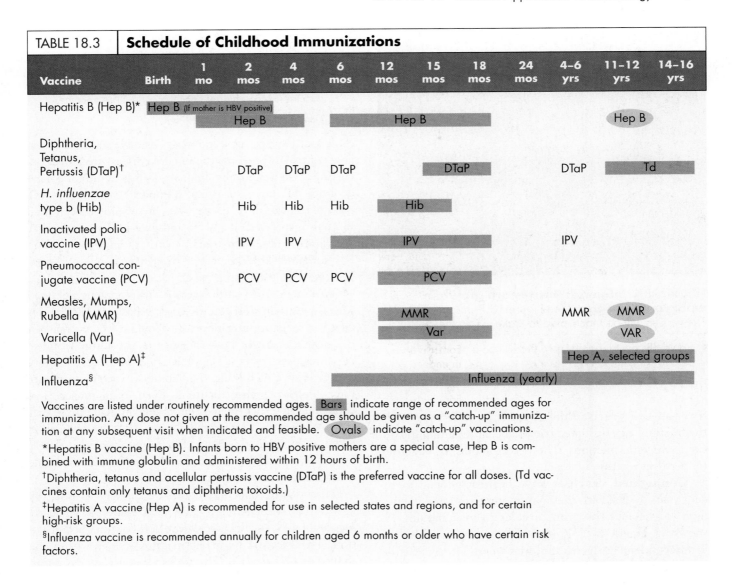

TABLE 18.3	Schedule of Childhood Immunizations											
Vaccine	Birth	1 mo	2 mos	4 mos	6 mos	12 mos	15 mos	18 mos	24 mos	4–6 yrs	11–12 yrs	14–16 yrs
Hepatitis B (Hep B)*	Hep B (If mother is HBV positive)	Hep B				Hep B					Hep B	
Diphtheria, Tetanus, Pertussis (DTaP)†		DTaP	DTaP	DTaP			DTaP			DTaP	Td	
H. influenzae type b (Hib)		Hib	Hib	Hib		Hib						
Inactivated polio vaccine (IPV)		IPV	IPV		IPV					IPV		
Pneumococcal conjugate vaccine (PCV)		PCV	PCV	PCV		PCV						
Measles, Mumps, Rubella (MMR)						MMR				MMR	MMR	
Varicella (Var)						Var					VAR	
Hepatitis A (Hep A)‡										Hep A, selected groups		
Influenza§					Influenza (yearly)							

Vaccines are listed under routinely recommended ages. Bars indicate range of recommended ages for immunization. Any dose not given at the recommended age should be given as a "catch-up" immunization at any subsequent visit when indicated and feasible. Ovals indicate "catch-up" vaccinations.

*Hepatitis B vaccine (Hep B). Infants born to HBV positive mothers are a special case, Hep B is combined with immune globulin and administered within 12 hours of birth.

†Diphtheria, tetanus and acellular pertussis vaccine (DTaP) is the preferred vaccine for all doses. (Td vaccines contain only tetanus and diphtheria toxoids.)

‡Hepatitis A vaccine (Hep A) is recommended for use in selected states and regions, and for certain high-risk groups.

§Influenza vaccine is recommended annually for children aged 6 months or older who have certain risk factors.

Attenuated vaccines are not recommended for people whose immune systems are compromised. If available, inactivated vaccines are substituted.

Inactivated whole-agent vaccines use microbes that have been killed, usually by formalin or phenol. Inactivated virus vaccines used in humans include those against rabies (animals sometimes receive a live vaccine considered too hazardous for humans), influenza (Figure 18.1), and polio (the Salk polio vaccine). Inactivated bacterial vaccines include those for pneumococcal pneumonia and cholera. Several long-used inactivated vaccines that are being replaced for most uses by newer, more effective types are those for pertussis (whooping cough) and typhoid.

Toxoids, which are inactivated toxins, are vaccines directed at the toxins produced by a pathogen. The tetanus and diphtheria toxoids have long been part of the standard childhood inoculation series. They require a series of injections for full immunity, followed by boosters every 10 years. Many older adults have not received boosters; they are likely to have low levels of protection.

Subunit vaccines use only those antigenic fragments of a microorganism that best stimulate an immune response. Subunit vaccines that are produced by genetic engineering techniques, meaning that other microbes are programmed to produce the desired antigenic fraction, are called **recombinant vaccines.** For example, the vaccine against the hepatitis B virus consists of a portion of the viral protein coat that is produced by a genetically engineered yeast.

Subunit vaccines are inherently safer because they cannot reproduce in the recipient. They also contain little or no extraneous material and therefore tend to produce

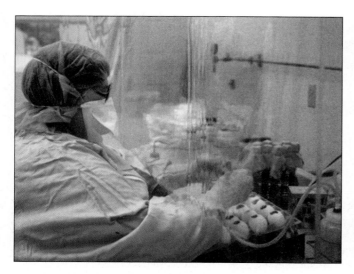

FIGURE 18.1 **Influenza viruses are grown in embryonated eggs.** (See Figure 13.7 on page 385.) The viruses will be inactivated to make a vaccine.

■ Could this method of virus cultivation be a problem for persons allergic to eggs?

fewer adverse effects. Similarly, it is possible to separate the fractions of a disrupted bacterial cell, retaining the desired antigenic fractions. The newer **acellular vaccines** for pertussis use this approach.

Conjugated vaccines have been developed in recent years to deal with the poor immune response of children to vaccines based on capsular polysaccharides. As shown in Figure 17.17 (page 501), polysaccharides are T-independent antigens; children's immune systems do not respond well to these antigens until the age of 15–24 months. Therefore, the polysaccharides are combined with proteins such as diphtheria toxoid; this approach has led to the very successful vaccine for *Haemophilus influenzae* type b, which gives significant protection even at 2 months.

Nucleic acid vaccines, or DNA vaccines, are among the newest and most promising vaccines, although at this time they have not yet resulted in any commercial vaccine for humans. Experiments with animals show that plasmids of "naked" DNA injected into muscle results in the production of the protein encoded in the DNA. (The "gene gun" method for injecting nucleic acids into plant cells is described in Chapter 9, Figure 9.6 on page 257.) These proteins persist and stimulate an immune response.

A problem with this type of vaccine is that the DNA remains effective only until it is degraded. Indications are that RNA, which could replicate in the recipient, might be a more effective agent.

The Development of New Vaccines

Learning Objectives

- *Compare and contrast the production of whole-agent vaccines, recombinant vaccines, and DNA vaccines.*

- *Define adjuvant.*

An effective vaccine is the most desirable method of disease control. It prevents the targeted disease from ever occurring at all in an individual, and it is generally the most economical. This is especially important in developing parts of the world. A "dream" vaccine would be swallowed instead of injected. It also would give lifelong immunity with a single dose, remain stable without refrigeration, and be affordable. This dream is currently far from realization, however. See the box on page 513.

Although interest in vaccine development declined with the introduction of antibiotics, it has intensified in recent years. Fear of litigation contributed to the decrease in the development of new vaccines in the United States. However, passage of the National Childhood Vaccine Injury Act in 1986, which limits the liability of vaccine manufacturers, helped reverse this trend. Even so, pharmaceutical companies find that the most profitable drugs are those that must be taken daily for extended periods, for example, drugs taken for high blood pressure. In contrast, a vaccine that is required for a few injections or even only once in a lifetime is inherently less attractive.

Historically, vaccines could be developed only by growing the pathogen in usefully large amounts. The early successful viral vaccines were developed by animal cultivation. The vaccinia virus for smallpox was grown on the shaved bellies of calves. The rabies virus used by Pasteur in his vaccine over 100 years ago was grown in the central nervous system of rabbits.

The introduction of vaccines against polio, measles, mumps, and a number of other viral diseases that would not grow in anything but a living human awaited the development of cell culture techniques. Cell cultures from human sources, or more often from animals such as monkeys that are closely related to humans, enabled growth of these viruses on a large scale. A convenient animal that will grow many viruses is the chick embryo (see Figure 13.7 on page 385). Viruses for several vaccines (influenza, for example) are grown this way (see Figure 18.1). Interestingly, the first vaccine against hepatitis B virus used viral antigens extracted from the blood of chronically infected humans because no other source was available.

Recombinant vaccines and DNA vaccines do not need a cell or animal host to grow the vaccine's microbe. This avoids a major problem with certain viruses that so

Why Not Vaccinate Against Everything?

Vaccines were responsible for eradicating smallpox and hold promise for eradicating polio and measles within the next 5 to 10 years. Research in, development of, and use of vaccines over the next decade are expected to increase. Vaccines are under development for contraception and cancer, and even a vaccine for cocaine addiction has been proposed. It appears that there is no absolute limit to the number of vaccines that can be given to a person. All standard vaccines recommended for children and adults can be given to the same person, at separate anatomical sites, on the same day or weeks apart, with no risk of interference of one vaccine by another and no potential adverse effects. A few exceptions exist; for example, live vaccines should not be given to immunosuppressed or pregnant individuals, and live bacterial vaccines should not be given within 24 hours after administering antibiotics.

So why not use more vaccines? For example, the vaccines against rabies, botulism, plague, and yellow fever are not routinely used. In the United States in 2002, there were 2 cases of rabies, 1 case of plague, and 14 cases of botulism; yellow fever has not occurred in the United States since 1924. One might argue that these cases of rabies and plague, few as there were, would have been prevented by vaccination. However, the plague vaccine affords only limited protection, and society must balance the

cost of manufacturing and distributing a vaccine against the overall benefit. The *primary* purpose of vaccination against an infectious disease is to provide herd immunity against it; if herd immunity already exists in a society, it is not usually necessary to undertake efforts at widespread vaccination of all individuals. Why don't we vaccinate against tuberculosis? Like the plague vaccine, the TB vaccine affords variable protection. This presents a risk; if people assume they are immune to a disease, they will not practice precautions.

Even in the United States, where the vaccination of children against diseases should be routine, not all children get vaccinated. In 1993, the U.S. government initiated the Childhood Immunization Initiative (CII) to increase vaccination coverage levels to at least 90% among 2-year-old-children by 1996 (for hepatitis B, the objective was set for 1998). By 1999, childhood immunization rates for the complete series of vaccinations were at an all-time high of 78%; however, 1 million children under age 2 still had not received all their immunizations. During 2000, the national rate decreased to 76%.

While efforts to vaccinate children in the United States are increasing, the World Health Organization is working to extend protection against diseases preventable with vaccine to children of all nations. Measles still accounts for 10% of global mortality among children aged

less than 5 years (approximately 1 million deaths annually). New vaccines are needed against diseases that affect millions of people in developing nations. Unfortunately, these diseases are not usually candidates for vaccine development. The affected populations cannot afford even minimal costs. Moreover, if electricity and therefore refrigeration are not available, many vaccines cannot be made available.

Recent attempts to develop a malaria vaccine provide an example of the clinical and technical difficulties involved in developing a vaccine. Some 300–500 million new cases of malaria, with 2 million deaths, occur every year throughout the world. Preliminary human trials of a vaccine against a malaria sporozoite antigen showed that human volunteers did develop antibodies against the antigen. However, the antibodies must kill the sporozoites within 30 minutes after the mosquito bites its victim, before the infective sporozoites enter the protection of the liver cells. A separate vaccine is needed to provoke an immune response against the later stage merozoites, which destroy red blood cells and cause the typical symptoms of malaria.

The CII has established ambitious goals for enhancing present vaccines and for developing new ones. The ultimate vaccine is an affordable, heat-stable, orally administered, multiple-antigen, single immunization given at birth.

far have not been grown in cell culture—hepatitis B, for example.

Plants are also a potential source for vaccines. They can be genetically altered to produce almost any desired antigen in large amounts, which theoretically would cause production of antibodies against that antigen. Also, the thick cell walls of plants help to protect the cellular contents from degradation by stomach acids. Furthermore, release of the antigens directly into the intestinal tract represents an ideal location for introduction of vaccines against many diseases. These would include the diarrheal diseases, in which the pathogen is ingested and immunity depends pri-

marily on defenses in the intestinal tract (see the discussion of M cells and Peyer's patches on page 496 in Chapter 17). Diarrheal diseases are a major cause of mortality for infants in the underdeveloped countries, where costs and distribution of vaccines also pose special problems. For example, a vaccine that must be constantly refrigerated can be nearly useless in underdeveloped countries that lack reliable electrical service. Edible, plant-derived vaccines of several types are undergoing clinical trials. Bananas are an especially attractive plant because they are eaten raw, usually relished by children, and can be grown in the tropics, where they are most likely to be needed.

The so-called golden age of immunology occurred from about 1870 to 1910, when most of the basic elements of immunology were discovered and several important vaccines were developed. We may soon be entering another golden age in which new technologies are brought to bear on emerging infectious diseases and problems arising from the decreasing effectiveness of antibiotics. It is remarkable that there are no useful vaccines against chlamydias, fungi, protozoa, or helminthic parasites of humans. Moreover, vaccines for some diseases, such as cholera and tuberculosis, are not reliably protective. At present, vaccines for at least 75 diseases are under development, ranging from those for prominent deadly diseases such as AIDS and malaria to such commonplace conditions as earaches. But we will probably find that the easy vaccines have already been made.

Infectious diseases are not the only possible target of vaccines. Researchers are investigating vaccines' potential for treating and preventing cocaine addiction, Alzheimer's disease, and cancer, and for contraception.

Work is underway to improve effectiveness of antigens that may not be very effective when injected by themselves. For example, chemicals added for this purpose, called **adjuvants,** greatly improve the effectiveness of many antigens. Only alum has been approved as an adjuvant for human use; others might be found. Injecting antigens attached to microscopic particles has been shown experimentally to improve their effectiveness.

Currently, nearly 20 separate injections are recommended for infants and children, sometimes requiring 3 or more at one appointment. Developing additional multiple combinations of vaccines, such as the MMR vaccine, would be of some help. Several such combination vaccines that are not yet approved in the United States are already in use in Canada and Europe. Delivery in ways other than by needle would also be a desirable advance. Perhaps you have already received injection by high-pressure "guns," which are commonly used for mass inoculations. Experiments are also underway to test delivery by intranasal aerosols, skin patches, and, as we have discussed, vaccines produced in edible foods and ingested.

Safety of Vaccines

Learning Objective

- *Explain the value of vaccines, and discuss acceptable risks for vaccines.*

We have seen how variolation, the first attempt to provide immunity to smallpox, sometimes caused the disease it was intended to prevent. At the time, however, the risk was considered very worthwhile. Even today the oral polio vaccine on rare occasions may *cause* the disease. In 1999, a vaccine to prevent infant diarrhea caused by rotaviruses was withdrawn from the market because several recipients developed a life-threatening intestinal obstruction. But public reaction to such risks has changed. Most parents have never seen a case of polio or measles and therefore tend to view the risk of these diseases as a remote abstraction. Moreover, reports or rumors of harmful effects often lead people to avoid certain vaccines for themselves or their children. Various vaccines have been publicized as being responsible for multiple sclerosis, inflammatory bowel disease, and diabetes, to name a few. In particular, a possible connection between the MMR vaccine and autism has received widespread publicity. Autism is a poorly understood developmental condition that causes a child to withdraw from reality. Because autism is usually diagnosed at the age of 18 to 30 months, about the time vaccine immunization schedules are nearing completion, some people have attempted to make a cause-and-effect connection. Medically, however, most experts agree that autism is a condition with a major genetic component and begins before birth. Extensive scientific surveys have provided no evidence to support a connection between the usual childhood vaccines and autism or any other disease condition. Some experts even maintain that again introducing the rotavirus vaccine, which has been withdrawn in the United States, would be well justified on a risk-versus-benefit calculus in much of the underdeveloped world. No vaccine will ever be perfectly safe or perfectly effective; neither is any antibiotic, for that matter. Nevertheless, vaccines still remain the safest and most effective means of preventing infectious disease in children.

Diagnostic Immunology

Learning Objective

- *Explain how antibodies are used to diagnose diseases.*

Throughout most of history, diagnosing a disease was essentially a matter of observing a patient's signs and symptoms. The writings of ancient and medieval physicians left descriptions of many diseases that are recognizable even today.

Knowledge of the high specificity of the immune system soon suggested that this might be used in diagnosing diseases. In fact, an accidental observation in the course of research into vaccines against tuberculosis led to one of the first diagnostic tests for an infectious disease. More than 100 years ago, Robert Koch was trying to develop a vaccine against tuberculosis. He observed that when

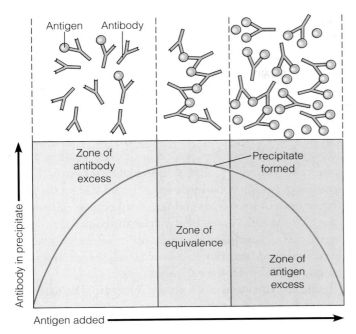

FIGURE 18.2 A precipitation curve. The curve is based on the ratio of antigen to antibody. The maximum amount of precipitate forms in the zone of equivalence, where the ratio is roughly equivalent.

■ Precipitation reactions involve soluble antigens.

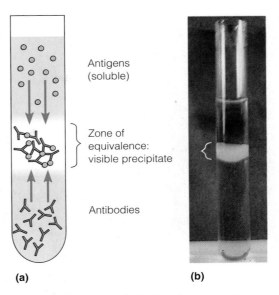

FIGURE 18.3 The precipitin ring test. (a) This drawing shows the diffusion of antigens and antibodies toward each other in a small-diameter test tube. Where they reach equal proportions, in the zone of equivalence, a visible line or ring of precipitate is formed. **(b)** A photograph of a precipitin ring.

■ What causes the visible line?

guinea pigs with the disease were injected with a suspension of *Mycobacterium tuberculosis*, the site of the injection became red and slightly swollen a day or two later. You may recognize this symptom as the positive result of the test to determine whether a person has been infected by the tuberculosis pathogen. This test is widely used today; almost all of us have been subjected to the skin test for tuberculosis. Koch, of course, had no idea of the mechanism of cell-mediated immunity that caused this phenomenon, nor did he know of the existence of antibodies.

Since the time of Robert Koch, immunology has given us many other invaluable diagnostic tools, most of which are based on interactions of humoral antibodies with antigens. In this section we will discuss several of the most important techniques used to detect antigens and antibodies.

One problem our diagnostic tools must overcome is that antibodies cannot be seen directly. Even at magnifications of well over 100,000×, they appear only as fuzzy, ill-defined particles. Therefore, we must establish their presence indirectly through a variety of reactions. In most instances the presence of antibodies is of primary interest, but the procedures can be reversed so that known antibodies can be used to detect antigens.

Precipitation Reactions

Learning Objective

• Explain how precipitation reactions and immunodiffusion tests work.

Precipitation reactions involve the reaction of soluble antigens with IgG or IgM antibodies to form large, interlocking molecular aggregates called lattices.

Precipitation reactions occur in two distinct stages. First, the antigens and antibodies rapidly form small antigen–antibody complexes. This interaction occurs within seconds and is followed by a slower reaction, which may take minutes to hours, in which the antigen–antibody complexes form lattices that precipitate from solution. Precipitation reactions normally occur only when the ratio of antigen to antibody is optimal. Figure 18.2 shows that no visible precipitate forms when either component is in excess. The optimal ratio is produced when separate solutions of antigen and antibody are placed adjacent to each other and allowed to diffuse together. In a **precipitin ring test** (Figure 18.3), a cloudy line of precipitation (ring) appears in the area in which the optimal ratio has been reached (the *zone of equivalence*).

Immunodiffusion tests are precipitation reactions carried out in an agar gel medium, on either a Petri plate or

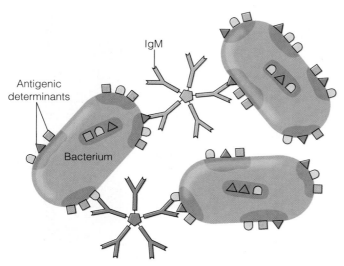

FIGURE 18.4 **An agglutination reaction.** When antibodies react with antigenic determinant sites on antigens carried on neighboring cells, such as these bacteria (or red blood cells), the particulate antigens (cells) agglutinate. IgM, the most efficient immunoglobulin for agglutination, is shown here, but IgG also participates in agglutination reactions.

■ Agglutination reactions involve particulate antigens.

a microscope slide. A line of visible precipitate develops between the wells at the point where the optimal antigen–antibody ratio is reached.

Other tests use electrophoresis to speed up the movement of antigen and antibody in a gel, sometimes in less than an hour, with this method. The techniques of immunodiffusion and electrophoresis can be combined in a procedure called **immunoelectrophoresis.** The procedure is used in research to separate proteins in human serum and is the basis of certain diagnostic tests. It is an essential part of the Western blot test used in AIDS testing (see Figure 10.12 on page 292).

Agglutination Reactions

Learning Objectives

- *Differentiate direct from indirect agglutination tests.*
- *Differentiate agglutination from precipitation tests.*
- *Define hemagglutination.*

Whereas precipitation reactions involve *soluble* antigens, agglutination reactions involve either *particulate* antigens (particles such as cells that carry antigenic molecules) or soluble antigens adhering to particles. These antigens can be linked together by antibodies to form visible aggregates, a reaction called **agglutination** (Figure 18.4). Agglutination reactions are very sensitive, relatively easy to read (see

Figure 10.10 on page 291), and available in great variety. Agglutination tests are classified as either direct or indirect.

Direct Agglutination Tests

Direct agglutination tests detect antibodies against relatively large cellular antigens, such as those on red blood cells, bacteria, and fungi. At one time they were carried out in a series of test tubes, but now they are usually done in plastic *microtiter plates,* which have many shallow wells that take the place of the individual test tubes. The amount of particulate antigen in each well is the same, but the amount of serum that contains antibodies is diluted, so that each successive well has half the antibodies of the previous well. These tests are used, for example, to test for brucellosis and to separate *Salmonella* isolates into serovars, types defined by serological means.

Clearly, the more antibody we start with, the more dilutions it will take to lower the amount to the point where there is not enough antibody for the antigen to react with. This is the measure of **titer,** or concentration of serum antibody (Figure 18.5). For infectious diseases in general, the higher the serum antibody titer, the greater the immunity to the disease. However, the titer alone is of limited use in diagnosing an existing illness. There is no way to know whether the measured antibodies were generated in response to the immediate infection or to an earlier illness. For diagnostic purposes, *a rise in titer* is significant; that is, the titer is higher later in the course of the disease than at its outset. Also, if it can be demonstrated that the person's blood had no antibody titer before the illness but has a significant titer while the disease is progressing, this change, called **seroconversion,** is also diagnostic. This situation is frequently encountered with HIV infections.

Some diagnostic tests specifically identify IgM antibodies. As discussed in Chapter 17, short-lived IgM is more likely to reflect a response to a current disease condition.

Indirect (Passive) Agglutination Tests

Antibodies against soluble antigens can be detected by agglutination tests if the antigens are adsorbed onto particles such as bentonite clay, or most often, minute latex spheres, each about one-tenth of the diameter of a bacterium. Such tests, known as *latex agglutination tests,* are commonly used for the rapid detection of serum antibodies against many bacterial and viral diseases. In such **indirect (passive) agglutination tests,** the antibody reacts with the soluble antigen adhering to the particles (Figure 18.6). The particles then agglutinate with one another, much as particles do in the direct agglutination tests. The same principle can be applied in reverse by using particles coated with antibodies to detect the antigens against which they are specific. This

1:20   1:40   1:80   1:160   1:320   1:640   Control

Top view
of wells

(a)

Enlarged
photo of
wells

Side view
of wells

(b) Agglutinated   **(c)** Nonagglutinated

(a) Each well in this microtiter plate contains, from left to right, only half the concentration of serum that is contained in the preceding well. Each well contains the same concentration of particulate antigens, in this instance red blood cells.

(b) In a positive (agglutinated) reaction, sufficient antibodies are present in the serum to link the antigens together, forming a mat of antigen–antibody complexes on the bottom of the well.

(c) In a negative (nonagglutinated) reaction, not enough antibodies are present to cause the linking of antigens. The particulate antigens roll down the sloping sides of the well, forming a pellet at the bottom. In this example, the antibody titer is 160 because the well with a 1:160 concentration is the most dilute concentration that produces a positive reaction.

FIGURE 18.5 Measuring antibody titer with the direct agglutination test.

■ What is meant by the term *antibody titer*?

approach is especially common in tests for the streptococci that cause sore throats. A diagnosis can be completed in about 10 minutes.

Hemagglutination

When agglutination reactions involve the clumping of red blood cells, the reaction is called **hemagglutination.** These reactions, which involve red blood cell surface antigens and their complementary antibodies, are used routinely in blood typing (see Table 19.2 on page 534) and in the diagnosis of infectious mononucleosis.

Certain viruses, such as those causing mumps, measles, and influenza, have the ability to agglutinate red blood cells without an antigen-antibody reaction; this process is called **viral hemagglutination** (Figure 18.7). This type of hemagglutination can be inhibited by antibodies that neutralize the agglutinating virus. Diagnostic tests based on such neutralization reactions are discussed in the next section.

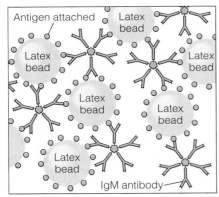

(a) Reaction in a positive indirect test for antibodies. When particles are coated with antigens, agglutination indicates the presence of antibodies, such as the IgM shown here.

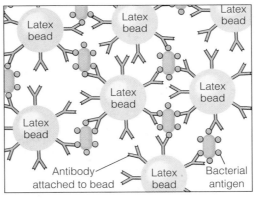

(b) Reaction in a positive indirect test for antigens. When particles are coated with monoclonal antibodies, agglutination indicates the presence of antigens.

FIGURE 18.6 Reactions in indirect agglutination tests. These tests are performed using antigens or antibodies coated onto particles such as minute latex spheres.

■ Differentiate direct from indirect agglutination tests.

FIGURE 18.7 **Viral hemagglutination.** Viral hemagglutination is not an antigen-antibody reaction.

■ Some viral diseases can be diagnosed by the virus's ability to agglutinate red blood cells.

Red blood cells   Viruses   Hemagglutination

Neutralization Reactions

Learning Objectives

• Explain how a neutralization test works.

• Differentiate precipitation from neutralization tests.

Neutralization is an antigen-antibody reaction in which the harmful effects of a bacterial exotoxin or a virus are blocked by specific antibodies. These reactions were first described in 1890, when investigators observed that immune serum could neutralize the toxic substances produced by the diphtheria pathogen, *Corynebacterium diphtheriae*. Such a neutralizing substance, which is called an antitoxin, is a specific antibody produced by a host as it responds to a bacterial exotoxin or its corresponding toxoid (inactivated toxin). The antitoxin combines with the exotoxin to neutralize it (Figure 18.8a). Antitoxins produced in an animal can be injected into humans to provide passive immunity against a toxin. Antitoxins

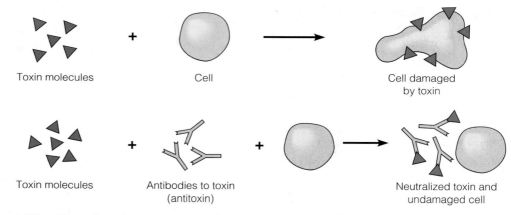

Toxin molecules   +   Cell   →   Cell damaged by toxin

Toxin molecules   +   Antibodies to toxin (antitoxin)   +   →   Neutralized toxin and undamaged cell

(a) The effects of a toxin on a susceptible cell and neutralization of the toxin by antitoxin.

Red blood cells   +   Antiviral antibodies from serum   +   Viruses   →   Viruses neutralized and hemagglutination inhibited

(b) Viral hemagglutination test to detect antibodies to a virus. These viruses will normally cause hemagglutination when mixed with red blood cells. If antibodies to the virus are present, as shown here, they neutralize and inhibit hemagglutination.

FIGURE 18.8 **Reactions in neutralization tests.**

■ Why does hemagglutination indicate that a patient does not have a specific disease?

from horses are routinely used to prevent or treat diphtheria and botulism; tetanus antitoxin is usually of human origin.

These therapeutic uses of neutralization reactions have led to their use as diagnostic tests. Viruses that exhibit their cytopathic (cell-damaging) effects in cell culture or embryonated eggs can be used to detect the presence of neutralizing viral antibodies (see Chapter 15 on page 450). If the serum to be tested contains antibodies against the particular virus, the antibodies will prevent that virus from infecting cells in the cell culture or eggs, and no cytopathic effects will be seen. Such tests, known as in vitro neutralization tests, can thus be used to both identify a virus and ascertain the viral antibody titer. In vitro neutralization tests are comparatively complex to carry out and are becoming less common in modern clinical laboratories.

A more frequently used neutralization test is the **viral hemagglutination inhibition test.** This test is used in the diagnosis of influenza, measles, mumps, and a number of other infections caused by viruses that can agglutinate red blood cells. If a person's serum contains antibodies against these viruses, these antibodies will react with the viruses and neutralize them (Figure 18.8b). For example, if hemagglutination occurs in a mixture of measles virus and red blood cells but does not occur when the patient's serum is added to the mixture, this result indicates that the serum contains antibodies that have bound to and neutralized the measles virus.

Complement-Fixation Reactions

Learning Objective

• Explain the basis for the complement-fixation test.

In Chapter 16 (pages 472–478) we discussed a group of serum proteins collectively called complement. During most antigen-antibody reactions, complement binds to the antigen–antibody complex and is used up, or fixed. This process of **complement fixation** can be used to detect very small amounts of antibody. Antibodies that do not produce a visible reaction, such as precipitation or agglutination, can be demonstrated by the fixing of complement during the antigen-antibody reaction. Complement fixation was once used in the diagnosis of syphilis (Wasserman test) and is still used to diagnose certain viral, fungal, and rickettsial diseases. Execution of the complement-fixation test requires great care and good controls, one reason the trend is to replace it with newer, simpler tests. The test is performed in two stages: complement fixation and indicator (Figure 18.9).

Fluorescent-Antibody Techniques

Learning Objective

• Compare and contrast direct and indirect fluorescent-antibody tests.

Fluorescent-antibody (FA) techniques can identify microorganisms in clinical specimens and can detect the presence of a specific antibody in serum (Figure 18.10). These techniques combine fluorescent dyes such as fluorescein isothiocyanate (FITC) with antibodies to make them fluoresce when exposed to ultraviolet light (see Figure 3.6 on page 62). These procedures are quick, sensitive, and very specific; the FA test for rabies can be performed in a few hours and has an accuracy rate close to 100%.

Fluorescent-antibody tests are of two types, direct and indirect. **Direct FA tests** are usually used to identify a microorganism in a clinical specimen (Figure 18.10a). During this procedure, the specimen containing the antigen to be identified is fixed onto a slide. Fluorescein-labeled antibodies are then added, and the slide is incubated briefly. Next the slide is washed to remove any antibody not bound to antigen and is then examined under the fluorescence microscope for yellow-green fluorescence.

Indirect FA tests are used to detect the presence of a specific antibody in serum following exposure to a microorganism (Figure 18.10b). They are often more sensitive than direct tests. During this procedure, a known antigen is fixed onto a slide. The test serum is then added, and, if antibody that is specific to that microbe is present, it reacts with the antigen to form a bound complex. In order for the antigen-antibody complex to be seen, fluorescein-labeled **antihuman immune serum globulin (anti-HISG),** an antibody that reacts specifically with *any* human antibody, is added to the slide. Anti-HISG will be present only if the specific antibody has reacted with its antigen and is therefore present as well. After the slide has been incubated and washed (to remove unbound antibody), it is examined under a fluorescence microscope. If the known antigen fixed to the slide appears fluorescent, the antibody specific to the test antigen is present.

An especially interesting adaptation of fluorescent antibodies is the **fluorescence-activated cell sorter (FACS).** In Chapter 17, we learned that T cells carry antigenically specific receptors such as CD4 and CD8 on their surface, and these are characteristic of certain groups of T cells. The depletion of CD4 T cells is used to follow the progression of AIDS; their populations can be determined with a FACS.

The FACS is a modification of a *flow cytometer,* in which a suspension of cells leaves a nozzle as droplets containing no more than one cell each (see page 293). A laser

FIGURE 18.9 **The complement-fixation test.** This test is used to indicate the presence of antibodies to a known antigen. Complement will combine (be fixed) with an antibody that is reacting with an antigen. If all the complement is fixed in the complement-fixation stage, then none will remain to cause hemolysis of the red blood cells in the indicator stage.

■ Why does red blood cell lysis indicate that the patient does not have a specific disease?

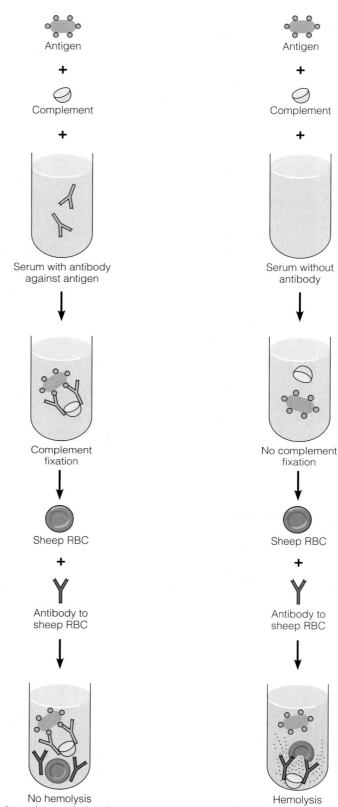

Antigen
+
Complement
+

Serum with antibody against antigen

Complement fixation

Sheep RBC
+

Antibody to sheep RBC

No hemolysis (complement tied up in antigen–antibody reaction)

(a) Positive test. All available complement is fixed by the antigen–antibody reaction; no hemolysis occurs, so the test is positive for the presence of antibodies.

Antigen
+
Complement
+

Serum without antibody

No complement fixation

Sheep RBC
+

Antibody to sheep RBC

Hemolysis (uncombined complement available)

(b) Negative test. No antigen–antibody reaction occurs. The complement remains, and the red blood cells are lysed in the indicator stage, so the test is negative.

Complement-fixation stage

Indicator stage

(a) Reactions in a positive direct fluorescent-antibody test

Group A streptococci from patient's throat

Fluorescent dye–labeled antibodies to group A streptococci

Fluorescent streptococci

LM | 4 μm

(b) Reactions in a positive indirect fluorescent-antibody test

T. pallidum from laboratory stock

Specific antibodies in serum of patient

Antibodies bind to T. pallidum

Fluorescent dye–labeled anti human immune serum globulin (This will react with any immunoglobulin)

Fluorescent spirochetes

FIGURE 18.10 Fluorescent-antibody (FA) techniques. (a) A direct FA test to identify group A streptococci. **(b)** In an indirect FA test such as that used in the diagnosis of syphilis, the fluorescent dye is attached to antihuman immune serum globulin, which reacts with any human immunoglobulin (such as the *Treponema pallidum*–specific antibody) that has previously reacted with the antigen. The reaction is viewed through a fluorescence microscope and the antigen with which the dye-tagged antibody has reacted—fluoresces (glows) in the ultraviolet illumination.

■ A direct FA test detects the presence of antigen in the patient. An indirect FA test detects antibodies in the patient.

beam strikes each cell-containing droplet and is then received by a detector that determines certain characteristics such as size (Figure 18.11). If the cells carry FA markers to identify them as CD4 or CD8 T cells, the detector can measure this fluorescence. As the laser beam detects a cell of a preselected size or fluorescence, an electrical charge, either positive or negative, can be imparted to it. As the charged droplet falls between electrically charged plates, it is attracted to one receiving tube or another, effectively separating cells of different types. Mil-

lions of cells can be separated in an hour with this process, all under sterile conditions, which allows them to be used in experimental work.

An interesting application of the flow cytometer is sorting sperm cells to separate male (Y-carrying) and female (X-carrying) sperm. The female sperm (meaning that it will result in a female embryo when it fertilizes the egg) contains more DNA, 2.8% more in humans, 4% in animals. When the sperm is stained with a fluorescent dye specific for DNA, the female sperm, with more DNA,

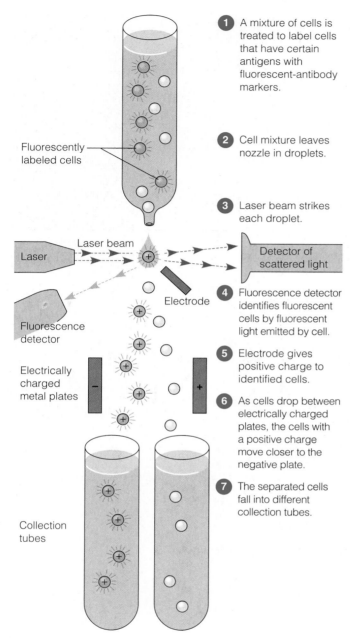

① A mixture of cells is treated to label cells that have certain antigens with fluorescent-antibody markers.

Fluorescently labeled cells

② Cell mixture leaves nozzle in droplets.

③ Laser beam strikes each droplet.

Laser beam

Laser

Detector of scattered light

Electrode

Fluorescence detector

Electrically charged metal plates

④ Fluorescence detector identifies fluorescent cells by fluorescent light emitted by cell.

⑤ Electrode gives positive charge to identified cells.

⑥ As cells drop between electrically charged plates, the cells with a positive charge move closer to the negative plate.

⑦ The separated cells fall into different collection tubes.

Collection tubes

FIGURE 18.11 **The fluorescence-activated cell sorter (FACS).** This technique can be used to separate different classes of T cells. A fluorescence-labeled antibody reacts with, for example, the CD4 receptor on a T cell.

■ The FACS is used to count T cells to follow progress of HIV infection.

glows more brightly when illuminated by the laser beam and can be separated out. The technique was developed for agricultural purposes. However, it has received medical approval for use in humans where couples carry genes for inherited diseases that affect only boys.

Enzyme-Linked Immunosorbent Assay (ELISA)

Learning Objective
• *Explain how direct and indirect ELISA tests work.*

The **enzyme-linked immunosorbent assay (ELISA)** is the most widely used of a group of tests known as *enzyme immunoassay (EIA)*. There are two basic methods. The *direct ELISA* detects antigens, and the *indirect ELISA* detects antibodies. A microtiter plate with numerous shallow wells is used in both procedures (see Figure 10.11a on page 291). Variations of the test exist; for example, the reagents can be bound to tiny latex particles rather than to the surfaces of the microtiter plates. ELISA procedures are popular primarily because they require little interpretive skill to read; the results tend to be clearly positive or clearly negative.

Direct ELISA

The direct ELISA method is shown in Figure 18.12.

❶ In the first step, the antibody specific for the antigen to be detected is adsorbed to the surface of the wells of the microtiter plate.

❷ A patient's sample containing unidentified antigen is then added to each well. If the antigen reacts specifically with the antibodies adsorbed to the well, the antigen (a drug in a urine test, for example) will be retained there when the well is washed free of unbound antigen.

❸ A second antibody specific for the antigen is then added. If both the antibody adsorbed to the wall of the well and the antibody known to be specific for the antigen have reacted with the antigen, a "sandwich" will be formed, with the antigen between two antibody molecules. This reaction is visible only because the second added antibody is linked to an enzyme, such as horseradish peroxidase or alkaline phosphatase. Unbound enzyme-linked antibody is washed from the well, and then

❹ the enzyme's substrate is added to it. Enzymatic activity is indicated by a color change that can be visually detected. The test will be positive if the antigen has reacted with adsorbed antibodies in the first step. If the test antigen was not specific for the antibody adsorbed to the wall of the well, the test will be negative because the unbound antigen will have been washed away.

A common use of the direct ELISA test is to detect the presence of drugs in urine. The availability of monoclonal antibodies has been essential to the widespread use of this type of test.

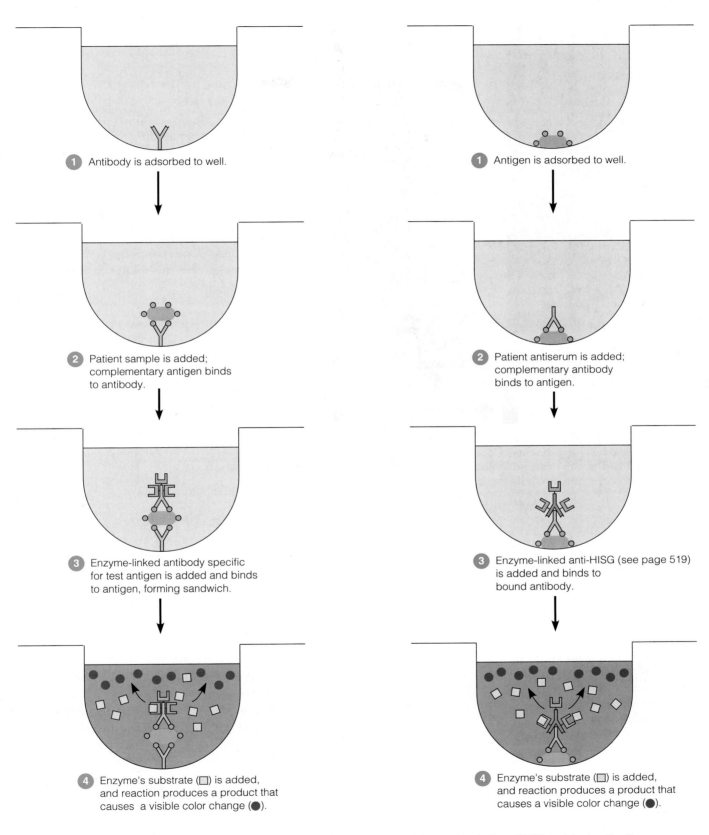

1 Antibody is adsorbed to well.

2 Patient sample is added; complementary antigen binds to antibody.

3 Enzyme-linked antibody specific for test antigen is added and binds to antigen, forming sandwich.

4 Enzyme's substrate (□) is added, and reaction produces a product that causes a visible color change (●).

(a) A positive direct ELISA to detect antigens

1 Antigen is adsorbed to well.

2 Patient antiserum is added; complementary antibody binds to antigen.

3 Enzyme-linked anti-HISG (see page 519) is added and binds to bound antibody.

4 Enzyme's substrate (□) is added, and reaction produces a product that causes a visible color change (●).

(b) A positive indirect ELISA to detect antibodies

FIGURE 18.12 **The ELISA method.** The components are usually contained in small wells of a microtiter plate.

■ Differentiate a direct from an indirect ELISA test.

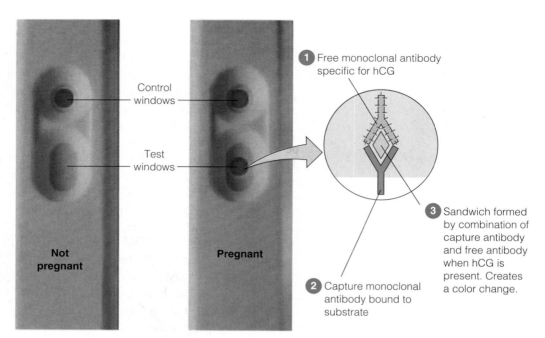

FIGURE 18.13 The use of monoclonal antibodies in a home pregnancy test.
Home pregnancy tests detect a hormone called human chorionic gonadotropin (hCG) that is
excreted only in the urine of a pregnant woman. The test area of the dipstick contains two
types of monoclonal antibodies: ❶ free monoclonal antibodies and ❷ capture monoclonal
antibodies, bound to the substrate in the test window. The free antibodies are specific for
hCG and are color-labeled. The capture antibodies are specific for the hCG/free antibody
complex. When the dipstick comes in contact with urine that contains hCG, the free antibodies bind the hCG and are then bound in turn by the capture antibodies, forming a
"sandwich." ❸ The color-labeled antibodies then create a visible color change in the test
window. (Other antibodies create the color change in the control window.)

■ What is the antigen in the home pregnancy test?

Indirect ELISA

In the first step of the indirect ELISA method,

❶ a known antigen from the laboratory, rather than the
antibody, is adsorbed to the shallow wells on the plate
(Figure 18.12b). To see whether a serum sample contains antibodies against this antigen,

❷ the antiserum is added to the well. If the serum contains antibody specific to the antigen, the antibody
will bind to the adsorbed antigen. All unreacted
antiserum is washed from the well. Anti-HISG (see
page 519) is then allowed to react with the
antigen–antibody complex.

❸ The anti-HISG, which has been linked with an enzyme, reacts with the antibodies that are bound to the
antigens in the well.

❹ Finally, all unbound anti-HISG is rinsed away, and
the correct substrate for the enzyme is added. A
colored enzymatic reaction occurs in the wells in
which bound antigen has combined with antibody in
the serum sample. This procedure resembles that of
the indirect FA test, except that the anti-HISG of the
indirect assay is linked with an enzyme rather than
with a fluorescing dye.

Many ELISA tests are available for clinical use in the
form of commercially prepared kits. Procedures are often
highly automated, with the results read by a scanner and
printed out by computer (see Figure 10.11 on page 291).
Some tests based on this principle are also available for use
by the public; one example is a commonly available home
pregnancy test (Figure 18.13). ELISA tests are also used to
screen blood for antibodies to HIV.

The Future of Diagnostic Immunology

Learning Objective

- *Explain the importance of monoclonal antibodies.*

The introduction of monoclonal antibodies has revolutionized diagnostic immunology by making available large, economical amounts of specific antibodies. This has led to many newer diagnostic tests that are more sensitive, specific, rapid, and simpler to use. For example, tests to diagnose sexually transmitted chlamydial infections and certain protozoan-caused intestinal parasitic diseases are coming into common use. These tests had previously required relatively difficult cultural or microscopic methods for diagnosis. At the same time, the use of many of the classic serological tests, such as complement-fixation tests, is declining. Most newer tests will require less human judgment to read, and they require fewer highly trained personnel.

The use of certain *nonimmunological* tests, such as the PCR and DNA probes that were discussed in Chapter 10 on page 295, is increasing. Some of these tests will become automated to a significant degree. For example, a DNA chip (see Figure 10.17 on page 297) containing over 50,000 DNA probes for genetic information expected in possible pathogens can be exposed to a test sample. This chip is scanned and its data automatically analyzed. PCR tests are also becoming highly automated. Federal approval has already been granted to instruments that measure quantities of several viral pathogens. At the other end of the technical spectrum, efforts are underway to provide low-cost diagnostic methods for developing nations, where the annual medical expenditure per person is only a few dollars. Interest in methods less invasive than blood sampling, such as urinalysis and oral swabs, is also increasing.

The tests described in this chapter are most often used to detect existing disease. In the future, diagnostic testing will probably also be directed at preventing disease. Food and water, for example, are frequently contaminated with pathogens. Controlling diseases spread by ingesting them is a significant public health problem. At present, detecting and identifying such pathogens usually require that the microorganism be selectively isolated and cultivated before diagnostic tests can be applied. This time-consuming process poses significant problems when the source of infection is a perishable consumer product. Much current research in diagnostic methods is now aimed at enabling health workers to detect and identify pathogenic organisms directly, and quickly, in a food sample without the need for isolation and cultivation.

STUDY OUTLINE

Vaccines (pp. 508–514)

1. Edward Jenner developed the modern practice of vaccination when he inoculated people with cowpox virus to protect them against smallpox.

Principles and Effects of Vaccination (pp. 509–510)

1. Herd immunity results when most of a population is immune to a disease.

Types of Vaccines and Their Characteristics (pp. 510–512)

1. Attenuated whole-agent vaccines consist of attenuated (weakened) microorganisms; attenuated virus vaccines generally provide lifelong immunity.
2. Inactivated whole-agent vaccines consist of killed bacteria or viruses.
3. Toxoids are inactivated toxins.
4. Subunit vaccines consist of antigenic fragments of a microorganism; these include recombinant vaccines and acellular vaccines.
5. Conjugated vaccines combine the desired antigen with a protein that boosts the immune response.

6. Nucleic acid vaccines, or DNA vaccines, are being developed. These cause the recipient to make the antigenic protein associated with class I MHC.

The Development of New Vaccines (pp. 512–514)

1. Viruses for vaccines may be grown in animals, cell cultures, or chick embryos.
2. Recombinant vaccines and nucleic acid vaccines do not need to be grown in cells or animals.
3. Genetically engineered plants may someday provide edible vaccines.
4. Adjuvants improve the effectiveness of some antigens.

Safety of Vaccines (p. 514)

1. Vaccines are the safest and most effective means of controlling infectious diseases.

Diagnostic Immunology (pp. 514–525)

1. Many tests based on the interactions of antibodies and antigens have been developed to determine the presence of antibodies or antigens in a patient.

Precipitation Reactions (pp. 515–516)

1. The interaction of soluble antigens with IgG or IgM antibodies leads to precipitation reactions.

2. Precipitation reactions depend on the formation of lattices and occur best when antigen and antibody are present in optimal proportions. Excesses of either component decrease lattice formation and subsequent precipitation.

3. The precipitin ring test is performed in a small tube.

4. Immunodiffusion procedures are precipitation reactions carried out in an agar gel medium.

5. Immunoelectrophoresis combines electrophoresis with immunodiffusion for the analysis of serum proteins.

Agglutination Reactions (pp. 516–517)

1. The interaction of particulate antigens (cells that carry antigens) with antibodies leads to agglutination reactions.

2. Diseases may be diagnosed by combining the patient's serum with a known antigen.

3. Diseases can be diagnosed by a rising titer or seroconversion (from no antibodies to the presence of antibodies).

4. Direct agglutination reactions can be used to determine antibody titer.

5. Antibodies cause visible agglutination of soluble antigens affixed to latex spheres in indirect or passive agglutination tests.

6. Hemagglutination reactions involve agglutination reactions using red blood cells. Hemagglutination reactions are used in blood typing, the diagnosis of certain diseases, and the identification of viruses.

Neutralization Reactions (pp. 518–519)

1. In neutralization reactions, the harmful effects of a bacterial exotoxin or virus are eliminated by a specific antibody.

2. An antitoxin is an antibody produced in response to a bacterial exotoxin or a toxoid that neutralizes the exotoxin.

3. In a virus neutralization test, the presence of antibodies against a virus can be detected by the antibodies' ability to prevent cytopathic effects of viruses in cell cultures.

4. Antibodies against certain viruses can be detected by their ability to interfere with viral hemagglutination in viral hemagglutination inhibition tests.

Complement-Fixation Reactions (p. 519)

1. Complement-fixation reactions are serological tests based on the depletion of a fixed amount of complement in the presence of an antigen-antibody reaction.

Fluorescent-Antibody Techniques (pp. 519–522)

1. Fluorescent-antibody techniques use antibodies labeled with fluorescent dyes.

2. Direct fluorescent-antibody tests are used to identify specific microorganisms.

3. Indirect fluorescent-antibody tests are used to demonstrate the presence of antibody in serum.

4. A fluorescence-activated cell sorter can be used to detect and count cells labeled with fluorescent antibodies.

Enzyme-Linked Immunosorbent Assay (ELISA) (pp. 522–524)

1. ELISA techniques use antibodies linked to an enzyme, such as horseradish peroxidase or alkaline phosphatase.

2. Antigen-antibody reactions are detected by enzyme activity. If the indicator enzyme is present in the test well, an antigen-antibody reaction has occurred.

3. The direct ELISA is used to detect antigens against a specific antibody bound in a test well.

4. The indirect ELISA is used to detect antibodies against an antigen bound in a test well.

The Future of Diagnostic Immunology (p. 525)

1. The use of monoclonal antibodies will continue to make new diagnostic tests possible.

STUDY QUESTIONS

Review

1. Classify the following vaccines by type. Which could cause the disease it is supposed to prevent?
   a. attenuated measles virus
   b. dead *Rickettsia prowazekii*
   c. *Vibrio cholerae* toxoid
   d. hepatitis B antigen produced in yeast cells
   e. purified polysaccharides from *Streptococcus pyogenes*
   f. *Haemophilus influenzae* polysaccharide bound to diphtheria toxoid
   g. a plasmid containing genes for influenza A protein

2. Explain the effects of excess antigen and antibody on the precipitation reaction. How is the precipitin ring test different from an immunodiffusion test?

3. How does the antigen in an agglutination reaction differ from that in a precipitation reaction?

4. Define the following terms, and give an example of how each reaction is used diagnostically:
   a. viral hemagglutination
   b. hemagglutination inhibition
   c. passive agglutination

5. Explain the fluorescent-antibody techniques.

6. Describe the direct and indirect FA tests in the following situations. Sketch how the reactions might appear in these tests if the antigens and antibodies could be seen.
   a. Rabies can be diagnosed postmortem by mixing fluorescent-labeled antibodies with brain tissue.
   b. Syphilis can be diagnosed by adding the patient's serum to a slide fixed with *Treponema pallidum*. Antihuman immune serum globulin tagged with a fluorescent dye is added.

7. Explain the ELISA techniques.

8. Describe the direct and indirect ELISA tests in the following situations:
   a. Respiratory secretions to detect respiratory syncytial virus.
   b. Blood to detect human immunodeficiency virus antibodies.

   Which of these tests provides definitive proof of disease?

9. Match the following serological tests to the descriptions.
   _____ Precipitation
   _____ Immuno-electrophoresis
   _____ Agglutination
   _____ Complement fixation
   _____ Neutralization
   _____ ELISA

   a. Occurs with particulate antigens
   b. Uses an enzyme for the indicator
   c. Uses red blood cells for the indicator
   d. Uses antihuman immune serum globulin
   e. Occurs with a free soluble antigen
   f. Used to determine the presence of antitoxin

10. Match each of the following tests to its positive reaction.
    _____ Agglutination
    _____ Complement fixation
    _____ ELISA
    _____ FA test
    _____ Neutralization
    _____ Precipitation

    a. Peroxidase activity
    b. Harmful effects of agents not seen
    c. No hemolysis
    d. Cloudy white line
    e. Cell clumping
    f. Fluorescence

Multiple Choice

Use the following choices to answer questions 1 and 2:
   a. hemolysis
   b. hemagglutination
   c. hemagglutination-inhibition
   d. no hemolysis
   e. precipitin ring forms

1. Patient's serum, influenza virus, sheep red blood cells, and antisheep red blood cells are mixed in a tube. What happens if the patient has antibodies against influenza?

2. Patient's serum, *Chlamydia*, guinea pig complement, sheep red blood cells, and antisheep red blood cells are mixed in a tube. What happens if the patient has antibodies against *Chlamydia*?

3. The examples in questions 1 and 2 are
   a. direct tests.
   b. indirect tests.

Use the following choices to answer questions 4 and 5:
   a. anti-*Brucella*
   b. *Brucella*
   c. substrate for the enzyme

4. Which is the third step in a direct ELISA test?

5. Which item is from the patient in an indirect ELISA test?

6. In an immunodiffusion test, a strip of filter paper containing diphtheria antitoxin is placed on a solid culture medium. Then bacteria are streaked perpendicular to the filter paper. If the bacteria are toxigenic,
   a. the filter paper will turn red.
   b. a line of antigen-antibody precipitate will form.
   c. the cells will lyse.
   d. the cells will fluoresce.
   e. none of the above

Use the following choices to answer questions 7–9.
   a. direct fluorescent antibody
   b. indirect fluorescent antibody
   c. rabies immune globulin
   d. killed rabies virus
   e. none of the above

7. Treatment given to a person bitten by a rabid bat.

8. Test used to identify rabies virus in the brain of a dog.

9. Test used to detect the presence of antibodies in a patient's serum.

10. In an agglutination test, eight serial dilutions to determine antibody titer were set up: Tube #1 contained a 1:2 dilution; tube #2, a 1:4, and so on. If tube #5 is the last tube showing agglutination, what is the antibody titer?
    a. 5
    b. 1:5
    c. 32
    d. 1:32

Critical Thinking

1. What problems are associated with the use of attenuated whole-agent vaccines?

2. The World Health Organization has announced the complete eradication of smallpox and is working toward the eradication of measles and polio. Why would vaccination be more likely to eradicate a viral disease than a bacterial disease?

3. Many of the serological tests require a supply of antibodies against pathogens. For example, to test for *Salmonella*, anti-*Salmonella* antibodies are mixed with the unknown bacterium. How are these antibodies obtained?

4. A test for antibodies against *Treponema pallidum* uses an antigen called cardiolipin and the patient's serum (suspected of having antibodies). Why do the antibodies react with cardiolipin? What is the disease?

Clinical Applications

1. Which of the following is proof of a disease state? Why doesn't the other situation confirm a disease state? What is the disease?
   a. *Mycobacterium tuberculosis* is isolated from a patient.
   b. Antibodies against *M. tuberculosis* are found in a patient.

2. Streptococcal erythrogenic toxin is injected into a person's skin in the Dick test. What results are expected if a person has antibodies against this toxin? What type of immunological reaction is this? What is the disease?

3. The following data were obtained from FA tests for anti-*Legionella* in four people. What conclusions can you draw? What is the disease?

	Antibody Titer			
	Day 1	**Day 7**	**Day 14**	**Day 21**
Patient A	128	256	512	1024
Patient B	0	0	0	0
Patient C	256	256	256	256
Patient D	0	0	128	512

4. Maria decided against the relatively new chickenpox vaccine and used her parents' method: She wanted her children to get chickenpox in order to develop natural immunity. Her two children did get chickenpox. Her son had slight itching and skin vesicles, but her daughter was hospitalized for months with streptococcal cellulitis and underwent several skin grafts before recovering. Maria's housekeeper contracted chickenpox from the children and subsequently died. Almost half of the deaths due to chickenpox occur in adults.
   a. What responsibilities do parents have for their children's health?
   b. What rights do individuals have? Should vaccination be required by law?
   c. What responsibilities do individuals (e.g., parents) have for the health of society?
   d. Vaccines are given to healthy people, so what risks are acceptable?

LEARNING WITH TECHNOLOGY

The Microbiology Place CD-ROM/website includes many resources to help you study and succeed in this course. In addition to self-quizzes, web links, and flashcards, you'll find the following activities for this chapter in the "Activities" section of The Microbiology Place:

Diagnostic Immunology: Precipitation Test

Diagnostic Immunology: Indirect Agglutination Test

Diagnostic Immunology: Direct Agglutination Test

Diagnostic Immunology: Hemagglutination Inhibition Test

Diagnostic Immunology: Complement Fixation Test

Diagnostic Immunology: ELISA Test

Diagnostic Immunology: Fluorescent Antibody Test

Case Study: *Pseudomonas*

Chapter 19

Disorders Associated with the Immune System

CYTOTOXIC T CELLS DESTROY CANCER CELLS. *Enzymes from the small cytotoxic T cell have made a hole in the larger cancer cell.*

*I*n this chapter, we will see that not all immune system responses produce a desirable result, such as immunity to disease. Occasionally the reactions of the immune system are harmful. A familiar example is hay fever, which results from repeated exposure to plant pollen. Most of us also know that a blood transfusion will be rejected if the blood of the donor and the blood of the recipient are not compatible, and that rejection is also a potential problem with transplanted organs. The body rejects transfusions or transplants because the immune system identifies the donated cells as nonself and attacks them. One's own tissue may also be mistakenly attacked by the immune system, causing diseases we classify as autoimmune.

We may also be harmed if the immune system fails to function properly. The obvious example is that we may be the victim of some pathogen if a vaccine, or recovery from a previous illness, has failed to provide immunity to it. Cancer can be another, less apparent, indication of a failure of the immune system.

Certain antigens provoke such a drastic immune response that they are termed **superantigens** (see Chapter 15, page 446). Superantigens are relatively nonspecific antigens and indiscriminately activate many T cell receptors at once, resulting in a damaging reaction.

Some people are born with a defective immune system, and in all of us the effectiveness of our immune system declines with age. In addition, many people have their immune systems deliberately crippled (immunosuppressed) to minimize the potential for the rejection of transplanted organs. Other people suffer from the effects of HIV, which attacks the immune system.

TABLE 19.1	**Types of Hypersensitivity**			
Type of Reaction	**Time Before Clinical Signs**	**Characteristics**	**Examples**	
Type I (anaphylactic)	<30 min	IgE binds to mast cells or basophils; causes degranulation of mast cell or basophil and release of reactive substances such as histamine	Anaphylactic shock from drug injections and insect venom; common allergic conditions, such as hay fever, asthma	
Type II (cytotoxic)	5–12 hours	Antigen causes formation of IgM and IgG antibodies that bind to target cell; when combined with action of complement, destroys target cell	Transfusion reactions, Rh incompatibility	
Type III (immune complex)	3–8 hours	Antibodies and antigens form complexes that cause damaging inflammation	Arthus reactions, serum sickness	
Type IV (cell-mediated or delayed-type)	24–48 hours	Antigens activate T_C that kill target cells.	Rejection of transplanted tissues; contact dermatitis, such as poison ivy; certain chronic diseases, such as tuberculosis	

Hypersensitivity

Learning Objective

• *Define hypersensitivity.*

The term **hypersensitivity** refers to an antigenic response beyond that which is considered normal; the term **allergy** is more familiar and is essentially synonymous. Hypersensitivity responses occur in individuals who have been *sensitized* by previous exposure to an antigen, which in this context is sometimes called an **allergen.** When an individual who was previously sensitized is exposed to that antigen again, his or her immune system reacts to it in a damaging manner. The four principal types of hypersensitivity reactions, summarized in Table 19.1, are anaphylactic, cytotoxic, immune complex, and cell-mediated, or delayed-type, reactions.

Type I (Anaphylactic) Reactions

Learning Objectives

• *Describe the mechanism of anaphylaxis.*
• *Compare and contrast systemic and localized anaphylaxis.*
• *Explain how allergy skin tests work.*
• *Define desensitization and blocking antibody.*

Type I, or anaphylactic, reactions often occur within 2 to 30 minutes after a person sensitized to an antigen is reexposed to that antigen. Anaphylaxis means "the opposite of protected," from the prefix *ana-*, against, and the Greek

phylaxis, protection. **Anaphylaxis** is an inclusive term for the reactions caused when certain antigens combine with IgE antibodies. Anaphylactic responses can be *systemic reactions*, which produce shock and breathing difficulties and are sometimes fatal, or *localized reactions*, which include common allergic conditions such as hay fever, asthma, and hives (slightly raised, often itchy and reddened areas of the skin).

The IgE antibodies produced in response to an antigen, such as insect venom or plant pollen, bind to the surfaces of cells such as mast cells and basophils. These two cell types are similar in morphology and in their contribution to allergic reactions. **Mast cells** are especially prevalent in the connective tissue of the skin and respiratory tract and in surrounding blood vessels. The name is from the German word *mastzellen*, meaning "well fed"; they are packed with granules that at one time were mistakenly thought to have been ingested (Figure 19.1). **Basophils** circulate in the bloodstream, where they constitute fewer than 1% of the leukocytes. Both are packed with granules containing a variety of chemicals called *mediators*.

Mast cells and basophils can have as many as 500,000 sites for IgE attachment. The Fc (stem) region of an IgE antibody (see Figure 17.5 on page 486) can attach to one of these specific receptor sites on such a cell, leaving two antigen-binding sites free. Of course, the attached IgE monomers will not all be specific for the same antigen. But when an antigen such as plant pollen encounters two adjacent antibodies of the same appropriate specificity, it can bind to one antigen-binding site on each antibody, bridging the space between them. This bridge triggers the mast cell or basophil to undergo **degranulation,** which releases the granules inside these cells and also the mediators they contain (Figure 19.1).

These mediators cause the unpleasant and damaging effects of an allergic reaction. The best-known mediator is **histamine.** The release of histamine increases the permeability and distension of blood capillaries, resulting in edema (swelling) and erythema (redness). Other effects include increased mucus secretion (a runny nose, for example) and smooth muscle contraction, which in the respiratory bronchi results in breathing difficulty.

Other mediators include **leukotrienes** of various types and **prostaglandins.** These mediators are not preformed and stored in the granules but are synthesized by the antigen-triggered cell. Because leukotrienes tend to cause prolonged contractions of certain smooth muscles, their action contributes to the spasms of the bronchial tubes that occur during asthmatic attacks. Prostaglandins affect smooth muscles of the respiratory system and cause increased mucus secretion.

Collectively, all these mediators serve as chemotactic agents that, in a few hours, attract neutrophils and eosinophils to the site of the degranulated cell. They then activate various factors that cause inflammatory symptoms, such as distension of the capillaries, swelling, increased secretion of mucus, and involuntary contractions of smooth muscles.

Systemic Anaphylaxis

At the turn of the century, two French biologists studied the responses of dogs to the venom of stinging jellyfish. Large doses of venom usually killed the dogs, but sometimes a few survived the injections. These surviving dogs were used for repeat experiments with the venom, and the results were surprising. Even a very tiny dose of the venom, one that should have been almost harmless, killed the dogs. They suffered difficulty in respiration, entered shock as their cardiovascular systems collapsed, and quickly died. This phenomenon was called *anaphylactic shock.*

Systemic anaphylaxis (or *anaphylactic shock*) can result when an individual sensitized to an antigen is exposed to it again. Injected antigens are more likely to cause a dramatic response than antigens introduced via other portals of entry. The release of mediators causes peripheral blood vessels throughout the body to enlarge, resulting in a drop in blood pressure (shock). This reaction can be fatal within a few minutes. There is very little time to act once someone develops systemic anaphylaxis. Treatment usually involves self-administration with a preloaded syringe of epinephrine, a drug that constricts blood vessels and raises the blood pressure. In the United States, 50–60 people die each year from anaphylactic shock caused by insect stings.

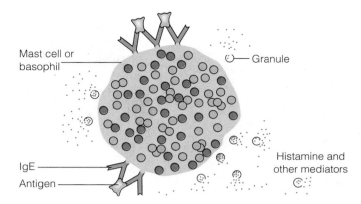

(a) IgE antibodies, produced in response to an antigen, coat mast cells and basophils. When an antigen bridges the gap between two adjacent antibody molecules of the same specificity, the cell undergoes degranulation and releases histamine and other mediators.

(b) A degranulated mast cell that has reacted with an antigen and released granules of histamine and other reactive mediators.

FIGURE 19.1 **The mechanism of anaphylaxis.**

■ IgE and an antigen interact to cause common allergic reactions.

You may be acquainted with someone who reacts to penicillin in this way. In these individuals, the penicillin, which is a hapten (it cannot induce antibody formation by itself; see Chapter 17), combines with a carrier serum protein. Only then is penicillin immunogenic. Penicillin allergy probably occurs in about 2% of the population. The most reliable test for allergy to penicillin will still fail to

(a) SEM | 10 μm   **(b)** SEM | 500 μm

FIGURE 19.2 **Localized anaphylaxis. (a)** A micrograph of pollen grains. **(b)** A micrograph of a house dust mite.

■ Localized anaphylaxis is usually associated with antigens that are ingested or inhaled.

identify about 7% of those at risk. Allergy to penicillin also includes risk from exposure to some related drugs, such as carbapenem (page 569).

Localized Anaphylaxis

Whereas sensitization to injected antigens is a common cause of systemic anaphylaxis, **localized anaphylaxis** is usually associated with antigens that are ingested (foods) or inhaled (pollen)(Figure 19.2a). The symptoms depend primarily on the route by which the antigen enters the body.

In allergies involving the upper respiratory system, such as hay fever (allergic rhinitis), sensitization usually involves mast cells in the mucous membranes of the upper respiratory tract. Reexposure to the airborne antigen, which might be a common environmental material such as plant pollen, fungal spores, feces of house dust mites (Figure 19.2b), or animal dander.* The typical symptoms are itchy and teary eyes, congested nasal passages, coughing, and sneezing. Antihistamine drugs, which compete for histamine receptor sites, are often used to treat these symptoms.

Asthma is an allergic reaction that mainly affects the lower respiratory system. Symptoms such as wheezing and shortness of breath are caused by the constriction of smooth muscles in the bronchial tubes.

For unknown reasons, asthma is becoming a near epidemic, affecting about 10% of children in Western society, although they often outgrow it in time. It is speculated that lack of childhood exposure in the developed world to many childhood infections, the so-called *hygiene hypothesis*, is a factor in the increase in the incidence of asthma. Mental or emotional stress can also be a contributing factor in precipitating an attack. Symptoms of asthma are usually controlled by aerosol inhalants, which, unfortunately, may be difficult for very small children to use.

Antigens that enter the body via the gastrointestinal tract can also sensitize an individual. Many of us may know someone who is allergic to a particular food. Frequently, so-called food allergies may not be related to hypersensitivity at all and are more accurately described as *food intolerances*. For example, many people are unable to digest the lactose in milk because they lack the enzyme that breaks down this disaccharide milk sugar. The lactose enters the intestine, where it osmotically retains fluid, causing diarrhea.

Gastrointestinal upset is a common symptom of food allergies, but it can also result from many other factors. Hives are more characteristic of a true food allergy, and ingestion of the antigen may result in systemic anaphylaxis. Death has even resulted when a person sensitive to fish ate french fries that had been prepared in oil previously used to fry fish. Skin tests are not reliable indicators for the diagnosis of food-related allergies, and completely controlled tests for hypersensitivity to ingested foods are very difficult to perform. Only eight foods are responsible for 97% of food-related allergies: eggs, peanuts, tree-grown nuts, milk,

---

*Dander is a general term for microscopic particles from the fur or skin of animals. A cat, for example, carries about 100 mg of dander on its coat and sheds about 0.1 mg a day. This accumulates in upholstery and carpeting. People with allergies to mice, gerbils, and similar small animals are more likely to be allergic to components of urine accumulating in cages.

soy, fish, wheat, and peas. Most children exhibiting allergies to milk, egg, wheat, and soy develop tolerance as they age, but reactions to peanuts, tree nuts, and seafood tend to persist. Sulfites, to which many people have an allergy, are a frequent problem. Their use is widespread in foods and beverages, and although food labels should indicate their presence, these foods may be difficult to avoid in practice. A food product may have come into contact with a food allergen through processing machinery or cookware previously used for other foods. In one Food and Drug Administration report, 25% of bakery, ice cream, and candy products tested positive for peanut allergens even though peanuts were not listed on the required product labels. In the United States it is estimated that 150 persons a year die of severe allergic reactions to foods.

Prevention of Anaphylactic Reactions

Avoiding contact with the sensitizing antigen is the most obvious way to prevent allergic reactions. Unfortunately, avoidance is not always possible. Some individuals experience an allergic reaction after eating an assortment of foods. In such cases, they may not know exactly what antigen they are sensitive to. In some cases, skin tests might be of use in diagnosis (Figure 19.3). These tests involve inoculating small amounts of the suspected antigen just beneath the epidermis of the skin. Sensitivity to the antigen is indicated by a rapid inflammatory reaction that produces redness, swelling, and itching at the inoculation site. This small affected area is called a *wheal*.

Once the responsible antigen has been identified, the person can either try to avoid contact with it or undergo **desensitization.** This procedure consists of a series of dosages of the antigen carefully injected beneath the skin. The objective is to cause the production of IgG rather than IgE antibodies in the hope that the circulating IgG antibodies will act as *blocking antibodies* to intercept and neutralize the antigens before they can react with cell-bound IgE. Desensitization is not a routinely successful procedure, but it is effective in 65–75% of individuals whose allergies are induced by inhaled antigens, and in a reported 97% for insect venom.

Type II (Cytotoxic) Reactions

Learning Objectives

- *Describe the mechanism of cytotoxic reactions and how drugs can induce them.*
- *Describe the basis of the ABO and Rh blood group systems.*
- *Explain the relationship between blood groups, blood transfusions, and hemolytic disease of the newborn.*

FIGURE 19.3 **A skin test to identify allergens.** Drops of fluid containing test substances are placed on the skin. A light scratch is made with a needle to allow the substances to penetrate the skin. Reddening and swelling at the site identify the substance as a probable cause of an allergic reaction.

- In skin tests, sensitivity to a substance is indicated by a rapid inflammation to the inoculated antigen.

Type II (cytotoxic) reactions generally involve the activation of complement by the combination of IgG or IgM antibodies with an antigenic cell. This activation stimulates complement to lyse the affected cell, which might be either a foreign cell or a host cell that carries a foreign antigenic determinant (such as a drug) on its surface. Additional cellular damage may be caused within 5 to 8 hours by the action of macrophages and other cells that attack antibody-coated cells.

The most familiar cytotoxic hypersensitivity reactions are *transfusion reactions*, in which red blood cells are destroyed as a result of reacting with circulating antibodies. These involve blood group systems that include the ABO and Rh antigens.

The ABO Blood Group System

In 1901, Karl Landsteiner discovered that human blood could be grouped into four principal types, which were designated A, B, AB, and O. This method of classification is called the **ABO blood group system.** Since then, other blood group systems, such as the Lewis system and the MN system, have been discovered, but our discussion will be limited to two of the best known, the ABO and the Rh systems. The main features of the ABO blood group system are summarized in Table 19.2.

A person's ABO blood type depends on the presence or absence of carbohydrate antigens located on the cell membranes of red blood cells (RBCs). Cells of blood type O lack both A and B antigens. Table 19.2 shows that the

| TABLE 19.2 | **The ABO Blood Group System** | | | | | | |

Blood Group	Erythrocyte or Red Blood Cell Antigens	Illustration	Plasma Antibodies	Blood That Can Be Received	Frequency (% U.S. Population) White	Black	Asian
AB	A and B		Neither anti-A nor anti-B antibodies	A, B, AB, O Universal recipient	3	4	5
B	B		Anti-A	B, O	9	20	27
A	A		Anti-B	A, O	41	27	28
O	None		Anti-A and Anti-B	O Universal donor	47	49	40

plasma of individuals with a given blood type, such as A, have antibodies against the alternative blood type, anti-B antibody. These antibodies arise in response to bacteria that have antigenic determinants very similar to blood group antigens. Individuals with type AB cells have plasma with no antibodies to either A or B antigens and therefore can receive cells with either A or B blood without a reaction. Type O individuals have antibodies against both A and B antigens. Lacking antigens, their blood cells can be transfused without difficulty to others, but they can receive transfusions of type O blood only. (The anti-A and anti-B antibodies contained in donated type O blood or blood plasma are quickly degraded by the recipient and cause no significant adverse reaction.)

When a transfusion is incompatible, as when type B blood is transfused into a person with type A blood, the antigens on the type B blood cells will react with anti-B antibodies in the recipient's serum. This antigen-antibody reaction activates complement, which in turn causes lysis of the donor's RBCs as they enter the recipient's system.

The Rh Blood Group System

In the 1930s, researchers discovered the presence of a different surface antigen on human red blood cells. Soon after they injected rabbits with RBCs from rhesus monkeys, the rabbit serum contained antibodies that were directed against the monkey blood cells but that would also agglutinate some human RBCs. This indicated that a common antigen was present on both human and monkey red blood

cells. The antigen was named the **Rh factor** (Rh for Rhesus monkey). The roughly 85% of the population whose cells possess this antigen are called Rh^+; those lacking this RBC antigen (about 15%) are Rh^-. Antibodies that react with the Rh antigen do not occur naturally in the serum of Rh^- individuals, but exposure to this antigen can sensitize their immune systems to produce anti-Rh antibodies.

Blood Transfusions and Rh Incompatibility If blood from an Rh^+ donor is given to an Rh^- recipient, the donor's RBCs stimulate the production of anti-Rh antibodies in the recipient. If the recipient then receives Rh^+ RBCs in a subsequent transfusion, a rapid, serious hemolytic reaction will develop.

Hemolytic Disease of the Newborn Blood transfusions are not the only way in which an Rh^- person can become sensitized to Rh^+ blood. When an Rh^- woman and an Rh^+ man produce a child, there is a 50% chance that the child will be Rh^+ (Figure 19.4). If the child is Rh^+, the Rh^- mother can become sensitized to this antigen during birth, when the placental membranes tear and the fetal Rh^+ RBCs enter the maternal circulation, causing the mother's body to produce anti-Rh antibodies of the IgG type. If the fetus in a subsequent pregnancy is Rh^+, her anti-Rh antibodies will cross the placenta and destroy the fetal RBCs. The fetal body responds to this immune attack by producing large numbers of immature RBCs called erythroblasts. Thus, the term *erythroblastosis fetalis* was

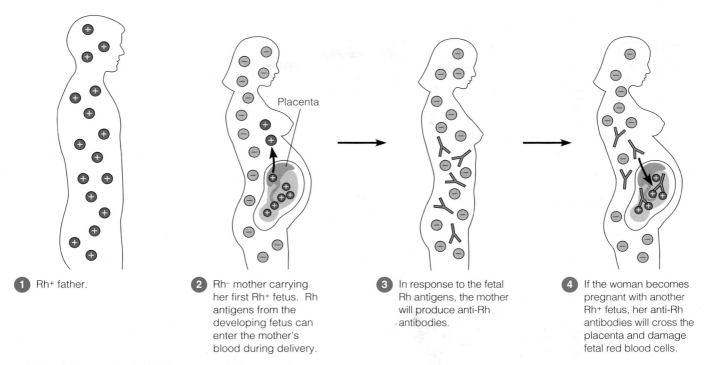

① Rh⁺ father.

② Rh⁻ mother carrying her first Rh⁺ fetus. Rh antigens from the developing fetus can enter the mother's blood during delivery.

③ In response to the fetal Rh antigens, the mother will produce anti-Rh antibodies.

④ If the woman becomes pregnant with another Rh⁺ fetus, her anti-Rh antibodies will cross the placenta and damage fetal red blood cells.

FIGURE 19.4 **Hemolytic disease of the newborn.**

■ Hemolytic disease of the newborn is caused by a mother's antibodies to the fetus crossing the placenta.

once used to describe what is now called **hemolytic disease of the newborn (HDNB).** Before the birth of a fetus with this condition, the maternal circulation removes most of the toxic by-products of fetal RBC disintegration. After birth, however, the fetal blood is no longer purified by the mother, and the newborn develops jaundice and severe anemia.

HDNB is usually prevented today by passive immunization of the Rh⁻ mother at the time of delivery of any Rh⁺ infant with anti-Rh antibodies, which are available commercially (Rhogam). These anti-Rh antibodies combine with any fetal Rh⁺ RBCs that have entered the mother's circulation, so it is much less likely that she will become sensitized to the Rh antigen. If the disease is not prevented, the newborn's Rh⁺ blood, contaminated with maternal antibodies, may have to be replaced by transfusion of uncontaminated blood.

Drug-Induced Cytotoxic Reactions

Blood platelets (thrombocytes) are minute cell-like bodies that are destroyed by drug-induced cytotoxic reactions in the disease called **thrombocytopenic purpura.** The drug molecules are usually haptens because they are too small to be antigenic by themselves, but in the situation illustrated in Figure 19.5, a platelet has become coated with molecules of a drug (quinine is a familiar example), and the

combination is antigenic. Both antibody and complement are needed for lysis of the platelet. Because platelets are necessary for blood clotting, their loss results in hemorrhages that appear on the skin as purple spots (purpura).

Drugs may bind similarly to white or RBCs, causing local hemorrhaging and yielding symptoms described as "blueberry muffin" skin mottling. Immune-caused destruction of granulocytic white cells is called **agranulocytosis,** and it affects the body's phagocytic defenses. When RBCs are destroyed in the same manner, the condition is termed **hemolytic anemia.**

Type III (Immune Complex) Reactions

Learning Objective

• *Describe the mechanism of immune complex reactions.*

Type III reactions involve antibodies against soluble antigens circulating in the serum. (In contrast, type II immune reactions are directed against antigens located on cell or tissue surfaces.) The antigen–antibody complexes are deposited in organs and cause inflammatory damage.

Immune complexes form only when certain ratios of antigen and antibody occur. The antibodies involved are usually IgG. A significant excess of antibody leads to the formation of complement-fixing complexes that are rapidly removed from the body by phagocytosis. When there is a

1. Drug binds to platelet, forming hapten-platelet complex.

2. Complex induces formation of antibodies against hapten.

Hapten-platelet complex

Anti-hapten antibody

3. Action of antibodies and complement causes platelet destruction.

Complement

FIGURE 19.5 Drug-induced thrombocytopenic purpura. Molecules of a drug such as quinine accumulate on the surface of a platelet and stimulate an immune response that destroys the platelet.

■ What actually destroys the platelets in thrombocytopenic purpura?

significant excess of antigen, soluble complexes form that do not fix complement and do not cause inflammation. However, when a certain antigen–antibody ratio exists, usually with a slight excess of antigen, the soluble complexes that form are small and escape phagocytosis.

Figure 19.6 illustrates the consequences. These complexes circulate in the blood, pass between endothelial cells of the blood vessels, and become trapped in the basement membrane beneath the cells. In this location, they may activate complement and cause a transient inflammatory reaction: attracting neutrophils that release enzymes. Repeated introduction of the same antigen can lead to more serious inflammatory reactions, causing damage to the basement membrane's endothelial cells within 2 to 8 hours. **Glomerulonephritis** is an immune complex condition that causes inflammatory damage to the kidney glomeruli, which are sites of blood filtration.

Type IV (Cell-Mediated) Reactions

Learning Objective

• Describe the mechanism of cell-mediated reactions, and name two examples.

Up to this point we have discussed humoral immune responses involving IgE, IgG, or IgM. Type IV reactions involve cell-mediated immune responses and are caused mainly by T cells. Instead of occurring within a few minutes or hours after a sensitized individual is again exposed to an antigen, these delayed reactions are not apparent for a day or more. A major factor in the delay is the time required for the participating T cells and macrophages to migrate to and accumulate near the foreign antigens. The T cells involved in **delayed-type hypersensitivity** reactions are primarily T_D cells. In some types of hypersensitivities resulting in tissue damage, T_C cells may also participate.

Causes of Type IV Reactions

Sensitization for type IV hypersensitivity reactions occurs when certain foreign antigens, particularly those that bind to tissue cells, are phagocytized by macrophages and then presented to receptors on the T-cell surface. Contact between the antigenic determinant sites and the appropriate T cell causes the T cell to proliferate into mature differentiated T cells and memory cells.

When a person sensitized in this way is reexposed to the same antigen, a cell-mediated hypersensitivity reaction might result. Memory cells from the initial exposure activate T cells, which release destructive cytokines in their interaction with the target antigen. In addition, some cytokines contribute to the inflammatory reaction to the foreign antigen by attracting macrophages to the site and activating them (see Chapter 17, Figures 17.15 and 17.19 on pages 499 and 503).

Cell-Mediated Hypersensitivity Reactions of the Skin

We have seen that hypersensitivity symptoms are frequently displayed on the skin. One cell-mediated hypersensitivity reaction that involves the skin is the familiar skin test for tuberculosis. Because *Mycobacterium tuberculosis* is often located within macrophages, this organism can stimulate a cell-mediated immune response. As a screening test, protein components of the bacteria are injected into the skin. If the recipient has (or has had) a previous infection by tuberculosis bacteria, an inflammatory reaction to the injection of these antigens will appear on the skin in 1–2 days; this interval is typical of delayed hypersensitivity reactions.

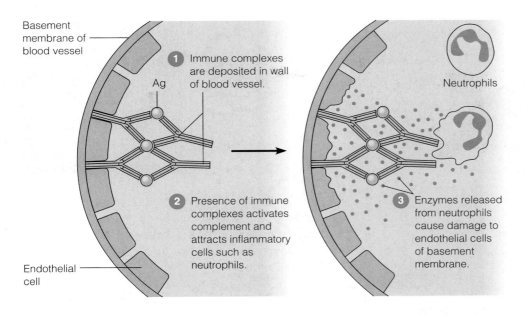

FIGURE 19.6 **Immune complex–mediated hypersensitivity.**
❶ Immune complexes are deposited on the basement membrane of the wall of a blood vessel, where they ❷ activate complement and attract inflammatory cells such as neutrophils to the site. ❸ The neutrophils discharge enzymes as they react with the immune complexes, resulting in damage to tissue cells.

- Name one immune complex disease.

Allergic contact dermatitis, another common manifestation of type IV hypersensitivity, is usually caused by haptens that combine with proteins in the skin of some people (particularly the amino acid lysine) to produce an immune response. Reactions to poison ivy (Figure 19.7), cosmetics, and the metals in jewelry (especially nickel) are familiar examples of these allergies.

The increasing exposure to latex in condoms, in certain catheters, and in gloves used by health care workers has led to a greater awareness of hypersensitivity to latex. Death from anaphylactic shock can also occur. During a recent 4-year period, 15 fatalities among patients were caused by exposure to latex tubing used for enemas or, during abdominal surgery, to the latex gloves of the surgeon. Many hospitals now even restrict entry of toy balloons. Among physicians and nurses, 3% or more have reported this type of hypersensitivity to surgical gloves. Delayed-type hypersensitivity reactions are not to latex itself, but to chemicals used in its manufacture. The result is a contact dermatitis such as that shown in Figure 19.8.

The powder in latex gloves is another source of problems; it absorbs allergens, becomes airborne, and can be inhaled.

Because of the growing problem of hypersensitivity to latex, non-latex gloves are now marketed, and latex gloves that are free of powder have been introduced. Many of the non-latex gloves now have tactile properties similar to latex, and some powder-free latex gloves are actually easier to put on than the powdered versions. However, these alternatives are often significantly more expensive.

Latex paint, however, does not pose a threat of hypersensitivity reactions. Despite its name, latex paint contains no natural latex, but only synthetic nonallergenic chemical polymers.

The identity of the environmental factor causing the dermatitis can usually be determined by a *patch test.* Samples of suspected materials are taped to the skin; after 48 hours, the area is examined for inflammation.

Autoimmune Diseases

Learning Objectives

- *Describe a mechanism for self-tolerance.*
- *Give an example of type I, type II, type III, and type IV autoimmune diseases.*

When the action of the immune system is in response to self-antigens and causes damage to one's own organs, the result is an **autoimmune disease.** Autoimmune diseases are relatively rare, but overall they affect about 5% of the population in the developed world. One prominent immunologist has observed, "We are all constantly teetering on the brink of autoimmune disease." About 75% of the cases of autoimmune disease selectively affect women. Treatments for autoimmune diseases are improving as knowledge of the mechanisms controlling immune reactions improves.

Autoimmune diseases occur when there is a loss of **self-tolerance,** the immune system's ability to discriminate self from nonself. In the generally accepted model by which T cells become capable of distinguishing self from nonself, the cells acquire this ability during their passage through the thymus. As we saw in Chapter 17 (page 491), any T cells that will target host cells are either eliminated

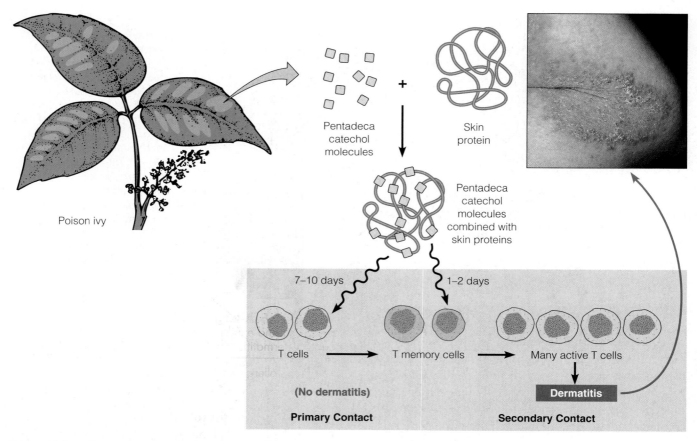

FIGURE 19.7 The development of an allergy (allergic contact dermatitis) to catechols from the poison ivy plant. Pentadecacatenol is a mixture of catechols, which are oils secreted by the plant that dissolve easily in skin oils and penetrate the skin. In the skin the catechols function as haptens, that is, they combine with skin proteins to become antigenic and provoke an immune response. The first contact with poison ivy sensitizes the susceptible person and subsequent exposure results in contact dermatitis (shown in the photo inset of a human arm and diagram).

■ How does a hapten cause an allergic reaction?

by *clonal deletion* or inactivated during this period. This requirement makes it unlikely that the T cell will attack its own tissue cells.

In autoimmune diseases, the loss of self-tolerance leads to the production of antibodies or a response by sensitized T cells against a person's own tissue antigens. Autoimmune reactions, and the diseases they cause, can be cytotoxic, immune complex, or cell-mediated in nature.

Type I Autoimmunity

Type I autoimmunity involves antibodies that attack self. These antibodies may be made in response to an infectious agent such as a virus but sequence similarities between viral and self-proteins may result in the antibodies attacking self cells. Hepatitis C virus may be responsible for autoimmune hepatitis by this mechanism.

Type II (Cytotoxic) Autoimmune Reactions

Graves' disease and myasthenia gravis are two examples of disorders caused by type II autoimmune reactions. Both diseases involve antibody reactions to cell-surface antigens, although there is no cytotoxic destruction of the cells.

Graves' disease is caused by antibodies called long-acting thyroid stimulators. These antibodies attach to receptors on thyroid gland cells that are the normal target cells of the thyroid-stimulating hormone produced by the pituitary gland. The result is that the thyroid gland is stimulated to produce increased amounts of thyroid hormones and becomes greatly enlarged. The most striking signs of the disease are goiter, a disfiguring swelling of the thyroid gland, and markedly bulging, staring eyes.

Myasthenia gravis is a disease in which muscles become progressively weaker. It is caused by antibodies that coat the acetylcholine receptors at the junctions at which nerve impulses reach the muscles. Eventually, the muscles controlling the diaphragm and the rib cage may fail to receive the necessary nerve signals, and respiratory arrest and death result.

Type III (Immune Complex) Autoimmune Reactions

Systemic lupus erythematosus is a systemic autoimmune disease, involving immune complex reactions, that mainly affects women. The etiology of the disease is not completely understood, but afflicted individuals produce antibodies directed at components of their own cells, including DNA, which is probably released during the normal breakdown of tissues, especially the skin. The most damaging effects of the disease result from deposits of immune complexes in the kidney glomeruli.

Crippling **rheumatoid arthritis** is a disease in which immune complexes of IgM, IgG, and complement are deposited in the joints. In fact, immune complexes called *rheumatoid factors* may be formed by IgM binding to the Fc region of normal IgG. These factors are found in 70% of individuals suffering from rheumatoid arthritis. The chronic inflammation caused by this deposition eventually leads to severe damage to the cartilage and bone of the joint.

Type IV (Cell-Mediated) Autoimmune Reactions

Multiple sclerosis is one of the more common autoimmune diseases, affecting mostly younger adults. Most individuals with multiple sclerosis are whites living in northern latitudes; females are twice as likely to have the disease. It is a neurological disease in which T cells and macrophages attack the myelin sheath of nerves. Symptoms range from only fatigue and weakness to, in some cases, eventual severe paralysis. Progression of the disease is slow, measured over many years. New attacks that worsen the condition are often separated by long periods of remission. There is considerable evidence of genetic susceptibility, probably not from a single gene, but from several genes that interact. The etiology of multiple sclerosis is unknown, but epidemiological evidence indicates that it probably involves some infective agent or agents acquired during early adolescence. The Epstein-Barr virus is frequently mentioned as a prime suspect. No cure for the condition exists, but treatments with interferons and several drugs that interfere with immune processes can significantly slow progression of symptoms.

FIGURE 19.8 **Allergic contact dermatitis.** This person's hand exhibits a severe case of delayed-type contact dermatitis from wearing latex surgical gloves.

■ What is allergic contact dermatitis?

Hashimoto's thyroiditis is a result of the destruction of the thyroid gland, primarily by T cells of the cell-mediated immune system. It is a fairly common disorder and is often found in related family members. **Insulin-dependent diabetes mellitus** is a familiar condition caused by immunological destruction of insulin-secreting cells of the pancreas. T cells are clearly implicated in this disease; animals that are genetically likely to develop diabetes fail to do so when their thymus is removed in infancy.

Reactions Related to the Human Leukocyte Antigen (HLA) Complex

Learning Objective

- Define HLA complex, and explain its importance in disease susceptibility and tissue transplants.

The inherited genetic characteristics of individuals are expressed not only in the color of their eyes and the curl of their hair but also in the composition of the self molecules on their cell surfaces. Some of these are called **histocompatibility antigens.** The genes controlling the production of the most important of these self molecules are known as the **major histocompatibility complex (MHC).** In humans, these genes are called the **human leukocyte antigen (HLA) complex.** We encountered these self molecules in Chapter 17, where we saw that most antigens

TABLE 19.3	**Diseases Related to Specific HLAs**	
Disease	**Increased Risk of Occurrence with Specific HLA***	**Description**
Inflammatory Diseases		
Multiple sclerosis	5 times	Progressive inflammatory disease affecting nervous system
Rheumatic fever	4–5 times	Cross-reaction with antibodies against streptococcal antigen
Endocrine Diseases		
Addison's disease	4–10 times	Deficiency in production of hormones by adrenal gland
Graves' disease	10–12 times	Antibodies attached to certain receptors in the thyroid gland cause it to enlarge and produce excessive hormones
Malignant Disease		
Hodgkin's disease	1.4–1.8 times	Cancer of lymph nodes

*Compared to the general population.

can stimulate an immune reaction only if they are associated with an MHC molecule.

A process called *HLA typing* is used to identify and compare HLAs. Certain HLAs are related to an increased susceptibility to specific diseases; one medical application of HLA typing is to identify such susceptibility. A few of these relationships are summarized in Table 19.3.

Another important medical application of HLA typing is in transplant surgery, in which the donor and the recipient must be matched by *tissue typing*. The serological technique shown in Figure 19.9 is the one most often used. In serological tissue typing, the laboratory uses standardized antisera or monoclonal antibodies that are specific for particular HLAs.

❶ Lymphocytes from the person being tested are incubated with a selected specific antiserum.

❷ Complement and a dye, such as trypan blue, are then added.

❸ If antibodies in the antiserum have reacted specifically with the lymphocyte, the cell is damaged. The damaged cell will take up the dye (undamaged cells will not), thus indicating that the lymphocyte possesses a certain antigen. This method is simple and rapid.

A promising new technique for analyzing HLA is the use of *PCR*, the *polymerase chain reaction*, to amplify the cell's DNA (see Figure 9.4 on page 255). If this is done for both donor and recipient, a match between donor DNA and recipient DNA can then be made. Such a match should result in a much higher success rate in transplant surgery.

Two main classes of HLAs are important in such matches: *class I* (HLA-A, -B, and -C) and *class II* (HLA-DR, -DP, and -DQ). The genes that control the production of these self molecules are located close together on chromosome 6. Matching donors and recipients for class I self molecules, which are on all cells in the body, has long been a standard procedure. These self molecules stimulate a strong immune response by the antibodies and T_C cells

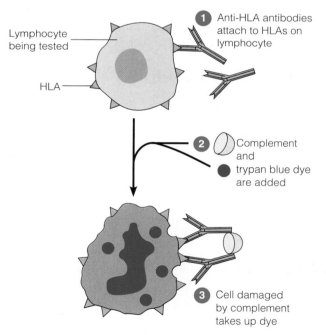

FIGURE 19.9 **Tissue typing, a serological method.** Lymphocytes from the person being tested are incubated with laboratory test stocks of anti-HLA antibodies specific for a particular HLA. If the antibodies react with the antigens on a lymphocyte, then complement damages the lymphocyte and dye can enter the cell. Such a positive test result indicates that the person has the particular HLA being tested for.

■ Why is tissue typing done?

mediating transplant rejection. However, matching for class II self molecules, found primarily on the surfaces of specialized immune system cells, might well be more important, especially for tissue from an unrelated person. If the donor and the recipient do not share these self molecules, the transplant will probably be rejected. The donor and the recipient must also be of the same ABO blood type.

Other factors may be involved in the success of a transplant, however. In Chapter 17, we briefly introduced a hypothesis that the body's reaction to transplanted foreign tissue may be a response to surgery-damaged cells. In other words, tissue rejection may result from a learned reaction to the danger signal posed by damaged cells, rather than a learned reaction to nonself.

Reactions to Transplantation

Learning Objectives

- *Explain how the rejection of a transplant occurs.*
- *Define privileged site.*
- *Define autograft, isograft, allograft, and xenotransplant.*
- *Explain how graft-versus-host disease occurs.*

In sixteenth-century Italy, crimes were often punished by cutting off the offender's nose. A surgeon of the time, in his attempts to repair this mutilation, observed that if skin was taken from the patient, it healed properly, but if it was taken from another person, it did not. He called this a manifestation of "the force and power of individuality."

We now know the principles behind this phenomenon. Transplants recognized as nonself are rejected—attacked by T cells that directly lyse the grafted cells, by macrophages activated by T cells, and, in certain cases, by antibodies, which activate the complement system and injure blood vessels supplying the transplanted tissue. However, transplants that are not rejected can add many healthy years to a person's life.

Since the first kidney transplant was performed in 1954, this particular type of transplant has become a nearly routine medical procedure. Other types of transplants that are now feasible include bone marrow, thymus glands, heart, liver, and cornea. Tissues and organs for transplant are usually taken from recently deceased individuals, although one of a pair of organs, such as a kidney, occasionally comes from a living donor.

Privileged Sites and Privileged Tissue

Some transplants or grafts do not stimulate an immune response. A transplanted cornea, for example, is rarely rejected, mainly because antibodies usually do not circulate into that portion of the eye, which is therefore considered an immunologically **privileged site.** (However, rejections do occur, especially when the cornea has developed many blood vessels from corneal infections or damage.) The brain is also an immunologically privileged site, probably because it does not have lymphatic vessels and because the walls of the blood vessels in the brain differ from blood vessel walls elsewhere in the body. Someday it may even be possible to graft foreign nerves to replace damaged nerves in the brain and spinal cord. Encouraging results have been obtained with these types of grafts in experiments with rats.

It is possible to transplant **privileged tissue** that does not stimulate an immune rejection. An example is replacing a person's damaged heart valve with a valve from a pig's heart. However, privileged sites and tissues are more the exception than the rule.

How animals tolerate pregnancy without rejecting the fetus is only partially understood. The uterus is not a privileged site; yet during pregnancy, the tissues of two genetically different individuals are in direct contact.

A development that promises to transform transplantation medicine is the use of **stem cells.** These cells can give rise to more stem cells but can also generate cell types such as nerve, blood, or liver cells (see Figure 17.6, on page 488); that is, they are pluripotent. Stem cells first appear in the human embryo as *embryonic stem cells.* In principle, these pluripotent cells could be cultured indefinitely and generate useful amounts of almost any tissue for transplantation.

As the fetus develops, embryonic stem cells migrate to the liver, becoming fetal stem cells and, later, adult stem cells. Blood-forming (*hematopoietic*) stem cells (HSC) in the bone marrow produce red blood cells, platelets, and the various white blood cells essential for the immune system. (This is the logical basis for bone marrow transplants for treatment of leukemia, etc.). Some HSC also seem to exhibit considerable plasticity, the ability to change course and give rise to cells of tissues such as the nervous or gastrointestinal system. At present, such pluripotent cells are present only in very small numbers and grow poorly in culture. It would be desirable if adult stem cells such as HSC could be used to produce most other tissue cells. This would avoid the political and ethical considerations concerning the use of embryos. Currently the prospects for this are uncertain.

The eventual goal of work with stem cells is **therapeutic cloning,** where the genetic material of a patient with a disease is used to create stem cell lines that can be used to treat that disease. Scientists hope eventually to routinely make an embryonic clone of an adult patient, remove the pluripotent stem cells, and then use the stem cells to grow tissues whose genetic material is the same as the patient's. This approach would avoid immunological rejection and would ideally be a way of treating conditions such as diabetes, Parkinson's disease, Alzheimer's disease, and spinal cord injuries. Perhaps even functional organs such as the liver could be grown from a patient's own tissue, entirely avoiding rejection.

Grafts

When one's own tissue is grafted to another part of the body, as is done in burn treatment or in plastic surgery, the graft is not rejected. Recent technology has made it possible to use a few cells of a burn patient's uninjured skin to culture extensive sheets of new skin. This new skin is an example of an **autograft.** Identical twins have the same genetic makeup; therefore, skin or organs such as kidneys may be transplanted between them without provoking an immune response. Such a transplant is called an **isograft.**

Most transplants, however, are made between people who are not identical twins, and these transplants do trigger an immune response. Attempts are made to match the HLAs of the donor and recipient as closely as possible so that the chances of rejection are reduced. Because HLAs of close relatives are most likely to match, blood relatives, especially siblings, are the preferred donors. Grafts between people who are not identical twins are called **allografts.**

Because of the shortage of available organs, medical researchers hope to increase the success of **xenotransplantation products** (formerly called **xenografts**), which are tissues or organs that have been transplanted from animals. However, the body tends to mount an especially severe immune assault on such transplants. Unsatisfactory attempts have been made to use organs from baboons and other nonhuman primates. Research interest is high in genetically engineering pigs—an animal that is in plentiful supply, is of the right size, and generates relatively little public sympathy—to make acceptable donors of organs. The primary concern about xenotransplantation products is the possibility of transferring harmful animal viruses.

Preliminary research is underway that may eventually allow some bones and organs to be grown from the host's own tissue cells.

In order to be successful, xenotransplantation products must overcome **hyperacute rejection,** caused by the development in early infancy of antibodies against all distantly related animals such as pigs. With the aid of complement, these antibodies attack the transplanted animal tissue and destroy it within an hour. Pigs are now being bred in which the cell-surface antigens that cause hyperacute rejection have been eliminated. These "knockout pigs" are considered an important step toward xenotransplantation. Hyperacute rejection occurs in human-to-human transplants only when antibodies have been preformed because of previous transfusions, transplantations, or pregnancies.

Bone Marrow Transplants

Transplants of bone marrow are frequently in the news. The recipients are usually individuals who lack the capacity to produce B cells and T cells vital for immunity, or who are suffering from leukemia. Recall from Chapter 17 that bone marrow stem cells give rise to red blood cells and immune system lymphocytes. The goal of bone marrow transplants is to enable the recipient to produce such vital cells. However, such transplants can result in **graft-versus-host (GVH) disease.** The transplanted bone marrow contains immunocompetent cells that mount primarily a cell-mediated immune response against the tissue into which they have been transplanted. Because the recipients lack effective immunity, GVH disease is a serious complication and can even be fatal.

An extremely promising technique for avoiding this problem is the use of *umbilical cord blood* instead of bone marrow. This blood is harvested from the placenta and umbilical cords of newborns, material that would otherwise be discarded. It is very rich in the stem cells (Figure 17.6, page 488) found in bone marrow. Not only do these cells proliferate into the variety of cells required by the recipient; but because stem cells from this source are younger and less mature, the "matching" requirements are also less stringent than with bone marrow. As a result, GVH disease is less likely to occur.

Immunosuppression

Learning Objective

• *Explain how rejection of a transplant is prevented.*

To keep the problem of transplant rejection in perspective, it is useful to remember that the immune system is simply doing its job and has no way of recognizing that its attack against the transplant is not helpful. In an attempt to prevent rejection, the recipient of an allograft usually receives treatment to suppress this normal immune response against the graft.

In transplantation surgery, it is generally desirable to suppress cell-mediated immunity, the most important factor in transplant rejection. If humoral (antibody-based) immunity, is not suppressed, much of the ability to resist microbial infection will remain. In 1976, the drug *cyclosporine* was isolated from a mold. (Curiously, this fungus has a sexual cycle in a dung beetle that stimulates the beetle to climb high on vegetation and die, ensuring more efficient airborne distribution of fungal spores.)

The successful transplantation of organs such as hearts and livers generally dates from the discovery of cyclosporine. The secretion of interleukin-2 (IL-2) is suppressed by cyclosporine, disrupting cell-mediated immunity by cytotoxic T cells. Following the success of this drug, other immunosuppressant drugs soon followed. *Tacrolimus*

(FK506) has a mechanism similar to cyclosporine and is a frequent alternative, although both have many serious side effects. Neither cyclosporine nor tacrolimus has much effect on antibody production by the humoral immune system.

Some newer drugs, such as *sirolimus* (rapamycin) are among those that inhibit both cell-mediated and humoral immunity. This can be an advantage if chronic or hyperacute rejection by antibodies is a consideration. Sirolimus blocks the action of IL-2. Drugs such as *mycophenolate mofetil* inhibit the proliferation of T cells and B cells. In a similar approach, in 1998 the FDA approved two chimeric monoclonal antibodies (see page 495), *basiliximab* and *daclizumab* that block IL-2. Immunosuppressive drugs are usually administered in combinations.

Several other drugs are being studied, all of which will probably improve the already impressive success rate of transplantation medicine. The ideal solution to tissue rejection would be to induce tolerance, in which the immune system treats the transplanted tissue as self rather than foreign. Transplant surgeons sometimes attempt to promote tolerance by dosing transplant recipients with bone marrow from the same donors as the transplanted tissues. Occasionally, an organ recipient is weaned from immunosuppressive drugs and still avoids rejection. A more complete understanding of the reasons for this would be an important advance in the science of transplantation.

Immune Deficiencies

Learning Objective

- *Compare and contrast congenital and acquired immune deficiencies.*

The absence of a sufficient immune response is called an **immune deficiency.** Immune deficiencies can be either congenital or acquired.

Congenital Immune Deficiencies

Some people are born with a defective immune system. Defects in, or the absence of, a number of inherited genes can result in **congenital immune deficiencies.** For example, individuals with a certain recessive trait, DiGeorge syndrome, do not have a thymus gland and therefore lack cell-mediated immunity. An animal equivalent, which is extremely valuable for research in transplantation science, is the nude (hairless) mouse (Figure 19.10). These mice have no thymus (the coincidental hairlessness is controlled by the same gene) and therefore do not produce T cells and do not reject transplanted tissue. Even chicken skin, complete with feathers, is readily accepted as a graft.

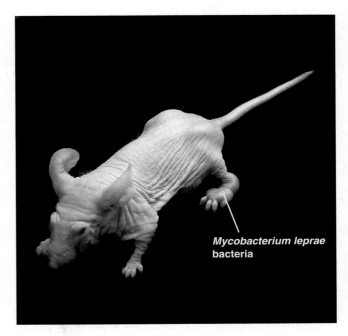

Mycobacterium leprae bacteria

FIGURE 19.10 **A nude mouse infected with *M. leprae* in the hind foot.** These animals also play important roles in leprosy research.

Acquired Immune Deficiencies

A variety of drugs, cancers, or infectious agents can result in **acquired immune deficiencies.** For example, Hodgkin's disease (a type of cancer) lowers the cell-mediated response. Many viruses are capable of infecting and killing lymphocytes, lowering the immune response. Removal of the spleen decreases humoral immunity. Table 19.4 summarizes several of the better known immune deficiency conditions, including AIDS.

The Immune System and Cancer

Learning Objective

- *Describe the immune responses to cancer and how cells evade immune responses.*

Like an infectious disease, cancer represents a failure of the body's defenses, including the immune system. One of the most promising avenues for effective cancer therapy makes use of immunological techniques.

One long-held hypothesis is that cancer cells arise by mutation or virus-induced changes, and that the immune system constantly patrols and eliminates these cells before they become established tumors. This process is called *immunological surveillance*, and it has even been suggested that the cell-mediated immune system may have arisen primarily for this purpose. This concept is supported by the observation that cancers occur most often in older adults,

TABLE 19.4	**Immune Deficiencies**	
Disease	**Cells Affected**	**Comments**
Acquired immunodeficiency syndrome (AIDS)	T cells (virus destroys T_H [CD4] cells)	Allows cancer and bacterial, viral, fungal, and protozoan diseases; caused by HIV infection
Selective IgA immunodeficiency	B, T cells	Affects about 1 in 700, causing frequent mucosal infections; specific cause uncertain
Common variable hypogammaglobulinemia	B, T cells (decreased immunoglobulins)	Frequent viral and bacterial infections; second most common immune deficiency, affecting about 1 in 70,000; inherited
Reticular dysgenesis	B, T, and stem cells (a combined immunodeficiency; deficiencies in B and T cells and neutrophils)	Usually fatal in early infancy; very rare; inherited; bone marrow transplant a possible treatment
Severe combined immunodeficiency	B, T, and stem cells (deficiency of both B and T cells)	Affects about 1 in 100,000; allows severe infections, inherited; treated with bone marrow, fetal thymus transplants; gene therapy treatment is promising
Thymic aplasia (DiGeorge syndrome)	T cells (defective thymus causes deficiency of T cells)	Absence of cell-mediated immunity; usually fatal in infancy from *Pneumocystis* pneumonia or viral or fungal infections; due to failure to develop in embryo
Wiskott-Aldrich syndrome	B, T cells (few platelets in blood, abnormal T cells)	Frequent infections by viruses, fungi, protozoa; eczema, defective blood clotting; usually causes death in childhood; inherited on X chromosome
X-linked infantile (Bruton's) agammaglobulinemia	B cells (decreased immunoglobulins)	Frequent extracellular bacterial infections; affects about 1 in 200,000; the first immunodeficiency disorder recognized (1952); inherited on X chromosome

whose immune systems are becoming less efficient, or in the very young, whose immune systems may not have developed fully or properly. Also, individuals who are immunosuppressed by either natural or artificial means are more susceptible to cancer, although these are usually restricted to cancers of the immune system. However, there are many observations that cannot be explained by the surveillance hypothesis.

A cell becomes cancerous when it undergoes transformation and begins to proliferate without control (see Chapter 13, page 400). Because of transformation, the surfaces of tumor cells may acquire tumor-associated antigens that mark them as nonself to the immune system. Activated T_C cells respond to these nonself antigens by attaching to and lysing the cancerous cells that carry them (Figure 19.11).

Cancer sometimes occurs even in people with a presumably healthy immune system. Initially, the cancer is an individual cell that arises from mutations, perhaps resulting from exposure to chemicals or radiation. These changes usually require multiple "hits" (constituting damage to genes, and often separated by considerable time), to complete the change from normal to cancerous.

Viral infection can also change a normal cell to a cancerous cell.

Individual cancer cells that arise are attacked by the immune system, much as it responds to transplanted foreign tissue. Once the cancerous cell attaches to tissue and becomes vascularized (connected with the body's blood supply), it develops rapidly into a tumor that becomes resistant to immune rejection. At present, a number of mechanisms have been identified. Tumor cells often lack surface molecules necessary to activate T_C cells, or lack parts of their structure that in other ways stimulate the immune system.

Tumor cells sometimes actively suppress the immune system; they produce factors that reduce the effectiveness of T_C cells, and some cancer cells induce immune cells to undergo apoptosis. Once established, tumor cells may even reproduce so rapidly that they exceed the capacity of an effective immune response.

It is encouraging to know that occasionally some malignancies spontaneously disappear, presumably because of a successful immune response. The phenomenon of cancer's resistance to the immune system is of interest because a better understanding could lead to ways to defeat this resistance, as will be discussed in the next section on immunotherapy.

FIGURE 19.11 **The interaction between a cytotoxic T (T_C) cell and a cancer cell.**
The small T_C cell (left panel) has already made a perforation in the cancer cell. At right,
shown at a lower magnification, only the skeleton of the cancer cell remains.

■ T_C cells can lyse cancer cells. How do they do this? (*Hint:* See Figure 17.14.)

Immunotherapy

Learning Objective

- *Define immunotherapy, and give two examples.*

At the turn of the twentieth century, William B. Coley, a
physician at a New York City hospital, observed that if
cancer patients contracted typhoid fever, their cancers often diminished noticeably. Following this lead, Coley
made vaccines of killed gram-negative bacteria, called Coley's toxins, to simulate a bacterial infection. Some of this
work was very promising, but its results were inconsistent,
and advances in surgery and radiation treatment caused it
to be nearly forgotten.

We now know that endotoxins from such bacteria are
powerful stimulants for the production of tumor necrosis
factor (TNF) by macrophages. TNF is a small protein that
interferes with the blood supply of cancers in animals. There
is now also much interest in cancer therapy with TNF and
other cytokines such as interleukin-2 (see Chapter 17, page
495) and interferons (see Chapter 16, page 476).

The treatment of cancer by immunological means is
termed **immunotherapy,** an approach that will probably
be used in the future. One promising technique involves
immunotoxins. Recall from Chapter 17 that an immunotoxin (page 494) is a combination of a monoclonal antibody and a toxic agent, such as the drug ricin or a
radioactive compound. The monoclonal antibody is directed against a particular type of tumor-associated antigen. The antibody selectively locates the cancer cell, and
the attached toxic agent destroys the cell but causes little
or no damage to healthy tissue. One such monoclonal antibody, *Herceptin,* is currently undergoing limited use in
certain applications in the treatment of breast cancer.

A promising new target for therapeutic antibodies is to
neutralize cytokines that furnish the blood supply for tumor cell proliferation. This therapy would prevent the formation of connective tissue that makes up most of the
mass of a tumor; thus, the cancerous cells would not develop into a tumor.

A vaccine against cancers also remains a possibility.
For many years, for example, there has been an effective
vaccine for preventing Marek's disease, a type of cancer in
poultry. A cancer vaccine will require isolation and cultivation of suitable tumor antigens. This would be only the
first step in a lengthy and expensive process of developing
a practical vaccine against human cancers.

Acquired Immunodeficiency Syndrome (AIDS)

In 1981, a cluster of cases of *Pneumocystis* pneumonia appeared in the Los Angeles area. This extremely

rare disease usually only occurred in immunosuppressed individuals. Investigators soon correlated the appearance of this disease with an unusual incidence of a rare form of cancer of the skin and blood vessels called Kaposi's sarcoma. The people affected were all young homosexual men, and all showed a loss of immune function. By 1983, the pathogen causing the loss of immune function had been identified as a virus that selectively infects helper T cells. This virus is now known as human immunodeficiency virus (HIV) (see Figure 1.1e).

The Origin of AIDS

Learning Objective

- *Give two examples of how emerging infectious diseases arise.*

HIV is now believed to have arisen by the mutation of a virus that had been endemic in some areas of central Africa for many years. Researchers speculate that a relatively benign virus infecting monkeys and chimpanzees entered the human population when the animals were skinned and butchered for food. Mathematical models of the supposed evolution of HIV, by Bette Korber of the Los Alamos National Laboratory, calculate that the virus probably made the transition to humans around 1930. The disease may have smoldered with little notice as long as transmission was limited to small villages where rates of sexual promiscuity were lower. The virus could not have killed or incapacitated its hosts quickly; otherwise, it could not have been maintained in the village population. With the sudden end of European colonialism, the social structure of sub-Saharan Africa was disrupted. The population became urbanized; the developments that result from urbanization, such as an increase in prostitution and the growth of highway transportation, which contributed to an increase in sexual promiscuity, are believed to be responsible for the spread of the disease. The earliest documented case of AIDS is from a patient in Leopoldville, Belgian Congo (now Kinshasa, capital of the Democratic Republic of the Congo). This man died in 1959; preserved samples of his blood contain antibodies to HIV. In the Western world, the first confirmed case of AIDS was the death of a Norwegian sailor in 1976, who probably was infected in 1961 or 1962 by contacts in western Africa.

HIV Infection

Learning Objectives

- *Explain the attachment of HIV to a host cell.*
- *List two ways in which HIV avoids the host's antibodies.*
- *Describe the stages of HIV infection.*
- *Describe the effects of HIV infection on the immune system.*

One of the most common misconceptions is that HIV infection is synonymous with AIDS. But AIDS denotes only the final stage of a long infection.

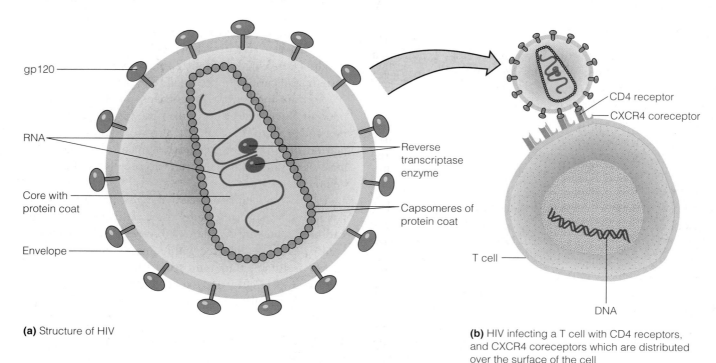

(a) Structure of HIV

(b) HIV infecting a T cell with CD4 receptors, and CXCR4 coreceptors which are distributed over the surface of the cell

FIGURE 19.12 **HIV structure and attachment to receptors on target T cell.**

■ Why does HIV preferentially infect CD4 cells?

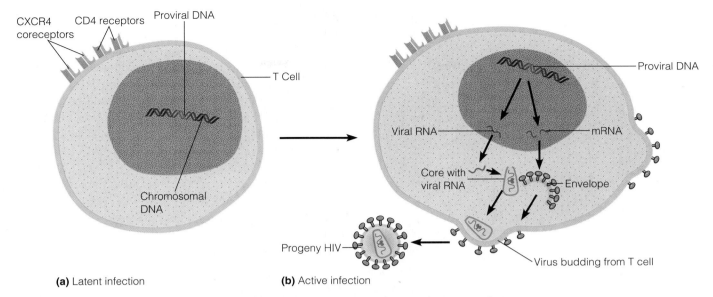

FIGURE 19.13 Latent and active HIV infection in CD4 T cells. (a) In latent infections, no viral genes are being transcribed and no new viruses are being formed. **(b)** In active infections, proviral DNA controls the synthesis of new viruses, which bud from the host cell.

■ CD4 T cells may be latently or actively infected with HIV.

The Structure of HIV

HIV, of the genus *Lentivirus*, is a retrovirus (see Figure 13.19 on page 397). It has two identical strands of RNA, the enzyme reverse transcriptase, and an envelope of phospholipid (Figure 19.12a). The envelope has spikes termed gp120 (the notation for a glycoprotein with a molecular weight of 120,000).

The Infectiveness and Pathogenicity of HIV

The spikes enable the virus to attach to the CD4 receptor on the host cells (Figure 19.12b). CD4 receptors are found on helper T cells, macrophages, and dendritic cells—the main targets of HIV infection. A CD4 receptor by itself is not sufficient for HIV infection. Certain coreceptors, which are receptors for chemokines, are also required. The two best known chemokine coreceptors, originally called fusins, are named CCR5 and CXCR4. This intimidating nomenclature is based upon the beginning amino acid sequence of these proteins. The term CCR5 indicates that the beginning sequence consists of cysteines, thus CC. If some other amino acid is located between the first two cysteines, this is shown by changing to CXC. The letter R is a convention for the balance of the chemical molecule. Simplistically, CCR5 is most important for the infection of macrophages and appears in the earlier stages of the infection. CXCR4 is involved mainly in the infection of T cells and predominates in later stages.

Attachment of the virus is followed by entry into the host cell (Figure 19.12b). In the host cell, viral RNA is released and transcribed into DNA by the enzyme reverse transcriptase. This viral DNA then becomes integrated into the chromosomal DNA of the host cell. The DNA may control the production of an active infection in which new viruses bud from the host cell, as shown in Figure 19.13b.

Alternatively, this integrated DNA may not produce new HIV but remains hidden in the host cell's chromosome as a *provirus* (Figures 19.13a and 19.14a). As a provirus, it is not detected by the immune system. HIV produced by a host cell is not necessarily released from the cell but may remain as *latent virions* in vacuoles within the cell, as shown in Figure 19.14b. In fact, a subset of the HIV-infected cells, instead of being killed, become long-lived memory T cells in which the reservoir of latent HIV can persist for decades. This ability of the virus to remain as a provirus or latent virus within host cells is an important reason why anti-HIV antibodies developed by infected individuals fail to inhibit progression of the infection. Another way HIV evades the immune system is by cell–cell fusion, by which the virus moves from an infected cell to an adjacent uninfected cell.

The virus also evades immune defenses by undergoing rapid antigenic changes. Retroviruses, with the reverse transcriptase enzyme step, have a high mutation rate compared to DNA viruses. They also lack the corrective "proofreading"

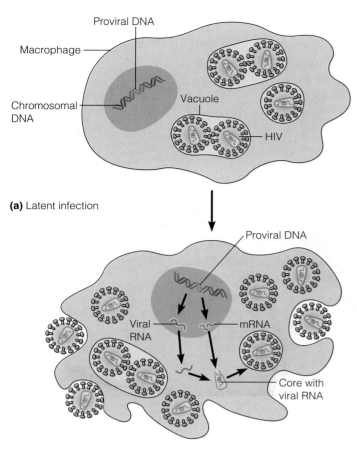

(a) Latent infection

(b) Active infection

FIGURE 19.14 **Latent and active HIV infection in macrophages. (a)** In latent infections, the newly made viruses remain in vacuoles. They are also present in the form of a provirus in the cell's chromosome. If the macrophage (not the infecting virus) is activated, these viruses will be released. **(b)** In active infections, new viruses are being produced; some of these remain in vacuoles, while others are released. Also, notice the proviruses latent in the chromosome.

■ How does an active infection differ from a latent infection?

capacity of DNA viruses. As a result, a mutation is probably introduced at every position in the HIV genome many times each day in an infected person. This may amount to an accumulation of 1 million variants of the virus in an asymptomatic person and 100 million variants during the final stages of the infection. These dramatic numbers illustrate the potential problems of drug resistance and obstacles to the development of vaccines and diagnostic tests.

Clades (Subtypes) of HIV

Worldwide, HIV is beginning to separate into distinctive groups called *clades* (Greek for branches), or subtypes. Viruses may vary by 15–20% within a clade; between clades, they may vary by 30% or more. Currently, *HIV-1*, the most common major type of HIV, has 11 such clades.

In the United States, about 90% of the cases are caused by HIV-1, clade B. Generally, this clade is also the most common elsewhere in the developed Western world. In sub-Saharan Africa, the predominant clade is C; in the Far East and Asia, clade E is the most common.

A second major HIV type, *HIV-2*, is found primarily in western Africa and is rare in the United States. The progression from infection to AIDS is much longer with HIV-2. In fact, most people infected with HIV-2 have a life span that is normal for the area.

The Stages of HIV Infection

The CDC classification divides the progress of HIV infection in adults into three clinical stages, or categories (Figure 19.15):

1. *Category A.* At this stage, the infection may be asymptomatic or cause persistent lymphadenopathy (swollen lymph nodes).

2. *Category B.* This stage is characterized by persistent infections by the yeast *Candida albicans*, which can appear in the mouth, throat, or vagina. Other conditions may include shingles, persistent diarrhea and fever, whitish patches on the oral mucosa (hairy leukoplakia), and certain cancerous or precancerous conditions of the cervix.

3. *Category C.* This stage is clinical AIDS. Important AIDS indicator conditions are *Candida albicans* infections of the esophagus, bronchi, and lungs; cytomegalovirus eye infections; tuberculosis; *Pneumocystis* pneumonia; toxoplasmosis of the brain; and Kaposi's sarcoma (caused by human herpesvirus 8).

The CDC also classifies the progress of HIV infections based on T cell populations. The purpose is primarily to furnish guidance for treatment, such as when to administer certain drugs. The normal population of a healthy individual is 800–1000 CD4 T cells/mm^3. In the United States, a count below 200/mm^3 is considered diagnostic for AIDS, regardless of the clinical category observed.

The progression from initial HIV infection to AIDS typically takes about 10 years in adults. This figure is typical in industrialized countries; in Africa, it is often about half this. Cellular warfare on an immense scale occurs during this time. At least 100 billion HIVs are generated every day, each with a remarkably short half-life of about 6 hours. These viruses must be cleared by the body's defenses, which include antibodies, cytotoxic T cells, and macrophages. Almost all HIVs, at least 99%, are produced by infected CD4 T cells, which survive for only about 2 days (T cells normally live for several years). Every day, an average of about 2 billion CD4 T cells are produced in an

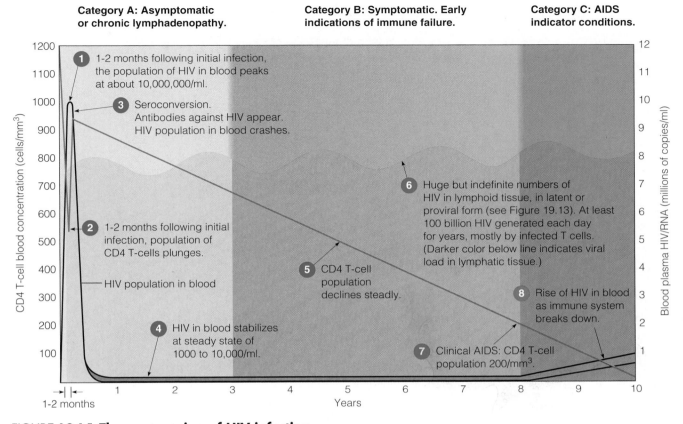

Category A: Asymptomatic or chronic lymphadenopathy.

Category B: Symptomatic. Early indications of immune failure.

Category C: AIDS indicator conditions.

1. 1-2 months following initial infection, the population of HIV in blood peaks at about 10,000,000/ml.

2. 1-2 months following initial infection, population of CD4 T-cells plunges.

3. Seroconversion. Antibodies against HIV appear. HIV population in blood crashes.

HIV population in blood

4. HIV in blood stabilizes at steady state of 1000 to 10,000/ml.

5. CD4 T-cell population declines steadily.

6. Huge but indefinite numbers of HIV in lymphoid tissue, in latent or proviral form (see Figure 19.13). At least 100 billion HIV generated each day for years, mostly by infected T cells. (Darker color below line indicates viral load in lymphatic tissue.)

7. Clinical AIDS: CD4 T-cell population 200/mm³.

8. Rise of HIV in blood as immune system breaks down.

FIGURE 19.15 **The progression of HIV infection.**

■ What causes the HIV population to drop in the first few months following infection?

attempt to compensate for losses. Over time, however, there is a daily net loss of at least 20 million CD4 T cells, one of the main markers for the progression of HIV infection. The most recent studies show that the decrease in CD4 T cells is not due entirely to direct viral destruction of the cells; rather, it is caused primarily by shortened life of the cells and the body's failure to compensate by increasing production of replacement T cells.

Survival with HIV Infection

HIV infection devastates the immune system, which is then unable to respond effectively to pathogens. The diseases or conditions most commonly associated with HIV infection and AIDS are summarized in Table 19.5. Success in treating these conditions has extended the lives of many HIV-infected people.

As the historical record of the AIDS epidemic lengthens, it has become apparent that not all HIV-positive individuals progress inexorably to AIDS and death. A significant group—about 5% of all those infected—are *long-term nonprogressors*. These infected individuals remain free of AIDS and even symptoms; and their CD4 T-cell counts remain

stable. Survival exceeding 25 years is predicted. In some cases, the virus seems less virulent; but in most, the immune system, especially T_C cells, is apparently more effective.

The age of the infected person can also be an important factor. Older adults are less able to replace antiviral T cell populations. Infants and younger children have an immune system that is not fully developed. They are much more susceptible to opportunistic infections.

Infants born to HIV-positive mothers are not always infected—in fact, only about 20% are. Infants who are most seriously infected survive less than 18 months.

Another surprising aspect of the AIDS epidemic is that a few people have been subjected to multiple HIV exposures and yet never become infected at all. The evidence is that their CD4 T cells are innately resistant.

Diagnostic Methods

Learning Objective

• Describe how HIV infection is diagnosed.

Generally speaking, it is simpler and less expensive to detect antibodies against HIV than to detect the virus itself.

TABLE 19.5	**Some Common Diseases Associated with AIDS**
Pathogen or Disease	**Disease Description**
Protozoa	
Cryptosporidium parvum	Persistent diarrhea
Toxoplasma gondii	Encephalitis
Isospora belli	Gastroenteritis
Viruses	
Cytomegalovirus	Fever, encephalitis, blindness
Herpes simplex virus	Vesicles of skin and mucous membranes
Varicella-zoster virus	Shingles
Bacteria	
Mycobacterium tuberculosis	Tuberculosis
M. avium-intracellulare	May infect many organs; gastroenteritis and other highly variable symptoms
Fungi	
Pneumocystis jiroveci	Life-threatening pneumonia
Histoplasma capsulatum	Disseminated infection
Cryptococcus neoformans	Disseminated, but especially meningitis
Candida albicans	Overgrowth on oral and vaginal mucous membranes (category B stage of HIV infection)
C. albicans	Overgrowth in esophagus, lungs (category C stage of HIV infection)
Cancers or Precancerous Conditions	
Kaposi's sarcoma	Cancer of skin and blood vessels (caused by human herpesvirus 8)
Hairy leukoplakia	Whitish patches on mucous membranes; commonly considered precancerous
Cervical dysplasia	Abnormal cervical growth

Therefore, most tests for HIV infection have been developed to detect HIV antibodies. The most commonly used screening tests are versions of the ELISA test (see Figure 18.12 on page 523). Positive ELISA tests are confirmed with the Western blot test (see Figure 10.12 on page 292). A problem with antibody-type testing is the window of time between infection and the appearance of detectable antibodies, or **seroconversion.** This interval, which can be as long as 3 months, is illustrated in Figure 19.15, where seroconversion follows the peak number of viruses in circulation. Because of this delay, the recipient of an organ transplant or a blood transfusion can become infected with HIV even though antibody tests did not show the presence of the virus. Improvements in testing have gradually narrowed the window to 21–25 days.

In contrast to antibody tests, **plasma viral load tests** can detect and quantify HIV circulating in the blood. Viral load tests detect viral RNA by methods such as PCR (see page 254) or nucleic acid hybridization (see page 295.) Despite their obvious advantages over antibody tests, however, viral load tests are costly and require 48–72 hours to complete.

Tests that detect the virus are also useful for instances in which antibody-detection tests cannot be used. For example, newborns of HIV-infected mothers have circulating maternal antibodies that interfere with conventional antibody-detection tests. To ensure safety of the blood supply as much as possible, the Red Cross has introduced PCR testing for HIV at all their regional collection centers. Pooled samples of 120 donations are tested, and if HIV is found, each individual sample is tested until the infected sample is found. This will allow blood banks to discontinue antibody-detecting tests and narrow the window of nondetection to about 12 days. Viral load tests are also used to follow the progress of chemotherapy treatment of HIV infections.

A caution to keep in mind in testing for HIV is that current tests may not reliably detect all of the myriad variants of rapidly mutating HIV, especially subtypes that are not normally present in a population. Furthermore, plasma viral load tests sample only the virions circulating in the blood. Large numbers of HIV in latent form (see Figures 19.13 and 19.14) remain in lymphoid tissue. These are constantly being released into peripheral circulation.

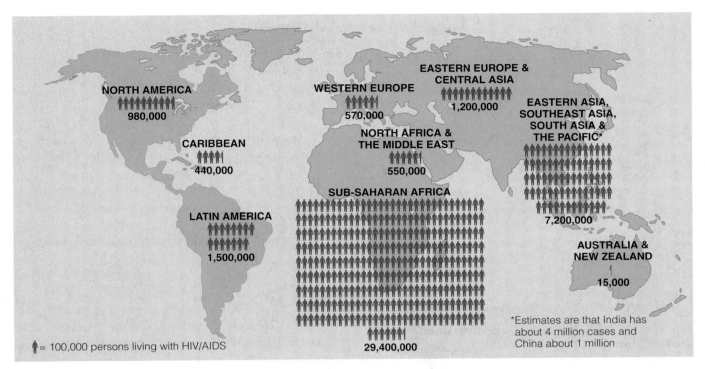

FIGURE 19.16 Distribution of HIV infection and AIDS in regions of the world.
Each figure represents 100,000 persons living with HIV infection or AIDS.
SOURCE: UNAIDS; World Health Organization

■ Where do you think the most accurate figures would be available?

HIV Transmission

Learning Objective

• List the routes of HIV transmission.

The transmission of HIV requires the transfer of, or direct contact with, infected body fluids. The most important of these is blood, which contains 1000–100,000 infective viruses per milliliter, and semen, which contains about 10–50 viruses per milliliter. The viruses are often located within cells in these fluids, especially in macrophages. HIV can survive more than 1.5 days inside a cell but only about 6 hours outside a cell.

Routes of HIV transmission include intimate sexual contact, breast milk, transplacental infection of a fetus, blood-contaminated needles, organ transplants, artificial insemination, and blood transfusion. Probably the most dangerous form of sexual contact is anal-receptive intercourse. Vaginal intercourse is much more likely to transmit HIV from male to female than vice versa, and transmission either way is much greater when genital lesions are present. Although rare, transmission can occur by oral-genital contact.

HIV is not transmitted by insects or casual contact such as hugging or sharing household items. Saliva generally contains less than 1 virus per milliliter, and kissing is not known to transmit the virus. In developed countries, transmission by transfusion is unlikely because blood is tested for HIV or HIV antibodies. However, there will always be a slight risk, as discussed earlier.

AIDS Worldwide

Learning Objective

• Identify geographic patterns of HIV transmission.

Only about two decades after AIDS was first recognized, HIV infection has become a global pandemic. An estimated 20 million AIDS deaths have occurred worldwide, and over 40 million adults, children, and infants are infected with HIV. It is the leading cause of death in sub-Saharan Africa and also in many of the largest cities in the United States and Europe (Figure 19.16). Worldwide, probably 14,000 new HIV infections occur each day.

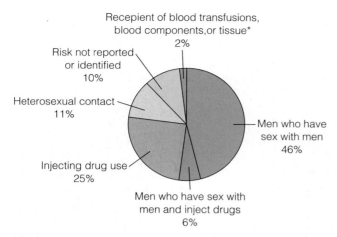

Recepient of blood transfusions,
blood components, or tissue*
2%

Risk not reported
or identified
10%

Heterosexual contact
11%

Injecting drug use
25%

Men who have
sex with men
46%

Men who have sex with
men and inject drugs
6%

* Most cases of AIDS resulting from transfusion of blood or tissue occurred before proper testing methods for HIV were in place. Hemophiliacs, who require regular infusion of blood-clotting factors, were an early group contracting HIV infections. Transfused blood, blood products, and transplanted tissue are now rigorously tested for HIV contamination and the risk is very low.

FIGURE 19.17 **Modes of HIV transmission in the United States.** In most of the world, transmission is primarily by heterosexual sex. In the United States, this form of transmission is much lower but is growing rapidly. Transmission in western Europe is similar to that in the United States.

■ Worldwide, the primary mode of HIV transmission is heterosexual sex.

Three basic epidemiological patterns have emerged:

• In the United States, Canada, western Europe, Australia, northern Africa, and certain parts of South America, HIV has primarily affected injecting drug users (IDUs) and homosexual and bisexual males. Eastern Europe and Russia have seen an accelerating growth in the number of cases, mostly in IDUs. In western Europe and North America, the incidence of heterosexual spread is increasing rapidly. It is expected that women will soon be infected at the same rate as men.

• In sub-Saharan Africa, HIV transmission has been almost entirely from heterosexual contact, and males and females are affected in about equal numbers. In some countries, more than 25% of the population is infected.

• The infection has been introduced more recently in eastern Europe, the Middle East, and, most importantly, in southeast Asia. The World Health Organization (WHO) believes that Asia's epidemic, which potentially includes the immense populations of India and China, may ultimately dwarf all others in scope and impact. In these areas, the epidemic involves IDUs, commercial sex workers, and young heterosexual males.

The HIV clades in Asia may be better adapted to heterosexual transmission through mucous membranes. In India, the WHO currently estimates that there are as many as 4 million infected adults. Female sex workers in urban areas are infected at rates sometimes approaching 50%, and the disease is increasingly appearing in rural villages. India, with its population of about 1 billion, may soon have more AIDS cases than any other country. HIV-infection statistics in China are murky, but the WHO estimates it at about 1 million.

Figure 19.17 compares modes of HIV transmission in the United States with other parts of the world.

The Prevention and Treatment of AIDS

Learning Objective

• List the current methods of preventing and treating HIV infection.

At present, for most of the world the only practical means of control is by minimizing transmission. This requires education programs to promote the use of condoms, as well as discouraging sexual promiscuity. In high-income countries, the availability of medications has made HIV infection no longer a certain death sentence. Unfortunately, improvements in managing HIV infection have resulted in a relaxed attitude toward safer-sex practices. The fact tends to be overlooked that the available drugs only delay the progress of the infection, they are not a cure. Attempts to prevent the use of contaminated needles among IDUs is also important. To be effective, educational programs often require fundamental social changes that are not easy to accomplish, but they have slowed the infection rate in some areas, such as Thailand.

HIV Vaccines

The rapid mutation rate of HIV makes it difficult to develop a vaccine that is effective against all forms of the virus. Another obstacle is the variety of routes by which HIV can be transmitted. An effective vaccine would have to protect against transmission by different mucosal routes, which is proving to be an elusive goal in tests with simian immunodeficiency virus (SIV) that have been performed in monkeys. (SIV is a virus closely related to HIV, especially HIV-2. Many AIDS researchers think it probable that HIV evolved from SIV.) Additionally, because the virus occurs both as free viruses and in virus-containing cells, a vaccine would have to be effective in both instances. In poorer parts of the world, affordability is an important concern. In summary, a vaccine for AIDS will be a much more difficult challenge to develop than any existing vaccine, and the prospects are uncertain (see the box on

New Weapons Against AIDS

The ability of HIV to destroy, evade, and even hide inside cells of the human immune system makes it extremely difficult to vaccinate against or to treat HIV infection. Because the virus is so unusual and so deadly, it has stimulated intense research into a variety of approaches to vaccination and chemotherapy to complement the highly active antiretroviral therapy (HAART) cocktail.

Looking for Drugs in New Places

A reverse transcriptase inhibitor, calmolide A, is produced by a Malaysian tree. This compound is effective against HIV-1, but not against HIV-2.

Sharon Hill of the University of Pittsburgh has identified two lactic acid–producing bacteria, *Lactobacillus crispatus* and *L. jensenii*, in the vaginal microbiota that produce bacteriocins that inhibit HIV. Studies have shown a correlation between the lack of these bacteria and transmission of HIV by sexual intercourse.

Virus Decoys

The idea behind a virus decoy is that if the decoy is present in numbers that are high enough, the virus will bind to the decoy instead of to a cell. Researchers at Trimeris Corporation in North Carolina are investigating a peptide called T-20 that should align itself with part of gp41 to prevent viral fusion with the cell. Progenics Corporation in New York is exploring a similar approach with a recombinant antibody–like molecule, PRO 542, a recombinant protein that is essentially an IgG antibody with the variable regions replaced by part of the CD4 molecule. The idea is that HIV will bind with this molecule instead of the CD4 molecules on a cell. Clinical trials on both of these drugs have shown tenfold decreases in HIV levels in patients.

Boosting Immunity

The conventional anti-HIV drugs work by inhibiting enzymes. HIV patients taking HAART still need to increase their CD4 cells; otherwise, they are at risk for opportunistic infections. Cyclosporine, given with HAART, is being tested in HIV patients to determine whether it can increase CD4 cells. Other approaches to increasing the T cell response include: a thymic hormone being tested by Hollis-Eden Pharmaceuticals in San Diego, California; HE2000, an analog of an adrenal hormone that stimulates T cells; and a synthetic compound, AS101 that enhances CSF and IL-2.

In a test conducted at the National Institute of Allergy and Infectious Diseases (NIAID), HIV-infected people with moderate immune suppression were given IL-2 over 8 months. The IL-2 caused a 50% increase in the number of CD4 cells in the test group. In 1999, clinical trials of IL-2 began in the United States and several other countries.

Vaccines

It is believed that antibodies are essential to preventing HIV from contacting a potential host cell. Unfortunately, vaccines designed to promote antibodies have been disappointing. Several vaccines in clinical trials are showing promise for slowing the progression of the disease by eliciting a T cell response. The vaccine trials taking place in the United States and several other countries are using the HIV clade that occurs in the region where the trials are being conducted. For example, clade B is primarily responsible for AIDS in North America and Europe; clade C predominates in sub-Saharan Africa; and clades B, C, and E are seen in south and southeast Asia. Chiron Corporation in California is trying to develop a multiclade vaccine.

A whole-cell vaccine that uses *Salmonella* bacteria has been engineered to remove the virulence gene and to insert the gp120 gene. Subunit vaccine preparations using HIV genes expressed in *Saccharomyces cerevisiae* or hamster cells promote an antibody response against the gp120 on HIV and may prevent infection. Animal studies suggest that the best way to control HIV is to target events that occur early in the lytic cycle of the virus. Therefore, some vaccines are being designed to stimulate immune responses against the virus proteins that are produced within a few hours after infection, such as gag and tat. Gag is one of the capsid proteins, and tat is a transcription factor. One such vaccine, called Alvac, uses a harmless canarypox virus that has some HIV capsid protein genes. The idea is that canarypox virus will infect cells, causing them to make these HIV proteins, which bud from the cell as "pseudovirions." The pseudovirions present HIV proteins in a natural shape and induce a T_C cell response in vaccine recipients.

Naked DNA vaccines consisting of either the *tat* or *gag* genes are being tested by Merck Pharmaceuticals. Cells take up the naked DNA and produce viral protein. The resulting protein stimulates both humoral and cellular immunity.

* * *

Anthony Fauci, director of the NIAID, cautions that one vaccine will not protect from all HIV clades and may not protect against all the viruses in one clade. Moreover, vaccines that work by stimulating cell-mediated immunity will probably require booster injections and will probably not provide protection indefinitely.

page 553). Some experts believe that no HIV vaccine is possible that will confer nearly complete protection, such as those for smallpox or measles. It is thought that a more practical goal may be to develop a vaccine that will stimulate cell-mediated immunity in already infected persons and help the patient's immune system to clear the virus.

Chemotherapy

Much progress has been made in the use of chemotherapy to inhibit HIV infections. Drugs that are available for chemotherapy of HIV are discussed on page 577 and in the box on page 553. The first target of anti-HIV drugs was the enzyme reverse transcriptase (page 397), an enzyme not present in human cells. These *nucleoside reverse transcriptase inhibitors* are analogs of nucleosides and cause the termination of the synthesis of viral DNA by competitive inhibition (page 118). There are other drugs that inhibit reverse transcription but are not analogs of nucleic acids; these are the so-called *non-nucleoside reverse transcriptase inhibitors*.

To reproduce, the virus makes use of certain protease enzymes that cut proteins into pieces, which are then reassembled into the coat of new HIV particles. Drugs called *protease inhibitors* inhibit this enzyme and are now in use. Some new drugs are discussed in the box on page 553.

Because of the increasing number of drugs that control reproduction of the virus, at least temporarily, HIV infection is almost at the stage where it can be considered a treatable chronic disease—assuming that the treatment is affordable. The rapid reproductive rate and frequent occurrence of drug-resistant mutations dictates that multiple drugs, given simultaneously, must be used. The current treatment is termed **highly active antiretroviral therapy (HAART).** This therapy consists of administering drug combinations; one of the most common combinations is two nucleoside analog reverse transcriptase inhibitors plus either a non-nucleoside transcriptase inhibitor or a protease inhibitor.

Patients are often required to take as many as 40 pills a day on a complex schedule, which must be adhered to rigorously because the virus is unforgiving. Even so, resistant strains of the virus are likely to emerge. As one scientist put it, drug resistance in HIV is driven by selective pressures that Darwin never imagined. Experience has also shown that eliminating all viruses in latent form in lymphoid tissue is especially difficult. The number of HIVs in circulation is often reduced to fewer than can be detected, but this is not necessarily the same as eradication. The more practical goal has been to reduce the viral load to a level where it can be controlled by the immune system. A caution is that a patient with a viral load at an undetectable level might still be infective. Frequent testing for viruses (not antibodies; see the discussion of diagnostic tests on page 549–550) is required to follow effectiveness of the treatment. A rise in numbers probably indicates development of a resistant population.

One clearly successful application of chemotherapy has been to reduce the chance of HIV transmission from an infected mother to her newborn. The administration of even one nucleoside reverse transcriptase drug alone reduces the incidence.

The AIDS Epidemic and the Importance of Scientific Research

The AIDS epidemic gives clear evidence of the value of basic scientific research. Without the advances in molecular biology of the past few decades, we would have been unable even to identify the causative agent of AIDS. We would not have been able to develop the tests for screening donated blood, to identify points in the viral life cycle for which selectively toxic drugs could be developed, or even to monitor the course of the infection. In the lifetime of most of us, we will have the opportunity to witness medical history being made as the struggle with this deadly and elusive virus continues.

STUDY OUTLINE

Introduction (p. 529)

1. Hay fever, transplant rejection, and autoimmunity are examples of harmful immune reactions.
2. Infection and immunosuppression are examples of failure of the immune system.
3. Superantigens activate many T-cell receptors, resulting in the release of excessive amounts of cytokines that can cause adverse host responses.

Hypersensitivity (pp. 530–537)

1. Hypersensitivity reactions represent immunological responses to an antigen (allergen) that lead to tissue damage rather than immunity.
2. Hypersensitivity reactions occur when a person has been sensitized to an antigen.
3. Hypersensitivity reactions can be divided into four classes: types I, II, and III are immediate reactions based on

humoral immunity, and type IV is a delayed reaction based on cell-mediated immunity.

Type I (Anaphylactic) Reactions (pp. 530–533)

1. Anaphylactic reactions involve the production of IgE antibodies that bind to mast cells and basophils to sensitize the host.

2. The binding of two adjacent IgE antibodies to an antigen causes the target cell to release chemical mediators, such as histamine, leukotrienes, and prostaglandins, which cause the observed allergic reactions.

3. Systemic anaphylaxis may develop in minutes after injection or ingestion of the antigen; this may result in circulatory collapse and death.

4. Localized anaphylaxis is exemplified by hives, hay fever, and asthma.

5. Skin testing is useful in determining sensitivity to an antigen.

6. Desensitization to an antigen can be achieved by repeated injections of the antigen, which leads to the formation of blocking (IgG) antibodies.

Type II (Cytotoxic) Reactions (pp. 533–535)

1. Type II reactions are mediated by IgG or IgM antibodies and complement.

2. The antibodies are directed toward foreign cells or host cells. Complement fixation may result in cell lysis. Macrophages and other cells may also damage the antibody-coated cells.

The ABO Blood Group System (pp. 533–534)

1. Human blood may be grouped into four principal types, designated A, B, AB, and O.

2. The presence or absence of two carbohydrate antigens designated A and B on the surface of the red blood cell determines a person's blood type.

3. Naturally occurring antibodies are present in serum against the opposite AB antigen.

4. Incompatible blood transfusions lead to the complement-mediated lysis of the donor red blood cells.

The Rh Blood Group System (pp. 534–535)

1. Approximately 85% of the human population possesses another blood group antigen, designated the Rh antigen; these individuals are designated Rh$^+$.

2. The absence of this antigen in certain individuals (Rh$^-$) can lead to sensitization upon exposure to it.

3. An Rh$^+$ person can receive Rh$^+$ or Rh$^-$ blood transfusions.

4. When an Rh$^-$ person receives Rh$^+$ blood, that person will produce anti-Rh antibodies.

5. Subsequent exposure to Rh$^+$ cells will result in a rapid, serious hemolytic reaction.

6. An Rh$^-$ mother carrying an Rh$^+$ fetus will produce anti-Rh antibodies.

7. Subsequent pregnancies involving Rh incompatibility may result in hemolytic disease of the newborn.

8. The disease may be prevented by passive immunization of the mother with anti-Rh antibodies.

Drug-Induced Cytotoxic Reactions (p. 535)

1. In the disease thrombocytopenic purpura, platelets are destroyed by antibodies and complement.

2. Agranulocytosis and hemolytic anemia result from antibodies against one's own blood cells coated with drug molecules.

Type III (Immune Complex) Reactions (pp. 535–536)

1. Immune complex diseases occur when IgG antibodies and soluble antigen form small complexes that lodge in the basement membranes of cells.

2. Subsequent complement fixation results in inflammation.

3. Glomerulonephritis is an immune complex disease.

Type IV (Cell-Mediated) Reactions (pp. 536–537)

1. Delayed-type hypersensitivity reactions are due primarily to T_D cell proliferation.

2. Sensitized T cells secrete cytokines in response to the appropriate antigen.

3. Cytokines attract and activate macrophages and initiate tissue damage.

4. The tuberculin skin test and allergic contact dermatitis are examples of delayed hypersensitivities.

Autoimmune Diseases (pp. 537–539)

1. Autoimmunity results from a loss of self-tolerance.

2. Self-tolerance occurs during fetal development; T cells that will target host cells are eliminated (clonal deletion) or inactivated.

3. Type I autoimmunity may be due to antibodies against infectious agents.

4. Graves' disease and myasthenia gravis are type II autoimmune reactions in which antibodies react to cell-surface antigens.

5. Systemic lupus erythematosus and rheumatoid arthritis are type III autoimmune reactions in which the deposition of immune complexes results in tissue damage.

6. Multiple sclerosis, Hashimoto's disease, and insulin-dependent diabetes mellitus are type IV autoimmune reactions mediated by T cells.

Reactions Related to the Human Leukocyte Antigen (HLA) Complex (pp. 539–543)

1. Histocompatibility self molecules located on cell surfaces express genetic differences among individuals; these antigens are coded for by MHC or HLA gene complexes.

2. To prevent the rejection of transplants, HLA and ABO blood group antigens of the donor and recipient are matched as closely as possible.

Reactions to Transplantation (pp. 541–542)

1. Transplants recognized as foreign antigens may be lysed by T cells and attacked by macrophages and complement-fixing antibodies.

2. Transplantation to a privileged site (such as the cornea) or of a privileged tissue (such as pig heart valves) does not cause an immune response.

3. Pluripotent stem cells differentiate into a variety of tissues that may provide tissues for transplant.

4. Four types of transplants have been defined on the basis of genetic relationships between the donor and the recipient: autografts, isografts, allografts, and xenotransplantation.

5. Bone marrow (with immunocompetent cells) can cause graft-versus-host disease.

6. Successful transplant surgery often requires immunosuppressant drugs to prevent an immune response to the transplanted tissue.

Immune Deficiencies (p. 543)

1. Immune deficiencies can be congenital or acquired.

2. Congenital immune deficiencies are due to defective or absent genes.

3. A variety of drugs, cancers, and infectious diseases can cause acquired immune deficiencies.

The Immune System and Cancer (pp. 543–545)

1. Cancer cells are normal cells that have undergone transformation, divide uncontrollably, and possess tumor-associated antigens.

2. The response of the immune system to cancer is called immunological surveillance.

3. T_C cells recognize and lyse cancerous cells.

4. Cancer cells can escape detection and destruction by the immune system.

5. Cancer cells may suppress T cells or grow faster than the immune system can respond.

Immunotherapy (p. 545)

1. Tumor necrosis factor (TNF) and other cytokines are being tested as cancer treatments.

2. Immunotoxins are chemical poisons linked to a monoclonal antibody; the antibody selectively locates the cancer cell for release of the poison.

3. A vaccine consisting of tumor antigens has been effective in controlling one type of cancer in poultry.

Acquired Immunodeficiency Syndrome (AIDS) (pp. 545–554)

The Origin of AIDS (p. 546)

1. HIV is thought to have originated in central Africa and was brought to other countries by modern transportation and unsafe sexual practices.

HIV Infection (pp. 546–549)

1. AIDS is the final stage of HIV infection.

2. HIV is a retrovirus with single-stranded RNA, reverse transcriptase, and a phospholipid envelope with gp120 spikes.

3. HIV spikes attach to CD4 and coreceptors on host cells; the CD4 receptor is found on helper T cells, macrophages, and dendritic cells.

4. Viral RNA is transcribed to DNA by reverse transcriptase. The viral DNA becomes integrated into the host chromosome to direct synthesis of new viruses or to remain latent as a provirus.

5. HIV evades the immune system in latency, in vacuoles, by using cell-cell fusion, and by antigenic change.

6. Genetically distinct groups of HIV are classified into clades.

7. HIV infection is categorized by symptoms: Category A (asymptomatic) and Category B (selected symptoms) are reported as AIDS if CD4 T cells fall below 200 cells/mm^3; Category C (AIDS indicator conditions) is reported as AIDS.

8. HIV infection is also categorized by CD4 T cell numbers: 200 CD4 cells/mm^3 is reported as AIDS.

9. The progression from HIV infection to AIDS takes about 10 years.

10. The life of an AIDS patient can be prolonged by the proper treatment of opportunistic infections.

Diagnostic Methods (pp. 549–550)

1. HIV antibodies are detected by ELISA. HIV antigens are detected by Western blotting.

2. Plasma viral load tests detect viral nucleic acid and are used to quantify HIV in blood.

HIV Transmission (p. 551)

1. HIV is transmitted by sexual contact, breast milk, contaminated needles, transplacental infection, artificial insemination, and blood transfusion.

2. In developed countries, blood transfusions are not a likely source of infection because blood is tested for HIV antibodies.

AIDS Worldwide (pp. 551–552)

1. In the United States, Canada, western Europe, Australia, northern Africa, and parts of South America, transmission has been by injecting drug use (IDU) and male-to-male sexual contact. Heterosexual transmission is increasing.

2. In sub-Saharan Africa, transmission is primarily heterosexual contact.

3. In eastern Europe and Asia, transmission is by IDU and heterosexual contact.

The Prevention and Treatment of AIDS (pp. 552–554)

1. The use of condoms and sterile needles prevents the transmission of HIV.

2. Vaccine development is difficult because the virus remains inside host cells.

3. Current chemotherapeutic agents target the virus enzymes, reverse transcriptase and protease.

STUDY QUESTIONS

Review

1. Define hypersensitivity.

2. List three mediators released in anaphylactic hypersensitivities, and explain their effects.

3. Discuss the roles of antibodies and antigens in an incompatible tissue transplant.

4. What happens to the recipient of an incompatible blood type?

5. Explain how hemolytic disease of the newborn develops and how this disease might be prevented.

6. Explain what happens when a person develops a contact sensitivity to poison oak.
  a. What causes the observed symptoms?
  b. How did the sensitivity develop?
  c. How might this person be desensitized to poison oak?

7. Which type of graft (autograft, isograft, allograft, or xenotransplantation) is most compatible? Least compatible?

8. Which of the following blood transfusions is compatible? Explain your answers.

Donor	Recipient
(a) AB, Rh−	AB, Rh+
(b) B, Rh+	B, Rh−
(c) A, Rh+	O, Rh+

9. Define autoimmunity. Propose a hypothesis that could explain autoimmune responses.

10. Differentiate type II, type III, and type IV autoimmune diseases. Name an example of each type.

11. Summarize the causes of immune deficiencies. What is the effect of an immune deficiency?

12. In what ways do tumor cells differ antigenically from normal cells? Explain how tumor cells may be destroyed by the immune system.

13. If tumor cells can be destroyed by the immune system, how does cancer develop? What does immunotherapy involve?

14. Differentiate HIV infection from AIDS. How is HIV transmitted? How can HIV infection be prevented?

Multiple Choice

1. Desensitization to prevent an allergic response can be accomplished by injecting small, repeated doses of
  a. IgE antibodies.
  b. the antigen (allergen).
  c. histamine.
  d. IgG antibodies.
  e. antihistamine.

2. In vitro, recipient serum type B will agglutinate donor cells type A; while in vivo, recipient serum type B will lyse donor cells type A. The in vivo response is due to
  a. T cells.
  b. complement.
  c. antibodies.
  d. autoimmunity.
  e. none of the above

3. Cytotoxic autoimmunity differs from immune complex autoimmunity in that cytotoxic reactions
  a. involve antibodies.
  b. do not involve complement.
  c. are caused by T cells.
  d. do not involve IgE antibodies.
  e. none of the above

4. Worldwide, the primary method of transmitting HIV is
  a. homosexual sex.
  b. heterosexual sex.
  c. use of injecting drugs.
  d. blood transfusions.
  e. kissing.

5. Which of the following is *not* the cause of a natural immune deficiency?
  a. a recessive gene resulting in lack of a thymus gland
  b. a recessive gene resulting in few B cells
  c. HIV infection
  d. immunosuppressant drugs
  e. none of the above

6. Which antibodies will be found naturally in the serum of a person with blood type A, Rh$^+$?
  a. anti A, anti B, anti Rh
  b. anti A, anti Rh
  c. anti A
  d. anti B, anti Rh
  e. anti B

Use the following choices to answer questions 7 through 10.
- **a.** type I hypersensitivity
- **b.** type II hypersensitivity
- **c.** type III hypersensitivity
- **d.** type IV hypersensitivity
- **e.** all of the above

7. Localized anaphylaxis.

8. Allergic contact dermatitis.

9. Due to immune complexes.

10. Reaction to an incompatible blood transfusion.

Critical Thinking

1. When and how does our immune system discriminate between self and nonself antigens?

2. The first preparations used for artificially acquired passive immunity were antibodies in horse serum. A complication that resulted from the therapeutic use of horse serum was immune complex disease. Why did this occur?

3. Do people with AIDS make antibodies? If so, why are they said to have an immune deficiency?

4. What are the methods of action of anti-AIDS drugs?

Clinical Applications

1. Fungal infections such as athlete's foot are chronic. These fungi degrade skin keratin but are not invasive and do not produce toxins. Why do you suppose that many of the symptoms of a fungal infection are due to hypersensitivity to the fungus?

2. After working in a mushroom farm for several months, a worker develops these symptoms: hives, edema, and swelling lymph nodes.
   - **a.** What do these symptoms indicate?
   - **b.** What mediators cause these symptoms?
   - **c.** How may sensitivity to a particular antigen be determined?
   - **d.** Other employees do not appear to have any immunological reactions. What could explain this?
   (*Hint:* The allergen is conidiospores from molds growing in the mushroom farm.)

3. Physicians administering live, attenuated mumps and measles vaccines prepared in chick embryos are instructed to have epinephrine available. Epinephrine will not treat these viral infections. What is the purpose of keeping this drug on hand?

4. A woman with blood type A+ once received a transfusion of AB+ blood. When she carried a type B+ fetus, the fetus developed hemolytic disease of the newborn. Explain why this fetus developed this condition when another type B+ fetus in a type A+ mother was normal.

LEARNING WITH TECHNOLOGY

The Microbiology Place CD-ROM/website includes many resources to help you study and succeed in this course. In addition to self-quizzes, web links, and flashcards, you'll find the following activities for this chapter in the "Activities" section of The Microbiology Place:

Case Study: HIV Transmission

Chapter 20

Antimicrobial Drugs

This bacterium is lysing because an antibiotic has prevented cell wall synthesis.

When the body's normal defenses cannot prevent or overcome a disease, it is often treated with **chemotherapy**. In this chapter we focus on antimicrobial drugs, the class of chemotherapeutic agents used to treat infectious diseases. Like the disinfectants discussed in Chapter 7, **antimicrobial drugs** act by interfering with the growth of microorganisms. Unlike disinfectants, however, they must often act within the host. Therefore, their effects on the cells and tissues of the host are important. The ideal antimicrobial drug kills the harmful microorganism without damaging the host; this is the principle of **selective toxicity**.

The History of Chemotherapy

Learning Objectives

- *Identify the contributions of Paul Ehrlich and Alexander Fleming to chemotherapy.*
- *Name the microbes that produce most antibiotics.*

The birth of modern chemotherapy is credited to the efforts of Paul Ehrlich in Germany during the early part of the twentieth century. While attempting to stain bacteria without staining the surrounding tissue, he speculated about some "magic bullet" that would selectively find and destroy pathogens but not harm the host. This idea provided the basis for chemotherapy, a term he coined.

In 1928, Alexander Fleming observed that the growth of the bacterium *Staphylococcus aureus* was inhibited in the area surrounding the colony of a mold that had contaminated a Petri plate (Figure 20.1). The mold was identified as *Penicillium notatum*, and its active compound, which was isolated a short time later, was named penicillin. Similar inhibitory reactions between colonies on solid media are commonly observed in microbiology, and the mechanism of inhibition is called *antibiosis*. From this word comes the term **antibiotic,** a substance produced by microorganisms that in small amounts inhibits another microorganism. Therefore, the wholly synthetic sulfa drugs, for example, technically are not antibiotics. However, this distinction is often ignored in practice.

FIGURE 20.1 **The discovery of penicillin.** Alexander Fleming took this photograph in 1928. The colony of *Penicillium* mold accidentally contaminated the plate and is inhibiting nearby bacterial growth.

- New antibiotic-producing organisms are discovered by looking for areas of inhibition around colonies.

TABLE 20.1	**Representative Sources of Antibiotics**
Microorganism	**Antibiotic**
Gram-Positive Rods	
Bacillus subtilis	Bacitracin
Bacillus polymyxa	Polymyxin
Actinomycetes	
Streptomyces nodosus	Amphotericin B
Streptomyces venezuelae	Chloramphenicol
Streptomyces aureofaciens	Chlortetracycline and tetracycline
Streptomyces erythraeus	Erythromycin
Streptomyces fradiae	Neomycin
Streptomyces griseus	Streptomycin
Micromonospora purpureae	Gentamicin
Fungi	
Cephalosporium spp.	Cephalothin
Penicillium griseofulvum	Griseofulvin
Penicillium notatum	Penicillin

In 1940, a group of scientists at Oxford University headed by Howard Florey and Ernst Chain succeeded in the first clinical trials of penicillin. Intensive research in the United States then led to the isolation of especially productive *Penicillium* strains for use in the mass production of the antibiotic. (The most famous of these high-producing strains was originally isolated from a cantaloupe bought at a market in Peoria, Illinois.)

Antibiotics are actually rather easy to discover, but few are of medical or commercial value. Some are used commercially other than for treating disease—for example, as a supplement in animal feed (see the box on page 581). Many antibiotics are toxic to humans or lack any advantage over antibiotics already in use.

More than half of our antibiotics are produced by species of *Streptomyces,* filamentous bacteria that commonly inhabit soil. A few antibiotics are produced by bacteria of the genus *Bacillus,* and others are produced by molds, mostly of the genera *Penicillium* and *Cephalosporium* (sef-ä-lō-spô′rē-um). See Table 20.1 for the sources of many antibiotics in use today—a surprisingly limited group of organisms. One study screened 400,000 microbial cultures that yielded only three useful drugs. It is especially interesting to note that practically all antibiotic-producing microbes have some sort of sporulation process.

The Spectrum of Antimicrobial Activity

Learning Objectives

- *Describe the problems of chemotherapy for viral, fungal, protozoan, and helminthic infections.*

- *Define the following terms: spectrum of activity, broad-spectrum antibiotic, superinfection.*

It is comparatively easy to find or develop drugs that are effective against prokaryotic cells and that do not affect the eukaryotic cells of humans. These two cell types differ substantially in many ways, such as in the presence or absence of cell walls, the fine structure of their ribosomes, and details of their metabolism. Thus, selective toxicity has numerous targets. The problem is more difficult when the pathogen is a eukaryotic cell, such as a fungus, protozoan, or helminth. At the cellular level, these organisms resemble the human cell much more closely than a bacterial cell does. We will see that our arsenal against these types of pathogens is much more limited than our arsenal of antibacterial drugs. Viral infections are particularly difficult to treat because the pathogen is within the human host's cells, and the genetic information of the virus is directing the human cell to make viruses rather than to synthesize normal cellular materials.

TABLE 20.2	**The Spectrum of Activity of Antibiotics and Other Antimicrobial Drugs**							
Prokaryotes				**Eukaryotes**				
Mycobacteria*	Gram-Negative Bacteria	Gram-Positive Bacteria	Chlamydias, Rickettsias†	Fungi	Protozoa	Helminths	Viruses

Penicillin

Ketoconazole

Niclosamide (tapeworms)

Streptomycin

Mefloquine (malaria)

Acyclovir

Praziquantel (flukes)

Tetracycline

Isoniazid

*Growth of these bacteria frequently occurs within macrophages or tissue structures.
†Obligately intracellular bacteria.

Some drugs have a narrow **spectrum of microbial activity,** or range of different microbial types they affect. Penicillin G, for example, affects gram-positive bacteria but very few gram-negative bacteria. Antibiotics that affect a broad range of gram-positive or gram-negative bacteria are therefore called **broad-spectrum antibiotics.**

A primary factor involved in the selective toxicity of antibacterial action lies in the lipopolysaccharide outer layer of gram-negative bacteria and the porins that form water-filled channels across this layer (see Figure 4.13c on page 85). Drugs that pass through the porin channels must be relatively small and preferably hydrophilic. Drugs that are lipophilic (having an affinity for lipids) or especially large do not enter gram-negative bacteria readily.

Table 20.2 summarizes the spectrum of activity of a number of chemotherapeutic drugs. Because the identity of the pathogen is not always immediately known, a broad-spectrum drug would seem to have an advantage in treating a disease by saving valuable time. The disadvantage is that many normal microbiota of the host are destroyed by broad-spectrum drugs. The normal microbiota ordinarily compete with and check the growth of pathogens or other microbes. If certain organisms in the normal microbiota are not destroyed by the antibiotic and their competitors are destroyed, the survivors may flourish and become opportunistic pathogens. An example that sometimes occurs is overgrowth by the yeastlike fungus *Candida albicans,* which is not sensitive to bacterial antibiotics. This overgrowth is called a **superinfec-** **tion,** a term that is also applied to growth of a target pathogen that has developed resistance to the antibiotic. In this situation, such an antibiotic-resistant strain replaces the original sensitive strain, and the infection continues.

The Action of Antimicrobial Drugs

Learning Objective

- Identify five modes of action of antimicrobial drugs.

Antimicrobial drugs are either **bactericidal** (they kill microbes directly) or **bacteriostatic** (they prevent microbes from growing). In bacteriostasis, the host's own defenses, such as phagocytosis and antibody production, usually destroy the microorganisms. The major modes of action are summarized in Figure 20.2.

The Inhibition of Cell Wall Synthesis

Recall from Chapter 4 that the cell wall of a bacterium consists of a macromolecular network called peptidoglycan. Peptidoglycan is found only in bacterial cell walls. Penicillin and certain other antibiotics prevent the synthesis of intact peptidoglycan; consequently, the cell wall is greatly weakened, and the cell undergoes lysis (Figure 20.3 on page 563). Because penicillin targets the synthesis process, only actively growing cells are affected by these antibiotics. And, because human cells do not have peptidoglycan cell walls, penicillin has very little toxicity for host cells.

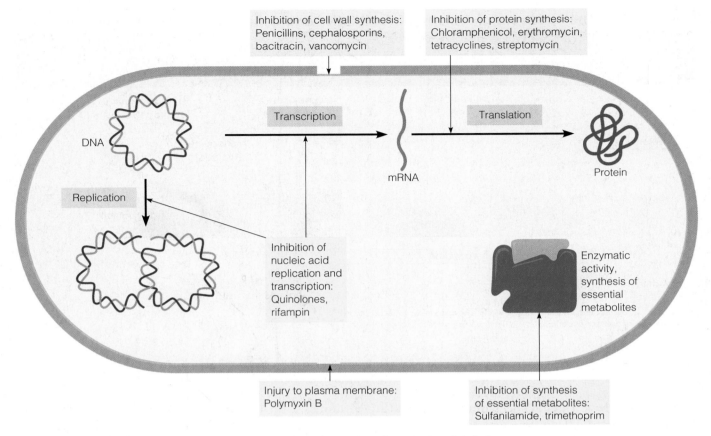

Inhibition of cell wall synthesis:
Penicillins, cephalosporins,
bacitracin, vancomycin

Inhibition of protein synthesis:
Chloramphenicol, erythromycin,
tetracyclines, streptomycin

Transcription

Translation

DNA

mRNA

Protein

Replication

Inhibition of
nucleic acid
replication and
transcription:
Quinolones,
rifampin

Enzymatic
activity,
synthesis of
essential
metabolites

Injury to plasma membrane:
Polymyxin B

Inhibition of synthesis
of essential metabolites:
Sulfanilamide, trimethoprim

FIGURE 20.2 A summary of the major modes of action of antimicrobial drugs.
This illustration shows these actions as they might affect a highly diagrammatic
representation of a bacterial cell.

■ Which modes of action are effective against prokaryotic cells? Against eukaryotic cells?

The Inhibition of Protein Synthesis

Because protein synthesis is a common feature of all cells, whether prokaryotic or eukaryotic, it would seem an unlikely target for selective toxicity. One notable difference between prokaryotes and eukaryotes, however, is the structure of their ribosomes. As discussed in Chapter 4 on page 94, eukaryotic cells have 80S ribosomes; prokaryotic cells have 70S ribosomes. (The 70S ribosome is made up of a 50S and a 30S unit. The S stands for Svedberg unit, which describes the relative rate of sedimentation in a high-speed centrifuge.) The difference in ribosomal structure accounts for the selective toxicity of antibiotics that affect protein synthesis. However, mitochondria (important eukaryotic organelles) also contain 70S ribosomes similar to those of bacteria. Antibiotics targeting the 70S ribosomes can therefore have adverse effects on the cells of the host. Among the antibiotics that interfere with protein synthe-

sis are chloramphenicol, erythromycin, streptomycin, and the tetracyclines (Figure 20.4).

Reacting with the 50S portion of the 70S prokaryotic ribosome, chloramphenicol inhibits the formation of peptide bonds in the growing polypeptide chain. Erythromycin also reacts with the 50S portion of the 70S prokaryotic ribosome. Most drugs that inhibit protein synthesis have a broad spectrum of activity; erythromycin is an exception. Because it does not penetrate the gram-negative cell wall, it affects mostly gram-positive bacteria.

Some other antibiotics react with the 30S portion of the 70S prokaryotic ribosome. The tetracyclines interfere with the attachment of the tRNA carrying the amino acids to the ribosome, preventing the addition of amino acids to the growing polypeptide chain. Tetracyclines do not interfere with mammalian ribosomes because they do not penetrate very well into intact mammalian cells. However, at least small amounts are able to enter the host

(a) SEM | 0.5 μm

(b) SEM | 0.5 μm

FIGURE 20.3 **The inhibition of bacterial cell synthesis by penicillin. (a)** Rod-shaped bacterium before penicillin. **(b)** The bacterial cell is lysing as penicillin weakens the cell wall.

(a) Three-dimensional detail of the protein synthesis site showing the 30S and 50S subunit portions of the 70S prokaryotic ribosome.

Growing peptide chain

Protein synthesis site

Tunnel

50S

5′

30S

mRNA

3′

Growing polypeptide

Chloramphenicol
Binds to 50S portion and inhibits formation of peptide bond

50S portion

Protein sythesis site

Erythromycin
Binds to 50S portion, prevents translocation–movement of ribosome along mRNA

tRNA

Messenger RNA

30S portion

70S prokaryotic ribosome

Translation

Tetracyclines
Interfere with attachment of tRNA to mRNA-ribosome complex

Streptomycin
Changes shape of 30S portion, causes code on mRNA to be read incorrectly

Direction of ribosome movement

(b) In the diagram the black arrows indicate the different points at which chloramphenicol, erythromycin, the tetracyclines, and streptomycin exert their activities.

FIGURE 20.4 **The inhibition of protein synthesis by antibiotics. (a)** The inset shows how the 70S prokaryotic ribosome is assembled from two subunits, 30S and 50S. Note how the growing peptide chain passes through a tunnel in the 50S subunit from the site of protein synthesis. **(b)** In the diagram, the black arrows indicate the different points at which chloramphenicol, erythromycin, the tetracyclines, and streptomycin exert their activities.

■ Why do antibiotics that inhibit protein synthesis affect bacteria and not human cells?

SEM |———| 10 mm

FIGURE 20.5 **Injury to the plasma membrane of a yeast cell caused by an antifungal drug.** The cell releases its cytoplasmic contents as the plasma membrane is disrupted by the antifungal drug miconazole.

■ Many antifungal drugs combine with sterols in the plasma membrane.

cell, as is apparent from the fact that the intracellular pathogenic rickettsias and chlamydias are sensitive to tetracyclines. The selective toxicity of the drug in this case is due to a greater sensitivity of the bacteria at the ribosomal level.

Aminoglycoside antibiotics, such as streptomycin and gentamicin, interfere with the initial steps of protein synthesis by changing the shape of the 30S portion of the 70S prokaryotic ribosome. This interference causes the genetic code on the mRNA to be read incorrectly.

Injury to the Plasma Membrane

Certain antibiotics, especially polypeptide antibiotics, bring about changes in the permeability of the plasma membrane; these changes result in the loss of important metabolites from the microbial cell. For example, polymyxin B causes disruption of the plasma membrane by attaching to the phospholipids of the membrane.

Some antifungal drugs, such as amphotericin B, miconazole, and ketoconazole, are effective against a considerable range of fungal diseases. Such drugs combine with sterols in the fungal plasma membrane to disrupt the membrane (Figure 20.5). Because bacterial plasma membranes generally lack sterols, these antibiotics do not act

on bacteria. However, the plasma membranes of animal cells do contain sterols, and amphotericin B and ketoconazole can be toxic to the host. Fortunately, animal cell membranes have mostly *cholesterol*, and fungal cells have mostly *ergosterol*, against which the drug is most effective, so that the balance of the toxicity is tilted against the fungus.

The Inhibition of Nucleic Acid Synthesis

A number of antibiotics interfere with the processes of DNA replication and transcription in microorganisms. Some drugs with this mode of action have an extremely limited usefulness because they interfere with mammalian DNA and RNA as well. Others, such as rifampin and the quinolones, are more widely used in chemotherapy because they are more selectively toxic.

Inhibiting the Synthesis of Essential Metabolites

In Chapter 5, we mentioned that a particular enzymatic activity of a microorganism can be *competitively inhibited* by a substance (antimetabolite) that closely resembles the normal substrate for the enzyme (see Figure 5.7 on page 118). An example of competitive inhibition is the relationship between the antimetabolite sulfanilamide (a sulfa drug) and **para-aminobenzoic acid (PABA).** In many microorganisms, PABA is the substrate for an enzymatic reaction leading to the synthesis of folic acid, a vitamin that functions as a coenzyme for the synthesis of the purine and pyrimidine bases of nucleic acids and many amino acids. In the presence of sulfanilamide, the enzyme that normally converts PABA to folic acid combines with the drug instead of with PABA. This combination prevents folic acid synthesis and stops the growth of the microorganism. Because humans do not produce folic acid from PABA (they obtain it as a vitamin in ingested foods), sulfanilamide exhibits selective toxicity—it affects microorganisms that synthesize their own folic acid but does not harm the human host. Other chemotherapeutic agents that act as antimetabolites are the sulfones and trimethoprim.

A Survey of Commonly Used Antimicrobial Drugs

Learning Objective
• Explain why the drugs described in this section are specific for bacteria.

Tables 20.3 and 20.4 summarize the commonly used antimicrobial drugs.

TABLE 20.3	**Antibacterial Drugs**
Drugs by Mode of Action	**Comments**

Inhibitors of Cell Wall Synthesis

Natural Penicillins

Penicillin G	Against gram-positive bacteria, requires injection
Penicillin V	Against gram-positive bacteria, oral administration

Semisynthetic Penicillins

Oxacillin	Resistant to penicillinase
Ampicillin	Broad spectrum
Amoxicillin	Broad spectrum; combined with inhibitor of penicillinase
Aztreonam	A monobactam; effective for gram-negative bacteria, including *Pseudomonas* spp.
Imipenem	A carbapenem; very broad spectrum

Cephalosporins

Cephalothin	First-generation cephalosporin; activity similar to penicillin; requires injection
Cefixime	Third-generation cephalosporin; oral administration

Polypeptide Antibiotics

Bacitracin	Against gram-positive bacteria; topical application
Vancomycin	A glycopeptide type; penicillinase resistant; against gram-positive bacteria

Antimycobacterial Antibiotics

Isoniazid	Inhibits synthesis of mycolic acid component of cell wall of *Mycobacterium* spp.
Ethambutol	Inhibits incorporation of mycolic acid into cell wall of *Mycobacterium* spp.

Inhibitors of Protein Synthesis

Chloramphenicol	Broad spectrum, potentially toxic

Aminoglycosides

Streptomycin	Broad spectrum, including mycobacteria
Neomycin	Topical use, broad spectrum
Gentamicin	Broad spectrum, including *Pseudomonas* spp.

Tetracyclines

Tetracycline, oxytetracycline, chlortetracycline	Broad spectrum, including chlamydias and rickettsias; animal feed additives

Macrolides

Erythromycin	Alternative to penicillin

Streptogramins

Quinupristin and dalfopristin (Synercid)	Alternative for treating vancomycin-resistant gram-positive bacteria

Oxazolidinones

Linezolid	Useful primarily against penicillin-resistant gram-positive bacteria

Injury to the Plasma Membrane

Polymyxin B	Topical use, gram-negative bacteria, including *Pseudomonas* spp.

TABLE 20.3	**Antibacterial Drugs** *(continued)*
Drugs by Mode of Action	**Comments**

Inhibitors of Nucleic Acid Synthesis

Rifamycins

Rifampin	Inhibits synthesis of mRNA; treatment of tuberculosis

Quinolones and Fluoroquinolones

Nalidixic acid, nofloxacin, ciprofloxacin	Inhibit DNA synthesis; broad spectrum; urinary tract infections

Competitive Inhibitors of the Synthesis of Essential Metabolites

Sulfonamides

Trimethoprim-sulfamethoxazole	Broad spectrum; combination is widely used

TABLE 20.4	**Antifungal, Antiviral, Antiprotozoan, and Antihelminthic Drugs**	
	Mode of Action	**Comments**

Antifungal Drugs

Agents Affecting Fungal Sterols (Plasma Membrane)

Polyenes

Amphotericin B	Injury to plasma membrane	Systemic fungal infections; fungicidal

Azoles

Antifungal drug	Mode of action	Comments
Clotrimazole, miconazole	Inhibit synthesis of plasma membrane	Topical use
Ketoconazole	Inhibits synthesis of plasma membrane	Can be taken orally for systemic fungal infections
Voriconazole	Inhibits synthesis of plasma membrane	Can penetrate blood-brain barrier to treat aspergillosis of the central nervous system

Allylamines

Terbinafine, naftifine	Inhibits synthesis of plasma membrane	New class of antifungals frequently used to treat diseases resistant to azoles

Agents Affecting Fungal Cell Walls

Echinocandins

Caspofungin (Cancidas)	New class of antifungals that inhibit synthesis of cell wall	

Agents Inhibiting Nucleic Acids

Flucytocine	Inhibits synthesis of RNA and therefore protein synthesis	

Other Antifungal Drugs

Griseofulvin	Inhibition of mitotic microtubules	Fungal infections of the skin
Tolnaftate	Unknown	Athlete's foot

TABLE 20.4	(continued)		
		Mode of Action	**Comments**

Antiviral Drugs

   Nucleoside and Nucleotide Analogs

Acyclovir, ganciclovir, ribavirin, lamivudine	Inhibit DNA or RNA synthesis	Used primarily against herpesviruses
Cidofovir	Inhibits DNA or RNA synthesis	Possibly effective against smallpox
Zidovudine, didanosine, zalcitabine, tenofovir	Antiretrovirals that block DNA synthesis by reverse transcriptase	Used for HIV chemotherapy

   Other Enzyme Inhibitors

Indinavir, saquinavir	Inhibit viral protease	Used for HIV chemotherapy

   Attachment and Uncoating

Zanamivir, oseltamivir phosphate	Inhibit neuraminidase on influenza virus	Treatment of influenza
Amantadine, zimantadine	Inhibit uncoating	Treatment of influenza

   Interferons

α-interferon	Inhibits spread of virus to new cells	Viral hepatitis

Antiprotozoan Drugs

Chloroquine	Inhibits DNA synthesis	Malaria; effective against red blood cell stage only
Diiodohydroxyquin	Unknown	Amoebic infections; amoebicidal
Metronidazole	Damages DNA	Protozoans, *Entamoeba*, *Trichomonas*, anaerobic bacteria; amoebicidal and trichomonicidal

Antihelminthic Drugs

Niclosamide	Prevents ATP generation in mitochondria	Tapeworm infections; kills tapeworms
Praziquantel	Alters permeability of plasma membranes	Tapeworm and fluke infections; kills flatworms
Pyantel pamoate	Neuromuscular block	Intestinal roundworms; kills roundworms
Mebendazole, albendazole	Inhibit absorption of nutrients	Intestinal roundworms
Ivermectin	Paralyzes worm	Intestinal roundworms primarily; occasional use for scabies mite, and lice

Antibacterial Antibiotics: Inhibitors of Cell Wall Synthesis

Learning Objectives

- List the advantages of each of the following over penicillin: semisynthetic penicillins, cephalosporins, and vancomycin.

- Explain why INH and ethambutol are antimycobacterial agents.

Penicillin

The term **penicillin** refers to a group of over 50 chemically related antibiotics (Figure 20.6). All penicillins have a common core structure containing a β-lactam ring called the nucleus. Penicillin molecules are differentiated by the chemical side chains attached to their nuclei. Penicillins can be produced either naturally or semisynthetically.

FIGURE 20.6 **The structure of penicillins, antibacterial antibiotics.** The portion that all penicillins have in common—which contains the β-lactam ring—is shaded in purple. The unshaded portions represent the side chains that distinguish one penicillin from another.

■ Semisynthetic penicillins include the β-lactam ring made by *Penicillium* with various side chains to extend the spectrum of activity or resist penicillinase.

Penicillin G (Requires injection)

Penicillin V (Can be taken orally)

β-lactam ring

(a) Natural penicillins

Oxacillin (Resistant to penicillinase)

β-lactam ring

Ampicillin (Extended spectrum)

(b) Semisynthetic penicillins

FIGURE 20.7 **Retention of penicillin G.** Penicillin G is normally injected (solid red line); when administered by this route, the drug is present in high concentrations in the blood but disappears quickly. Taken orally (dotted red line), penicillin G is destroyed by stomach acids and is not very effective. It is possible to improve retention of penicillin G by combining it with compounds such as procaine and benzathine (blue and black lines). However, the blood concentration reached is low, and the target bacterium must be extremely sensitive to the antibiotic.

Penicillins prevent the cross-linking of the peptidoglycans, which interferes with the final stages of the construction of the cell wall; see Figure 4.13a on page 85.

Natural Penicillins Penicillin extracted from cultures of the mold *Penicillium* exists in several closely related forms. These are the so-called **natural penicillins.** The prototype compound of all the penicillins is *penicillin G*. It has a narrow but useful spectrum of activity and is often the drug of choice against most staphylococci, streptococci, and several spirochetes. When injected intramuscularly, penicillin G is rapidly excreted from the body in 3–6 hours (Figure 20.7). When taken orally, the acidity of the digestive fluids in the stomach diminishes its concentration. *Procaine penicillin*, a combination of the drugs procaine and penicillin G, is retained at detectable concentrations for up to 24 hours; concentration peaks at about 4 hours. Still longer retention times can be achieved with *benzathine penicillin*, a combination of benzathine and penicillin G. Although retention times of as long as 4 months can be obtained, the concentration of the drug is so low that the organisms must be very sensitive to it. Penicillin V, which is stable in stomach acids β-lactam ring

Penicillin → Penicillinase → Penicilloic acid

FIGURE 20.8 **The effect of penicillinase on penicillins.** Bacterial production of this enzyme, which is shown breaking the β-lactam ring, is by far the most common form of resistance to penicillins. R is an abbreviation for the chemical side groups that differentiate similar or otherwise-identical compounds.

and can be taken orally, and penicillin G are the natural penicillins most often used.

Natural penicillins have some disadvantages. Chief among them are their narrow spectrum of activity and their susceptibility to penicillinases. *Penicillinases* are enzymes produced by many bacteria, especially *Staphylococcus* species, that cleave the β-lactam ring of the penicillin molecule (Figure 20.8). Because of this characteristic, penicillinases are sometimes called *β-lactamases*.

Semisynthetic Penicillins A large number of **semisynthetic penicillins** have been developed in attempts to overcome the disadvantages of natural penicillins (see Figure 20.6b). Scientists develop these penicillins in either of two ways. First, they can interrupt synthesis of the molecule by *Penicillium* and obtain only the common penicillin nucleus for use. Second, they can remove the side chains from the completed natural molecules and then chemically add other side chains that make them more resistant to penicillinase, or the scientists can give them an extended spectrum. Thus the term *semisynthetic*: Part of the penicillin is produced by the mold, and part is added synthetically.

Penicillinase-Resistant Penicillins The first semisynthetic penicillin designed to evade the action of penicillinases was *methicillin*. Eventually, so many strains of staphylococci developed resistance to methicillin that the abbreviation MRSA, for methicillin-resistant *Staphylococcus aureus*, made its appearance. Resistance became so prevalent that methicillin has been discontinued in the United States. Replacement antibiotics similar to methicillin, such as *oxacillin* and *nafcillin*, have been developed.

Extended-Spectrum Penicillins To overcome the problem of the narrow spectrum of activity of natural penicillins, broader-spectrum semisynthetic penicillins have been developed. These new penicillins are effective against many gram-negative bacteria as well as gram-positive ones, although they are not resistant to penicillinases. The first such penicillins were the aminopenicillins, such as *ampi-*

cillin and *amoxicillin*. When bacterial resistance to these became more common, the carboxypenicillins were developed. Members of this group, such as *carbenicillin* and *ticarcillin*, have even greater activity against gram-negative bacteria and have the special advantage of activity against *Pseudomonas aeruginosa*. Among the more recent additions to the penicillin family are the ureidopenicillins, such as *mezlocillin* and *azlocillin*. These broader-spectrum penicillins are modifications of the structure of ampicillin. The search for even more effective modifications of penicillin continues.

Penicillins plus β-lactamase Inhibitors A different approach to the proliferation of penicillinase is to combine penicillins with *potassium clavulanate (clavulanic acid)*, a product of a streptomycete. Potassium clavulanate is a noncompetitive inhibitor of penicillinase with essentially no antimicrobial activity of its own. It has been combined with some new broader-spectrum penicillins, such as *amoxicillin* (the combination is called by its trade name Augmentin). The same antibiotics are sold worldwide under numerous different trade names.

Carbapenems The **carbapenems** are a class of β-lactam antibiotics that substitute a carbon atom for a sulfur atom and add a double bond to the penicillin nucleus. These antibiotics, which inhibit cell wall synthesis, have an extremely broad spectrum of activity. Representative of this group is Primaxin, a combination of *imipenem* and *cilastin*. The cilastin has no antimicrobial activity but prevents degradation of the combination in the kidneys. Tests have demonstrated that Primaxin is active against 98% of all organisms isolated from hospital patients.

Monobactams Another method of avoiding the effects of penicillinase is shown by *aztreonam*, which is the first member of a new class of antibiotics. It is a synthetic antibiotic that has only a single ring and is therefore known as a **monobactam,** rather than the conventional β-lactam double ring. Aztreonam's spectrum of activity is remarkable for a penicillin-related compound—this antibiotic, which has unusually low toxicity, affects only certain gram-negative bacteria, including pseudomonads and *E. coli*.

Cephalosporins

In structure, the nuclei of **cephalosporins** resemble those of penicillin (Figure 20.9). Cephalosporins inhibit cell wall synthesis in essentially the same way as do penicillins. However, cephalosporins differ from penicillin in that they are resistant to penicillinases and are effective against more gram-negative organisms than the natural penicillins. However, the cephalosporins are susceptible to a separate group of β-lactamases.

The number of second-, third-, and even fourth-generation cephalosporins has proliferated in recent years; there are now more than 70 versions. Each generation tends to be more effective against gram-negatives and has a broader spectrum of activity than the previous generation. Some first-generation cephalosporins are *cephalothin*, *cefamandole*, and *cefotaxime*.

Oral administration is preferred by patients, and several newer cephalosporins allow this. *Cefpodoxime* and *cefixime* are two cephalosporins that have been approved for oral administration.

Polypeptide Antibiotics

Bacitracin Bacitracin (the name is derived from its source, a *Bacillus* isolated from a wound on a girl named Tracy) is a polypeptide antibiotic effective primarily against gram-positive bacteria, such as staphylococci and streptococci. Bacitracin inhibits the synthesis of cell walls at an earlier stage than penicillins and cephalosporins. It interferes

Cephalosporin nucleus

Penicillin nucleus

FIGURE 20.9 The nuclear structures of cephalosporin and penicillin compared.

■ Why are cephalosporins resistant to penicillinase?

with the synthesis of the linear strands of the peptidoglycans. (See Figure 4.13a on page 84.) Its use is restricted to topical application for superficial infections.

Vancomycin Vancomycin (optimistically named from the word *vanquish*) is one of a small group of glycopeptide antibiotics derived from a species of *Streptomyces* found in the jungles of Borneo. Originally, toxicity of vancomycin was a serious problem, but improved purification procedures in its manufacture have largely corrected this. Although it has a very narrow spectrum of activity, which is based on inhibition of cell wall synthesis, vancomycin has been extremely important in addressing the problem of MRSA. Vancomycin has been considered the last line of antibiotic defense for treatment of *Staphylococcus aureus* infections that are resistant to other antibiotics. The recent appearance of strains of *S. aureus* and other important pathogens, such as certain enterococci (see box on page 581), that are resistant to vancomycin, is considered a medical emergency.

Antimycobacterial Antibiotics

The cell wall of members of the genus *Mycobacterium* differs from the cell wall of most other bacteria. It incorporates mycolic acids that are a factor in their staining properties, causing them to stain as acid-fast (page 69). The genus includes important pathogens, such as those that cause leprosy and tuberculosis.

Isoniazid (INH) is a very effective synthetic antimicrobial drug against *Mycobacterium tuberculosis*. The primary effect of INH is to inhibit the synthesis of mycolic acids, which are components of cell walls only of the mycobacteria. It has little effect on nonmycobacteria. When used to treat tuberculosis, INH is usually administered simultaneously with other drugs, such as rifampin or ethambutol. This minimizes development of drug resistance. Because the tubercle bacillus is usually found only within macrophages or walled off in tissue, any antitubercular drug must be able to penetrate into such sites.

Ethambutol is effective only against mycobacteria. The drug apparently inhibits incorporation of mycolic acid into the cell wall. It is a comparatively weak antitubercular drug; its principal use is as the secondary drug to avoid resistance problems.

Inhibitors of Protein Synthesis

Learning Objective

• *Describe how each of the following inhibits protein synthesis: aminoglycosides, tetracyclines, chloramphenicol, macrolides.*

FIGURE 20.10 **The structure of the antibacterial antibiotic chloramphenicol.** Notice the simple structure, which makes synthesizing this drug less expensive than isolating it from *Streptomyces*.

- What effect does the binding of chloramphenicol to the 50S portion of the ribosomes have on a cell?

Chloramphenicol

Chloramphenicol is a broad-spectrum antibiotic with serious toxicity problems. Because of its relatively simple structure (Figure 20.10), it is less expensive for the pharmaceutical industry to synthesize it chemically than to isolate it from *Streptomyces*. Because it is relatively inexpensive, chloramphenicol is often used where low cost is essential. Its relatively small molecular size promotes its diffusion into areas of the body that are normally inaccessible to many other drugs. However, chloramphenicol has serious adverse effects; most important is the suppression of bone marrow activity. This suppression affects the formation of blood cells. In about 1 in 40,000 users, the drug appears to cause aplastic anemia, a potentially fatal condition; the normal rate for this condition is only about 1 in 500,000 individuals. Physicians are advised not to use the drug for trivial conditions or ones for which suitable alternatives are available.

Aminoglycosides

Aminoglycosides are a group of antibiotics in which amino sugars are linked by glycoside bonds. They were among the first antibiotics to have significant activity against gram-negative bacteria. Probably the best-known aminoglycoside is *streptomycin*, which was discovered in 1944. Streptomycin is still used as an alternative drug in the treatment of tuberculosis, but rapid development of resistance and serious toxic effects have diminished its usefulness.

Aminoglycosides are bactericidal and inhibit protein synthesis. They can affect hearing by causing permanent damage to the auditory nerve, and damage to the kidneys has also been reported. Because of this, their use has been declining. *Neomycin* is present in many nonprescription topical preparations. *Gentamicin*, spelled with an "i" to reflect its source, the filamentous bacterium *Micromonospora*, is especially useful against *Pseudomonas* infections. Pseudomonads are a major problem for persons suffering from cystic fibrosis. The aminoglycoside *tobramycin* is administered in an aerosol to aid in the control of infections that occur in patients with cystic fibrosis.

Tetracycline

FIGURE 20.11 **The structure of the antibacterial antibiotic tetracycline.** Other tetracycline-type antibiotics share the four-cyclic-ring structure of tetracycline and closely resemble it.

- Tetracyclines inhibit protein synthesis.

Tetracyclines

Tetracyclines are a group of closely related broad-spectrum antibiotics produced by *Streptomyces* spp. Tetracyclines inhibit protein synthesis. They not only are effective against gram-positive and gram-negative bacteria, but also penetrate body tissues well and are especially valuable against the intracellular rickettsias and chlamydias. Three of the more commonly used tetracyclines are *oxytetracycline* (Terramycin), *chlortetracycline* (Aureomycin), and tetracycline itself (Figure 20.11). Some semisynthetic tetracyclines, such as *doxycycline* and *minocycline*, are available. They have the advantage of longer retention in the body.

Tetracyclines are used to treat many urinary tract infections, mycoplasmal pneumonia, and chlamydial and rickettsial infections. They are also frequently used as alternative drugs for such diseases as syphilis and gonorrhea. Tetracyclines often suppress the normal intestinal microbiota because of their broad spectrum, causing gastrointestinal upsets and often leading to superinfections, particularly by the fungus *Candida albicans*. They are not advised for children, who might experience a brownish discoloration of the teeth, or for pregnant women, in whom they might cause liver damage. Tetracyclines are among the most common antibiotics added to animal feeds, where their use results in significantly faster weight gain; however, some human health problems can also result (see the box on page 581).

Macrolides

Macrolides are a group of antibiotics named for the presence of a macrocyclic lactone ring. The best-known macrolide in clinical use is *erythromycin* (Figure 20.12). Its mode of action is the inhibition of protein synthesis. However, erythromycin is not able to penetrate the cell walls of most gram-negative bacilli. Its spectrum of activity is therefore similar to that of penicillin G, and it is a frequent alternative drug to penicillin. Because it can be administered orally, an orange-flavored preparation of erythromycin is a frequent penicillin substitute for the treatment

FIGURE 20.12 **The structure of the antibacterial antibiotic erythromycin, a representative macrolide.** All macrolides have the macrocyclic lactone ring shown here.

■ Macrolides inhibit protein synthesis.

of streptococcal and staphylococcal infections in children. Erythromycin is the drug of choice for the treatment of legionellosis, mycoplasmal pneumonia, and several other infections.

Other macrolides now available include *azithromycin* and *clarithromycin*. Compared to erythromycin, they have a broader antimicrobial spectrum and penetrate tissues better. This is especially important in the treatment of conditions caused by intracellular bacteria such as *Chlamydia*, a frequent cause of sexually transmitted infection.

A new generation of semisynthetic macrolides, the **ketolides,** is being developed to cope with increasing resistance to other macrolides. The prototype of this generation is *telithromycin* (Ketek).

Streptogramins

We mentioned previously that the appearance of vancomycin-resistant pathogens constitutes a serious medical problem. One answer may be a unique group of antibiotics, the **streptogramins.** The first of these drugs to be released, Synercid, is a combination of two cyclic peptides, *quinupristin* and *dalfopristin*, which are distantly related to the macrolides. They block protein synthesis by attaching to the 50S portion of the ribosome, as do other antibiotics such as the macrolides and chloramphenicol. Synercid, however, acts at uniquely different points on the ribosome. Dalfopristin blocks an early step in protein synthesis, and quinupristin blocks a later step. The combination causes incomplete peptide chains to be released and is synergistic in its action. Synercid is effective against a broad range of gram-positive bacteria that are resistant to other antibiotics. This makes Synercid especially valuable, even though it is expensive and has a high incidence of adverse side effects.

Oxazolidinones

The oxazolidinones are another new class of antibiotics developed in response to vancomycin resistance. When the FDA approved this class of antibiotic in 2001, it represented the first new class of antibiotics approved in 25 years. Like several other antibiotics that inhibit protein synthesis, oxazolidinone antibiotics act on the ribosome (see Figure 20.4). However, they are unique in their target, binding to the 50S ribosomal subunit close to the point where it interfaces with the 30S subunit. These drugs are totally synthetic, which may make resistance slower to develop. Like vancomycin, they have no usefulness against gram-negative bacteria, but they are active against certain enterococci that are not sensitive to Synercid. One member of this antibiotic group is *linezolid* (Zyvox), used mainly to combat MRSA.

Injury to the Plasma Membrane

Learning Objective

- Compare the mode of action of polymyxin B, bacitracin, and neomycin.

Polymyxin B is a bactericidal antibiotic effective against gram-negative bacteria. For many years, it was one of very few drugs used against infections by gram-negative *Pseudomonas*. The mode of action of polymyxin B is to injure plasma membranes. Polymyxin B is seldom used today except in the topical treatment of superficial infections.

Both *bacitracin* and *polymyxin B* are available in non-prescription antiseptic ointments, in which they are usually combined with *neomycin*, a broad-spectrum aminoglycoside. In a rare exception to the rule, these antibiotics do not require a prescription.

Inhibitors of Nucleic Acid (DNA/RNA) Synthesis

Learning Objective

- Describe how rifamycins and quinolones kill bacteria.

Rifamycins

The best known derivative of the **rifamycin** family of antibiotics is *rifampin*. These drugs are structurally related to the macrolides and inhibit the synthesis of mRNA. By far the most important use of rifampin is against mycobacteria in the treatment of tuberculosis and leprosy. A valuable characteristic of rifampin is its ability to penetrate tissues and reach therapeutic levels in cerebrospinal fluid and abscesses. This characteristic is probably an important factor in its antitubercular activity, because the tuberculosis pathogen is usually located inside tissues or macrophages. An unusual side effect of rifampin is the appearance of orange-red urine, feces, saliva, sweat, and even tears.

Quinolones and Fluoroquinolones

In the early 1960s, the synthetic drug *nalidixic acid* was developed—the first of the **quinolone** group of antimicrobials. It exerted a unique bactericidal effect by selectively inhibiting an enzyme (DNA gyrase) needed for the replication of DNA. Although nalidixic acid found only limited use (its only application being for urinary tract infections), it led to the development in the 1980s of a prolific group of new antibiotics, the **fluoroquinolones.**

The most widely used fluoroquinolones are *norfloxacin* and *ciprofloxacin*. The latter is better known under its trade name of Cipro and gained widespread publicity for its use against anthrax infections. Although they are relatively safe for adults, fluoroquinolones adversely affect the development of cartilage, and their use is limited among children, adolescents, and pregnant women.

Synthetic versions of the fluoroquinolones are also being developed. Two of these are *moaxifloxacin* and *gatifloxacin*. They are broader in spectrum and may be taken orally. Their use is not indicated for children; they also interfere with the activity of certain drugs used to control heart rhythm.

Competitive Inhibitors of the Synthesis of Essential Metabolites

Learning Objective

* Describe how sulfa drugs inhibit microbial growth.

Sulfonamides

As noted earlier, **sulfonamides,** or **sulfa drugs,** were among the first synthetic antimicrobial drugs used to treat microbial diseases. Antibiotics have diminished the importance of sulfa drugs in chemotherapy. But they continue to be used to treat certain urinary tract infections and have other specialized uses, as in the combination drug *silver sulfadiazine*, used to control infections in burn patients. Sulfonamides are bacteriostatic; as mentioned, their action is due to their structural similarity to para-aminobenzoic acid (PABA) (see Figure 5.7 on page 118).

Probably the most widely used sulfa drug today is a combination of *trimethoprim* and *sulfamethoxazole* (TMP-SMZ). This combination is an excellent example of drug **synergism.** When used in combination, only 10% of the concentration is needed, compared to when each drug is used alone. The combination also has a broader spectrum of action and greatly reduces the emergence of resistant strains. (Synergism is discussed more fully later in the chapter; see Figure 20.22.)

Figure 20.13 illustrates how the two drugs interfere with different steps of a metabolic sequence leading to the synthesis of precursors of proteins, DNA, and RNA.

❶ Sulfamethoxazole inhibits the synthesis of dihydrofolic acid from PABA, so that the amount of dihydrofolic acid made is much less than usual.

❷ Then trimethoprim (which is not classified as a sulfa drug) inhibits the conversion of any dihydrofolic acid present into tetrahydrofolic acid. TMP-SMZ has a broad spectrum of activity, except against pseudomonads, and can be taken orally. It is even used for the control of *Pneumocystis* pneumonia, which is caused by a fungus and often occurs as a complication of AIDS. TMP-SMZ is very effective in penetrating the brain and cerebrospinal fluid.

Antifungal Drugs

Learning Objective

* Explain the modes of action of currently used antifungal drugs.

Eukaryotes, such as fungi, use the same mechanisms to synthesize proteins and nucleic acids as higher animals. Therefore it is more difficult to find a point of selective toxicity in eukaryotes than in prokaryotes. Moreover, fungal infections are becoming more frequent because of their role as opportunistic infections in immunosuppressed individuals, especially those with AIDS.

Agents Affecting Fungal Sterols

Many antifungal drugs target the sterols in the plasma membrane. In fungal membranes, the principal sterol is ergosterol; in animal membranes, cholesterol. When the biosynthesis of ergosterol in a fungal membrane is interrupted, the membrane becomes excessively permeable, killing the cell. Inhibition of ergosterol biosynthesis is the basis for the selective toxicity of many antifungals, which include members of the polyene, azole, and allylamine groups.

Polyenes Amphotericin B is the most commonly used member of the antifungal **polyene antibiotics** (Figure 20.14). For many years amphotericin B, produced by *Streptomyces* species of soil bacteria, has been a mainstay of clinical treatment for systemic fungal diseases such as histoplasmosis, coccidioidomycosis, and blastomycosis. The drug's toxicity, particularly to the kidneys, is a strongly limiting factor in these uses. Administering the drug encapsulated in lipids (liposomes) appears to minimize toxicity.

Azoles Some of the most widely used antifungal drugs are represented among the **azole antibiotics.** Before they

FIGURE 20.13 **Actions of the antibacterial synthetics trimethoprim and sulfamethoxazole.** TMP-SMZ works by inhibiting different stages in the synthesis of precursors of DNA, RNA, and proteins. Together the drugs are synergistic. ❶ Sulfamethoxazole, a sulfonamide that is a structural analog of PABA, competitively inhibits the synthesis of dihydrofolic acid from PABA. ❷ Then trimethoprim, a structural analog of a portion of dihydrofolic acid, competitively inhibits the synthesis of tetrahydrofolic acid.

■ In some instances, two drugs act synergistically to increase the effectiveness of each and decrease the potential for resistance.

made their appearance, the only drugs available for systemic fungal infections were amphotericin B and flucytocine (discussed below). The first azoles were **imidazoles,** such as *clotrimazole* and *miconazole* (Figure 20.15), which are now sold without a prescription for topical application for treatment of cutaneous mycoses, such as athlete's foot and vaginal yeast infections. An important addition to this group was *ketoconazole*, which has an unusually broad spectrum of activity among fungi. Ketoconazole, taken orally, is often used as a less toxic alternative to amphotericin B for many systemic fungal infections, although occasional liver damage has been reported. Ketoconazole topical oint-

ments are used to treat dermatomycoses. A promising new broad spectrum antifungal, *voriconazole*, is expected to replace amphotericin B for treatment of many systemic antifungal infections. It has a special advantage in treatment of aspergillosis of the central nervous system because it is able to penetrate the blood-brain barrier.

The use of ketoconazole diminished sharply upon the introduction of the **triazole antibiotics.** Drugs of this group, such as *fluconazole* and *itraconazole*, are less toxic and have other advantages as well. Unlike ketoconazole, they are very water soluble, which makes their administration for systemic infections easier and more effective.

Amphotericin B

FIGURE 20.14 The structure of the antifungal drug amphotericin B, representative of the polyenes.

- Why do polyenes injure fungal plasma membranes and not bacterial membranes?

Allylamines The **allylamines** represent a recently developed class of antifungals that inhibit the biosynthesis of ergosterols in a manner that is functionally distinct. *Terbinafine* and *naftifine*, examples of this group, are frequently used when resistance to azole-type antifungals arises.

Agents Affecting Fungal Cell Walls

The fungal cell wall contains compounds that are unique to these organisms. A primary target for selective toxicity among these compounds is β-glucan. Inhibition of the biosynthesis of this glucan results in an incomplete cell wall and results in lysis of the fungal cell. The first of a new class of antifungal drugs (the first in 40 years) is the **echinocandins,** which inhibit the biosynthesis of glucans. A member of the echinocandin group, *caspofungin (Cancidas)* is now available commercially. This new antifungal agent is expected to become especially valuable for combating systemic *Aspergillus* infections in persons whose immune system is compromised. It is also effective against important fungi such as *Candida* spp. and *Pneumocystis jiroveci*, which causes a pneumonia often seen in AIDS patients.

Agents Inhibiting Nucleic Acids

Flucytocine, an analog of the pyrimidine cytosine, interferes with the biosynthesis of RNA and therefore protein synthesis. The selective toxicity lies in the ability of the fungal cell to convert flucytocine into 5-fluorouracil, which is incorporated into RNA and eventually disrupts protein synthesis. Mammalian cells lack the enzyme to make this conversion of the drug. Flucytocine has a narrow spectrum of activity, and toxicity to the kidneys and bone marrow further limit its use.

Other Antifungal Drugs

Griseofulvin is an antibiotic produced by a species of *Penicillium*. It has the interesting property of being active against superficial dermatophytic fungal infections of the

Miconazole

FIGURE 20.15 The structure of the antifungal drug miconazole, representative of the imidazoles.

- Imidazoles injure fungal plasma membranes.

hair (tinea capitis, or ringworm) and nails, even though its route of administration is oral. The drug apparently binds selectively to the keratin found in the skin, hair follicles, and nails. Its mode of action is primarily to block microtubule assembly, which interferes with mitosis and thereby inhibits fungal reproduction.

Tolnaftate is a common alternative to miconazole as a topical agent for the treatment of athlete's foot. Its mechanism of action is not known. *Undecylenic acid* is a fatty acid that has antifungal activity against athlete's foot, although it is not as effective as tolnaftate or the imidazoles.

Pentamidine isethionate is used in the treatment of *Pneumocystis* pneumonia. The drug's mode of action is unknown, but it appears to bind DNA.

Antiviral Drugs

Learning Objective

- *Explain the modes of action of currently used antiviral drugs.*

In developed parts of the world, it is estimated that at least 60% of infectious illnesses are caused by viruses, and about 15% by bacteria. Every year, at least 90% of the U.S. population suffers from a viral disease. Yet relatively few antiviral drugs have been approved in the United States, and they are effective against only an extremely limited group of diseases. Many of the most recently developed antiviral drugs are directed at HIV. Targets for antiviral drugs are various points in viral reproduction (see Figures 13.16, 13.17, and 13.19), such as attachment to host cell, penetration, uncoating (which is the mechanism of action in the drugs used to reduce symptoms of influenza, *amantadine* and *rimantadine*), DNA or RNA syntheses or reproduction, or assembly of the new virion.

Nucleoside and Nucleotide Analogs

Several important antiviral drugs are analogs of nucleosides and nucleotides (page 47). Among the nucleoside analogs, *acyclovir* is the one more widely used (Figure 20.16). While best known for treating genital herpes, it is

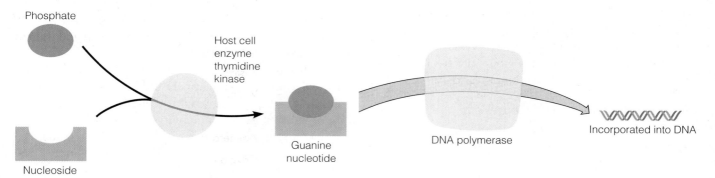

(a) Structural resemblance between acyclovir and guanine-containing nucleoside

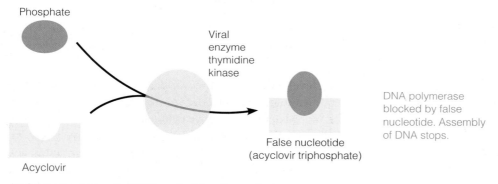

(b) Synthesis of normal viral DNA guanine nucleotide

(c) Synthesis of false viral DNA nucleotide with acyclovir

FIGURE 20.16 **The structure and function of the antiviral drug acyclovir.**
(a) Acyclovir structurally resembles the nucleoside 2'-deoxyguanosine. **(b)** The viral enzyme thymidine kinase combines phosphates with such nucleosides to form nucleotides, which are then incorporated into DNA. **(c)** Because acyclovir is a structural analog of the nucleoside, the enzyme uses it to assemble a false nucleotide that cannot be incorporated into DNA. The selective toxicity of acyclovir is based upon the fact that the phosphorylated form of the drug (acyclovir triphosphate) is a very strong inhibitor of DNA polymerase. The *viral enzyme* thymidine kinase is much more effective in forming acyclovir triphosphate than is the *host-cell* thymidine kinase.

■ Why are viral infections generally difficult to treat with chemotherapeutic agents?

TABLE 20.5	**Drugs for Chemotherapy of HIV**
Generic Name	**Alternative or Brand Name**
Nucleoside Reverse Transcriptase Inhibitors	
Abacavir	ABC, Ziagen
Didanosine	ddI, Videx
Lamivudine	3TC, Epivir
Stavudine	d4T, Zerit
Zalcitabine	ddC, Hivid
Zidovudine	AZT, ZDV, Retrovir
Nucleotide Reverse Transcriptase Inhibitor	
Tenofovir	Viread
Non-Nucleoside Reverse Transcriptase Inhibitors	
Delavirdine	Rescriptor
Elvirenz	Sustiva
Nevirapine	Viramune
Protease Inhibitors	
Amprenavir	Agenerase
Indinavir	Crixivan
Nelfinavir	Viracept
Ritonavir	Norvir
Saquinavir	Invirase, Fortovase
Lopinavir* plus ritonavir	Kaletra

*Available only as this combination.

generally useful for most herpesvirus infections, especially in immunosuppressed individuals. The antiviral drugs *famciclovir*, which can be taken orally, and *ganciclovir* are derivatives of acyclovir and have a similar mode of action. *Ribavirin* resembles the nucleoside guanine and accelerates the already high mutation rate of RNA viruses until the accumulation of errors reaches a crisis point, killing the virus. The nucleoside analog, *lamivudine*, is used to treat hepatitis B. Another nucleoside analog, *cidofovir*, is presently used for treatment of cytomegalovirus infections of the eye, but this drug is especially interesting because it shows promise as a possible treatment of smallpox.

Antiretrovirals The need for effective chemotherapy of HIV infections has led to the development of a relatively numerous group of antiviral drugs. HIV is an RNA virus, and its reproduction depends on the enzyme reverse transcriptase, controlling the synthesis of RNA from DNA.

Analogs of nucleotides or nucleosides are often a basis for drugs to block this essential step. In fact, the term **antiretroviral** currently implies that a drug is used to treat HIV infection. A well-known example of a nucleoside analog is *zidovudine*. Currently, the only nucleotide reverse transcriptase inhibitor is *tenofovir*. Only recently approved, it is often used when other regimens have failed. (Table 20.5 summarizes the current drugs available for HIV chemotherapy.)

Not all drugs that inhibit reverse transcriptase are nucleoside or nucleotide analogs. For example, several non-nucleoside agents, such as *nevirapine* (see Table 20.5), block RNA synthesis by other mechanisms.

Other Enzyme Inhibitors

Two inhibitors of the enzyme neuraminidase have recently been introduced for treatment of influenza. These are *zanamivir* (Relenza) and *oseltamivir phosphate* (Tamiflu).

Another approach to controlling HIV infections is to inhibit the enzymes that control the last stage of viral reproduction. When the host cell, at the direction of the infecting HIV, makes a new virus, it must begin by cutting up large proteins with protease enzymes. These fragments are used to assemble new viruses. The protease inhibitors are analogs of amino acid sequences in the large proteins and competitively interfere with the action of the proteases. The protease inhibitors *indinavir* and *saquinavir* (Table 20.5) have proved especially effective when combined with inhibitors of reverse transcriptase.

Interferons

Cells infected by a virus often produce interferon, which inhibits further spread of the infection. Interferons are classified as cytokines, discussed in Chapter 17. *Alpha-interferon* (see Chapter 16 on page 476) is currently the drug of choice for viral hepatitis infections. The production of interferons can be stimulated by a recently introduced antiviral, **imiquimod.** This drug is often prescribed to treat genital warts caused by a herpesvirus.

Antiprotozoan and Antihelminthic Drugs

Learning Objective

- *Explain the modes of action of currently used antiprotozoan and antihelminthic drugs.*

For hundreds of years, quinine from the Peruvian cinchona tree was the only drug known to be effective for the treatment of a parasitic infection (malaria). It was first introduced into Europe in the early 1600s and was known as "Jesuit's powder." There are now many antiprotozoan and antihelminthic drugs, although many of them are still considered experimental. This does not preclude their use, however, by qualified physicians. The CDC provides several of them on request when they are not available commercially.

Antiprotozoan Drugs

Quinine is still used to control the protozoan disease malaria, but synthetic derivatives, such as *chloroquine*, have largely replaced it. For preventing malaria in areas where the disease has developed resistance to chloroquine, the new drug *mefloquine* is recommended, although serious psychiatric side effects have been reported. *Quinacrine* is the drug of choice for treating the protozoan disease giardiasis. *Diiodohydroxyquin (iodoquinol)* is an important drug prescribed for several intestinal amoebic diseases, but its dosage must be carefully controlled to avoid optic nerve damage. Its mode of action is unknown.

Metronidazole (Flagyl) is one of the most widely used antiprotozoan drugs. It is unique in that it acts not only against parasitic protozoa but also against obligately anaerobic bacteria. For example, as an antiprotozoan agent, it is the drug of choice for vaginitis caused by *Trichomonas vaginalis*. It is also used in the treatment of giardiasis and amoebic dysentery. The mode of action is to interfere with anaerobic metabolism, which incidentally these protozoans share with certain obligately anaerobic bacteria, such as *Clostridium*.

Antihelminthic Drugs

With the increased popularity of sushi, a Japanese specialty often made with raw fish, the CDC began to notice an increased incidence of tapeworm infections. To estimate the incidence, the CDC documents requests for *niclosamide*, which is the usual first choice in treatment. The drug is effective because it inhibits ATP production under aerobic conditions. *Praziquantel* is about equally effective for the treatment of tapeworms; it kills worms by altering the permeability of their plasma membranes. Praziquantel has a broad spectrum of activity and is highly recommended for treating several fluke-caused diseases, especially schistosomiasis. It causes the helminths to undergo muscular spasms and apparently makes them susceptible to attack by the immune system. Apparently, its action exposes surface antigens, which antibodies can then reach.

Mebendazole and *albendazole* are broad-spectrum antihelminthics that have few side effects and have become the drugs of choice for treatment of many intestinal helminthic infections. The mode of action of both drugs is to inhibit the formation of microtubules in the cytoplasm, which interferes with the absorption of nutrients by the parasite. These drugs are also widely used in the livestock industry; for veterinary applications they are relatively more effective in ruminant animals.

Ivermectin is a drug that is used primarily in the livestock industry as an antihelminthic. Its exact mode of action is uncertain, but the final result is paralysis of the worm. This drug has also been occasionally used for treatment of several human parasitic diseases and even for treatment of scabies (a skin mite) and head lice.

Tests to Guide Chemotherapy

Learning Objective

- *Describe two tests for microbial susceptibility to chemotherapeutic agents.*

Different microbial species and strains have different degrees of susceptibility to different chemotherapeutic agents. Moreover, the susceptibility of a microorganism can change with time, even during therapy with a specific drug. Thus, a physician must know the sensitivities of the

FIGURE 20.17 **The disk-diffusion method for determining the activity of antimicrobials.** Each disk contains a different chemotherapeutic agent, which diffuses into the surrounding agar. The clear zones indicate inhibition of growth of the microorganism swabbed onto the plate surface.

■ Which agent is the most effective against the bacterium being tested?

FIGURE 20.18 **The E test (for epsilometer), a gradient diffusion method that determines antibiotic sensitivity and estimates minimal inhibitory concentration (MIC).** The plastic strip, which is placed on an agar surface inoculated with test bacteria, contains an increasing gradient of the antibiotic. The MIC is clearly shown.

■ What other factors must be considered before treatment is prescribed?

pathogen before treatment can be started. However, physicians often cannot wait for sensitivity tests and must begin treatment based on their "best guess" estimation of the most likely pathogen causing the illness.

Several tests can be used to indicate which chemotherapeutic agent is most likely to combat a specific pathogen. However, if the organisms have been identified—for example, *Pseudomonas aeruginosa*, beta-hemolytic streptococci, or gonococci—certain drugs can be selected without specific testing for susceptibility. Tests are necessary only when susceptibility is not predictable, or when antibiotic resistance problems develop.

The Diffusion Methods

Probably the most widely used, although not necessarily the best, method of testing is the **disk-diffusion method,** also known as the *Kirby–Bauer test* (Figure 20.17). A Petri plate containing an agar medium is inoculated ("seeded") uniformly over its entire surface with a standardized amount of a test organism. Next, filter paper disks impregnated with known concentrations of chemotherapeutic agents are placed on the solidified agar surface. During incubation, the chemotherapeutic agents diffuse from the disks into the agar. The farther the agent diffuses from the disk, the lower its concentration. If the

chemotherapeutic agent is effective, a **zone of inhibition** forms around the disk after a standardized incubation. The diameter of the zone can be measured; in general, the larger the zone, the more sensitive the microbe is to the antibiotic. The zone diameter is compared to a standard table for that drug and concentration, and the organism is reported as *sensitive, intermediate,* or *resistant.* For a drug with poor solubility, however, the zone of inhibition indicating that the microbe is sensitive will usually be smaller than for another drug that is more soluble and has diffused more widely. Results obtained by the disk-diffusion method are often inadequate for many clinical purposes. However, the test is simple and inexpensive and is most often used when more sophisticated laboratory facilities are not available.

A more advanced diffusion method, the **E test,** enables a lab technician to estimate the **minimal inhibitory concentration (MIC),** the lowest antibiotic concentration that prevents visible bacterial growth. A plastic-coated strip contains a gradient of antibiotic concentrations, and the MIC can be read from a scale printed on the strip (Figure 20.18).

Broth Dilution Tests

A weakness of the diffusion method is that it does not determine whether a drug is bactericidal and not just bacteriostatic. A **broth dilution test** is often useful in determining the MIC and the **minimal bactericidal concentration**

Doxycycline
(Growth in all wells, resistant)

Sulfamethoxazole
(Trailing end point; usually read where there
is an estimated 80% reduction in growth)

Streptomycin
(No growth in any well; sensitive at all
concentrations)

Ethambutol
(Growth in second wells;
equally sensitive to
ethambutol and kanamycin)

Kanamycin

Decreasing concentration of drug ⟶

FIGURE 20.19 A microdilution, or microtiter, plate used for testing for minimal inhibitory concentration (MIC) of antibiotics. Such plates contain as many as 96 shallow wells that contain measured concentrations of antibiotics. They are usually purchased frozen or freeze dried (page 169). The test microbe is added simultaneously, with a special dispenser, to all the wells in a row of test antibiotics. A button of growth appears if the antibiotic has no effect on the microbe; the microbe is recorded as not sensitive. If there is no growth in a well, the microbe is sensitive to the antibiotic at that concentration. To ensure that the microbe is capable of growth in the absence of the antibiotic, wells that contain no antibiotic are also inoculated (positive control). To ensure against contamination by unwanted microbes, wells that contain nutrient broth but no antibiotics or inoculum are also included (negative control).

■ The MIC is the lowest concentration that inhibits microbial growth.

(MBC) of an antimicrobial drug. The MIC is determined by making a sequence of decreasing concentrations of the drug in a broth, which is then inoculated with the test bacteria (Figure 20.19). The wells that do not show growth (higher concentration than the MIC) can be cultured in broth free of the drug. If growth occurs in this broth, the drug was not bactericidal, and the MBC can be determined. Determining the MIC and MBC is important because it avoids the excessive or erroneous use of expensive antibiotics and minimizes the chance of toxic reactions that larger-than-necessary doses might cause.

Dilution tests are often highly automated. The drugs are purchased already diluted into broth in wells formed in a plastic tray. A suspension of the test organism is prepared and inoculated into all the wells simultaneously by a special inoculating device. After incubation, the turbidity may be read visually, although clinical laboratories with high workloads may read the trays with special scanners that enter the data into a computer that provides a printout of the MIC.

Other tests are also useful for the clinician; a determination of the microbe's ability to produce β-lactamase is one example. One popular, rapid method makes use of a cephalosporin that changes color when its β-lactam ring is opened. In addition, a measurement of the *serum concentration* of an antimicrobial is especially important when toxic drugs are used. These assays tend to vary with the drug and may not always be suitable for smaller laboratories.

The Effectiveness of Chemotherapeutic Agents

Drug Resistance

Learning Objective

• *Describe the mechanisms of drug resistance.*

Bacteria become resistant to chemotherapeutic agents by four major mechanisms:

Antibiotics in Animal Feed Linked to Human Disease

Over 40 years ago, livestock growers began using antibiotics in the feed of closely penned animals that were being fattened for market. The drugs helped reduce the number of bacterial infections and controlled their spread in such ripe conditions. They also unexpectedly accelerated the animals' growth, an effect that led to wider and wider use. The growth-promoting effects are thought to be due to suppression of intestinal bacteria, such as *Clostridium* spp., that produce excessive gas and toxins, which may retard the animals' growth. Today, more than half the antibiotics used worldwide are given to farm animals to promote weight gain and treat infections.

Meat and milk that reach the consumer's table are not heavily laden with antibiotics because the U.S. Food and Drug Administration (FDA) has established limits for antibiotic residues in edible tissues. But the constant presence of antibiotics in these animals puts selective pressure on their normal microbiota, as well as on their pathogens, to develop drug resistance.

Salmonella

Salmonella can be transferred from animals to humans in meat or milk. The first known case of salmonellosis occurred in Germany in 1888, when 50 people became ill after eating ground beef from one cow. The use of antibiotics in animal feed preferentially allows the growth of strains of bacteria that are resistant to drugs commonly used to treat human infections. In the 1980s, researchers proved that antibiotic-resistant bacteria are being transferred from animals to humans directly in meat or milk. In 1984, Scott Holmberg and his associates at the CDC traced an antibiotic-resistant *Salmonella newport* infection from South Dakota beef cattle to 18 people in four states. Antibiotic-resistant *S. typhimurium* originating from

one Illinois dairy infected 16,000 people in seven states in 1985.

In 1985, nearly 1000 cases of infection by *S. newport* resistant to multiple antibiotics were reported in Los Angeles County, California. A radioactive DNA probe was used to identify an identical plasmid carrying multiple antibiotic resistance genes in 99% of the *S. newport* isolates. A plasmid profile analysis was used to track the bacterium from the sick people back to a slaughterhouse, to a meat-deboning plant, and finally to three farms.

Enterococcus

Vancomycin-resistant *Enterococcus* spp. (VRE) were first isolated in France in 1986, and were found in the United States in 1989. Vancomycin and another glycopeptide, avoparcin, are widely used in animal feed in Europe. Veterinary researchers in Denmark cultured vancomycin-resistant enterococci from horses, pigs, and poultry in the eight European countries that use avoparcin in animal feed. They did not find VRE in livestock in Sweden or the United States, which do not use avoparcin. Their conclusion is that use of avoparcin has created a reservoir for VRE in food animals. VRE are frequently present in food produced in European countries, and human VRE carriers are not uncommon.

In 1996, veterinary use of avoparcin was banned in Germany. After the ban, VRE-positive samples decreased from 100% to 25%, and the human carrier rate dropped from 12% to 3%. In 1997, all European Union countries banned the use of avoparcin.

Campylobacter

The FDA is expected to ban the use of fluoroquinolones in poultry because of the rise in human infections caused by fluoroquinolone-resistant *Campylobacter*. In 2002, several U.S. poultry farms stopped using fluoroquinolones.

Alternatives to Antibiotics

The U.S. Department of Agriculture (USDA) estimates that bacteria account for 3.6–7.1 million foodborne illnesses each year. Among the leading causes are *Salmonella, Campylobacter*, and *Escherichia coli* O157:H7, which colonize the intestines of farm animals and subsequently contaminate meat products during processing. *E. coli* O157:H7 colonize most cattle herds at one time or another, and 25% of raw chickens carry *Salmonella*. A variety of approaches may be necessary to reduce the possibility of illness: (1) prevent colonization in the animals at the farm, (2) reduce fecal contamination of meat during processing at the slaughterhouse, and (3) use proper storage and cooking methods.

Competitive exclusion (CE), the use of bacteria to prevent the colonization of undesirable microorganisms, is being investigated to eliminate undesirable bacteria from animals without using antibiotics. Desirable bacteria can be included with the animals' feed. When ingested, these bacteria will colonize the digestive system and compete for nutrients, compete for attachment sites, and produce antibacterial compounds. Current research has shown that a mixture of three strains of bacteria can inhibit *Campylobacter* from colonizing in poultry.

Researchers at the University of Wisconsin are investigating the use of an antibody to block a hormone involved in digestion. Chickens gained weight when the hormone was blocked because food was retained in the intestine longer and therefore more nutrients were absorbed.

The antibiotic virginiamycin has long been used in livestock. Similar to Synercid, it was recently licensed for human use. Perhaps use of virginiamycin should be discontinued because its use might lead to Synercid resistance.

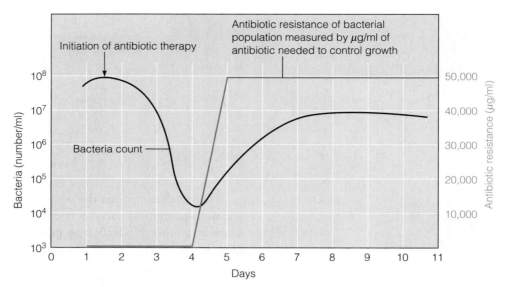

FIGURE 20.20 The development of an antibiotic-resistant mutant during antibiotic therapy. The patient, suffering from a chronic kidney infection caused by a gram-negative bacterium, was treated with streptomycin. The red line records the antibiotic resistance of the bacterial population. Until about the 4th day, essentially all of the bacterial population is sensitive to the antibiotic. At this time, resistant mutants that require 50,000 μg/ml of antibiotic (a very high amount) to control them appear, and their numbers increase rapidly. The black line records the bacterial population in the patient. After antibiotic therapy is begun, the population declines until the 4th day. At this time, mutants in the population that are resistant to streptomycin appear. The bacterial population in the patient rises as these resistant mutants replace the sensitive population.

SOURCE: From *Biology of Microorganisms,* 7th Ed. by T. D. Brock, M.T. Madigan, J. M. Martinko, and J. Parker, p. 410 © 1993. Adapted by permission of Prentice-Hall Inc., Upper Saddle River, New Jersey.

■ Sensitive bacteria are killed, and resistant mutants are able to grow in the presence of an antibiotic.

1. Destruction or inactivation of the drug (by β-lactamase, for example)

2. Prevention of penetration to the target site within the microbe (a frequent mechanism for tetracycline resistance)

3. Alteration of the drug's target sites (for example, a single amino acid change in the ribosome may be enough to make the microbe resistant to certain macrolide antibiotics)

4. Rapid efflux (ejection), which pumps the drug out of the cell before it can become effective.

Variations on these mechanisms also occur. For example, a microbe could become resistant to trimethoprim by synthesizing very large amounts of the enzyme against which the drug is targeted. Conversely, polyene antibiotics can become less effective when resistant organisms produce smaller amounts of the sterols against which the drug is effective. Of particular concern is the possibility that such *resistant mutants* will increasingly replace the susceptible normal populations (see the box on page 581).

Figure 20.20 shows how rapidly bacterial numbers increase during infection as resistance develops.

Hereditary drug resistance is often carried by plasmids, or by small segments of DNA called transposons that can jump from one piece of DNA to another (Chapter 8 on page 240). Some plasmids, including those called resistance (R) factors, can be transferred between bacterial cells in a population and between different but closely related bacterial populations (see Figure 8.27 on page 239). R factors often contain genes for resistance to several antibiotics.

Antibiotics have been a much misused product, nowhere more so than in the less developed areas of the world. Well-trained personnel are scarce, especially in rural areas, which is perhaps one reason why antibiotics can almost universally be purchased without prescription in these countries. A survey in rural Bangladesh, for example, showed that only 8% of antibiotics had been prescribed by a physician. In much of the world, antibiotics are sold to treat headaches and for other inappropriate uses (Figure 20.21). Even when the use of antibiotics is appropriate, dose regimens are usually shorter than needed to eradicate the infection. This encourages the

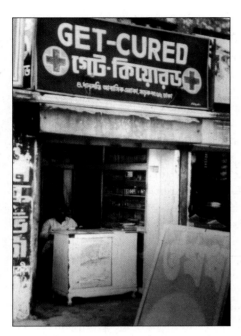

FIGURE 20.21 **Antibiotics are sold without prescriptions in much of the world.** This frequently leads to development of resistant strains of pathogenic bacteria.

survival of resistant strains of bacteria. Outdated and adulterated, and therefore weakened, antibiotics are also common. These problems have no easy solutions. Poverty is endemic and is complicated by political turmoil that disrupts already inadequate medical services.

The developed world is also contributing to the rise of antibiotic resistance. The CDC estimates that in the United States, 30% of antibiotic prescriptions for ear infections, 100% of prescriptions for the common cold, and 50% of prescriptions for sore throats were unnecessary. At least half of the antibiotics produced in the United States are used to promote animal growth, a practice many people feel should be curtailed. Patients also contribute to survival of antibiotic resistant microbes when they fail to finish the full regimen of the prescription or use leftover antibiotics prescribed for someone else.

Strains of bacteria that are resistant to antibiotics are particularly common among people who work in hospitals, where antibiotics are in constant use. Drugs should not be used indiscriminately, and the concentrations administered should always be of optimum strength to minimize the chances of survival of resistant mutants. Many hospitals have special monitoring committees to review the use of antibiotics for effectiveness and cost.

Another way to minimize the development of resistant strains is to administer two or more drugs simultaneously. If a strain is resistant to one of the drugs, the other might be effective against it. The probability that the organism will acquire resistance to both drugs is roughly the product of the two individual probabilities. However, multidrug-resistant tuberculosis is an emerging problem (see the box in Chapter 15, page 440). Bacteria that drugs are unable to control are the so-called *superbugs*.

It is hoped that if antibiotic use is controlled, the relative numbers of susceptible bacteria will rise. It is believed that they will reproduce in greater numbers than will the resistant bacteria because they do not have to expend energy to maintain the resistance genes.

There is concern that the use of antimicrobials (which are not antibiotics) such as triclosan (see page 195) in soaps and other products may promote the survival and growth of plasmid-carrying bacteria. These bacteria may also carry plasmids for antibiotic resistance.

Antibiotic Safety

In our discussions of antibiotics, we have occasionally mentioned side effects. These may be potentially serious, such as liver or kidney damage or hearing impairment. Administration of almost any drug involves an assessment of risks against benefits; this is called the *therapeutic index*. Sometimes, the use of another drug can cause toxic effects that do not occur when the drug is taken alone. One drug may also neutralize the intended effects of the other. For example, a few antibiotics have been reported to neutralize the effectiveness of contraceptive pills. Also, some individuals may have hypersensitivity reactions, for example, to penicillins.

A pregnant woman should take only those antibiotics that are classified by the U.S. Food and Drug Administration as presenting no evidence of risk to the fetus.

Effects of Combinations of Drugs

Learning Objective
• Compare and contrast synergism and antagonism.

The chemotherapeutic effect of two drugs given simultaneously is sometimes greater than the effect of either given alone (Figure 20.22). This phenomenon, called synergism, was introduced earlier. For example, in the treatment of bacterial endocarditis, penicillin and streptomycin are much more effective when taken together than when either drug is taken alone. Damage to bacterial cell walls by penicillin makes it easier for streptomycin to enter.

Other combinations of drugs can show **antagonism.** For example, the simultaneous use of penicillin and tetracycline is often less effective than when either drug is used alone. By stopping the growth of the bacteria, the bacteriostatic drug tetracycline interferes with the action of penicillin, which requires bacterial growth.

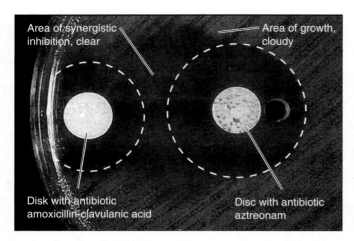

FIGURE 20.22 **An example of synergism between two different antibiotics.** The photograph shows the surface of a Petri plate seeded with bacteria. The paper disc at the left contains the antibiotic amoxicillin plus clavulanic acid. The disc on the right contains the antibiotic aztreonam. The dashed circles drawn over the photo show the clear areas surrounding each disc where bacterial growth would have been inhibited if there had been no synergy. The additional clear area between these two areas and outside the drawn circles, illustrates inhibition of bacterial growth through the effects of synergy.

The Future of Chemotherapeutic Agents

Learning Objective

- Identify three areas of research on new chemotherapeutic agents.

Antibiotics are clearly one of the greatest triumphs of medical science, but they rely upon a limited range of targets (see Figure 20.2). Microbes are now developing resistance to these modes of action. Unfortunately, the problem of antibiotic resistance has now reached a crisis point. Dealing with this crisis will require both intensified educational campaigns to promote the wise use of existing antibiotics and redoubled efforts toward the design of new ones.

A first step will be to modify existing drugs to extend their spectrum and prevent their destruction by bacterial enzymes of resistance. This is the approach that has been used in the penicillin group. Research is under way to prevent the resistance strategy of rapid antibiotic efflux. Investigators are also working to identify new targets for antimicrobial activity other than those summarized in Figure 20.2.

Antimicrobial Peptides

Most of our present antibiotics are the products of other microorganisms. Interest is now also focusing on antibiotics produced by plants and animals that often exhibit extraordinary resistance to microbial infections. Multicellular animals, plants, and insects are able to survive successfully in a world full of potentially harmful microbes. Plant seeds survive and reproduce successfully in a soil seething with bacteria and fungi. Insects do not have the animal-type defenses that include phagocytes and antibodies, nor do octopuses or starfish (sea stars). What plants, insects, and many animals do have are powerful broad-spectrum **antimicrobial peptides** (sometimes called *cationic peptides*). More than 300 such peptides have been identified. Their mechanism of action is not completely clear, but they seem to have targeted some feature in the plasma membrane of many microbes that distinguishes them from multicellular plants and animals. The peptides form disruptive channels in the plasma membrane, eventually killing the microbe. Resistant mutants do not seem to have appeared in nature; apparently, this would require a complete and excessively difficult redesign of the membrane.

Examples of antimicrobial peptides include *magainin* (from Hebrew for shield), which is found in the skin of certain frogs; *squalamine*, an easily synthesized peptide found in the tissue of the spiny dogfish shark; and *protegrins*, isolated from pigs. Several such peptides are in advanced clinical trials.

Apparently, human skin and epithelial cells of the tongue and trachea also are sources of antimicrobial peptides. *Nisin*, which we have mentioned for its use as a food preservative, resembles magainin in its mode of action on membranes. It is the subject of considerable research to see whether nisin derivatives might replace vancomycin for use against gram-positive pathogens.

Insects lack a mammalian-type immune system and produce an array of antimicrobial peptides to defend against infecting bacteria and fungi. One such compound is *cecropin*, which is produced by certain moths.

Antisense Agents

Another promising new approach is the use of short synthetic strands of DNA, called **antisense agents.** (Because antisense agents involve nucleic acids, persons investigating them whimsically call them "nubiotics.") The principle is to identify sites on the DNA or RNA of the pathogen that are responsible for the pathogenic effects. Segments of DNA are then synthesized that will selectively recognize and bind to the target site, a process that blocks the biosynthesis of the target protein. This approach has a great advantage: It prevents the production of a pathogenic protein, rather than trying to selectively neutralize it once it is made. An antiviral based upon antisense principles, *fomivirsen*, has been approved for treatment of the eye disease cytomegalovirus retinitis.

A peculiar modification of this idea is to synthesize a molecule in which the nucleic acid strands have a backbone that differs from the normal alternating sugar and phosphate units (Figure 2.16). The molecule now has its nucleic acid units attached to a strand of modified amino acids resembling a protein, called an antisense **peptide nucleic acid (PNA).** This unnatural chemical is stable to the action of enzymes that might confer resistance to a microbe. The main problem encountered has been poor penetration into microbial cells.

Most of these techniques are still in the investigative stages. Their primary potential targets are viral infections and cancer. There is special interest in such antiviral drugs, as well as in antifungal and antiparasitic drugs, because our arsenal in these categories is especially limited.

The past few chapters have shown how profoundly science has changed the effects of infectious disease on human mortality and life span. At the turn of the twentieth century, the most common causes of death were infectious diseases. Most of these, including tuberculosis, typhoid fever, and diphtheria, were caused by bacteria. At the beginning of the twenty-first century, viral diseases such as influenza, viral pneumonias, and AIDS, are the only infectious diseases among the ten leading causes of death in the United States. These facts bear testimony to the effectiveness of sanitation, vaccines, and (as discussed in this chapter) the discovery and use of antibiotics.

* * *

STUDY OUTLINE

Introduction (p. 559)

1. An antimicrobial drug is a chemical substance that destroys pathogenic microorganisms with minimal damage to host tissues.

2. Chemotherapeutic agents include chemicals that combat disease in the body.

The History of Chemotherapy (pp. 559–560)

1. Paul Ehrlich developed the concept of chemotherapy to treat microbial diseases; he predicted the development of chemotherapeutic agents, which would kill pathogens without harming the host.

2. Sulfa drugs came into prominence in the late 1930s.

3. Alexander Fleming discovered the first antibiotic, penicillin, in 1929; its first clinical trials were done in 1940.

The Spectrum of Antimicrobial Activity (pp. 560–561)

1. Antibacterial drugs affect many targets in a prokaryotic cell.

2. Fungal, protozoan, and helminthic infections are more difficult to treat because these organisms have eukaryotic cells.

3. Narrow-spectrum drugs affect only a select group of microbes—gram-positive cells, for example; broad-spectrum drugs affect a large number of microbes.

4. Small, hydrophilic drugs can affect gram-negative cells.

5. Antimicrobial agents should not cause excessive harm to normal microbiota.

6. Superinfections occur when a pathogen develops resistance to the drug being used or when normally resistant microbiota multiply excessively.

The Action of Antimicrobial Drugs (pp. 561–564)

1. General action is either by directly killing microorganisms (bactericidal) or by inhibiting their growth (bacteriostatic).

2. Some agents, such as penicillin, inhibit cell wall synthesis in bacteria.

3. Other agents, such as chloramphenicol, erythromycin, tetracyclines, and streptomycin, inhibit protein synthesis by acting on 70S ribosomes.

4. Agents such as polymyxin B cause injury to plasma membranes.

5. Rifampin and the quinolones inhibit nucleic acid synthesis.

6. Agents such as sulfanilamide act as antimetabolites by competitively inhibiting enzyme activity.

A Survey of Commonly Used Antimicrobial Drugs (pp. 564–578)

Antibacterial Antibiotics: Inhibitors of Cell Wall Synthesis (pp. 564–570)

1. All penicillins contain a β-lactam ring.

2. Natural penicillins produced by *Penicillium* are effective against gram-positive cocci and spirochetes.

3. Penicillinases (β-lactamases) are bacterial enzymes that destroy natural penicillins.

4. Semisynthetic penicillins are made in the laboratory by adding different side chains onto the β-lactam ring made by the fungus.

5. Semisynthetic penicillins are resistant to penicillinases and have a broader spectrum of activity than natural penicillins.

6. Carbapenems are broad-spectrum antibiotics that inhibit cell wall synthesis.

7. The monobactam aztreonam affects only gram-negative bacteria.

8. Cephalosporins inhibit cell wall synthesis and are used against penicillin-resistant strains.

9. Polypeptides such as bacitracin inhibit cell wall synthesis primarily in gram-positive bacteria.

10. Vancomycin inhibits cell wall synthesis and may be used to kill penicillinase-producing staphylococci. Streptogramins are bactericidal agents that inhibit protein synthesis and may be used to kill vancomycin-resistant bacteria.

11. Isoniazid (INH) inhibits mycolic acid synthesis in mycobacteria. INH is administered with rifampin or ethambutol to treat tuberculosis.

12. The antimetabolite ethambutol is used with other drugs to treat tuberculosis.

Inhibitors of Protein Synthesis (pp. 570–572)

1. Chloramphenicol, aminoglycosides, tetracyclines, macrolides, and streptogramins inhibit protein synthesis at 70S ribosomes.

2. Oxazolidinones prevent formation of 70S ribosomes.

Injury to the Plasma Membrane (p. 572)

1. Polymyxin B and bacitracin cause damage to plasma membranes.

Inhibitors of Nucleic Acid (DNA/RNA) Synthesis (pp. 572–573)

1. Rifamycin inhibits mRNA synthesis; it is used to treat tuberculosis.

2. Quinolones and fluoroquinolones inhibit DNA gyrase for treatment of urinary tract infections.

Competitive Inhibitors of the Synthesis of Essential Metabolites (p. 573)

1. Sulfonamides competitively inhibit folic acid synthesis.

2. TMP-SMZ competitively inhibits dihydrofolic acid synthesis.

Antifungal Drugs (pp. 573–575)

1. Polyenes, such as nystatin and amphotericin B, combine with plasma membrane sterols and are fungicidal.

2. Azoles and allylamines interfere with sterol synthesis and are used to treat cutaneous and systemic mycoses.

3. Echinocandins interfere with fungal cell wall synthesis.

4. The antifungal agent flucytosine is an antimetabolite of cytosine.

5. Griseofulvin interferes with eukaryotic cell division and is used primarily to treat skin infections caused by fungi.

Antiviral Drugs (pp. 575–578)

1. Nucleoside and nucleotide analogs, such as acyclovir, AZT, ddI, and ddC, inhibit DNA or RNA synthesis.

2. Inhibitors of viral enzymes are used to treat influenza and HIV infection.

3. Attachment and uncoating are other targets of antiviral drugs.

4. Alpha-interferons inhibit the spread of viruses to new cells.

Antiprotozoan and Antihelminthic Drugs (p. 578)

1. Chloroquine, quinacrine, diiodohydroxyquin, pentamidine, and metronidazole are used to treat protozoan infections.

2. Antihelminthic drugs include niclosamide, mebendazole, praziquantel, and piperazine.

3. Mebendazole disrupts microtubules.

4. Pyantel pamoate and ivermectin paralyze intestinal roundworms.

Tests to Guide Chemotherapy (pp. 578–580)

1. These tests are used to determine which chemotherapeutic agent is most likely to combat a specific pathogen.

2. These tests are used when susceptibility cannot be predicted or when drug resistance arises.

The Diffusion Methods (p. 579)

1. In this test, also known as the Kirby-Bauer test, a bacterial culture is inoculated on an agar medium, and filter paper disks impregnated with chemotherapeutic agents are overlaid on the culture.

2. After incubation, the absence of microbial growth around a disk is called a zone of inhibition.

3. The diameter of the zone of inhibition, when compared with a standardized reference table, is used to determine whether the organism is sensitive, intermediate, or resistant to the drug.

4. MIC is the lowest concentration of drug capable of preventing microbial growth; MIC can be estimated using the E test.

Broth Dilution Tests (pp. 579–580)

1. In a broth dilution test, the microorganism is grown in liquid media containing different concentrations of a chemotherapeutic agent.

2. The lowest concentration of a chemotherapeutic agent that kills bacteria is called the minimum bactericidal concentration (MBC).

The Effectiveness of Chemotherapeutic Agents (pp. 580–585)

Drug Resistance (pp. 580–583)

1. Resistance may be due to enzymatic destruction of a drug, prevention of penetration of the drug to its target site, or cellular or metabolic changes at target sites.

2. Hereditary drug resistance (R) factors are carried by plasmids and transposons.

3. Resistance can be minimized by the discriminating use of drugs in appropriate concentrations and dosages.

Antibiotic Safety (p. 583)

1. The risk (e.g., side-effects) versus the benefit (e.g., curing an infection) must be evaluated prior to use of antibiotics.

Effects of Combinations of Drugs (p. 583)

1. Some combinations of drugs are synergistic; they are more effective when taken together.

2. Some combinations of drugs are antagonistic; when taken together, both drugs become less effective than when taken alone.

The Future of Chemotherapeutic Agents (pp. 584–585)

1. Many bacterial diseases, previously treatable with antibiotics, have become resistant to antibiotics.

2. Chemicals produced by plants and animals are providing new antimicrobial agents called antimicrobial peptides.

3. New antimicrobial drugs include DNA that is complementary to specific genes in a pathogen; the DNA will bind to the pathogen's DNA or mRNA and inhibit protein synthesis.

STUDY QUESTIONS

Review

1. Fill in the following table:

Antimicrobial Agent	Synthetic or Antibiotic	Method of Action	Principal Use
Isoniazid			
Sulfonamides			
Ethambutol			
Trimethoprim			
Fluoroquinolones			
Penicillin, natural			
Penicillin, semisynthetic			
Cephalosporins			
Carbapenems			
Aminoglycosides			
Tetracyclines			
Macrolides			
Polypeptides			
Vancomycin			
Rifamycins			
Polyenes			
Griseofulvin			
Amantadine			
Zidovudine			
Niclosamide			
Albendazole			

2. Define chemotherapeutic agent. Distinguish a synthetic drug from an antibiotic.

3. Discuss the contributions to chemotherapy made by Ehrlich and Fleming.

4. List and explain five criteria used to identify an effective antimicrobial agent.

5. What similar problems are encountered with antiviral, antifungal, antiprotozoan, and antihelminthic drugs?

6. Identify four modes of action of antiviral drugs. Give an example of a currently used antiviral drug for each mode of action.

7. Compare and contrast the broth dilution and disk-diffusion tests. Identify at least one advantage of each.

8. Describe the disk-diffusion test for microbial susceptibility. What information can you obtain from this test?

9. Define drug resistance. How is it produced? What measures can be taken to minimize drug resistance?

10. List the advantages of using two chemotherapeutic agents simultaneously to treat a disease. What problem can be encountered using two drugs?

11. Why does a cell die from the following antimicrobial actions?
    a. Colistimethate binds to phospholipids.
    b. Kanamycin binds to 70S ribosomes.

12. How is translation inhibited by each of the following?
    a. chloroamphenicol d. streptomycin
    b. erythromycin e. oxazolidinone
    c. tetracycline f. streptogramin

13. Dideoxyinosine (ddI) is an antimetabolite of guanine. The −OH is missing from carbon 3 in ddI. How does ddI inhibit DNA synthesis?

14. Compare the method of action of the following pairs:
    a. Penicillin and echinocandin
    b. Imidazole and polymyxin B

Multiple Choice

1. Which of the following pairs is mismatched?
   a. antihelminthic—inhibition of oxidative phosphorylation
   b. antihelminthic—inhibition of cell wall synthesis
   c. antifungal—injury to plasma membrane
   d. antifungal—inhibition of mitosis
   e. antiviral—inhibition of DNA synthesis

2. All of the following are modes of action of antiviral drugs except
   a. inhibition of protein synthesis at 70S ribosomes.
   b. inhibition of DNA synthesis.
   c. inhibition of RNA synthesis.
   d. inhibition of uncoating.
   e. none of the above

3. Which of the following modes of action would not be fungicidal?
   a. inhibition of peptidoglycan synthesis
   b. inhibition of mitosis
   c. injury to the plasma membrane
   d. inhibition of nucleic acid synthesis
   e. none of the above

4. An antimicrobial agent should meet all of the following criteria except
   a. selective toxicity.
   b. the production of hypersensitivities.
   c. a narrow spectrum of activity.
   d. no production of drug resistance.
   e. none of the above

5. The most selective antimicrobial activity would be exhibited by a drug that
   a. inhibits cell wall synthesis.
   b. inhibits protein synthesis.
   c. injures the plasma membrane.
   d. inhibits nucleic acid synthesis.
   e. all of the above

6. Antibiotics that inhibit translation have side effects
   a. because all cells have proteins.
   b. only in the few cells that make proteins.
   c. because eukaryotic cells have 80S ribosomes.
   d. at the 70S ribosomes in eukaryotic cells.
   e. none of the above

7. Which of the following will *not* affect eukaryotic cells?
   a. inhibition of the mitotic spindle
   b. binding with sterols
   c. binding to 80S ribosomes
   d. binding to DNA
   e. none of the above

8. Cell membrane damage causes death because
   a. the cell undergoes osmotic lysis.
   b. cell contents leak out.
   c. the cell plasmolyzes.
   d. the cell lacks a wall.
   e. none of the above

9. A drug that intercalates into DNA has the following effects. Which one leads to the others?
   a. It disrupts transcription.
   b. It disrupts translation.
   c. It interferes with DNA replication.
   d. It causes mutations.
   e. It alters proteins.

10. Chloramphenicol binds to the 50S portion of a ribosome, which will interfere with
    a. transcription in prokaryotic cells.
    b. transcription in eukaryotic cells.
    c. translation in prokaryotic cells.
    d. translation in eukaryotic cells.
    e. DNA synthesis.

Critical Thinking

1. Which of the following can affect human cells? Explain why or why not.
   a. penicillin
   b. indinavir
   c. erythromycin
   d. polymyxin

2. Why is idoxuridine effective if host cells also contain DNA?

3. Some bacteria become resistant to tetracycline because they don't make porins. Why can a porin-deficient mutant be detected by its inability to grow on a medium containing a single carbon source such as succinic acid?

4. The following data were obtained from a disk-diffusion test.

Antibiotic	Zone of Inhibition
A	15 mm
B	0 mm
C	7 mm
D	15 mm

   a. Which antibiotic was most effective against the bacteria being tested?
   b. Which antibiotic would you recommend for treatment of a disease caused by this bacterium?
   c. Was antibiotic A bactericidal or bacteriostatic? How can you tell?

5. Why do you suppose *Streptomyces griseus* produces an enzyme that inactivates streptomycin? Why is this enzyme produced early in metabolism?

6. The following results were obtained from a broth dilution test for microbial susceptibility.

Antibiotic Concentration	Growth	Growth in Subculture
200 μg/ml	–	–
100 μg/ml	–	–
50 μg/ml	–	+
25 μg/ml	+	+

a. The MIC of this antibiotic is _____.
b. The MBC of this antibiotic is _____.

Clinical Applications

1. Penicillin and streptomycin can be used together under certain circumstances, but penicillin and tetracycline cannot be used together. Offer an explanation for this.

2. A patient with a urinary bladder infection took nalidixic acid, but her condition did not improve. Explain why her infection disappeared when she switched to a sulfonamide.

3. A patient with streptococcal sore throat takes penicillin for 2 days of a prescribed 10-day regimen. Because he feels better, he then saves the remaining penicillin for some other time. After 3 more days, he suffers a relapse of the sore throat. Discuss the probable cause of the relapse.

LEARNING WITH TECHNOLOGY

The Microbiology Place CD-ROM/website includes many resources to help you study and succeed in this course. In addition to self-quizzes, web links, and flashcards, you'll find the following activities for this chapter in the "Activities" section of The Microbiology Place:

Enzyme Inhibitors

Case Study: Determining the Method of Action

Appendix A

Classification of Bacteria According to *Bergey's Manual**

Domain: Archaea
  Phylum I: Crenarchaeota
    Class I: Thermoprotei
      Order I: Thermoproteales
        Family I: Thermoproteaceae
          Caldivirga
          Pyrobaculum
          Thermocladium
          Thermoproteus
        Family II: Thermofilaceae
          Thermofilum
      Order II: Desulfurococcales
        Family I: Desulfurococcaceae
          Acidolobus
          Aeropyrum
          Desulfurococcus
          Igniococcus
          Staphylothermus
          Stetteria
          Sulfophobococcus
          Thermodiscus
          Thermosphaera
        Family II: Pyrodictiaceae
          Hyperthermus
          Pyrodictium
          Pyrolobus
      Order III: Sulfolobales
        Family I: Sulfolobaceae
          Acidianus
          Metallosphaera
          Stygiolobus
          Sulfolobus
          Sulfurisphaera
          Sulfurococcus
  Phylum II: Euryarchaeota
    Class I: Methanobacteria
      Order I: Methanobacteriales
        Family I: Methanobacteriaceae
          Methanobacterium
          Methanobrevibacter
          Methanosphaera
          Methanothermobacter

        Family II: Methanothermaceae
          Methanothermus
    Class II: Methanococci
      Order I: Methanococcales
        Family I: Methanococcaceae
          Methanococcus
          Methanothermococcus
        Family II: Methanocaldococcaceae
          Methanocaldococcus
          Methanotorris
      Order II: Methanomicrobiales
        Family I: Methanomicrobiaceae
          Methanoculleus
          Methanogenium
          Methanolacinia
          Methanomicrobium
          Methanoplanus
          Methanofollis
        Family II: Methanocorpusculaceae
          Methanocorpusculum
        Family III: Methanospirillaceae
          Methanospirillum
        Genera incertae sedis
          Methanocalculus
      Order III: Methanosarcinales
        Family I: Methanosarcinaceae
          Methanococcoides
          Methanohalobium
          Methanohalophilus
          Methanolobus
          Methanomicrococcus
          Methanosarcina
          Methanosalsum
        Family II: Methanosaetaceae
          Methanosaeta
    Class III: Halobacteria
      Order I: Halobacteriales
        Family I: Halobacteriaceae
          Haloarcula
          Halobacterium
          Halobaculum
          Halococcus
          Haloferax
          Halogeometricum
          Halorhobdus
          Halorubrum
          Haloterrigena

**Bergey's Manual of Systematic Bacteriology*, 2nd ed., 5 vols. (2002), is the reference for classification. *Bergey's Manual of Determinative Bacteriology*, 9th ed. (1994), should be used for identification of culturable bacteria and archaea.

Natrialba
Natrinema
Natronobacterium
Natronococcus
Natronomonas
Natronorubrum
Class IV: Thermoplasmata
   Order I: Thermoplasmatales
      Family I: Thermoplasmataceae
         Thermoplasma
      Family II: Picrophilaceae
         Picrophilus
      Family III: Ferroplasmatacea
         Ferroplasma
Class V: Thermococci
   Order II: Thermococcales
      Family I: Thermococcaceae
         Palaeococcus
         Pyrococcus
         Thermococcus
Class VI: Archaeoglobi
   Order I: Archaeoglobales
      Family I: Archaeoglobaceae
         Archaeoglobus
         Ferroglobus
Class VII: Methanopyri
   Order I: Methanopyrales
      Family I: Methanopyraceae
         Methanopyrus

Domain: Bacteria
  Phylum I: Aquificae
    Class I: Aquificae
       Order I: Aquificales
          Family I: Aquificaceae
             Aquifex
             Calderobacterium
             Hydrogenobacter
             Thermocrinis
          Genera incertae sedis
             Desulfurobacterium
  Phylum II: Thermotogae
    Class I: Thermotogae
       Order I: Thermotogales
          Family I: Thermotogaceae
             Fervidobacterium
             Geotoga
             Petrotoga
             Thermosipho
             Thermotoga
  Phylum III: Thermodesulfobacteria
       Order I: Thermodesulfobacteriales
          Family I: Thermodesulfobacteriaceae
             Thermodesulfobacterium

Phylum IV: Deinococcus-Thermus
  Class I: Deinococci
    Order I: Deinococcales
       Family I: Deinococcaceae
          Deinococcus
    Order II: Thermales
       Family I: Thermaceae
          Meiothermus
          Thermus
Phylum V: Chrysiogenetes
  Class I: Chrysiogenetes
    Order I: Chrysiogenales
       Family I: Chrysiogenaceae
          Chrysiogenes
Phylum VI: Chloroflexi
  Class I: Chloroflexi
    Order I: Chloroflexales
       Family I: Chloroflexaceae
          Chloroflexus
          Chloronema
          Heliothrix
       Family II: Oscillochloridaceae
          Oscillochloris
    Order II: Herpetosiphonales
       Family I: Herpetosiphonaceae
          Herpetosiphon
Phylum VII: Thermomicrobia
  Class I: Thermomicrobia
    Order I: Thermomicrobiales
       Family I: Thermomicrobiaceae
          Thermomicrobium
Phylum VIII: Nitrospira
  Class I: Nitrospira
    Order I: Nitrospirales
       Family I: Nitrospiraceae
          Leptospirillum
          Magnetobacterium
          Nitrospira
          Thermodesulfovibrio
Phylum IX: Deferribacteres
  Class I: Deferribacteres
    Order I: Deferribacterales
       Family I: Deferrivacter
          Deferribacter
          Denitrovibrio
          Flexistipes
          Geovibrio
       Genera incertae sedis
          Synergistes
Phylum X: Cyanobacteria
  Class I: Cyanobacteria
    SubSection I
         Chamaesiphon
         Chroococcus
         Cyanobacterium
         Cyanobium

Cyanothece
Dactylococcopsis
Gloeobacter
Gloeocapsa
Gloeothece
Microcystis
Prochlorococcus
Prochloron
Synechococcus
Synechocystis
SubSection II
    Chroococcidiopsis
    Cyanocystis
    Dermocarpella
    Myxosarcina
    Pleurocapsa
    Stanieria
    Xenococcus
SubSection III
    Arthrospira
    Borzia
    Crinalium
    Geitlerinema
    Halospirulina
    Leptolyngbia
    Limnothrix
    Lyngbya
    Microcoleus
    Oscillatoria
    Planktothrix
    Prochlorothrix
    Pseudoanabaena
    Spirulina
    Starria
    Symploca
    Trichodesmium
    Tychonema
Subsection IV
    Anabaena
    Anabaenopsis
    Aphanizomenon
    Calothrix
    Cyanospira
    Cylindrospermum
    Nodularia
    Nostoc
    Rivularia
    Scytonema
    Tolypothrix
SubSection V
    Chlorogloeopsis
    Fischerella
    Geitleria
    Iyengariella
    Nostochopsis
    Stigonema

Phylum XI: Chlorobi
Class I: Chlorobia
    Order I: Chlorobiales
        Family I: Chlorobiaceae
            Ancalochloris
            Chlorobium
            Chloroherpeton
            Pelodictyon
            Prosthecochloris
Phylum XII: Proteobacteria
Class I: Alpha-Proteobacteria
    Order I: Rhodospirillales
        Family I: Rhodosprillaceae
            Azospirillum
            Magnetospirillum
            Phaeospirillum
            Rhodocista
            Rhodospira
            Rhodospirillum
            Rhodothalassium
            Rhodovibrio
            Roseospira
            Shermanella
        Family II: Acetobacteraceae
            Acetobacter
            Acidiphilium
            Acidosphaera
            Acidocella
            Acidomonas
            Asaia
            Craurococcus
            Gluconacetobacter
            Gluconobacter
            Paracraurococcus
            Rhodopila
            Roseococcus
            Stella
            Zavarzinia
    Order II: Rickettsiales
        Family I: Rickettsiaceae
            Orientia
            Rickettsia
            Wolbachia
        Family II: Ehrlichiaceae
            Aegyptianella
            Anaplasma
            Cowdria
            Ehrlichia
            Neorickettsia
            Xenohaliotis
        Family III: Holosporaceae
            Caedibacter
            Holospora
            Lyticum

Odyssella
Polynucleobacter
Pseudocaedibacter
Symbiotes
Tectibacter
Order III: Rhodobacterales
    Family I: Rhodobacteraceae
        Ahrensia
        Amaricoccus
        Antarctobacter
        Gemmobacter
        Hirschia
        Hyphomonas
        Maricaulis
        Methylarcula
        Octadecabacter
        Paracoccus
        Rhodobacter
        Rhodovulum
        Roseobacter
        Roseibium
        Roseinatronobacter
        Roseovarius
        Roseovivax
        Rubrimonas
        Ruegeria
        Sagittula
        Staleya
        Stappia
        Sulfitobacter
Order IV: Sphingomonadales
    Family I: Sphingomonodaceae
        Blastomonas
        Erythrobacter
        Erythromicrobium
        Erythromonas
        Porphyrobacter
        Rhizomonas
        Sandaracinobacter
        Sphingomonas
        Zymomonas
Order V: Caulobacterales
    Family I: Caulobacteraceae
        Asticcacaulis
        Brevundimonas
        Caulobacter
        Phenylobacterium
Order VI: Rhizobiales
    Family I: Rhizobiaceae
        Agrobacterium
        Carbophilus
        Chelatobacter
        Ensifer
        Rhizobium
        Sinorhizobium

Family II: Bartonellaceae
    Bartonella
Family III: Brucellaceae
    Brucella
    Mycoplana
    Ochrobactrum
Family IV: Phyllobacteriaceae
    Allorhizobium
    Aminobacter
    Aquamicrobium
    Defluvibacter
    Mesorhizobium
    Phyllobacteriaceae
    Pseudaminobacter
Family V: Methylocystaceae
    Methylocystis
    Methylopila
    Methylosinus
Family VI: Beijerinckiaceae
    Beijerinckia
    Chelatococcus
    Derxia
Family VII: Bradyrhizobiaceae
    Afipia
    Agromonas
    Blastobacter
    Bosea
    Bradyrhizobium
    Nitrobacter
    Oligotropha
    Rhodopseudomonas
Family VIII: Hyphomicrobiaceae
    Ancalomicrobium
    Ancylobacter
    Angulomicrobium
    Aquabacter
    Azorhizobium
    Blastochloris
    Devosia
    Dichotomicrobium
    Filomicrobium
    Gemmiger
    Hyphomicrobium
    Labrys
    Methylorhabdus
    Pedomicrobium
    Prosthecomicrobium
    Rhodomicrobium
    Rhodoplanes
    Seliberia
    Starkya
    Xanthobacter
Family IX: Methylobacteriaceae
    Methylobacterium
    Protomonas
    Roseomonas

Family X: Rhodobiaceae
  Rhodobium
Class II: Beta-Proteobacteria
  Order I: Burkholderiales
    Family I: Burkholderiaceae
      Burkholderia
      Cupriavidus
      Lautropia
      Pandoraea
      Thermothrix
    Family II: Ralstoniaceaee
      Ralstonia
    Family III: Oxalobacteraceae
      Duganella
      Herbaspirillum
      Janthinobacterium
      Massilia
      Oxalobacter
      Telluria
    Family IV: Alcaligenaceae
      Achromobacter
      Alcaligenes
      Bordetella
      Pelistega
      Sutterella
      Taylorella
    Family V: Comamonadaceae
      Acidovorax
      Aquabacterium
      Brachymonas
      Comamonas
      Delftia
      Hydrogenophaga
      Ideonella
      Leptothrix
      Polaromonas
      Rhodoferax
      Rubrivivax
      Sphaerotilus
      Tepidimonas
      Thiomonas
      Variovorax
  Order II: Hydrogenophilales
    Family I: Hydrogenophilus
      Hydrogenophaga
      Thiobacillus
  Order III: Methylophilales
    Family I: Methylophilaceae
      Methylobacillus
      Methylophilus
      Methylovorus
  Order IV: Neisseriales
    Family I: Neisseriaceae
      Alysiella
      Aquaspirillum

Catenococcus
Chromobacterium
Eikenella
Formivibrio
Iodobacter
Kingella
Microvirgula
Neisseria
Prolinoborus
Simonsiella
Vitreoscilla
Vogesella
  Order V: Nitrosomonadales
    Family I: Nitrosomonadaceae
      Nitrosomonas
      Nitrosospira
    Family II: Spirillaceae
      Spirillum
    Family III: Gallionellacea
      Gallionella
  Order VI: Rhodocyclales
    Family I: Rhodocyclaceae
      Azoarcus
      Azonectus
      Azospira
      Azovibrio
      Propionibacter
      Propionivibrio
      Rhodocyclus
      Thauera
      Zoogloea
Class III: Gammaproteobacteria
  Order I: Chromatiales
    Family I: Chromatiaceae
      Allochromatium
      Amoebobacter
      Chromatium
      Halochromatium
      Halothiobacillus
      Isochromatium
      Lamprobacter
      Lamprocystis
      Marichromatium
      Nitrosococcus
      Pfennigia
      Rhabdochromatium
      Thermochromatium
      Thioalkalicoccus
      Thiocapsa
      Thiococcus
      Thiocystis
      Thiodictyon
      Thiohalocapsa
      Thiolamprovum
      Thiopedia
      Thiorhodococcus

Thiorhodovibrio
Thiospirillum
Family II: Ectothiorhodospiraceae
Arhodomonas
Ectothiorhodospira
Halorhodospira
Nitrococcus
Thiorhodospira
Order II: Acidothiobacillales
Family 1: Acidothiobacillaceae
Acidothiobacillus
Order III: Xanthomonadales
Family I: Xanthomonadaceae
Frateuria
Lureimonas
Lyssobacter
Nevskia
Pseudoxanthomonas
Rhodanobacer
Stenotrophomonas
Xanthomonas
Xylella
Order IV: Cardiobacteriales
Family I: Cardiobacteriaceae
Cardiobacterium
Dichelobacter
Suttonella
Order V: Thiotrichales
Family I: Thiotrichaceae
Achromatium
Beggiatoa
Leucothrix
Macromonas
Thiobacterium
Thiomargarita
Thioploca
Thiospira
Thiothrix
Family II: Piscirickettsiaceae
Cycloclasticus
Hydrogenovibrio
Methylophaga
Piscirickettsia
Thiomicrospira
Family III: Francisellaceae
Francisella
Order VI: Legionellales
Family I: Legionellaceae
Legionella
Family II: Coxiellaceae
Coxiella
Rickettsiella
Order VII: Methylococcales
Family I: Methylococcaceae
Methylobacter
Methylocaldum

Methylococcus
Methylomicrobium
Methylomonas
Methylosphaera
Order VIII: Oceanospirillales
Family I: Oceanospirillaceae
Balneatrix
Fundibacter
Marinomonas
Marinospirillum
Neptunomonas
Oceanospirillum
Family II: Halomonadaceae
Alcanivorax
Carnimonas
Chromohalobacter
Deleya
Halomonas
Zymobacter
Order IX: Pseudomonadales
Family I: Pseudomonadaceae
Azomonas
Azotobacter
Cellvibrio
Chryseomonas
Flavimonas
Lampropedia
Mesophilobacter
Morococcus
Oligella
Pseudomonas
Rhizobacter
Rugamonas
Serpens
Thermoleophilum
Xylophilus
Family II: Moraxellaceae
Acinetobacter
Moraxella
Psychrobacter
Order X: Alteromonadales
Family I: Alteromonadaceae
Allishewanella
Alteromonas
Colwellia
Ferrimonas
Idiomarina
Marinobacter
Marinobacterium
Microbulbifer
Moritella
Pseudoalteromonas
Shewanella
Order XI: Vibrionales
Family I: Vibrionaceae
Allomonas

Enhydrobacter
Listonella
Photobacterium
Salvinivibrio
Vibrio
Order XII: Aeromonadales
    Family I: Aeromonadaceae
      Aeromonas
      Oceanomonas
      Tolumonas
    Family II: Succinivibrionaceae
      Anaerobiospirillumm
      Ruminobacter
      Succinomonas
      Succinivibrio
Order XIII: Enterobacteriales
    Family I: Enterobacteriaceae
      Alterococcus
      Arsenophonus
      Brenneria
      Buchnera
      Budvicia
      Buttiauxella
      Calymmatobacterium
      Cedecea
      Citrobacter
      Edwardsiella
      Enterobacter
      Erwinia
      Escherichia
      Ewingella
      Hafnia
      Klebsiella
      Kluyvera
      Leclercia
      Leminorella
      Moellerella
      Morganella
      Obesumbacterium
      Pantoea
      Pectobacterium
      Photorhabdus
      Plesiomonas
      Pragia
      Proteus
      Providencia
      Rahnella
      Saccharobacter
      Salmonella
      Serratia
      Shigella
      Sodalis
      Tatumella
      Trabulsiella
      Wigglesworthia
      Xenorhabdus

Yersinia
Yokenella
Order XIV: Pasteurellales
    Family I: Pasteurellaceae
      Actinobacillus
      Haemophilus
      Lonepinella
      Pasteurella
      Mannheimia
      Phocoenobacter
Class VI: Deltaproteobacteria
    Order I: Desulfurales
      Family I: Desulfurellaceae
        Desulfurella
        Hippea
    Order II: Desulfovibrionales
      Family I: Desulfovibrionaceae
        Bilophila
        Desulfovibrio
        Lawsonia
      Family II: Desulfomicrobiaceae
        Desulfomicrobium
      Family III: Desulfonalobiaceae
        Desulfohalobium
        Desulfomonaas
        Desulfonatronovibrio
    Order III: Desulfobacterales
      Family I: Desulfobacteraceae
        Desulfobacter
        Desulfobacterium
        Desulfobacula
        Desulfococcus
        Desulfofaba
        Desulfofrigus
        Desulfonema
        Desulfosarcina
        Desulfospira
        Desulfocella
        Desulfotalea
        Desulfotignum
      Family II: Desulfobulbaceae
        Desulfobulbus
        Desulfocapsa
        Desulfofustis
        Desulforhopalus
      Family III: Nitrospinaceae
        Nitrospina
        Desulfobacca
        Desulfomonile
    Order IV: Desulfuromonadales
      Family I: Desulfomonadaceae
        Desulfuromonas
        Desulfuromusa
      Family II: Geobacteraceae
        Geobacter

Family III: Pelobacteriaceae
Pelobacter
Melonomonas
Trichlorobacter
Order V: Syntrophobacterales
Family I: Syntrophobacteraceae
Desulfacinum
Syntrophobacter
Desulforhabdus
Desulfovirga
Thermodesulforhabdus
Family II: Syntrophaceae
Smithella
Syntrophus
Order VI: Bdellovibrionales
Family VI: Bdellovibrionaceae
Bacteriovorax
Bdellovibrio
Micavibrio
Vampirovibrio
Order VII: Myxococcales
Family I: Myxococcaceae
Angiococcus
Myxococcus
Family II: Archangiaceae
Archangium
Family III: Cystobacteraceae
Cystobacter
Melittangium
Stigmatella
Family IV: Polyangiaceae
Chondromyces
Nannocystis
Polyangium
Class V: Epsilonproteobacteria
Order I: Campylobacterales
Family I: Campylobacteraceae
Arcobacter
Campylobacter
Sulfurospirillum
Thiovulum
Family II: Helicobacteraceae
Helicobacter
Wolinella
Phylum XIII: Firmicutes
Class I: Clostridia
Order I: Clostridiales
Family I: Clostridiaceae
Acetivibrio
Acidaminobacter
Anaerobacter
Caloramator
Clostridium
Coprobacillus
Natronineola
Oxobacter

Sarcina
Sporobacter
Thermobrachium
Thermohalobacter
Tindallia
Family II: Lachnospiraceae
Acetitomaculum
Anaerofilum
Butyrivibrio
Catenibacterium
Catonella
Coprococcus
Johnsonella
Lachnospira
Pseudobutyrivibrio
Roseburia
Ruminococcus
Sporobacterium
Family III: Peptostreptococcace
Filifactor
Fusibacter
Helcococcus
Micromonas
Peptostreptococcus
Tissierella
Family IV: Eubacteriaceae
Acetobacterium
Anaerovorax
Eubacterium
Mogibacterium
Pseudoramibacter
Family V: Peptococcaceae
Anaeroaarcus
Anaerosinus
Anaerovibrio
Carboxydothermus
Centipeda
Dehalobacter
Dendrosporobacter
Desulfitobacterium
Desulfonispora
Desulfosporosinus
Desulfotomaculum
Mitsuokella
Peptococcus
Propionispira
Succinispira
Syntrophobotulus
Thermoterrabacterium
Family VI: Heliobacteriaceae
Heliobacterium
Heliobacillus
Heliophilum
Heliorestis
Family VII: Acidaminocaccaceae
Acetonema

Anaeromonas
Acidaminococcus
Dialister
Megasphaera
Papillibacter
Pectinatus
Phascolarctobacterium
Quinella
Schwartzia
Selenomonas
Sporomusa
Succiniclasticum
Veillonella
Zymophilus
Family VII: Syntrophomonadaceae
   Acetogenium
   Aminobacterium
   Aminomonas
   Anaerobaculum
   Anaerobranca
   Caldicellulosiruptor
   Dethiosulfovibrio
   Pelospora
   Syntrophomonas
   Syntrophospora
   Syntrophothermus
   Thermoaerobacter
   Thermoaerovibrio
   Thermohydrogenium
   Thermosyntropha
Order II: Thermoanaerobactriales
   Family I: Thermoanaerobacteriaceae
      Ammonifex
      Carboxydobrachium
      Coprothermobacter
      Moorella
      Sporotomaculum
      Thermacetogenium
      Thermoanaerobacter
      Thermoanaerobacterium
      Thermoanaerobium
Order III: Haloanaerobiales
   Family I: Haloanaerobiaceae
      Haloanaerobium
      Halocella
      Halothermothrix
      Natroniella
   Family II: Halobacteroidaceae
      Acetohalobium
      Haloanaerobacter
      Halobacteroides
      Orenia
      Sporohalobacter
Class II: Mollicutes
   Order I: Mycoplasmatales
      Family I: Mycoplasmataceae

Mycoplasma
Eperythrozoon
Haemobartonella
Ureaplasma
Order II: Entoplasmatales
   Family I: Entoplasmatales
      Entomoplasma
      Mesoplasma
   Family II: Spiroplasmataceae
      Spiroplasma
Order III: Acholeplasmataceae
   Family I: Anaeroplasmataceae
      Acholeplasma
Order IV: Anaeroplasmatales
      Anaeroplasma
      Asteroleplasma
Order V: Incertae sedis
   Family V: Erysipelothrichaeceae
      Bulleidia
      Erysipelothrix
      Holdemania
      Solobacterium
Class III: Bacilli
   Order I: Bacillales
      Family I: Bacillaceae
         Amphibacillus
         Anoxybacillus
         Bacillus
         Exiguobacterium
         Gracilibacillus
         Halobacillus
         Saccharococus
         Salibacillus
         Virgibacillus
      Family II: Planococcaceae
         Filibacter
         Kurthia
         Planococcus
         Sporosarcina
      Family III: Caryophanaceae
         Caryophanon
      Family IV: Listeriaceae
         Brochothrix
         Listeria
      Family V: Staphylococcaceae
         Gemella
         Macrococcus
         Salinicoccus
         Staphylococcus
      Family VI: Sporolactobacillaceae
         Marinococcus
         Sporolactobacillus
      Family VII: Paenibacillaceae
         Ammoniphilus
         Aneurinibacillus
         Brevibacillus

Oxalophagus
Paenibacillus
Thermicanus
Thermobacillus
Family VIII: Alicyclobacillaceae
Alicyclobacillus
Pasteuria
Sulfobacillus
Family VII: Thermoactinomycetaceae
Thermoactinomyces
Order II: Lactobacillales
Family I: Lactobacillaceae
Lactobacillus
Paralactobacillus
Pediococcus
Family II: Aerococcaceae
Abiotrophia
Aerococcus
Dolosicoccus
Eremococcus
Facklamia
Globicatella
Tetragenococcus
Ignavigranum
Family III: Carnobacteriaceae
Agitococcus
Alloiococcus
Carnobacterium
Desemzia
Dolosigranulum
Granulicatella
Lactosphaera
Trichococcus
Family IV: Enterococcaceae
Atopobacter
Enterococcus
Melissococcus
Tetragenococcus
Vagococcus
Family V: Leuconostocaceae
Leuconostoc
Oenococcus
Weissela
Family VI: Streptococcaceae
Lactococcus
Streptococcus
Family VII
Acetoanaerobium
Oscillospira
Syntrophococcus
Phylum XIV: The Actinobacteria
Class I: Actinobacteria
Order I: Acidimicrobiales
Family I: Acidimicrobiaceae
Acidimicrobium

Order II: Rubrobacterales
Family I: Rubrobacteraceae
Rubrobacter
Order III: Coriobacteridae
Family I: Coriobacteriaceae
Atophobium
Collinsella
Coriobacterium
Cryptobacterium
Denitrobacterium
Eggerthella
Slakia
Order IV: Sphaerobacterales
Family I: Sphaerobacteraceae
Sphaerobacter
Order V: Actinomycetales
Suborder: Actinomycineae
Family I: Actinomycetaceae
Actinobaculum
Actinomyces
Arcanobacterium
Mobiluncus
Suborder: Micrococcineae
Family I: Micrococcaceae
Arthrobacter
Kocuria
Micrococcus
Nesterenkonia
Renibacterium
Rothia
Stomatococcus
Family II: Bogoricellaceae
Bogoriella
Family III: Rarobacteriaceae
Rarobacter
Family IV: Sanguibacteraceae
Sanguibacter
Family V: Brevibacteriaceae
Brevibacterium
Family VI: Cellulomonadaceae
Cellulomonas
Oerskovia
Family VII: Dermabacteraceae
Brachybacterium
Dermabacter
Family VIII: Dermatophilaceae
Dermatophilus
Family IX: Dermacoccaceae
Dermacoccus
Demetria
Kytococcus
Family X: Intrasporangiaceae
Intrasporangium
Janibacter
Ornithinicoccus
Ornithinimicrobium

Terrabacter
Terracoccus
Tetrasphaera
Family XI: Jonesiaceae
  Jonesia
Family XII: Microbacteriaceae
  Agrococcus
  Agromyces
  Aureobacterium
  Clavibacter
  Cryobacterium
  Curtobacterium
  Frigoribacterium
  Leifsonia
  Leucobacter
  Microbacterium
  Rathayibacter
  Subtercola
Family XIII: Beutenbergiaceae
  Beutenbergia
Family XIV: Promicromonosporaceae
  Promicromonospora
Suborder: Corynebacterineae
  Family I: Corynebacteriaceae
    Corynebacterium
  Family II: Dietziaceae
    Dietzia
  Family III: Gordoniaceae
    Gordonia
    Skermania
  Family IV: Mycobacteriaceae
    Mycobacterium
  Family V: Nocardiaceae
    Nocardia
    Rhodococus
  Family VI: Tsukamurellaceae
    Tsukamurella
  Family VII: Williamsiaceae
    Williamsia
Suborder: Micromonosporaceae
  Family I: Micromonosporaceae
    Actinoplanes
    Catellatospora
    Catenuloplanes
    Couchioplanes
    Dactylosporangium
    Micromonospora
    Pilimelia
    Spirilliplanes
    Verrucosispora
Suborder: Propionibacterineae
  Family I: Propionibacteriaceae
    Luteococcus
    Microlunatus
    Propionibacterium

    Propioniferax
    Tessaracoccus
  Family II: Nocardioidaceae
    Aeromicrobium
    Friedmanniella
    Hongia
    Kribbella
    Micropruina
    Marmoricola
    Nocardiodes
Suborder: Pseudonocardineae
  Family I: Pseudonocardiaceae
    Actinoalloteichus
    Actinopolyspora
    Amycolatopsis
    Kibdelosporangium
    Kutzneria
    Prauserella
    Pseudonocardia
    Saccharomonospora
    Saccharopolyspora
    Streptoalloteichus
    Thermobispora
    Thermocrispum
  Family II: Actinosynnemataceae
    Actinokineospora
    Actinosynnema
    Lentzea
    Saccharothrix
Suborder: Streptomycineae
  Family I: Streptomycetaceae
    Streptomyces
    Streptoverticillium
Suborder: Streptosporangineae
  Family I: Streptosporangiaceae
    Acrocarpospora
    Herbidospora
    Microbispora
    Microtetraspora
    Nonomuraea
    Planobispora
    Planomonospora
    Planopolyspora
    Planotetraspora
    Streptosporangium
  Family II: Nocardiopsaceae
    Nocardiopsis
    Thermobifida
  Family III: Thermomonosporaceae
    Actinomadura
    Spirillospora
    Thermomonospora
Suborder: Frankineae
  Family I: Frankiaceae
    Frankia

Family II: Geodermatophilaceae
 Blastococcus
 Geodermatophilus
 Modestobacter
Family III: Microsphaeraceae
 Microsphaera
Family IV: Sporichthyaceae
 Sporichthya
Family V: Acidothermaceae
 Acidothermus
Family VI: Kineosporaceae
 Cryptosporangium
 Kineococcus
 Kineosporia
Suborder XI: Glycomycineae
Family I: Glycomycetaceae
 Glycomyces
Order VI: Bifidobacteriales
Family I: Bifidobacteriaceae
 Bifidobacterium
 Falcivibrio
 Gardnerella
Family II: Unknown Affiliation
 Actinobispora
 Actinocorallia
 Excellospora
 Pelczaria
 Turicella

Phylum XV: Planctomycetes
Order I: Planctomycetales
Family I: Planctomycetaceae
 Gemmata
 Isosphaera
 Pirellula
 Planctomyces

Phylum XVI: Chlamydiae
Order I: Chlamydiales
Family I: Chlamydiaceae
 Chlamydia
 Chlamydophila
Family II: Parachlydiaceae
 Parachlamydia
Family III: Simkaniaceae
 Simkania
Family IV: Waddliaceae
 Waddlia

Phylum XII: Spirochetes
Class I: Spirochaetes
Order I: Spirochaetales
Family I: Spirochaetaceae
 Borrelia
 Brevinema
 Clevelandina
 Cristispira
 Diplocalyx

Hollandina
Pillotina
Spirochaeta
Treponema
Family II: Serpulinaceae
 Brachyspira
 Serpulina
Family III: Leptospiraceae
 Leptonema
 Leptospira

Phylum XVIII: Fibrobacteres
Class I: Fibrobacteres
Family I: Fibrobacteraceae
 Fibrobacter

Phylum XIX: Actinobacter
Family I: Acidobacteriaceae
 Acidobacterium
 Geothrix
 Holophaga

Phylum XX: Bacteroidetes
Class I: Bacteroidetes
Order I: Bacteroidales
Family I: Bacteroidaceae
 Acetofilamentum
 Acetomicrobium
 Acetothermus
 Anaerorhabdus
 Bacteroides
 Megamonas
Family II: Rikenellaceae
 Marinilabilia
 Rikenella
Family III: Porphyromonadaceae
 Dysgonomonas
 Porphyromonas
Family IV: Prevotellaceae
 Prevotella
Class II: Flavobacteria
Order I: Flavobacteriales
Family I: Flavobacteriaceae
 Bergeyella
 Capnocytophaga
 Cellulophaga
 Chryseobacterium
 Coenonia
 Empedobacter
 Flavobacterium
 Gelidibacter
 Ornithobacterium
 Polaribacter
 Psychroflexus
 Psychroserpens
 Riemerella
 Saligentibacter
 Weeksella

Family II: Myroideaceae
  Myroides
  Psychromonas
Family III: Blattabacteriaceae
  Blattabacterium
Class III: Sphingobacteria
  Order I: Sphingobacteriales
    Family I: Sphingobacteriaceae
      Pedobacter
      Sphingobacterium
    Family II: Saprospiraceae
      Haliscomenobacter
      Lewinella
      Saprospira
    Family III: Flexibacteraceae
      Cyclobacterium
      Cytophaga
      Dyadobacter
      Flectobacillus
      Flexibacter
      Hymenobacter
      Meniscus
      Microscilla
      Runella
      Spirosoma
      Sporocytophaga
    Family IV: Flammeovirgaceae
      Flammeovirga
      Flexithrix
      Persicobacter
      Thermonema

Family V: Crenotrichaceae
  Chitinophaga
  Crenothrix
  Rhodothermus
  Toxothrix
Phylum XXI: Fusobacteria
  Class I: Fusobacteria
    Order I: Fusobacteriales
      Family I: Fusobacteriaceae
        Fusobacterium
        Ilyobacter
        Leptotrichia
        Propionigenium
        Sebaldella
        Streptobacillus
      Genera incertae sedis
        Cetobacterium
Phylum XXII: Verrucomicrobia
  Class I: Verrucomicrobiae
    Order I: Verrucomicrobiales
      Family I: Verrucomicrobiaceae
        Prosthecobacter
        Verrucomicrobium
      Family II: Xiphinematobacteriaceae
        Xiphinematobacter
Phylum XXIII: Dictyoglomas
  Class I: Dictyoglomi
    Order I: Dictyoglomales
      Family I: Dictyoglomaceae
        Dictyoglomus

Appendix B

Methods for Taking Clinical Samples

To diagnose a disease, it is often necessary to obtain a sample of material that may contain the pathogenic microorganism. Samples must be taken aseptically. The sample container should be labeled with the patient's name, room number (if hospitalized), date, time, and medications being taken. Samples must be transported to the laboratory immediately for culture. Delay in transport may result in the growth of some organisms, and their toxic products may kill other organisms. Pathogens tend to be fastidious and die if not kept in optimum environmental conditions.

In the laboratory, samples from infected tissues are cultured on differential and selective media in an attempt to isolate and identify any pathogens or organisms that are not normally found in association with that tissue.

Universal Precautions

The following procedures should be used by all health care workers, including students, whose activities involve contact with patients or with blood or other body fluids. These procedures were developed to minimize the risk of transmitting HIV or AIDS in a health care environment, but adherence to these guidelines will minimize the transmission of *all* nosocomial infections.

1. Gloves should be worn when touching blood and body fluids, mucous membranes, and nonintact skin and when handling items or surfaces soiled with blood or body fluids. Gloves should be changed after contact with each patient.

2. Hands and other skin surfaces should be washed immediately and thoroughly if contaminated with blood or other body fluids. Hands should be washed immediately after gloves are removed.

3. Masks and protective eyewear or face shields should be worn during procedures that are likely to generate droplets of blood or other body fluids.

4. Gowns or aprons should be worn during procedures that are likely to generate splashes of blood or other body fluids.

5. To prevent needlestick injuries, needles should not be recapped, purposely bent or broken, or otherwise manipulated by hand. After disposable syringes and needles, scalpel blades, and other sharp items are used, they should be placed in puncture-resistant containers for disposal.

6. Although saliva has not been implicated in HIV transmission, mouthpieces, resuscitation bags, and other ventilation devices should be available for use in areas in which the need for resuscitation is predictable. Emergency mouth-to-mouth resuscitation should be minimized.

7. Health care workers who have exudative lesions or weeping dermatitis should refrain from all direct patient care and from handling patient-care equipment.

8. Pregnant health care workers are not known to have a greater risk of contracting HIV infection than health care workers who are not pregnant; however, if a health care worker develops HIV infection during pregnancy, the infant is at risk of infection. Because of this risk, pregnant health care workers should be especially familiar with, and strictly adhere to, precautions to minimize the risk of HIV transmission.

Instructions for Specific Procedures

Wound or Abscess Culture

1. Cleanse the area with a sterile swab moistened in sterile saline.

2. Disinfect the area with 70% ethanol or iodine solution.

3. If the abscess has not ruptured spontaneously, a physician will open it with a sterile scalpel.

4. Wipe the first pus away.

5. Touch a sterile swab to the pus, taking care not to contaminate the surrounding tissue.

6. Replace the swab in its container, and properly label the container.

Source: Centers for Disease Control and Prevention and National Institutes of Health. *Biosafety in Microbiological and Biomedical Laboratories,* 4th ed. Washington, D. C.: US Government Printing Office, 1999.

Ear Culture

1. Clean the skin and auditory canal with 1% tincture of iodine.
2. Touch the infected area with a sterile cotton swab.
3. Replace the swab in its container.

Eye Culture

This procedure is often performed by an ophthalmologist.

1. Anesthetize the eye with topical application of a sterile anesthetic solution.
2. Wash the eye with sterile saline solution.
3. Collect material from the infected area with a sterile cotton swab. Return the swab to its container.

Blood Culture

1. Close the room's windows to avoid contamination.
2. Clean the skin around the selected vein with 2% tincture of iodine on a cotton swab.
3. Remove dried iodine with gauze moistened with 80% isopropyl alcohol.
4. Draw a few milliliters of venous blood.
5. Aseptically bandage the puncture.

Urine Culture

1. Provide the patient with a sterile container.
2. Instruct the patient to collect a midstream sample by first voiding a small volume from the urinary bladder before collection. This washes away extraneous bacteria of the skin microbiota.
3. A urine sample may be stored under refrigeration (4°–6°C) for up to 24 hours.

Fecal Culture

For bacteriological examination, only a small sample is needed. This may be obtained by inserting a sterile swab into the rectum or feces. The swab is then placed in a tube of sterile enrichment broth for transport to the laboratory. For examination for parasites, a small sample may be taken from a morning stool. The sample is placed in a preservative (polyvinyl alcohol, buffered glycerol, saline, or formalin) for microscopic examination for eggs and adult parasites.

Sputum Culture

1. A morning sample is best because microorganisms will have accumulated while the patient is sleeping.
2. The patient should rinse his or her mouth thoroughly to remove food and normal microbiota.
3. The patient should cough deeply from the lungs and expectorate into a sterile glass wide-mouth jar.
4. Care should be taken to avoid contaminating health care workers.
5. In cases such as tuberculosis in which there is little sputum, stomach aspiration may be necessary.
6. Infants and children tend to swallow sputum. A fecal sample may be of some value in these cases.

Appendix C

Metabolic Pathways

FIGURE C.1 **The Calvin-Benson cycle for photosynthetic carbon metabolism.** ❶–❸ The initial fixation and reduction of carbon occurs, generating the three-carbon compounds glyceraldehyde 3-phosphate and dihydroxyacetone phosphate, ❹ which are interconvertible. Ⓐ–Ⓓ On average, 2 of every 12 three-carbon molecules are used in the synthesis of glucose. ❺ Ten of every 12 three-carbon molecules are used to generate ribulose 5-phosphate by a complex series of reactions. ❻ The ribulose 5-phosphate is then phosphorylated at the expense of ATP, forming ribulose 1,5-diphosphate, the acceptor molecule with which the sequence began. (See Figure 5.25, p. 140, for a simplified version of the Calvin-Benson cycle.)

Glucose (6C)

1 Hexokinase

ATP
ADP

**Glucose
6-phosphate (6C)**

2 Phosphoglucoisomerase

**Fructose
6-phosphate (6C)**

3 Phosphofructokinase

ATP
ADP

**Fructose
1,6-diphosphate (6C)**

4 Aldolase

**Dihydroxyacetone
phosphate (3C)**

5 Isomerase

**Glyceraldehyde
3-phosphate (3C)**

To step **6**

FIGURE C.2 **Glycolysis (Embden-Meyerhof pathway).** Each of the ten steps of glycolysis is catalyzed by a specific enzyme, which is named under each step number. (See Figure 5.12, p. 125, for a simplified version of glycolysis.)

1 Glucose enters the cell and is phosphorylated by the enzyme hexokinase, which transfers a phosphate group from ATP to the number 6 carbon of the sugar. The product of the reaction is glucose 6-phosphate. The electrical charge of the phosphate group traps the sugar in the cell because of the impermeability of the plasma membrane to ions. Phosphorylation of glucose also makes the molecule more chemically reactive. Although glycolysis is supposed to *produce* ATP, in step **1**, ATP is actually consumed—an energy investment that will be repaid with dividends later in glycolysis.

2 Glucose 6-phosphate is rearranged to convert it to its isomer, fructose 6-phosphate. Isomers have the same number and types of atoms but in different structural arrangements.

3 In this step, still another molecule of ATP is invested in glycolysis. An enzyme transfers a phosphate group from ATP to the sugar, producing fructose 1,6-diphosphate.

4 This is the reaction from which glycolysis gets its name ("sugar splitting"). An enzyme cleaves fructose 1,6-diphosphate into two different three-carbon sugars: glyceraldehyde 3-phosphate and dihydroxyacetone phosphate. These two sugars are isomers.

5 The enzyme isomerase interconverts the three-carbon sugars. The next enzyme in glycolysis uses only glyceraldehyde 3-phosphate as its substrate. This pulls the equilibrium between the two three-carbon sugars in the direction of glyceraldehyde 3-phosphate, which is removed as fast as it forms.

6

Triose phosphate dehydrogenase

2 NAD⁺

2 NADH + H⁺

2 Pᵢ

1,3-diphosphoglyceric acid (3C)
(2 molecules)

7

Phospho-glycerokinase

2 ADP

2 ATP

3-phosphoglyceric acid (3C)
(2 molecules)

8

Phospho-glyceromutase

2-phosphoglyceric acid (3C)
(2 molecules)

9

Enolase

H₂O

Phosphoenolpyruvic acid (PEP) (3C)
(2 molecules)

10

Pyruvate kinase

2 ADP

2 ATP

6 An enzyme now catalyzes two sequential reactions while it holds glyceraldehyde 3-phosphate in its active site. First, the sugar is oxidized at the number one carbon and NAD⁺ is reduced, resulting in the formation of NADH + H⁺. Second, the enzyme couples this reaction to the creation of a high-energy phosphate bond at the number one carbon of the oxidized substrate. The source of the phosphate is inorganic phosphate, which is always present in the cell. As products, the enzyme releases NADH + H⁺ and 1,3-diphosphoglyceric acid. Notice in the figure that the new phosphate bond is symbolized with a high-energy bond (~), which indicates that the bond is at least as energetic as the phosphate bonds of ATP.

7 At this step, glycolysis produces ATP. The phosphate group, with its high-energy bond, is transferred from 1,3-diphosphoglyceric acid to ADP. For each glucose molecule that began glycolysis, step **7** produces two molecules of ATP, because every product after the sugar-splitting step (step **4**) is doubled. Of course, two ATPs were invested to get sugar ready for splitting. The ATP ledger now stands at zero. By the end of step **7**, glucose has been converted to two molecules of 3-phosphoglyceric acid.

8 Next, an enzyme relocates the remaining phosphate group of 3-phosphoglyceric acid to form 2-phosphoglyceric acid. This prepares the substrate for the next reaction.

9 An enzyme forms a double bond in the substrate by extracting a water molecule from 2-phosphoglyceric acid to form phosphoenolpyruvic acid. This results in the electrons of the substrate being arranged in such a way that the remaining phosphate bond becomes very unstable.

10 The last reaction of glycolysis produces another molecule of ATP by transferring the phosphate group from phosphoenolpyruvic acid to ADP. Because this step occurs twice for each glucose molecule, the ATP ledger now shows a net gain of two ATPs. Thus, the glycolysis of one molecule of glucose results in two molecules of pyruvic acid, two molecules of NADH + H⁺, and two molecules of ATP. Each molecule of pyruvic acid can now undergo respiration or fermentation.

Pyruvic acid (3C)
(2 molecules)

NAD⁺   NADH + H⁺   CoA   CO₂

Aerobic conditions (respiration)

Acetyl CoA (2C)

To Krebs cycle

NADH + H⁺   NAD⁺

Anaerobic conditions (fermentation)

CO₂ +

Ethanol (2C) and CO₂ (1C)   OR   **Lactic acid (3C)**

FIGURE C.3 **The pentose phosphate pathway.** This pathway, which operates simultaneously with glycolysis, provides an alternate route for the oxidation of glucose and plays a role in the synthesis of biological molecules, depending on the needs of the cell. Possible fates of the various intermediates are shown in blue. (See Chapter 5, p. 124.)

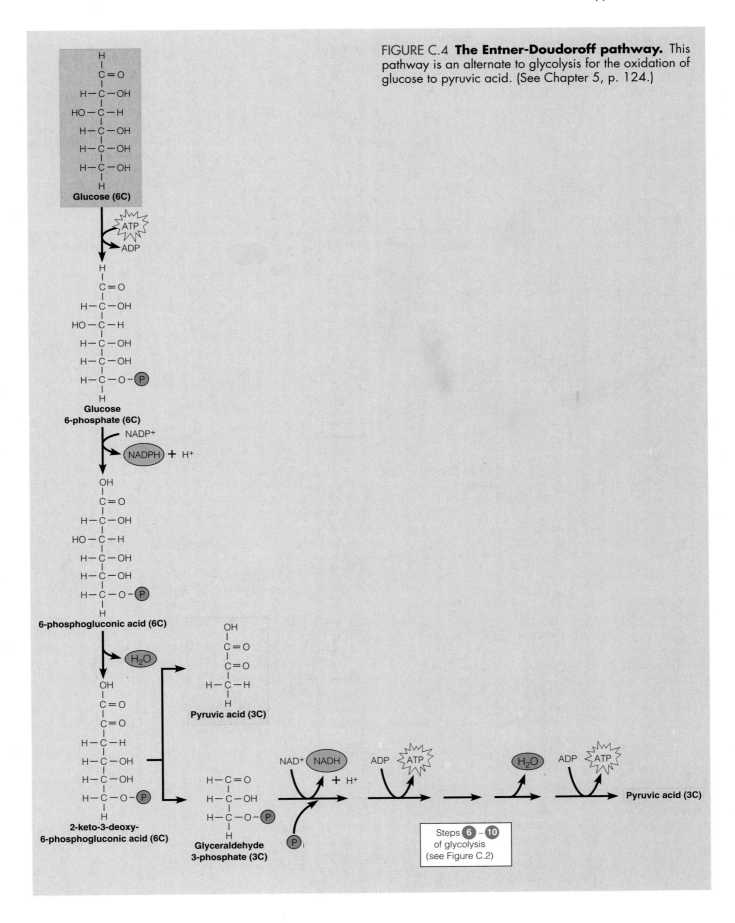

FIGURE C.4 **The Entner-Doudoroff pathway.** This pathway is an alternate to glycolysis for the oxidation of glucose to pyruvic acid. (See Chapter 5, p. 124.)

FIGURE C.5 The Krebs cycle. See Figure 5.13, p. 126, for a simplified version.

From glycolysis ➝

Pyruvic acid (3C)

NAD^+ — CoA

$NADH + H^+$

CO_2

Acetyl~CoA (2C)

8 The last oxidative step produces another molecule of $NADH + H^+$ and regenerates oxaloacetic acid, which accepts a two-carbon fragment from acetyl CoA for another turn of the cycle.

8

Oxaloacetic acid (4C)

NAD^+

$NADH + H^+$

Malic acid (4C)

1 Acetyl CoA adds its two-carbon acetyl fragment (*pink*) to oxaloacetic acid, a four-carbon compound. The unstable bond of acetyl CoA is broken as oxaloacetic acid displaces the coenzyme and attaches to the acetyl group. The product is the six-carbon citric acid. CoA is then free to prime another two-carbon fragment derived from pyruvic acid.

1 CoA

2 A molecule of water is removed, and another is added back. The net result is the conversion of citric acid to its isomer, isocitric acid.

2 H_2O

Citric acid (6C)

7 Bonds in the substrate are rearranged in this step by the addition of a water molecule.

7 H_2O

Fumaric acid (4C)

To electron transport chain

Krebs cycle

$FADH_2$

FAD

Isocitric acid (6C)

NAD^+

$NADH + H^+$

3 CO_2

3 The substrate loses a CO_2 molecule (*pale violet*), and the remaining five-carbon compound is oxidized, reducing NAD^+ to $NADH + H^+$.

6 In another oxidative step, two hydrogens are transferred to FAD to form $FADH_2$. The function of this coenzyme is similar to that of $NADH + H^+$, but $FADH_2$ stores less energy.

6

Succinic acid (4C)

$NADH + H^+$

CoA

NAD^+

5

GTP

$GDP + P_i$

$P_i + ADP$

ATP

Succinyl~CoA (4C)

4 CO_2

α-Ketoglutaric acid

4 CO_2 (*pale violet*) is lost; the remaining four-carbon compound is oxidized by the transfer of electrons to NAD^+ to form $NADH + H^+$ and is then attached to CoA by an unstable bond.

5 Substrate-level phosphorylation occurs in this step. CoA is displaced by a phosphate group, which is then transferred to GDP to form guanosine triphosphate (GTP). GTP is similar to ATP, which is formed when GTP donates a phosphate group to ADP.

Appendix D

Exponents, Exponential Logarithms, and Generation Time

Exponents and Exponential Notation

Very large and very small numbers, such as 4,650,000,000 and 0.00000032, are cumbersome to work with. It is more convenient to express such numbers in exponential notation—that is, as a power of 10. For example, 4.65×10^9 is in standard exponential notation, or **scientific notation:** 4.65 is the *coefficient*, and 9 is the power or *exponent*. In standard exponential notation, the coefficient is always a number between 1 and 10, and the exponent can be positive or negative.

To change a number into exponential notation, follow two steps. First, determine the coefficient by moving the decimal point so there is only one nonzero digit to the left of it. For example,

$$0.0000003\,2$$

The coefficient is 3.2. Second, determine the exponent by counting the number of places you moved the decimal point. If you moved it to the left, the exponent is positive. If you moved it to the right, the exponent is negative. In the example, you moved the decimal point seven places to the right, so the exponent is -7. Thus

$$0.00000032 = 3.2 \times 10^{-7}$$

Now suppose you are working with a larger number instead of a very small number. The same rules apply, but the exponential value will be positive rather than negative. For example,

$$4,650,000,000 = 4.65 \times 10^{-9}$$
$$= 4.65 \times 10^9$$

To multiply numbers written in exponential notation, multiply the coefficients and *add* the exponents. For example,

$$(3 \times 10^4) \times (2 \times 10^3) =$$
$$(3 \times 2) \times (10^{4+3}) = 6 \times 10^7$$

To divide, divide the coefficient and *subtract* the exponents. For example,

$$\frac{3 \times 10^4}{2 \times 10^3} = \frac{3}{2} \times 10^{4-3} = 1.5 \times 10^1$$

Microbiologists use exponential notation in many situations. For instance, exponential notation is used to describe the number of microorganisms in a population. Such numbers are often very large (see Chapter 6). Another application of exponential notation is to express concentrations of chemicals in a solution—chemicals such as media components (Chapter 6), disinfectants (Chapter 7), or antibiotics (Chapter 20). Such numbers are often very small. Converting from one unit of measurement to another in the metric system requires multiplying or dividing by a power of 10, which is easiest to carry out in exponential notation.

Logarithms

A **logarithm (log)** is the power to which a base number is raised to produce a given number. Usually we work with logarithms to the base 10, abbreviated $\log_{10}$. The first step in finding the $\log_{10}$ of a number is to write the number in standard exponential notation. If the coefficient is exactly 1, the $\log_{10}$ is simply equal to the exponent. For example

$$\log_{10} 0.00001 = \log_{10}(1 \times 10^{-5})$$
$$= -5$$

If the coefficient is not 1, as is often the case, the logarithm function on a calculator must be used to determine the logarithm.

Microbiologists use logs for calculating pH levels and for graphing the growth of microbial populations in culture (see Chapter 6).

Calculating Generation Time

As a cell divides, the population increases exponentially. Numerically this is equal to 2 (because one cell divides into two) raised to the number of times the cell divided (generations);

$$2^{\text{number of generations}}$$

To calculate the final concentration of cells:

Initial number of cells $\times 2^{\text{number of generations}} =$

number of cells

For example, if 5 cells were allowed to divide 9 times, this would result in

$$5 \times 2^9 = 2560 \text{ cells}$$

To calculate the number of generations a culture has undergone, cell numbers must be converted to logarithms. Standard logarithm values are based on 10. The log of 2 (0.301) is used because one cell divides into two.

$$\text{No. of generations} = \frac{\text{log number of cells (end)} - \text{log number of cells (beginning)}}{0.301}$$

To calculate the generation of time for a population:

$$\frac{60 \text{ min} \times \text{hours}}{\text{number of generations}} = \text{minutes/generation}$$

As an example, we will calculate the generation time if 100 bacterial cells growing for 5 hours produced 1,720,320 cells:

$$\frac{\log 1,720,320 - \log 100}{0.301} = 14 \text{ generations}$$

$$\frac{60 \text{ min} \times 5 \text{ hr}}{14 \text{ generation}} = 21 \text{ minutes/generation}$$

A practical application of the calculation is determining the effect of a newly developed food preservative on the culture. Suppose 900 of the same species were grown under the same conditions as the previous example, except that the preservative was added to the culture medium. After 15 hours, there were 3,276,800 cells. Calculate the generation time, and decide whether the preservative inhibited growth.

Answer: 75 min/generation. The preservative did inhibit growth.

Appendix E

Pronunciation of Scientific Names

Rules of Pronunciation

The easiest way to learn new material is to talk about it, and that requires saying scientific names. Scientific names may look difficult at first glance, but keep in mind that generally every *syllable* is pronounced. The primary requirement in saying a scientific name is to communicate it.

The rules for the pronunciation of scientific names depend, in part, on the derivation of the root word and its vowel sounds. We have provided some general guidelines here. Pronunciations frequently do not follow the rules because a common usage has become "accepted," or the derivation of the name cannot be determined. And for many scientific names there are alternative correct pronunciations.

Vowels

Pronounce all the vowels in scientific names. Two vowels written together and pronounced as one sound are called a *diphthong* (for example, the *ou* in *sound*). A special comment is needed about the pronunciation of the vowel endings *-i* and *-ae:* There are two alternative ways to pronounce each of these. In this book, we usually give the pronunciation of a long *e* (*ē*) to the *-i* ending, and a long *i* (*ī*) to the *-ae* ending. However, the reverse pronunciations are also correct and in some cases are preferred. For example, *coli* is usually pronounced kō′lī.

Consonants

When *c* or *g* is followed by *ae, e, oe, i,* or *y,* it has a soft sound. When *c* or *g* is followed by *a, o, oi,* or *u,* it has a hard sound. When a double *c* is followed by *e, i,* or *y,* it is pronounced as *ks* (e.g., cocci).

Accent

The accented syllable is usually the next-to-last or third-to-last syllable.

1. The accent is on the next-to-last syllable:

   a. When the name contains only two syllables. Example: pes′tis.

   b. When the next-to-last syllable is a diphthong. Example: a-kan-thä-mē′bä.

   c. When the vowel of the next-to-last syllable is long. Example: tre-pō-nē′mä. The vowel in the next-to-last syllable is long in words ending in the following suffixes:

Suffix	Example
-ales	Orders such as Eubacteriales
-ina	Sarcina
-anus, anum	pasteurianum
-uta	diminuta

   d. When the word ends in one of the following suffixes:

Suffix	Example
-atus, atum	caudatum
-ella	Salmonella

2. The accent is on the third-to-last syllable in family names. Families end in *-aceae,* which is always pronounced -ā′sē-ē.

Pronunciation of Microorganisms in This Text

Pronunciation key:

a hat	ē see	o hot	th thin
ā age	ė term	ō go	u cup
ã care	g go	ô order	u̇ put
ä father	i sit	oi oil	ü rule
ch child	ī ice	ou out	ū use
e let	ng long	sh she	zh seizure

Acanthamoeba polyphaga   a-kan-thä-mē′bä pol′if-ä-gä
Acetobacter   a-sē′tō-bak-tėr
Achinanthes minutissima   ā′kin-an-thēs min′ü-tis-sē-mä
Acinetobacter   a-si-ne′tō-bak-tėr
Actinomyces israelii   ak-tin-ō-mī′sēs is-rā′lē-ē
Aedes aegypti   ā′ē-dēz ē-jip′tē
A. albopictus   al-bō-pik′tus
Aeromonas hydrophilia   ār′ō-mō-nas hī′dro-fil-ē-ä
Afipia felis   ä-fi′pē-ä fē′lis
Agrobacterium tumefaciens   ag′rō-bak-ti′rē-um tü′me-fāsh-enz
Ajellomyces   ä-jel-lō-mī′sēs
Alcaligenes   al′kä-li-gen-ēs
Alexandrium   äl-eg-zan′drē-um

615

Amanita phalloides   am-an-ī′ta fal-loi′dēz
Anabaena   an-ä-bē′nä
Ancylostoma duodenale   an-sil-os′tō-mä dü′o-den-al-ē
Anopheles   an-of′e-l-ēz
Aquaspirillum bengal   ä-kwä-spī-ril′lum ben′gal
A. graniferum   gra-ni′fér-um
A. magnetotacticum   mag-ne-tō-tak′ti-kum
Arthroderma   är-thrō-dér′mä
Ascaris lumbricoides   as′kar-is lum-bri-koi′dēz
Ashbya gossypii   ash′bē-ä gos-sip′e-ē
Aspergillus flavus   a-spér-jil′lus flā′vus
A. fumigatus   fü-mi-gä′tus
A. niger   nī′jér
A. oryzae   ô′ri-zī
A. soyae   soi′ī
Azolla   ā-zō′lä
Azomonas   ā-zō-mō′nas
Azospirillum   ā-zō-spī′ril-lum
Azotobacter   ä-zo′tō-bak-tér
Babesia microti   ba-bē′sē-ä mī-krō′tē
Bacillus anthracis   bä-sil′lus an-thrā′sis
B. brevis   brev′is
B. cereus   se′rē-us
B. circulans   sér′ku-lans
B. coagulans   kō-ag′ū-lanz
B. licheniformis   lī-ken-i-fôr′mis
B. megaterium   meg-ä-tér′ē-um
B. polymyxa   po-lē-miks′ä
B. sphaericus   sfe′ri-kus
B. stearothermophilus   ste-rō-thér-mä′fil-us
B. subtilis   su′til-us
B. thuringiensis   thúr-in-jē-en′sis
Bacteroides hypermegas   bak-té-roi′dēz hī-pér′meg-äs
Balamuthia   bal′am-üth-ē-ä
Balantidium coli   bal-an-tid′ē-um kō′lī (or kō′lē)
Bartonella henselae   bär′tō-nel-lä hen′sel-ī
Bdellovibrio bacteriovorus   del-lō-vib′rē-ō bak-té-rē-o′vô-rus
Beauveria   bō-vär′ē-a
Beggiatoa   bej′jē-ä-tō-ä
Beijerinckia   bī-yé-rink′ē-ä
Bifidobacterium globosum   bī-fi-dō-bak-ti′rē-um glob-ō′sum
B. pseudolongum   sū-dō-lông′um
Blastomyces dermatitidis   blas-tō-mī′sēz dér-mä-tit′i-dis
Blattabacterium   blat-tä-bak-ti′rē-um
Boletus edulis   bō-lē′tus e′dū-lis
Bordetella pertussis   bór-de-tel′lä pér-tus′sis
Borrelia burgdorferi   bór-rel′ē-ä burg-dôr′fér-ē
Bradyrhizobium   brad-ē-rī-zō′bē-um
Brucella abortus   brü′sel-lä ä-bôr′tus
B. melitensis   me-li-ten′sis
B. suis   sü′is
Burkholderia   berk′hōld-ér-ē-ä
Byssochlamys fulva   bis-sō-klam′is fül′vä
Campylobacter fetus   kam′pi-lō-bak-tér fē′tus
C. jejuni   jē-jū′nē
Candida albicans   kan′did-ä al′bi-kanz
C. krusei   krūs′ā-ē
C. oleophila   ō-lē-of′il-ä
C. utilis   ū′til-is
Canis familiaris   kānis fa-mil′ē-är-is
Carpenteles   kär-pen′tel-ēz
Caulobacter   kô-lō-bak′tér
Cephalosporium   sef-ä-lō-spô′rē-um
Ceratocystis ulmi   sē-rä-tō-sis′tis ul′mē
Chilomastix   kē′lō-ma-sticks
Chlamydia pneumoniae   kla-mi′dē-ä nü-mō′nē-ī
C. psittaci   sit′tä-sē

C. trachomatis   trä-kō′mä-tis
Chlamydomonas   klam-i-dō-mō′näs
Chlorobium   klô-rō′bē-um
Chloroflexus   klô-rō-flex′us
Chromatium vinosum   krō-mä′tē-um vi-nō′sum
Chroococcus turgidus   krō-ō-kok′kus tér′ji-dus
Chrysops   krī′sops
Citrobacter   sit′rō-bak-tér
Claviceps purpurea   kla′vi-seps púr-pù-rē′ä
Clonorchis sinensis   klo-nôr′kis si-nen′sis
Clostridium acetobutylicum   klôs-tri′dē-um a-sē-tō-bū-til′li-kum
C. botulinum   bo-tū-lī′num
C. butyricum   bü-ti′ri-kum
C. difficile   dif′fi-sil-ē
C. pasteurianum   pas-tyér-ē-ā′num
C. perfringens   pér-frin′jens
C. sporogenes   spô-rä′jen-ēz
C. tetani   te′tan-ē
C. thermosaccharolyticum   thér-mō-sak-kär-ō-li′ti-kum
Coccidioides immitis   kok-sid-ē-oi′dēz im′mi-tis
Corynebacterium diphtheriae   kôr′ī-nē-bak-ti-rē-um dif-thi′rē-ī
C. glutamicum   glü-tam′i-kum
C. xerosis   ze-rō′sis
Coxiella burnetii   käks′ē-el-lä bér-ne′tē-ē
Cristispira pectinis   kris-tē-spī′rä pek′tin-is
Crucibulum vulgare   krü-si-bū′lum vul′gär-ē
Cryphonectria parasitica   kri-fō-nek′trē-ä par-ä-si′ti-kä
Cryptococcus neoformans   krip′tō-kok-kus nē-ō-fôr′manz
Cryptosporidium parvum   krip′tō-spô-ri-dē-um pär′vum
Culex   kū′leks
Culiseta   kū-li′se-tä
Cyclospora cayetanensis   sī′klō-spô-rä kī′ē-tan-en-sis
Cytophaga   sī-täf′äg-ä
Daptobacter   dap′to-bak-tér
Dermacentor andersoni   dér-mä-sen′tôr an-dér-sōn′ē
D. variabilis   vār-ē-a′bil-is
Desulfotomaculum nigrificans   dē′sul-fō-to-ma-kū-lum nī′gri-fi kans
Desulfovibrio desulfuricans   dē′sul-fō-vib-rē-ō dē-sul-fér′i-kans
Didinium nasutum   dī-di′nē-um nä-sūt′um
Diphyllobothrium latum   dī-fil-lō-bo′thrē-um lā′tum
Dracunculus medininsis   dra-kun′ku-lus med-in′in-sis
Echinococcus granulosus   ē-kīn-ō-kok′kus gra-nū-lō′sus
Ectothiorhodospira mobilis   ek-tō-thī-ō-rō-dō-spī′rä mō′bil-is
Emericella   em′ér-ē-sel-lä
Emmonsiella   em′mon-sē-el-lä
Entamoeba histolytica   en-tä-mē′bä his-tō-li′ti-kä
Enterobacter aerogenes   en-te-rō-bak′tér ā-rä′jen-ēz
E. cloacae   klō-ā′kī
Enterobius vermicularis   en-te-rō′bē-us ver-mi-kū-lar′is
Enterococcus faecalis   en-te-rō-kok′kus fē-kä′lis
Entomorphaga   en-tō-môr′fag-ä
Epidermophyton   ep-i-dér-mō-fī′ton
Epulopiscium   ep′ū-lō-pis-sē-um
Ehrlichia   ér′lik-ē-ä
Erwinia dissolvens   ér-wi′nē-ä dis-solv′ens
Erysipelothrix rhusiopathiae   ār-i-si-pel′ō-thrix rus-ē-ō-path′ē-ī
Erysiphe grammis   ār′i-sīf gram′mis
Escherichia coli   esh-ér-i′kē-ä kō′lī (or kō′lē)
Euglena   ū-glē′nä
Eupenicillium   ū-pen-i-sil′lē-um
Eurotium   yér-ō′tē-um
Filobasidiella   fi-lō-ba-si-dē-el′lä
Fonsecaea pedrosoi   fon-se′kē-ä pe-drō′sō-ē
Francisella tularensis   fran′sis-el-lä tü′lä-ren-sis
Frankia   frank′ē-ä
Fusobacterium   fü-sō-bak-ti′rē-um
Gambierdiscus toxicus   gam′bē-ér-dis-kus toks′i-kus

Gardnerella vaginalis gärd-né-rel′lä va-jin-al′is
Gelidium jel-id′ē-um
Geotrichum jē-ō-trik′um
Giardia lamblia jē-är′dē-ä lam′lē-ä
Gloeocapsa glē-ō-kap′sä
Glomus fasciculatus glo′mus fa-sik-ū-lä′tus
Glossina gläs-sē′nä
Gluconobacter glü′kon-ō-bak-tér
Gymnoascus jim-nō-as′kus
Gymnodynium breve jim-nō-din′ē-um brev′ē
Haemophilus ducreyi hē-mä′fil-us dü-krā′ē
Haloarcula hā′lō-är-kū-lä
Halobacterium halobium ha-lō-bak-ti′rē-um hal-ō′bē-um
Halococcus hā′lō kok-kus
Helicobacter pylori hē′lik-ō-bak-tér pī′lô-rē
H. influenzae in-flü-en′zī
H. vaginalis va-jin-al′is
Histoplasma capsulatum his-tō-plaz′mä kap-su-lä′tum
Holospora hō-lo′spô-rä
Homo sapiens hō′mō sā′pē-ens
Hydrogenomonas hī-drō-je-nō-mō′näs
Hyphomicrobium hī-fō-mī-krō′bē-um
Isospora ī-so′spô-rä
Ixodes scapularis iks-ō′dēs scap-ū-lär′is
I. pacificus pas-i′fi-kus
Klebsiella pneumoniae kleb-sē-el′lä nü-mō′nē-ī
Lactobacillus bulgaricus lak-tō-bä-sil′lus bul-gā′ri-kus
L. delbrueckii del-brúk′ē-ē
L. leichmannii līk-man′nē-ē
L. plantarum plan-tä′rum
L. sanfrancisco sän-fran-sis′kō
Lactococcus lactis lak-tō-kok′kus lak′tis
Laminaria japonica lam′i-när-e-ä ja-pon′i-kä
Legionella pneumophila lē-jä-nel′lä nü-mō′fi-lä
Leishmania braziliensis lish′mä-nē-ä brä-sil′ē-en-sis
L. donovani don′ō-van-ē
L. tropica trop′i-kä
Leptospira interrogans lep-tō-spī′rä in-tér′rä-ganz
Leuconostoc mesenteroides lü-kō-nos′tok mes-en-ter-oi′dēz
Licmorphora lik-môr′fôr-ä
Listeria monocytogenes lis-te′rē-ä mo-nō-sī-tô′je-nēz
Meniscus mē-nis′kus
Methanobacterium meth-a-nō-bak-tér′ē-um
Methanosarcina meth-a-nō-sär′sī-nä
Methylophilus methylotrophus meth-i-lo′fi-lus meth-i-lō-trôf′us
Microcladia mī-krō-klād′ē-ä
Micrococcus luteus mī-krō-kok′kus lū′tē-us
Micromonospora purpureae mī-krō-mo-nä′spô-rä púr-pú-rē′ī
Microsporum gypseum mī-krō-spô′rum jip′sē-um
Mixotricha mix-ō-trik′ä
Moraxella catarrhalis mô-raks-el′lä ka-tär′al-is
M. lacunata la-kü-nä′tä
M. osloensis os′lō-en-sis
Mucor rouxii mū-kôr rō′ē-ē
Mycobacterium avium-intracellulare mī-kō-bak-ti′rē-um ā′vē-um-in′trä-cel-ū-lä-rē
M. bovis bō′vis
M. leprae lep′rī
M. smegmatis smeg-ma′tis
M. tuberculosis tü-bér-kū-lō′sis
Mycoplasma hominis mī-kō-plaz′mä ho′mi-nis
M. pneumoniae nu-mō′nē-ī
Myxococcus fulvus micks-ō-kok′kus ful′vus
Naegleria fowleri nī-gle′rē-ä fou′lér-ē
Nannizia nan′nē-zē-ä
Necator americanus ne-kā′tôr ä-me-ri-ka′nus
Neisseria gonorrhoeae nī-se′rē-ä go-nôr-rē′ī

N. meningitidis me-nin-ji′ti-dis
Nereocystis nē-rē-ō-sis′tis
Nitrobacter nī-trō-bak′tér
Nitrosomonas nī-trō-sō-mō′näs
Nocardia asteroides nō-kär′dē-ä as-tér-oi′dēz
Oocystis ō-ō-sis′tis
Opisthorchis sinensis [opisthorchis] si-nen′sis
Ornithodorus ôr-nith-ō′dô-rus
Paracoccus denitrificans pār-ä-kok′kus dē-nī-tri′fi-kanz
Paragonimus westermani pār-ä-gōn′e-mus we-stér-ma′nē
Paramecium multimicronucleatum pār-ä-mē′sē-um mul-tē-mī-krō-nü-klē-ä′tum
Pasteurella haemolytica pas-tyér-el′lä hēm-ō-lit′i-kä
P. multocida mul-tō′si-dä
Pediculus humanus corporis ped-ik′ū-lus hü′ma-nus kôr′pô-ris
Pediococcus cerevisiae pe-dē-ō-kok′kus se-ri-vis′ē-ī
Penicillium chrysogenum pen-i-sil′lē-um krī-so′gen-um
P. griseofulvum gri-sē-ō-fúl′vum
P. notatum nō-tä′tum
Peridinium per-i-din′ē-um
Petriellidium pet-rē-el-li′dē-um
Pfiesteria piscicida fes-tér′ē-ä pi′si-sid-ä
Phormidium luridum fôr-mi′dē-um ler′i-dum
Phytophthora infestans f ī-tof′thô-rä in-fes′tans
Pichia pik′ē-ä
Pityrosporum ovale pit-i-ros′pô-rum ovale
Plasmodium falciparum plaz-mō′dē-um fal-sip′är-um
P. malariae mä-lā′rē-ī
P. ovale ō-vä′lē
P. vivax vī′vaks
Plesiomonas shigelloides ple-sē-ō-mō′nas shi-gel-loi′des
Pneumocystis carinii nü-mō-sis′tis kär-i′nē-ē
P. jiroveci ye-rō′vet-zē
Porphyromonas pôr′f ī-rō-mō-nas
Prevotella intermedia prev′ō-tel-la in′tér-mē-dē-ä
Propionibacterium acnes prō-pē-on′ē-bak-ti-rē-um ak′nēz
P. freudenreichii froi-den-rīk′ē-ē
Proteus mirabilis prō′tē-us mi-ra′bi-lis
P. vulgaris vul-ga′ris
Prototheca prō-tō-thä′kä
Pseudomonas aeruginosa sū-dō-mō′nas ā-rü-ji-nō′sä
P. andropogonis än′dro-po-gō-nis
P. cepacia se-pā′sē-ä
P. fluorescens flôr-es′ens
P. haemolytica hē-mō-lit′i-kä
P. multocida mul′tō-sid-ä
P. syringae sér-in′jī
Pyridictum abyssi pir-i′dik-tum a-bis′sē
Quercus kwer′kus
Rhizobium japonicum rī-zō′bē-um jap-on′i-kum
R. meliloti mel-i-lot′ē
Rhizopus nigricans rī′zō-pùs nī′gri-kans
R. oryzae ô′rī-zī
Rhodococcus bronchialis rō-dō-kok′kus bron-kē′al-is
R. erythropolis er-i-throp′ō-lis
Rhodopseudomonas rō-dō-su-dō-mō′nas
Rhodospirillum rubrum rō-dō-spī-ril′um rūb′rum
Rickettsia prowazekii ri-ket′sē-ä prou-wä-ze′kē-ē
R. rickettsii ri-ket′sē-ē
R. typhi tī′fē
Riftia rift′ē-ä
Rosa multiflora rō-sä mul-ti-flô′rä
Saccharomyces beticus sak-ä-rō-mī′sēs bet′i-kus
S. bayanus bā′an-us
S. cerevisiae se-ri-vis′ē-ī
S. ellipsoideus ē-lip-soi′dē-us
S. exiguus egz-ij′ū-us

S. fragilis fra′jil-is
Salmonella bongori sal-mōn-el′lä bon′gôr-ē
S. choleraesuis kol-ėr-ä-sü′is
S. enterica en-ter′i-kä
S. enteritidis en-tėr-ī′ti-dis
S. typhi tī′fē
S. typhimurium tī-fi-mür′ē-um
Sarcoptes sär-kop′tēs
Sargassum sär-gas′sum
Sartorya sär-tô′rē-ä
Schistosoma shis-tō-sō′mä or skis-tō-sō′mä
Schizosaccharomyces skiz-ō-sak-ä-rō-mī′sēs
Serratia marcescens ser-rä′tē-ä mär-ses′sens
Shigella boydii shi-gel′lä boi′dē-ē
S. dysenteriae dis-en-te′rē-ī
S. flexneri fleks′nėr-ē
S. sonnei sōn′ne-ē
Sphaerotilus natans sfe-rä′ti-lus nā′tans
Spirillum ostreae spī′ril-lum ost′rä-ī
S. volutans vō′lū-tans
Spiroplasma spī-rō-plaz′mä
Spirosoma spī-rō-sō′mä
Spirulina spī-rü-lī′nä
Spongomorphora spon′jō-môr-fô-rä
Sporothrix schenkii spô-rō′thriks shen′kē-ē
Stachybotrys stak′ē-bo-tris
Staphylococcus aureus staf-i-lō-kok′kus ô′rē-us
S. epidermidis e-pi-der′mi-dis
S. saprophyticus sap-rō-fit′i-kus
Stella stel′lä
Stigmatella aurantiaca stig-mä′tel-lä ô-rän-tē′ä-kä
Streptobacillus strep-tō-bä-sil′lus
Streptococcus agalactiae strep-tō-kok′kus ā′gal-act-ē-ī
S. mutans mū′tans
S. pneumoniae nü-mō′nē-ī
S. pyogenes pī-äj′en-ēz
S. thermophilus thėr-mo′fil-us
Streptomyces aureofaciens strep-tō-mī′sēs ô-rē-ō-fa′si-ens
S. erythraeus ā-rith′rē-us
S. fradiae frā′dē-ī
S. griseus gri′sē-us
S. nodosus nō-dō′sus
S. noursei nôr′sē-ī
S. olivoreticuli ō-liv-ō-re-tik′ū-lē
S. venezuelae ve-ne-zü-e′lī
Sulfolobus sul′fō-lō-bus

Taenia saginata te′nē-ä sa-ji-nä′tä
T. solium sō′lē-um
Talaromyces ta-lä-rō-mī′sēs
Tetrahymena tet-rä-hī′me-nä
Thermoactinomyces vulgaris thėr-mō-ak-tin-ō-mī′sēs vul-ga′ris
Thermoplasma thėr-mō-plaz′mä
Thermotoga maritima thėr′mō-tō-gä mar-it′ē-mä
Thiobacillus ferrooxidans thī-ō-bä-sil′lus fer-rō-oks′i-danz
T. thiooxidans thī-ō-oks′i-danz
Thiocapsa floridana thī-ō-kap′sä flôr′i-dä-nä
Thiomargarita namibiensis thī′ō-mär-gär-ē-tä na′mi-bē-n-sis
Thunnus tün′nus
Toxoplasma gondii toks-ō-plaz′mä gon′dē-ē
Trachelomonas trä-kel-ō-mōn′as
Treponema pallidum tre-pō-nē′mä pal′li-dum
Triatoma trī-ä-tō′ma
Tribonema vulgare trī′bō-nē-mä vul′gär-ē
Trichinella spiralis trik-in-el′lä spī-ra′lis
Trichoderma trik′ō-dėr-mä
Trichomonas vaginalis trik-ō-mōn′as va-jin-al′is
Trichonympha sphaerica trik-ō-nimf′ä sfe′ri-kä
Trichophyton trik-ō-fī′ton
Trichosporon trik-ō-spôr′on
Trichuris trichiura trik′ėr-is trik-ē-yėr′a
Tridacna trī-dak′nä
Tropheryma trō-fer-ē′mä
Trypanosoma brucei gambiense tri-pa′nō-sō-mä brüs′ē gam-bē-ens′
T. brucei rhodesiense rō-dē-sē-ens′
T. cruzi kruz′ē
Ureaplasma urealyticum ū-rē-ä-plaz′mä ū-rē-ä-lit′i-kum
Usnea üs′nē-ä
Veillonella vī-lo-nel′lä
Vibrio cholerae vib′rē-ō kol′ėr-ī
V. parahaemolyticus pa-rä-hē-mō-li′ti-kus
Wolbachia wol-ba′kē-ä
Wuchereria bancrofti vū-kėr-ār′ē-ä ban-krof′tē
Xenopsylla cheopis ze-nop′sil-lä kē-ō′pis
Yersinia enterocolitica yėr-sin′ē-ä en′tėr-ō-kōl-it-ik-ä
Y. pestis pes′tis
Y. pseudotuberculosis sū-dō-tü-bėr-kū-lō′sis
Zoogloea zō′ō-glē-ä
Zygosaccharomyces rouxii zī′gō-sak-ä-rō-mī-ses rō′ē-ē

Multiple Choice

1. c	**6.** c
2. b	**7.** a
3. b	**8.** a
4. e	**9.** b
5. b	**10.** c

Chapter 3

Review

1. $1 \ \mu m = 10^{-6} \ m$
$1 \ nm = 10^{-9} \ m$
$1 \ \mu m = 10^3 \ nm$

2. a. Ocular lens
  b. Objective lens
  c. Diaphragm
  d. Condensor
  e. Illuminator

3.

Ocular lens magnification	×	Oil immersion lens magnification	=	Total magnification of speciman
10×		100×		1000×

4. a. Compound light microscope
  b. Darkfield microscope
  c. Phase-contrast microscope
  d. Fluorescence microscope
  e. Electron microscope
  f. Differential interference contrast microscope

5. . . . that <u>a beam of electrons</u> focused by <u>magnets</u> . . . on a <u>television-like screen or photographic plate.</u>

6.

Type of Microscope	Maximum Magnification	Resolution
Compound light	2,000×	0.2 μm
Electron	100,000×	0.0025 μm

7. Bacterial cells have a slightly negative charge, and the colored positive ion of a basic dye is attracted to the negative charge of the cell. Acid dyes do not stain bacterial cells because the negatively charged colored ion is repelled by the like charge of the cell.

8. a. A simple stain is used to determine cell shape and arrangement.
  b. A differential stain is used to distinguish kinds of bacteria based on their reaction to the differential stain.
  c. A negative stain does not distort the cell and is used to determine cell shape, size, and the presence of a capsule.
  d. A flagella stain is used to determine the number and arrangement of flagella.

9. In a Gram stain, the mordant combines with the basic dye to form a complex that will not wash out of gram-positive cells. In a flagella stain, the mordant accumulates on the flagella so that they can be seen with a light microscope.

10. A counterstain stains the colorless non–acid-fast cells so that they are easily seen through a microscope.

11. In the Gram stain, the decolorizer removes the color from gram-negative cells. In the acid-fast stain, the decolorizer removes the color from non–acid-fast cells.

12. Endospore: safranin is the *counterstain.*
  Gram: safranin is the *counterstain.*

13.

Steps	Appearance after this step of Gram-positive cells	Gram-negative cells
Crysytal violet	Purple	Purple
Iodine	Purple	Purple
Alcohol-acetone	Purple	Colorless
Safranin	Purple	Red

Multiple Choice

1. e	**6.** e
2. c	**7.** d
3. b	**8.** b
4. a	**9.** a
5. a	**10.** c

Chapter 4

Review

1.

2. Endospore formation is called <u>sporogenesis</u>. It is initiated by <u>certain adverse environmental conditions</u>. Formation of a new cell from an endospore is called <u>germination</u>. This process is triggered by <u>favorable growth conditions</u>.

3.

a. ⁓⁓⁓

d.

b.

e.

c.

f.

4. Matching
d	Cell wall
f	Endospore
a	Fimbriae
c	Flagella
a, e	Glycocalyx
i	Pili
b, h	Plasma membrane
g	Ribosomes

5. An endospore is called a resting structure because it is a method of one cell "resting," or surviving, as opposed to growing and reproducing. The protective endospore wall allows a bacterium to withstand adverse conditions in the environment.

6. **a.** Both allow materials to cross the plasma membrane from a high concentration to a low concentration without expending energy. Facilitated diffusion requires carrier proteins.
   b. Both require enzymes to move materials across the plasma membrane. In active transport, energy is expended.
   c. Both move materials across the plasma membrane with an expenditure of energy. In group translocation, the substrate is changed after it crosses the membrane.

7. Mycoplasmas do not have cell walls.

8. **a.** Both lack cell walls. A spheroplast is a gram-negative cell whose wall has beeen destroyed by lysozyme. An L form is a cell that is not synthesizing a complete wall.
   b. Both lack cell walls. Mycoplasmas do not normally (genetically) make walls. L forms do not make walls because of environmental reasons, e.g., penicillin.

9. **a.** Diagram (a) refers to a gram-positive bacterium because the lipopolysaccharide–phospholipids–lipoprotein layer is absent.
   b. The gram-negative bacterium initially retains the violet stain, but it is released when the outer membrane is dissolved by the decolorizing agent. After the dye–iodine complex enters, it becomes trapped by the peptidoglycan of gram-positive cells.
   c. The outer layer of the gram-negative cells prevents penicillin from entering the cells.
   d. Essential molecules diffuse through the gram-positive wall. Porins and specific channel proteins in the gram-negative outer membrane allow passage of small water-soluble molecules.
   e. Gram-negative.

10. An extracellular enzyme (amylase) hydrolyzes starch into disaccharides (maltose) and monosaccharides (glucose). A carrier enzyme (maltase) hydrolyzes maltose and moves one glucose into the cell. Glucose can be transported by group translocation as glucose-6-phosphate.

11. **Matching**
c	Centriole
d	Chloroplasts
g	Golgi complex
a	Lysosomes
f	Mitochondria
b	Peroxisomes
e	Rough ER

12. A mitochondrion is an example of an organelle that resembles a prokaryotic cell. The inner membrane of a mitochondrion is arranged in folds similar to mesosomes. ATP is generated on this membrane just as it is in prokaryotic plasma membranes. Mitochondria can reproduce by binary fission, and they contain circular DNA and 70S ribosomes.

13. Phagocytosis. Pinocytosis.

14. Erythromycin inhibits protein synthesis in a prokaryotic cell; it will inhibit protein synthesis in mitochondria and chloroplasts.

Multiple Choice

1. e	**6.** e
2. d	**7.** b
3. b	**8.** e
4. a	**9.** a
5. d	**10.** b

Chapter 5

Review

1. Metabolism is the sum of all chemical reactions that occur within a living organism.

2. Catabolic reactions break down organic compounds and release energy, while anabolic reactions use the products of catabolism and energy to build cell material.

3.

Enzyme (a) (b) (c) Substrate Competitive inhibitor Noncompetitive inhibitor (d)

4. **a.** When the enzyme and substrate combine, the substrate molecule will be transformed.
   b. When the competitive inhibitor binds to the enzyme, the enzyme will not be able to bind with the substrate.
   c. When the noncompetitive inhibitor binds to the enzyme, the active site of the enzyme will be changed so the enzyme cannot bind with the substrate.
   d. The noncompetitive inhibitor.

5. **(a)** is the Calvin–Benson cycle, **(b)** is glycolysis, and **(c)** is the Krebs cycle.

6. Glycerol is catabolized by pathway (b) as dihydroxyacetone phosphate. Fatty acids by pathway (c) as acetyl groups.

7. In pathway (c) at ∝-ketoglutaric acid.

8. Glyceraldehyde-3-phosphate from the Calvin–Benson cycle enters glycolysis. Pyruvic acid from glycolysis is decarboxylated to produce acetyl for the Krebs cycle.

9. In (a), between glucose and glyceraldehyde-3-phosphate.

10. The conversion of pyruvic acid to acetyl, isocitric acid to ∝-ketoglutaric acid, and ∝-ketoglutaric acid to succinyl~CoA.

11. By pathway (c) as acetyl groups.

12.

	Uses	Produces
Calvin–Benson cycle	6 NADPH	
Glycolysis		2 NADH
Pyruvic acid → acetyl		1 NADH
Isocitric acid → ∝-ketoglutaric acid		1 NADH
∝-ketoglutaric acid → Succinyl~CoA		1 NADH
Succinic acid → Fumaric acid		1 FADH2
Malic acid → Oxaloacetic acid		1 NADH

13. Dihydroxyacetone phosphate; acetyl; oxaloacetic acid; ∝-ketoglutaric acid.

14. The optimum temperature for an enzyme is one that favors movement of molecules so the enzyme can "find" its substrate. Lower temperatures will decrease the rate of collisions and the rate of reactions. Increased temperatures will denature the enzyme.

15. Ethyl alcohol, lactic acid, butyl alcohol, acetone, and glycerol are some of the possible products. Refer to Table 5.4 and Figure 5.18b.

16.

Organism	Carbon Source	Energy Source
Photoautotroph	CO_2	Light
Photoheterotroph	Organic molecules	Light
Chemoautotroph	CO_2	Inorganic molecules
Chemoheterotroph	Organic molecules	Organic molecules

17.

ATP generated by	Reaction
Photophosphorylation	An electron, liberated from chlorophyll by light, is passed down an electron transport chain.
Oxidative phosphorylation	Cytochrome *c* passes two electrons to cytochrome *a*.
Substrate-level phosphorylation	(structural diagram) Phosphoenolpyruvic acid → Pyruvic acid

18. **a.** Oxidation–reduction: A coupled reaction in which one substance is oxidized and one is reduced.

b. The final electron acceptor in aerobic respiration is molecular oxygen; in anaerobic respiration, it is another inorganic molecule.

c. In cyclic photophosphorylation, electrons are returned to chlorophyll. In noncyclic photophosphorylation, chlorophyll receives electrons from hydrogen atoms.

19. The pentose phosphate pathway produces pentoses for the synthesis of nucleic acids, precursors for the synthesis of glucose by photosynthesizing organisms, precursors in the synthesis of certain amino acids, and NADPH.

20. Oxidation

21. Reactions requiring ATP are coupled with reactions that produce ATP.

22.

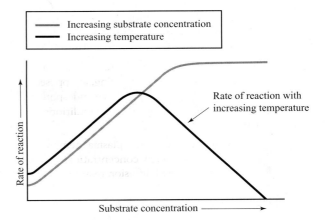

The reaction rate will increase until the enzymes are saturated.

Multiple Choice

1. a	3. b	5. c	7. b	9. c
2. d	4. c	6. b	8. a	10. b

Chapter 6
Review

1. In binary fission, the cell elongates and the chromosome replicates. Next, the nuclear material is evenly divided. The plasma membrane invaginates toward the center of the cell. The cell wall thickens and grows inward between the membrane invaginations; two new cells result.

2. Refer to Figure 6.14. The period of no cell division is called lag phase. During lag phase, the bacteria are synthesizing enzymes that are necessary for growth. In log phase, the cells are dividing at the maximum rate under the conditions provided. The number of cells dividing equals the number of cells dying in stationary phase. When the number of deaths exceeds the number of divisions, death phase is observed.

3. Carbon (C) is required for synthesis of molecules that make up a living cell. Carbon-containing compounds also are required as an energy source for heterotrophs.

4. Most bacteria grow best between pH 6.5 and 7.5.

5. The addition of salt or sugar to foods increases the osmotic pressure for microorganisms on the food. The resulting hypertonic environment causes plasmolysis of the microbial cells.

6. **a.** Catalyzes the breakdown of H_2O_2 to O_2 and H_2O.
   b. H_2O_2; peroxide ion is O_2^{2-}.
   c. Catalyzes the breakdown of H_2O_2;
   $$NADH + H^+ + H_2O_2 \xrightarrow{\text{Peroxidase}} NAD^+ + 2H_2O$$
   d. O_2^-; this diatom has one unpaired electron.
   e. Converts superoxide to O_2 and H_2O_2;
   $$2O_2^- \cdot + 2H^+ \xrightarrow{\text{Superoxide dismutase}} O_2 + H_2O_2$$
   The enzymes are important in protecting the cell from the strong oxidizing agents, peroxide and superoxide, that form during respiration.

7. Both environments prevent molecular oxygen from reaching the bacterial cells. In reducing media, thioglycolate combines with dissolved oxygen, thereby removing it from the medium. In an anaerobic incubator, air is replaced with an atmosphere of CO_2 (and N_2). *Clostridium* is an obligate anaerobe that lacks superoxide dismutase and catalase. Consequently, the accumulation of superoxides and peroxides will kill the cell in an aerobic environment.

8. Direct methods are those in which the microorganisms are seen and counted. Direct methods are direct count, standard plate count, filtration, and most probable number.

9. The growth rate of bacteria slows down with decreasing temperatures. Mesophilic bacteria will grow slowly at refrigeration temperatures and will remain dormant in a freezer. Bacteria will not spoil food quickly in a refrigerator.

10. Number of cells $\times$ 2^n generations = Total number of cells
    $$6 \quad \times \quad 2^7 \quad = \quad 768$$

11. Petroleum can meet the carbon and energy requirements for an oil-degrading bacterium; however, nitrogen and phosphate are usually not available in large quantities. Nitrogen and phosphate are essential for making proteins, phospholipids, nucleic acids, and ATP.

12. A chemically defined medium is one in which the exact chemical composition is known. A complex medium is one in which the exact chemical composition is not known.

13.
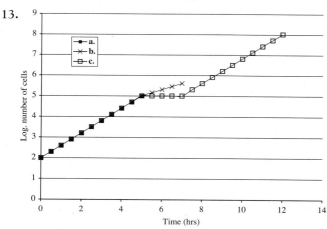

Multiple Choice

1. c   **6.** d
2. a   **7.** e
3. c   **8.** a
4. a   **9.** b
5. c   **10.** b

Chapter 7

Review

1. **a.** Lysis.
   b. Altered permeability and leakage of cell contents.
   c. Destruction of enzymes and structural proteins such as those in the plasma membrane.
   d. Interference with protein synthesis and cell division.

2. Autoclave. Due to the high specific heat of water, moist heat is readily transferred to cells.

3. Most organisms that cause disease or rapid spoilage of food are destroyed by pasteurization.

4. Variables that affect determination of the thermal death point are
   a. The innate heat resistance of the strain of bacteria.
   b. The past history of the culture, whether it was freeze-dried, wetted, etc.
   c. The clumping of the cells during the test.
   d. The amount of water present.
   e. The organic matter present.
   f. Media and incubation temperature used to determine viability of the culture after heating.

5. **a.** Ionizing radiation can break DNA directly. However, due to the high water content of cells, the formation of free radicals (H· and OH·) that break DNA strands is likely to occur.
   b. Ultraviolet radiation damages DNA by the formation of thymine dimers.

6. Microorganisms tend to die at a constant rate over a period of time. The constant rate is indicated by the straight line after exposure to the bactericidal compound.

7.

Sterilization Method	Temp.	Time	Type	Preferred Use	Mechanism of Action
Autoclaving	121°C	15 min	Moist	Media, equipment	Protein denaturation
Hot air	170°C	2 hr	Dry	Glassware	Oxidation
Pasteurization	72°C	15 sec	Moist	Milk, alcoholic drinks	Protein denaturation

8. All three processes kill microorganisms; however, as moisture and/or temperatures are increased, less time is required to achieve the same result.

9. Salts and sugars create a hypertonic environment. Salts and sugars (as preservatives) do not directly affect cell structures or metabolism; instead, they alter the osmotic pressure. Jams and jellies are preserved with sugar; meats are usually preserved with salt. Molds are more capable of growth in high osmotic pressure than bacteria.

10. 1. Acts rapidly.
   2. Attacks all, or a wide range of, microbes.
   3. Is able to penetrate.
   4. Readily mixes with water.
   5. Is not hampered by organic matter.
   6. Stable.
   7. Does not stain or corrode.
   8. Nontoxic.
   9. Pleasant odor.
   10. Economical.
   11. Safe to transport.

11.

Method of Action	Standard Use
a. Disrupts plasma membrane	Skin surfaces
b. Inhibits protein function	Antiseptic
c. Oxidation	Disinfect water
d. Denatures proteins, destroys lipids	Skin surfaces
e. Oligodynamic	AgNO₃ to prevent gonococcal eye infections
f. Inactivation of proteins	Chemical sterilizer
g. Denatures proteins	Chemical sterilizer
h. Oxidation	Antiseptic

12. Disinfectant B is preferable because it can be diluted more and still be effective.

13. Quaternary ammonium compounds are most effective against gram-positive bacteria. Gram-negative bacteria that were stuck in cracks or around the drain of the tub would not have been washed away when the tub was cleaned. These gram-negative bacteria could survive the washing procedure. Some pseudomonads can grow on quats that have accumulated.

Multiple Choice

1. d	6. b
2. b	7. b
3. d	8. a
4. d	9. a
5. b	10. b

Chapter 8

Review

1. DNA consists of a strand of alternating sugars (deoxyribose) and phosphate groups with a nitrogenous base attached to each sugar. The bases are adenine, thymine, cytosine, and guanine. DNA exists in a cell as two strands twisted together to form a double helix. The two strands are held together by hydrogen bonds between their nitrogenous bases. The bases are paired in a specific, complementary way: A-T and C-G. The information held in the sequence of nucleotides in DNA is the basis for synthesis of RNA and proteins in a cell.

2. a.

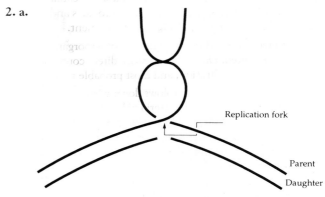

Replication fork

Parent

Daughter

   b. DNA polymerases synthesize a complementary strand of DNA from a DNA template. RNA polymerase starts each fragment of the lagging strand with an RNA primer.
   c. Each new double-stranded DNA molecule contains one original strand and one new strand.

3. a. ATAT<u>TACTTTGCATGGACT</u>.
   b. met-lys-arg-thr-(end).
   c. TATAATGAAACGTTCCTGA.
   d. No change.
   e. Cysteine substituted for arginine.
   f. Proline substituted for threonine (missense mutation).
   g. Frameshift mutation.
   h. Adjacent thymines might polymerize.
   i. ACT.

4. One end of the mRNA molecule becomes associated with a ribosome. Ribosomes are composed of rRNA and protein. The anticodon of a tRNA with its activated amino acid pairs with the mRNA codon at the ribosome.

5. A mutant is isolated by direct selection because it grows on a particular medium. The colonies on an antibiotic-containing medium can be identified as resistant to that antibiotic.
   A mutant is isolated by indirect selection because it does not grow on a particular medium. Replica plating

could be employed to inoculate an antibiotic-containing medium. Colonies that did not grow on this medium can be isolated from the original plate and are antibiotic sensitive.

6. **Matching**

b	A mutagen that is incorporated into DNA in place of a normal base.
d	A mutagen that causes the formation of highly reactive ions.
c	A mutagen that alters adenine so that it base-pairs with cytosine.
a	A mutagen that causes insertions.
e	A mutagen that causes the formation of pyrimidine dimers.

7. The basis for the Ames test is that a mutated cell can revert to a cell resembling the original, nonmutant cell by undergoing another mutation. The reversion rate of histidine auxotrophs of *Salmonella* in the presence of a mutagen will be higher than the spontaneous rate (in the absence of a mutagen).

8. Plasmids are small, self-replicating circles of DNA that are not associated with the chromosome. The F plasmid can be integrated into the chromosome. The F plasmid can be transferred from a donor to a recipient cell in conjugation. When the F plasmid becomes integrated into the chromosome, the cell is called an Hfr. During conjugation between an Hfr and an F cell, the chromosome of the Hfr cell, with its integrated F factor, replicates, and the new copy of the chromosome is transferred to the recipient cell.

9. a. . . . a <u>repressor</u> protein must be bound tightly to the <u>operator</u> site . . . it will bind to the <u>repressor</u> so that <u>transcription</u> can occur.
   b. . . . called a <u>corepressor</u>, causes the <u>repressor</u> to bind to the <u>operator</u>. Derepression is by removal of the <u>corepressor</u>, C in this case, when the <u>corepressor</u> is needed in the cell.
   c. None; constitutive enzymes are produced at certain necessary levels regardless of the amount of substrate or end-product.

10. Light repair; dark repair; proofreading by DNA polymerase.

11. a. The genetic makeup of an organism.
    b. The external manifestations of the genotype.
    c. Rearrangement of genes to form new combinations; in nature, this usually occurs between members of the same species; in vitro, recombinant DNA is made from genes of different species.

12. CTTTGA. Endospores and pigments offer protection against UV radiation. Additionally, repair mechanisms can remove and replace thymine polymers.

13. a. Culture 1 will remain the same. Culture 2 will convert to F^+ but will have its original genotype.
    b. The donor and recipient cells' DNA can recombine to form combinations of $A^+B^+C^+$ and $A^-B^-C^-$. If the F plasmid also is transferred, the recipient cell may become F^+.

14. Semiconservative replication ensures the offspring cell will have one correct strand of DNA. Any mutations that may have occurred during DNA replication have a greater chance of being correctly repaired.

15. Mutation and recombination provide genetic diversity. Environmental factors select for the survival of organisms through natural selection. Genetic diversity is necessary for the survival of some organisms through the processes of natural selection. Organisms that survive may undergo further genetic change, resulting in the evolution of the species.

Multiple Choice

1. c	6. b
2. d	7. a
3. c	8. c
4. d	9. d
5. c	10. a

Chapter 9

Review

1. Recombinant DNA (rDNA) is DNA that is combined from different sources. In nature, rDNA results from conjugation, transduction, and tranformation. Genetic engineering is the artifical making of rDNA.

2. a. Both are DNA. cDNA is a segment of DNA made by RNA-dependent DNA polymerase. It is not necessarily a gene; a gene is a transcribable unit of DNA that codes for protein or RNA.
   b. Both are DNA. A restriction fragment is a segment of DNA produced when a restriction endonuclease hydrolyzes DNA. It is not usually a gene; a gene is a transcribable unit of DNA that codes for protein or RNA.
   c. Both are DNA. A DNA probe is a short, single-stranded piece of DNA. It is not a gene; a gene is a transcribable unit of DNA that codes for protein or RNA.
   d. Both are enzymes. DNA polymerase synthesizes DNA one nucleotide at a time using a DNA template; DNA ligase joins pieces (strands of nucleotides) together.
   e. Both are DNA. Recombinant DNA results from joining DNA from two different sources; cDNA results from copying a strand of RNA.
   f. The proteome is the expression of the genome. An organism's genome is one complete copy of its genetic information. The proteins encoded by this genetic material comprise the proteome.

3. a. A desired gene can be spliced into a plasmid and inserted into a cell by transformation.
   b. A desired gene can be spliced into a viral genome and inserted into a cell by transduction.
   c. Antibiotic resistance genes are used as markers or labels on plasmids so that the cell containing the plasmid can be found by direct selection on an antibiotic-containing medium.
   d. A genetically engineered bacterium should be producing a new protein product. Radioactively labeled antibodies against a specific protein can be used to locate the bacterial colony producing the protein.

4. Restriction fragments from one source can be cloned in microbial cells to make a gene library. Synthetic DNA is made in a lab.

5. In protoplast fusion, two wall-less cells fuse together to combine their DNA. A variety of genotypes can result from this process. In b, c, and d, specific genes are inserted directly into the cell.

6. *Bam*H I, *Eco*R I, and *Hind* III make sticky ends. Fragments of DNA produced with the same restriction enzyme will spontaneously anneal to each other at their sticky ends.

7. The gene can be spliced into a plasmid and inserted into a bacterial cell. As the cell grows, the number of plasmids will increase. The polymerase chain reaction can make copies of a gene using DNA polymerase in vitro.

8. In a eukaryotic cell, RNA polymerase copies DNA; RNA processing removes the introns, leaving the exons in the mRNA. cDNA can be made from the mRNA by reverse transcriptase.

9. See Tables 9.1 and 9.2.

10. You probably used a few plant cells in a Petri plate for your experiment. How will you select the plant cells that actually have the new Ti plasmid? You can grow these cells on plant-cell culture media with tetracycline. Only the cells with the new plasmid will grow.

Multiple Choice

1. b   6. d
2. b   7. c
3. b   8. b
4. b   9. e
5. c   10. a

Chapter 10

Review

1. Taxonomy is the science of classifying organisms to establish the relatedness between groups of organisms.

2. The three distinct chemical types of cells (see Table 10.1).

3. Living organisms cannot be grouped into two groups. For example, plant and animal is not acceptable because if fungi are grouped with plants, the definition of plants can't include *cellulose* and *photosynthesis*. If fungi are grouped with animals, the definition of animals can't include *no cell wall* and *ingestive*. The goal is to look for a "natural" scheme; that is, what criteria can be used to characterize all organisms.

4. Fungi: Unicellular or multicellular organisms that absorb organic nutrients; noncellulose cell walls; lack flagella.
Plantae: Multicellular eukaryotes with tissue formation; cellulose cell walls; generally photosynthetic.
Animalia: Multicellular eukaryotes with tissue formation; develop from an embryo (gastrula); lacking cell walls; ingest organic nutrients through a mouth of some kind.

5. **a.** Both are prokaryotic. They differ in composition of their cell walls, plasma membranes, and rRNAs.
   b. Both are bound by ester-linked plasma membranes. Eukarya have membrane-bound organelles.
   c. Both use methionine as the start signal. Eukarya have membrane-bound organelles and ester-linked membranes.

6. Binomial nomenclature is the system of assigning a genus and specific epithet to each organism.

7. Common names are not specific and can be misleading. According to the rules of scientific nomenclature, each organism has only one binomial.

8. The genus name must be written out so the reader knows what organism is being discussed, since the abbreviation for both of these species is *E. coli*.

9. Domain, kingdom, phylum, class, order, family, genus, species.

10. Domain: Bacteria
Phylum: Firmicutes
Class: Bacilli
Order: Bacillales
It is more related to *Gemella*. Family: Staphylococcaceae

11. A eukaryotic species is a group of closely related organisms having limited geographical distribution that interbreeds but does not breed with other species. Species can be distinguished morphologically. Because of the distinct differences between eukaryotic organisms and bacteria, a bacterial species is defined as a population of cells with similar characteristics. A viral species is a population of viruses with similar characteristics that occupies a particular ecological niche.

12. (See Table 10.5)
Used primarily for identification:
   morphological characteristics
   differential staining
   biochemical tests
   serology
   phage typing
   fatty acid profiles
Used primarily for taxonomic classification:
   flow cytometry
   DNA base composition
   DNA fingerprinting
   rRNA sequencing
   PCR
   nucleic acid hybridization
Data obtained from laboratory tests employing any (or all) of these twelve techniques can be assimilated using numerical taxonomy to provide information on classification.

13. Most microorganisms do not contain structures that are readily fossilized, making it difficult to obtain information on the evolution of microorganisms. Recent developments in molecular biology have provided techniques for determining evolutionary relationships amongst bacteria.

14. A and D appear to be most closely related because they have similar G-C moles %. No two are the same species.

Multiple Choice

1. b	**6.** a
2. e	**7.** a
3. d	**8.** e
4. b	**9.** a
5. e	**10.** b

Chapter 11

Review

1. I.   Gram-positive
    A. Endospore-forming rod
        1. *Clostridium*
        2. *Bacillus*
    B. Nonendospore-forming
        1. Cells are rods
           a. *Streptomyces*
           b. *Mycobacterium*
        2. Cells are cocci
           a. *Streptococcus*
           b. *Staphylococcus*
  II.   Gram-negative
    A. Cells are helical or curved
        1. *Treponema*
        2. *Spirillum*
    B. Cells are rods
        1. *Pseudomonas*
        2. *Escherichia*
  III.   Lack of cell walls
    A. *Mycoplasma*
    B. *Chlamydia*
  IV.   Obligate intracellular parasites
    A. *Rickettsia*
    B. *Coxiella*

2. a. Both are oxygenic photoautotrophs. Cyanobacteria are prokaryotes; algae are eukaryotes.

  b. Both are chemoheterotrophs capable of forming mycelia; some form conidia. Actinomycetes are prokaryotes; fungi are eukaryotes.

  c. Both are large rod-shaped bacteria. *Bacillus* forms endospores, *Lactobacillus* is a fermentative non-endospore-forming rod.

  d. Both are small rod-shaped bacteria. *Pseudomonas* has an oxidative metabolism; *Escherichia* is fermentative. *Pseudomonas* has polar flagella; *Escherichia* has peritrichous flagella.

  e. Both are helical bacteria. *Leptospira* (a spirochete) has an axial filament. *Spirillum* has flagella.

  f. Both are gram-negative, rod-shaped bacteria. *Escherichia* are facultative anaerobes, and *Bacteroides* are anaerobes.

  g. Both are obligatory intracellular parasites.

  h. Both lack peptidoglycan cell walls. *Ureaplasma* are archaea; *Mycoplasma* are bacteria (see Table 10.2).

3. Matching

Nitrogen-fixing	d	*Frankia*
Anoxygenic	i	Purple bacteria
Oxygenic	a	Cyanobacteria
Oxidize NO_2^-	h	*Nitrobacter*
Reduce CO_2	f	Methanogens
Slime layer	b	*Cytophaga* and
	g	Myxobacteria
Myxocysts	g	Myxobacteria
Anaerobic	c	*Desulfovibrio*
Thermophilic	k	*Sulfolobus*
Filaments	j	*Sphaerotilus*
Projections	e	*Hyphomicrobium*

Multiple Choice

1. d	**6.** d
2. b	**7.** e
3. e	**8.** b
4. a	**9.** b
5. b	**10.** a

Chapter 12

Review

1. Conidiospores are asexual spores formed by the aerial mycelia of one organism. Ascospores are sexual spores resulting from the fusion of the nuclei of two opposite mating strains of the same species of fungus.

2.

	Spore Type(s)	
Phylum	**Asexual**	**Sexual**
Zygomycota	Zygospore	Sporangiospore
Ascomycota	Blastoconidia, Arthrospore	Ascospore
Basidiomycota	Blastoconidia	Basidiospore

3.

Genus	**Mycosis**
Blastomyces	Systemic
Sporothrix	Subcutaneous
Microsporum	Cutaneous
Trichosporon	Superficial
Aspergillus	Systemic

4. a. *E. coli*
  b. *P. chrysogenum*

5. The alga produces carbohydrates for the lichen, and the fungus provides both the holdfast and protection from desiccation.

6. As the first colonizers on newly exposed rock or soil, lichens are responsible for the chemical weathering of large inorganic particles and the consequent accumulation of soil.

Appendix F **629**

7.

Phylum	Cell wall composition	Special features
Oomycotes	Chitin	Heterotrophic; flagellated spores.
Dinoflagellates*	Cellulose and silica	Some produce neurotoxins
Diatoms*	Pectin and silica	Produce much O_2; fossilized inclusions form petroleum
Red algae	Cellulose	Grow in deep water; source of agar
Brown algae	Cellulose and alginic acid	Source of algin
Green algae	Cellulose	Plantlike

*Unicellular

The green algae (Chlorophyta) could be placed in the plant kingdom. They have chlorophyll b, as do land plants, and have colonial forms. In the most advanced colonial form (*Volvox*), groups of *Chlamydomonas*-like cells live together; some are specialized for reproductive functions, which suggests a possible evolutionary route for the formation of plant tissue. The other algae are most often classified as protists.

8. Cellular slime molds exist as individual amoeboid cells. Plasmodial slime molds are multinucleate masses of protoplasm. Both survive adverse environmental conditions by forming spores.

9. Complete the following table.

Phylum	Method of Motility	One Human Parasite
Archaezoa	Flagella	*Giardia*
Microsporidia	None	*Nosema*
Rhizopoda	Pseudopods	*Entamoeba*
Apicomplexa	None	*Plasmodium*
Ciliophora	Cilia	*Balantidium*
Euglenozoa	Flagella	*Trypanosoma*

10. *Trichomonas* cannot survive for long outside of a host because it does not form a protective cyst. *Trichomonas* must be transferred from host to host quickly.

11. Asexual reproduction occurs in the human host and sexual reproduction takes place in the mosquito. The definitive host and the vector are the mosquito.

12. Ingestion.

13. This is a cestode. The encysted larva is called a cysticercus. Tapeworms are dorsoventrally flattened and have an incomplete digestive system.

14. The male reproductive organs are in one individual, and the female reproductive organs in another. Nematodes belong to the Phylum Aschelminthes.

15.

Vector Type	Example	Disease
Mechanical	Housefly	Salmonellosis
Suitable for reproduction of parasite	*Ixodes*	Lyme disease
As a host	*Anopheles*	Malaria

Multiple Choice

1. c
2. b
3. b
4. a
5. d

6. b
7. a
8. c
9. a
10. d

Chapter 13

Review

1. The term filterable describes the property of passing through filters that retain bacteria. Viruses are too small to be seen with a light microscope, but their presence is known because material passed through a filter is still capable of causing a disease.

2. Viruses absolutely require living host cells to multiply.

3. A virus:
   a. Contains DNA or RNA;
   b. Has a protein coat surrounding the nucleic acid;
   c. Multiplies inside a living cell using the synthetic machinery of the cell; and
   d. Causes the synthesis of virions. A virion is a fully developed virus particle that transfers the viral nucleic acid to other cells and initiates multiplication.

4. The capsid of a helical virus is a hollow cylinder with a helical shape, which surrounds the nucleic acid (see Figure 13.4). An example of a helical virus is tobacco mosaic virus. Polyhedral viruses are many-sided (Figure 13.2). A polyhedral virus in the shape of an icosahedron is adenovirus. Polyhedral or helical viruses surrounded by an envelope are called enveloped viruses. An example of an enveloped helical virus is *Influenzavirus* (Figure 13.3), and herpes simplex is an enveloped polyhedral virus.

5. A sample of bacteriophage is mixed with host bacteria and melted nutrient agar. The mixture is then poured over a layer of nutrient agar in a Petri plate. Each phage infects a bacterium, multiplies, and releases new phages. These newly produced phages infect other bacteria, and more new

viruses are produced. Following multiplication, the bacteria are destroyed. This produces a number of clearings or plaques in the layer of bacteria. The number of phages in the original sample can be estimated by counting the number of plaques.

6. Primary cell lines tend to die after a few generations. Continuous cell lines can be maintained through an indefinite number of generations. Continuous cell lines then allow long-term observations of viruses. Continuous cell lines are transformed cells.

7. A prophage gene codes for the cholera toxin. When phage DNA is incorporated into the cell's DNA, toxin can be produced.

8. **Adsorption**: The virus attaches to the cell membrane by means of spikes located on its envelope.
Penetration: The virus gains entrance by piriocytosis, or its envelope may fuse with the plasma membrane of the host cell, allowing the virus to enter the cell.
Uncoating: Uncoating refers to the separation of the capsid from the viral DNA.
Biosynthesis: Viral DNA is released into the cell's nucleus, and transcription and translation from viral DNA occur. Viral DNA is synthesized.
Maturation: Capsids form around strands of viral DNA.
Release: The assembled capsid-containing nucleic acid pushes through the plasma membrane; a portion of the plasma membrane adheres to the capsid, thus forming the envelope.

9. **a.** Viruses cannot easily be observed in host tissues. Viruses cannot easily be cultured in order to be inoculated into a new host. Additionally, viruses are specific for their hosts and cells, making it difficult to substitute a laboratory animal for the third step of Koch's postulates.
b. Some viruses can infect cells without inducing cancer. Cancer may not develop until long after infection. Cancers do not seem to be contagious.

10. Subacute sclerosing panencephalitis . . . common viruses . . . Students will have to suggest a mechanism to fill in the last blank; some examples are latent, in an abnormal tissue.

11. Provirus
TSTA appear on the host cell surface, or T antigens appear in the nucleus. Transformed cells do not exhibit contact inhibition.
RNA-containing oncogenic viruses produce a double-stranded DNA molecule using reverse transcriptase. The DNA is integrated into the host cell's DNA as a provirus. The provirus may transform the host cell into a tumor cell.

12. Prions are infectious proteins that appear to lack any nucleic acid. Viroids are infectious RNAs that do not have a protein coat. A prion causes CJD. A viroid causes potato spindle tuber viroid disease.

13. Of the rigid cell walls . . . vectors such as sap-sucking insects . . . plant protoplasts and insect cell cultures.

Multiple Choice

1. e	6. e
2. c	7. c
3. b	8. d
4. c	9. d
5. b	10. c

Chapter 14

Review

1. **a.** Etiology is the study of the cause of a disease, whereas pathogenesis is the manner in which the disease develops.
b. Infection refers to the colonization of the body by a microorganism. Disease is any change from a state of health. A disease may, but does not always, result from infection.
c. A communicable disease is a disease that is spread from one host to another, whereas a noncommunicable disease is not transmitted from one host to another.

2. Microorganisms that reside more or less permanently on the body are called normal microbiota. Microorganisms that are present for a few days or weeks are transient microbiota.

3. Symbiosis refers to different organisms living together. Commensalism is a symbiotic relationship in which one of the organisms is benefited and the other is unaffected. Corynebacteria living on the surface of the eye are commensals. Mutualism is a symbiosis in which both organisms are benefited. *E. coli* receives nutrients and a constant temperature in the large intestine and produces vitamin K and certain B vitamins that are useful for the human host. In parasitism, one organism benefits while the other is harmed. *Salmonella enterica* receives nutrients and warmth in the large intestine, and the human host experiences gastroenteritis or typhoid fever.

4. A reservoir of infection is a source of continual infection.

Matching

b	Influenza
c	Rabies
a	Botulism

5. Koch's postulates establish the etiology of an infectious disease because the microorganism is removed from a sick organism, grown in a laboratory culture, and introduced into a healthy, susceptible organism. This shows that the microorganism caused the disease—not contact with a sick individual or environmental conditions.

Some organisms are not easily seen in a host. Some microorganisms cannot be cultured on laboratory media. And some microorganisms are specific for one host. In a human, these pathogens will give rise to a group of signs and symptoms, but they will not cause the same disease in laboratory animals.

6. **a.** In transmission by direct contact, some kind of body contact between an infected individual and a susceptible host is required.

b. Pathogens are transmitted from one host to another by fomites via indirect contact.

c. Arthropod vectors can transmit pathogens mechanically where the pathogen is carried on external body parts. When an arthropod ingests a pathogen and the pathogen reproduces in the vector, it is called biological transmission.

d. Droplet transmission also is a method of contact transmission. Pathogens are transmitted by droplets of saliva or mucus.

e. Pathogens can be transmitted to a large number of individuals by food or water. This is called common-vehicle transmission.

f. Airborne transmission refers to the spread of pathogens by droplet nuclei or dust.

7. Nutrition, fatigue, age, habits, lifestyle, occupation, preexisting illness, chemotherapy.

8. a. Acute
   b. Chronic
   c. Subacute

9. Hospital patients may be in a weakened condition and therefore predisposed to infection. Pathogenic microorganisms are generally transmitted to patients by contact and airborne transmission. The reservoirs of infection are the hospital staff, visitors, and other patients.

10. A disease constantly present in a population is an endemic disease. When many people acquire the disease in a relatively short time, it is an epidemic disease.

11. Epidemiology is the science dealing with when and where diseases occur and how they are transmitted in the human population. The Centers for Disease Control and Prevention (CDC) is a central source of epidemiological information.

12. Changes in body function felt by the patient are called symptoms. Symptoms such as weakness or pain are not measurable by a physician. Objective changes that the physician can observe and measure are called signs.

13. When microorganisms causing a local infection enter a blood or lymph vessel and are spread throughout the body, a systemic infection can result.

14. Mutualistic microorganisms are providing a chemical or environment that is essential for the host. In a commensal relationship, the microorganisms are obtaining nutrients from sloughed-off cells and secretions, which benefits the host by removing materials that might be invaded by pathogens. These organisms, however, are not essential; another microorganism might serve the function as well.

15. Incubation period, prodromal period, period of illness, period of decline (may be crisis), period of convalescence.

Multiple Choice

1. a	6. c
2. b	7. d
3. a	8. a
4. d	9. c
5. b	10. b

Chapter 15
Review

1. Mucous membranes: Microorganisms can adhere to and then penetrate mucous membranes. Skin: Microorganisms can penetrate unbroken skin through hair follicles and sweat ducts. Parenteral route: Pathogens can be introduced into tissues beneath the skin and mucous membranes by punctures, injections, bites, and cuts.

2. The ability of a microorganism to produce a disease is called pathogenicity. The degree of pathogenicity is virulence.

3. a. Would prevent adherence by making the mannose attachment site unavailable.
   b. Would prevent adherence of *N. gonorrhoeae*.
   c. *S. pyogenes* would not be able to attach to host cells and would be more susceptible to phagocytosis.

4. Cytopathic effects are observable changes produced in cells infected by viruses. Five examples are:
   a. Cessation of mitosis.
   b. Autolysis.
   c. The presence of inclusion bodies.
   d. Cell fusion producing syncytia.
   e. Transformation.

5.

	Exotoxin	Endotoxin
Bacterial source	Gram +	Gram −
Chemistry	Proteins	Lipid A
Toxigenicity	High	Low
Pharmacology	Destroy certain cell parts or physiological functions	Systemic, fever, weakness, aches, and shock
Example	Botulinum toxin	Salmonellosis

6. Encapsulated bacteria can resist phagocytosis and continue growing. *Streptococcus pneumoniae* and *Klebsiella pneumoniae* produce capsules that are related to their virulence. M protein found in the cell walls of *Streptococcus pyogenes* and A protein in the cell walls of *Staphylococcus aureus* help these bacteria resist phagocytosis.

7. Hemolysins are enzymes that cause the lysis of red blood cells; hemolysis might supply nutrients for bacterial growth. Leukocidins destroy neutrophils and macrophages that are active in phagocytosis; this decreases host resistance to infection. Coagulase is an enzyme that causes the fibrinogen in blood to clot; the clot may protect the bacterium from phagocytosis and other host defenses. Bacterial kinases break down fibrin; kinases can destroy a clot that was made to isolate the bacteria, thus allowing the bacteria to spread. Hyaluronidase dissolves the hyaluronic acid that binds cells together; this could allow the bacteria to spread through tissues. Siderophores take iron from host iron-transport proteins, thus allowing bacteria to get iron for growth. IgA proteases destroy IgA antibodies; IgA antibodies protect mucosal surfaces.

8. Pathogenic fungi do not have specific virulence factors; capsules, metabolic products, toxins, and allergic responses

contribute to the virulence of pathogenic fungi. Some fungi produce toxins that, when ingested, produce disease. Protozoa and helminths elicit symptoms by destroying host tissues and producing toxic metabolic wastes.

9. *Legionella*.

10. Botulinum toxin is more potent than *Salmonella* toxin. A much smaller amount of botulinum toxin will kill 50 percent of the inoculated hosts.

11. Food infection refers to a disease that results from pathogens entering through the gastrointestinal route. The pathogens infect the gastrointestinal tract and produce endotoxins while they are growing. Food intoxication results from ingestion of a toxin formed in food. Pathogens grow in the food and excrete an exotoxin. The pathogens do not infect the host; symptoms are due to the toxin.

12. Viruses avoid the host's immune response by growing inside host cells; some can remain latent in a host cell for prolonged periods. Some protozoa avoid the immune response by mutations that change their antigens.

Multiple Choice

1. e	**3.** d	**5.** c	**7.** b	**9.** d
2. c	**4.** d	**6.** a	**8.** a	**10.** c

Chapter 16

Review

1. **a.** The ability of the human body to ward off diseases.
   b. The lack of resistance to an infectious disease.
   c. Host defenses that tend to protect the body from any kind of pathogen.

2.

	Mechanical	Chemical
Skin	Dry, packed cells	Sebum
Eyes	Tears	Lysozyme
Digestive tract	Movement out	HCl
Respiratory tract	Ciliary escalator	
Urinary tract	Movement out	
Genital tract	Movement out	Acidic in female

3. See Table 16.1.

4. Phagocytosis is the ingestion of a microorganism or any foreign particulate matter by a cell.

5. Granulocytes have granules in the cytoplasm. Among the granulocytes, neutrophils have the most prominent phagocytic activity. Monocytes are agranulocytes (without granules) that develop into macrophages.
   When an infection occurs, granulocytes migrate to the infected area. Monocytes follow the granulocytes to the infected tissue. During migration, monocytes enlarge and develop into actively phagocytic cells called macrophages. Macrophages phagocytize dead or dying bacteria.

6. Phagocytic cells that migrate to the infected area are called wandering macrophages. Fixed macrophages remain in certain tissues and organs.

7. Refer to Figures 16.8 and 16.9.

8. Inflammation is the body's response to tissue damage. The characteristic symptoms of inflammation are redness, pain, heat, and swelling.

9. The functions of inflammation are:
   1. To destroy the injurious agent, if possible, and to remove it and its by-products from the body;
   2. If destruction is not possible, to confine or wall off the injurious agent and its by-products by forming an abscess;
   3. To repair or replace tissues damaged by the injurious agent or its by-products.

10. Leukocytic pyrogen, released from phagocytic granulocytes, has the ability to raise body temperature. The higher temperature is believed to inhibit the growth of some microorganisms. The higher temperature speeds up body reactions and may help body tissues to repair themselves more quickly.

11. The chill is an indication that body temperature is rising. Shivering and cold skin are mechanisms for increasing internal temperature. Crisis indicates body temperature is falling. The skin becomes warm as circulation is returned to it when the body attempts to dissipate extra heat.

12. Complement is a group of proteins found in normal blood serum. The classical pathway is activated by an antigen-antibody complex and C1 (see Figure 16.13). The alternative pathway is activated by microbial lipid-carbohydrate complex and factor B, factor D, and factor P (see Figure 16.14). The lectin pathway is initiated by carbohydrate-binding proteins (lectins) binding the mannose on microbes (see Figure 16.15).

13. Activation of complement can result in immune adherence and phagocytosis, local inflammation, and cell lysis.

14. Endotoxin binds C3b, which activates C5–C9 to cause cell lysis. This can result in free cell wall fragments, which bind more C3b, resulting in C5–C9 damage to host cell membranes.

15. Interferons are antiviral proteins produced by infected cells in response to viral infections. Alpha-IFN and Beta-IFN induce uninfected cells to produce antiviral proteins. Gamma-IFN is produced by lymphocytes and activates neutrophils to kill bacteria.

Multiple Choice

1. a	**3.** c	**5.** b	**7.** c	**9.** d
2. d	**4.** d	**6.** a	**8.** b	**10.** e

Chapter 17

Review

1. The ability to produce antibodies against microorganisms and their toxins provides a type of resistance called immunity.

2. **a.** Acquired immunity is the resistance to infection obtained during the life of the individual. Acquired immunity results from the production of antibodies. Innate resistance refers to the resistance of species or individuals to certain diseases that is not dependent on antigen-specific immunity such as antibodies.

   b. Humoral immunity is due to antibodies (and B cells). Cell-mediated immunity is due to T cells.

   c. Active immunity refers to antibodies produced by the individual who carries them. Passive immunity refers to antibodies produced by another source and then transferred to the individual who needs the antibodies.

   d. T_H1 cells produce cytokines that activate T cells. Cytokines produced by T_H2 cells activate B cells.

   e. Natural immunity is acquired naturally, i.e., from mother to newborn, or following an infection. Artificial immunity is acquired from medical treatment, i.e., by injection of antibodies or by vaccination.

   f. T-dependent antigens: Certain antigens must combine with self-antigens to be recognized by T_H cells and then by B cells. T-independent antigens can elicit an antibody response without T cells.

   g. T cells can be classified by their surface antigens: T_H cells possess the CD4 antigen; T_C and T_S cells have the CD8 antigen.

3. **a.** Artificially acquired active immunity.

   b. Naturally acquired active immunity.

   c. Naturally acquired passive immunity.

   d. Artificially acquired passive immunity.

4. An antigen is a chemical substance that causes the body to produce specific antibodies and can combine with these antibodies. A hapten is a low-molecular-weight substance that is not antigenic unless it is attached to a carrier molecule. Once an antibody has been formed against the hapten, the hapten alone will react with the antibodies independently of its carrier.

5. An antibody is a protein produced by the body in response to the presence of an antigen; it is capable of combining specifically with that antigen. Antibodies are proteins and usually consist of four polypeptide chains. Two of the chains are identical and are called heavy (H) chains. The other two chains are identical to each other but are of lower molecular weight and are called light (L) chains. The variable portions of the H and L chains are where antigen binding occurs. The variable portion is different for each kind of antibody. The remaining constant portions of each chain are identical for all of the antibodies in one class of antibody. Refer to Figure 17.5 for the structure of IgG antibodies.

6. Each person has a population of B cells with receptors for different antigens. When the appropriate antigen contacts the antigen receptor on a B cell, the cell proliferates to produce a clone of cells. Plasma cells in this clone produce antibodies specific to the antigen that caused their formation.

7. See Figures 17.8, 17.13, 17.14, 17.16, 17.18, and 17.19.

8. Cytotoxic T cells (T_C) destroy target cells upon contact. Delayed hypersensitivity T cells (T_D) produce lymphokines. Helper T cells (T_H) interact with an antigen to "present" it to a B cell for antibody formation. Suppressor T cells (T_S) inhibit the conversion of B cells into plasma cells. Lymphokines cause an inflammatory response. An example of a cytokine is macrophage chemotactic factor, which attracts macrophages to the infection site. See Table 17.2 for functions of other cytokines.

9. **a.** Area *a* shows the primary response to the antigen. Area *b* shows the anamnestic response, in which the antibody titer is greater and remains high longer than in the primary response. The booster dose stimulated the memory cells to respond to the antigen.

10. Neutralize toxins, inactivate viruses, fix complement to initiate cytolysis.

11. Surface recognition sites for antigen peptides and MHC proteins.

12. NK cells lyse target cells (usually tumor cells and virus-infected cells) on contact.

13. Both would prevent attachment of the pathogen; (a) interferes with the attachment site on the pathogen and (b) interferes with the pathogen's receptor site.

14. See Figure 17.11.

15. The person recovered because s/he produced antibodies against the pathogen. The memory response will continue to protect the person against that pathogen.

16. Human gamma globulin is the fraction of human serum in which antibodies are found. If antibodies against hepatitis are in the gamma globulin, this would be artificially acquired passive immunity.

Multiple Choice

1. d	**3.** b	**5.** e	**7.** d	**9.** a
2. e	**4.** d	**6.** c	**8.** d	**10.** d

Chapter 18

Review

1. **a.** Whole-agent. Live, avirulent virus that can cause the disease if it mutates back to its virulent state.

   b. Whole-agent; (heat-) killed bacteria.

   c. Subunit; (heat- or formalin-) inactivated toxin.

   d. Subunit

   e. Subunit

   f. Conjugated

   g. Nucleic acid

2. If excess antibody is present, an antigen will combine with several antibody molecules. Excess antigen will result in an antibody combining with several antigens. Refer to Figure 18.2.

3. Particulate antigens react in agglutination reactions. The antigens can be cells or soluble antigens bound to synthetic particles. Soluble antigens take part in precipitation reactions.

4. **a.** Some viruses are able to agglutinate red blood cells. This is used to detect the presence of large numbers of virions capable of causing hemagglutination (e.g., *Influenzavirus*).
   b. Antibodies produced against viruses that are capable of agglutinating red blood cells will inhibit the agglutination. Hemagglutination inhibition can be used to detect the presence of antibodies against these viruses.
   c. This is a procedure to detect antibodies that react with soluble antigens by first attaching the antigens to insoluble latex spheres. This procedure may be used to detect the presence of antibodies that develop during certain mycotic or helminthic infections.

5. See Figure 18.10.

6. **a.** Direct test (see Figure 18.10a)
   b. Indirect test (see Figure 18.10b)

7. An indirect ELISA test is used to detect the presence of antibodies. A known antigen is fixed to a small well, and the patient's serum is added. Patient's antibodies will react with the antigen in the well. Antihuman immunoglobulins bound to an enzyme are added to the well. The antihuman immunoglobulins will bind to the patient's antibodies. Substrate for the enzyme is then added and a positive reaction indicating presence of the antibody in the patient's serum is shown by the enzyme–substrate reaction.

   A direct ELISA test is used to detect the presence of an antigen. Antibodies are fixed to a small well, and the unknown antigen is added. If the antigen reacts with the antibodies, the antigen will be bound to the well. Antibodies specific for the antigen are then added to the well. This second layer of antibodies is bound to an enzyme. Substrate for the enzyme is then added, and a positive reaction indicating the identity of the antigen is shown by the enzyme–substrate reaction (see Figure 18.12).

8. **a.** Direct test
   b. Indirect test
   The direct test provides definitive proof.

9. **Matching**

e	Precipitation
d, f	Immunoelectrophoresis
a	Agglutination
c	Complement fixation
f	Neutralization
b, d	ELISA

10.
e	Agglutination
c	Complement fixation
a	ELISA
f	FA
b	Neutralization
d	Precipitation

Multiple Choice

1. c	3. b	5. a	7. c	9. b
2. d	4. a	6. b	8. a	10. c

Chapter 19

Review

1. The immune state that results in altered immunologic reactions leading to pathogenic changes in tissue.

2.
Mediator	Function
Histamine	Increases blood capillary permeability, mucus secretion, and smooth muscle contraction.
Leukotrienes	Increase blood capillary permeability and smooth muscle contraction.
Prostaglandins	Increase smooth muscle contraction and mucus secretion.

3. Recipient's antibodies will react with donor's tissues.

4. The recipient will experience symptoms due to lysis of the donor RBCs. Hemolysis occurs because the antigen (donor RBCs)–antibody reaction fixes complement.

5. This condition develops when an Rh2 mother becomes sensitized to the Rh1 antigens of her fetus. The mother's anti–Rh antibodies (IgG) can cross the placenta and react with fetal RBCs, causing their destruction. This condition can be prevented by passive immunization of the Rh2 mother with anti-Rh antibodies shortly after birth. These anti-Rh antibodies combine with fetal Rh1 RBCs, which may have entered maternal circulation, and enhance their clearance, thereby reducing the sensitization of the mother's immune system to this antigen.

6. Refer to Figure 19.7.
   a. The observed symptoms are due to lymphokines.
   b. When a person contacts poison oak initially, the antigen (catechols on the leaves) binds to tissue cells, is phagocytized by macrophages, and is presented to receptors on the surface of T cells. Contact between the antigen and the appropriate T cell stimulates the T cell to proliferate and become sensitized. Subsequent exposure to the antigen results in sensitized T cells releasing lymphokines, and a delayed hypersensitivity occurs.
   c. Small repeated doses of the antigen are believed to cause the production of IgG (blocking) antibodies.

7. Autografts and isografts are the most compatible. Xenotransplants are the least compatible.

8. **a.** Compatible. There are no Rh antigens on the donor's RBCs.
   b. Incompatible. The recipient will produce anti-Rh antibodies. If the recipient receives Rh+ RBCs in a subsequent transfusion, a hemolytic reaction will develop.
   c. Incompatible. The recipient has anti-A antibodies that will result in lysis of the donor's RBCs.

9. Autoimmunity is a humoral (types I, II, and III) or cell-mediated (type IV) immune response against a person's own tissue antigen. During development, T cells that recognize self may not be eliminated. During adulthood, inactive T cells may become active or antibodies could cross-react with host cell antigens. New or altered antigens may be formed on the surface of host cells. These antigens may result from the use of certain drugs, or from infections by certain viruses.

Antibodies to cell-membrane antigens of certain group A streptococci cross-react with human heart tissue. Severe, recurrent infections caused by β-hemolytic group A streptococci sometimes lead to the development of rheumatoid arthritis long after the streptococcal infection has subsided.

10. Type I Antibodies against microbes react with self
 Type II Antibodies react with self.
 Type III Antibody–complement complexes deposit in tissues.
 Type IV T cells destroy self cells.
 See Table 19.3.

11. Natural
 Inherited
 Viral infections, most notably HIV

 Artificial
 Induced by immunosuppression drugs
 Result: Increased susceptibility to various infections depending on the type of immune deficiency.

12. Tumor cells have tumor-specific antigens such as TSTA and T antigen. Sensitized T_C cells may react with tumor-specific antigens, initiating lysis of the tumor cells.

13. Some malignant cells can escape the immune system by antigen modulation or immunological enhancement. Immunotherapy might trigger immunological enhancement. The body's defense against cancer is cell-mediated and not humoral. Transfer of lymphocytes could cause graft-versus-host disease.

14. AIDS is the last stage of an HIV infection. HIV is transmitted by sexual contact, by intravenous drug use, across the placenta, and in mother's milk. HIV is prevented by using condoms for hetero- and homosexual intercourse and oral and anal copulation, and by not re-using needles.

Multiple Choice

1. b	3. b	5. d	7. a	9. c
2. b	4. b	6. e	8. d	10. b

Chapter 20

Review

1.

Antimicrobial Agents	Synthetic or Antibiotic	Method of Action	Principal Use
Isoniazid	Synthetic	Vitamin B_6 analog	Tuberculosis
Sulfonamides	Synthetic	Inhibit folic acid synthesis	Gram-negative bacteria
Ethambutol	Synthetic	Competitive inhibitor	Tuberculosis
Trimethoprim	Synthetic	Inhibits folic acid synthesis	*Pneumocystis*
Fluoroquinolones	Synthetic	Inhibit DNA synthesis	Urinary tract infections
Penicillin, natural	Antibiotic	Inhibits cell wall synthesis	Gram-positive bacteria
Penicillin, semisynthetic	Antibiotic	Inhibits cell wall synthesis	Broad spectrum; penicillin-resistant bacteria
Cephalosporins	Antibiotic	Inhibit cell wall synthesis	Penicillin-resistant bacteria
Carbapenems	Antibiotic	Inhibit cell wall synthesis	Broad spectrum
Aminoglycosides	Antibiotic	Inhibit protein synthesis	Gram-negative bacteria
Tetracyclines	Antibiotic	Inhibit protein synthesis	Broad spectrum
Macrolides	Antibiotic	Inhibit protein synthesis	Gram-negative bacteria
Polypeptides	Antibiotic	Inhibit cell wall synthesis; injure plasma membrane	Gram-positive bacteria; gram-negative bacteria
Vancomycin	Antibiotic	Inhibits cell wall synthesis	Penicillin-resistant *Staphylococcus*
Rifamycins	Antibiotic	Inhibit mRNA synthesis	Tuberculosis
Polyenes	Antibiotic	Injure plasma membrane	Fungicide
Griseofulvin	Antibiotic	Inhibits mitosis	Antifungal
Amantadine	Synthetic	Blocks viral entry or uncoating	Influenza A
Zidovudine	Synthetic	Inhibits DNA synthesis	AIDS
Niclosamide	Synthetic	Inhibits oxidative phosphorylation	Tapeworms
Albenazole	Synthetic	Inhibits microtubule formation	Antihelminthic, antiprotozoal

2. A chemotherapeutic agent is a substance taken into the body to combat disease. A synthetic chemotherapeutic agent is prepared in a laboratory, whereas antibiotics are produced naturally by bacteria and some fungi.

3. **a.** Ehrlich discovered the first chemotherapeutic agent (salvarsan, which was used to treat syphilis).
   b. Fleming discovered the antibiotic penicillin.

4. The drug should be toxic to the undesired microorganisms and not harmful to the host (selective toxicity).
   The drug should be active against many microorganisms (broad spectrum).
   The drug should not produce hypersensitivity in the host.
   The drug should not produce drug resistance in the host.
   The drug should not harm normal microbiota.

5. Because a virus uses the host cell's metabolic machinery, it is difficult to damage the virus without damaging the host. Fungi, protozoa, and helminths possess eukaryotic cells. Therefore, antiviral, antifungal, antiprotozoan, and antihelminthic drugs must also affect eukaryotic cells.

6. Pyrimidine (idoxuridine) and purine (acyclovir) analogs. Prevent release of nucleic acid from viruses into the host cell (amantadine).
   Inhibition of infection of cells (interferon).
   Enzyme inhibitors (indinavir).

7. In the broth dilution test, a series of cultures is prepared in a microtiter plate. To each well of liquid medium, the test organism and a different concentration of chemotherapeutic agent are added. The plate is incubated for 16–20 hours and observed for the presence of microbial growth.
   The minimal inhibitory concentration (MIC) is the lowest concentration of chemotherapeutic agent capable of preventing growth of the test organism. The lowest concentration of the agent that results in no growth in a subculture is the minimal bactericidal concentration (MBC).
   In the agar dilution method, bacterial colonies are replica-plated onto nutrient media plus varying concentrations of antimicrobial agents. The MIC is determined by measuring the colony growth. In the tube dilution test, both the MIC and the MBC can be determined. The agar dilution method has the advantage of ease of inoculation and media preparation.

8. In the disk-diffusion test, filter paper disks impregnated with chemotherapeutic agents are overlaid on an inoculated agar medium. During incubation, the agents diffuse from the disk and a zone of inhibition is observed in the area immediately around the disks. The zone of inhibition indicates susceptibility of the test organism to the agent tested.

9. Drug resistance is the lack of susceptibility of a microorganism to a chemotherapeutic agent. Drug resistance may develop when microorganisms are constantly exposed to an antimicrobial agent. The development of drug-resistant microorganisms can be minimized by judicious use of antimicrobial agents; following directions on the prescription; or by administering two or more drugs simultaneously.

10. **a.** Prevention of resistant strains of microorganisms;
    b. Take advantage of the synergistic effect;
    c. Provide therapy until a diagnosis is made; and
    d. Lessen the toxicity of individual drugs by reducing the dosage of each in combination.

11. **a.** Like polymyxin B, causes leaks in the plasma membrane.
    b. Interferes with translation.

12. **a.** Inhibits formation of peptide bond.
    b. Prevents translocation of ribosome along mRNA.
    c. Interferes with attachment of tRNA to mRNA-ribosome complex.
    d. Changes shape of 30S portion of ribosome, resulting in misreading mRNA.
    e. Prevents 70S ribosomal subunits from forming.
    f. Prevents release of peptide from ribosome.

13. DNA polymerase adds bases to the 3′–OH.

14. **a.** Penicillin inhibits bacterial cell wall synthesis. Echinocandin inhibits fungal cell wall synthesis.
    b. Imidazole interferes with fungal plasma membrane synthesis. Polymyxin B disrupts any plasma membrane.

Multiple Choice

1. b	6. d
2. a	7. e
3. a	8. b
4. b	9. c
5. a	10. d

Glossary

ABO blood group system The classification of red blood cells based on the presence or absence of A and B carbohydrate antigens.

abscess A localized accumulation of pus.

A-B toxin Bacterial exotoxins consisting of two polypeptides.

acellular vaccine A vaccine consisting of antigenic parts of cells.

acetyl group

$$H_3C - \overset{\overset{\displaystyle O}{\|}}{C} - $$

acid A substance that dissociates into one or more hydrogen ions (H^+) and one or more negative ions.

acid-fast stain A differential stain used to identify bacteria that are not decolorized by acid-alcohol.

acidic dye A salt in which the color is in the negative ion; used for negative staining.

acidophile A bacterium that grows below pH 4.

acquired immune deficiency The inability, obtained during the life of an individual, to produce specific antibodies or T cells, due to drugs or disease.

acquired immunity The ability, obtained during the life of the individual, to produce specific antibodies.

activated macrophage A macrophage that has increased phagocytic ability and other functions after exposure to mediators released by T cells after stimulation by antigens.

activated sludge system A process used in secondary sewage treatment in which batches of sewage are held in highly aerated tanks; to ensure the presence of microbes efficient in degrading sewage, each batch is inoculated with portions of sludge from a precious batch.

activation energy The minimum collision energy required for a chemical reaction to occur.

active site A region on an enzyme that interacts with the substrate.

active transport Net movement of a substance across a membrane against a concentration gradient; requires the cell to expend energy.

acute disease A disease in which symptoms develop rapidly but last for only a short time.

acute-phase proteins Serum proteins whose concentration changes by at least 25% during inflammation.

adenocarcinoma Cancer of glandular epithelial tissue.

adenosine diphosphate (ADP) The substance formed when ATP is hydrolyzed and energy is released.

adenosine triphosphate (ATP) An important intracellular energy source.

adherence Attachment of a microbe or phagocyte to another's plasma membrane or other surface.

adhesin A carbohydrate-specific binding protein that projects from prokaryotic cells; used for adherence, also called a ligand.

adjuvant A substance added to a vaccine to increase its effectiveness.

aerobe An organism requiring molecular oxygen (O_2) for growth.

aerobic respiration Respiration in which the final electron acceptor in the electron transport chain is molecular oxygen (O_2).

aerotolerant anaerobe An organism that does not use molecular oxygen (O_2) but is not affected by its presence.

aflatoxin A carcinogenic toxin produced by *Aspergillus flavus*.

agar A complex polysaccharide derived from a marine alga and used as a solidifying agent in culture media.

agglutination A joining together or clumping of cells.

agranulocyte A leukocyte without visible granules in the cytoplasm; includes monocytes and lymphocytes.

agranulocytosis Destruction of granulocytic white blood cells.

AIDS (acquired immunodeficiency syndrome) An infectious disease caused by the human immunodeficiency virus (HIV) in which the virus infects CD4 cells.

airborne transmission The spread of pathogens farther than 1 meter in air from reservoir to susceptible host.

alcohol An organic molecule with the functional group—OH.

alcohol fermentation A catabolic process, beginning with glycolysis, that produces ethyl alcohol to reoxidize NADH.

aldehyde An organic molecule with the functional group

alga (plural: **algae**) A photosynthetic eukaryote; may be unicellular, filamentous, or multicellular but lack the tissues found in plants.

algal bloom An abundant growth of microscopic algae producing visible colonies in nature.

algin A sodium salt of mannuronic acid ($C_6H_8O_6$); found in brown algae.

allergen An antigen that evokes a hypersensitivity response.

allergy *See* hypersensitivity.

allograft A tissue graft that is not from a genetically identical donor (i.e., not from self or an identical twin).

allosteric inhibition The process in which an enzyme's activity is changed because of binding to the allosteric site.

allosteric site The site on an enzyme at which a noncompetitive inhibitor binds.

allyamines Antifungal agents that interfere with sterol synthesis.

amanitin A polypeptide mushroom toxin that causes liver and nerve damage.

Ames test A procedure using bacteria to identify potential carcinogens.

amination The addition of an amino group.

amino acid An organic acid containing an amino group and a carboxyl group.

aminoglycoside An antibiotic consisting of amino sugars and an aminocyclitol ring; for example, streptomycin.

amino group —NH_2.

ammonification The release of ammonia from nitrogen-containing organic matter by the action of microorganisms.

amoeba A unicellular eukaryote that moves by means of pseudopods.

amphibolic pathway A pathway that is both anabolic and catabolic.

amphitrichous Having tufts of flagella at both ends of a cell.

anabolism All synthesis reactions in a living organism; the building of complex organic molecules from simpler ones.

anaerobe An organism that does not require molecular oxygen (O_2) for growth.

anaerobic respiration Respiration in which the final electron acceptor in the electron transport chain is an inorganic molecule other than molecular oxygen (O_2); for example, a nitrate ion or CO_2.

anaerobic sludge digester Anaerobic digestion used in secondary sewage treatment.

anal pore A site in certain protozoa for elimination of waste.

analytical epidemiology Comparison of a diseased group and a healthy group to determine the cause of the disease.

anamnestic response *See* memory response.

anamorph Ascomycete fungi that have lost the ability to reproduce sexually; the asexual stage of a fungus.

anaphylaxis A hypersensitivity reaction involving IgE antibodies, mast cells, and basophils.

animalia The kingdom composed of multicellular eukaryotes lacking cell walls.

anion An ion with a negative charge.

anoxygenic Not producing molecular oxygen; typical of cyclic photophosphorylation.

antagonism Active opposition; for example, beyond two drugs or two microbes.

antibiotic An antimicrobial agent, usually produced naturally by a bacterium or fungus.

antibody A protein produced by the body in response to an antigen, and capable of combining specifically with that antigen.

antibody-dependent cell-mediated cytotoxicity (ADCC) The killing of antibody-coated cells by natural killer cells and phagocytes.

antibody titer The amount of antibody in serum.

anticodon The three nucleotides by which a tRNA recognizes an mRNA codon.

antigen Any substance that causes antibody formation and reacts only with its specific antibody; also called immunogen.

antigen-antibody complex The combination of an antigen with the antibody that is specific for it; the basis of immune protection and many diagnostic tests.

antigen-binding sites A site on an antibody that binds to an antigenic determinant.

antigenic determinant A specific region on the surface of an antigen against which antibodies are formed; also called epitope.

antigenic drift A minor variation in the antigenic makeup of influenza viruses that occurs with time.

antigenic shift A major genetic change in influenza viruses causing changes in H and N antigens.

antigenic variation Changes in surface antigens that occur in a microbial population.

antigen-presenting cell (APC) A macrophage or dendritic cell that engulfs an antigen and presents fragments to T cells.

antigen receptors An antibody-like molecule on B and T cells that enables them to recognize and bind to their specific antigen.

antihuman immune serum globulin (anti-HISG) An antibody that reacts specifically with human antibodies.

antimicrobial drug A chemical that destroys pathogens without damaging body tissues.

antimicrobial peptide An antibiotic that is bactericidal and has a broad spectrum of activity; *see* bacteriocin.

antisense DNA DNA that is complementary to the DNA encoding a protein; the antisense RNA transcript will hybridize with the mRNA encoding the protein and inhibit synthesis of the protein.

antisense strand (− strand) Viral RNA that cannot act as mRNA.

antisepsis A chemical method for disinfection of the skin or mucous membranes; the chemical is called an antiseptic.

antiserum A blood-derived fluid containing antibodies.

antitoxin A specific antibody produced by the body in response to a bacterial exotoxin or its toxoid.

antiviral protein (AVP) A protein made in response to interferon that blocks viral multiplication.

apicomplexa Unicellular eukaryotes that are obligate intracellular parasites; possess a special organelle at the tip of the cell.

apoenzyme The protein portion of an enzyme, which requires activation by a coenzyme.

apoptosis The natural programmed death of a cell; the residual fragments are disposed of by phagocytosis.

aquatic microbiology The study of microorganisms and their activities in natural waters.

arbuscule Fungal mycelia in plant root cells.

archaea Prokaryotic cells lacking peptidoglycan; one of the three domains.

archaezoa Primitive, unicellular eukaryotes that lack mitochondria, such as *Giardia*.

arthrospore An asexual fungal spore formed by fragmentation of a septate hypha.

Arthus reaction Inflammation and necrosis at the site of injection of foreign serum, due to immune complex formation.

artificially acquired active immunity The production of antibodies by the body in response to a vaccination.

artificially acquired passive immunity The transfer of humoral antibodies formed by one individual to a susceptible individual, accomplished by the injection of antiserum.

artificial selection Choosing one organism from a population to grow because of its desirable traits.

ascospore A sexual fungal spore produced in an ascus, formed by the ascomycetes.

ascus A saclike structure containing ascospores; found in the ascomycetes.

asepsis The absence of contamination by unwanted organisms.

aseptic packaging Commercial food preservation by filling sterile containers with sterile food.

aseptic surgery Techniques used in surgery to prevent microbial contamination of the patient.

aseptic techniques Laboratory techniques used to minimize contamination.

asexual spore A reproductive cell produced by mitosis and cell division (eukaryotes) or binary fission (actinomycetes).

atom The smallest unit of matter that can enter into a chemical reaction.

atomic number The number of protons in the nucleus of an atom.

atomic weight The total number of protons and neutrons in the nucleus of an atom.

attenuated whole-agent vaccine A vaccine containing live, attenuated (weakened) microorganisms.

atomic force microscopy *See* scanned-probe microscopy.

autoclave Equipment for sterilization by steam under pressure, usually operated at 15 psi and 121°C.

autograph A tissue graft from one's self.

autoimmune disease Damage to one's own organs due to action of the immune system.

autotroph An organism that uses carbon dioxide (CO_2) as its principal carbon source. *See also* chemoautotroph, photoautotroph.

auxotroph A mutant microorganism with a nutritional requirement that is absent in the parent.

axial filament The structure for motility found in spirochetes; also called endoflagellum.

azoles Antifungal agents that interfere with sterol synthesis.

bacillus (plural: bacilli) (1) Any rod-shaped bacterium. (2) When written as a genus (*Bacillus*) refers to rod-shaped, endospore-forming, facultatively anaerobic, gram-positive bacteria.

bacteremia A condition in which there are bacteria in the blood.

bacteria Kingdom of prokaryotic organisms, characterized by peptidoglycan cell walls; **bacterium** (singular) when referring to a single organism.

bacteriology The scientific study of prokaryotes, including bacteria and archaea.

bacterial growth curve A graph indicating the growth of a bacterial population over time.

bactericide A substance capable of killing bacteria.

bacteriocin An antimicrobial peptide produced by bacteria that kills other bacteria.

bacteriophage (phage) A virus that infects bacterial cells.

bacteriostasis A treatment capable of inhibiting bacterial growth.

base A substance that dissociates into one or more hydroxide ions (OH^-) and one or more positive ions.

base pairs The arrangement of nitrogenous bases in nucleic acids based on hydrogen bonding; in DNA, base pairs are A-T and G-C; in RNA, base pairs are A-U and G-C.

base substitution The replacement of a single base in DNA by another base, causing a mutation; also called point mutation.

basic dye A salt in which the color is in the positive ion; used for bacterial stains.

basidiospore A sexual fungal spore produced in a basidium, characteristic of the basidiomycetes.

basidium A pedestal that produces basidiospores; found in the basidiomycetes.

basophil A granulocyte (leukocyte) that readily takes up basic dye and is not phagocytic; has receptors for IgE Fc regions.

batch flow An industrial process in which cells are grown for a period of time after which the product is collected.

B cell A type of lymphocyte; differentiates into antibody-secreting plasma cells and memory cells.

BCG vaccine A live, attenuated strain of *Mycobacterium bovis* used to provide immunity to tuberculosis.

beer Alcoholic beverage produced by fermentation of starch.

benthic zone The sediment at the bottom of a body of water.

Bergey's Manual *Bergey's Manual of Systematic Bacteriology*, the standard taxonomic reference on bacteria; also refers to *Bergey's Manual of Determinative Bacteriology*.

beta oxidation The removal of two carbon units from a fatty acid to form acetyl CoA.

biguanide A group of antimicrobial chemicals, including chlorhexidine, especially useful on skin and mucous membranes.

binary fission Prokaryotic cell reproduction by division into two daughter cells.

binomial nomenclature The system of having two names (genus and specific epithet) for each organism; also called scientific nomenclature.

bioaugmentation The use of pollutant-acclimated microbes or genetically engineered microbes for bioremediation.

biochemical oxygen demand (BOD) A measure of the biologically degradable organic matter in water.

biocide A substance capable of killing microorganisms.

bioconversion Changes in organic matter brought about by the growth of microorganisms.

biofilm A microbial community that usually forms as a slimy layer on a surface.

biogenesis The theory that living cells arise only from preexisting cells.

biogeochemical cycle The recycling of chemical elements by microorganisms for use by other organisms.

bioinformatics The science of determining the function of genes through computer-assisted analysis.

biological transmission The transmission of a pathogen from one host to another when the pathogen reproduces in the vector.

bioluminescence The emission of light from the electron transport chain; requires the enzyme luciferase.

biomass Organic matter produced by living organisms and measured by weight.

bioreactor A fermentation vessel with controls for environmental conditions, e.g., temperature and pH.

bioremediation The use of microbes to remove an environmental pollutant.

biotechnology The industrial application of microorganisms, cells, or cell components to make a useful product.

biotype *See* biovar.

biovar A subgroup of a serovar based on biochemical or physiological properties; also called biotype.

bisphenol A group of antimicrobial chemicals composed of two phenolic groups; includes triclosan.

blade A flat leaflike structure of multicellular algae.

blastoconidium An asexual fungal spore produced by budding from the parent cell.

blood-brain barrier Cell membranes that allow some substances to pass from the blood to the brain but restrict others.

brightfield microscope A microscope that uses visible light for illumination; the specimens are viewed against a white background.

broad-spectrum antibiotic An antibiotic that is effective against a wide range of both gram-positive and gram-negative bacteria.

broth dilution test A method of determining the minimal inhibitory concentration by using serial dilutions of an antimicrobial drug.

bubo An enlarged lymph node caused by inflammation.

budding (1) Asexual reproduction beginning as a protuberance from the parent cell that grows to become a daughter cell. (2) Release of an enveloped virus through the plasma membrane of an animal cell.

budding yeast Following mitosis, a yeast cell that divides unevenly to produce a small cell (bud) from the parent cell.

buffer A substance that tends to stabilize the pH of a solution.

bulking A condition arising when sludge floats rather than settles in secondary sewage treatment.

bullae (singular: bulla) Large serum-filled vesicles in the skin.

burst size The number of newly synthesized bacteriophage particles released from a single cell.

burst time The time required from bacteriophage attachment to release.

Calvin-Benson cycle The fixation of CO_2 into reduced organic compounds; used by autotrophs.

capnophile A microorganism that grows best at relatively high CO_2 concentrations.

capsid The protein coat of a virus that surrounds the nucleic acid.

capsomere A protein subunit of a viral capsid.

capsule An outer, viscous covering on some bacteria composed of a polysaccharide or polypeptide.

carbapenems Antibiotics that contain a β-lactam antibiotic and cilastatin.

carbohydrate An organic compound composed of carbon, hydrogen, and oxygen, with the hydrogen and oxygen present in a 2:1 ratio; carbohydrates include starches, sugars, and cellulose.

carbon cycle The series of processes that converts CO_2 to organic substances and back to CO_2 in nature.

carbon fixation The synthesis of sugars by using carbons from CO_2. *See also* Calvin-Benson cycle.

carbon skeleton The basic chain or ring of carbon atoms in a molecule; for example,

$$-\overset{|}{\underset{|}{C}}-\overset{|}{\underset{|}{C}}-\overset{|}{\underset{|}{C}}-$$

carboxyl group

$$-C\overset{\displaystyle O}{\underset{\displaystyle OH}{}}$$

carboxysome A prokaryotic inclusion containing ribulose 1,5-diphosphate carboxylase.

carcinogen Any cancer-causing substance.

carrier Organisms (usually refers to humans) that harbor pathogens and transmit them to others.

casein Milk protein.

catabolism All decomposition reactions in a living organism; the breakdown of complex organic compounds into simpler ones.

catabolite repression Inhibition of the metabolism of alternate carbon sources by glucose.

catalase An enzyme that catalyzes the breakdown of hydrogen peroxide to water and oxygen.

catalyst A substance that increases the rate of a chemical reaction but is not altered itself.

cation A positively charged ion.

CD (cluster of determination) Number assigned to an epitope on a single antigen, for example, CD4 protein, which is found on helper T cells.

cDNA (complementary DNA) DNA made in vitro from an mRNA template.

cell culture Eukaryotic cells grown in culture media; also called tissue culture.

cell-mediated immunity An immune response that involves T cells binding to antigens presented on infected cells; T cells then differentiate into several types of effector T cells, including helper and cytotoxic.

cell theory All living organisms are composed of cells and arise from preexisting cells.

cellular respiration *See* respiration.

cell wall The outer covering of most bacterial, fungal, algal, and plant cells; in bacteria, it consists of peptidoglycan.

Centers for Disease Control and Prevention (CDC) A branch of the U.S. Public Health Service that serves a central source of epidemiological information.

central nervous system (CNS) The brain and the spinal cord. *See also* peripheral nervous system.

centriole A structure consisting of nine microtubule triplets, found in eukaryotic cells.

centrosome Region in a eukaryotic cell consisting of a pericentriolar area (protein fibers) and a pair of centrioles; involved in formation of the mitotic spindle.

cephalosporin An antibiotic produced by the fungus *Cephalosporium* that inhibits the synthesis of gram-positive bacterial cell walls.

cercaria A free-swimming larva of trematodes.

chancre A hard sore, the center of which ulcerates.

chemical bond An attractive force between atoms forming a molecule.

chemical element A fundamental substance composed of atoms that have the same atomic number and behave the same way chemically.

chemical energy The energy of a chemical reaction.

chemically defined medium A culture medium in which the exact chemical composition is known.

chemical reaction The process of making or breaking bonds between atoms.

chemiosmosis A mechanism that uses a proton gradient across a cytoplasmic membrane to generate ATP.

chemistry The science of the interactions between atoms and molecules.

chemoautotroph An organism that uses an inorganic chemical as an energy source and CO_2 as a carbon source.

chemoheterotroph An organism that uses organic molecules as a source of carbon and energy.

chemokine A cytokine that induces, by chemotaxis, the migration of leukocytes into infected areas.

chemotaxis Movement in response to the presence of a chemical.

chemotherapeutic agent A chemical (drug) used in the treatment of disease. *See also* chemotherapy.

chemotherapy Treatment with chemical substances.

chemotroph An organism that uses oxidation-reduction reactions as its primary energy source.

chimeric monoclonal antibody A genetically engineered antibody made of human constant regions and mouse variable regions.

chlamydospore An asexual fungal spore formed within a hypha.

chloramphenicol A broad-spectrum bacteriostatic chemical.

chloroplast The organelle that performs photosynthesis in photoautotrophic eukaryotes.

chlorosome Plasma membrane folds in green sulfur bacteria containing bacteriochlorophylls.

chromatin Threadlike, uncondensed DNA in an interphase eukaryotic cell.

chromatophore An infolding in the plasma membrane where bacteriochlorophyll is located in photoautotrophic bacteria; also known as thylakoids.

chromosome The structure that carries hereditary information, chromosomes contain genes.

chronic disease An illness that develops slowly and is likely to continue or recur for long periods.

ciliary escalator Ciliated mucosal cells of the lower respiratory tract that move inhaled particulates away from the lungs.

cilium (plural: cilia) A relatively short cellular projection from some eukaryotic cells, composed of nine pairs plus two microtubules. *See* flagellum.

ciliate A member of the protozoan phylum Ciliophora that uses cilia for locomotion.

cistern A flattened membranous sac in endoplasmic reticulum and the Golgi complex.

cladogram A dichotomous phylogenetic tree that branches repeatedly, suggesting the classification of organisms based on the time sequence in which evolutionary branches arose.

class A taxonomic group between phylum and order.

clonal deletion The elimination of B and T cells that react with self.

clonal selection The development of clones of B and T cells against a specific antigen.

clone A population of cells arising from a single parent cell.

clue cells Sloughed-off vaginal cells covered with *Gardnerella vaginalis*.

coagulase A bacterial enzyme that causes blood plasma to clot.

coccobacillus (plural: **coccobacilli**)   A bacterium that is an oval rod.

coccus (plural: **cocci**)   A spherical or ovoid bacterium.

codon   A sequence of three nucleotides in mRNA that specifies the insertion of an amino acid into a polypeptide.

coenocytic hypha   A fungal filament that is not divided into uninucleate cell-like units because it lacks septa.

coenzyme   A nonprotein substance that is associated with and that activates an enzyme.

coenzyme A (CoA)   A coenzyme that functions in decarboxylation.

coenzyme Q   *See* ubiquinone.

cofactor   (1) The nonprotein component of an enzyme. (2) A microorganism or molecule that acts with others to synergistically enhance or cause disease.

coliforms   Aerobic or facultatively anaerobic, gram-negative, nonendospore-forming, rod-shaped bacteria that ferment lactose with acid and gas formation within 48 hours at 35°C.

collagenase   An enzyme that hydrolyzes collagen.

collision theory   The principle that chemical reactions occur because energy is gained as particles collide.

colony   A visible mass of microbial cells arising from one cell or from a group of the same microbes.

colony hybridization   The identification of a colony containing a desired gene by using a DNA probe that is complementary to that gene.

colony-stimulating factor (CSF)   A substance that induces certain cells to proliferate or differentiate.

commensalism   A symbiotic relationship in which two organisms live in association and one is benefited while the other is neither benefited nor harmed.

commercial sterilization   A process of treating canned goods aimed at destroying the endospores of *Clostridium botulinum.*

communicable disease   Any disease that can be spread from one host to another.

competence   The physiological state in which a recipient cell can take and incorporate a large piece of donor DNA.

competitive inhibitor   A chemical that competes with the normal substrate for the active site of an enzyme. *See also* noncompetitive inhibitor.

complement   A group of serum proteins involved in phagocytosis and lysis of bacteria.

complementary DNA (cDNA)   DNA made in vitro from an mRNA template.

complement fixation   The process in which complement combines with an antigen–antibody complex.

complex medium   A culture medium in which the exact chemical composition is not known.

complex virus   A virus with a complicated structure, such as a bacteriophage.

composting   A method of solid waste disposal, usually plant material, by encouraging its decomposition by microbes.

compound   A substance composed of two or more different chemical elements.

compound light microscope (LM)   An instrument with two sets of lenses that uses visible light as the source of illumination.

compromised host   A host whose resistance to infection is impaired.

condensation reaction   A chemical reaction in which a molecule of water is released; also called dehydration synthesis.

condenser   A lens system located below the microscope stage that directs light rays through the specimen.

confocal microscopy   A light microscope that uses fluorescent stains and laser to make two- and three-dimensional images.

congenital   Refers to a condition existing at birth; may be inherited or acquired in utero.

congenital immune deficiency   The inability, due to an individual's genotype, to produce specific antibodies or T cells.

conidiophore   An aerial hypha bearing conidiospores.

conidium   An asexual spore produced in a chain from a conidiophore.

conjugated vaccine   A vaccine consisting of the designed antigen and other proteins.

conjugation   The transfer of genetic material from one cell to another involving cell-to-cell contact.

conjugative plasmid   A prokaryotic plasmid that carries genes for sex pili and for transfer of the plasmid to another cell.

constitutive enzyme   An enzyme that is produced continuously.

contact inhibition   The cessation of animal cell movement and division as a result of contact with other cells.

contact transmission   The spread of disease by direct or indirect contact or via droplets.

contagious disease   A disease that is easily spread from one person to another.

continuous cell line   Animal cells that can be maintained through an indefinite number of generations in vitro.

continuous flow   An industrial fermentation in which cells are grown indefinitely with continual addition of nutrients and removal of waste and products.

corepressor   A molecule that binds to a repressor protein, enabling the repressor to bind to an operator.

cortex   The protective fungal covering of a lichen.

counterstain   A second stain applied to a smear, provides contrast to the primary stain.

covalent bond   A chemical bond in which the electrons of one atom are shared with another atom.

crisis   The phase of a fever characterized by vasodilation and sweating.

crista (plural: **cristae**)   Folding of the inner membrane of a mitochondrion.

crossing over   The process by which a portion of one chromosome is exchanged with a portion of another chromosome.

culture   Microorganisms that grow and multiply in a container of culture medium.

culture medium   The nutrient material prepared for growth of microorganisms in a laboratory.

curd   The solid part of milk that separates from the liquid (whey) in the making of cheese, for example.

cutaneous mycosis   A fungal infection of the epidermis, nails, or hair.

cuticle   The outer covering of helminths.

cyanobacteria   Oxygen-producing photoautotrophic prokaryotes.

cyclic AMP (cAMP)   A molecule derived from ATP, in which the phosphate group has a cyclic structure; acts as a cellular messenger.

cyclic photophosphorylation   The movement of an electron from chlorophyll through a series of electron acceptors and back to chlorophyll; anoxygenic; purple and green bacterial photophosphorylation.

cyst   A sac with a distinct wall containing fluid or other material; also, a protective capsule of some protozoa.

cysticercus   An encysted tapeworm larva.

cystitis   Inflammation of the urinary bladder.

cytochrome oxidase   An enzyme that oxidizes cytochrome *c*.

cytochrome A protein that functions as an electron carrier in cellular respiration and photosynthesis.

cytokine A small protein released from human cells in response to bacterial infection; directly or indirectly may induce fever, pain, or T-cell proliferation.

cytolysis The destruction of cells, resulting from damage to their cell membrane, that causes cellular contents to leak out.

cytopathic effect (CPE) A visible effect on a host cell, caused by a virus, that may result in host cell damage or death.

cytoplasm In a prokaryotic cell, everything inside the plasma membrane; in a eukaryotic cell, everything inside the plasma membrane and external to the nucleus.

cytoplasmic streaming The movement of cytoplasm in a eukaryotic cell.

cytoskeleton Microfilaments, intermediate filaments, and microtubules that provide support and movement for eukaryotic cytoplasm.

cytosol The fluid portion of cytoplasm.

cytostome The mouthlike opening in some protozoa.

cytotoxic T (T_C) cells A specialized T cell that destroys infected cells presenting antigens.

cytotoxin A bacterial toxin that kills host cells or alters their functions.

darkfield microscope A microscope that has a device to scatter light from the illuminator so that the specimen appears white against a black background.

dark (light-independent) reaction The process by which electrons and energy from ATP are used to reduce CO_2 to sugar. *See also* Calvin-Benson cycle.

deamination The removal of an amino group from an amino acid to form ammonia. *See also* ammonification.

death phase The period of logarithmic decrease in a bacterial population; also called logarithmic decline phase.

debridement Surgical removal of necrotic tissue.

decarboxylation The removal of CO_2 from an amino acid.

decimal reduction time (DRT) The time (in minutes) required to kill 90% of a bacterial population at a given temperature; also called D value.

decolorizing agent A solution used in the process of removing a stain.

decomposition reaction A chemical reaction in which bonds are broken to produce smaller parts from a large molecule.

deep-freezing Preservation of bacterial cultures at −50°C to −95°C.

definitive host An organism that harbors the adult, sexually mature form of a parasite.

degeneracy Redundancy of the genetic code; that is, most amino acids are encoded by several codons.

degerming The removal of microorganisms in an area; also called degermation.

degranulation The release of contents of secretory granules from mast cells or basophils during anaphylaxis.

dehydration synthesis *See* condensation reaction.

dehydrogenation The loss of hydrogen atoms from a substrate.

delayed-type hypersensitivity Cell-mediated hypersensitivity.

delayed hypersensitivity T (T_D) cells A specialized T cell that produces lymphokines in type IV hypersensitivities.

denaturation A change in the molecular structure of a protein, usually making it nonfunctional.

dendritic cell A type of antigen-presenting cell characterized by long fingerlike extensions; found in lymphatic tissue and skin.

denitrification The reduction of nitrogen in nitrate to nitrite or nitrogen gas.

dental plaque A combination of bacterial cells, dextran, and debris adhering to the teeth.

deoxyribonucleic acid (DNA) The nucleic acid of genetic material in all cells and some viruses.

deoxyribose A five-carbon sugar contained in DNA nucleotides.

dermatomycosis A fungal infection of the skin; also known as tinea or ringworm.

dermatophyte A fungus that causes a cutaneous mycosis.

dermis The inner portion of the skin.

descriptive epidemiology The collection and analysis of all data regarding the occurrence of a disease to determine its cause.

desensitization The prevention of allergic inflammatory responses.

desiccation The removal of water.

Deuteromycota A fungus taxon that is no longer used.

diapedesis *See* emigration.

dichotomous key An identification scheme based on successive paired questions; answering one question leads to another pair of questions, until an organism is identified.

differential interference contrast (DIC) microscope An instrument that provides a three-dimensional, magnified image.

differential medium A solid culture medium that makes it easier to distinguish colonies of the desired organism.

differential stain A stain that distinguishes objects on the basis of reactions to the staining procedure.

differential white blood cell count The number of each kind of leukocyte in a sample of 100 leukocytes.

dimorphism The property of having two forms of growth.

dioecious Referring to organisms in which organs of different sexes are located in different individuals.

diplobacilli (singular: diplobacillus) Rods that divide and remain attached in pairs.

diplococci (singular: diplococcus) Cocci that divide and remain attached in pairs.

diploid cell A cell having two sets of chromosomes; diploid is the normal state of a eukaryotic cell.

diploid cell line Eukaryotic cells grown in vitro.

direct agglutination test The use of known antibodies to identify an unknown cell-bound antigen.

direct contact transmission A method of spreading infection from one host to another through some kind of close association between the hosts.

direct FA test A fluorescent-antibody test to detect the presence of an antigen.

direct microscopic count Enumeration of cells by observation through a microscope.

disaccharide A sugar consisting of two simple sugars, or monosaccharides.

disease An abnormal state in which part or all of the body is not properly adjusted or is incapable of performing normal functions; any change from a state of health.

disinfection Any treatment used on inanimate objects to kill or inhibit the growth of microorganisms; a chemical used is called a disinfectant.

disk-diffusion method An agar-diffusion test to determine microbial susceptibility to chemotherapeutic agents; also called Kirby-Bauer test.

D-isomer A stereoisomer.

dissimilation A metabolic process in which nutrients are not assimilated but are excreted as ammonia, hydrogen sulfide, and so on.

dissimilation plasmid A plasmid containing genes encoding production of enzymes that trigger the catabolism of certain unusual sugars and hydrocarbons.

dissociation The separation of a compound into positive and negative ions in solution. *See also* ionization.

disulfide bond A covalent bond that holds together two atoms of sulfur.

division A phylum; used in botany.

DNA base composition The moles-percentage of guanine plus cytosine in an organism's DNA.

DNA fingerprinting Analysis of DNA by electrophoresis of restriction enzyme fragments of the DNA.

DNA gyrase *See* topoisomerase II.

DNA ligase An enzyme that covalently bonds a carbon atom of one nucleotide with the phosphate of another nucleotide.

DNA polymerase Enzyme that synthesizes DNA by copying a DNA template.

DNA probe A short, labeled, single strand of DNA or RNA used to locate its complementary strand in a quantity of DNA.

DNA sequencing A process by which the nucleotide sequence of DNA is determined.

domain A taxonomic classification based on rRNA sequences; above the kingdom level.

domoic acid intoxication Diarrhea and memory loss caused by domoic acid, produced by diatoms.

donor cell A cell that gives DNA to a recipient cell during genetic recombination.

DPT vaccine A combined vaccine used to provide active immunity, containing diphtheria and tetanus toxoids and killed *Bordetella pertussis* cells or cell fragments.

droplet transmission The transmission of infection by small liquid droplets carrying microorganisms.

D value *See* decimal reduction time.

dysentery A disease characterized by frequent, watery stools containing blood and mucus.

echinocandins Antifungal agents that interfere with cell wall synthesis.

eclipse period The time during viral multiplication when complete, infective virions are not present.

ecology The study of the interrelationships between organisms and their environment.

edema An abnormal accumulation of interstitial fluid in body parts or tissues, causing swelling.

electron A negatively charged particle in motion around the nucleus of an atom.

electron acceptor An ion that picks up an electron that has been lost from another atom.

electron donor An ion that gives up an electron to another atom.

electronic configuration The arrangement of electrons in shells or energy levels in an atom.

electron microscope A microscope that uses electrons instead of light to produce an image.

electron shell A region of an atom where electrons orbit the nucleus, corresponding to an energy level.

electron transport chain, electron transport system A series of compounds that transfer electrons from one compound to another, generating ATP by oxidative phosphorylation.

electroporation A technique by which DNA is inserted into a cell using an electrical current.

elementary body An infectious form of *Chlamydia*.

ELISA (enzyme-linked immunosorbent assay) A group of serological tests that use enzyme reactions as indicators.

emerging infectious disease (EID) A new or changing disease that is increasing or has the potential to increase in incidence in the near future.

emigration The process by which phagocytes move out of blood vessels; also called diapedesis.

Embden-Meyerhof pathway *See* glycolysis.

emulsification The process of mixing two liquids that do not dissolve in each other.

encephalitis Infection of the brain.

encystment Formation of a cyst.

endemic disease A disease that is constantly present in a certain population.

endergonic reaction A chemical reaction that requires energy.

endocarditis Infection of the lining of the heart (endocardium).

endocytosis The process by which material is moved into a eukaryotic cell.

endoflagellum *See* axial filament.

endolith An organism that lives inside rock.

endoplasmic reticulum (ER) A membranous network in eukaryotic cells connecting the plasma membrane with the nuclear membrane.

endospore A resting structure formed inside some bacteria.

endosymbiotic theory A model for the evolution of eukaryotes which states that organelles arose from prokaryotic cells living inside a host prokaryote.

endotoxic shock Loss of blood pressure caused by the endotoxin component of a gram-negative cell wall. *See* septic shock.

endotoxin Part of the outer portion of the cell wall (lipid A) of most gram-negative bacteria; released on destruction of the cell.

end-product inhibition *See* feedback inhibition.

energy level Potential energy of an electron in an atom. *See also* electron shell.

enrichment culture A culture medium used for preliminary isolation that favors the growth of a particular microorganism.

enteric The common name for a bacterium in the family Enterobacteriaceae.

enterotoxin An exotoxin that causes gastroenteritis, such as those produced by *Staphylococcus*, *Vibrio*, and *Escherichia*.

Entner-Doudoroff pathway An alternate pathway for the oxidation of glucose to pyruvic acid.

envelope An outer covering surrounding the capsid of some viruses.

enzyme A molecule that catalyzes biochemical reactions in a living organism, usually a protein. *See also* ribozyme.

enzyme induction The process by which a substance can cause the synthesis of an enzyme.

enzyme-linked immunosorbent assay *See* ELISA.

enzyme–substrate complex A temporary union of an enzyme and its substrate.

eosinophil A granulocyte whose granules take up the stain eosin.

epidemic disease A disease acquired by many hosts in a given area in a short time.

epidemiology The science that studies when and where diseases occur and how they are transmitted.

epidermis The outer portion of the skin.

epitope *See* antigenic determinant.

equilibrium The point of even distribution.

equivalent treatments   Different methods that have the same effect on controlling microbial growth.

ergot   A toxin produced in sclerotia by the fungus *Claviceps purpurea* that causes ergotism.

E test   An agar diffusion test to determine antibiotic sensitivity using a plastic strip impregnated with varying concentrations of an antibiotic.

ethambutol   A synthetic antimicrobial agent that interferes with the synthesis of RNA.

ethanol

$$H-\underset{\underset{H}{\displaystyle|}}{\overset{\overset{H}{\displaystyle|}}{C}}-\underset{\underset{H}{\displaystyle|}}{\overset{\overset{H}{\displaystyle|}}{C}}-OH$$

etiology   The study of the cause of a disease.

eukarya   All eukaryotes (animals, plants, fungi, and protists); members of the Domain Eukarya.

eukaryote   A cell having DNA inside a distinct membrane-enclosed nucleus.

eukaryotic species   A group of closely related organisms that can interbreed.

eutrophication   The addition of organic matter and subsequent removal of oxygen from a body of water.

exchange reaction   A chemical reaction that has both synthesis and decomposition components.

exergonic reaction   A chemical reaction that releases energy.

exon   A region of a eukaryotic chromosome that encodes a protein.

exotoxin   A protein toxin released from living, mostly gram-positive bacterial cells.

experimental epidemiology   The study of a disease using controlled experiments.

exponential growth phase   *See* log phase.

extracellular polysaccharide (EPS)   A glycocalyx, composed of sugars, that permits bacteria to attach to various surfaces.

extreme halophile   An organism that requires a high salt concentration for growth.

extreme thermophile   *See* hyperthermophile.

extremophile   A microorganism that lives in environmental extremes of temperature, acidity, alkalinity, salinity, or pressure.

eyespot   A pigmented area in a cell, capable of detecting the presence of light.

facilitated diffusion   The movement of a substance across a plasma membrane from an area of higher concentration to an area of lower concentration, mediated by transporter proteins.

facultative anaerobe   An organism that can grow with or without molecular oxygen (O_2).

facultative halophile   An organism capable of growth in, but not requiring, 1–2% salt.

FAD   Flavin adenine dinucleotide; a coenzyme that functions in the removal and transfer of hydrogen ions (H^+) and electrons from substrate molecules.

family   A taxonomic group between order and genus.

feedback inhibition   Inhibition of an enzyme in a particular pathway by the accumulation of the end-product of the pathway; also called end-product inhibition.

fermentation   The enzymatic degradation of carbohydrates in which the final electron acceptor is an organic molecule, ATP is synthesized by substrate-level phosphorylation, and O_2 is not required.

fermentation test   Method used to determine whether a bacterium or yeast ferments a specific carbohydrate; usually performed in a peptone broth containing the carbohydrate, a pH indicator, and an inverted tube to trap gas.

fever   An abnormally high body temperature.

F factor (fertility factor)   A plasmid found in the donor cell in bacterial conjugation.

fibrinolysin   A kinase produced by streptococci.

filtration   The passage of a liquid or gas through a screenlike material; a 0.45-μm filter removes most bacteria.

fimbria (plural: fimbriae)   An appendage on a bacterial cell used for attachment.

fission yeast   Following mitosis, a yeast cell that divides evenly to produce two new cells.

fixed macrophage   A macrophage that is located in a certain organ or tissue (e.g., liver, lungs, spleen, or lymph nodes); also called a histiocyte.

fixing   (1) In slide preparation, the process of attaching a specimen to a slide. (2) Regarding chemical elements, combining elements so that a critical element can enter the food chain. *See also* Calvin-Benson cycle; nitrogen fixation.

flaccid paralysis   Loss of muscle movement, loss of muscle tone.

flagellum (plural: flagella)   A thin appendage from the surface of a cell; used for cellular locomotion; composed of flagellin in prokaryotic cells, composed of nine pairs plus two microtubules in eukaryotic cells.

flaming   The process of sterilizing an inoculating loop by holding it in an open flame.

flat sour spoilage   Thermophilic spoilage of canned goods not accompanied by gas production.

flatworm   An animal belonging to the phylum Platyhelminthes.

flavin adenine dinucleotide   *See* FAD.

flavin mononucleotide   *See* FMN.

flavoprotein   A protein containing the coenzyme flavin; functions as an electron carrier in electron transport chains.

flocculation   The removal of colloidal material during water purification by adding a chemical that causes colloidal particles to coalesce.

flow cytometry   A method of counting cells using a flow cytometer, which detects cells by the presence of a fluorescent tag on the cell surface.

fluid mosaic model   A way of describing the dynamic arrangement of phospholipids and proteins comprising the plasma membrane.

fluke   A flatworm belonging to the class Trematoda.

fluorescence   The ability of a substance to give off light of one color when exposed to light of another color.

fluorescence-activated cell sorter (FACS)   A modification of a flow cytometer that counts and sorts cells labeled with fluorescent antibodies.

fluorescence microscope   A microscope that uses an ultraviolet light source to illuminate specimens that will fluoresce.

fluorescent-antibody (FA) technique   A diagnostic tool using antibodies labeled with fluorochromes and viewed through a fluorescence microscope; also called immunofluorescence.

fluoroquinolone   A synthetic antibacterial agent that inhibits DNA synthesis.

FMN   Flavin mononucleotide; a coenzyme that functions in the transfer of electrons in the electron transport chain.

focal infection   A systemic infection that began as an infection in one place.

folliculitis   An infection of hair follicles, often occurring as pimples.

fomite   A nonliving object that can spread infection.

forespore   A structure consisting of chromosome, cytoplasm, and endospore membrane inside a bacterial cell.

frameshift mutation   A mutation caused by the addition or deletion of one or more bases in DNA.

freeze-drying   *See* lyophilization.

FTA-ABS test   An indirect fluorescent-antibody test used to detect syphilis.

fulminating   A condition that develops quickly and rapidly increases in severity.

functional group   An arrangement of atoms in an organic molecule that is responsible for most of the chemical properties of that molecule.

fungus (plural: **fungi**)   An organism that belongs to the Kingdom Fungi; a eukaryotic absorptive chemoheterotroph.

furuncle   An infection of a hair follicle.

fusion   The merging of plasma membranes of two different cells, resulting in one cell containing cytoplasm from both original cells.

gamete   A male or female reproductive cell.

gametocyte   A male or female protozoan cell.

gamma globulin   *See* immune serum globulin.

gastroenteritis   Inflammation of the stomach and intestine.

gas vacuole   A prokaryotic inclusion for buoyancy compensation.

gel electrophoresis   The separation of substances (such as serum proteins or DNA) by their rate of movement through an electrical field.

gene   A segment of DNA (a sequence of nucleotides in DNA) encoding a functional product.

gene library   A collection of cloned DNA fragments created by inserting restriction enzyme fragments in a bacterium, yeast, or phage.

gene therapy   Treating a disease by replacing abnormal genes.

generalized transduction   The transfer of bacterial chromosome fragments from one cell to another by a bacteriophage.

generation time   The time required for a cell or population to double in number.

genetic code   The mRNA codons and the amino acids they encode.

genetic engineering   Manufacturing and manipulating genetic material in vitro; also called recombinant DNA technology.

genetic recombination   The process of joining pieces of DNA from different sources.

genetics   The science of heredity and gene function.

genetic screening   Techniques for determining which genes are in a cell's genome.

genome   One complete copy of the genetic information in a cell.

genomics   The study of genes and their function.

genotype   The genetic makeup of an organism.

genus (plural: **genera**)   The first name of the scientific name (binomial); the taxon between family and species.

germicide   *See* biocide.

germination   The process of starting to grow from a spore or endospore.

germ theory of disease   The principle that microorganisms cause disease.

global warming   Retention of solar heat by gases in the atmosphere.

globulin   The class of proteins that includes antibodies. *See also* immunoglobulin.

glycocalyx   A gelatinous polymer surrounding a cell.

glycolysis   The main pathway for the oxidation of glucose to pyruvic acid; also called Embden-Meyerhof pathway.

gnotobiotic   Germfree animals colonied with known microorganisms.

Golgi complex   An organelle involved in the secretion of certain proteins.

graft-versus-host (GVH) disease   A condition that occurs when a transplanted tissue has an immune response to the tissue recipient.

gram-negative bacteria   Bacteria that lose the crystal violet color after decolorizing by alcohol; they stain red after treatment with safranin.

gram-negative cell wall   The cell wall of gram-negative bacteria, a peptidoglycan layer surrounded by a lipopolysaccharide outer membrane.

gram-positive bacteria   Bacteria that retain the crystal violet color after decolorizing by alcohol; they stain dark purple.

gram-positive cell wall   The cell wall of most gram-positive bacteria, consisting of peptidoglycan and teichoic acids.

Gram stain   A differential stain that classifies bacteria into two groups, gram-positive and gram-negative.

granulocyte   A leukocyte with visible granules in the cytoplasm; includes neutrophils, basophils, and eosinophils.

green nonsulfur bacteria   Gram-negative, nonproteobacteria; anaerobic and phototrophic; use reduced organic compounds as electron donors for CO_2 fixation.

green sulfur bacteria   Gram-negative, nonproteobacteria; strictly anaerobic and phototrophic; no growth in dark; use reduced sulfur compounds as electron donors for CO_2 fixation.

griseofulvin   A fungistatic antibiotic.

group translocation   In prokaryotes, active transport in which a substance is chemically altered during transport across the plasma membrane.

gumma   A rubbery mass of tissue characteristic of tertiary syphilis.

HAART (highly active antiretroviral therapy)   A combination of drugs used to treat HIV infection.

halogen   One of the following elements: fluorine, chlorine, bromine, iodine, or astatine.

haploid cell   A eukaryotic cell or organism with one of each type of chromosome.

hapten   A substance of low molecular weight that does not cause the formation of antibodies by itself but does so when combined with a carrier molecule.

helminth   A parasitic roundworm or flatworm.

helper T (T_H) cell   A specialized T cell that often interacts with an antigen before B cells interact with the antigen.

hemagglutination   The clumping of red blood cells.

hemoflagellate   A parasitic flagellate found in the circulatory system of its host.

hemolysin   An enzyme that lyses red blood cells.

herd immunity   The presence of immunity in most of a population.

hermaphroditic   Having both male and female reproductive capacities.

heterocyst   A large cell in certain cyanobacteria; the site of nitrogen fixation.

heterolactic   Describing an organism that produces lactic acid and other acids or alcohols as end-products of fermentation; e.g., *Escherichia*.

heterotroph   An organism that requires an organic carbon source; also called organotroph.

Hfr cell   A bacterial cell in which the F factor has become integrated into the chromosome; Hfr stands for high frequency of recombination.

high-efficiency particulate air (HEPA) filter A screenlike material that removes particles larger than 0.3 μm from air.

high-temperature short-time (HTST) pasteurization Pasteurizing at 72°C for 15 seconds.

histamine A substance released by tissue cells that causes vasodilation, capillary permeability, and smooth muscle contraction.

histocompatibility antigen An antigen on the surface of human cells.

histone A protein associated with DNA in eukaryotic chromosomes.

holdfast The branched base of an algal stipe.

haloenzyme An enzyme consisting of an apoenzyme and a cofactor.

homolactic Describing an organism that produces only lactic acid from fermentation; e.g., *Streptococcus*.

horizontal gene transfer Transfer of genes between two organisms in the same generation. *See also* vertical gene transfer.

host An organism infected by a pathogen. *See also* definitive host; intermediate host.

host range The spectrum of species, strains, or cell types that a pathogen can infect.

hot-air sterilization Sterilization by the use of an oven at 170°C for approximately 2 hours.

human leukocyte antigen (HLA) complex Human cell surface antigens. *See also* major histocompatibility complex.

humoral immunity Immunity produced by antibodies dissolved in body fluids, mediated by B cells; also called antibody-mediated immunity.

hyaluronidase An enzyme secreted by certain bacteria that hydrolyzes hyaluronic acid and helps spread microorganisms from their initial site of infection.

hybridoma A cell made by fusing an antibody-producing B cell with a cancer cell.

hydrogen bond A bond between a hydrogen atom covalently bonded to oxygen or nitrogen and another covalently bonded oxygen or nitrogen atom.

hydrolysis A decomposition reaction in which chemicals react with the H^+ and OH^- of a water molecule.

hydroxyl radical A toxic form of oxygen (OH•) formed in cytoplasm by ionizing radiation and aerobic respiration.

hyperacute rejection Very rapid rejection of transplanted tissue, usually in the case of tissue from nonhuman sources.

hyperbaric chamber An apparatus to hold materials at pressures greater than 1 atmosphere.

hypersensitivity An altered, enhanced immune reaction leading to pathological changes; also called allergy.

hyperthermophile An organism whose optimum growth temperature is at least 80°C; also called extreme thermophile.

hypertonic (hyperosmotic) solution A solution that has a higher concentration of solutes than an isotonic solution.

hypha A long filament of cells in fungi or actinomycetes.

hypotonic (hypoosmotic) solution A solution that has a lower concentration of solutes than an isotonic solution.

ID$_{50}$ The number of microorganisms required to produce a demonstrable infection in 50% of the test host population.

idiophase The period in the production curve of an industrial cell population in which secondary metabolites are produced; a period of stationary growth following the phase of rapid growth. *See also* trophophase.

IgA The class of antibodies found in secretions.

IgD The class of antibodies found on B cells.

IgE The class of antibodies involved in hypersensitivities.

IgG The most abundant class of antibodies in serum.

IgM The first class of antibodies to appear after exposure to an antigen.

immune complex A circulating antigen-antibody aggregate capable of fixing complement.

immune deficiency The absence of an adequate immune response; may be congenital or acquired.

immune serum globulin The serum fraction containing immunoglobulins (antibodies); also called gamma globulin.

immunity The body's defense against particular pathogenic microorganisms; also called specific resistance.

immunization *See* vaccination.

immunodiffusion test A test consisting of precipitation reactions carried out in an agar gel medium.

immunoelectrophoresis The identification of proteins by electrophoretic separation followed by serological testing.

immunofluorescence *See* fluorescent-antibody technique.

immunogen *See* antigen.

immunoglobulin (Ig) A protein (antibody) formed in response to an antigen and can react with that antigen. *See also* globulin.

immunological escape Resistance of cancer cells to the immune response.

immunological surveillance The body's immune response to cancer.

immunology The study of a host's specific defenses to a pathogen.

immunosuppression Inhibition of the immune response.

immunotherapy Making use of the immune system to attack tumor cells, either by enhancing the normal immune response or by using toxin-bearing specific antibodies. *See also* immunotoxin.

immunotoxin An immunotherapeutic agent consisting of a poison bound to a monoclonal antibody.

inapparent infection *See* subclinical infection.

incidence The fraction of the population that contracts a disease during a particular period of time.

inclusion Material held inside a cell, often consisting of reserve deposits.

inclusion body A granule or viral particle in the cytoplasm or nucleus of some infected cells; important in the identification of viruses that cause infection.

incubation period The time interval between the actual infection and first appearance of any signs or symptoms of disease.

indicator organism A microorganism, such as a coliform, whose presence indicates conditions such as fecal contamination of food or water.

indirect (passive) agglutination test An agglutination test using soluble antigens attached to latex or other small particles.

indirect contact transmission The spread of pathogens by fomites (nonliving objects).

indirect FA test A fluorescent-antibody test to detect the presence of specific antibodies.

inducer A substance that initiates transcription of a gene.

induction The process that turns on the transcription of a gene.

infection The invasion or growth of microorganisms in the body.

infectious disease A disease in which pathogens invade a susceptible host and carry out at least part of their life cycle in the host.

inflammation A host response to tissue damage characterized by redness, pain, heat, and swelling; and sometimes loss of function.

innate (nonspecific) immunity Host defenses that afford protection against any kind of pathogen. *See also* specific resistance.

innate resistance The resistance of an individual to diseases that affect other species and other individuals of the same species.

inoculum A culture medium in which microorganisms are implanted.

inorganic compound A small molecule that does not contain carbon and hydrogen.

insertion sequence (IS) The simplest kind of transposon.

interferon (IFN) An antiviral protein produced by certain animal cells in response to a viral infection.

interleukin A chemical that causes T-cell proliferation. *See also* cytokine.

intermediate host An organism that harbors the larval or asexual stage of a helminth or protozoan.

intoxication A condition resulting from the ingestion of a microbially produced toxin.

intron A region in a eukaryotic gene that does not code for a protein or mRNA.

invasin A surface protein produced by *Salmonella typhimurium* and *Escherichia coli* that rearranges nearby actin filaments in the cytoskeleton of a host cell.

iodophor A complex of iodine and a detergent.

ion A negatively or positively charged atom or group of atoms.

ionic bond A chemical bond formed when atoms gain or lose electrons in the outer energy levels.

ionization The separation (dissociation) of a molecule into ions.

ionizing radiation High-energy radiation with a wavelength less than 1nm; causes ionization. X rays and gamma rays are examples.

ischemia Localized decreased blood flow.

isograft A tissue graft from a genetically identical source (i.e., from an identical twin).

isomer One or two molecules with the same chemical formula but different structures.

isoniazid (INH) A bacteriostatic agent used to treat tuberculosis.

isotonic (isosmotic) solution A solution in which, after immersion of a cell, osmotic pressure is equal across the cell's membrane.

isotope A form of a chemical element in which the number of neutrons in the nucleus is different from the other forms of that element.

karyogamy Fusion of the nuclei of two cells; occurs in the sexual stage of a fungal life cycle.

kelp A multicellular brown alga.

keratin A protein found in epidermis, hair, and nails.

kinase (1) An enzyme that removes a ℗ from ATP and attaches it to another molecule. (2) A bacterial enzyme that breaks down fibrin (blood clots).

kingdom A taxonomic classification between domain and phylum.

kinin A substance released from tissue cells that causes vasodilation.

Kirby-Bauer test *See* disk-diffusion method.

Koch's postulates Criteria used to determine the causative agent of infectious diseases.

Koji A microbial fermentation on rice; usually *Aspergillus oryzae*; used to produce amylase.

Krebs cycle A pathway that converts two-carbon compounds to CO_2, transferring electrons to NAD^+ and other carriers; also called tricarboxylic acid (TCA) cycle or critic acid cycle.

lactic acid fermentation A catabolic process, beginning with glycolysis, that produces lactic acid to reoxidize NADH.

lagging strand During DNA replication, the daughter strand that is synthesized discontinuously.

lag phase The time interval in a bacterial growth curve during which there is no growth.

larva The sexually immature stage of a helminth or arthropod.

latent disease A disease characterized by a period of no symptoms when the pathogen is inactive.

latent infection A condition in which a pathogen remains in the host for long periods without producing disease.

LD_{50} The lethal dose for 50% of the inoculated hosts within a given period.

leading strand During DNA replication, the daughter strand that is synthesized continuously.

lectin Carbohydrate-binding proteins on a cell.

lepromin test A skin test to determine the presence of antibodies to *Mycobacterium leprae*, the cause of leprosy.

leukocidins Substances produced by some bacteria that can destroy neutrophils and macrophages.

leukocyte A white blood cell.

leukotriene A substance produced by mast cells and basophils that causes increased permeability of blood vessels and helps phagocytes attach to pathogens.

L-form Prokaryotic cells that lack a cell wall; can return to walled state.

lichen A mutualistic relationship between a fungus and an alga or a cyanobacterium.

ligand *See* adhesin.

light-independent (dark) reactions The process by which electrons and energy from ATP are used to reduce CO_2 to sugar. *See also* Calvin-Benson cycle.

light (light-dependent) reaction The process by which light energy is used to convert ADP and phosphate to ATP. *See also* photophosphorylation.

light-repair enzyme An enzyme that splits thymine dimers in the presence of visible light.

limnetic zone The surface zone of an inland body of water away from the shore.

***Limulus* amoebocyte lysate (LAL) assay** A test to detect the presence of bacterial endotoxins.

lipase An enzyme that breaks down triglycerides into their component glycerol and fatty acids.

lipid A non–water soluble organic molecule, including triglycerides, phospholipids, and sterols.

lipid A A component of the gram-negative outer membrane; endotoxin.

lipid inclusion *See* inclusion.

lipopolysaccharide (LPS) A molecule consisting of a lipid and a polysaccharide, forming the outer membrane of gram-negative cell walls.

L-isomer A stereoisomer.

lithotroph *See* autotroph.

littoral zone The region along the shore of the ocean or a large lake where there is considerable vegetation and where light penetrates to the bottom.

local infection An infection in which pathogens are limited to a small area of the body.

localized anaphylaxis An immediate hypersensitivity reaction that is restricted to a limited area of skin or mucous membrane; for example, hayfever, a skin rash, or asthma. *See also* systemic anaphylaxis.

logarithmic decline phase *See* death phase.

log phase The period of bacterial growth or logarithmic increase in cell numbers; also called exponential growth phase.

lophotrichous   Having two or more flagella at one end of a cell.

luciferase   An enzyme that accepts electrons from flavoproteins and emits a photon of light in bioluminescence.

lymphangitis   Inflammation of lymph vessels.

lymphocyte   A leukocyte involved in specific immune responses.

lyophilization   Freezing a substance and sublimating the ice in a vacuum; also called freeze-drying.

lysogenic conversion   The acquisition of new properties by a host cell infected by a lysogenic phage.

lysogenic cycle   Stages in viral development that result in the incorporation of viral DNA into host DNA.

lysis   (1) Destruction of a cell by the rupture of the plasma membrane, resulting in a loss of cytoplasm. (2) In disease, a gradual period of decline.

lysogeny   A state in which phage DNA is incorporated into the host cell without lysis.

lysosome   An organelle containing digestive enzymes.

lysozyme   An enzyme capable of hydrolyzing bacterial cell walls.

lytic cycle   A mechanism of phage multiplication that results in host cell lysis.

macrolide   An antibiotic that inhibits protein synthesis; for example, erythromycin.

macromolecule   A large organic molecule.

macrophage   A phagocytic cell; a mature monocyte.

macrophage activation factor   Increases macrophages' efficiency at destroying ingested cells.

macrophage chemotactic factor   Attracts macrophages to infection site.

macrophage migration-inhibiting factor   Prevents macrophages from leaving infection site.

macule   A flat, reddened skin lesion.

magnetosome   An iron oxide inclusion, produced by some gram-negative bacteria, that acts like a magnet.

major histocompatibility complex (MHC)   The genes that code for histocompatibility antigens; also known as human leukocyte antigen (HLA) complex.

malolactic fermentation   The conversion of malic acid to lactic acid by lactic acid bacteria.

malt   Germinated barley grains containing maltose, glucose, and amylase.

malting   The germination of starchy grains resulting in glucose and maltose production.

margination   The process by which phagocytes stick to the lining of blood vessels.

mast cell   A type of cell found throughout the body that contains histamine and other substances that stimulate vasodilation.

maximum growth temperature   The highest temperature at which a species can grow.

M cell   Intestinal cells that take up and transfer antigens to lymphocytes.

mechanical transmission   The process by which arthropods transmit infections by carrying pathogens on their feet and other body parts.

medulla   A lichen body consisting of algae (or cyanobacteria) and fungi.

meiosis   A eukaryotic cell replication process that results in cells with half the chromosome number of the original cell.

membrane attack complex   Complement proteins C5–C9, which together make lesions in cell membranes that lead to cell death.

membrane filter   A screenlike material with pores small enough to retain microorganisms; a 0.45-μm filter retains most bacteria.

memory cells   A long-lived B or T cell responsible for the memory, or secondary, response.

memory response   A rapid rise in antibody titer following exposure to an antigen after the primary response to that antigen; also called anamestic response or secondary response.

meningitis   Inflammation of the meninges, the three membranes covering the brain and spinal cord.

merozoite   A trophozoite of *Plasmodium* found in red blood cells or liver cells.

mesophile   An organism that grows between about 10°C and 50°C; a moderate-temperature–loving microbe.

mesosome   An irregular fold in the plasma membrane of a prokaryotic cell that is an artifact of preparation for microscopy.

messenger RNA (mRNA)   The type of RNA molecule that directs the incorporation of amino acids into proteins.

metabolic pathway   A sequence of enzymatically catalyzed reactions occurring in a cell.

metabolism   The sum of all the chemical reactions that occur in a living cell.

metacercaria   The encysted stage of a fluke in its final intermediate host.

metachromatic granule   A granule that stores inorganic phosphate and stains red with certain blue dyes; characteristic of *Corynebacterium diphtheriae*. Collectively known as volutin.

methane   The hydrocarbon CH_4, a flammable gas formed by the microbial decomposition of organic matter; natural gas.

methylate   Addition of a methyl group ($-CH_3$) to a molecule; methylated cytosine is protected from digestion by restriction enzymes.

microaerophile   An organism that grows best in an environment with less molecular oxygen (O_2) than is normally found in air.

microbial antagonism   *See* antagonism.

microinjection   The introduction of material, such as DNA, directly into a cell, usually with a glass pipette.

micrometer (μm)   A unit of measurement equal to 10^{-6} m.

microorganism   A living organism too small to be seen with the naked eye; includes bacteria, fungi, protozoa, and microscopic algae; also includes viruses.

microsporidia   Eukaryotic organisms that lack mitochondria and microtubules; obligate intracellular parasites.

microtubule   A hollow tube made of the protein tubulin; the structural unit of eukaryotic flagella and centrioles.

microwave   Electromagnetic radiation with wavelength between 10^{-1} and 10^{-3} m.

minimal bactericidal concentration (MBC)   The lowest concentration of chemotherapeutic agent that will kill test microorganisms.

minimal inhibitory concentration (MIC)   The lowest concentration of a chemotherapeutic agent that will prevent growth of the test microorganisms.

minimum growth temperature   The lowest temperature at which a species will grow.

miracidium   The free-swimming, ciliated larva of a fluke that hatches from the egg.

missense mutation   A mutation that results in the substitution of an amino acid in a protein.

mitochondrion (plural: mitochondria)   An organelle containing Krebs cycle enzymes and the electron transport chain.

mitosis   A eukaryotic cell replication process in which the chromosomes are duplicated; followed by division of the cytoplasm of the cell.

MMWR *Morbidity and Mortality Weekly Report;* a CDC publication containing data on notifiable diseases and topics of special interest.

mole An amount of a chemical equal to the atomic weights of all the atoms in a molecule of the chemical.

molecular biology The science dealing with DNA and protein synthesis of living organisms.

molecular weight The sum of the atomic weights of all atoms making up a molecule.

molecule A combination of atoms forming a specific chemical compound.

monobactam A synthetic antibiotic with a β-lactam ring that is monocyclic in structure, in contrast to the bicyclic β-lactam structure of the penicillins and cephalosporins.

monoclonal antibody A specific antibody produced in vitro by a clone of B cells hybridized with cancerous cells.

monocyte A leukocyte that is the precursor of a macrophage.

monoecious Having both male and female reproductive capacities.

monomer A small molecule that collectively combines to form polymers.

mononuclear phagocytic system A system of fixed macrophages located in the spleen, liver, lymph nodes, and red bone marrow.

monosaccharide A simple sugar consisting of 3–7 carbon atoms.

monotrichous Having a single flagellum.

morbidity (1) The incidence of a specific disease. (2) The condition of being diseased.

morbidity rate The number of people affected by a disease in a given period of time in relation to the total population.

mordant A substance added to a staining solution to make it stain more intensely.

mortality The number of deaths from a specific notifiable disease.

mortality rate The number of deaths resulting from a disease in a given period of time in relation to the total population.

most probable number (MPN) method A statistical determination of the number of coliforms per 100 ml of water or 100 g of food.

motility The ability of an organism to move by itself.

M protein A heat- and acid-resistant protein of streptococcal cell walls and fibrils.

mucous membranes Membranes that line body openings, including the intestinal tract, open to the exterior; also called mucosa.

mutagen An agent in the environment that brings about mutations.

mutation Any change in the nitrogenous base sequence of DNA.

mutation rate The probability that a gene will mutate each time a cell divides.

mutualism A type of symbiosis in which both organisms or populations are benefited.

mycelium A mass of long filaments of cells that branch and intertwine, typically found in molds.

mycology The scientific study of fungi.

mycorrhiza A fungus growing in symbiosis with plant roots.

mycosis A fungal infection.

mycotoxin A toxin produced by a fungus.

NAD⁺ A coenzyme that functions in the removal and transfer of hydrogen ion (H^+) and electrons from substrate molecules.

NADP⁺ A coenzyme similar to NAD^+.

nanometer (nm) A unit of measurement equal to 10^{-9} m, 10^{-3} μm.

natural killer (NK) cell A lymphoid cell that destroys tumor cells and virus-infected cells.

natural penicillins Penicillin molecules made by *Penicillium* spp.; penicillins G and V are examples. *See also* semisynthetic penicillins.

naturally acquired active immunity Antibody production in response to an infectious disease.

naturally acquired passive immunity The natural transfer of humoral antibodies, for example, transplacental transfer.

necrosis Tissue death.

negative (indirect) selection The process of identifying mutations by selecting cells that do not grow using replica plating.

negative staining A procedure that results in colorless bacteria against a stained background.

neurotoxin An exotoxin that interferes with normal nerve impulse conduction.

neutralization An antigen–antibody reaction that inactivates a bacterial exotoxin or virus.

neutron An uncharged particle in the nucleus of an atom.

neutrophil A highly phagocytic granulocyte; also called polymorphonuclear leukocyte (PMN) or polymorph.

nicotinamide adenine dinucleotide *See* NAD⁺.

nicotinamide adenine dinucleotide phosphate *See* NADP⁺.

nitrification The oxidation of nitrogen in ammonia to produce nitrate.

nitrogen cycle The series of processes that converts nitrogen (N_2) to organic substances and back to nitrogen in nature.

nitrogen fixation The conversion of nitrogen (N_2) into ammonia.

nitrosamine A carcinogen formed by the combination of nitrite and amino acids.

nomenclature The system of naming things.

noncommunicable disease A disease that is not transmitted from one person to another.

noncompetitive inhibitor An inhibitory chemical that does not compete with the substrate for an enzyme's active site. *See also* allosteric inhibition; competitive inhibitor.

noncyclic photophosphorylation The movement of an electron from chlorophyll to NAD^+; plant and cyanobacterial photophosphorylation.

nonionizing radiation Short-wavelength radiation that does not cause ionization; ultraviolet (UV) radiation is an example.

nonsense codon A codon that does not encode any amino acid.

nonsense mutation A base substitution in DNA that results in a nonsense codon.

nonspecific resistance Host defenses that afford protection against any kind of pathogen. *See also* specific resistance.

normal microbiota The microorganisms that colonize a host without causing disease; also called normal flora.

nosocomial infection An infection that develops during the course of a hospital stay and was not present at the time the patient was admitted.

notifiable disease A disease that physicians must report to the U.S. Public Health Service; also called reportable disease.

N (neuraminidase) spikes Antigenic projections from the outer lipid bilayer of *Influenzavirus*.

nuclear envelope The double membrane that separates the nucleus from the cytoplasm in a eukaryotic cell.

nuclear pore An opening in the nuclear envelope through which materials enter and exit the nucleus.

nucleic acid A macromolecule consisting of nucleotides; DNA and RNA are nucleic acids.

nucleic acid hybridization The process of combining single complementary strands of DNA.

nucleic acid vaccine A vaccine made up of DNA, usually in the form of a plasmid; also called DNA vaccine.

nucleoid   The region in a bacterial cell containing the chromosome.

nucleolus (plural: nucleoli)   An area in a eukaryotic nucleus where rRNA is synthesized.

nucleoside   A compound consisting of a purine or pyrimidine base and a pentose sugar.

nucleoside analog   A chemical that is structurally similar to the normal nucleosides in nucleic acids but with altered base-pairing properties.

nucleotide   A compound consisting of a purine or pyrimidine base, a five-carbon sugar, and a phosphate.

nucleotide excision repair (NER)   The repair of DNA involving removal of defective nucleotides and replacement with functional ones.

nucleus   (1) The part of an atom consisting of the protons and neutrons. (2) The part of a eukaryotic cell that contains the genetic material.

numerical identification   Bacterial identification schemes in which test values are assigned a number.

nutrient agar   Nutrient broth containing agar.

nutrient broth   A complex medium made of beef extract and peptone.

objective lenses   In a compound light microscope, the lenses closest to the specimen.

obligate aerobe   An organism that requires molecular oxygen (O_2) to live.

obligate anaerobe   An organism that does not use molecular oxygen (O_2) and is killed in the presence of O_2.

obligate halophile   An organism that requires high osmotic pressures such as high concentrations of NaCl.

ocular lens   In a compound light microscope, the lens closest to the viewer; also called the eyepiece.

oligodynamic action   The ability of small amounts of a heavy metal compound to exert antimicrobial activity.

oligoadenylate synthetase   A eukaryotic antiviral protein that degrades viral RNA.

oligosaccharide   A carbohydrate consisting of 2 to approximately 20 monosaccharides.

oncogene   A gene that can bring about malignant transformation.

oncogenic virus   A virus that is capable of producing tumors; also called oncovirus.

oocyst   An encysted apicomplexan zygote in which cell division occurs to form the next infectious stage.

Opa   A bacterial outer membrane protein; cells with Opa form opaque colonies.

operator   The region of DNA adjacent to structural genes that controls their transcription.

operon   The operator and promoter sites and structural genes they control.

opportunistic pathogen   A microorganism that does not ordinarily cause a disease but can become pathogenic under certain circumstances.

opsonization   The enhancement of phagocytosis by coating microorganisms with certain serum proteins (opsonins); also called immune adherence.

optimum growth temperature   The temperature at which a species grows best.

order   A taxonomic classification between class and family.

organelle   A membrane-enclosed structure within eukaryotic cells.

organic compound   A molecule that contains carbon and hydrogen.

organic growth factor   An essential organic compound that an organism is unable to synthesize.

osmosis   The net movement of solvent molecules across a selectively permeable membrane from an area of lower solute concentration to an area of higher solute concentration.

osmotic lysis   Rupture of the plasma membrane resulting from movement of water into the cell.

osmotic pressure   The force with which a solvent moves from a solution of lower solute concentration to a solution of higher solute concentration.

oxidation   The removal of electrons from a molecule.

oxidation pond   A method of secondary sewage treatment by microbial activity in a shallow standing pond of water.

oxidation-reduction   A coupled reaction in which one substance is oxidized and one is reduced; also called redox reaction.

oxidative phosphorylation   The synthesis of ATP coupled with electron transport.

oxygenic   Producing oxygen, as in plant and cyanobacterial photosynthesis.

PABA   Para-aminobenzoic acid; a precursor for folic acid synthesis.

pandemic disease   An epidemic that occurs worldwide.

papule   Small, solid elevation of the skin.

parasite   An organism that derives nutrients from a living host.

parasitism   A symbiotic relationship in which one organism (the parasite) exploits another (the host) without providing any benefit in return.

parasitology   The scientific study of parasites (protozoa and parasitic worms).

parenteral route   A portal of entry for pathogens by deposition directly into tissues beneath the skin and mucous membranes.

pasteurization   The process of mild heating to kill particular spoilage microorganisms or pathogens.

pathogen   A disease-causing organism.

pathogenesis   The manner in which a disease develops.

pathogenicity   The ability of a microorganism to cause disease by overcoming the defenses of a host.

pathology   The scientific study of disease.

pellicle   (1) The flexible covering of some protozoa. (2) Scum on the surface of a liquid medium.

penicillins   A group of antibiotics produced either by *Penicillium* (natural penicillins) or by adding side chains to the β-lactam ring (semisynthetic penicillins).

pentose phosphate pathway   A metabolic pathway that can occur simultaneously with glycolysis to produce pentoses and NADH without ATP production; also called hexose monophosphate shunt.

peptide bond   A bond joining the amino group of one amino acid to the carboxyl group of a second amino acid with the loss of a water molecule.

peptidoglycan   The structural molecule of bacterial cell walls consisting of the molecules N-acetylglucosamine, N-acetylmuramic acid, tetrapeptide side chain, and peptide side chain.

perforin   Protein that makes a pore in a target cell membrane, released by T_C cells.

pericarditis   Inflammation of the pericardium, the sac around the heart.

period of convalescence   The recovery period, when the body returns to its predisease state.

peripheral nervous system (PNS)   The nerves that connect the outlying parts of the body with the central nervous system.

periplasm   The region of a gram-negative cell wall between the outer membrane and the cytoplasmic membrane.

peritrichous Having flagella distributed over the entire cell.

peroxidase An enzyme that breaks down hydrogen peroxide.

peroxide anion An oxygen anion consisting of two atoms of oxygen.

peroxisome Organelle that oxidizes amino acids, fatty acids, and alcohol.

peroxygen A class of oxidizing-type sterilizing disinfectants.

persistent viral infection A disease process that occurs gradually over a long period.

pH The symbol for hydrogen ion (H^+) concentration; a measure of the relative acidity or alkalinity of a solution.

phage *See* bacteriophage.

phage conversion Genetic change in the host cell resulting from infection by a bacteriophage.

phage typing A method of identifying bacteria using specific strains of bacteriophages.

phagocyte A cell capable of engulfing and digesting particles that are harmful to the body.

phagocytosis The ingestion of solids by eukaryotic cells.

phagolysosome A digestive vacuole.

phagosome A food vacuole of a phagocyte; also called a phagocytic vesicle.

phase-contrast microscope A compound light microscope that allows examination of structures inside cells through the use of a special condenser.

phenol OH Also called carbolic acid.

phenolic A synthetic derivative of phenol used as a disinfectant.

phenotype The external manifestations of an organism's genotype, or genetic makeup.

phosphate group A portion of a phosphoric acid molecule attached to some other molecule, Ⓟ,

$$PO_4^{3-}, \quad {}^-O-\overset{\overset{O}{\|}}{\underset{\underset{O^-}{|}}{P}}-O^-$$

phospholipid A complex lipid composed of glycerol, two fatty acids, and a phosphate group.

phosphorous cycle The various solubility stages of phosphorus in the environment.

phosphorylation The addition of a phosphate group to an organic molecule.

photoautotroph An organism that uses light as its energy source and carbon dioxide (CO_2) as its carbon source.

photoheterotroph An organism that uses light as its energy source and an organic carbon source.

photophosphorylation The production of ATP in a series of redox reactions; electrons from chlorophyll initiate the reactions.

photosynthesis The conversion of light energy from the sun into chemical energy; the light-fueled synthesis of carbohydrate from carbon dioxide (CO_2).

phototaxis Movement in response to the presence of light.

phototroph An organism that uses light at its primary energy source.

phylogeny The evolutionary history of a group of organisms; phylogenetic relationships are evolutionary relationships.

phylum A taxonomic classification between kingdom and class.

phytoplankton Free-floating photoautotrophs.

pilus (plural: **pili**) An appendage on a bacterial cell used for the transfer of genetic material during conjugation.

pinocytosis The engulfing of fluid by infolding of the plasma membrane, in eukaryotes.

plankton Free-floating aquatic organisms.

plantae The kingdom composed of multicellular eukaryotes with cellulose cell walls.

plaque A clearing in a bacterial lawn resulting from lysis by phages. *See also* dental plaque.

plaque-forming units (pfu) Visible viral plaques counted.

plasma The liquid portion of blood in which the formed elements are suspended.

plasma cell A cell that an activated B cell differentiates into; plasma cells manufacture specific antibodies.

plasma (cytoplasmic) membrane The selectively permeable membrane enclosing the cytoplasm of a cell; the outer layer in animal cells, internal to the cell wall in other organisms.

plasmid A small circular DNA molecule that replicates independently of the chromosome.

plasmodium (1) A multinucleated mass of protoplasm, as in plasmodial slime molds. (2) When written as a genus, refers to the causative agent of malaria.

plasmogamy Fusion of the cytoplasm of two cells; occurs in the sexual stage of a fungal life cycle.

plasmolysis Loss of water from a cell in a hypertonic environment.

plate count A method of determining the number of bacteria in a sample by counting the number of colony-forming units on a solid culture medium.

pleomorphic Having many shapes, characteristic of certain bacteria.

pneumonia Inflammation of the lungs.

point mutation *See* base substitution.

polar molecule A molecule with an unequal distribution of charges.

polyene antibiotic An antimicrobial agent that alters sterols in eukaryotic plasma membranes and contains more than four carbon atoms and at least two double bonds.

polymer A molecule consisting of a sequence of similar molecules, or monomers.

polymerase chain reaction (PCR) A technique using DNA polymerase to make multiple copies of a DNA template in vitro. *See also* cDNA.

polymorphonuclear leukocyte (PMN) *See* neutrophil.

polypeptide (1) A chain of amino acids. (2) A group of antibiotics.

polysaccharide A carbohydrate consisting of 8 or more monosaccharides joined through dehydration synthesis.

porins A type of protein in the outer membrane of gram-negative cell walls that permits the passage of small molecules.

portal of entry The avenue by which a pathogen gains access to the body.

portal of exit The route by which a pathogen leaves the body.

positive (direct) selection A procedure for picking out mutant cells by growing them.

pour plate method A method of inoculating a solid nutrient medium by mixing bacteria in the melted medium and pouring the medium into a Petri dish to solidify.

precipitation reaction A reaction between soluble antigens and multivalent antibodies to form visible aggregates.

precipitin ring test A precipitation test performed in a capillary tube.

predisposing factor Anything that makes the body more susceptible to a disease or alters the course of a disease.

prevalence The fraction of a population having a specific disease at a given time.

primary cell line Human tissue cells that grow for only a few generations in vitro.

primary infection An acute infection that causes the initial illness.

primary metabolite A product of an industrial cell population produced during the time of rapid logarithmic growth. *See also* secondary metabolite.

primary producer An autotrophic organism, either chemotroph or phototroph, that converts carbon dioxide into organic compounds.

primary response Antibody production in response to the first contact with an antigen. *See also* memory response.

primary sewage treatment The removal of solids from sewage by allowing them to settle out and be held temporarily in tanks or ponds.

prion An infectious agent consisting of a self-replicating protein, with no detectable nucleic acids.

privileged site (tissue) An area of the body (or a tissue) that does not elicit an immune response.

probiotics Microbes inoculated into a host to occupy a niche and prevent growth of pathogens.

prokaryote A cell whose genetic material is not enclosed in a nuclear envelope.

prokaryotic species A population of cells that share certain rRNA sequences; in conventional biochemical testing, it is a population of cells with similar characteristics.

prodromal period The time following the incubation period when the first symptoms of illness appear.

profundal zone The deeper water under the limnetic zone in an inland body of water.

proglottid A body segment of a tapeworm containing both male and female organs.

promoter The starting site on a DNA strand for transcription of RNA by RNA polymerase.

prophage Phage DNA inserted into the host cell's DNA.

prostaglandin A hormonelike substance that is released by damaged cells, intensifies inflammation.

prostheca A stalk or bud protruding from a prokaryotic cell.

protease inhibitor A chemical that inhibits proteases (proteolytic enzymes).

protein A large molecule containing carbon, hydrogen, oxygen, and nitrogen (and sulfur); some proteins have a helical structure and others are pleated sheets.

protein kinase An enzyme that activates another protein by adding a Ⓟ from ATP.

proteobacteria Gram-negative, chemoheterotrophic bacteria that possess a signature rRNA sequence.

proteomics The science of determining all of the proteins expressed in a cell.

protist Term used for unicellular and simple multicellular eukaryotes; usually protozoa and algae.

proton A positively charged particle in the nucleus of an atom.

protoplast A gram-positive bacterium or plant cell treated to remove the cell wall.

protoplast fusion A method of joining two cells by first removing their cell walls; used in genetic engineering.

protozoan (plural: protozoa) Unicellular eukaryotic organisms; usually chemoheterotrophic.

provirus Viral DNA that is integrated into the host cell's DNA.

pseudohypha A short chain of fungal cells that results from the lack of separation of daughter cells after budding.

pseudopod An extension of a eukaryotic cell that aids in locomotion and feeding.

psychrophile An organism that grows best at about 15°C and does not grow above 20°C; a cold-loving microbe.

pscyhrotroph An organism that is capable of growth between about 0°C and 30°C.

purines The class of nucleic acid bases that includes adenine and guanine.

purple nonsulfur bacteria Alpha-proteobacteria; strictly anaerobic and phototrophic; grow on yeast extract in dark; use reduced organic compounds as electron donors for CO_2 fixation.

purple sulfur bacteria Gamma-proteobacteria; strictly anaerobic and phototrophic; use reduced sulfur compounds as electron donors for CO_2 fixation.

pus An accumulation of dead phagocytes, dead bacterial cells, and fluid.

pustule A small pus-filled elevation of skin.

pyocyanin A blue-green pigment produced by *Pseudomonas aeruginosa*.

pyrimidines The class of nucleic acid bases that includes uracil, thymine, and cytosine.

quaternary ammonium compound (quat) A cationic detergent with four organic groups attached to a central nitrogen atom; used as a disinfectant.

quinine An antimalarial drug derived from the cinchona tree and effective against schizogomy stage in red blood cells.

quinolone An antibiotic whose mode of action is to inhibit DNA replication by interfering with the enzyme DNA gyrase.

R Used to represent nonfunctional groups of a molecule. *See also* resistance factor.

radioimmunoassay (RIA) A method of measuring the amount of a compound using radioactively labeled antibodies.

random shotgun sequencing A technique for determining the nucleotide sequence in an organism's genome.

rapid plasma reagin (RPR) test A serological test for syphilis.

r-determinant A group of genes for antibiotic resistance carried on R factors.

receptor An attachment for a pathogen on a host cell.

recipient cell A cell that receives DNA from a donor cell during genetic recombination.

recombinant DNA A DNA molecule produced by recombination.

recombinant DNA technology *See* genetic engineering.

recombinant vaccine A vaccine made by recombinant DNA techniques.

redia A trematode larval stage that reproduces asexually to produce cercariae.

redox reaction *See* oxidation-reduction.

red tide A bloom of planktonic dinoflagellates.

reducing medium A culture medium containing ingredients that will remove dissolved oxygen from the medium to allow the growth of anaerobes.

reduction The addition of electrons to a molecule.

refractive index The relative velocity with which light passes through a substance.

rennin An enzyme that forms curds as part of any dairy fermentation product; originally from calves' stomachs, now produced by molds and bacteria.

replica plating A method of inoculating a number of solid minimal culture media from an original plate to produce the same pattern of colonies on each plate.

replication fork   The point where DNA strands separate and new strands will be synthesized.

repression   The process by which a repressor protein can stop the synthesis of a protein.

repressor   A protein that binds to the operator site to prevent transcription.

reservoir of infection   A continual source of infection.

resistance   The ability to ward off diseases through nonspecific and specific defenses.

resistance (R) factor   A bacterial plasmid carrying genes that determine resistance to antibiotics.

resistance transfer factor   A group of genes for replication and conjugation on the R factor.

resolution   The ability to distinguish fine detail with a magnifying instrument; also called resolving power.

respiration   A series of redox reactions in a membrane that generates ATP; the final electron acceptor is usually an inorganic molecule.

restriction enzyme   An enzyme that cuts double-stranded DNA at specific sites between nucleotides.

retort   A device for commercially sterilizing canned food by using steam under pressure; operates on the same principle as an autoclave but is much larger.

reverse transcriptase   An RNA-dependent DNA polymerase; an enzyme that synthesizes a complementary DNA from an RNA template.

reversible reaction   A chemical reaction in which the end-products can readily revert to the original molecules.

RFLP   Restriction fragment length polymorphism; a fragment resulting from restriction-enzyme digestion of DNA.

Rh factor   An antigen on red blood cells of rhesus monkeys and most humans; possession makes the cells Rh^+.

Rh blood group system   The classification of red blood cells based on the presence or absence of Rh antigens.

rhizine   A rootlike hypha that anchors a fungus to a surface.

rhizopoda   Eukaryotic organisms that move using pseudopods; amoebas.

ribonucleic acid (RNA)   The class of nucleic acids that comprises messenger RNA, ribosomal RNA, and transfer RNA.

ribose   A five-carbon sugar that is part of ribonucleotide molecules and RNA.

ribosomal RNA (rRNA)   The type of RNA molecule that forms ribosomes.

ribosomal RNA (rRNA) sequencing   Determination of the order of nucleotide bases in rRNA.

ribosome   The site of protein synthesis in a cell, composed of RNA and protein.

ribozyme   An enzyme consisting of RNA that specifically acts on strands of RNA to remove introns and splice together the remaining exons.

rifamycin   An antibiotic that inhibits bacterial RNA synthesis.

ring stage   A young *Plasmodium* trophozoite that looks like a ring in a red blood cell.

RNA primer   A short strand of RNA used to start synthesis of the lagging strand of DNA, and to start the polymerase chain reaction.

root nodule   A tumorlike growth on the roots of certain plants containing symbiotic nitrogen-fixing bacteria.

rotating biological contactor   A method of secondary sewage treatment in which large disks are rotated while partially submerged in a sewage tank exposing sewage to microorganisms and aerobic conditions.

rough ER   Endoplasmic reticulum with ribosomes on its surface.

roundworm   An animal belonging to the phylum Nematoda.

S (Svedberg unit)   Notes the relative rate of sedimentation during ultra-high-speed centrifugation.

salt   A substance that dissolves in water to cations and anions, neither of which is H^+ or OH^-.

sanitization   The removal of microbes from eating utensils and food preparation areas.

saprophyte   An organism that obtains its nutrients from dead organic matter.

sarcina (plural: sarcinae)   (1) A group of eight bacteria that remain in a packet after dividing. (2) When written as a genus, refers to gram-positive, anaerobic cocci.

sarcoma   A cancer of fleshy, nonepithelial tissue or connective tissue.

saturation   The condition in which the active site on an enzyme is occupied by the substrate or product at all times.

saxitoxin   A neurotoxin produced by some dinoflagellates.

scanned probe microscopy   Microscopic technique used to obtain images of molecular shapes, to characterize chemical properties, and to determine temperature variations within a specimen.

scanning electron microscope (SEM)   An electron microscope that provides three-dimensional views of the specimen magnified $1000-10,000\times$.

scanning tunneling microscopy   *See* scanned-probe microscopy.

schizogony   The process of multiple fission, in which one organism divides to produce many daughter cells.

scientific nomenclature   *See* binomial nomenclature.

sclerotia   The compact mass of hardened mycelia of the fungus *Claviceps purpurea* that fills infected rye flowers; produces the toxin ergot.

scolex   The head of a tapeworm, containing suckers and possibly hooks.

secondary infection   An infection caused by an opportunistic microbe after a primary infection has weakened the host's defenses.

secondary metabolite   A product of an industrial cell population produced after the microorganism has largely completed its period of rapid growth and is in a stationary phase of the growth cycle. *See also* primary metabolite.

secondary response   *See* memory response.

secondary sewage treatment   Biological degradation of the organic matter in wastewater following primary treatment.

secretory vesicle   A membrane-enclosed sac produced by the ER; transports synthesized material into cytoplasm.

selective medium   A culture medium designed to suppress the growth of unwanted microorganisms and encourage the growth of desired ones.

selective permeability   The property of a plasma membrane to allow certain molecules and ions to move through the membrane while restricting others.

selective toxicity   The property of some antimicrobial agents to be toxic for a microorganism and nontoxic for the host.

self-tolerance   The ability of an organism to recognize and not make antibodies against self.

semiconservative replication   The process of DNA replication in which each double-stranded DNA molecule contains one original strand and one new strand.

semisynthetic penicillins   Modifications of natural penicillins by introducing different side chains that extend the spectrum of antimicrobial activity and avoid microbial resistance.

sense codon   A codon that codes for an amino acid.

sense strand (+ strand)   Viral RNA that can act as mRNA.

sepsis   A toxic condition resulting from the growth and spread of bacteria in blood and tissue.

septa   A cross-wall in a fungal hypha.

septate hypha   A hypha consisting of uninucleate cell-like units.

septicemia   The proliferation of bacteria in the blood, accompanied by fever; sometimes causes organ damage.

septic shock   A sudden drop in blood pressure induced by bacterial toxins, usually endotoxins from gram-negative bacteria.

serial dilution   The process of diluting a sample several times.

seroconversion   A change in a person's response to an antigen in a serological test.

serological testing   Techniques for identifying a microorganism based on its reaction with antibodies.

serology   The branch of immunology that studies blood serum and antigen–antibody reactions in vitro.

serotype   *See* serovar.

serovar   A variation within a species; also called serotype.

serum   The liquid remaining after blood plasma is clotted; contains antibodies (immunoglobulins).

sexual dimorphism   The distinctly different appearance of adult male and female organisms.

sexual spore   A spore formed by sexual reproduction.

Shiga toxin   An exotoxin produced by *Shigella dysenteriae* and enterohemorrhagic *E. coli.*

shock   Any life-threatening loss of blood pressure. *See also* septic shock.

shuttle vector   A plasmid that can exist in several different species; used in genetic engineering.

siderophore   Bacterial iron-binding proteins.

sign   A change due to a disease that a person can observe and measure.

simple diffusion   The net movement of molecules or ions from an area of higher concentration to an area of lower concentration.

simple stain   A method of staining microorganisms with a single basic dye.

singlet oxygen   Highly reactive molecular oxygen (O_2).

site-directed mutagenesis   Techniques used to modify a gene in a specific location to produce the desired polypeptide.

slide agglutination test   A method of identifying an antigen by combining it with a specific antibody on a slide.

slime layer   A glycocalyx that is unorganized and loosely attached to the cell wall.

slime mold   A funguslike protist.

sludge   Solid matter obtained from sewage.

smear   A thin film of material containing microorganisms, spread over the surface of a slide.

smooth ER   Endoplasmic reticulum without ribosomes.

solute   A substance dissolved in another substance.

solvent   A dissolving medium.

Southern blotting   A technique that uses DNA probes to detect the presence of specific DNA in restriction fragments separated by electrophoresis.

specialized transduction   The process of transferring a piece of cell DNA adjacent to a prophage to another cell.

species   The most specific level in the taxonomic hierarchy. *See also* bacterial species; eukaryotic species; viral species.

specific epithet   The second or species name in a scientific binomial. *See also* species.

specific resistance   *See* immunity.

spectrum of microbial activity   The range of distinctly different types of microorganisms affected by an antimicrobial drug; a wide range is referred to as a broad spectrum of activity.

spheroplast   A gram-negative bacterium treated to damage the cell wall, resulting in a spherical cell.

spicule   One of two external structures on the male roundworm used to guide sperm.

spike   A carbohydrate-protein complex that projects from the surface of certain viruses.

spiral   *See* spirillum and spirochete.

spirillum (plural: spirilla)   (1) A helical or corkscrew-shaped bacterium. (2) When written as a genus, refers to aerobic, helical bacteria with clumps of polar flagella.

spirochete   A corkscrew-shaped bacterium with axial filaments.

splicing   A process by which an RNA molecule cuts out introns and rejoins exons to produce a molecule of mRNA.

spontaneous generation   The idea that life could arise spontaneously from nonliving matter.

spontaneous mutation   A mutation that occurs without a mutagen.

sporadic disease   A disease that occurs occasionally in a population.

sporangiophore   An aerial hypha supporting a sporangium.

sporangiospore   An asexual fungal spore formed within a sporangium.

sporangium   A sac containing one or more spores.

spore   A reproductive structure formed by fungi and actinomycetes.

sporogenesis   *See* sporulation.

sporozoite   A trophozoite of *Plasmodium* found in mosquitoes, infective for humans.

sporulation   The process of spore and endospore formation; also called sporogenesis.

spread plate method   A plate count method in which inoculum is spread over the surface of a solid culture medium.

staining   Colorizing a sample with a dye to view through a microscope or to visualize specific structures.

staphylococci (singular: staphylococcus)   Cocci in a grapelike cluster or broad sheet.

stationary phase   The period in a bacterial growth curve when the number of cells dividing equals the number dying.

stem cell   An undifferentiated embryonic cell that can mature into a variety of specialized cells.

stereoisomers   Two molecules consisting of the same atoms, arranged in the same manner but differing in their relative positions; mirror images; also called D-isomer and L-isomer.

sterile   Free of microorganisms.

sterilization   The removal of all microorganisms, including endospores.

steroid   A specific group of lipids, including cholesterol and hormones.

stipe   A stemlike supporting structure of multicellular algae and basidiomycetes.

storage vesicle   Organelles that form from the Golgi complex; contain proteins made in the rough ER and processed in the Golgi complex.

strain   Genetically different cells within a clone. *See* serovar.

streak plate method   A method of isolating a culture by spreading microorganisms over the surface of a solid culture medium.

streptobacilli (singular: streptobacillus)   Rods that remain attached in chains after cell division.

streptococci (singular: streptococcus)   (1) Cocci that remain attached in chains after cell division. (2) When written as a genus, refers to gram-positive, catalase-negative bacteria.

streptogramin   Antibiotics that block protein synthesis at 70S ribosomes.

structural gene   A gene that determines the amino acid sequence of a protein.

subacute disease   A disease with symptoms that are intermediate between acute and chronic.

subclinical infection   An infection that does not cause a noticeable illness; also called inapparent infection.

subcutaneous mycosis   A fungal infection of tissue beneath the skin.

substrate   Any compound with which an enzyme reacts.

substrate-level phosphorylation   The synthesis of ATP by direct transfer of a high-energy phosphate group from an intermediate metabolic compound to ADP.

subunit vaccine   A vaccine consisting of an antigenetic fragment.

sulfa drug   A bacteriostatic compound that interferes with folic acid synthesis by competitive inhibition.

sulfhydryl group   —SH.

sulfur cycle   The various oxidation and reduction stages of sulfur in the environment, mostly due to the action of microorganisms.

sulfur granule   *See* inclusion.

superantigen   An antigen that activates many different T cells, thereby eliciting a large immune response.

superficial mycosis   A fungal infection localized in surface epidermal cells and along hair shafts.

superinfection   The growth of a pathogen that has developed resistance to an antimicrobial drug being used; the growth of an opportunistic pathogen.

superoxide dismutase (SOD)   An enzyme that destroys superoxide free radicals.

superoxide free radical   A toxic form of oxygen ($O_2^{\bullet}$) formed during aerobic respiration.

suppressor T cell (T_S)   A T cell that is thought to end an immune response after an antigen is no longer present.

surface-active agent   Any compound that decreases the tension between molecules lying on the surface of a liquid; also called surfactant.

susceptibility   The lack of resistance to a disease.

symbiosis   The living together of two different organisms or populations.

symptom   A change in body function that is felt by a patient as a result of a disease.

syncytium   A multinucleated giant cell resulting from certain viral infections.

syndrome   A specific group of signs or symptoms that accompany a disease.

synergism   (1) The effect of two microbes working together that is greater than the effect of either acting alone. (2) The principle whereby the effectiveness of two drugs used simultaneously is greater than that of either drug used alone.

synthesis reaction   A chemical reaction in which two or more atoms combine to form a new, larger molecule.

synthetic drug   A chemotherapeutic agent that is prepared from chemicals in a laboratory.

systematics   The science organizing groups of organisms into a hierarchy.

systemic anaphylaxis   A hypersensitivity reaction causing vasodilation and resulting in shock; also called anaphylactic shock.

systemic (generalized) infection   An infection throughout the body.

systemic mycosis   A fungal infection in deep tissues.

tachyzoite   A rapidly growing trophozoite form of a protozoan.

T antigen   An antigen in the nucleus of a tumor cell.

tapeworm   A flatworm belonging to the class Cestoda.

target cell   An infected body cell to which defensive cells of the immune system bind.

taxa   Subdivisions used to classify organisms, e.g., domain, kingdom, phylum.

taxis   Movement in response to an environmental stimulus.

taxonomy   The science of the classification of organisms.

T cell   A type of lymphocyte, which develops from a stem cell processed in the thymus gland, that is responsible for cell-mediated immunity.

T-dependent antigen   An antigen that will stimulate the formation of antibodies only with the assistance of helper T cells. *See also* T-independent antigen.

teichoic acid   A polysaccharide found in gram-positive cell walls.

teleomorph   The sexual stage in the life cycle of a fungus; also refers to a fungus that produces both sexual and asexual spores.

temperature abuse   Improper food storage at a temperature that allows bacteria to grow.

terminator   The site on a DNA strand at which transcription ends.

tertiary sewage treatment   A method of waste treatment that follows conventional secondary sewage treatment; nonbiodegradable pollutants and mineral nutrients are removed, usually by chemical or physical means.

tetracycline   Broad-spectrum antibiotics that interfere with protein synthesis.

tetrad   A group of four cocci.

thallus   The entire vegetative structure or body of a fungus, lichen, or alga.

thermal death point (TDP)   The temperature required to kill all the bacteria in a liquid culture in 10 minutes.

thermal death time (TDT)   The length of time required to kill all bacteria in a liquid culture at a given temperature.

thermoduric   Heat resistant.

thermophile   An organism whose optimum growth temperature is between 50°C and 60°C; a heat loving microbe.

thermophilic anaerobic spoilage   Spoilage of canned foods due to the growth of thermophilic bacteria.

thylakoid   A chlorophyll-containing membrane in a chloroplast. A bacterial thylakoid is also known as a chromatophore.

tincture   A solution in aqueous alcohol.

T-independent antigen   An antigen that will stimulate the formation of antibodies without the assistance of helper T cells. *See also* T-dependent antigen.

tinea   Fungal infection of hair, skin, or nails.

Ti plasmid   An *Agrobacterium* plasmid carrying genes for tumor induction in plants.

titer   An estimate of the amount of antibodies or viruses in a solution; determined by serial dilution and expressed as the reciprocal of the dilution.

topoisomerase II   Enzyme that supercoils DNA thus allowing it to fit compactly in a cell.

total magnification   The magnification of a microscopic specimen, determined by multiplying the ocular lens magnification by the objective lens magnification.

toxemia   The presence of toxins in the blood.

toxigenicity   The capacity of a microorganism to produce a toxin.

toxin   Any poisonous substance produced by a microorganism.

toxoid   An inactivated toxin.

trace element   A chemical element required in small amounts for growth.

transamination   The transfer of an amino group from an amino acid to another organic acid.

transcription The process of synthesizing RNA from a DNA template.

transduction The transfer of DNA from one cell to another by a bacteriophage. *See also* generalized transduction; specialized transduction.

transferrin A human iron-binding protein that reduces iron available to a pathogen.

transfer RNA (tRNA) The type of RNA molecule that brings amino acids to the ribosomal site where they are incorporated into proteins.

transfer vesicle Membrane-bound sacs that move proteins from the Golgi complex to specific areas in the cell.

transformation (1) the process in which genes are transferred from one bacterium to another as "naked" DNA in solution. (2) The changing of a normal cell into a cancerous cell.

transient microbiota The microorganisms that are present in an animal for a short time without causing a disease.

translation The use of mRNA as a template in the synthesis of protein.

transmembrane channel A pore in a target cell membrane produced by complement.

transmission electron microscope (TEM) An electron microscope that provides high magnifications ($10,000–100,000\times$) of thin sections of a specimen.

transport vesicle Membrane-bound sacs that move proteins from the rough ER to the Golgi complex.

transporter protein A carrier protein in the plasma membrane.

transposon A small piece of DNA that can move from one DNA molecule to another.

trickling filter A method of secondary sewage treatment in which sewage is sprayed out of rotating arms onto a bed of rocks or similar materials, exposing the sewage to highly aerobic conditions and microorganisms.

triglyceride A simple lipid consisting of glycerol and three fatty acids.

triplex agent A short segment of DNA that binds to a target area on a double strand of DNA blocking transcription.

trophophase The period in the production curve of an industrial cell population in which the primary metabolites are formed; a period of rapid, logarithmic growth. *See also* idiophase.

trophozoite The vegetative form of a protozoan.

tuberculin skin test A skin test used to detect the presence of antibodies to *Mycobacterium tuberculosis*.

tumor necrosis factor (TNF) A polypeptide released by phagocytes in response to bacterial endotoxins; induces shock; also called cachectin.

tumor-specific transplantation antigen (TSTA) A viral antigen on the surface of a transformed cell.

turbidity The cloudiness of a suspension.

turnover number The number of substrate molecules acted on per enzyme molecule per second.

12D treatment A sterilization process that would result in a decrease of the number of *Clostridium botulinum* endospores by 12 logarithmic cycles.

ubiquinone A low–molecular weight, nonprotein carrier in an electron transport chain; also called coenzyme Q.

UDP-N-acetylglucosamine (UDPNAc) A compound necessary for the biosynthesis of peptidoglycan.

ultra-high-temperature (UHT) treatment A method of treating food with high temperatures ($140–150°C$) for very short times to make the food sterile so that it can be stored at room temperature.

uncoating The separation of viral nucleic acid from its protein coat.

undulating membrane A highly modified flagellum on some protozoa.

use-dilution test A method of determining the effectiveness of a disinfectant using serial dilutions.

vaccination The process of conferring immunity by administering a vaccine; also called immunization.

vaccine A preparation of killed, inactivated, or attenuated microorganisms or toxoids to induce artificially acquired active immunity.

vacuole An intracellular inclusion, in eukaryotic cells, surrounded by a plasma membrane; in prokaryotic cells, surrounded by a proteinaceous membrane.

valence The combining capacity of an atom or a molecule.

vancomycin An antibiotic that inhibits cell wall synthesis.

variolation An early method of vaccination using infected material from a patient.

vasodilation Dilation or enlargement of blood vessels.

VDRL test A rapid screening test to detect the presence of antibodies against *Treponema pallidum*. (VDRL stands for Venereal Disease Research Laboratory.)

vector (1) A plasmid or virus used in genetic engineering to insert genes into a cell. (2) An arthropod that carries disease-causing organisms from one host to another.

vegetative Referring to cells involved with obtaining nutrients, as opposed to reproduction.

vehicle transmission The transmission of a pathogen by an inanimate reservoir.

V factor NAD^+ or $NADP^+$.

vertical gene transfer Transfer of genes from an organism or cell to its offspring.

vesicle (1) A small serum-filled elevation of the skin. (2) Smooth oval bodies formed in plant roots by mycorrhizae.

vibrio (1) A curved or comma-shaped bacterium. (2) When written as a genus, a gram-negative, motile, facultatively anaerobic curved rod.

vibroid A curved, helical bacterium that does not make a complete turn or twist.

viral hemagglutination The ability of certain viruses to cause the clumping of red blood cells in vitro.

viral hemagglutination inhibition test A neutralization test in which antibodies against particular viruses prevent the viruses from clumping red blood cells in vitro.

viral species A group of viruses sharing the same genetic information and ecological niche.

viremia The presence of viruses in the blood.

virion A complete, fully developed viral particle.

viroid Infectious RNA.

virology The scientific study of viruses.

virulence The degree of pathogenicity of a microorganism.

virus A submicroscopic, parasitic, filterable agent consisting of a nucleic acid surrounded by a protein coat.

volutin Stored inorganic phosphate in a prokaryotic cell. *See also* metachromatic granule.

wandering macrophage A macrophage that leaves the blood and migrates to infected tissue.

Western blotting A technique that uses antibodies to detect the presence of specific proteins separated by electrophoresis.

wheal   A small, burning or itching swelling of the skin, such as that resulting from a skin test or a mosquito bite.

whey   The fluid portion of milk that separates from curd.

xenobiotics   Synthetic chemicals that are not readily degraded by microorganisms.

xenodiagnosis   A method of diagnosis based on exposing a parasite-free normal host to the parasite and then examining the host for parasites.

xenotransplantation product   A tissue graft from another species; also called xenotransplant.

X factor   Substances from the heme fraction of blood hemoglobin.

yeast   Nonfilamentous, unicellular fungi.

yeast infection   Disease caused by growth of certain yeasts in a susceptible host.

zone of inhibition   The area of no bacterial growth around an antimicrobial agent in the disk-diffusion method.

zoonosis   A disease that occurs primarily in wild and domestic animals but can be transmitted to humans.

zoospore   An asexual algal spore; has two flagella.

zygospore   A sexual fungal spore characteristic of the zygomycetes.

zygote   A diploid cell produced by the fusion of two haploid gametes.

Credits

Photo Credits

11.1b Bergey's Manual Trust, Michigan State University
11.2 ©Biology Media/Science Source/Photo Researchers, Inc.
11.3a ©R. Calentine/Visuals Unlimited
11.3 ©Biological Photo Service
11.4 ©John D. Cunningham/Visuals Unlimited
11.5 ©Chris Bjornberg/Photo Researchers, Inc.
11.5 ©Science VU/Visuals Unlimited
11.7 ©Dr. Linda Stannard, UCT/Photo Researchers, Inc.
11.8 ©SPL/Custom Medical Stock Photo
11.9 ©Institut Pasteur/CNRI/Phototake
11.10a Wolgemuth, C., Hoiczyk, E., Kaiser, D., and Oster, G., How Myxobacteria Glide, (cover image), Current Biology, March 5, 2002
11.10b ©J. Eisenback/Phototake
11.11 ©A.B. Dowsett/SPL/Photo Researchers, Inc.
11.12b Photo by Dennis Strete
11.12c Courtesy of Susan M. Barnes, Los Alamos National Laboratory
11.13 ©Paul Johnson/Biological Photo Service
11.14 ©Cabisco/Visuals Unlimited
11.15 ©Charles O'Rear/Westlight
11.16a From C.L. Hannay and P. Titz-James, Canadian Journal of Microbiology 1:694, 1955
11.16b From R.E. Strange and J.R. Hunter, in G.W. Gould and A. Hurst (eds), The Bacterial Spore, p. 461, Fig. 4 (Orlando, FL:Academic Press, 1969)
11.17 ©Dr. Tony Brain/SPL/Custom Medical Stock
11.18 ©David M. Phillips/Visuals Unlimited
11.19a Krause, D. C. and D. Taylor-Robinson. 1992. Mycroplasmas which infect humans, Page 419, Figure 1. In J. Maniloff, R. N. McElhaney, L. R. Finch, and J. B. Baseman (ed.), Mycoplasmas: Molecular Biology and Pathogenesis. American Society for Microbiology, Washington D.C.
11.19b ©Michael Gabridge/Visuals Unlimited
11.20 © Frederick P. Mertz/Visuals Unlimited
11.21 ©David M. Phillips/Visuals Unlimited
11.22b Courtesy of Kurt Reed, M.D., Marshfield Medical Research Foundation
11.23 ©J.A. Breznak and H.S. Pankratz/Biological Photo Service
11.24 Courtesy of Sydney M. Finegold
11.25 Courtesy of Karl O. Stetter
11.26 Helde Schultz/Max Planck Institute
11.Box 1 Reproduced by permission of The NutraSweet Kelco Company, a unit of the Monsanto Company, San Diego

CO 12 ©M.F. Brown
12.2a Photo by Dennis Strete
12.2b Photo by Dennis Strete
12.3 Budding yeast David Scharf/Peter Arnold, Inc.
12.4 Courtesy of Christine Case
12.5a Condiospores David M. Phillips/Visuals Unlimited
12.5b Arthrospores M.F. Brown/Visuals Unlimited
12.5c Blastospores David M. Phillips/Visuals Unlimited

12.5d Chlamydospores Candida David M. Phillips/Visuals Unlimited
12.5e Sporangiospores R. Kessel & G. Shih/Visuals Unlimited
12.6.1 ©G. Shih-R. Kessel/Visuals Unlimited, Inc.
12.6.2 Photo by Dennis Strete
12.7 Manfred Kage/Peter Arnold, Inc.
12.7 M.F. Brown/Visuals Unlimited
12.7 ©Keith A. Seifert/Agriculture and Agri-food, Canada
12.8 ©Biophoto Associates/Photo Researchers, Inc.
12.9a Courtesy of Christine Case
12.9b Courtesy of Christine Case
12.10a Courtesy of Christine Case
12.11b ©David Wright
12.11c Courtesy of Christine Case
12.12a ©E.R. Degginger/Bruce Coleman
12.13a ©Manfred Kage/Peter Arnold, Inc.
12.15 Courtesy of Davy Reynolds
12.16 ©M. Abbey/Visuals Unlimited, Inc.
12.17c Apple Trees Productions
12.17d ©E. Koneman/Visuals Unlimited, Inc.
12.18b ©Manfred Kage/Peter Arnold, Inc.
12.20b ©E.R. Degginger/Photo Researchers, Inc.
12.21 Euglena ScienceSource/Photo Researchers, Inc.
12.22 Slime mold showing spore caps Cabisco/Visuals Unlimited
12.23 (right) Courtesy of Christine Case
12.23 (left) Courtesy of Christine Case
12.24 Courtesy of Dr. David Hoppe from Rosenberry, D.O., 2001, Malformed frogs in Minnesota: an update: U.S. Geological Survey Water Fact Sheet FS-043-01, 4p.
12.25b Photo by Dennis Strete
12.27 ©Stanley Flegler/Visuals Unlimited
12.28a Photo by Dennis Strete
12.28b ©R. Calentine/Visuals Unlimited
12.29 Photo by Dennis Strete
12.30 ©David Scharf/Peter Arnold, Inc.
12.31a ©Hans Pfletschinger/Peter Arnold, Inc.
12.32 ©Dr. Robert Lane, UC Berkeley
12.Box 1 Pfiesteria Joann Burkholder/North Carolina State University

CO 13 ©George Musil/Visuals Unlimited
13.2b R.C. Valentine and H.G. Pereira, J. Mol. Biol./BPS
13.3b ©G. Murti/Visuals Unlimited
13.4b ©Science Source/Photo Researchers, Inc.
13.5a Courtesy of Robley Williams
13.5b ©Hans Gelderblom/Visuals Unlimited
13.6 Courtesy of Christine Case
13.9a, b ©S. Martin/Visuals Unlimited
13.14a–d ©Chris Bjornberg/Photo Researchers, Inc.
13.15b ©George Musil/Visuals Unlimited
13.16a C. Garon and J. Rose, National Institute of Allergy and Infectious Diseases, CDC
13.16b ©George Musil/Visuals Unlimited, Inc.
13.18a ©Alfred Pasieka/Science Photo
13.18b ©K.G. Murti/Visuals Unlimited
13.18c ©Dr. Linda Stannard, UCT/Photo Researchers, Inc.

13.20b ©Science VU/Visuals Unlimited, Inc.
13.22 Courtesy of T.O. Diener, USDA

CO 14 ©David M. Phillips/Photo Researchers, Inc.
14.1a ©David M. Phillips/Visuals Unlimited
14.1b ©Stanley Flegler/Visuals Unlimited
14.1c ©David M. Phillips/Photo Researchers, Inc.
14.6a ©Michael P. Gadomski 1988/Photo Researchers, Inc.
14.6b ©Jeff Greenberg/Peter Arnold, Inc.
14.6c ©Steve Allen/The Image Bank
14.7 ©Mitch Kezar/Phototake, NYC
14.8 ©Kent Wood/Photo Researchers, Inc.

CO 15 Photo courtesy of Stanley Falkow
15.1b ©SPL/Photo Researchers, Inc.
15.1c ©David M. Phillips/Visuals Unlimited, Inc.
15.2 Photo courtesy of Stanley Falkow
15.7a ©Edward C. Klatt MD/Webpath/Florida State University Research Foundation, Tallahassee, Florida, USA
15.7b ©Diana Hardie, University of Cape Town Medical School, South Africa
15.8 ©John P. Bader/Biological Photo Service

CO 16 ©Lennart Nilsson, Boehringer Ingelheim International, GmbH
16.2 ©Ed Reschke/Peter Arnold, Inc.
16.4 ©CNRI/SPL/Photo Researchers, Inc.
16.5.1 Courtesy of Barbara Safiejko-Mroczka and Paul B. Bell Jr./Department of Zoology, University of Oklahoma
16.5.2 Courtesy of Barbara Safiejko-Mroczka and Paul B. Bell Jr./Department of Zoology, University of Oklahoma
16.5.3 Courtesy of Barbara Safiejko-Mroczka and Paul B. Bell Jr./Department of Zoology, University of Oklahoma
16.5.4 Courtesy of Barbara Safiejko-Mroczka and Paul B. Bell Jr./Department of Zoology, University of Oklahoma
16.5.5 Courtesy of Nivaldo Medeiros, MD, www.hematologyatlas.com
16.5.6 Courtesy of Barbara Safiejko-Mroczka and Paul B. Bell Jr./Department of Zoology, University of Oklahoma
16.7 ©Lennart Nilsson, Boehringer Ingelheim International, GmbH
16.8 ©Jurgen Berger/Photo Researchers, Inc.
16.11.1 © Journal of Experimental Medicine
16.11.2 © Journal of Experimental Medicine
T16.1 Courtesy of Barbara Safiejko-Mroczka and Paul B. Bell Jr./Department of Zoology, University of Oklahoma

CO 17 ©Boehringer Ingelheim International GmbH, photo by Lennart Nilsson
17.5d Courtesy of Arthur Olson, The Scripps Research Institute
17.7 ©Dr. Gopal Murti/SPL/Photo Researchers, Inc.
17.11 A.K. Szakal et al., Ann. Rev. Immunol. 1989, 7:91–109
17.12 ©David Scharf/Peter Arnold, Inc.

Index